T0304779

Biorefineries and Chemical Processes

Biorefineries and Chemical Processes

Design, Integration and Sustainability Analysis

JHUMA SADHUKHAN

Centre for Environmental Strategy, University of Surrey, UK

KOK SIEW NG

Centre for Process Integration, University of Manchester, UK

ELIAS MARTINEZ HERNANDEZ

Centre for Environmental Strategy, University of Surrey, UK

WILEY

Library of Congress Cataloging-in-Publication Data

Sadhukhan, Jhuma, author.
  Biorefineries and chemical processes : design, integration and sustainability analysis / Jhuma Sadhukhan, Kok Siew Ng, Elias Martinez Hernandez.
     pages cm
  Includes index.
  ISBN 978-1-119-99086-4 (paperback)
1. Biomass–Refining. 2. Biomass chemicals–Technological innovations. 3. Biomass chemicals industry. 4. Biomass energy industries–Environmental aspects. I. Ng, Kok Siew, author. II. Hernandez, Elias Martinez. III. Title.
  TP155.2.E58S34 2014
  662′.88–dc23

2013049101

A catalogue record for this book is available from the British Library.

ISBN: 9781119990864

Set in 9/11pt Times by Aptara Inc., New Delhi, India

Contents

V   ONLINE RESOURCES

Preface

Transformation of fossil resources, coal, crude oil and natural gas into useful products was arguably what allowed chemical engineering to emerge as an essential discipline for industrial and socioeconomic development. Fossil derived fuels, petrochemicals and products have decided the lifestyle and comfort of our civilization. Although the life of fossil resources may be extended by the discovery of shale oil and gas reserves, they will always be finite and will continue to get depleted due to the increasing consumption by a growing world population. The chemical engineering discipline continues to evolve, as scientific and engineering fundamentals are discovered for renewable carbon feedstock: biomass and its conversion processes. This evolution is needed not only because the renewable resources are different from the fossil resources but also because of concerns about sustainability of new supply chains and emergence of a bio- and renewable economy. In the same way as traditional chemical engineering helped to create highly efficient and integrated crude oil refineries, the newly evolved discipline will be key to the advancement of more sustainable biorefineries, in order to supply daily life goods and services. Chemical engineering as a result will extend its reach to cross-disciplinary tools from physical, environmental, biological and material sciences, mathematics, economics and social science.

Some of the major questions that the newly evolved discipline needs to respond to include: How can the renewable resources and biomass be effectively used without depleting land, water, mineral and fossil resources? What are the processes to be integrated to economically produce chemicals and energy with least environmental impact? How can the sustainability of a biorefinery system be achieved? Even more importantly, how can chemical engineers be trained to deal with so many complex challenges? This book and the companion website (referred collectively here as textbook) provides modern multidisciplinary tools and methods to equip chemical engineers with analytical and synthesis skills essential for tackling sustainable design challenges. Holistic and integrative approaches enriched with life cycle thinking are presented towards novel and effective solutions to the above questions.

This textbook is intended for educators, postgraduate and final year undergraduate students in chemical engineering, as well as researchers and practitioners in industry. The main feature of this textbook is the presentation of process modeling and integration and whole system life cycle analysis tools for the synthesis, design, operation and sustainable development of biorefinery and chemical processes.

The book has the following structure.

Part I: Introduction

This gives an opening introduction on the concept, development and tools for sustainable design of biorefineries and utilization of biomass. Biomass can be used for the generation of liquid transportation fuels, that is, biofuels; gaseous fuels; chemicals – commodity as well as specialty; and materials – polymers; alongside combined heat and power (CHP) generation. The integration between biomass feedstocks, processes and products results in complex site configurations for biorefineries. Renewable carbon resources are the waste and lignocellulosic materials. Lignocellulose conversion processes into products need to be synthesized using fundamental chemistry, physics and engineering principles and tools. Insights and suggestions for research into cross-disciplinary areas are also given.

Part II: Tools

This section includes the fundamental methods for economic and environmental impact analyses; combined economic value and environmental impact (EVEI) analysis; life cycle assessment (LCA); multicriteria analysis; heat integration and utility system design; value analysis; mathematical programming based optimization and genetic algorithms. These tools can be taught as individual topics as well as part of computer-aided process design subjects.

Chapters 2 to 8 focus on acquiring fundamental engineering skills. Models for estimating the economics of biomass conversion processes and ways to reduce the costs are shown with example problems. Heat integration using pinch analysis, a well-established methodology in process engineering, is shown not only in the traditional form but also with the adaptation for practical and workable solutions. Life cycle assessment (LCA) considers whole system environmental impacts from element and primary energy resource extraction through conversion to end of life ('cradle to grave'). Following the ISO standards 14040, 14041 and 14044, LCA can be done for the cradle to grave whole system environmental impact analysis, much beyond a chemical plant's boundary. A wide range of biorefinery and energy system LCA problem formulations and solution approaches are shown. Data uncertainty analysis, sensitivity analysis and Monte Carlo simulations, the essential tools in LCA studies to minimize error in estimation, to identify more important indicators, and emission hotspots, and to present results in an accessible and transparent way, are discussed.

Designing a biorefinery according to both economic and environmental objectives is a challenging task due to the wide range of alternatives. In practice, it is essential to be able to prioritize the most promising process pathways for integration into a biorefinery process network. The value analysis approach shown is a powerful tool for differential marginal analysis of process networks. It enables evaluation and graphical presentation of the cost of production (COP), value on processing (VOP) and margins of individual components in a network. Equivalent to COP and VOP, the environmental impact (EI) cost and credit value can be evaluated to analyze the environmental performance of new products and processes quantitatively and select an optimal biorefinery configuration. The EVEI analysis tool gives a fresh perspective to multicriteria analysis.

This section also discusses optimization as a core activity in chemical process design. The linear programming (LP), nonlinear programming (NLP), mixed integer LP (MILP), MINLP and stochastic search methodologies are discussed with example problems. Problem formulations and recognition of the type of algorithm and solution strategy to solve a problem are important. For this, the search techniques need to be understood.

Part III: Process Synthesis and Design

Conventional unit operations face the challenge of slow reactions, lower conversions and purity of desired products when processing lignocellulose as feedstocks. Engineers need to be trained to use a whole range of fundamental tools, such as reaction thermodynamics, reaction engineering, product and process synthesis, unit operations and multiscale simulation, to be able to design products and processes with desired performances. This section gives teaching materials on chemical, polymer, biorefinery and energy systems, reaction engineering and unit operations, synthesis and design subjects.

In this part, biomass based products identified as potential building blocks for chemical synthesis are shown from process synthesis perspectives. Modern process technologies, such as pyrolysis, gasification, catalytic fast pyrolysis, fermentation, crystallization, membrane, membrane filtration, electrodialysis, extraction, reactive extraction, reactive distillation, crystallization and multifunctional catalytic reaction processes, etc., for production of such fuels, chemicals and their subsequent derivatives, are shown alongside design, modeling and simulation frameworks. The exercise problems range from simple flowsheeting through reaction thermodynamic, kinetic and reactor modeling to GHG emission balances and economic analysis. A whole range of biopolymer synthesis processes is discussed from reactor, separator and combined reactor and separator process design, modeling, integration and LCA perspectives. This part also features robust, insightful and accessible process modeling frameworks for CO_2 capture by absorption, adsorption, refrigeration and chemical looping combustion processes. Modern, innovative process flowsheets are illustrated. Examples of economic and life cycle assessments are included, where an opportunity arises to reinforce and encourage life cycle and integrative thinking.

Part IV: Biorefinery Systems

The first chapters in this section deal with the synthesis of bio-oil and algae-based biorefineries as well as hybrid systems integrating fuel cells and renewable energy. The last chapter in this section shows material pertaining to various aspects of multiscale modeling of heterogeneously catalyzed reaction systems, taking transesterification and esterification reactions as an example. Modeling of the intrinsic kinetics, multicomponent diffusion coefficients, effective diffusivities and dynamic multiscale simulation framework is shown. This section can be taught as part of several postgraduate and final year undergraduate level courses including design project, computer-aided process design, process synthesis, reaction engineering, biorefinery engineering and sustainability analysis modules.

Part V: Interacting Systems of Biorefineries (available on the companion website)

This looks at minimizing waste and emissions (Web Chapter 1), energy storage systems, materials and process control (Web Chapter 2) and the optimization and reuse of water (Web Chapter 3). The texts give insights into research in these areas. Sustainability of biorefinery systems needs to draw upon the interactions with other energy and material systems. The companion website discusses the important interacting systems of biorefineries. Emphasis is placed on the mitigation of climate change and environmental impacts by prudent selection of raw materials, products, unit operations and systems.

Case Studies (available on the companion website)

For the chapters in the book, additional exercises and examples can be found on the companion website. Life cycle based problem solving approaches are shown using four case studies on the companion website.

- Case Study 1: Biomass combined heat and power (CHP) system
- Case Study 2: Epoxy resin production from biomass
- Case Study 3: Waste water sludge based CHP system
- Case Study 4: Solar organic photovoltaic cells manufacturing

These case studies are suitable as graduate and postgraduate design and research projects.

This textbook contains exercises (complete with worked solutions) on industrial processes and systems. More than 150 exercise problems are provided to practice the application of the analytical tools shown and encourage systematic thinking. It provides teaching materials with spreadsheet based calculations, conceptual analysis and process simulation based evaluations. Insightful doctorate and postdoctoral research ideas are given throughout the textbook. Chapters feature multidisciplinary problems using Aspen Plus, GAMS, MATLAB and the LCA software GaBi. It is important that, instead of building from scratch, engineers make use of the market leading software in innovative ways.

This textbook is designed to bridge a gap between engineering design and sustainability assessment, for advanced students and practicing process designers and engineers. It is envisaged that this platform of tools will propel a generation of ideas for integrated solutions to complex problems arising from interactions between physical, biological and chemical systems and economic, environmental and policy criteria. The complexity is reduced by step-by-step decision making, from fundamental conceptual unit operation, heat recovery network synthesis, through intra- and interprocess integration, operability and control studies, to life cycle assessment of systems. It is intended to ease integrated designs of sustainable process systems and develop a community of engineers for whom integrated thinking, across the scales, is their distinctive and defining attribute.

Acknowledgments

The authors believe that a cross-disciplinary book of new science such as this is a collective contribution of many previous works from a variety of streams of science and engineering. We wish to acknowledge all those who made contributions in the creation of this book.

The idea for writing this book germinated from a gap in current engineering education literature by coupling process synthesis and life cycle assessment principles. We are indebted to Ms Sarah Higginbotham (Hall) from John Wiley & Sons, Ltd who approached Dr Jhuma Sadhukhan in 2010 in order to fill this gap and for her cooperation throughout the writing phase of the book. Special acknowledgement also goes to Ms Sarah Keegan (Tilley) for her inputs on the structure and materials of the book and Ms Rebecca Ralf for bearing with us for the several iterations requested for the cover page design.

We are grateful to Ms Jasmine Kao (Wiley, Singapore) and Mr Shubham Dixit (Aptara Inc.) and team for making an excellent production exercise, as well as useful communications and feedbacks throughout the production process.

We are indebted to various funding bodies, especially EPSRC, UK, for supporting important research projects, the outputs from which have contributed to the body of knowledge of this book.

Our acknowledgement goes to Khor Chen, a PhD Graduate of Imperial College for writing some of the optimization problems for Chapter 8, and Pedro Manuel Arcelus Arrillaga, another PhD researcher of Imperial College for drawing many diagrams. Special acknowledgement goes to Merari Garcia-Perez for drawing and formatting. Figure 9.1 uses images from http://commons.wikimedia.org/wiki/Commons:Reusing_content_outside_Wikimedia. We are grateful to its author, HaSee at the German Wikipedia project, who made it possible to use the images.

Dr Jhuma Sadhukhan: I would like to express my gratitude to a number of people who have helped in the preparation of the book. It is of special note that Dr Grant Campbell (The University of Manchester), Professor Matthew Leach (The University of Surrey) and Professor Nilay Shah (Imperial College) have provided support throughout the process of writing this book over the last three years. Their vision and valuable insights have been particularly helpful and inspirational. Dr Grant Campbell had stimulated my interest in biorefineries during my time of Lectureship in The University of Manchester, UK (formerly UMIST). At the time, there was very little understanding of how process integration tools could be applied to design biorefineries. Dr Campbell envisaged that application of process integration tools is inevitable for successful deployment of biorefineries. Dr Campbell's encouragement, support and constructive comments helped us to extend our reach beyond the traditional process integration boundaries. Professors Nilay Shah, Nigel Brandon and Matthew Leach shared valuable experiences and gave access to numerous ground-breaking research activities, fuel cells, electrochemistry, polygeneration, micro and distributed systems, environmental and energy systems, and LCA, during my sabbatical research with their respective groups in 2009.

Professor Colin Webb at The University of Manchester, who brought the best out of me as an academic, encouraged me to write this book. Professor Robin Smith's book *Chemical Process: Design and Integration* has been tremendously inspirational to the creation of this book. His pioneering book in the field set an example for us and showed how to systematically communicate an apparently complex subject. My sincere thanks also go to Professor Philippa Browning of The University of Manchester.

The Centre for Environmental Strategy (CES) in The University of Surrey has been particularly supportive to me during the process of writing. I am grateful to Professor Matthew Leach (Director of CES), my Line Manager, who made special arrangements and adjustments in my work load to help me complete this book. He also involved me in important pedagogical activities that gave me useful insights into how students absorb new material, and this has influenced my style of writing. Materials from many of the chapters have been trialled in my classroom teaching. I am indebted to Dr Jonathan Chenoweth, Professor Richard Murphy, Mrs Marilyn Ellis, Barbara Millington and Penny Savill for providing the support needed for the successful completion of this book.

Special thanks go to: Professors Jonathan Seville (Dean of the Faculty of Engineering and Physical Science, University of Surrey), Rex Thorpe (Head of Chemical Engineering Department, University of Surrey), Roland Clift (CBE, Founder of CES and Distinguished Professor of Environmental Technology), Steven Morse, Marios Chryssanthopoulos, Craig Underwood, John Varcoe, Bob Slade, Chris France, Adel Sharif, Drs Aidong Yang, Franjo Cecelja, Devendra Saroj, Walter Wehrmeyer, Yacob Mulugetta, Lucia Elghali, Angela Druckman, Muhammad Ali Imran, Prashant Kumar, Seyed Ali Hosseini and Ahmet Özgür Yazaydin (UCL), Mr Tony Rachwal and Mr Nigel Hartley (University of Surrey); Professors Ferda Mavituna, Adisa Azapagic, Roger Davey, Drs Kostas Theodoropoulos, Nan Zhang and Mr Simon Perry (The University of Manchester); Professors Karen Wilson, Adam Lee, Tony Bridgwater (Universities of Cardiff/Aston/Warwick); Keith Scott, Ian Metcalfe, Adam Harvey, Ian Head, Tom Curtis and Dr Eileen Yu (University of Newcastle); Yingru Zhao; Lauren Basson; Professor Thokozani Majozi (University of Witwatersrand); Drs Murni M. Ahmed (Universiti Teknologi PETRONAS), Denny K.S. Ng (University of Nottingham, Malaysia), Professors Mahmoud El-Halwagi (University of Texas), Raymond Tan (De La Salle University), Dominic Foo (University of Nottingham, Malaysia), Larry Biegler and Ignacio Grossmann (Carnegie Mellon University), Antonis Kokossis (National Technical University of Athens), Rafiqul Gani (Technical University of Denmark), Richard Venditti (North Caroline State University), Mr Howard Simons (previously in MW Kellogg Ltd), Dr M.O. Garg (Indian Institute of Petroleum), Mr Venkat Ramanujam (GTC Technology), Mr Amit Neogi (Degremont), Mr Nigel Ridgway (Arup), Geoffrey Drage and Kenneth Day (Biosep Ltd), and Graduate Researchers and Research Engineers Drs Mian Xu, Ankur Kapil (Johnson Matthey), Shrikant Bhat (ABB), Mustafa Mustafa and Fernan Mateos-Salvador (SABIC), Mr Chinedu Okoli, Mr Abhijeet Parvatker (BITS Pillani, India), Mr Naqqash Todai (Atkins), Mr Ashwin Nagarajan, Mr Thomas Lau, Ms Sophie Persons (National Physical Laboratory), Mr Elliott Seath, Mr Polyxeni Tofalou, Mr Nick Mills (Thames Water), Mr Freddie Pask (3M), Ms Stefanie Nikamp (TWI Ltd) and Mr Jamal Miah (Nestle) for their contributions to research and the body of knowledge in the area, and hence to the book.

My earliest teachers of Chemical Engineering at Jadavpur University, Kolkata, and Indian Institute of Technology, Kanpur, India, have had a major role in my understanding of the subject. My heartiest gratitude goes out to Professors Kajari Kargupta, Abhijeet Bhawal, Late Shyamal K. Sanyal, Pinaki Bhattacharya and Chiranjeeb Bhattacharya for transferring fundamental knowledge in undergraduate classrooms that forms the foundation of my understanding of this subject.

An endeavour of this scale would not have been possible without the love and support from family and friends. Over the last three years my husband and boys, Sohum and Ritam of age 11 and 5, have been extremely patient in excusing me from joining them in family weekends and holiday fun activities for the noble cause of *"mum writing a book."* Sohum has enthusiastically produced some spreadsheet based graphs for the book as part of achieving his own Excel learning goal.

I am grateful to my brother Mr Deepayan Sadhukhan for providing many useful tips on water systems engineering and interactions with other industries, and research results on chemical looping combustion. Personal acknowledgment goes to my parents, Mrs Pratima Sadhukhan and Mr Tushar Kanti Sadhukhan, my brother, Mr Deepayan Sadhukhan, and Deepadi (Mrs Deepa Mukherjee) for providing me with endless and unconditional love and trust, Gorada (Professor Subrata Mukherjee, Delhi Univeristy), Sanskriti Reading, UK, and Jadavpur University, IIT Kanpur, UMIST, and my school BMGS friends and many others, without whose encouragement and support I could not have completed the book.

Dr Kok Siew Ng: A chemical engineer plays an important role in building society and improving the human lifestyle in various aspects. As academics, we believe that the inheritance of knowledge to the next generation of chemical engineers is one of the utmost important tasks of all. It is indeed a great honour for me to write a textbook like this during my early professional career, transferring the theoretical and practical knowledge and experience that I have gained throughout these years to the engineering and science communities.

My life would not have been so meaningful (and of course challenging!) without the responsibility of writing this book. I am indebted to Dr Jhuma Sadhukhan for giving me such an invaluable opportunity to join the team and to communicate knowledge through writing. I also thank her for the encouragement, guidance and support throughout my university education. I faithfully thank Dr Elias Martinez Hernandez, another co-author of this book, for his enormous contribution in every sense. I would like to express my sincerest gratitude to Professor Robin Smith for providing me with the opportunity to join Process Integration Limited (PIL) and to participate in a number of important projects in the company. I always love to pick his brain for anything related to research and design in chemical engineering. I have gained substantial industrial project experience by working as a consultant in PIL. I have been learning from the experts in the company including Professor Robin Smith, Dr Nan Zhang and Dr Steve Hall. The knowledge and practical experience that I have gained through consultancy projects have been helpful in writing this book. My sincere gratitude goes to Dr Grant Campbell, Professor Colin Webb, Dr Nan Zhang and Dr Steve Hall for their guidance and motivation throughout my studies and professional career. Special thanks go to all the staff and colleagues within the Centre for Process Integration, Process Integration Limited, and the School of Chemical Engineering and Analytical Science at The University of Manchester for fostering me from undergraduate, postgraduate and, until now, as a professional. It has been a great pleasure to be acquainted with Dr Denny Kok Sum Ng and Professor Dominic Foo from The University of Nottingham, Malaysia. Their enthusiasm and devotion to chemical engineering have vastly inspired and stimulated me. Special acknowledgement goes to Dr Carol Lin from City University of Hong Kong who was one of my favourite tutors during my undergraduate studies. She has been very helpful all the time when I was in need.

My heartfelt thanks to all my best friends, Way Sheong, Seong Keat, See Beng, Yet Th'ng, Elladora, Thomas, Sonia Farrokhpanah, Pasika, Enn Swan, Tzyy Shyang, Wei Ning, Tze Howe, Seong Chuin, Nan Jia, Yuhang, Kunpeng, Li Sun, Xuesong and many others, for their company and precious friendships. The whole book writing and editing process that took nearly three years to complete would not have been an easy journey without their support.

I owe my deepest gratitude to my parents, Mr Kweng Chuan Ng and Mdm Eng Tho Woo, for their selfless love and care. I am also thankful to my sister, Chui Mei Ng, for being with me all the time with joy and sorrow, acting as a role model since I was young. I would also like to convey my appreciation to all my cousins and relatives, Chui Lim, Kok Chuan, Kok Siong, Alice and many others.

Dr Elias Martinez Hernandez: I would like to express infinite gratitude to Dr Jhuma Sadhukhan for her support and advice during the last five years, first in The University of Manchester and now in The University of Surrey. I thank her for the invitation to write this book, one of the most unique opportunities in my career and my life. I have enjoyed discussions on the preparation and presentation of the material. Writing this book together has been a highly rewarding learning experience for me. I also thank co-author Dr Kok Siew Ng for his help and advice during doctoral studies, and his pertinent suggestions and efforts to improve the quality of presentation of this book. I would like to thank Dr Grant Campbell for his intriguing and inspirational advice on how to adapt and apply process integration tools to biorefinery design. His vision has much influenced the aims of this book.

I would like to thank Drs Kostas Theodoropoulos, Nan Zhang and Megan Jobson from The Centre for Process Integration (The University of Manchester) for allowing me to help them as teaching assistant for their lectures, which have been helpful for the preparation of this book. Special thanks go to Dr Aidong Yang from The University of Oxford and Professor Matthew Leach from The University of Surrey for their support and comprehension when I took time off from our research project during the proofreading stage.

I wish to acknowledge Professors Celestino Montiel and Antonio Ortiz in the Faculty of Chemistry at the National Autonomous University of Mexico (UNAM), who have greatly contributed to my engineering experience while we worked on PEMEX projects. The experience on reaction kinetics and reactor engineering shared by Professors Tatiana Klimova and Martin Hernandez Luna (also at UNAM) greatly helped in the preparation of material on these topics. I specially appreciate the guidance from Mr Ernesto Rios Montero (Consultant in CONSUTEC, Grupo Mexico). His brilliant career as both practitioner and teacher in Chemical Engineering has been inspiring. I am indebted to him for giving me the opportunity to contribute in setting up standards and procedures for a project on process design of a crude oil produced water treatment plant. Those documents I prepared with his assistance and this book have a purpose in common, to support the training and guidance of current and future chemical engineers.

Special personal acknowledgements go to my parents, Luisa and Herminio, and all my family members who miss my presence at home and gave all their support to pursue a worthy job of writing this book. I wish to thank Merari for her support and comprehension during the last few years and also for helping me with the proofreading during her summer holidays in 2013. My heartfelt thanks go to Professor Celestino, because his encouragement to pursue doctoral studies has eventually led me to write this book. Personal thanks to my friend Pedro Arcelus for his company and help since undergraduate studies. I am thankful to my friends in Mexico and fellows at The Centre for Process Integration for their support throughout the last five years.

Finally, I wish to thank the chemical engineers in the Faculty of Chemistry (UNAM), who took my lab classes for Reactor Chemical Engineering. Your enquiries and troubles in trying to understand and translate theory into practice have definitely enriched my experience, which has influenced my way of communication in this book. This book is mainly dedicated to people like you, engineering students and practitioners, who aim not only how to make life on this planet more comfortable but also how to make this planet more sustainable. In some way, this underlying message has been conveyed in the design of the book's front cover page.

About the Authors

Dr Jhuma Sadhukhan is Senior Lecturer of Sustainable Resources in the Centre for Environmental Strategy at the University of Surrey, a Visiting Academic in Imperial College, London, and was previously Lecturer in The University of Manchester in the UK. She has extensive industrial experience with MW Kellogg Ltd and Technip (formerly Kinetics Technology India), in oil, gas, energy and process engineering design, synthesis and optimization. She received a BEng degree in Chemical Engineering from Jadavpur University and an MTech degree in Chemical Engineering from IIT Kanpur in India, and a PhD in Process Integration from UMIST, now The University of Manchester in the UK. She has published widely in biorefinery design, integration and technoeconomic and life cycle assessments. Her research and teaching interest is in integrated engineering design and life cycle assessments. In 1999, her work on gasification received the first prize in the IChemE International Conference on *Gasification for the Future*. In 2006, she was awarded the IChemE *Hanson Medal* for contribution to an article on biorefinery engineering education. In 2011, she was jointly awarded the IChemE *Junior Moulton Medal* for the best publication by the IChemE with Dr Kok Siew Ng.

Dr Kok Siew Ng is Researcher in the Centre for Process Integration, School of Chemical Engineering and Analytical Science, at The University of Manchester in the UK. He received an MEng in Chemical Engineering with Chemistry and PhD degrees from The University of Manchester in the UK. He has contributed extensively in the field of chemical and biorefinery engineering, and particularly in process integration subjects. Formerly, he was a consultant in Process Integration Limited and had extensive experience of developing process integration tools and carrying out consultancy projects for a diverse range of systems including refrigeration systems for the LNG process, upgrading of gasoline into chemicals through the hydrocracking process and water systems integration in refineries and steel plants, in the UK, Europe and China. He extensively worked on data collection, verification and reconciliation, conceptual process design and synthesis, process modification and integration, simulation modeling, economic analysis and optimization. He had contributed to teaching a number of graduate and postgraduate modules in The University of Manchester in the UK, including process control, enquiry based learning, energy systems as well as reaction and separation systems. His research interests are in biorefinery, clean coal technologies, cogeneration and polygeneration and engineering solutions to CO_2 emission reduction through reuse. He has published widely in these fields. In 2011, he was jointly awarded the IChemE *Junior Moulton Medal* for the best publication with Dr Jhuma Sadhukhan.

Dr Elias Martinez Hernandez is Research Fellow in the Centre for Environmental Strategy and the Department of Chemical and Process Engineering at The University of Surrey in the UK. He received a PhD in Process Integration from The University of Manchester in the UK. Formerly, he was consultant in the Faculty of Chemistry of the National Autonomous University of Mexico (UNAM), from which he graduated as a Chemical Engineer. He has been involved in a wide range of industrial projects including technoeconomic and environmental feasibility analyses for revamping projects for PEMEX refineries, conceptual designs of LNG regasification processes, simulation and rating of refinery fired heaters, and front end engineering design projects for refineries. He contributed to the Reactor Engineering Laboratory Unit in UNAM, biorefinery engineering, reaction–separation systems, computer-aided process design, process control, distillation systems design and advanced mass transfer methods in The University of Manchester and life cycle assessment in the University of Surrey in the UK. He has extensively contributed to the development of process integration and life cycle assessment tools for sustainable design of biorefineries and methodologies and tools for sustainable design of local production systems using renewable resources. He regularly posts latest methodological and software developments in the area to http://biorefinerydesign.webs.com/. He has widely published in the fields of chemical process design and integration, biofuel production, biorefinery engineering, life cycle assessment and sustainable chemical engineering.

Companion Website

This textbook is supported by a website which contains a variety of supplementary material:

http://www.wiley.com/go/sadhukhan/biorefineries

On the website you will find the following resources:

Web Chapters:
Three additional online chapters covering the following topics:

- Web Chapter 1: Waste and Emission Minimization
- Web Chapter 2: Energy Storage and Control Systems
- Web Chapter 3: Water Reuse, Footprint and Optimization Analysis

Case Studies
Four case studies covering LCA-based problem solving approaches:

- Case Study 1: Biomass CHP Plant Design Problem – LCA and Cost Analysis
- Case Study 2: Comparison between Epoxy Resin Productions from Algal or Soya Oil – An LCA Based Problem Solving Approach
- Case Study 3: Waste Water Sludge Based CHP and Agricultural Application System – An LCA Based Problem Solving Approach
- Case Study 4: LCA Approach for Solar Organic Photovoltaic Cells Manufacturing

Exercises and examples
Additional exercises and examples referred to within each chapter of the book can be found online.

Figures from the book as PowerPoint slides.
Copies of all the figures in the book are provided in PowerPoint format for viewing or downloading.

Nomenclature

Chapter 2

Symbols

a	cost of the first unit in Equation (2.12)
b	parameter in Equation (2.12)
C_D	total direct cost
C_f	cash flow in a particular year in Equation (2.4)
C_{ID}	total indirect cost
C_o	original cost in Equation (2.1)
$COST_{size1}$	cost of the base system in Equation (2.2)
$COST_{size2}$	cost of the system after scaling up/down in Equation (2.2)
C_{pr}	present cost in Equation (2.1)
C_{TCI}	total capital investment
FV	future value in Equation (2.3)
i	annual interest rate
I_o	original index value in Equation (2.1)
I_{pr}	present index value in Equation (2.1)
I_T	total interest paid in Equation (2.6)
M	monthly instalment in Equation (2.5)
n	number of years of investment in Equation (2.3)
N	number of payments in total in Equation (2.5)
P	principal of the loan in Equation (2.5)
PV	present value in Equation (2.3)
q	number of payments in a year in Equation (2.5)
r	discount rate or interest rate in Equation (2.3)
R	scaling factor in Equation (2.2)
$SIZE_1$	capacity of the base system in Equation (2.2)
$SIZE_2$	capacity of the system after scaling up/down in Equation (2.2)
T_{PL}	plant life in Equation (2.4)
x	cumulative number of units in Equation (2.12)
y	cost of xth unit in Equation (2.12)

Chapter 3

Symbols

CP	heat capacity flow rate
C_p	specific heat capacity
h_1	specific enthalpy at point 1 in Figure 3.20
$h_{2,act}$	actual specific enthalpy at point 2 in Figure 3.20
$h_{2,is}$	theoretical specific enthalpy at point 2 in Figure 3.20
h_{sup}	specific enthalpy of superheated steam
h_v	specific enthalpy of saturated steam
LHV_{fuel}	lower heating value of fuel
m_1	initial amount of steam generated in Figure 3.10
m_2	amount of steam generated after composite curve shifting in Figure 3.10
m_{BFW}	mass flow rate of boiler feed water in Figure 3.19
m_{fuel}	mass flow rate of fuel required in the boiler in Figure 3.19
m_{steam}	mass flow rate of steam generated from the boiler in Figure 3.19
$Q_{c,min}$	minimum cold utility load
$Q_{h,min}$	minimum hot utility load
Q_R	potential heat recovery
$\dot{Q}_{steam}$	heat required for steam generation
$\dot{Q}_{fuel}$	heat supplied from fuel
$\dot{Q}_{saturate}$	heat required for generating saturated steam
$\dot{Q}_{sensible}$	heat required for heating BFW
$\dot{Q}_{superheat}$	heat required for superheating steam
T^*	shifted temperature
T_{BFW}	temperature of boiler feed water in Figure 3.19
T_S	supply temperature of a stream
T_S^*	shifted supply temperature
T_{sat}	saturation temperature of steam
T_T	target temperature of a stream
T_T^*	shifted target temperature
W	power generated from steam turbine in Figure 3.19
W_{act}	actual work done
W_T	theoretical work done

Greek letters

ΔH	heat load from a heat transfer process
Δh_{vap}	enthalpy of vaporization of water
ΔT_{min}	minimum temperature difference
$\eta_{turbine}$	isentropic efficiency of steam turbine in Figure 3.19

Chapter 6

Sets

UNIT	set of process units $\{i/\text{process units}\}$
ED	set of downstream elements $\{e'(i \in UNIT)/\text{downstream elements of process unit } i\}$
EU	set of upstream elements $\{e(i \in UNIT)/\text{upstream elements to process unit } i\}$

Symbols

ACC	annualised capital cost
CI	capital investment of a plant in Section 6.2.1
CI_i	capital cost of process unit i per unit flow rate of feed stream
F	unit market price of feedstock in Figure 6.2
$F_{1-2-3-\cdots-i}$	market price of feed $F1-2-3-\cdots-i$
$(F_{1-2-3-\cdots-i})_{COP}$	cost of production of feed $F1-2-3-\cdots-i$
$(F_{1-2-3-\cdots-i})_{VOP}$	value on processing of feed $F1-2-3-\cdots-i$
F_{e-i}	market price of feed to unit i from path e
$(F_{e-i})_{COP}$	cost of production of feed to unit i from upstream element $e \in EU(i)$
$(F_{e-i})_{VOP}$	value on processing of feed to unit i from upstream element $e \in EU(i)$
$(F)_{VOP}$	value on processing of feed to a process unit
m^F	flow rate of feed to a process unit
$m^F_{1-2-3-\cdots-i}$	flow rate of feed $F1-2-3-\cdots-i$
m^F_{e-i}	flow rate of feed to unit i from upstream element $e \in EU(i)$
m^P	flow rate of product from a process unit
m^P_1	flow rate of end product from unit 1
$\hat{o}$	annualized cost (operating plus capital) of process unit per unit flow rate of feed
$\hat{o}_i$	annualized cost (operating plus capital) of process unit i per unit flow rate of feed
O	total cost (operating plus capital) of a unit
O_i	total cost (operating plus capital) of a unit i
O_i'	total cost (operating plus capital) of a unit i considering the cost of a recycled stream
P	unit market price of end product in Figure 6.2
$(P)_{COP}$	cost of production of product from a process unit
TF_i	tolling fee of process unit i equal to the *ACC* times CI_i
yr	year
Y	payback time in years (after plant start-up) based on discounted cash flow
Y^c	number of years to build a plant before its start-up

Greek letters

Δ	economic margin of a process unit
Δ_i	economic margin of unit i
Δe	economic margin of a stream in Figure 6.2
$\Delta_i(F_{1-2-3-\cdots-i})$	economic margin of unit i incurred from stream $F_{1-2-3-\cdots-i}$

Chapter 7

Symbols

A	tank surface area in Exercise 3 of Chapter 7
$\bar{A}_k$	single-column vector of flow rates of auxiliary raw materials
B_f	CO_2 binding by photosynthesis
CC_k	annualized capital cost of process unit k
C_f	economic cost of a feedstock f
CI_k	annualized impact from the materials of construction of process unit k
C_m	economic cost for emission control and treatment of emissions/wastes
$\bar{C}$	vector containing the "costs" (COP and ICP) of a stream
$\bar{C}_p$	vector containing the "costs" (COP and ICP) of a product or outlet stream p from a process unit
$\bar{C}_{a,k}$	single-row vector of unit economic costs of auxiliary raw materials
$\bar{C}_{m,k}$	single-row vector of unit economic costs of emissions/wastes treatment
$\bar{C}_{u,k}$	single-row vector of unit economic costs of utilities
D_p	EI credit value of a product
f	subscript for feed (inlet) streams in Chapter 7
F_f	mass flow rates of a feed or inlet stream
g	number of feeds or inlet streams to a process unit (excluding auxiliary raw materials) in Equation (7.6)
G_f	EI from production or growing a biomass feedstock
I_a	EI cost of auxiliary raw materials
I_{end}	impact from end use or end of life of a product
I_f	EI cost of a biomass feedstock
I_m	EI cost from emissions/wastes
I_{peq}	EI cost of an existing equivalent product to be displaced or substituted
I_u	EI cost of utilities
$\bar{I}_{a,k}$	single-row vector containing the EI costs of auxiliary raw materials
$\bar{I}_{m,k}$	single-row vector containing the EI costs of emissions/wastes
$\bar{I}_{u,k}$	single-row vector containing the EI costs of utilities
k	subscript for process units in Chapter 7
L	tank length in Exercise 3 of Chapter 7
M\$	million dollar
$\bar{M}_k$	single-column vector of flow rates of emissions/wastes
O_{Ik}	total EI costs (operating plus annualized EI from construction) associated with a process unit k
O_k	total economic costs (operating plus capital) associated with a process unit k
$\bar{O}_k$	vector of total costs (O_k and O_{Ik}) of a process unit k
p	subscript for product (outlet) streams in Chapter 7
P_p	mass flow rates of a product or outlet stream
q	number of products or outlet streams from a process unit (excluding emissions/wastes) in Equation (7.6)
r	tank radius in Exercise 3 of Chapter 7
s_p	relative percentage of EI saving of a product p
T_f	EI from transportation of a feedstock f

T_p	EI from transportation of a product p
$\bar{U}_k$	single-column vector of flow rates of utilities
V	volume
$\bar{V}$	vector containing the "values" (VOP and CVP) of a stream
$\bar{V}_f$	vector containing the "values" (VOP and CVP) of a feed or inlet stream f to a process unit

Greek letters

α'_p	final allocation factor of an end product
α	allocation factor of a stream as defined in Equation (7.7)
β	equivalency factor in Chapter 7
Δe_p	economic margin of a product stream
Δi_p	EI saving margin of a product stream
Δi_{target}	EI saving margin to achieve an EI target
ρ_{steel}	steel density
ℓ	tank wall thickness

Chapter 10

Symbols

a_{ij}	number of atoms of the element j in component i
A_j	total number of atoms of element j in the feed
C	weight percent of carbon, determined in the ultimate analysis
C_p	heat capacity in Equation (10.6)
$C_{p,mean}$	mean heat capacity
$C_{p,product}$	specific heat capacity of product
$C_{p,reactant}$	specific heat capacity of reactant
FC	fixed carbon in proximate analysis
G	Gibbs free energy
G_T	total Gibbs free energy of a system
H	weight percent of hydrogen, determined in the ultimate analysis
HHV	higher heating value of fuel
HHV_{dry}	higher heating value on dry basis
h_{vap}	latent heat of vaporization of water at standard condition
K_{eq}	equilibrium constant
K_P	equilibrium constant of a reaction at constant pressure
k	rate constant
L	Lagrange function
LHV	lower heating value of fuel
LHV_{dry}	lower heating value on dry basis
m_{air}	mass of air
m_B	mass fraction of biomass

m_C	mass fraction of char
m_{COND}	mass of water formed
m_{fuel}	mass of fuel
m_G	mass fraction of gas
m_O	mass fraction of oil
N	number of components in Equation (10.42)
n	stoichiometric number of moles
n_i	number of moles of component i
n_P	total number of products
n_R	total number of reactants
n_T	total number of moles; see Table 10.10
O	weight percent of oxygen, determined in the ultimate analysis
P	total pressure of the system
p_i	partial pressure of product component i
p_j	partial pressure of reactant component j
R	universal gas constant
R_{AF}	air-to-fuel ratio defined in Equation (10.22)
T	temperature in Equation (10.6)
T_1	initial temperature of combustion reaction
T_2	final temperature of combustion reaction
T_{AFT}	adiabatic flame temperature
VM	volatile matters in proximate analysis
x	number of moles of steam in Equation (10.30)
X_C	carbon conversion
y	number of moles of air in Equation (10.29) and (10.30)
y_i	mole fraction of product component i in gaseous phase
y_j	mole fraction of reactant component j in gaseous phase

Greek letters

$\Delta G°$	standard Gibbs free energy change
$\Delta \overline{G}°_{f,i}$	standard Gibbs energy of formation
ΔH_C	heat of combustion of a reaction
$\Delta H_C°$	standard heat of combustion
$\Delta H°_{f,product}$	standard heat of formation of product
$\Delta H°_{f,reactant}$	standard heat of formation of reactant
$\Delta H°_f$	standard heat of formation
$\Delta H°_{f,1}$	standard heat of formation of reaction 1 (Reactant(s) → Intermediate Product(s))
$\Delta H°_{f,2}$	standard heat of formation of reaction 2 (Intermediate Product(s) → Product(s))
ΔH_P	enthalpy change due to heating of product
ΔH_R	enthalpy change due to cooling of reactant
η_{CG}	cold gas efficiency
η_{HG}	hot gas efficiency

λ	stoichiometric ratio defined in Equation (10.23)
λ_j	Lagrange multiplier
μ_i	chemical potential of component i
Π	equivalence ratio defined in Equation (10.26)

Chapter 11

Symbols

a to f	exponent or power of concentrations in kinetic reaction rate equations; see Section 11.4
A_M	membrane area
A^-	conjugate base of the acid HA
B	extractant species
B-HA	complex formed between extractant B and acid HA
C^*	concentration or solubility of a solute at saturation
C_i	concentration of component i
C_B	total concentration of extractant B in Chapter 11
C_C	cellulose concentration
C_{CO_2}	total dissolved CO_2 concentration
$C_{CO_2,carb}$	CO_2 concentration due to dissolution of carbonate salts
C_F	cell biomass concentration
C_{FA}	formic acid concentration
C_F^{max}	maximum attainable biomass concentration in fermentation
C_h	humins concentration
$C_{F,i}$	concentration of component i in the feed
C_G	glucose concentration
C_{H+}	concentration of hydrogen cations in Equations (11.25) and (11.26)
$C_{H_2SO_4}$	sulfuric acid concentration
$C_{HA(aq)}$	total concentration of an acid HA in the aqueous phase
$C_{HA(org)}$	total concentration of an acid HA in the organic phase
C_{HMF}	5-hydroxymethylfurfural concentration
C_{in}	inlet concentration
$C_{in,i}$	inlet concentration of component i
C_{LA}	levulinic acid concentration
C_{out}	outlet concentration
$C_{out,i}$	outlet concentration of component i
C_P	product concentration, final product concentration in Equation (11.6)
$C_{P,i}$	concentration of component i in the permeate
C_{P0}	initial product concentration
C_P^{max}	maximum product concentration above which bacteria do not grow
$C_{R,i}$	concentration of component i in the retentate
C_{SA}	succinic acid concentration
D	flow rate of organic phase feed (diluent only)

D_{HA}	diffusivity of lactic acid in the diluent
$e(\%)$	recovery or extraction percentage
E	flow rate of extract (diluent only)
E_a	activation energy
f'	dimensionless parameter related to the inhibitory effect of cell biomass concentration
F_i	flow rate of a stream or component i
F_{in}	inlet flow rate in Exercise 12 of Chapter 11
F_L	liquid flow rate in Exercise 12 of Chapter 11
F_v	vapor flow rate in Exercise 12 of Chapter 11
g	exponent or power of the catalytic activity of the acid in Equation (11.25)
G_m	mean crystal growth rate
H	Henry's constant in fermentation broth
H'	Henry's constant in pure water = 4320 kPa L mol^{-1}
HA	acid species
h_v	latent heat of vaporization of water
$[H^+]$	concentration of H$^+$
$[HA]_{0,aq}$	initial acid concentration in the aqueous phase
$[i]$	concentration of species i
J_i	flux or rate of transport of a component i across a membrane
k_1	first order rate constant in Equation (11.82)
k_i	reaction constant of reaction i
k_o	reaction constant at the reference temperature
K_{1,H_2CO_3}	first dissociation constant for carbonic acid, including the equilibrium between CO$_2$ and H$_2$CO$_3$
K_{2,H_2CO_3}	second dissociation constant for carbonic acid, corresponding to the dissociation of HCO$_3^-$
K_a	equilibrium constant corresponding to the ionization of an acid
K_{a2}	second dissociation constant of H$_2$SO$_4$
K_c	equilibrium constant for complex formation
$K_{c,n}$	equilibrium constant for complex formation involving n molecules of solute
K_d	distribution coefficient or equilibrium constant in extraction process
$K_{d,overall}$	overall distribution coefficient
K_{dim}	equilibrium constant of dimerization reaction
K_S	saturation constant in Monod kinetic equation (Equation (1.1))
L_m	mean crystal size
m_S	substrate maintenance coefficient
m_{SA}	mass flow rate of succinic acid crystals
n	number of ions formed when the solute molecule dissociates, e.g., $n = 2$ for NaCl
N_s	number of moles of solute
p'	dimensionless parameter related to the inhibitory effect of product concentration
p_{CO_2}	partial pressure of CO$_2$
p_i	partial pressure of component i
pK_a	equal to $-\log K_a$
$P_{M,i}$	permeability of component i
$\bar{P}_{M,i}$	permeance of component i

Q	heat flow rate in Equations (11.93) and (11.94) and Figure 11.40
Q	ion exchange equilibrium quotient in Equation (11.61)
Q_F	volumetric flow rate of the feed
Q_P	volumetric flow rate of the permeate
Q_R	volumetric flow rate of the retentate
r_i	rate of reaction i
R	universal gas constant in Equations (11.25) and (11.58)
R	flow rate of raffinate (water only) in Exercise 10, Chapter 11
RX	resin R with ion X bond in Equations (11.60) and (11.61)
RZ	resin R with ion Z bond in Equations (11.60) and (11.61)
S_G	glucose selectivity
t	time
T_o	reference temperature
V	reactor volume
V	volume of pure solvent in Equation (11.58)
W	flow rate of aqueous phase feed (water only)
x	concentration in the aqueous phase in Equation (11.62); see also Exercise 10
x_R	solute concentration in raffinate, kg solute kg^{-1} water
x_W	solute concentration in aqueous feed, kg solute kg^{-1} water
X_i	conversion of reactant i
$X^{\pm}$	ion originally bound to an ion exchange resin
y	concentration in the organic phase in Equation (11.62)
y_D	solute concentration in organic feed, kg solute kg^{-1} diluent
y_E	concentration of solute in the extract, kg solute kg^{-1} diluent
Y_P	dimensionless product yield coefficient
Y_{SA}	succinic acid production rate
$Z^{\pm}$	ion to be recovered by the ion exchange resin

Greek letters

α	growth-associated product formation coefficient in Equation (11.4)
$\alpha 1, \alpha 2$	orders of reaction in Exercise 8 of Chapter 11
$\alpha i,j$	membrane separation factor or selectivity
β	non-growth-associated product formation coefficient in Equation (11.4)
δ_M	membrane thickness
ΔP	pressure difference generated across a membrane in Section 11.5.2
$\Delta \pi$	pressure difference between the osmotic pressure of the feed solution and that of the permeate solution
θ	cut fraction, the fraction of feed permeated in a membrane
μ	specific growth rate
μ_{max}	maximum achievable growth rate
π	osmotic pressure
τ	space–time or residence time

Chapter 12

Symbols

a	fraction of propagating chains terminated by combination
$[A]$	concentration or activity of carboxyl groups (—COOH) in Equations (12.31) to (12.37)
$[A]_0, [B]_0$	initial concentrations of A and B, respectively, at time $t = 0$
$[B]$	concentration of hydroxyl groups (—OH) in Equations (12.31) to (12.37)
$\bar{C}$	dimensionless concentration quantity
$C_{A,f}$	concentration in the bulk fluid
$C_{A,s}$	concentration of reactant A at the particle external surface
$C_{CL,1}$	concentration of caprolactam in the outlet stream from CSTR 1 to CSTR 2
$C_{CL,2}$	concentration of caprolactam in the outlet from CSTR 2
$C_{CL,3}$	concentration of caprolactam at certain point in the PFR section of the VK column
$C_{CL,R'}$	concentration of caprolactam in the reflux stream
C_{DB}	concentration of double bonds in the oil
C_i	concentration of species or component i
C_{in}	concentration in the inlet stream
$C_{in,A}$	concentration of A in the inlet stream
$C_{in,CL}$	concentration of caprolactam in the inlet stream
$C_{in,S}$	concentration of substrate in the inlet stream
$C_{in,W}$	concentration of water in the inlet stream
C_{oil}	concentration of oil
C_{out}	concentration in the outlet stream
C_p	specific heat capacity of reaction mixture
$C_{W,1}$	concentration of water in the outlet from CSTR 1 to CSTR 2
$C_{W,2}$	concentration of water in the outlet from CSTR 2 to PFR
$C_{W,3}$	concentration of water at certain point in the PFR section of the VK column
$C_{W,R'}$	concentration of water in the reflux stream
d_p	catalyst particle diameter
$D_{A,fluid}$	diffusivity of reactant A in the bulk fluid
D_{eff}	effective diffusivity
E	elimination product from step-growth polymerization in Equations (12.26) to (12.29)
E	total energy in Section 12.4.4
E_{in}	inlet energy in units of (energy) (mass)$^{-1}$
E_{out}	outlet energy in units of (energy) (mass)$^{-1}$
f	initiator efficiency in Equation (12.8)
F	flow rate
F_1	flow rate from the CSTR 1 to CSTR 2
F_2	outlet flow rate from the CSTR 2 to PFR
F_3	mass flow rate of the polymer product leaving the PFR section of the VK column
F_{in}	feed flow rate to the VK column
F_{v1}	vapor flow rate from CSTR 1 to the top condenser

F_{v2}	vapor flow rate from CSTR 2 to CSTR 1
H	enthalpy in Section 12.4.4
$\bar{H}_i$	partial molar enthalpy of species i in Section 12.4.4
$[i]$	concentration of species i
I	initiator or catalyst species in a polymerization reaction in Section 12.3.1
I_0, I_1	Bessel functions of order 1 and order zero, respectively, in Equation (12.103)
k'	overall polymerization reaction constant including the effect of the catalyst in Equations (12.31) to (12.37)
k	reaction rate constant
k_1	forward reaction rate constant in Equations (12.28) and (12.29)
k_{-1}	backward reaction rate constant in Equations (12.28) and (12.29)
k_d	rate constant of decomposition of an initiator
k_i	rate constant of initiation reaction
k_{ij}	reaction constant of reaction between species from monomer i and species from monomer j; see Section 12.3.3.
k_m	mass transfer coefficient
k_p	rate constant of propagation reaction
k_t	overall rate constant of termination reaction
k_{tc}	rate constant of a termination reaction by combination
k_{td}	reaction constant of a termination reaction by disproportionation
K	polymer-specific constant in Equation (12.3)
K	kinetic energy in Equation (12.109)
K_{eq}	equilibrium constant
l	kinetic chain length
L	characteristic length of a catalyst particle
L'	mass flow rate of the liquid from the top condenser
m_{in}	inlet mass flow rate
m_{out}	outlet mass flow rate
M_0	molar mass of the repeat unit of a polymer
$M, M1, M2$	monomer molecules
M^*	chain-initiating radical of monomer molecule M
M_n^*, M_m^*	monomer growing chain radicals
M_n	number average molar mass
M_{pol}	molar mass of a polymer in the Mark–Houwink equation
M_w	weight-average molar mass
n	polymer chain length indicating the number of repeat units in a polymer molecule
n	order of reaction
nc	number of components in Exercise 11 of Chapter 12
n_i	number of moles of species i
$n_{in,T}$	total inlet molar flow rate
$n_{in,i}$	inlet molar flow rate of component i
n_r	total number of reactions
N	number of repeat units in the polymeric molecules coming from the initial monomer molecules
N_0	initial number of monomer molecules in Equation (12.35)

N_A	molar flux of species A
$N_{DB,i}$	number of double bonds in the oil component i
N_i	number of moles of polymer specie i with a molar mass M_i
P	polymer molecule in Section 12.3
P	total gas pressure in Section 12.4.1
PE	potential energy in Equation (12.109), Section 12.4.4
p_i	partial pressure of component i
P_n, P_m	growing polymer molecules
Q	heat flow in units of [energy] [time]$^{-1}$ in Section 12.4.4
r	radius at certain point in the catalyst particle in Section 12.4.3
$\bar{r}$	dimensionless radius quantity
r_A	reaction rate of A
$r_{A,cat}$	overall consumption rate of A in the catalyst particle
r_{CL}	rate of consumption of caprolactam
r_{epox}	reaction rate of epoxidation of vegetable oil in Exercise 11 of Chapter 12
r_i	reaction rate of consumption or production of i
r_{ij}	reaction rate of the reaction between species from monomer i with species from monomer j; see Section 12.3.3
r_{is}	the rate of initiation step reaction in Equation (12.8)
r_k	rate of reaction k
r_{pol}	rate of polymerization
r_t	reaction rate of termination reaction
r_W	reaction rate of water
R	radius of a catalyst particle
R'	reflux rate from the top condenser
$R*$	free radical from decomposition of an initiator in addition polymerization reaction
S, S_i	selectivity of species i
T	temperature
T_g	glass transition temperature
T_g^∞	glass transition temperature of a hypothetical polymer having an infinite molecular weight
T_{in}	inlet temperature
u	empirical parameter in the Mark–Houwink equation
U	internal energy in Section 12.4.4
U_h	overall heat transfer coefficient in Exercise 11 of Chapter 12
v	transformation variable for $\bar{C}$; see Equation (12.89)
v_i	stoichiometric coefficient of reactant or product i in Equation (12.59)
$v_{i,k}$	stoichiometric coefficient of component i in reaction k in Equation (12.122)
V	reactor volume
V_j	volume of section j of the VK column
W	work in units of (energy) (time)$^{-1}$ in Section 12.4.4
W_c	catalyst weight in Equation (12.100)
W_f	work done by the flow of the streams
W_s	shaft work applied to a system (e.g., by stirrers)

W_v	work due to change in volume
X, X_i	conversion of species or component i
x	in polymer structures indicate the number of repeat units in a polymer molecule
x	mass fraction of a component in polymer blend in Exercise 3 of Chapter 12
x_1	weight fraction of a single polymer with glass transition temperature T_{g1} in Equation (12.4)
x_i	molar composition of the oil component i
y	in polymer structures indicate the number of repeat units in a polymer molecule
y	transformation variable for integration, see Equation (12.94)
y_i	molar fraction of component i in gas phase
$y_{CL,1}$	caprolactam fraction in the vapour flow from CSTR 1 to the top condenser
$y_{CL,2}$	caprolactam fraction in the vapour flow from CSTR 2 to CSTR 2
$y_{in,i}$	molar fraction of i in the inlet
$y_{W,1}$	caprolactam fraction in the vapour flow from CSTR 1 to the top condenser
$y_{W,2}$	water fraction in the vapour flow from CSTR 2 to CSTR 1

Greek letters

α	coefficient of expansion of the reaction mixture in Section 12.4.4
δ	thickness of the fluid film around a catalyst particle
Δn	difference in stoichiometric coefficients defined in Equation (12.59)
Δr	infinitesimal radius increment from radius r
η	effectiveness factor defined in Equation (12.96)
Θ	empirical parameter in Mark–Houwink equation
Λ	intrinsic viscosity in Mark–Houwink equation
ρ	density of reaction mixture
ρ_b	density of the catalyst bed
ρ_p	density of the catalyst particle
Π	dimensionless Thiele modulus

Chapter 13

Symbols

A	absorption factor
A	correlation constant in Antoine equation
b_{CO_2}	Langmuir model constant for CO_2 adsorption
B, C	correlation coefficients in Antoine equation
C_{ik}^f	feed concentration of species i in column k
C_{ik}	concentration of species i in column k
C_i^{ic}	initial concentration of species i
C_p	heat capacity of a refrigerant at constant pressure
Cp_g	heat capacity of gas
Cp_s	heat capacity of solid

C_v	heat capacity of a refrigerant at constant volume
d_p	diameter of particle
d_t	diameter of tube
D_L	axial diffusion coefficient
D_m	molecular diffusivity
D_p	diffusivity inside particle
E	stage efficiency
E_j^{act}	energy of activation in Arrhenius equation
G	mass (or molar) flow rate of insoluble gas
G_{in}	inlet mass (or molar) flow rate of insoluble gas
G_{out}	outlet mass (or molar) flow rate of insoluble gas
$k_{CO_2}^{ads}$	mass transfer coefficient of CO_2 in adsorbed phase
keq_i	equilibrium constant in Arrhenius equation
kit_j	pre-exponential factor of reaction j
K	equilibrium constant
L	mass (or molar) flow rate of nonvolatile solvent
L_{in}	inlet mass (or molar) flow rate of nonvolatile solvent
L_{out}	outlet mass (or molar) flow rate of nonvolatile solvent
m_{CO_2}	Langmuir model coefficient for CO_2 adsorption
MW_L	molecular weight of liquid
n_s	isentropic efficiency of compressor
N	theoretical number of stages
P_{CO_2}	partial pressure of CO_2 in the gas phase
P^{cond}	saturation vapor pressure for condensation of a refrigerant
P^{evap}	saturation vapor pressure for evaporation of a refrigerant
P^{exit}	exit pressure of turbine or compressor
P^{inlet}	inlet pressure to turbine or compressor
P^{sat}	saturation vapor pressure
$\bar{q}_{CO_2}$	concentration of CO_2 in adsorbed phase
$q_{CO_2}^*$	solid phase concentration of CO_2 in equilibrium
r_p	particle radius
R	molar % of carbon dioxide removal
S	solvent to feed gas molar flow ratio
T^{cond}	saturation temperature for condensation of a refrigerant at P^{cond}
T^{evap}	saturation temperature for evaporation of a refrigerant at P^{evap}
T^f	feed temperature
T^{inlet}	inlet temperature to turbine or compressor
T^{sat}	saturation temperature
T_w	wall temperature
u	linear axial velocity
U	overall heat transfer coefficient
v_{ij}	stoichiometric coefficient of species i in jth reaction
$W^{compression}$	work required to drive a compressor

$W^{expansion}$	work generated from an expander
X_{in}	mass (or moles) of solute (absorbed component) in the inlet nonvolatile solvent per inlet mass (or molar) flow rate of nonvolatile solvent
X_{out}	mass (or moles) of solute (absorbed component) in the outlet nonvolatile solvent per outlet mass (or molar) flow rate of nonvolatile solvent
Y_{in}	mass (or moles) of solute (absorbing component) in the inlet insoluble gas per inlet mass (or molar) flow rate of insoluble gas
Y_{out}	mass (or moles) of solute (absorbing component) in the outlet insoluble gas per outlet mass (or molar) flow rate of insoluble gas

Greek letters

$\gamma = C_p/C_v$	heat capacity ratio
$\Delta H^{ads}_{CO_2}$	heat of adsorption of CO_2
ΔH^R_j	heat of reaction in reaction j
ε	porosity of catalyst bed
ε_p	particle void fraction
η_j	effectiveness factor of jth reaction
λ_z	thermal conductivity
μ_L	viscosity of liquid
ρ_L	density of liquid
ρ_{ads}	density of adsorbent
ρ_{cat}	density of catalyst
ρ_p	density of particle

Chapter 16

Symbols

A	area of the cell for ionic and electronic transfer
F	Faraday's constant, $96\,485$ C mol^{-1} (C is coulomb)
i	current density
$i_{0,anode/cathode}$	anode/cathode exchange current density
$i_{l,anode/cathode}$	limiting current density of the anode/cathode
L_k	thickness, $\forall k \in \{\text{electrolyte, anode, cathode, interconnect}\}$
n_{air-in}	inlet molar flow rate of air to the cell
n_e	number of electrons transferred in reactions
$n_{exhaust}(j)$	molar flow rate of component j in the exhaust gas from the combustor, $\forall j \in \{H_2O, CO_2, N_2, O_2\}$
n_{in}	molar flow rate of fuel intake to the cell
$n_{in}(j)$	molar flow rate of fuel j to the cell, $j \in \{H_2, CO, CH_4, H_2O, CO_2\}$
$n_{out}(j)$	molar outlet flow rate of component j from the cell, $j \in \{CO_2, H_2, H_2O, N_2, O_2\}$
P_j	partial pressure of component j
U_{air}	utilization factor of inlet oxygen on molar basis

U_f	fuel utilization factor on molar basis
$V_{activation,anode/cathode}$	voltage loss due to activation overpotentials in anode/cathode
$V_{concentration,anode/cathode}$	voltage loss due to concentration overpotentials in anode/cathode
V_{ohmic}	ohmic overpotentials
$W_{fuel\ cell}$	net power generation from SOFC
x_j	molar fraction of fuel j intake to the cell, $j \in \{H_2, CO, CH_4, H_2O, CO_2\}$

Greek letters

$\Delta g^o (T)$	molar Gibbs energy change as a function of T at a pressure of 1 atm
$\Delta g (T, P)$	molar Gibbs energy change as a function of temperature (T) and pressure (P)
σ_k	electronic or ionic conductivity in k, $\forall k \in \{$electrolyte, anode, cathode, interconnect$\}$
$\omega_{air\ blower}$	ratio of power consumption by air blower to power generation by fuel cell

Chapter 17

Symbols

a	parameter relating the photon flux with the C fixation per amount of chlorophyll, g C (mol^{-1} photons) m^2 g^{-1} chlorophyll, in Equation (17.4)
A	area
A_p	cross-sectional area of a particle in the direction of flow in Equation (17.35)
$C_{alg}F\big|_z$	algae biomass flow rate entering the element of volume ΔV at length z
$C_{alg}F\big|_{z+\Delta z}$	algae biomass flow rate leaving the element of volume ΔV at length $z+\Delta z$
C_{alg}	algae biomass concentration
C_D	drag coefficient in Equation (17.35)
$C_{in,alg}$	algae concentration in the inlet liquid stream
C_j	algae biomass concentration in the stream $j = 1,2$ in Exercise 10 of Chapter 17
$C_{out,alg}$	algae concentration in the outlet liquid stream
$[CO_2]$	concentration of CO_2 dissolved in the liquid medium
$CWAP$	sum of the costs per kg of wet algae paste
d_p	diameter of a settling particle in Equation (17.37)
D	dilution factor in Equation (17.12)
E_s	specific centrifugation energy consumption
$f(CO_2)$	function describing the influence of CO_2
$f(I)$	function describing the influence of light irradiation
$f(N)$	function describing the influence of nutrient concentration
$f(T)$	function describing the influence of the temperature
F	flow rate
F_c	pond circulation flow rate
F_{in}	inlet liquid flow rate
F_{out}	outlet liquid flow rate

g	gravitational acceleration in Equation (17.35)
G_{in}	inlet gas flow rate
G_{out}	outlet gas flow rate
h	fraction of area actually occupied by water
I	light irradiance, W m^{-2}, or photon flux (mol photons) m^{-2} d^{-1}, in Equation (17.3)
I_{op}	optimum irradiance for maximum algae
I_s	light irradiance on the pond liquid surface
$I(z)$	light irradiance as function of pond depth
K	empirical constant in Equation (17.5)
K_{CO2}	CO_2 saturation constant
K_e	light extinction coefficient in Equation (17.15)
K_{e1}	correlation constant in Equation (17.15)
K_{e2}	correlation constant in Equation (17.15)
K_I	saturation constant for light irradiance
m_{paste}	mass of algae paste
P_{AS}	price of aluminum sulfate
$P_{C,max}$	maximum carbon-specific photosynthesis rate at light saturation
P_E	price of electricity
r	radius
r_{alg}	algae biomass production rate
r_m	maintenance metabolic coefficient equal to the respiration rate at $\mu = 0$
R	alkyl chain in triglyceride in Figure 17.10
R'	alkyl chain in alcohol in Figure 17.10
Re	dimensionless Reynolds number
$\% R_{CFG}$	algae biomass recovery in centrifugation
$\%R_{C-F}$	recovery of biomass in coagulation–flocculation process
s	fraction of particles that settles and can be recovered in sedimentation process
T	temperature
T_d	temperature limit for algae growth
T_{op}	optimal temperature for maximum algae growth
v_c	critical velocity
v_p	particle velocity
v_s	settling velocity in Equation (17.35)
V_p	volume of a settling particle in Equation (17.35)
V_{pond}	pond volume
V_{PBR}	photobioreactor volume
WSI	water stress index
x_C	constant carbon content in algae biomass in solution to Exercise 6
x_{alg}	algae mass fraction in the paste
$x_{in,C}$	carbon content in the algae biomass in the inlet stream (mol C kg^{-1} biomass)
$x_{out,C}$	carbon content in the algae biomass in the outlet stream (mol C kg^{-1} biomass)
Y_{alg}	algae biomass yield per land area used, kg m^{-2} d^{-1}
$y_{in,CO2}$	CO_2 mole fraction in the inlet gas stream

$y_{in,CO2}$	CO_2 mole fraction in the outlet gas stream
z	depth

Greek letters

β	correlation parameter in Equation (17.6)
θ	mass ratio of chlorophyll to carbon, g chlorophyll g^{-1} C
θ_{max}	maximum mass ratio of chlorophyll to carbon, g chlorophyll g^{-1} C
μ	specific growth or algal biomass production in Chapter 17
$\mu_{(light)}$	specific algal biomass production as a function of light irradiance
$\mu_{(T),max}$	maximum specific growth rate at optimum temperature
μ_f	viscosity of the fluid medium in Equation (17.37)
μ_{max}	maximum algal production under optimum conditions
ρ_f	density of the fluid medium in Equation (17.35)
ρ_p	density of a settling particle in Equation (17.35)
τ	residence time in Section 17.1.2

Chapter 18

Symbols

C_i	concentration of species i in the bulk
C_i^p	concentration of species i inside catalyst particle
C_i^{pore}	concentration of species i inside pore, inside a catalyst particle
D_L	axial diffusion coefficient
d_i^{eff}	diffusivity of species i inside particle
$Ð_i^{eff}$	effective molecular diffusivity of species i, cm^2 s^{-1}
D_{ij}	axial diffusion coefficient
k_j, $k(j)$	rate constant of reaction j
K_j, $K_{eq}(j)$, $keq(j)$	equilibrium constant for reaction j
t	time
T	temperature
u	velocity

Greek letters

ε	bed void fraction
ε_p	void fraction inside particle
τ	tortuosity
σ	constriction factor
μ	viscosity

List of Abbreviations

aq	aqueous phase (subscript)
ABE	acetone–butanol–ethanol
AD	anaerobic digestion
AFC	alkaline fuel cell
AP	acidification potential
APR	annual percentage rate
ASF	Anderson–Schulz–Flory distribution model for FT (Fischer–Tropsch) reaction
ASU	air separation unit
BCL	Batelle Columbus
BDO	1,4-butanediol
BeFS	best first search
BES	bioelectrochemical system
BFS	breadth first search
BFW	boiler feed water
BGCC	biomass gasification combined cycle
BGFC	biomass gasification fuel cell
BOD	biological oxygen demand
BOIG	bio-oil integrated gasification
BTL	biomass to liquid
BTX	benzene, toluene and xylenes
BUWAL	Bundesamt für Umwelt, Wald und Landschaft
CAPEX	capital cost
CCS	carbon capture and storage
CEPCI	Chemical Engineering's Plant Cost Index
CFC	chlorofluorocarbon
CFP	catalytic fast pyrolysis
CHP	combined heat and power
CL	caprolactam
CLC	chemical looping combustion
CML	Institute of Environmental Sciences, Leiden University, Netherlands
COP	cost of production
CPE	cumulative fossil primary energy
CSTR	continuous stirred tank reactor
CTL	coal-to-liquid
CTUh	comparative toxic unit for human
CVP	credit value on processing
DCB	1,4-dichlorobenzene
DCF	discounted cash flow
DCFRR	discounted cash flow rate of return
DCO	decarboxylation
DDGS	dried distillers grains with solubles

DEA	diethanolamine
DEAE	diethylaminoethyl
DES	diethyl succinate
DFS	depth first search
DMEPEG	dimethyl ethers of polyethylene glycol
DMSO	dimethyl sulfoxide
DOE	US Department of Energy
DP	degree of polymerization
DP	degree of polymerization
DPC	direct production costs
ED	electrodialysis
EDBM	electrodialysis with bipolar membrane
EDF	electrodialysis fermentation
EI	environmental impact
ELCD	European Reference Life Cycle Database
EM	economic margin
EMI	Equated Monthly Instalment
ENR	Engineering New-Record Construction Cost Index
EP	eutrophication potential
EP	economic potential in Chapter 2
ER	Eley–Rideal
ETP	effluent treatment plant
EU	European Union
f.o.b.	free-on-board
FAETP	freshwater ecotoxicity potential
FAME	fatty acid methyl ester
FCC	fluid catalytic cracking
FDA	US Food and Drug Administration
FT	Fischer–Tropsch
GA	genetic algorithm
GBL	γ-butyrolactone
GHG	greenhouse gas
GRG	generalized reduced gradient
GT	gas turbine
GTL	gas-to-liquid
GVL	γ-valerolactone
GWP	global warming potential or global warming potential 100-year horizon
h	humins (subscript)
H/C	hydrogen-to-carbon atomic ratio
HC	hydrocracking
HDO	hydrodeoxygenation
HDPE	high density polyethylene
HETP	height equivalent of a theoretical plate

HHV	higher heating value
HMF	hydroxymethylfurfural
5-HMF	5-hydroxymethylfurfural
HOAc	acetic acid
HP	high pressure (used to indicate steam pressure)
HRSG	heat recovery steam generator
HT	hydrotreating in Chapter 15
HTFT	high temperature Fischer–Tropsch
HTP	human toxicity potential
HTWGS	high temperature water gas shift reactor
HV	heating value
ICP	impact cost of production
ICP_{limit}	ICP limit for EI saving target
IBGCC	integrated biomass gasification combined cycle
IEA	International Energy Agency
IGCC	integrated gasification combined cycle
IGT	Institute of Gas Technology
ILCD	International Reference Life Cycle Data System
int	interface (subscript)
IPCC	Intergovernmental Panel on Climate Change
IRR	internal rate of return
ISBL	inside battery limits
ISO	International Organization for Standardization
KKT	Karush–Kuhn–Tucker *necessary* and *sufficient* conditions for a solution to be optimal
LA	levulinic acid
LCA	life cycle assessment
LCC	life cycle cost
LCI	life cycle inventory
LCIA	life cycle impact assessment
LCT	life cycle thinking
LDPE	low density polyethylene
LEGS	lime enhanced gasification sorption
LHHW	Langmuir–Hinshelwood–Hougen–Watson
LHSV	liquid hourly space velocity
LHV	lower heating value
LNG	liquefied natural gas
LP	linear programming in Chapter 8 and in optimization contexts
LP	low pressure (used to indicate steam pressure)
LTFT	low temperature Fischer–Tropsch
LTWGS	low temperature water gas shift reactor
M&S	Marshall and Swift Equipment Cost Index
MAETP	marine aquatic ecotoxicity potential
MCLCA	Monte Carlo simulation combined with LCA

MDEA	methyldiethanolamine
MEA	monoethanolamine
MEC	microbial electrolysis cell
MeOAc	methyl acetate
MeOH	methanol
MFC	microbial fuel cell
MIBK	methyl isobutyl ketone
MIEC	mixed ionic electronic conducting
MILP	mixed integer linear programming
MINLP	mixed integer nonlinear programming
MP	medium pressure (used to indicate steam pressure)
MTBE	methyl *tert*-butyl ether
MTO	methanol-to-olefins (process)
NF	Nelson–Farrar Refinery Construction Cost Index
NLP	nonlinear programming
NMR	nuclear magnetic resonance
NMVOC	non-methane volatile organic compounds
NO_x	nitrogen oxides
NPV	net present value
NREL	US National Renewable Energy Laboratory
NRTL	nonrandom two liquid
NSGA	nondominated sorting genetic algorithm
O/C	oxygen-to-carbon atomic ratio
ODP	ozone depletion potential
OECD	Organization for Economic Co-operation and Development
OPEX	operating cost
OPV	organic photovoltaic
org	organic phase (subscript)
OSBL	outside battery limits
PA	polyamides
PAN	peroxyacetyl nitrate
PBAT	poly(butylene adipate terephthalate)
PBR	photobioreactor
PBS	poly(butylene succinate)
PBSA	poly(butylene succinate adipate)
PBST	poly(butylene succinate terephthalate)
PBT	poly(butylene terephthalate)
PC	polycaprolactam
PCE	per capita equivalents
PCL	polycaprolactone
PE	polyethylene
PEAA	poly(ethylene-co-acrylic acid)
PEF	poly(ethylene furanoate)

PEMFC	proton exchange membrane fuel cell
PET	poly(ethylene terephthalate)
PFR	plug flow reactor
PHAs	polyhydroxyalkanoates
PHB	polyhydroxybutyrate
PHV	polyhydroxyvalerate
PLA	poly(lactic acid)
PMT	financial function in Excel
PNNL	Pacific Northwest National Laboratory
POCP	photochemical ozone creation potential or photochemical oxidant creation potential
POX	partial oxidation
PP	polypropylene
PPC	poly(propylene carbonate)
PS	polystyrene
PSA	pressure swing adsorption
PTT	poly(trimethylene terephthalate)
PURs	polyurethanes
PVC	polyvinyl chloride
PWBs	printed wiring boards
R^2	coefficient of determination equal to $1 - SSQ$/summation of square (SSQ) of variance
RDF	refuse-derived fuel
RO	reverse osmosis
ROI	return on investment
RSSQ	residual sum of square errors
SA	succinic acid
SCF	simple cash flow
SESMR	sorption enhanced steam methane reforming
SETAC	Society of Environmental Toxicology and Chemistry
SEWGS	sorption enhanced water gas shift reactor
SLCA	social LCA (life cycle assessment)
SMR	steam methane reforming
SOFC	solid oxide fuel cell
SO_x	sulfur oxides
SSQ	summation of square
TETP	terrestrial ecotoxicity potential
Tetraglyme	tetraethylene glycol dimethyl ether
THF	tetrahydrofuran
TPS	thermoplastic starch
TS	total solids
TSA	temperature swing adsorption
UNEP	United Nations Environmental Programme
US	United States of America
VK	Vereinfacht Kontinuierliches Rohr (column reactor)

VOC	volatile organic compound
VOP	value on processing
VS	volatile solids
WGS	water gas shift reactor/reaction
WHSV	weight hourly space velocity

Part I
Introduction

1

Introduction

Much has been learnt about the detrimental effects of finite fossil resources on the environment, society and economy, making their current exploitation to satisfy human needs unsustainable. This has renewed the surge of biomass as a low carbon source of energy, combined heat and power (CHP), liquid transportation fuels known as biofuels, gaseous fuels, chemicals (commodity and specialty) and materials (polymers and elements). The complex site configurations arising from integration between biomass feedstocks, processes and products are known as biorefineries. Biorefineries have brought opportunities of using biomass feedstocks in an efficient way to ensure benefits to the environment and society as well as long-term economic viability against fossil based counterparts. The call for cost-effective and sustainable production of energy, chemical and material products from biomass gives light for the conception of biorefineries. In the most advanced sense, a biorefinery is a facility with integrated, efficient and flexible conversion of biomass feedstocks, through a combination of physical, chemical, biochemical and thermochemical processes, into multiple products. The concept was developed by analogy to the complex crude oil refineries adopting the process engineering principles applied in their designs, such as feedstock fractionation, multiple value-added productions, process flexibility and integration. For sustainable biorefinery design, the nature and range of alternatives for feedstocks, process technologies, intermediate platforms and products are important to know. In this chapter, the fundamental features and principles of biorefinery configurations are introduced alongside some research problems, concepts and tools to assess the sustainability of biorefineries.

1.1 Fundamentals of the Biorefinery Concept

1.1.1 Biorefinery Principles

The generation of products from biomass is not new. Biorefinery is a concept created for efficient processing of biomass coming from plant, animal and food wastes into energy, fuels, chemicals, polymers, food additives, etc. There are several definitions for biorefineries emphasizing the key elements of sustainability, integration and multiple value-added productions. The biorefinery concept has been developed by analogy to crude oil refineries. Biorefining must embrace the process engineering principles applied to crude oil refining for their successful development.

Table 1.1 shows the processing principles used in modern crude oil refineries and their adoption in the concept of biorefineries. As in petroleum refineries, biorefineries must follow the strategy of feedstock separation into more useful and treatable fractions, known as *platforms* or *precursors*. Then each fraction must create a production line to diversify their product slate and to increase profit and adaptability for low carbon pathways. A combination of various high throughput technologies allows conversion of the whole ton of biomass into commodity (e.g., biofuels, electricity) and specialty products (e.g., chemical building blocks replacing petrochemicals). These combinations create a complex system

Table 1.1   *Principles adopted in biorefinery from its analogy to a crude oil refinery.*

Oil Refinery	Biorefinery
Mature process technology (e.g., thermal and catalytic cracking, reforming, hydrotreatment)	Mature and innovative process technology (e.g., biomass gasification, pyrolysis, fermentation, anaerobic digestion, hydrocracking and bioseparations)
Use of every crude oil fraction	Use of every biomass fraction and components
Process flexibility and product diversification	Process flexibility and product diversification
Coproduction of valuable chemical building blocks	Coproduction of valuable and highly functionalized chemical building blocks
Cogeneration of heat and power	Cogeneration of heat and power
Process integration	Process integration and design for sustainability
Economy of scale	Scale according to biomass logistics but must be maximized to benefit from economy of scale

able to exchange material (waste streams, platforms and products) and energy streams to supply their requirements and achieve self-sufficiency. The complexity gives opportunities for process integration to increase energy efficiency, save water and reduce wastes and emissions that will contribute to the overall economic and environmental sustainability of the biorefinery. Thus, biorefinery design must be carried out by adopting process integration strategies and sustainability concepts in every stage.

The learning experience of the processing technologies, capability of processing various feedstocks, diversification of product portfolio, and application of process integration will make biorefineries into highly integrated, resource efficient, and flexible facilities. Although this is the final goal of the biorefinery concept for biomass processing, there is still a long way to go for biorefineries to reach such an advanced stage of development. New processing technologies and process engineering concepts are to be developed; there are barriers to overcome and lessons to learn in order to make the full biorefinery concept a reality.

1.1.2   Biorefinery Types and Development

The biorefinery principles have been practised to some degree in corn wet mill, pulp and paper and, more recently, biofuel plants by introducing additional production lines and process flexibility in the search for improved process economics. These facilities are considered as the precursors to biorefineries. Three types of biorefineries can be identified according to their phases of development defined by their degrees of complexity and flexibility, shown in Figure 1.1[1].

I.   Single feedstock, fixed process and no product diversification. Examples include dry-milling bioethanol plants using wheat or corn, and biodiesel plants using vegetable oils, which have no process flexibility and produce fixed amounts of fuels and coproducts.

II. Single feedstock, multiple, and flexible processes and product diversification. An example is a wet-milling plant using corn and various processes with the capability to adapt multiple productions depending on product demands and market prices.

III. Multiple, highly integrated and flexible processes allowing conversion of multiple feedstocks of a different or the same nature into a highly diverse portfolio of products. Flexible biorefineries will allow switching between feedstocks and blending of feedstocks for conversion into products.

The biorefinery Type III corresponds to the biorefinery concept in its broader extension. Various schemes for this type of biorefinery are under extensive research and development. The most workable one is the *lignocellulosic* feedstock based biorefinery for the processing of *agricultural residues, straw, wood; wastes such as sewage sludge and municipal solid wastes or refuse-derived fuels*, etc. Other developments include the two-platform biorefinery which combines biochemical with thermochemical processes and, more recently, the algae biorefinery. Materials and tools involved for sustainable designs of these processes are discussed in this book.

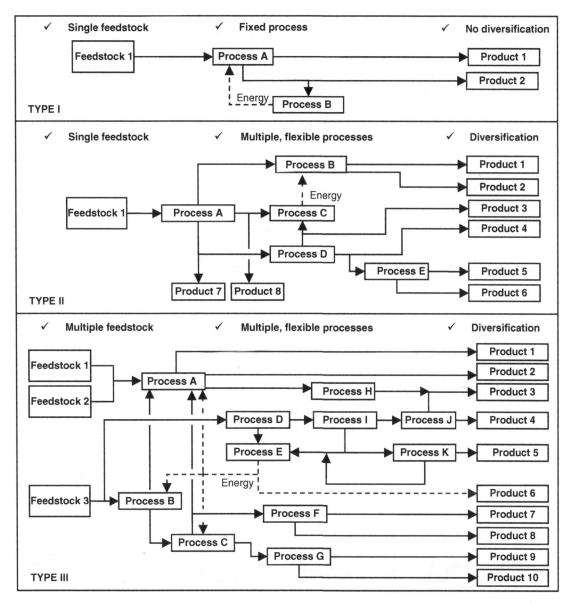

Figure 1.1 *Biorefinery types according to the phase of development. (Reproduced with permission from Kamm, Gruber, and Kamm (2006)[1]. Copyright © 2006 WILEY-VCH Verlag GmbH & Co. KGaA.)*

1.2 Biorefinery Features and Nomenclature

The basic features of a biorefinery can be grouped into *feedstock(s), processing technologies, platforms* and *products*, discussed in Sections 1.3 to 1.5. A biorefinery configuration is formed by a combination of at least one of each of those features. Thus, in a biorefinery, the biomass feedstock is first fractionated into *intermediate components* or *platforms*, which are further processed to produce a set of end products.

A systematic nomenclature would help to identify different biorefinery configurations. Attempts have been presented in the literature to name and classify biorefineries according to one of their features, that is, feedstock, platform or product. However, the lack of consistency in criteria can lead to ambiguity. A nomenclature system based on the four biorefinery basic features, accepted by the International Energy Agency (IEA) within the Bioenergy Task 42 "Biorefinery," is explained[2]. The features in a biorefinery configuration are identified and classified according to groups and subgroups in feedstocks, processes, platforms, and products. Then, the biorefinery is named following the structure:

Number of platforms (name(s)) + biorefinery for + products + from + feedstock (name(s))

In an example shown in Figure 1.2, the biorefinery has the following features: wheat as feedstock which is a starch crop, C6 is the platform followed in this case, and the products are bioethanol and animal feed (dried distillers grains with solubles, DDGS). This biorefinery system is then named as: one platform (C6 sugar) biorefinery for bioethanol and

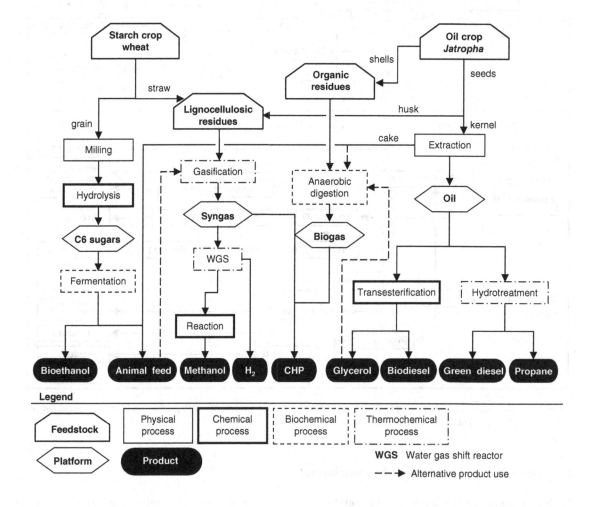

Figure 1.2   *Example of two biorefinery configurations. One platform (C6 sugar) biorefinery for bioethanol and animal feed from starch crops (wheat). One platform (oil) biorefinery for biodiesel, glycerol and animal feed from oil crops (Jatropha). Note that a syngas or biogas platform can also be integrated for the use of residues or cake.*

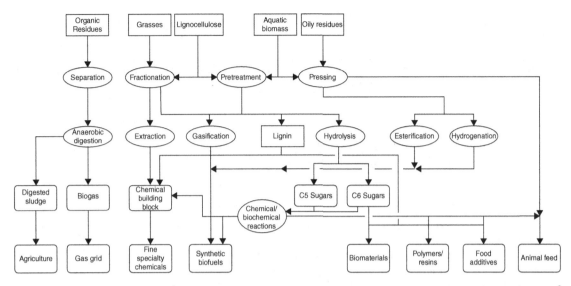

Figure 1.3   *Network of interlinked biorefinery configurations. (Reproduced with permission from Cherubini et al. (2009)[2]. Copyright © 2009 Society of Chemical Industry and John Wiley & Sons, Ltd.)*

animal feed from starch crops (wheat). For classification purposes, the biorefinery type corresponds to the description without the names between brackets, that is, "C6 sugar platform biorefinery for bioethanol and animal feed from starch crops." A biorefinery with the same process and products but using corn instead of wheat falls within this classification. As shown in Figure 1.2, a syngas or biogas platform can also be integrated for the use of residues or cake.

The various options for the biorefinery features forming alternative configurations can be combined into a major *superstructure* network for optimization. A network structure including all possible inter- and intraprocess connections is called a superstructure. A superstructure poses a great engineering and mathematical challenge to solve for optimal network configurations. Figure 1.3 shows a network including some possible configurations mentioned and their combinations for biorefineries[2]. It can be seen from this network that feedstocks can be converted to almost any platform in one or several steps. In addition, one particular product can be produced from different combinations of feedstocks and using different processing pathways. This facility can be integrated with utility systems to create a biorefinery design. The biorefinery system needs to be colocated with feedstock supply systems and synergistically integrated with product distribution systems to form a complete value chain.

Much research will have to go into the creation of a renewable future. It will be a shifting from a fossil energy era into a renewable energy era and into a biomass era – mainly for chemical and material production. Biomass is going to be the only renewable source of carbon. A timeline for the displacement of fossil fuel energy by biomass is shown in Figure 1.4. Current biofuel technologies include biodiesel and bioethanol production (known as *first generation biofuel*) from food crops. *Second generation biofuels* include lignocellulosic ethanol and biomass-to-liquid (BTL) fuels, which use the lignocellulosic crops such as miscanthus, wood, etc., rather than food crops as the feedstock. Advanced biorefineries can be lignocellulosic, green and multiplatform biorefineries. It is expected that renewable energy would widely and commercially be available on a large scale within the next few decades. The motivations and the rationale for this book are to help renewable energy deployment of substituting fossil resources, by training engineers in multidisciplinary areas.

1.3   Biorefinery Feedstock: Biomass

The biorefinery concept relies on the availability of lignocellulosic biomass as feedstock. The biomass feedstock is the starting point for planning and design of biorefineries. Its current and potential availability influences the scale and

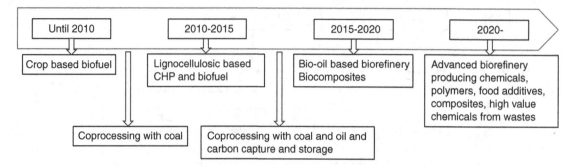

Figure 1.4   *Estimated timeline for biorefinery deployment.*

location of a biorefinery, whilst its nature decides the processes that can be used and the platforms and products that can be generated from it.

A comprehensive definition of *biomass*, including all the potential biomass as renewable feedstock, is: "*any organic matter that is available on a renewable or recurring basis, including dedicated energy crops and trees, agricultural residues, algae and aquatic plants, wood and wood residues, animal wastes, wastes from food and feed processing and other waste materials usable for the production of energy, fuels, chemicals, materials.*"[1] The biodegradable fraction of industrial and municipal waste is also included.

The biomass production phase reduces the global warming potential impact of the biorefinery products. Biomass from plants or algae or marine biomass is advantageous over fossil resources since they naturally capture CO_2 from the atmosphere during photosynthesis. For this reason, biomass is considered as carbon neutral or balanced feedstock and biorefinery is an effective way to alleviate climate change. However, food crops, i.e., corn, wheat, sugarcane, oil seeds, etc., to produce bioethanol and biodiesel have generated a sensible debate about deviating food and feed for biofuels and thus its socioeconomic consequences. As a response, alternative feedstocks are being explored, including lignocellulosic residues, *Jatropha curcas*, algae, etc. Lignocellulosic feedstocks are seen as promising options if the technological barriers for the breaking of their complex structure into functional products are overcome. Crop residues have the advantage to be cheap and do not need added land for production. However, lignocellulose processing pathways are yet to be fully developed to reduce the environmental footprint, by learning from the successes and failures with the first generation crops and fossil resources.

As the biorefinery processes and technologies reach a mature state of development and efficiency, the cost of biomass will dominate the economics of biorefineries. Cheap waste feedstocks, municipal waste, sludge and algae, etc., are essential to make the processes profitable and to deliver products that can be competitive with fossil based products. If the high crude oil prices trend remains, biomass processing may become economically attractive. A list of dedicated biomass feedstocks for biorefining is shown as follows.

1.   Agricultural and forestry residues and energy crops: wood, short rotation coppice, poplar, switchgrass and miscanthus.
2.   Grass: leaves, green plant materials, grass silage, empty fruit bunch, immature cereals.
3.   Oily crops and *Jatropha*.
4.   Oily residues: waste cooking oils and animal fat.
5.   Aquatic: algae and seaweed.
6.   Organic residues: municipal waste, manure, and sewage.

1.3.1   Chemical Nature of Biorefinery Feedstocks

In order to develop conversion processes for a biomass feedstock, it is important to understand its chemical nature. Biomass is composed of *cellulose, hemicellulose, lignin*, ash and a small amount of extractives with a wide range of chemical structures. The components of interest for biorefining are discussed as follows.

Table 1.2 *Chemical compositions of some lignocellulosic feedstocks*[3].

Feedstock	Cellulose	Hemicellulose	Lignin	Protein	Ash
Corn stover	26	38	23	5	6
Soybean straw	14	33	14	5	6
Wheat straw	29	38	24	5	6
Switchgrass	29	37	25	4	6
Miscanthus	24	43	19	3	6
Alfalfa	12	27	8	3	2
Sweet sorghum	14	23	11	17	9
Willow	49	14	20	–	5

Sugars. Sugar crops, for example, sugarcane and sugar beet, convert CO_2 and solar energy into the readily fermentable sugars: glucose and fructose ($C_6H_{12}O_6$, known as C6 sugars) mainly forming the disaccharide sucrose ($C_{12}H_{22}O_{11}$). Sugarcane in Brazil and sugar beet in the EU are used for bioethanol production.

Starch. Cereals, for example, corn, wheat, sorghum; roots and tubers, for example, cassava and sweet potato; store energy in the form of minute granules of starch ($(C_6H_{10}O_5)_n$). Starch is a polymer of glucose and consists of amylose and amylopectin as the main structural components, which can be easily broken down to fermentable C6 sugars by enzymatic hydrolysis. Wheat and corn are used as feedstock for bioethanol production in the EU and US, respectively. These cereals also contain protein, oil, and fibers that are recoverable as value-added products.

Lignocellulose. Lignocellulosic feedstocks include wood, grasses, and agricultural and forestry residues composed mainly of cellulose, hemicellulose and lignin forming a three-dimensional polymeric composite called lignocellulose. Table 1.2 provides typical chemical compositions for some lignocellulosic feedstocks. Cellulose is a polysaccharide having the generic formula $(C_6H_{10}O_6)_n$ yielding individual glucose monomers on hydrolysis. Hemicellulose is a macromolecular polysaccharide forming a mixture of straight and highly branched chains of both C5 and C6 sugars. Their hydrolysis produces the C6 sugars: glucose, mannose and galactose and the C5 sugars: xylose and arabinose. C5 hemicelluloses ($(C_5H_8O_4)_n$) include xylan, arabinan and mannan, and they can occur in large amounts (20 to 40%) in corncobs and corn stalks, straws and brans. The lignin fraction ($C_9H_{10}O_2(OCH_3)_n$) consists of complex phenolic polymers. The dominant monomeric units in the lignin polymers are benzene rings bearing methoxyl, hydroxyl and propyl groups that can be attached to other units. Lignin poses an obstacle to microbial digestion of structural carbohydrates because of its physical barrier and the depressing effect on microbial activity due to its content of phenolic compounds. Lignin has the potential to unlock the market for high functional material and chemical production, thus making biorefinery commercially practical.

Lipids. Most lipids in biomass are esters formed between one molecule of glycerol and one, two or three of fatty acids called monoglycerides, diglycerides and triglycerides, respectively. These water-insoluble esters are the main fraction of interest in the oily feedstocks including oil seeds, algae, animal fat, waste cooking oil and other food residues. Common triglycerides having the same saturated fatty acid in their structure include trilaurin ($C_{39}H_{74}O_6$), trimyristin ($C_{45}H_{86}O_6$) and tripalmitin ($C_{51}H_{98}O_6$). Triolein ($C_{57}H_{104}O_6$) and trilinolein ($C_{57}H_{98}O_6$) are triglycerides of the unsaturated oleic (one unsaturated bond) and linoleic acid (two unsaturated bonds), respectively. The content of free fatty acids, that is, not esterified, is critical especially in biodiesel production. The content of saturated and unsaturated fatty acids and triglycerides affects the properties of the biodiesel produced (via esterification and transesterification) and the amounts of auxiliary raw materials required for their processing.

Proteins and other components. Proteins are polymers of natural amino acids bonded by peptide linkages. Proteins are found in most biomass, but are particularly abundant in cereals and herbaceous, perennial species. They do not represent a viable feedstock for fuels, but may be useful for production of amino acids and other nutraceutical products. Protein can be extracted from feedstock or by-products as a food and feed additive. Other valuable biomass components include vitamins, dyes, flavoring, pesticides and pharmaceuticals, which can be extracted as value-added products before conversion.

Organic residues. Some residues such as fruit shells, plant pruning, food processing wastewater, animal manure, domestic food waste, etc., are rich in organic matter and nutrients. They may contain high amounts of water and can be exploited as substrate for anaerobic digestion to generate biogas or hydrogen.

1.3.2   Feedstock Characterization

Information about the composition and properties of a biomass feedstock is helpful to evaluate its suitability for a process technology. Chemical analyses and physical properties characterizing a biomass feedstock are described as follows.

Chemical composition. Chemical composition plays a major role in defining the processes for pretreatment and further processing of biomass. Feedstocks rich in sugars can be readily fermented, whilst starch rich feedstocks need an enzymatic pretreatment to release the sugars for fermentation. High protein content can be a negative factor for bioethanol production from wheat, thus needing pre-extraction by mechanical processes such as pearling before the saccharification and fermentation to bioethanol production. The processes of arabinoxylan extraction from wheat bran before fermentation to bioethanol production are shown in detail in the literature[4–6]. For oily feedstocks, free fatty acid content determines whether a feedstock needs to be treated by esterification using acid catalyst before transesterification to biodiesel production. The processes including mechanistic studies are shown in detail in later literature[7–9].

Compositional analysis is useful in biochemical processing of lignocellulosic feedstocks. For bioethanol production, a biomass feedstock with a high ratio of [(cellulose+hemicellulose)/lignin] is desirable for high yields. Other relevant analyses include fatty acid profile analysis in oil and fat processing, type and content of nutrients in biochemical processing of organic residues, proximate and ultimate analyses in thermochemical processing, etc.

Proximate and ultimate analyses. These analyses are relevant to combustion and other thermochemical processes. Proximate analysis shows the volatile matter and fixed carbon fractions. The volatile matter is the fraction (mainly organic matter) released in the form of gas when a biomass is heated to high temperature (950 °C) and indicates how easily the biomass can be combusted, gasified or partially oxidized. The remaining fraction is the fixed carbon, generally determined by the difference between the results of volatile matter, moisture and ash contents. Ultimate analysis shows the elemental composition (C, H, O, N, S and, sometimes, Cl) of biomass. Table 1.3 shows the proximate and ultimate analyses including higher heating values (HHV) of some biomass feedstocks. The main compositional difference between biomass and crude oil is the amount of oxygen, which can be up to 45% by mass in biomass, whilst oxygen is practically absent in crude oil. Biomass also has a lower carbon content than crude oil. As a result, biomass has a lower calorific value than fossil resources and produces unstable bio-oil upon pyrolysis due to the presence of oxygen in biomass. Bio-oil can be stabilized and can become an important platform compound for advanced biorefinery design (Figure 1.4), giving rise to transportation fuels, commodity chemicals and CHP[10–13]. The configurations are ready to capture carbon dioxide for an overall negative carbon footprint[13–16].

Ultimate analysis provides insights into the biomass quality as fuel. For example, high O/C and H/C ratios reduce the energy value of a fuel due to the lower energy contained in carbon–oxygen and carbon–hydrogen bonds than in carbon–carbon bonds[22]. Ultimate analysis can also be useful in identifying potential processing problems such as

Table 1.3   *Proximate and ultimate analyses of some biomass feedstocks.*

Component	Corn Stover[a]	Wheat Straw[b]	Rice Husk[c]	Rice Straw[c]	Switchgrass[d]
Proximate analysis (% weight)					
Moisture	11.75	4.10	8.20	10	8.38
Ash	4.63	6.04	13.17	10.40	7.33
Volatile matter	69.72	73.27	58.93	60.71	69.63
Fixed carbon	13.90	16.59	19.70	18.90	14.66
HHV (MJ kg^{-1})	18.6	18.9	16.6	17.2	18.3
Ultimate analysis (% weight)					
Carbon	42.37	45.36	39.05	39.13	42.6
Hydrogen	4.83	6.52	4.58	5.15	4.9
Nitrogen	1.93	0.77	0.18	0.60	0.6
Oxygen	34.40	36.92	34.75	36.24	36.11
Sulfur	0.09	0.29	0.04	0.09	0.08

[a] Sebesta Blomberg, 2002[17].
[b] Bridgeman et al., 2008[10].
[c] Jangsawang, et al., 2007[19], HHV calculated after Gaur and Reed, 1998[20].
[d] Carpenter et al., 2010[21], HHV calculated after Gaur and Reed, 1998[20].

NO_x, SO_x and H_2S emissions and corrosion. In general, sulfur concentrations are much lower in biomass than in fossil resources, decreasing the potential for SO_2 emissions. Generation of H_2S, however, can be a problem in gasification, anaerobic digestion and in particular downstream energy generation processes: fuel cell, Fischer–Tropsch, gas turbine and methanol synthesis, etc. Hence, sulfur components are removed using scrubbing and Claus processes from the gas produced.

Moisture content. For thermochemical processes, high moisture content in biomass feedstock indicates poor quality because energy is required for heating and evaporating the water, hence lowering the efficiency. Crop residues have the lowest moisture content with 4–18% on a weight basis, while aquatic biomass can have up to 85–97% on a weight basis. High moisture feedstock (e.g., sugarcane, marine biomass, manure, wet organic residues) can be better for biochemical processes that are carried out in the aqueous phase. However, moisture can affect feedstock during storage, degrading its quality, even for biochemical processing. Moisture is also an issue for biomass logistics since biorefineries will need a consistent supply of feedstock.

Ash content. The solid residue formed from the inorganic mineral matter in biomass after high temperature treatment is called *ash*. In comparison to carbon and crude oil, biomass has a low ash content. However, an ash chemistry leading to low melting point solids called *tar* represents operational challenges to thermochemical processing. Ash can be recovered as fertilizer or for mineral extraction after, for example, rotating cone filtration.

Heating value. The energy content of biomass is a crucial parameter for thermochemical processes producing heat and power. The heating value (HV) indicates the energy content of a substance and refers to the heat released when the substance is combusted. The *higher heating value (HHV)* includes the latent heat contained in the water vapor recoverable by condensation. The *lower heating value (LHV)* indicates the heat available excluding that heat. The form and the actual amount of energy recovered from a feedstock will depend on the conversion process applied. Energy efficiency of conversion is often reported on an HHV or LHV basis. The heating value is also affected by the nature and content of the biomass components, that is, cellulose, lignin, etc.

Density and particle size. The bulk density of biomass feedstock is an important characteristic with regard to transportation. The particle size and density also impact on the handling, feeding and storage system requirements. Particle size and size distribution may affect the fluid dynamics in biochemical and thermochemical processing.

Digestibility parameters. Parameters indicating the digestibility or biodegradability properties of organic wastes are relevant in biochemical processing, particularly in anaerobic digestion for biogas production. These parameters include the dry matter content as total solids (TSs), the biodegradable fraction or volatile solids (VSs) as a percentage of TS, the nutrient ratio C:N and chemical composition. Water content is also important in terms of equipment sizing and fluid dynamics of the mixture. Compositional analysis is required to identify likely inhibitory problems due to high ammonia content or the presence of toxic components like antibiotics, pesticides, heavy metals, etc. Carbohydrates and proteins are beneficial due to fast conversion rates. All the factors above affect the yield and methane content of the biogas produced.

Feedstock characterization is also useful to track quality variations and respond accordingly by manipulating process conditions. In addition, composition and properties of biomass components are essential input to perform process simulations under different scenarios. Wooley and Putsche (1996) developed a database for biomass components present in lignocellulosic feedstock and bioethanol production[23]. Chang and Liu (2010) developed models for property prediction of triglycerides and other components involved in biodiesel production[24]. Information on composition, energy content, and other properties for a wide range of biomass can be found in the following databases:

- IEA Task 32 biomass database (www.ieabcc.nl)
- Phyllis biomass database (www.ecn.nl/phyllis/)
- University of Technology of Vienna biomass database (www.vt.tuwien.ac.at/biobib)
- US Department of Energy (DOE) biomass feedstock composition and property database (http://www1.eere.energy.gov/biomass/printable_versions/feedstock_databases.html)

Table 1.4 shows the main feedstock characteristics and their relevance to different process types. Both biomass properties and process requirements must be evaluated simultaneously to develop a technically and economically feasible and environmentally sustainable biorefinery design.

Table 1.4 *Feedstock characteristics and their relevance to different process types. (Reproduced with permission from Klass (1998)[25]. Copyright © 1998, Elsevier.)*

Characteristic	Physical	Thermochemical	Biochemical	Chemical
Chemical composition		×	×	×
Proximate and ultimate analyses		×		
Moisture content	×	×	×	×
Ash content	×	×	×	
Energy content	×	×		×
Density	×			
Particle size/size distribution	×	×	×	×
Digestibility/biodegradability			×	
Nutrient type and content			×	

1.4   Processes and Platforms

Biomass refining through fractionation and upgrading allows an efficient use of biomass feedstock and generation of value-added products through valorization. As discussed in the earlier section, biomass is a complex feedstock made up of carbohydrate and phenolic polymers that need to be broken down to access to more treatable and versatile components. Biomass components need to be modified according to the type of products desired. For example, oxygen content needs to be reduced for biofuels since oxygen reduces their energy content and makes them polar, hydrophilic and unstable, which are problematic for storage, transportation and blending. On the other hand, oxygen provides functionality for chemical building blocks. Thus, various processes are necessary to extract, depolymerize, deoxygenate or modify functionality of biomass components to produce useful and valuable chemical and material products. Biorefinery processes can be classified as follows.

- *Mechanical/physical.* These processes are mainly used to perform size reduction (e.g., chopping, milling) and densification of feedstock (e.g., chipping, briquetting) or physical separation (e.g., mechanical fractionation, pressing, distillation, centrifugation, filtration, decantation, extraction, etc.) of components and products.
- *Biochemical.* These processes include anaerobic digestion, fermentation and other enzymatic conversions using microorganisms. Biochemical processes have the potential to convert substrates into final products in one or few steps and using mild reaction conditions (e.g., fermentation at 20–32 °C), which can lead to a more sustainable production due to less energy requirements and less waste generation.
- *Chemical.* These processes (e.g., hydrolysis, esterification and transesterification, deoxygenation, hydrodeoxygenation and decarboxylation, steam reforming, electrochemistry, Fischer–Tropsch and methanol synthesis, etc.) are used to change the chemical structure of a substrate. They may need high temperature and pressure. They need catalysts to keep the operating temperature and pressure at moderate levels and increase reaction conversion and desired product yield and purity.
- *Thermochemical.* Thermochemical processing is a special case of chemical processing, involving thermal decomposition, thermal oxidation, etc. In these processes (e.g., pyrolysis, gasification, combustion, and supercritical processing) feedstock is treated under medium to high temperature (350–1300 °C) and/or pressure with or without a catalyst.

The processes are able to generate products via a building block called *platform*. The following platforms deduced from biomass are established.

1. Syngas (using gasification)
2. Biogas (using anaerobic digestion)
3. Bio-oil (using pyrolysis)
4. C5 sugars (using fractionation into hemicellulose)
5. C6 sugars (using fractionation into cellulose)

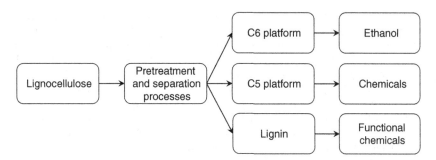

Figure 1.5 *Conceptual block diagram for various productions upon lignocellulose fractionation into cellulose, hemicellulose, and lignin.*

6. Lignin (using fractionation of lignocellulose)
7. Oils (using extraction)
8. Hydrogen (using chemical, thermochemical routes)

Biomass is best utilized for value-added chemical production (medium-to long-term solution) and has the least value-added from energy production (short-term solution). Biomass being the only alternative renewable carbon source, naturally occurring polymers from biomass and chemicals upon conversion of biomass can be extracted in various ways to replace a fossil resource.

Biomass-derived products broadly fall into the following categories: (1) energy and low molecular weight chemicals, (2) natural polymers, (3) monomers and aromatics. Their optimal production relies on how effectively functionalities, conversion, separation and purification steps are coupled. Thus, process integration inspired heuristics are needed to find synergies in driving forces to couple functionalities (e.g., reaction chemistry, electrochemistry and physical chemistry) to perform within one vessel. By this, the incentive is to save energy and capital costs as well as to increase productivity and selectivity. Furthermore, heuristics may be generated by applying the concepts of industrial symbiosis, where waste from one process becomes an essential feedstock to another. Once a conceptual process configuration is generated by applying such process synthesis heuristics, optimization is needed to find the best configuration (e.g., least cost and least environmental impacts) for desired products.

The most exploitable route is the depolymerization and conversion into chemicals. There are two ends of the product type from biomass: (1) energy and low molecular weight chemicals and (2) natural polymers. Figure 1.5 shows the conceptual block diagram for various productions from three major components of lignocellulose: C5 and C6 platforms and lignin. Figure 1.6 shows the essential difference between the hydrolysis process for decomposition of lignocellulose

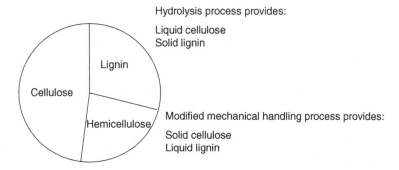

Figure 1.6 *Difference between hydrolysis process for decomposition into liquid cellulose and eventually sugar monomers and modified pulping process for recovery of lignin compounds.*

into liquid cellulose and eventually sugar monomers and solid lignin; and a modified pulping process for recovery of liquid lignin compounds and solid cellulose for production of energy commodities, respectively. Soft wood performs the best (ease of separation in the modified pulping process and greater recovery of the original hemicellulose and lignin present in the biomass) in Borregaard's process, synthesizing ethanol, C5 chemicals and vanillin (from lignin).

In conventional kraft pulping processes, most of the hemicelluloses from wood are degraded into oligomers or mono sugars and dissolved in black liquor along with dissolved lignin and the pulping chemicals (inorganic substances). The black liquor, even though with low calorific value, is usually combusted for steam and electricity generation, due to lack of other uses. However, with optimized pretreatment and separation process schemes, C6 compounds of cellulose and hemicellulose fractions can be converted into ethanol upon fermentation; hemicellulose is a platform for C5 chemicals production (xylose, arabinose and mannose); lignin can be a platform for functional chemicals (functional phenolics). For the production of ethanol, the removal of lignin and hemicelluloses is desired to improve the accessibility of cellulosic material to hydrolytic enzymes for high ethanol yield.

Biomass components can be transformed into intermediate *platforms* to offer processing flexibility to a biorefinery. As shown in Figure 1.3, a platform can be generated from a variety of feedstocks. At the same time, a platform allows shifting between processing pathways according to a desired product. Platforms are generated from biomass fractionation, pretreatment or conditioning, depending on the nature of the feedstock and the type of products to be produced. More than one platform can be present in a biorefinery configuration and the number of platforms is a sign of biorefinery complexity. The primary biorefinery platforms as shown before are discussed as follows.

- *Biogas* (mainly CH_4 and CO_2 and impurities such as H_2S) is produced from anaerobic digestion of organic residues. This biofuel after purification can be utilized for lighting, cooking and heat and power generation. Biogas has 50–70% methane but can be upgraded to 97% purity to be used as a natural gas substitute.
- *Syngas* (CO, CO_2, CH_4 and H_2 and other gases) is produced from biomass gasification. After cleaning, syngas can be used as fuel and for heat and power generation. Depending on the gasification conditions, the syngas produced can also be conditioned to undergo either thermochemical or biochemical processing for production of liquid biofuels and chemicals such as Fischer–Tropsch fuels, methanol, ethanol, dimethyl ether, isobutene, organic acids, ammonia, etc.
- *Hydrogen* (H_2) can be generated from the water gas shift reaction, steam reforming, water electrolysis and fermentation. H_2 can be used as fuel and as a chemical reactant for hydrotreatment of oils, hydrogenation of sugars, ammonia production, etc.
- *C6 sugars* (e.g., glucose, fructose, galactose: $C_6H_{12}O_6$) are released from hydrolysis of sucrose, starch, cellulose and hemicellulose. This platform is mainly used for the production of bioethanol and other chemicals with diverse functionality such as furfural, acetic acid, formic acid, etc.
- *C5 sugars* (e.g., xylose, arabinose: $C_5H_{10}O_5$) are released from hydrolysis of hemicellulose and food and feed side streams. These sugars can also be used for the production of biofuels and chemicals.
- *Lignin* is produced from the fractionation of lignocellulosic biomass. It can be applied as fuels or composite material. Further valorization of lignin includes the production of chemicals such as vanillin and phenolic based aromatic compounds.
- *Pyrolysis liquid* or *bio-oil,* a multicomponent mixture of oxygenated hydrocarbons, is produced from fast pyrolysis of biomass. Bio-oils can be separated into heavy (suitable for cocombustion in coal power stations) and light fractions, which can be upgraded via hydrodeoxygenation to liquid biofuels and chemicals (phenolic based aromatic compounds).
- *Oil and fat* (mainly containing lipids or triglycerides) can be present in oilseed crops, algae and oil based residues, and animal fat. These can be converted into fuels such as biodiesel, green diesel and jet fuel, etc. Due to their chemical functionalities, these can also serve as reactants for the production of surfactants or biodegradable lubricants.
- *Organic juice* is a liquid mixture of different biomass components produced after the pressing of wet biomass (e.g., grass juice, sugarcane bagasse).
- *Electricity and heat* can be internally produced and used to meet the energy needs of the biorefinery and any excess can be sold to the grid.

1.5 Biorefinery Products

Biorefinery products can be grouped into those used as energy carriers (biofuels, heat and electricity) and those used as materials for different industries or human needs including chemicals, nutraceuticals and food ingredients, pharmaceuticals, fertilizers, biodegradable plastics, surfactants, fibers, adhesives, enzymes, etc. Defining the product portfolio will depend on their potential to generate revenues and potential for avoided emission by replacing similar functionality fossil-derived products.

Biofuels, heat and power as renewable energy, or *bioenergy,* have been the most prominent products from biomass. Biofuels for transportation have received special attention due to the importance of such a sector in future energy demands and the avoidance of greenhouse gas emissions. Biofuels have been classified according to the type of feedstock used for their production as *first generation, second generation* and *advanced biofuels. First generation biofuels* are those produced from food crops and using well-established technologies, mainly ethanol from wheat, corn and sugarcane and biodiesel from edible vegetable oils. *Second generation biofuels* include biogas, methanol, dimethyl ether, ethyl *tert*-butyl ether, methyl *tert*-butyl ether, synthetic and green diesel, gasoline, jet fuel, biobutanol, biohydrogen, and Fischer–Tropsch liquids produced from alternative feedstocks including lignocellulosic materials, organic residues, algae and *Jatropha.* Production of these biofuels uses more advanced technologies including the Fischer–Tropsch process, gasification, hydrotreatment of oils, pyrolysis, etc.

Following the strategy from the crude oil and petrochemical industries, biorefineries aim to produce the chemical *building blocks* for industrial applications. Building blocks are simple but highly functionalized molecules using which further product diversification is possible by organic synthesis. An example of a building block, attracting research and industrial attention, is succinic acid, shown in Figure 1.7. Succinic acid can be produced by fermentation of glycerol coproduced with biodiesel. In fact, several derivatives can be produced from glycerol, making glycerol a building block itself[26].

The presence of oxygen, various functional groups, and bond types makes it possible to produce building blocks of a much wider range of chemicals from biomass than from the crude oil. Two attempts to find the more promising building block products in terms of economics and functionality are the study by Werpy and Petersen (2004) for the US National Renewable Laboratory (NREL) and Patel *et al.* (2006) for the EU BREW project[26,27]. The NREL study is based on the potential market and economic value of building blocks (both specialty and commodity chemicals). The BREW project focuses on the production of bio based bulk chemicals mainly as replacement or substitute products for fossil based products. Results of BREW project screening are summarized and compared to NREL study results in Table 1.5.

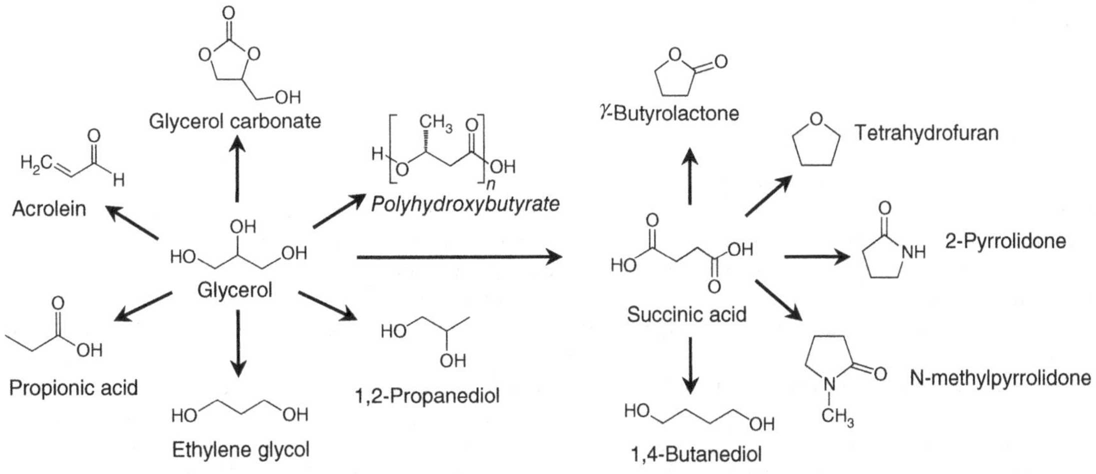

Figure 1.7   *Glycerol and its derivative succinic acid as examples of chemical building blocks.*

Table 1.5   *Promising chemical building blocks[26,27].*

Building Block	C No.	NREL	BREW	Building Block	C No.	NREL	BREW
Syngas (H$_2$ + CO)	C1	×	×	Aspartic acid	C4	×	
Ethanol	C2	×	×	Arabinitol	C5	×	
Acetic acid	C2	×	×	Furfural	C5	×	
Lactic acid	C3	×	×	Glutamic acid	C5	×	
Glycerol	C3	×		Itaconic acid	C5	×	
Malonic acid	C3	×		Levulonic acid	C5	×	
Serine	C3	×		Xylitol	C5	×	
Propionic acid	C3	×		Xylonic acid	C5	×	
3-Hydroxypropionic acid	C3	×		Glucaric and gluconic acid	C6	×	
1,3-Propanediol	C3		×	1-Butanol	C6	×	×
Acrylic acid	C3		×	1,4-Butanediol	C6	×	
Acrylamide	C4		×	Sorbitol	C6	×	×
Acetoin	C4	×		Adipic acid	C6	×	×
3-Hydroxybutryolactone	C4	×		Citric acid	C6	×	×
Malic acid	C4	×		Caprolactam	C6		×
Theonine	C4	×		Lysine	C6	×	×
Succinic acid	C4	×	×	Fat and oil derivatives	>C6		×
Fumaric acid	C4	×		Polyhydroxyalkanoates (PHA)	>C6		×

Further, US Department of Energy (DOE) studies suggest a list of organics from cellulose conversion, as follows[26,28].

Succinic, fumaric, and malic acids
2,5-Furandicarboxylic acid
3-Hydroxypropionic acid
Aspartic acid
Glucaric acid
Glutamic acid
Itaconic acid
Levulinic acid
3-Hydroxybutyrolactone
Glycerol
Sorbitol
Xylitol/arabinitol

The report embraced these screened products as a guide for research. The methodology for screening included factors such as known processes, economics, industrial viability, size of markets and the ability of a compound to serve as a platform for the production of derivatives. The list retains 2,5-furandicarboxylic acid and hydroxymethylfurfural in recent investigations because of their most efficient process and one-pot process development, while glutamic acid remained as an end product and not a building block, and has been dropped from the list.

Of highest importance is the need to extract lignin early from lignocellulose. Unless lignin fraction is routed to value-added chemical production, a biorefinery is unlikely to be economically and environmentally sustainable. Lignin value addition is an important stumbling block, unveiling of which can unlock the sustainability of biorefineries. The potential applications of lignin go from simple combustion for heat and power production to the production of aromatics typically produced from crude oil (phenols; benzene, toluene and xylene: BTX chemicals) or other highly functional chemicals used in the food and flavoring industry (e.g., vanillin). A study by PNNL shows the choices of priority chemicals, as in Figure 1.8[29]. The chemistries and markets are yet to be fully known for the blocks and arrows shown in dotted lines,

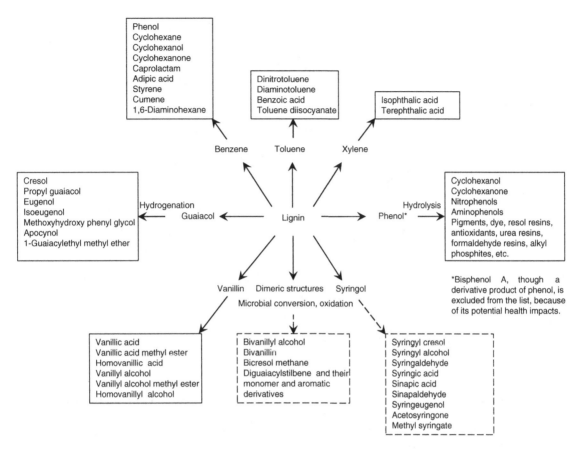

Figure 1.8   *Chemicals from lignin. The chemistries and markets are yet to be fully known for the blocks and arrows shown in dotted lines.*

whilst the productions of BTX derivatives are well established from the petrochemical industry. The sequence from the most mature technology to the furthest-from-market technology utilizing lignin is as follows:

BTX derivatives > phenol and guaiacol derivatives > vanillin derivatives > dimer and syringol derivatives

The revenue potentials have been estimated as US$ 80 billion for biofuels, US$ 10–15 billion for bio based bulk chemicals and polymers, and US$ 65 billion for heat and power by 2020 (King, 2010)[30]. The revenue potential from biofuel and heat and power production is higher compared to chemical production; the higher revenue potential is because of the volume produced, but not the market price of products. A biorefinery is most beneficial by producing one or several specialty chemicals to enhance profitability and a commodity product such as biofuel to ensure market penetration at present. Decision making attending the trade-off between low value–high volume and high value–low volume productions requires a market analysis, while also taking into account the technical feasibility of the processes needed to deliver a particular set of products.

A summary of biorefinery product categories and their market drivers are shown in Figure 1.9. The lowest value products are the energy products (however, due to high volume production, the overall revenue from their production may be higher compared to chemical production), followed by biofuel. This is followed by chemical (methanol and equivalent to primary petrochemical) and hydrogen production. Polymer, composites and food and pharmaceuticals are

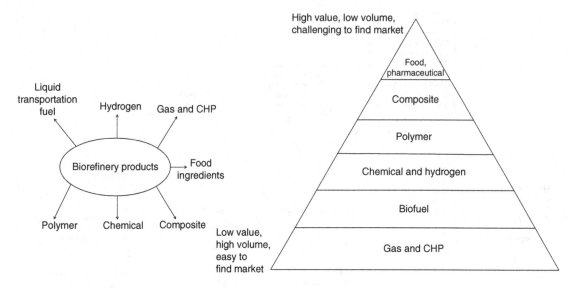

Figure 1.9   *Biorefinery products and their market drivers.*

consecutively the most value-added products. However, with lower extractable volumes in different or new molecular forms (with equal functionality), higher challenges are faced to find a market.

1.6   Optimization of Preprocessing and Fractionation for Bio Based Manufacturing

Biomass based manufacturing is urgently needed to replace fossil based manufacturing of carbon containing products: chemicals, polymers and materials. Electricity and heat though has a large market, but renewable energy resources are practical to fill the needs. Biomass CHP can be brought into action in place of fossil based power plants, to meet peak energy demands. However, biomass is the only carbon source. Amongst the biomass feedstock options, lignocellulosic biomass is the only workable biomass to replace crop based manufacturing causing depletion of land and water resources. The lignocellulose based manufacturing needs to be investigated from more fundamental perspectives: structural integrity and carbon balance of lignocellulose throughout the life cycle, process synthesis for least cost and environmental impacts and highest value generation. Sustainable biomanufacturing thus requires a "whole" systems optimization study.

Lignin encasing cellulose in cell walls provides cell wall rigidity, while encased cellulose microfibrils give tensile strength of cell walls. Extraction of hemicellulose and lignin reduces the cost of production of cellulosic products. This stage is called preprocessing, after which cellulose microfibrils can be easily converted into glucose by enzymatic hydrolysis, followed by ethanol and chemical productions. Alternatively, cellulose microfibrils can be used to form continuous phase (matrix) of composite materials. The discontinuous phase, also known as reinforcement, may consist of glass, carbon or Kevlar fibers. Together, these can form thin films with applications in electronics, for example, moulding transistors or printed circuit boards, construction, automotive as laminates, and marine construction industries.

The preprocessing of biomass at a broad level includes the physical (e.g., kraft processes), chemical (acid hydrolysis, alkali treatment) and biochemical (enzymatic hydrolysis) processes and fractionation to recover products. The concept avoids "end-of-pipe cleaning" of impurities from the main products, by separating the cellulose, hemicellulose and lignin materials upfront and diverting them to desired products, leading to environmentally benign process designs. The screening of lignocellulose preprocessing technologies and products (Figure 1.10) describes a more complex and challenging task than the development of sugar based bioethanol plant. However, to achieve the full potential of biorefineries, extensive extraction is needed for every component, to produce value-added products.

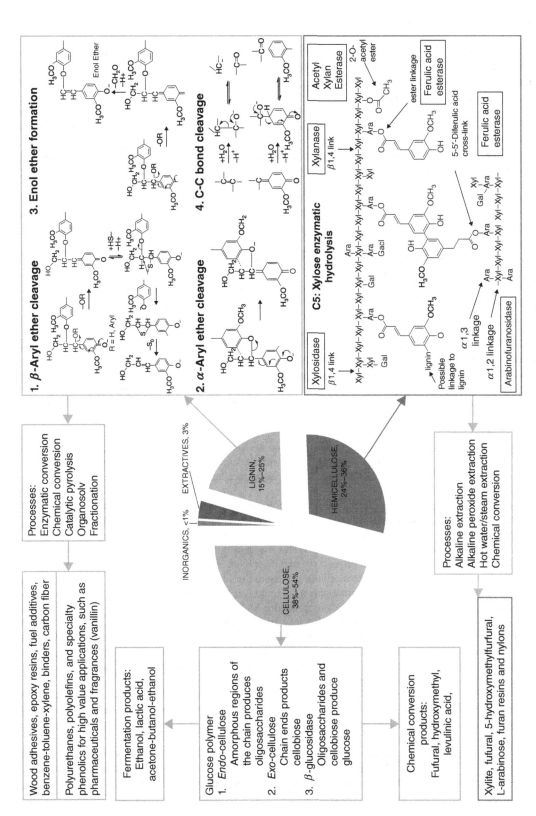

Figure 1.10 Preprocessing technologies, mechanisms and products.

On the right hand bottom corner in Figure 1.10, the hemicellulose bond cleavage mechanisms by enzymes are shown (Peng *et al.*, 2012; Reith *et al.*, 2009)[31,32]. The following enzymes are noted to act on the following cleavages. α-L-Arabinofuranosidase liberates the L-arabinose from positions 2 and 3 of the backbone xylan. Ferulic acid esterase hydrolyses the position 5 of arabinofuranoside residues, producing ferulic acid. Acetyl xylan esterase releases the *o*-acetyl of acetyl xylan. Xylanase randomly cleaves the main chain of xylan, producing a mixture of xylo-oligosaccharides. There are two types of Xylanase, *endo*-xylanase randomly breaks the xylan chain, while β-xylosidase releases monosaccharides and xylose from the nonreducing end of the short oligosaccharides.

Similarly, for galactoglucoglucomannans, β-mannosidases and *endo*-1,4-β-mannanses are used to break down into corresponding mono- and oligosaccharides. In order to produce products of desired properties, coordinated activities of cleavage of bonds are required, which must be assisted by genetic engineering and enzymatic pathway analysis.

The established processes for the above enzymatic treatments are as follows:

α-L-Arabinofuranosidase. Ultrafiltration and anion exchange chromatographic columns; pH: 5.8; temperature 45 °C; molecular weight: 160–210.

Ferulic acid esterase. Diethylaminoethyl cellulose chromatography and ammonium sulfate precipitation; pH: 7.5; temperature 45 °C; molecular weight: 20.

Acetyl xylan esterase. Ammonium sulfate precipitation, diethylaminoethyl cellulose chromatography, tertiary butyl hydrophobic interaction and hydroxyapatite chromatography, and Ni^{2+}–NTA agarose column chromatography; pH: 6.5–7; temperature 37–50 °C; molecular weight: 30–34.

Xylanase. Ultrafiltration, anion exchange chromatography, metal affinity resin and precipitation by gravity flow column, ultrafiltration and diethylaminoethyl cellulose chromatography, and a ammonium sulfate precipitation, ion exchange, and gel filtration chromatography; pH: 6–7; temperature 37–60 °C; molecular weight: 22–45.

β-Mannosidases. Diethylaminoethyl cellulose chromatography, anion exchange chromatography, and gel chromatography; pH: 3.5–7; temperature 30–53 °C; Molecular weight: 72–94.

Endo-1,4-β-mannanses. Ultrafiltration and chromatographic columns; pH: 6; temperature 60–65 °C; molecular weight: 38–40.

The main mechanism for cleavage of lignin bonds shown on the right hand top corner of Figure 1.10 includes β-aryl and α-aryl ether cleavages, enol ether formation and C—C bond breakage. Cellulose depolymerization mechanisms are well known, as shown in Figure 1.10. Each fraction, cellulose, hemicellulose and lignin can then be converted into a range of products.

Compositions of lignocellulosic feedstocks are often unpredictable when they are accumulated within a system boundary, maybe an urban system. Hence, which products can be generated and what are those flexible processes that can uptake the mixture of feedstocks and still achieve the desired product slate pose a huge challenge to solve. Unoptimized schemes would suffer from waste formation of low values that will ultimately be released to the atmosphere after combustion or cogeneration or to land from disposal to landfills or to an aquatic body by leaching. Most routes produce low value lignin residues. The lignin and hemicellulose fractions are increasingly recognized as a valuable source of value-added productions in the top of the pyramid shown in Figure 1.9. There are numerous possibilities, amongst which the niche areas need to be carefully selected based on market drives within a geographical region and policy context. The application principles encompass integrated feedstock management, conversion, use of end products, and reuse in cyclical and synergistic loops. Renewable feedstocks from a range of local activities can be converted into products to fulfill local needs and can reversely help produce biomass via anthropogenic activities to meet demands. With co-optimization of anthropogenic activities (agricultural, forestry, residential, industrial and commercial) and demand management, the impacts on land, soil, water and atmosphere will be reduced. A true sustainable biorefinery design calls for such a challenging solution for "whole systems." However, the complete fusion of an industrial symbiosis framework is a gradual process and requires years to stabilize from one state to another, for example, from a fossil based era to a renewable era.

Several routes are commercially established or under development. Table 1.6 shows the physical description and specifications of processes and associated challenges for research.

The economic performance of a biorefinery depends on adding value to the lignin and hemicellulose fractions. The most promising route for exploiting the hemicellulose fraction appears to be by extracting functional oligosaccharides, such as xylo-oligosaccharides, with a potential market as pharmacological supplements and food ingredients. Lignin

Table 1.6 *Preprocessing technologies for separation of cellulose, hemicellulose and lignin fractions and downstream purification technologies.*

Process	Physical Description and Specifications	Challenges
Preprocessing for separation of cellulose, hemicellulose and lignin fractions		
Alkali treatment	Alkali treatment is the most established method, amongst all, for lignocellulose separation into basic components: cellulose, hemicellulose and lignin. The process uses hazardous and expensive $NaClO_2$ for delignifying the lignocellulose. Replacing this with environmentally more benign aqueous alcohol treatment is desirable. Alkali type, concentration, time and temperature have an effect on product yields and qualities. Aqueous solutions of sodium, potassium, lithium, barium, calcium and ammonia (ammonium hydroxide) are commonly used. Some of the elements are abundant, sodium and potassium, compared to which lithium availability may be constrained, especially due to increased demand for lithium ion battery for the transport sector. Sodium hydroxide is more effective in terms of yield compared to potassium hydroxide, while the latter is more effective in terms of purity of products. The experimental conditions are generally within the following range: 1.0 M NaOH at 30 °C for 18 hours, to result > 80% recovery of the original amounts of lignin and hemicellulose present in agricultural residues[33]. Barium hydroxide has shown higher affinity for arabinoxylans extraction, from wheat bran fraction, for example. Alkaline peroxide has shown higher effectiveness in terms of delignification as well as solubilization of hemicellulose than alkaline solutions. Hydrogen peroxide (H_2O_2) in the presence of transition metal, such as iron, manganese and copper, readily decomposes into hydroxyl radicals (OH^-) and oxide anion radicals (O^{2-}). These radicals are thought to oxidize lignin, release carboxyl groups by the cleavage of bonds and eventually decompose lignin and hemicelluloses (by the cleavage of covalent bonds between lignin and hemicellulose and ester bonds) into C5 and phenolic compounds. 2% H_2O_2 at 48 °C for 16 hours at pH 12–12.5 proved to be more effective than alkaline solutions, resulting in almost pure activated lignin and hemicellulose products from agricultural residues[34]. The H_2O_2 stream can be recovered from cellulose fraction, yielding a cellulose rich insoluble residue that can be enzymatically converted into glucose and further to ethanol and chemicals.	Optimize operating conditions. Optimize alkali aqueous mixture not only for better yield and purity of products but also for the recovery of alkali solvent and to minimize make-up solvent requirement. It is essential to optimize solvent recirculation columns with heat integration. Further, microwave assisted columns can be designed for high efficiency and environmentally more friendly performance. Catalyst such as acid catalysts can also be introduced for further improvement. It is essential to analyze life cycle impacts and costs for the choice of raw materials, process configurations and production options.

Table 1.6 *(Continued)*

Process	Physical Description and Specifications	Challenges
Organosolv with DMSO	Uses organic solvent for pretreatment of lignocellulose and extraction of hemicellulose and lignin. It is possible to design a multistage extraction process by solvent in aqueous media with various concentrations. Most commonly used solvent is the dimethyl sulfoxide (DMSO). The solvent is more effective for low-branched heteroxylans, because the extraction mechanism does not include cleavage of acetyl ester and glycosidic linkages. This helps to carry out a hemicellulose and lignin structural study and pretreatment and access of wood polymers. However, the main problem is the cost and potential hazard in handling a large quantity.	Design of multistage extraction column. Minimize solvent to biomass ratio. Reduce energy cost. Reduce environmental impacts.
Organosolv with lower molecular weight solvent	The process has the advantage of energy efficient recovery of solvent and low environmental impact, when using low molecular weight organic solvents, such as ethanol, methanol, acetic acid and formic acid, etc. Various solvent mixes are possible. Hence, rigorous investigations must be made for an optimal mixture of solvent for highest value-added generation. The mixtures are methanol–water, ethanol–water, methanol–ethanol–water, acetic acid–water, formic acid–water, acetic acid–formic acid–water, etc. Lower molecular weight solvents are easy to recover, but the degree of recovery is affected and decomposition of hemicellulose and lignin polymers occurs.	Optimize solvent mix. Design of multistage extraction column. Minimize solvent to biomass ratio. Reduce energy cost. Reduce environmental impacts.
Organosolv with catalyst	Acid catalysts are used to increase the yield of hemicellulose and lignin, e.g., 0.1% HCl as catalyst, together with DMSO and aqueous KOH as solvent. To prevent depolymerization, neutral solvents, e.g., 90% dioxane and DMSO, can be used. To facilitate cleavage of acetyl ester and glycosidic bonds and saponification of the ester groups in the polymers, acidic dioxane or medium can be used. Sequential treatments with 80% acidic dioxane, DMSO and 8% KOH primarily give rise to arabinoxylans, especially effectively from crop residues, due to the absence of lignin.	Rigorous simulation for analyses of process configuration, costs and thermodynamic properties. Fully integrated and instrumented one-apparatus system. Saponification and downstream separation for homogeneous systems.

Table 1.6   *(Continued)*

Process	Physical Description and Specifications	Challenges
Organosolv with heterogeneous catalyst	Heterogeneous catalysts, such as solid acids (zeolites, clays, metal oxides, etc.), can give high conversion of hemicellulose into xylose, arabinose and furfural. The reaction conditions are varied by temperature, reaction time and solvent addition and phase separation. When solvent and heterogeneous solid catalysts are optimized for furfural production, the greatest carbon balance can be observed. *The heterogeneous reactive extraction process results in selective conversion into cellulose, hemicellulose and lignin.* The catalysts can be characterized by X-ray diffraction, temperature-programmed desorption of NH_3, inductively coupled plasma spectroscopy, elemental analysis and solid-state nuclear magnetic resonance (NMR) spectroscopy techniques.	Catalyst characterization and meso–nano porous structure. Reactor arrangement and configuration. Continuous reaction. Scaling up of the processes. Physicochemical properties of reaction mixture at various stages of reactive extraction.
Extrusion	The process combines mechanical and physical separation and reaction process using an extruder type twin screw reactor equipped with a filtration system. Functionalities integrated within one apparatus are the extrusion, cooking, liquid–solid extraction and liquid–solid separation (by filtration). Sodium hydroxide can be used as solvent. The sequential treatment with 80% aqueous dioxane containing 0.05 M HCl and DMSO at 85 °C for 4 hours and 8% NaOH at 50 °C for 3 hours results in over 85% recovery of each of the original hemicellulose and lignin present in crop residues.	Fully integrated, optimized and instrumented system. Scale-up risk. Flexible feedstock and process conditions. Catalyst deactivation. Solvent minimization. Solvent and catalyst optimization.
Steam explosion	This is one of the oldest technologies, still effective to fractionate lignocellulose into three major components: cellulose increasing its susceptibility to enzymes; hemicellulose easy to hydrolyze; depolymerized lignin. The basic principle is that biomass is first treated with high pressure steam, the pressure of which is suddenly reduced by expansion to decompose biomass extensively. The Mason process by flash decomposition was developed in 1928. Sequential extraction using steam followed by alcohol solvent (ethanol is often used) is carried out first to separate hemicellulose and then lignin. The controlling parameters are residence time, temperature and feedstock particulate size. The optimal conditions need to be determined to achieve ease of hemicellulose hydrolysis and separation from cellulose. Steam explosion, followed by alkaline peroxide solvent extraction are very effective. Though the processes are environmentally friendly, controlling the degradation of hemicellulose and lignin poses the biggest challenge.	Avoid gas production, but control operating conditions for functional chemical and material production. Controlling degradation of hemicellulose and lignin. Process sequencing and optimization.

(Continued)

Table 1.6 *(Continued)*

Process	Physical Description and Specifications	Challenges
Microwave irradiation	Comparatively newer efficient and environmentally friendly technique, this technique applies electromagnetic radiation to decompose lignocellulose. The process is more environmentally friendly, because less or no solvent is required, higher efficiency and better quality products are obtained. The characterization properties include molar mass and its distribution, degree of polymerization and degree of substitution. *The technology can be used in combination with steam explosion to produce external chemical (solvent) free or impurity free products.* Many studies have shown microwave irradiation as an effective technology for mannan production. Also, the technology enables design of microwave assisted alkali or acid pretreatment.	Controllability. Controlling the heat. Adjustment of electromagnetic waves to supply heat of absorption by materials. Good yield without extensive decomposition. Flexibility to change rate according to feedstocks to process and products to produce.
Ultrasonication	*Ultrasonication using sound waves to prefractionate lignocellulose is now showing promise in terms of energy effectiveness amongst all.* The technologies are mostly proprietary to biomass fractionation industries.	Yet to be proven.

Purification processes: the isolated fractions of hemicellulose and lignin contain numerous polymers need to be characterized and purified into functional products.

Membrane	Purification process involves multisteps such as filtration, ethanol precipitation, centrifuge, etc. Microfiltration, ultrafiltration, nanofiltration using forward, reverse and modified osmosis processes (uses reverse osmosis for purification followed by forward process for membrane clean-up and energy recovery) are effective for purification of lignocellulose products. Membrane processes can also be effectively integrated to other processes, catalytic and solvent treatment, twin screw extrusion and chromatographic separation, etc. Ultrafiltration can be applied to separate high purity arabinoxylans, glucuronoxylans and acetyl-galactoglucomannans of hemicellulose fraction using various cut-off ranges. $150–1000$ g mol^{-1} cut-off ranges economic options. Further, nanofiltration can also purify xylo-oligosaccharides from monosaccharide and low molar mass materials such as salts and lignin polymers from functional phenolics, both in high purity forms. The processes can be a good substitute for chromatographic separation for purification. Molar mass cut-off yields, degree of polymerization and degree of substitution are the key design parameters.	Energy recovery across membrane. Forward osmosis to recover energy and clean-up of membranes. Optimize membrane pore sizes. Optimize membrane modular arrangements. Optimize membrane and other pre-processing modular configurations. Cost, operability, controllability and yield optimization.

(Continued)

Table 1.6   *(Continued)*

Process	Physical Description and Specifications	Challenges
Chromatographic separation	The chromatographic or adsorption based separation technique is used to characterize materials as well as purify materials by precipitation and ion exchange. The ion exchange technique is more effective in terms of high purity fractions recovery. Diethylaminoethyl (DEAE) is a common chromatography agent to directly recover and fractionate hemicelluloses (e.g., into arabinose, xylose, fractions) from the plant materials. The main strength of the chromatographic method lies in the separation of activated lignin. While these techniques are quite common for crop based lignocellulose, they are less established for wood. The size based elution was applied to separate oligomers from monomers and low molecular weight salts, followed by steam explosion. The lignin is also recovered but to a lesser extent. The technique can find the molecular weight of polysaccharides.	Many columns are used; requires innovative techniques for cost-effective recovery. Generally accompanied by other recovery processes, such as alkali treatment, steam explosion upstream, for effective purity, especially for complex lignocellulose structures, such as wood. Simulated moving bed columns and other innovative sorption–reaction columns are needed.

recovery for specialty phenolics production for high value applications, such as pharmaceuticals and fragrances, is also promising. Within the context of an integrated biorefinery, xylo-oligosaccharides and function phenolics could be extracted economically to enhance the overall biorefinery economics.

Finding a niche market for hemicellulose and lignin based products will be the key to sustainable biorefinery development. The case of vanillin can be shown[35]. Rhodia SA dominates the vanillin market using the catechol-guaiacol process. Borregaard (Norway), the second largest vanillin producer, is one of the remaining producers of lignin based vanillin. Essentially, Borregaard supplies the European market and its vanillin production is almost exclusively for large-scale customers under long-term contracts. Since vanillin is mostly produced from guaiacol, vanillin prices are sensitive to the world oil market. However, lignin based vanillin is in high demand for certain market sectors, particularly for the perfume industry, European chocolate manufacturers, and Japanese market, and as such tends to command a price premium. The price of lignin based vanillin has been consistently maintained at about $100–200 kg^{-1} above that of guaiacol based vanillin. Systematic process integration and sustainability analysis will allow lignocellulose biorefineries to be a commercial success. Processes can be optimized for one or more of the following particular production objectives.

1. Enzymatic extraction of cellulose into ethanol, acetone–butanol–ethanol (ABE), xylonic acid, lactic acid, 5-hydroxymethylfurfural (HMF), levulinic acid, etc. Cellulosic polymers are the most abundant renewable polymers.
2. Hemicellulose hydrolysis for further processing of C5 into xylite, furfural, HMF, L-arabinose, etc. Hemicellulose is the second most abundant source of biopolymers, consisting of heteropolysaccharides, linked with cellulose and lignin, in cell walls by covalent and hydrogen bonds and by ionic and hydrophobic interactions. After the kraft pulping process, hemicelluloses are obtained along with dissolved lignin, as black liquor. Hemicelluloses in this mixture consist of oligomers or monosugars of C5 and C6: glucose, xylose, mannose, galactose, arabinose, rhamnose, glucuronic acid and galacturonic acid in various amounts depending upon the source.
3. High quality lignin stream recovery wood adhesives and resins, fuel additives, BTX, binders, carbon fiber, etc. Lignin is the third most abundant source of biopolymers and materials after cellulose and hemicellulose.
4. Lignin recovery for phenolic products, activated lignin, epoxy resins, polyurethanes, polyolefins and specialty phenolics for high value applications, such as pharmaceuticals and fragrances.

A biorefinery converting lignin and hemicellulose into functional products would only be economic if the commercial potential of all three fractions, cellulose, hemicellulose and lignin, is fully exploited. Now, it is clear how to add value

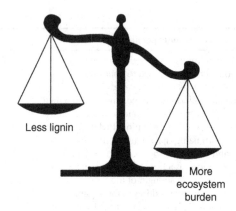

Figure 1.11   *Lignin value generation becomes threshold to sustainable manufacturing.*

to the starch fraction (e.g., through fermentation), but in particular the lignin fraction still needs to be investigated from fundamental synthesis perspectives.

1.6.1   Background of Lignin

Until the late 1990s research has been focused to repress the lignin content in transgenic trees, considering the postharvesting problems to separate it from cellulose. Lignin appears to be inhibiting the use of cellulose to produce biofuels. Lignin is problematic for the pulp and paper industry. Not only had the lignin separation technologies begun to evolve to ease cellulose conversion in these industries but suppression of lignin also became another main line of research in bioengineering.

On the other hand, lignin's resistance to microbial degradation increases plants' persistence to soil, while also retaining approximately 30% of the biogenic carbon captured. Retaining high quality and quantity of lignin in plants directly helps in enhancing plants' structural integrity, strength, rigidity, prolonged capture of biogenic carbon and enhanced ecosystem balance. Reduction of lignin on the contrary reduces the biogenic carbon capture and increases the overall biogenic carbon release to the atmosphere. Thus, lignin quality and quantity becomes a pivot for carbon balancing across the biomanufacturing sector. Figure 1.11 shows that less lignin in plants implies more ecosystem burden.

Lignin fractions are increasingly recognized as a valuable source for functional chemical and material production. A high quality lignin stream can be manufactured, such as vanillin, vanillic acid, dispersing agents, synthetic tannins, polymer filter sand, binding agents, activated carbon, ion-exchanger subtract, etc.

The main lignin compounds are polymers of *para*-hydroxyphenyl (H lignin), guaiacyl (G lignin) and syringyl (S lignin) alcohol, involving many pathways, enzymes, and cofactors (Figure 1.12)[36]. By suppressing some genes, by reducing particular enzymes, the quality and quantity of lignin can be controlled. For example, reduction of CAD enzyme leads to S lignin production provided there is up-regulation of F5H and COMT enzymes. Hence, down-regulation of F5H and COMT enzymes decreases S lignin content. Absence of HCT and C3H enzymes leads to H lignin prominence. It is seen that the structural strength of cell walls depends more prominently on the presence of H lignin. Up-regulating HCT, C3H and CAD enzymes leads to G lignin, a platform for vanillin production. However, studies have shown that perturbation of pathways affects all other pathways and expression of genes apparently seems to be unrelated. This means that research needs to address beyond genetic engineering to uncover molecular mechanisms that accompany lignin modifications, such that carbon can be preserved and lignocellulose cells can act as a bioreactor.

The lignin biosynthetic enzymes are:

PAL: phenylalanine ammonia-lyase
C4H: cinnamate 4-hydroxylase
4CL: 4-coumarate:CoA-ligase
C3H: *p*-coumarate 3-hydroxylase

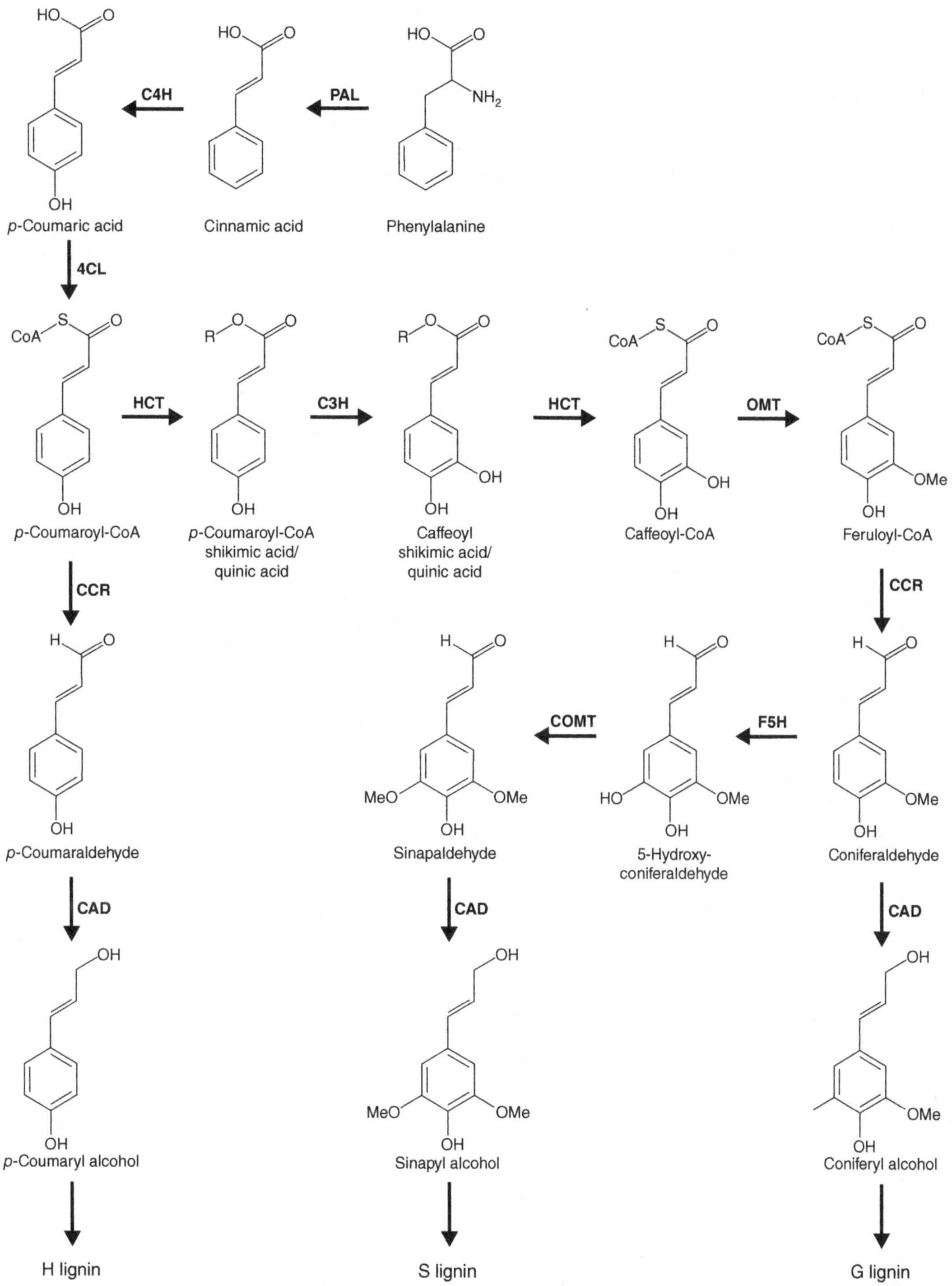

Figure 1.12 *Lignin biosynthetic pathways.*

OMT: *S*-adenosylmethione:caffeate/5-hydroxyferulate-O-methyltransferase
F5H: ferulate 5-hydroxylase
HCT: 4-hydroxycinnamoyl-CoA 3-hydroxylase
COMT: sadenosyl-methionine:caffeoyl-CoA/5-hydroxyferuloyl-CoA-O-methyltransferase
CCR: hydroxycinnamoyl-CoA:NADPH oxidoreductase
CAD: hydroxycinnamyl alcohol dehydrogenase; glucosyltransferase
UDP-Glc: coniferyl alcohol 4-O-glucosyltransferase; glucosidase, coniferin-specific 4-O-glucosidase.

The lignin chemical constituents vary greatly depending on the source of biomass and age, morphological location and growth environment (for agricultural and forestry residues). Lignin from softwoods is predominantly based on guaiacyl as the structural unit. Guaiacyl units are the precursors to vanillin based products. Hardwood lignin consists of syringil:guaiacyl with weight ratios in the range of 1.8–2.3.

Lignin extraction from lignocellulosic materials is the most basic step practiced over hundreds of years in the pulp making process and paper industry. The composition, property and quantity of lignin are highly influenced by the chosen extraction method. Sulfite, kraft and soda processes are most commonly used in the pulp and paper industry. The sulfite process uses salts (sodium, magnesium and ammonium) of sulfurous acid to perform three reaction steps to separate lignin: lignin and free sulfurous acid form lignosulfonic acid; lignosulfonates are formed by the reaction between lignin and cations; fragmentation of the lignosulfonates. The reaction operating conditions are as follows: 140–160 °C and pH: 1.5–2 for acidic and 4–5 for basic media respectively.

The kraft process uses sodium hydroxide and sodium sulfide under alkaline conditions to cleave the ether bonds in lignin in two main stages: lower temperature preheating and premixing at 150 °C and reaction at 170 °C. The reaction is mostly complete in the final stage. In the kraft process, β-1,4 links in cellulose are cleaved, extracting the lignin components. However, the lignin itself is also susceptible to alkali: ethers in lignin readily undergo base-induced hydrolysis under relatively mild conditions. Cleavage of α-aryl ether then takes place before β-aryl ether cleavage, particularly when free phenolic hydroxyl group is available in the para position. The black liquor is separated by light acid treatment, such as using sulfuric acid (more recently carbon dioxide) also to increase pH to 5–7.5 of the kraft lignin, which has a market value. Most of the process development has occurred for kraft lignin.

Most of the lignin separated from wood in the sulfite pulping and the kraft pulping processes is burned to avoid volatile organic matter emissions that could be toxic. With no carbon taxation or financial burden due to carbon dioxide emissions on these industries, burning the volatile matter for energy recovery seems to be an easy way. The remaining amount is isolated from the spent pulping liquors and sold for specialty applications, normally around 1 million tons per year worldwide. Lignin is a sticky material causing numerous operational problems in downstream processes. Lignin also gives rise to phenolic compound emissions with some of these being toxic. Preprocessing to extract lignin and any capture of organic vapor during processing of lignocellulose is highly desirable. The natural use of lignin is as binders in panels and boards. Phenol formaldehyde resins are used globally with a demand over 1 million tonnes per year for this purpose. Lignin can be extracted mechanically by modified kraft processes. The resulting products can be furan resins or modified lignin. These materials have 45% higher mechanical strength (e.g., 2.7 N mm^{-2} for modified lignin) than phenol formaldehyde (e.g., 1.6 N mm^{-2}). Hence, direct substitution of fossil-derived phenol formaldehyde by lignocellulose-derived modified karft lignin is possible.

The soda process being the first chemical extraction method for lignin was commercialized after patenting in 1845. The soda process using aqueous solution of sodium hydroxide (13–16 weight %) is used for lignin extraction from non-wood biomass: wheat straw, help, bagasse, etc. Amongst the three processes, sulfite, kraft and soda processes, the soda process is technically and environmentally more sustainable because there is no sulfur residual from the chemical reagents used in the resulting aqueous (cellulose) and organic (lignin) fractions. The presence of sulfur and its hydrophobic nature makes the lignin from sulfite and kraft processes very difficult to be accessible for value-added production. Therefore the main use of lignin is in energy production, still a usual feature in the pulp and paper industry.

With the push to produce cellulose for bioethanol or platform chemical production, the lignin fraction may as well be recovered at high quality so as to produce functional chemicals and materials from it. Such processes have not been semantically evolved but to link with lignocellulose production and product value chain. Organosolv with lower molecular weight solvents and a heterogeneous catalyst can be used, while keeping both the pulp and the liquid quality controlled at desired levels.

Supercritical carbon dioxide as an antisolvent agent can be used to precipitate a range of chemicals from nano- to microstructures, when used together with a chromatographic separation material (such as dimethyl sulfoxide in water). Carbon dioxide can be easily separated from the product at high purity and recirculated.

Ionic liquids have recently been researched to fractionate lignocellulosic materials. Ionic liquids having a large asymmetric organic cation and a small anion are an excellent solvent to depolymerize lignin for value-added productions. The arrangement is a special type of organosolv using ionic liquid to cleave the linkages between ether and alkyl, to yield monomeric methoxy phenols. Once depolymerized, the organosolv lignin can be catalytically converted into fuels or chemicals. Lewis–Brønsted acid zeolites and supported metal catalysts can be used alongside the ionic liquids as a solvent in organosolv to produce a high yield of propylguaiacol[37]. The stem woody tissue has shown a consistent mass ratio of syringyl to guaiacyl of 1.8–2.3. Depolymerization of these polymers for high value product generation loses the trapped bond energy, while direct extraction of biopolymers for functional materials, such as epoxy resins and composites, can retain the inherent properties of lignin, such as strength and rigidity. The process involves lignin polymer blending and reinforcement with cellulose based microfibrils.

Catalytic transfer hydrogenation is another concept applied to bio-oil hydrocracking that can be applied for lignin value-added chemical extraction. It is potentially attractive as it may not require a source of high pressure, high purity hydrogen gas and can be employed in relatively low pressure equipment whose capital costs are low. The expected outcome of this activity was selection of a process and catalyst that would result in lignin based products after decarboxylation with minimal oxygen removal, with no phase separation, with properties of functional phenolics, and also with a possibility of deriving guaiacol as the primary chemical to many products, shown in Figure 1.13. High throughput screening tests can be employed with a variety of conditions, donors and catalysts to select elementary reactions or reaction pathways to lead to effective product alternative evaluations. The final product can be anything in the product chain shown in Figure 1.13, but can be closely controlled by operating conditions and selective *in situ* removal.

Another alternative route is the hydrogenation to treat lignin with high pressure hydrogen in the presence of nickel-molybdenum or chromium oxide catalysts at 400 °C. The process produces monocyclic aromatics breaking down the lignin polymers. In terms of exergy loss in each conversion route of lignin into functional phenolic products, from the highest to the lowest, the following sequence is obtained:

Hydrogenation > Catalytic transfer hydrogenation > Ionic liquid catalytic organosolv > Mechanical/thermal/chemical blending and reinforcement of natural polymers. Hence, the mechanical/thermal/chemical blending and reinforcement of natural polymers is the most sustainable lignin conversion route.

Figure 1.14 shows the products from the various lignin conversion routes and their place in the market. World demand for phenolic derivatives is 8 million tonnes per year with a price of 1200 euros t^{-1}. In 2009, it was utilized as feedstock to produce bisphenol-A (48% of the world's phenol production), phenolic resin (25%), caprolactam (11%), alkyl phenols

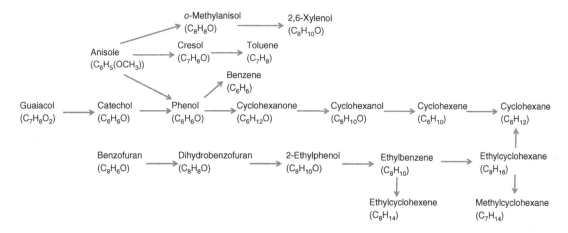

Figure 1.13   *Guaiacol derivatives.*

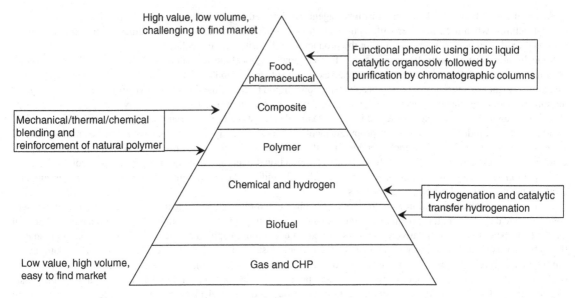

High value, low volume, challenging to find market

Food, pharmaceutical

Functional phenolic using ionic liquid catalytic organosolv followed by purification by chromatographic columns

Composite

Mechanical/thermal/chemical blending and reinforcement of natural polymer

Polymer

Chemical and hydrogen

Hydrogenation and catalytic transfer hydrogenation

Biofuel

Low value, high volume, easy to find market

Gas and CHP

Figure 1.14 *Products from various lignin conversion routes and their place in the market.*

(4%), xylenols (4%), aniline (2%) and others. However, soon after, the United States Food and Drug Administration (FDA) found out about possible hazards of bisphenol-A to fetuses, infants, and young children. In September 2010, Canada for the first time declared bisphenol-A as a toxic substance. The European Union and the United States afterwards have also banned bisphenol-A use in baby bottles. Epoxy resins, as long as they contain no polycarbonate, are a substitute for bisphenol-A in the application of water bottles, dental fillings sealants, eyeglass lenses, sports equipment, CDs, DVDs and lining for water pipes, etc. Another possible route for lignin value addition is the production of renewable carbon fiber, used in lightweight vehicle bodies, lithium ion batteries and semiconductors. The conversion process involved is the multiphase rotating disc contactor organosolv, commercially available for carbon pitch fiber, adaptable for bio based fiber production.

As shown in Figure 1.14, functional phenolics as high value chemicals with moderate thermodynamic performance can be produced using ionic liquid catalytic organosolv. The normal purification process with chromatographic columns and ultra- and nanofiltration membranes can be followed thereafter.

The most commonly used ionic liquid solvent is alkylmidazolium for dissolving cellulose, while 1-ethyl-3-methylimidazolium triflate and chloride (EMIM-TF and EMIM-Cl), butyl and octyl of methylimidazolium triflate and chloride (BMIM and OMIM) and EMIM with a mixture of alkylbenzenesulfonates and xylenesulfonate are the known ionic liquid solvents for lignin, respectively. In general, dialkylimidazolium ionic liquid solvents are known for lignin separation. In addition, a wide range of Lewis and Brønsted acid catalyst can be used. Their effects on the lignin conversion, product yields and composition vary by a wide range, as low as no reaction happening to about 75% by weight of product yield. The main products characterized by gas chromatography are 4-propylguaiacol (highest quantity) > isoeugenol > guaiacol (lowest quantity).

The mechanism for cellulose dissolution is the intramolecular hydrogen bond breakage within cellulose fibers, due to ionic bonds between the anions (e.g., chlorides) of the ionic liquid and the hydroxy groups of the cellulose. The mechanism for lignin breakage is much more complex than cellulose as lignin is structurally more complex. Lignin model compounds have been studied for the dealkylation reaction mechanism. However, the unknown functionality of alkyl substituents and the unknown mechanism of cleavage of ether linkages in lignin are a major stumbling block to design the solvent, catalyst, and experiments. Previously experimented dealkylation conditions for other systems are still used. There is a real risk that the experimental efforts may not reveal anything useful to carry forward. High throughput experimentation is a good strategy, though expensive, but is only practical to apply once the fundamental mechanism is modeled and optimal solvents and experimental conditions are determined by modeling.

Lignin completely dissolves in ionic liquid solvent that is good for separating the lignin from cellulose. However, no identifiable product was formed using gas chromatography of Agilent 6890 DB-column[37]. However, ionic liquids are difficult to separate from lignin for reuse. The potential processes yet to be investigated are precipitation of polymeric lignin based by-products by water and by ultrafiltration membrane. Bioseparation routes are yet to be engineered. Nevertheless, none of the routes discussed has been optimized from bio-origins to reuse, pointing to a promising area for research.

The addition of acid catalysts helps the lignin conversion. Amongst them, some have shown no effect or recoverable production. The failed acid catalysts are shown as follows. However, it must be remembered that the reaction conditions were not optimized and were maintained at the dealkylation conditions studied for other systems[37].

High surface area titania (Ti-0720 T)
Nb_2O_5 hydrate (6518–43–11)
Silica/alumina T-869
Montmorillonite K10
Sulfated ZrO_2
$CoCl_2$
$NiCl_2$
$CuCl_2$
$ZnCl_2$
$InCl_3$
$SnCl_4$
$CeCl_3$
$SmCl_3$
$InCl_3$
etc.

4-Propylguaiacol product was produced in a recoverable quantity (>75% by weight in the product mixture) by the following catalyst: 0.5%Pd/alumina (C3677); $RhCl_3$, in EMIM-TF and EMIM-Cl ionic liquids. The reaction temperature was 180 °C. With increasing temperature, completion to 4-propylguaiacol production may be seen, but with more difficulty in ionic liquid solvent recovery and reuse. With increasing temperature the problem of acid hydrolysis reoccurring may be overcome. Also, using heterogeneous acid catalyst is a promising way to separate the functional phenolic products from the reaction mixture by in situ adsorption or by membrane, as soon as the products form. Further, the top ten chemical productions from lignin can be found in DOE documentation[29]. However, the engineered process designs are yet to be developed.

The screening of products and processing technologies describes a complex task more challenging than the development of fossil based refineries that feature fixed feedstocks (oil or gas) and an established portfolio of products. Considering the importance to scope for alternative feedstocks, processes and products, the biorefinery design requires models with strong *synthesis capabilities* and with functions to optimize degrees of freedom. Multiple chemistries should be screened to produce the most profitable products in the most resource efficient and integrated manner. This book discusses the essential and most important tools to design sustainable biorefinery systems.

1.7 Electrochemistry Application in Biorefineries

The application of electrochemistry powered by electron harvesting from organic wastes and transfer via the electrodes and electrolyte can greatly enhance the sustainability of energy production from biomass. Such technologies discussed in this book include the *proton exchange membrane fuel cell* (PEMFC) processing biogas for electricity and heat generation; *allothermal gasification* (combustion of char and steam gasification of the remaining biomass are carried out in separate, but heat integrated vessels, through which heat is circulated) and *solid oxide fuel cell* (SOFC) integration for efficient energy production, *mixed ionic electronic conducting hollow fiber membrane, microbial fuel cell* (MFC) and *bio fuel cell*, etc. Figure 1.15 shows the various types of fuel cells named after the electrolytes used.

Design, integration and sustainability analysis and modeling of SOFC in particular are discussed in this book. Life cycle assessment (LCA) and energy analysis data are provided for a range microgeneration options using fuel cells.

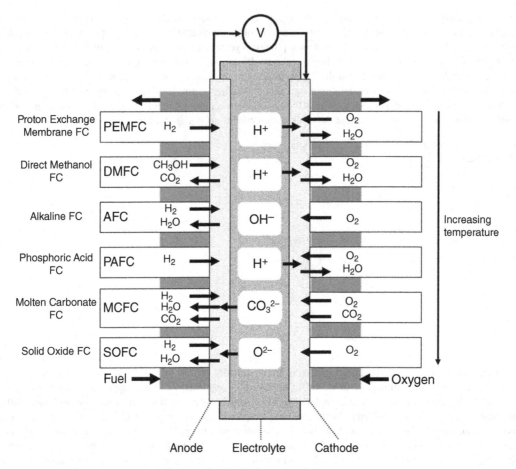

Figure 1.15   *Fuel cell types and reactions.*

MFC and bio fuel cells highlighted also hold promises. Bio fuel cells using glucose as substrate have the potential in the field of biomedical engineering, for example, as a pacemaker for the heart. There is great potential for organic waste disposal (including wastewater treatment) into electricity generation and metal recovery through MFC technology. A major challenge is the low power density compared to other types of fuel cells, as shown in Table 1.7. Though the amount of electric power generation is relatively low at present, future research is expected to improve the efficiency of these unique systems. These systems are a sustainable way to directly convert lignocellulosic biomass or wastewaters into useful energy and water using oxygen from air and hydrogen sourced from the biomass. An MFC can be single or double chambered. In MFC, microorganisms oxidize organic matter in the anode chamber (anaerobic conditions), producing electrons and protons. Electrons transfer via the external circuit to the cathode chamber, while protons transfer through the electrolyte. A solid proton exchange membrane electrolyte can be used to transfer protons from the anode to the cathode chamber. In the cathode chamber, electrons, protons and electron acceptor–oxygen–react together to produce water. Thus, if wastewater is the substrate used as a source of proton in the anode chamber, clean water can be produced from the cathode chamber by oxidizing protons. In a two-chamber setup, the anode and cathode compartments are separated by a proton selective membrane, allowing proton transfer from the anode to the cathode and preventing oxygen diffusion from the cathode to the anode. In a single-chamber MFC, the cathode is exposed directly to the air. Besides these two common designs, several adaptations can be made in MFC designs. The most common anode material is activated carbon cloth (wound or knitted) and brushes. Increasing the surface area per unit volume of anode helps in transferring proton

Table 1.7   *MFC systems research profile. (Reproduced with permission from Catal et al. (2008)[38]. Copyright © 2008, Elsevier.)*

Source Inoculum	Substrate and Concentration (g L^{-1})	MFC Anode Specification	MFC Cathode Specification	Power Density (A m^{-2})
Pre-acclaimed bacteria from MFC	Arabitol: 1.22	Non-wet proofed carbon cloth (2 cm^2 per 12 ml)	Wet proofed carbon cloth (7 cm^2 per 12 ml)	6.8
	Galactitol: 1.22	Non-wet proofed carbon cloth (2 cm^2 per 12 ml)	Wet proofed carbon cloth (7 cm^2 per 12 ml)	7.8
	Mannitol: 1.22	Non-wet proofed carbon cloth (2 cm^2 per 12 ml)	Wet proofed carbon cloth (7 cm^2 per 12 ml)	5.8
	Ribitol: 1.22	Non-wet proofed carbon cloth (2 cm^2 per 12 ml)	Wet proofed carbon cloth (7 cm^2 per 12 ml)	7.3
	Sorbitol: 1.22	Non-wet proofed carbon cloth (2 cm^2 per 12 ml)	Wet proofed carbon cloth (7 cm^2 per 12 ml)	6.2
	Xylitol: 1.22	Non-wet proofed carbon cloth (2 cm^2 per 12 ml)	Wet proofed carbon cloth (7 cm^2 per 12 ml)	7.1
Mixed bacterial culture maintained on sodium acetate (*Rhodococcus* and *Paracoccus*)	Glucose: 6.7 mmol	Non-wet proofed carbon cloth (2 cm^2 per 12 ml)	Wet proofed carbon cloth (7 cm^2 per 12 ml)	7
Mixed bacterial culture	Glucuronic acid: 6.7 mmol	Non-wet proofed carbon cloth (2 cm^2 per 12 ml)	Wet proofed carbon cloth (7 cm^2 per 12 ml)	11.8
Pure culture of *Geobacter sulfurreducens*	Sodium fumarate: 25 mmol	Ag/AgCl	Stainless steel (2.5 cm^2)	20.5

to the cathode that is also made up of activated carbon cloth, but with an embedded catalyst, for example, platinum. Reducing platinum quantity has the cost and environmental incentive, indicating that the surface area per unit volume of the anode must be increased with the anode brought closer to the cathode chamber to reduce the distance for proton transport. At the same time, oxygen must be prevented from penetrating through the cathode to the anode to increase the electron transfer rate externally generating power. Hence, proton exchange membrane material such as nafion (sulfonated tetrafluoroethylene based fluoropolymer–copolymer) is commonly used to facilitate proton transfer and a membrane is added on the oxygen side of the cathode chamber to prevent oxygen transport through the cathode.

The demand for water, mineral and energy resources is rapidly increasing due to the growing world population and rising economies. Currently mineral and energy resources are primarily obtained from nonrenewable geological deposits. These extraction processes need water. Though water is considered as a renewable resource, water will inevitably be depleted, beyond recovery, generating wastewater. The development of more sustainable routes to recover resources is paramount. Wastewater generating from agricultural, municipal, industrial and commercial sectors can be potential sources of metals that can be recovered using MFC. Their recovery is otherwise not possible using chemical processes, for example, precipitation, filtration and solvent extraction, etc., due to low metal selectivity and because of their low concentrations in a highly complex mixture. They can also cause secondary pollution from metal enriched sludge, from which it is not technically feasible to recover metals. MFC using bioelectrochemical technology to recover energy and minerals from wastewater offers the potential to overcome these problems. Elemental recovery from wastewaters can add economic and environmental value to the existing MFC technologies. The various electrochemical methods, such as electrodeposition, electrodialysis and electroprecipitation, can be used to recover Cu, Co, Ni, Pb, Cr, Sn and salt from scrap printed circuit boards and dilute wastewater streams.

Water electrolysis and microbial electrolysis cells (MECs) use electrical voltage (and work on the reverse of fuel cell principles) to produce hydrogen. Hydrogen production is an energy intensive process. Natural gas steam reforming (that is a highly endothermic chemical reaction) is the most common hydrogen production method. Renewable electrical energy (e.g., solar, wind, hydro, etc.) can be supplied in MEC to produce hydrogen, renewably. Electrohydrogenesis is an MEC process that requires application of voltage to produce hydrogen using acetate or glucose as a substrate in the anode. The cathode is sealed to keep the oxygen source away to produce hydrogen from a single chamber MEC without a membrane. The process delivers a higher hydrogen yield than fermentation of acetate and a higher energy efficiency than water electrolysis. However, hydrogen produced needs to be compressed and stored before further use as an energy vector. This adds to the economic and environmental life cycle costs. Also, more than a third of the cost is incurred from expensive platinum catalyst used on the cathode; in some cases its loading is as high as 0.5 mg cm^{-2}. To reduce the cost, alternative catalysts have been explored, but these resulted in high loading to achieve the same performance as Pt. A more cost-effective and environmentally sustainable route is to use hydrogen to produce products in the cathode chamber within the same electrochemical apparatus, generally MEC. Electromethanogenesis is a process to produce methane in the cathode chamber combining carbon dioxide, proton, and electron evolved from the anode chamber in a two-chamber MEC, by applying a slightly different electrical voltage needed for electrohydrogenesis. Based on thermodynamic calculations, methane could be produced electrochemically through carbon dioxide reduction at a voltage of 0.169 V under standard conditions, or −0.244 V under more biologically relevant conditions at a pH of 7[39], by primarily using Sabatier's reaction[13]. This has opened up a whole range of opportunities for carrying out environmentally friendly biorefining productions in bioelectrochemical systems (BESs) from organic wastes.

1.8   Introduction to Energy and Water Systems

While most chapters in this book discuss the subject of biorefineries and chemical processes, some features of energy and water systems modeling have been discussed in Web Chapters 2 and 3, because these systems are interacting systems of biorefineries and call for overall integration and analysis for sustainability. Web Chapter 1 includes waste and emission mitigation technologies relevant for biorefineries.

The main use of lignocellulose is for chemical and material productions, because there is no other alternative source of renewable carbon. There are options for renewable energy generation, other than biomass, such as wind, solar, tidal, geothermal and hydropower. Depending upon the availability of these resources, choices of local level renewable energy supplies can be made. The main stress should be on local availability of natural resources and how to make sustainable use of them, not only for technoeconomic viability but also to help grow an economy through education and job creation. Biorefinery engineering is just not about effective use of biomass, but also how whole supply chain systems can be synergistically designed using local resources delivering products of need, for socioeconomic welfare and ecosystem balance.

Renewable technologies need to be developed to mitigate global warming potential resulting from fossil based energy generation systems. However, a renewable energy supply is intermittent in nature. This intermittency has to be resolved by energy storage systems. Energy storage systems can be of various forms, including chemical, such as hydrogen and methanol. Biomass is a form to entrap renewable carbon and hydrogen and thus energy. Hence, to resolve supply intermittency of renewable energy systems listed above, biomass CHP systems can be operated when required. Largely, the renewable energy supply intermittency is addressed by fossil (coal, crude oil) based power plants, even today. However, the world is approaching an era of a lack of or prohibitively expensive fossil resources.

The other ways of energy storage are through the design of functional materials: physical, thermochemical, and electrochemical storage systems. The storage systems can be effectively designed to store water and heat (thermochemical: absorption and adsorption based) as well as water and electricity (fuel cells) simultaneously. Therefore, there are two ways to address the renewable energy resource intermittencies.

1. Biomass CHP generation. (However, remember that the best use of biomass is to produce products containing carbon as there is no other renewable carbon resource available. Excess biomass can be used for energy generation.)
2. Energy storage.

Under these broad levels of groupings, some options for energy supply and storage exist. Technologies need to be developed using abundant, cheap and environmentally friendly material in a scalable and reproducible process. Graphene (the lowest assembly of carbon atoms with one atom thickness) is an important material for energy storage and generation processes. Graphene has been discussed with novel applications in Web Chapter 2, because many fundamental theories in physics, electrochemistry and chemical physics can be generated investigating graphene.

The energy transfer model is at the core of design of materials and systems for energy storage and renewable energy generation; hence, it is the key focus of Web Chapter 2. Two strategies can be adopted to control energy transfer of a system.

1. Active thermal control strategy. An active thermal control strategy is based on process control theory. Processes are associated with unsteady state operations, in which the various properties of process streams and operating conditions change. In order to ensure safe operation of processes and achieve the greatest productivity and purity and least cost, etc., the processes need to operate at optimal operating conditions. Process control is needed to achieve optimal operating conditions during process operation. Hence, the subject of process control has been discussed with sufficient depth for biorefinery engineers in Web Chapter 2.
2. Passive thermal control strategy. A passive thermal control strategy deals with the design of materials to optimize physical properties so as to minimize external heat and power requirements or to maximize net heat and power generations from a system to the surrounding, without incorporation of any external fuel. This is the general philosophy for biorefinery design and is specifically addressed in Web Chapter 2.

Global greenhouse gas emissions result from direct consumption of fossil fuels by 58% and the balance from indirect electricity consumption. Figure 1.16 shows a generic trend of greenhouse gas emissions in percentage weight from end use sectors. The majority is contributed by the transport sector (33%), which can be decarbonized by biofuel production together with carbon capture and storage, followed by the industrial sector (28%) consisting of energy use by the sector, chemical and material production and extraction of metals, residential building (21%) and commercial building (18%). Decarbonization of the industrial sector will depend on how effectively biomass is utilized. Biomass is the only alternative source to chemical and material production, whilst renewable energy systems and energy storage systems can share some portions of the industrial energy use and energy required for metal extraction. Thus, only co-optimization of biorefinery, renewable energy and energy storage systems can fully decarbonize the industrial sector.

More than half of the industrial greenhouse gas emissions is caused by direct fossil resource consumption that must be avoided. At the point of use, a significant fraction of energy use is for heating water for domestic and industrial purposes. This water–energy nexus is the relationship between how energy is used to abstract, treat and transport water to users and water used in many energy conversion processes, waste heat recovery and cooling cycles. With increased insurgence

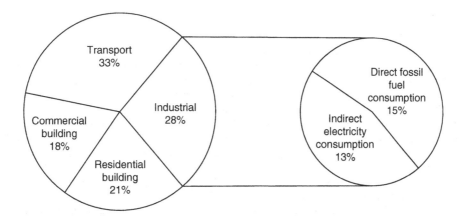

Figure 1.16 *Greenhouse gas emissions in weight percentages from end use sectors.*

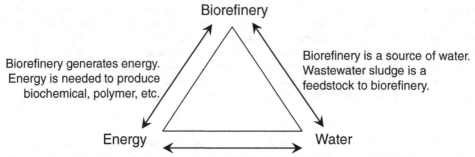

Biorefinery

Biorefinery generates energy.
Energy is needed to produce
biochemical, polymer, etc.

Biorefinery is a source of water.
Wastewater sludge is a
feedstock to biorefinery.

Energy   Water

Energy is needed throughout from abstraction through recovery reuse to deliver water.
Wastewater sludge is a source of energy. Water is an important medium for energy transmission.

Figure 1.17   *Biorefinery–energy–water nexus.*

of biorefineries, water footprint and thereby energy use is going to increase as biomass has more water content per unit of energy delivered compared to fossil resources. Energy is required to recover this water from biomass. In the biomass gasification route, the wet product gas is cooled to just below its dew point to separate noncondensable gas from water. While biomass is a source of water and energy, wastewater sludge is an important feedstock for a biorefinery. Thus, there are three-way interactions or nexus between the biorefinery, energy and water (Figure 1.17). Hence, the three systems must be designed synergistically for an urban system. Thus, in this book, the latter two are also addressed alongside in-depth learning materials for biorefinery systems. Clearly, engineers need to be aware of such cross-disciplinary nexus areas.

1.9   Evaluating Biorefinery Performances

1.9.1   Performance Indicators

The promise of biorefineries as contributors to sustainable development makes necessary the introduction of sustainability metrics for biorefinery planning and design. The complexity and multifaceted nature of the concept has led to the derivation of various indicators for the economic, environmental and social components. Indicators derived from concepts such as green chemistry and engineering, cleaner production, ecoefficiency, industrial ecology, ecodesign and life cycle analysis are used by industry and institutions such as the Center for Waste Reduction Technologies, the Institution of Chemical Engineers, among others. The various sets of indicators proposed often have similarities and discrepancies, but they can be useful if applied as complementary measures. Indicators continue evolving and new concepts and methodologies are proposed. The various indicators used to analyze sustainability must allow their evaluation from a life cycle approach and must be useful for the identification of hot spots and trade-offs. Environmental impacts of biorefineries have become important, such as climate change, eutrophication, water depletion, land use, depletion of energy and material resources and aquatic toxicity. Several aspects can be categorized according to the particularities of biorefineries.

Considering the significance of indicators for assessing biorefinery sustainability, the most prominent set of parameters are shown in Table 1.8. The metrics are grouped according to their relation to resources use, process efficiency, advantage over fossil-based products and economic performances, etc. The indicators are appropriate for comparison of biorefinery configuration alternatives and for identification of processing pathways for process integration. Not all the indicators are necessary at the same time because they may not differ from one configuration to another and they may be comparable with fossil based equivalent systems.

In addition, local versus global consequences are considered to adopt a biorefinery design configuration. Amongst the indicators shown in Table 1.8, the resource depletion, climate change impacts and aquatic toxicities have global consequences, whilst the acidification potential has regional consequences and the photochemical oxidant creation potential has local level consequences.

Table 1.8 *Indicators used throughout the present work to evaluate biorefinery performances*[40–42].

Indicators	Definition	Measured as: Better or Worse than Fossil Based Equivalents?
Process performance		
Energy conversion efficiency	Ratio between net energy output (as valuable products, including electricity, steam, fuels and other streams contributing to energy generation) to energy input (e.g., total biomass calorific value).	Fraction or percentage. Worse than fossil based equivalents.
Primary resources use		
Cumulative primary energy (CPE)	Primary energy resources (crude oil, natural gas, coal, etc.) consumed throughout the life cycle of a system, process or product.	Energy units. Saves fossil resource.
Abiotic resources use	Depletion of nonrenewable resources.	Mass of Sb equivalent. Saves abiotic resource.
Land use	Amount of land used per unit biomass feedstock production. Lignocellulosic feedstocks are chosen to reduce land use.	ha (hectare) Worse than fossil based equivalents. Lignocellulosic feedstocks are thus urgently sought.
Water use	Amount of freshwater depletion, beyond recovery.	m^3 Worse than fossil based equivalents.
Environment		
Greenhouse gases or global warming or climate change	Gases causing greenhouse effect, i.e., with global warming potential.	Mass of CO_2 equivalent. Avoid GHG emissions.
Eutrophication	Overenrichment of water by nutrients such as nitrogen and phosphorus which, among other effects, can destroy aquatic life in affected areas.	Mass of PO_4^{3-} equivalent. Worse than fossil based equivalents.
Acidification	Emissions of sulfur and nitrogen oxides cause acid rains that destroy vegetation.	Mass of SO_2 equivalent. Better than fossil based equivalents.
Photochemical oxidant creation potential	The volatile organic compounds and ozone in the lower atmosphere are responsible for urban smog and ground level ozone formation and are classified under photochemical oxidant creation potential.	Mass of ethylene equivalent. Worse than fossil based equivalents.
Aquatic toxicity	Toxic substance emissions to water bodies.	Mass of 1,4-dichlorobenzene equivalent. Worse than fossil based equivalents.
Economic performance		
Economic margin	Revenue potential of a biorefinery.	Currency units. Worse than fossil based equivalents.

1.9.2 Life Cycle Analysis

The metrics and tools for evaluating biorefinery sustainability are still evolving as the sustainability concept itself. There is a need for extensive quantitative tools to support decision making in biorefinery design from cradle to grave, that is, focusing not only on a product or process but the entire biorefinery system. Several tools have been developed for this purpose including LCA, the energy-based sustainability index (ESI), material flow analysis (MFA), ecological input/output analysis (EIOA), among others. LCA is the most suitable tool for comprehensive and quantitative assessments capturing the direct and indirect environmental impacts associated with a given product or process design. This tool standardized by the ISO 14040, 14041 and 14044 has been internationally accepted. LCA is playing a leading role for the sustainability analysis of biorefinery systems.

LCA is used to find the environmental burdens associated with a product, process, or activity by identifying and quantifying energy and material resources used and wastes and emissions released to the environment. The life cycle of a system or product includes extraction and processing of raw materials, manufacturing, transportation, distribution, and use of products, and reuse, maintenance and recycling of material of construction. The holistic approach of LCA is needed for sustainable biorefinery design and screening of options, discussed in detail with practical applications to biorefinery and energy systems in this book.

1.10 Chapters

Opening with an introduction on the concept and development of biorefinery, the book is then split into four parts:

Tools. Detailed analysis of economic, environmental and whole system impact, as well as combined economic value and environmental impact (EVEI) analysis. Life cycle assessment and heat integration and utility system design. Mathematical programming based optimization and genetic algorithms.

Process synthesis and design. Focuses on modern unit operations and innovative process flowsheets: reactors; electrochemical, membrane and combined reaction and separation processes with multifunctionalities. Production of chemicals and polymers from biomass. Thermochemical processing of biomass and biochemical processing of biomass. Processes for carbon dioxide capture.

Biorefinery systems. Biorefinery process synthesis examples using design, integration and sustainability analysis tools as appropriate. Bio-oil and algae biorefineries, integrated fuel cells and renewables, multiscale modeling of heterogeneously catalyzed reactions using the example of biodiesel.

Interacting systems of biorefineries. Looks at minimalizing waste and emissions, storing energy and the optimization and reuse of water.

Additional exercises and examples referred to within each chapter of the book can be found on a companion website. Four case studies are given on the companion website: LCA based problem solving approaches on biomass CHP plant design problems, epoxy resin production from biomass, wastewater sludge based CHP and the LCA approach for solar organic photovoltaic cells manufacturing. Figure 1.18 shows the structure of the book for lecture planning.

1.11 Summary

The introduction chapter provides overviews of biorefinery design options, including feedstock, product and platform chemical and process choices. Lignocellulosic biomass has been specifically covered for sustainable biorefinery systems. Lignin value addition holds the key to the commercial success of biorefineries. The background of lignin including process design options has been shown in great detail. The future biorefinery designs will feature electrochemical and bio-oil platforms to produce a range of products in a compact and efficient way. Co-design of lignocellulosic biorefineries, renewable energy systems and energy storage systems is necessary for overall sustainable development in a world without fossil resources. Economics, LCA and multicriteria indicators are applied for biorefinery design, integration and sustainability analysis.

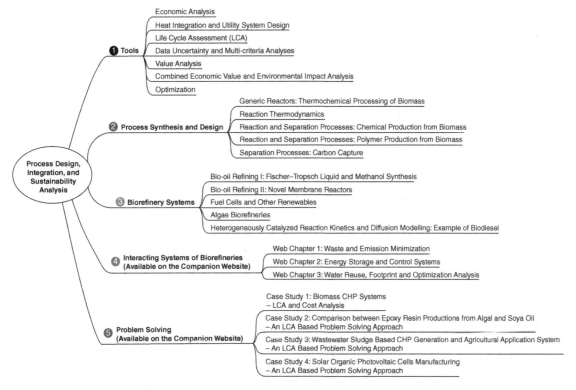

Figure 1.18 *Structure for lecture planning.*

References

1. B. Kamm, P.R. Gruber, M. Kamm, *Biorefineries – Industrial Processes and Products: Status Quo and Future Directions*, Wiley-VCH Verlag GmbH & Co, KGaA, Weinheim, Germany, 2006.

2. F. Cherubini, G. Jungmeier, M. Wellisch, T. Willke, I. Skiadas, R. Van Ree, E. Jong, Toward a common classification approach for biorefinery systems, *Biofuel Bioprod. Bior.*, **3**, 534–546 (2009).

3. D.K. Lee, V.N. Owens, A. Boe, P. Jeranyama, *Composition of Herbaceous Biomass Feedstocks*, Report prepared for the Sun Grant Initiative, North Central Sun Grant Center and South Dakota State University, SGINC1, 2007.

4. J. Sadhukhan, M.M. Mustafa, N. Misailidis, F. Mateos-Salvador, C. Du, G.M. Campbell, Value analysis tool for feasibility studies of biorefineries integrated with value added production, *Chem. Eng. Sci.*, **63**, 503–519 (2008).

5. C. Du, G.M. Campbell, N. Misailidis, F. Mateos-Salvador, J. Sadhukhan, M. Mustafa, R.M. Weightman, Evaluating the feasibility of commercial arabinoxylan production in the context of a wheat biorefinery principally producing ethanol. Part 1. Experimental studies of arabinoxylan extraction from wheat bran, *Chem. Eng. Res. Des.*, **87**, 1232–1238 (2009).

6. N. Misailidis, G.M. Campbell, C. Du, J. Sadhukhan, M. Mustafa, F. Mateos-Salvador, R.M. Weightman, Evaluating the feasibility of commercial arabinoxylan production in the context of a wheat biorefinery principally producing ethanol: Part 2. Process simulation and economic analysis, *Chem. Eng. Res. Des.*, **87**, 1239–1250 (2009).

7. T.J. Davison, C. Okoli, K. Wilson, A.F. Lee, A. Harvey, J. Woodford, J. Sadhukhan, Multiscale modeling of heterogeneously catalysed transesterification reaction process: an overview, *R. Soc. Chem. Adv.*, **3**, 6226–6240 (2013).

8. A. Kapil, A.F. Lee, K. Wilson, J. Sadhukhan, Kinetic modelling studies of heterogeneously catalyzed biodiesel synthesis reactions, *Ind. Eng. Chem. Res.*, **50**, 4818–4830 (2011).

9. A. Kapil, S.A. Bhat, J. Sadhukhan, Dynamic simulation of sorption enhanced simulated moving bed reaction processes for high purity biodiesel production, *Ind. Eng. Chem. Res.*, **49**, 2326–2335 (2010).

10. K.S. Ng and J. Sadhukhan, Techno-economic performance analysis of bio-oil based Fischer–Tropsch and CHP synthesis platform, *Biomass Bioenergy*, **35**, 3218–3234 (2011).
11. J. Sadhukhan and K.S. Ng, Economic and European Union environmental sustainability criteria assessment of bio-oil based biofuel systems: refinery integration cases, *Ind. Eng. Chem. Res.*, **50**, 6794–6808 (2011).
12. K.S. Ng and J. Sadhukhan, Process integration and economic analysis of bio-oil platform for the production of methanol and combined heat and power, *Biomass Bioenergy*, **35**, 1153–1169 (2011).
13. K.S. Ng, N. Zhang, J. Sadhukhan, Techno-economic analysis of polygeneration systems with carbon capture and storage and CO_2 reuse, *Chem. Eng. J.*, **219**, 96–108 (2013).
14. K.S. Ng, Y. Lopez, G.M. Campbell, J. Sadhukhan, Heat integration and analysis of decarbonised IGCC sites, *Chem. Eng. Res. Des.*, **88**, 170–188 (2010).
15. K.S. Ng, N. Zhang, J. Sadhukhan, A graphical CO_2 emission treatment intensity assessment for energy and economic analyses of integrated decarbonised production systems, *Comput. Chem. Eng.*, **45**, 1–14 (2012).
16. J. Sadhukhan, K.S. Ng, N. Shah, H.J. Simons, Heat integration strategy for economic production of CHP from biomass waste, *Energy Fuels*, **23**, 5106–5120 (2009).
17. S. Blomberg, *Biomass-to-Energy Feasibility Study*, Final Report, DOE Award Number: DE-FC26-01NT41352, Tulsa, OK (US), 2002.
18. T.G. Bridgeman, J.M. Jones, I. Shield, P.T. Williams, Torre faction of reed canary grass, wheat straw and willow to enhance solid fuel qualities and combustion properties, *Fuel*, **87**, 844–856 (2008).
19. W. Jangsawang, A.K. Gupta, K. Kitagawa, S.C. Lee, High temperature steam and air gasification of non-woody biomass wastes, *As. J. Energy Env.*, **08**, 601–609 (2007).
20. S. Gaur and T.B. Reed, *Thermal Data for Natural and Synthetic Fuels*, Marcel Dekker, New York, 1998.
21. D.L. Carpenter, R.L. Bain, R.E. Davis, A. Dutta, C.J. Feik, K.R. Gatson, W. Jablonski, S.D. Phillips, M.R. Nimlos, Pilot-scale gasification of corn stover, switchgrass, wheat straw, and wood: 1. Parametric study and comparison with literature, *Ind. Eng. Chem. Res.*, **49**, 1859–1871 (2010).
22. P. McKendry, Energy production from biomass (Part 1): overview of biomass, *Bioresour. Technol.*, **83**, 37–46 (2002).
23. R. Wooley and V. Putsche, *Development of an ASPEN PLUS Physical Property Database for Biofuel Components*, Technical Report NREL/MP-425-20685, National Renewable Energy Laboratory (NREL), US, 1996.
24. A. Chang and Y.A. Liu, Integrated process modeling and product design of biodiesel manufacturing, *Ind. Eng. Chem. Res.*, **49**, 1197–1213 (2010).
25. D.L. Klass, *Biomass for Renewable Energy, Fuels and Chemicals*, Academic Press, US, 1998.
26. T. Werpy and G. Petersen, *Top Value Added Chemicals from Biomass. Volume I – Results of Screening for Potential Candidates from Sugars and Synthesis Gas*, National Renewable Energy Laboratory, Golden, CO (US), 2004.
27. M. Patel, M. Crank, V. Dornburg, B. Hermann, L. Roes, B. Hüsing, L. Overbeek, F. Terragni, E. Recchia, *Medium and Long-Term Opportunities and Risks of the Biotechnological Production of Bulk Chemicals from Renewable Resources*, Department of Science, Technology and Society (STS)/Copernicus Institute, Utrecht University, Utrecht, Netherlands, 2006.
28. J.J. Bozell and G.R. Petersen, Technology development for the production of biobased products from biorefinery carbohydrates – the US Department of Energy's "Top 10" revisited, *Green Chem.*, **12**, 539–554 (2010).
29. J.J. Bozell, J.E. Holladay, D. Johnson, J.F. White, *Top Value Added Chemicals from Biomass. Volume II – Results of Screening for Potential Candidates from Biorefinery Lignin*, Pacific Northwest National Laboratory, Richland, WA, PNNL-16983, 2007.
30. D. King, *The Future of Industrial Biorefineries. World Economic Forum*, Switzerland. Available at: http://www3.weforum.org/docs/WEF_FutureIndustrialBiorefineries_Report_2010.pdf, 2010.
31. F. Peng, P. Peng, F. Xu, R.C. Sun, Fractional purification and bioconversion of hemicelluloses, *Biotechnol. Adv.*, **30**, 879–903 (2012).
32. J.H. Reith, R. van Ree, R.C. Campos, R.R. Bakker, P.J. de Wild, F. Monot, B. Estrine, A.V. Bridgwater, A. Agostini, *Lignocellulosic Feedstock Biorefinery for Co-production of Chemicals, Transportation Fuels, Electricity and Heat*, Energy Research Centre of the Netherlands, International Workshop on Biorefinery, Madrid, 2009.
33. B. Xiao, X.F. Sun, R.C. Sun, Chemical, structural, and thermal characterizations of alkali soluble lingins and hemicelluloses, and cellulose from maize stems, rye straw, and rice straw, *Polym. Degrad. Stab.*, **74**, 307–319 (2001).
34. R.C. Sun, J. Tomkinson, F.C. Mao, X.F. Sun, Physicochemical characterization of lignins from rice straw by hydrogen peroxide treatment, *J. Appl. Polym. Sci.*, **79**, 719–732 (2001).
35. E.A.B de Silva, M. Zabkova, J.D. Araújo, C.A. Cateto, M.F. Barreiro, M.N. Belgacem, A.E. Rodrigues, An integrated process to produce vanillin and lignin-based polyurethanes from kraft lignin, *Chem. Eng. Res. Des.*, **87**, 1276–1292 (2009).
36. M.M Campbell and R.R. Sederoff, Variation in lignin content and composition, *Plant Physiol.*, **110**, 3–13 (1996).
37. J.B. Binder, M.J. Gray, J.F. White, Z.C. Zhang, J.E. Holladay, Reactions of lignin model compounds in ionic liquids, *Biomass Bioenergy*, **33**, 1122–1130 (2009).
38. T. Catal, S. Xu, K. Li, H. Bermek, H. Liu, Electricity production from polyalcohols in single-chamber microbial fuel cells, *Biosens. Bioelectron.*, **24**, 855–860 (2008).

39. S. Cheng, D. Xing, D.F. Call, B.E. Logan, Direct biological conversion of electrical current into methane by electromethanogenesis, *Environ. Sci. Technol.*, **43**, 3953–3958 (2009).
40. E. Martinez-Hernandez, G.M. Campbell, J. Sadhukhan, Economic value and environmental impact (EVEI) analysis of biorefinery systems, *Chem. Eng. Res. Des.*, In press (2013).
41. E. Martinez-Hernandez, M.H. Ibrahim, M. Leach, P. Sinclair, G.M. Campbell, J. Sadhukhan, Environmental sustainability analysis of UK whole-wheat bioethanol and CHP systems, *Biomass Bioenergy*, **50**, 52–64 (2013).
42. J. Sadhukhan, N. Zhang, X.X. Zhu, Analytical optimisation of industrial systems and applications to refineries, petrochemicals. *Chem. Eng. Sci.*, **59**(20), 4169–4192 (2004).

Part II

Tools

2

Economic Analysis

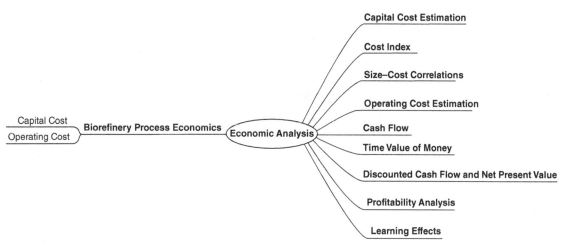

Structure for Lecture Planning

2.1 Introduction

Process economics is one of the crucial aspects for evaluation of process designs. It is often the main criterion in justifying the feasibility of a new design or modification of a plant. Many biomass based technologies are not yet widely employed. This is primarily attributed to the high capital investment and operating costs associated with these technologies, and thus they are less competitive compared to the fossil fuel based technologies. Therefore, it is highly essential to master the skills of performing a proper economic evaluation of a biorefinery plant design in order to gain deeper insights and achieve a better outcome in terms of economics and optimal designs.

This chapter discusses the fundamental concepts of economics germane to chemical engineering process design (Section 2.2) and the methodology for performing economic analysis of process technologies (Section 2.3). The correlation and cost data information of biorefinery design options (Section 2.4) are also shown.

Biorefineries and Chemical Processes: Design, Integration and Sustainability Analysis, First Edition.
Jhuma Sadhukhan, Kok Siew Ng and Elias Martinez Hernandez.
© 2014 John Wiley & Sons, Ltd. Published 2014 by John Wiley & Sons, Ltd.
Companion Website: http://www.wiley.com/go/sadhukhan/biorefineries

The learning outcomes from this chapter are as follows.

- To assimilate the fundamental economic terminologies and concepts.
- To perform a capital and operating costs evaluation using the standard techniques, equations, graphs, cost correlations and factors.
- To use economic criteria such as economic potential and netback to justify the economic viability of a design.

2.2 General Economic Concepts and Terminology

2.2.1 Capital Cost and Battery Limits

Capital cost is the cost for building a plant. It can be categorized into two parts: direct and indirect capital costs. Direct capital costs refer to the purchased and installation costs of equipment for constructing a plant. The cost data and correlation for standard equipment can be obtained from various sources including (1) website: *Matches' Process Equipment Cost Estimates*, www.matche.com; (2) reference books: *Coulson & Richardson's Chemical Engineering Design Volume 6, Product & Process Design Principles* and *Guide to Capital Cost Estimating* published by the Institute of Chemical Engineers (IChemE); (3) peer-reviewed archived journals. Most of the equipment costs provided in the literature are the free-on-board (f.o.b.) purchased cost. This means that the delivery cost of equipment is not included. The equipment cost is also influenced by various factors, namely, material of construction, pressure and temperature. These factors need to be considered for each piece of equipment by multiplying the correction factors. The geographical location of a plant is also a highly influential factor for the capital cost due to variations in local regulations, labor, taxes, cost of transportation, etc. A correction factor, if available, should be applied to account for such variations.

Battery limit is used to classify direct capital cost. Inside battery limits (ISBL) comprise the cost of purchasing and installation of major process equipment such as reactors, separators and gas turbines, etc. Other supporting facilities such as utilities and services are considered as outside battery limits (OSBL). Indirect capital costs refer to the design and engineering costs for building a site, contractor's fees and contingency allowances (costs forecasted for some unforeseen circumstances). These are estimated by taking a certain factor on top of the direct capital costs. Working capital should also be included in the capital cost evaluation. It is the cost required for the acquisition of raw materials during the initial start-up stage of a plant, until the plant becomes productive or makes money.

2.2.2 Cost Index

The cost data of equipment obtained for a year is only valid for that particular year. Costs vary with time. Therefore, the cost index method is applied, as shown below for updating the cost taken from previous years and to be used in the current cost analysis:

$$C_{pr} = C_o \left(\frac{I_{pr}}{I_o} \right) \tag{2.1}$$

where

C_{pr} is the present cost
C_o is the original cost
I_{pr} is the present index value
I_o is the original index value.

Many methods to estimate the cost index are available, such as the Chemical Engineering Plant Cost Index (CEPCI), the Marshall and Swift (M&S) Equipment Cost Index, the Nelson-Farrar–(NF) Refinery Construction Cost Index and the Engineering New-Record (ENR) Construction Cost Index. These include the cost index for equipment, labor, engineering and supervision, etc. The Chemical Engineering Plant Cost Index (CEPCI) and the Marshall and Swift (M&S) Equipment Cost Index are used for the cost estimation of biorefineries. Both cost indices are published monthly in *Chemical Engineering*. Figure 2.1 shows the annual CEPCI from year 1996 to 2010.

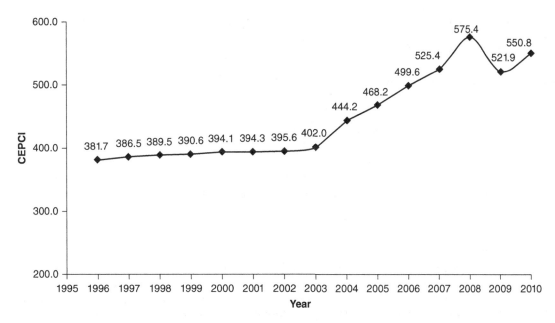

Figure 2.1 *Annual Chemical Engineering Plant Cost Index (CEPCI) from year 1996 to 2010.*

Exercise 1. Calculate the capital cost of two boiler units in year 2010 (CEPCI: 550.8), if the capital cost of one unit was $0.345 million in year 1998 (CEPCI: 389.5).

Solution to Exercise 1. Using Equation (2.1), the capital cost of one boiler in year 2010 is

$$C_{pr} = C_o \left(\frac{I_{pr}}{I_o} \right)$$

$$= 0.345 \times \left(\frac{550.8}{389.5} \right)$$

$$= \$0.488 \text{ million}$$

Hence, two boilers cost: $2 \times \$0.488$ million = **\$0.976 million** in year 2010.

2.2.3 Economies of Scale

A scaling factor R is applied to estimate the cost of a system based on the known cost of the system for a different size, as shown below. This relationship assumes that the equipment or unit operations can be scaled up/down. The maximum size limit is normally given and multiple units have to be taken into account if the size of the unit exceeds the maximum size.

$$\frac{COST_{size2}}{COST_{size1}} = \left(\frac{SIZE_2}{SIZE_1} \right)^R \tag{2.2}$$

where

$SIZE_1$ is the capacity of the base system
$COST_{size1}$ is the cost of the base system
$SIZE_2$ is the capacity of the system after scaling up/down
$COST_{size2}$ is the cost of the system after scaling up/down
R is the scaling factor.

Exercise 2. Calculate the capital cost of two water gas shift reactors in million $ for a flow rate of 942 kmol h^{-1} through each reactor. The known capital cost of $40.59 million was obtained for a flow rate of 15 600 kmol h^{-1} through one reactor. The scaling factor is 0.85.

Solution to Exercise 2. Using Equation (2.2), the capital cost of one water gas shift reactor in million $ is as follows:

$$\frac{COST_{size2}}{COST_{size1}} = \left(\frac{SIZE_2}{SIZE_1} \right)^R$$

$$COST_{size2} = 40.59 \times \left(\frac{942}{15600} \right)^{0.85} = \$3.73 \text{ million}$$

Hence, the capital cost of two water gas shift reactors is $2 \times 3.73 = $ **$7.46 million.**

2.2.4 Operating Cost

The operating costs can be classified into two main categories: fixed and variable operating costs. Fixed operating costs are independent of the production rate and quantity, in contrast to variable operating costs. These include the costs of maintenance, labor, taxation, insurance, royalties, etc. Fixed operating costs are estimated using factors that are normally based on indirect capital costs. Variable operating costs consist of the costs of raw materials, utilities, etc. The sum of fixed and variable operating costs is the direct production costs (DPCs) of a plant. Other costs such as the costs of research and development, sales expenses and general overheads are added as % of DPC to obtain the total operating cost.

Variable operating costs include the costs of raw materials (e.g., feedstock, catalyst, solvent, etc.) and utilities (e.g., electricity, steam, cooling water, etc.). The costs of raw materials and prices of products are highly volatile, vary with time and thus have the largest impact on the economic performance of a plant in most cases. These values can be obtained from business information providers such as *ICIS Pricing* and *IHS Chemical Week*. The costs of utilities also contribute to a major part of the variable operating costs. These costs vary across organizations. Thus to obtain relevant results, specific information must be collected from associated utility providers.

Table 2.1 shows the factors associated with the fixed cost and other DPC specifications.

Table 2.1 *Cost estimation of fixed operating cost. (Reproduced with permission from Sinnott (2006)[1]. Copyright © 2006, Elsevier: Butterworth-Heinemann.)*

No.	Specification	Cost Estimation
	Fixed Operating Costs	
1	Maintenance	5–10% of indirect capital cost
2	Personnel	See "labor cost" in Section 2.4.2
3	Laboratory costs	20–23% of (2)
4	Supervision	20% of (2)
5	Plant overheads	50% of (2)
6	Capital charges	10% of indirect capital cost
7	Insurance	1% of indirect capital cost
8	Local taxes	2% of indirect capital cost
9	Royalties	1% of indirect capital cost
	Direct Production Cost (DPC) = Variable + Fixed Operating Costs	
10	Sales expense	
11	General overheads	} 20–30% of DPC
12	Research and development	
	Total operating cost = 1.2 or 1.3 times the DPC	

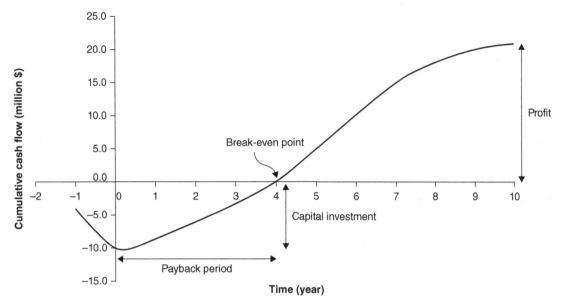

Figure 2.2   *Typical cash flow diagram.*

2.2.5   Cash Flows

An organization runs its everyday business by relying on a sustainable cash flow. It is thus crucial to understand the input and output of the money flows within an organization. A cash flow diagram, Figure 2.2, shows the capital investment during the initial start-up period (year −1 and 0) and the net cash flow. The net cash flow accounts for the earnings and expenditure over a project lifetime. The process begins at the end of year 0, with a variable yet increasing income until the profits stabilize at around a net cash flow. This relates to the equipment lifespan, increasing production costs and also the market value of the products. The cumulative cash flow remains negative until all the capital investment is reimbursed; this period of time is known as the *payback period*. The cumulative cash flow becomes positive as soon as the plant receives a net positive cash flow from selling products (starting from year 4 in the figure). Having a shorter payback period corresponds to an economically more viable design.

2.2.6   Time Value of Money

The "time value of money" reflects that the present value of money is worthier than the future value of money. For example, $100 cash that you receive today would have the same amount as the $100 you would receive after two years, but the "worthiness" of the money is different. If you invest $100 in a bank at present, you will eventually be getting more than $100 after two years, due to the addition of the interest paid by the bank. The relationship between present and future values is shown by

$$PV = \frac{FV}{(1+r)^n} \tag{2.3}$$

where

PV is the present value
FV is the future value
r is the discount rate or interest rate
n is the number of years of investment.

By taking an interest rate of 15% each year, an initial investment of $100 (present value) will become $100 \times (1 + 0.15)^2$ = $132.25 (future value) after two years. The present value is lower than the future value since the money at present is more valuable than in the future. This is the reason for using the discount rate to predict the "worthiness" of the future money in the present context.

> ## Did you know?
>
> If you win a lottery, it is better to redeem the whole lump sum of money now rather than receiving numerous payments over a few years. This is because the value of the money at present is greater than in the future and the money will depreciate over time. Your money will "shrink"!

2.2.7   Discounted Cash Flow Analysis and Net Present Value

A discounted cash flow (*DCF*) analysis is a method to evaluate the economic potential of an investment, where the projected future value of the cash flow based on capital investment is converted into its present value by applying a discount rate. This analysis considers the aforementioned time value of money.

The cumulative discounted cash flow is expressed as the net present value (*NPV*) in the *DCF* analysis. *NPV* is calculated using the following equation, where C_f is the cash flow in a particular year and T_{PL} is the plant life:

$$NPV = \sum_{n=0}^{n=T_{PL}} \frac{C_f}{(1 + r)^n} \qquad (2.4)$$

An example of *DCF* analysis is shown in Table 2.2. In this example, the capital investment is assumed to be $10 million and the cost is distributed over two years, that is, year −1 and 0, by 40% and 60% of the total capital investment, respectively.

In general, *NPV* also serves as an indicator for the profit of a project and thus deciding the feasibility of a particular project; for example, $NPV > 0$ means that the project can bring profits, $NPV < 0$ will result in a loss while $NPV = 0$ represents neither gain nor loss. See Equation (6.7) for a further discussion on calculation of the DCF analysis.

Table 2.2   *An example of DCF analysis for a plant life of 10 years and annual discount rate of 15%.*

Year	Cash Flow (million $)	Discounted Cash Flow (million $)	NPV (million $)
−1	−4.0	−4.0	−4.0
0	−6.0	−6.0	−10.0
1	1.5	1.3	−8.7
2	2.5	1.9	−6.8
3	2.8	1.8	−5.0
4	3.2	1.8	−3.1
5	5.0	2.5	−0.6
6	5.0	2.2	1.5
7	5.0	1.9	3.4
8	3.0	1.0	4.4
9	2.0	0.6	4.9
10	1.0	0.2	5.2

Did you know?

Discounted cash flow is strongly associated with our everyday lives. The calculations of the repayments of home mortgage, car loan and credit card are a few applications of discounted cash flow analysis.

Exercise 3. Andy is applying for a housing mortgage from a bank. He intends to borrow $300 000 where the current annual interest rate offered is 4.0%. The repayment period he has chosen is 30 years with a fixed rate scheme. How much does Andy have to pay monthly? The financial advisor of the bank suggested that he should choose a repayment period of 20 years. What is the benefit of having a shorter repayment period?

Solution to Exercise 3. There are 360 payments in total that need to be made over a 30 year period, that is, 1 year = 12 months. Monthly interest rate is 4/12 = 0.333%.

Method 1. By using the *DCF* method shown in Table 2.2, applying Equation (2.4) and setting *NPV* = 0 at the end of 30 years

or

Method 2. Using the PMT function in Excel, that is, = PMT(0.04/12, 360, −300 000)

Note that the PMT function is a Financial function in Excel that can be used for calculating the periodic payment for a loan based on constant payments and a constant interest rate.

or

Method 3. Using formula

$$\text{Monthly instalment}, M = \frac{P \times r \times (1+r)^N}{[(1+r)^N] - 1}$$

$$= \frac{P \times \left(\frac{i}{q}\right) \times \left(1 + \frac{i}{q}\right)^N}{\left[\left(1 + \frac{i}{q}\right)^N\right] - 1} \tag{2.5}$$

where

P is the principal of the loan
r = *i*/*q*, *i* is the annual interest rate and *q* is the number of payments a year
N is the number of payments in total.

$$\text{Monthly instalment}, M, \text{ for 30 years} = \frac{300\,000 \times \left(\frac{0.04}{12}\right) \times \left(1 + \frac{0.04}{12}\right)^{360}}{\left[\left(1 + \frac{0.04}{12}\right)^{360}\right] - 1} = \$1432.25$$

There are a total 240 payments that need to be made over the 20 years period, that is, 1 year = 12 months. The monthly interest rate is $4/12 = 0.333\%$.

$$\text{Monthly instalment, } M \text{ for 20 years} = \frac{300\,000 \times \left(\frac{0.04}{12}\right) \times \left(1 + \frac{0.04}{12}\right)^{240}}{\left[\left(1 + \frac{0.04}{12}\right)^{240}\right] - 1} = \$1817.94$$

The monthly repayment for the 30 years period is estimated to be **$1432.25**.
If the 20 years repayment period is chosen, the monthly repayment would be **$1817.94**.
Total interest paid can be determined using

$$\text{Total interest paid, } I_T = (M \times n \times q) - P \tag{2.6}$$

$$\text{Total interest paid, } I_T \text{ for 30 years} = (1432.25 \times 30 \times 12) - 300\,000 = \$215\,610$$

$$\text{Total interest paid, } I_T \text{ for 20 years} = (1817.94 \times 20 \times 12) - 300\,000 = \$136\,305.60$$

The total amount of interest paid over 30 years and 20 years are **$215 610** and **$136 305.60**, respectively. Therefore, the financial advisor was right! A shorter repayment period will incur less amount of interest. In this case, **$79 304.40** savings can be achieved.

Note that the monthly instalment is also known as the EMI (equated monthly instalment). Sometimes, processing fees are included, for example, for a mortgage payment. This should be accounted as part of the loan. The APR (annual percentage rate) is normally used as an indicator to compare among different schemes with different interest rates and processing fees.

2.2.8   Profitability Analysis

A profitability analysis is essential to justify the economic feasibility of a project. This can be carried out using various methods and measures, depending on the level of details required. During the preliminary stage of a project, the time value of money is not considered. The approximate quantitative indicators at the preliminary stage are payback time, return on investment and total annualized cost.

Payback time is the period from the start of a project until the time when all capital investment is recovered from selling of products (breakeven), shown in Figure 2.2. Intuitively, a shorter payback time is preferred. However, the usefulness of the payback time as an indicator is limited since it is only valid up to the break-even point. The economic performance after the break-even point cannot be measured using this indicator.

Return on investment (*ROI*) is a common profitability measure of a project, defined as the ratio of the annual income over a project life to the total capital investment, shown below. *ROI* gives a sense of the efficiency of an investment being made:

$$ROI = \frac{\text{Annual income}}{\text{Capital investment}} \times 100\% \tag{2.7}$$

The discounted cash flow rate of return (DCFRR) is essentially the interest rate that makes the *NPV* zero at the end of a project. The DCFRR is a way to measure the performance of utilizing a capital for projects, but does not give any indication of the profit, unlike the *NPV*. The DCFRR can be calculated using Equation (2.4) through the trial-and-error method (e.g., using the Solver function in Excel or the iteration method) or the graphical method.

The total annualized cost includes capital and operating costs in most of the cases, shown in the following equation. The annualized capital cost can be estimated using a fixed interest rate over the plant life. The operating cost is estimated by assuming operating hours in a year.

$$\text{Total annual cost} = \text{Annualized capital cost} + \text{Annual operating cost} \tag{2.8}$$

A more rigorous profitability analysis considering the time value of money is desirable in the detailed process design stage. Economic criteria such as economic potential, netback and cost of production can be applied.

Economic potential (*EP*) is the economic margin and can be evaluated using the following equation when values of products, feed, capital cost and operating cost are available. To obtain an annualized capital cost with the consideration of time value of money, the DCFRR needs to be calculated using Equation (2.4) by setting $NPV = 0$.

$$EP = \text{Value of products} - (\text{Value of feed} + \text{Annualized capital cost} + \text{Annual operating cost}) \quad (2.9)$$

Netback indicates the value of a feedstock from its products selling and can be determined using the following equation. Product prices, capital cost and operating cost except the feedstock cost are known. The market price or cost of the feedstock thus must be less than the netback to result in a positive economic margin.

$$\text{Netback} = \text{Value of products} - (\text{Annualized capital cost} + \text{Annual operating cost}) \quad (2.10)$$

Cost of production of a product is a meaningful indicator when comparing the economic viability between various production routes. It is used when the value of a product is not known, especially when a new product is synthesized or a conventional product is generated from a nonconventional feedstock. The cost of production is calculated from

$$\text{Cost of production} = \frac{\text{Value of feed} + \text{Annualized capital cost} + \text{Annual operating cost}}{\text{Production rate}} \quad (2.11)$$

2.2.9 Learning Effect

The cost of a new and developing technology such as fast pyrolysis of biomass and gasification of biomass is usually very high in the beginning of the development stage. The cost of the technologies decreases as more plants are built and productivity increases due to more experience gain by organizations. This effect is known as a *learning curve* or *experience curve* or *progress curve* or *learning by doing effect*, as shown in Figure 2.3.

The trend of the learning curve can be described by

$$y = ax^{-b} \quad (2.12)$$

where y is the cost of the xth unit, a is the cost of the first unit, x is the cumulative number of units and b is a parameter shown as $b = \log (\text{progress ratio})/\log 2$.

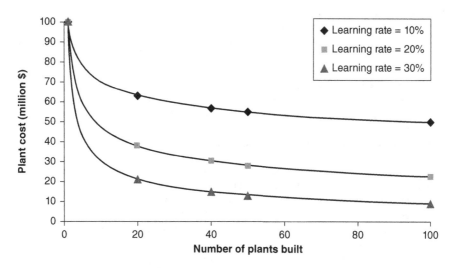

Figure 2.3 *Learning curve effect.*

The learning rate is shown by

$$\text{Learning rate} = 1 - (\text{Progress ratio}) \tag{2.13}$$

The rates of learning vary across different organizations. The factors influencing the rate of learning are crucial and particular attention has to be paid to enhance the performance and economics of the plant. A few factors have been identified as follows:

- Organization forgetting
- Employee turnover
- Transfer of knowledge
- Failure to control other factors such as economies of scale.

The learning curve effect and the learning rate can be easily understood by visualizing the teaching and learning environment in a classroom. If there are 30 students in a classroom, the amount of material that can be assimilated varies from individual to individual, assuming they have no background knowledge on a particular subject and the time of learning is the same. In this case, the method of learning, individual ability and attitude are variables.

2.3   Methodology

2.3.1   Capital Cost Estimation

There is extensive published literature reporting the correlations for estimating the cost of standard equipment, for example, heat exchangers, boilers, turbines, pumps, etc. This information is normally enough for carrying out a chemical plant economics evaluation. However, for a plant undergoing development, such as the biorefinery systems, much of the cost data is unavailable, not readily accessible or may be proprietarily to organizations. Therefore, the costs of equipment may need to be estimated based on existing methodologies or vendor quotes of similar equipment. The Aspen Icarus process evaluator is widely used for an equipment cost estimation in industry. The software may not be available with academic and small organizations; hence spreadsheet based economic evaluation is more convenient and still preferable.

The approach for estimating the equipment cost presented here has avoided using a cost chart that may lead to inconsistent results. A set of established information on equipment costs for biorefinery systems is used as the base cost and the suggested methodology is to scale up or down according to the desired capacity. The base costs and scale factors are collected from various sources[2-7].

The capital cost can be estimated using different methods, that is, order of magnitude estimate, study estimate, preliminary estimate, definitive estimate and detailed estimate. These methods need information at different levels of detail, hence leading to different levels of accuracy. A capital cost estimation by the order of magnitude results in an accuracy of 30–50%. The definitive and detailed estimates require Aspen Icarus software and also a vendor's quote, and thus are more complicated. For these reasons, study and preliminary estimates are preferable since they involve moderate complexity while giving accuracy at a reasonable level. The recommended capital cost estimation approaches are Guthrie's[8] and Lang's[9] methods, able to achieve an accuracy of 20–30%. These approaches should be used for a new design and more time should be allowed to complete the evaluation since many details are required, that is, a complete flowsheet with mass and energy balances, equipment sizing, etc.

The recommended approach for estimating capital costs is shown as follows:

1. Generate a list of equipment and estimate the size of each piece of equipment. The size information in terms of flow rates and power requirements (e.g., for pumps and compressors, etc.) can be obtained directly from simulation. However, some equipment require additional sizing procedures, such as weight of vessel and heat transfer area of the heat exchanger.
2. By applying the concept of economies of scale (Equation (2.2)), the base cost of equipment with a specific size is scaled up/down to obtain the cost of equipment for a desired size. Refer to Table 2.3 (gasification and hydrocracking systems) and Table 2.4 (bioethanol system) for information on base sizes, base costs and scale factors.
3. If the desired size of equipment exceeds the maximum size given, multiple units have to be assumed.

4. Apply the correction factors to the cost of equipment (where relevant) depending upon the material of construction, pressure and temperature.
5. The estimated cost of a piece of equipment can be obtained for different years. Therefore it has to be updated to the cost at the current year, by adopting Equation (2.1) and the cost index in Figure 2.1.
6. Estimate the total capital investment of the system, either based on the individual factor method (Guthrie's method) or the overall factor method (Lang's method).
    a. Guthrie's method. The cost of equipment is given in terms of the f.o.b. purchased cost. Apply individual installation factors for each unit operation to determine the total capital investment.
    b. Lang's method. The cost of equipment is given in terms of the f.o.b. purchased cost. Assume 10% delivery charges as a preliminary estimation. Calculate the total installed cost of equipment using an overall Lang factor based on the plant type, shown in Table 2.5.

For technologies at an early development stage, the estimated cost is normally very high. The learning curve effect (Equation (2.12)) can be taken into account to forecast the future cost of a system and to obtain a reasonable estimation of the cost. A reasonable reference point is the cost of building the tenth plant ($x = 10$ in Equation (2.12)) and a progress ratio of 0.8 can be used as a preliminary estimation.

Remarks. The Golden Rule: an estimated capital cost of a facility is unlikely to give its actual value. No matter how sophisticated the capital cost estimation method is, it will never be accurate. It is not possible to apply the same values for different locations and time. However, an estimated cost needs to be reasonable. Aspen Icarus is widely accepted and has been a standard in the chemical industry due to its strength in performing sizing and costing of equipment. Vendor's quotes are more up-to-date, yet it is still unreliable at some point due to several factors such as the seller-and-buyer relationship, contemporary supply and demand, etc. It is advisable to use the same basis throughout a capital cost evaluation of equipment in a plant so that the same degree of accuracy can be attained. A sensitivity analysis should be carried out to examine the effects of inaccuracies on results.

2.3.2   Profitability Analysis

The economic feasibility of a plant can be examined using the economic criteria mentioned in Section 2.2. A systematic approach is provided as a guideline to evaluate the economic performance of a plant up to a reasonably detailed level, as follows:

1. Evaluate the capital cost using the procedures given in Section 2.3.1. The annualized capital cost is calculated using the discounted cash flow (*DCF*) analysis (Section 2.2.7), by assuming a reasonable plant life.
2. Evaluate the annual operating cost. Variable operating costs such as biomass feedstock and transportation costs are shown in Tables 2.6 and 2.7, respectively, and the costs of chemicals and utilities can be obtained conveniently elsewhere. The cost factors for fixed operating costs are shown in Table 2.1.
3. Select an indicative economic measure, that is, economic potential, netback and cost of production, and determine the economic performance of the plant.

2.4   Cost Estimation and Correlation

The correlations for estimating the capital and operating costs of biorefinery systems are shown in Sections 2.4.1 and 2.4.2, respectively.

2.4.1   Capital Cost

The parameters required for estimating the equipment cost (base scales, base costs and scale factors) of gasification, hydrocracking and bioethanol systems[2–7] are shown in Tables 2.3 and 2.4, respectively. The installation factors for

Table 2.3 *Equipment cost correlation – gasification and hydrocracking systems[2–4,6,7].*

Item No.	Component	Base cost (million $)	Scale Factor	Base Size	Capacity	Maximum Size	Installation Factor	Base Year
A. Pretreatment								
A.1	Conveyers	0.35	0.8	33.5	wet t h^{-1} biomass feed	110	1.86	2001
A.2	Grinding	0.41	0.6	33.5	wet t h^{-1} biomass feed	110	1.86	2001
A.3	Storage	1.0	0.65	33.5	wet t h^{-1} biomass feed	110	1.86	2001
A.4	Dryer	7.6	0.8	33.5	wet t h^{-1} biomass feed	110	1.86	2001
A.5	Iron removal	0.37	0.7	33.5	wet t h^{-1} biomass feed	110	1.86	2001
A.6	Feeding system	0.41	1.0	33.5	wet t h^{-1} biomass feed	110	1.86	2001
B. Gasification System								
B.1	Gasifier BCL[a]	16.3	0.65	68.8	dry t h^{-1} biomass feed	83	1.69	2001
B.2	Gasifier IGT[b]	38.1	0.7	68.8	dry t h^{-1} biomass feed	75	1.69	2001
C. Syngas Cleaning								
C.1	Tar cracker	3.1	0.7	34.2	m^3 s^{-1} gas input	52	1.86	2001
C.2	Cyclones	2.6	0.7	34.2	m^3 s^{-1} gas input	180	1.86	2001
C.3	Gas cooling	6.99	0.6	39.2	kg s^{-1} steam generation		1.84	2001
C.4	Baghouse filter	1.6	0.65	12.1	m^3 s^{-1} gas input	64	1.86	2001
C.5	Condensing scrubber	2.6	0.7	12.1	m^3 s^{-1} gas input	64	1.86	2001
C.6	Hot gas cleaning	30	1.0	74.1	m^3 s^{-1} gas input		1.72	2001
D. Syngas Processing								
D.1	Steam reformer	9.4	0.6	1390	kmol h^{-1} gas input		2.3	2001
D.2	Autothermal reformer	4.7	0.6	1390	kmol h^{-1} gas input		2.3	2001
D.3	Shift reactor	36.9	0.85	15.6	Mmol h^{-1} CO+H$_2$ input		1.0	2001
D.4	Selexol CO$_2$ removal	54.1	0.7	9909	kmol h^{-1} CO$_2$ removed		1.0	2001
E. Product Synthesis and Upgrading								
E.1	Solid bed FT gas phase 60 bar	25.3	1.0	100	MW (HHV) FT produced		1.3	2001
E.2	Slurry phase FT 60 bar	36.5	0.72	131	MW (HHV) FT produced		1.0	2001
E.3	FT product upgrading	233	0.7	286	m^3 h^{-1} FT produced		1.0	2001
E.4	Gas phase methanol synthesis reactor	7	0.6	87.5	t h^{-1} MeOH produced		2.1	2001
E.5	Liquid phase methanol synthesis reactor	3.5	0.72	87.5	t h^{-1} MeOH produced		2.1	2001
E.6	Methanol product upgrading	15.1	0.7	87.5	t h^{-1} MeOH produced		2.1	2001
F. Combined Cycle								
F.1	Gas turbine + HRSG	18.9	0.7	26.3	MW$_e$ electrical output		1.86	2001
F.2	Steam turbine and steam system	5.1	0.7	10.3	MW$_e$ electrical output		1.86	2001
G. Common Process Machinery								
G.1	Compressor	11.1	0.85	13.2	MW$_e$ compression work		1.72	2001
H. Separation Unit								
H.1	Oxygen plant	44.2	0.85	41.7	t h^{-1} O$_2$ produced		1.0	2001
H.2	Pressure swing adsorption (PSA) unit[c]	28	0.7	9600	kmol h^{-1} throughput		1.69	2001
H.3	Membrane[c]	21.6	0.8	17	t h^{-1} H$_2$ recovered		1.0	2001
I. Pyrolysis Unit								
I.1	Pyrolyzer (circulating fluidized bed)	3.392	0.7	500	t d^{-1} biomass feed		2.47	2003
J. Refinery Unit								
J.1	Hydrocracker unit[d]	30	0.65	2250	bbl d^{-1} pyrolysis oil feed		2.47	2005
J.2	Separation[e]	2.28	0.65	2250	bbl d^{-1} pyrolysis oil feed		1.0	2007

[a] Gasifier BCL (Batelle Columbus) is an indirect, air-blown and atmospheric gasifier.

[b] Gasifier IGT (Institute of Gas Technology) is a direct, oxygen-blown and pressurized gasifier.

[c] Pressure swing adsorption unit and membrane are employed for hydrogen recovery process.

[d] Hydrocracker unit consists of fired heater, hydrocracker vessel, feed/product exchanger, air cooler, trim cooler, high and low pressure flash.

[e] Separation units include fractionators, splitters, reboilers, condensers and reflux drums.

Notes:

FT: Fischer–Tropsch (c.f. Chapter 14).

HRSG: heat recovery steam generator.

HHV: higher heating value (c.f. Section 10.2.2).

Table 2.4 Equipment cost correlation – bioethanol system[5].

Item No.	Component	Base Cost (million $)	Scale Factor	Base Size	Capacity	Maximum Size	Installation Factor	Base Year
A. Pretreatment								
A.1	Mechanical	4.44	0.67	83.3	dry t h^{-1} biomass feed	83.3	2.00	2003
A.2	Mill	0.37	0.7	50	wet t h^{-1} biomass feed		1.00	2003
A.3	Dilute acid	14.1	0.78	83.3	dry t h^{-1} biomass feed		2.36	2003
A.4	Steam explosion	1.41	0.78	83.3	dry t h^{-1} biomass feed		2.36	2003
A.5	Liquid hot water	5.62	0.78	83.3	dry t h^{-1} biomass feed		2.36	2003
A.6	Ion exchange	2.39	0.33	83.3	dry t h^{-1} biomass feed		1.88	2003
A.7	Overliming	0.77	0.46	83.3	dry t h^{-1} biomass feed		2.04	2003
B. Hydrolysis and Fermentation								
B.1	Cellulase production (SSF)	1.28	0.8	50	kg h^{-1} cellulase produced	50	2.03	2003
B.2	Seed fermenters (SSF+SSCF)	0.26	0.6	3.53	t h^{-1} ethanol produced	3.53	2.20	2003
B.3	C5 fermentation (SSF)	0.67	0.8	1.04	t h^{-1} ethanol produced	1.04	1.88	2003
B.4	Hydrolyze-fermentation (SSF)	0.67	0.8	1.04	t h^{-1} ethanol produced	1.04	1.88	2003
B.5	SSCF	0.67	0.8	1.04	t h^{-1} ethanol produced	1.04	1.88	2003
B.6	CBP	0.67	0.8	1.04	t h^{-1} ethanol produced	1.04	1.88	2003
C. Upgrading								
C.1	Distillation and purification	2.96	0.7	18.466	t h^{-1} ethanol produced	18.466	2.75	2003
C.2	Molecular sieve	2.92	0.7	18.466	t h^{-1} ethanol produced	18.466	1.00	2003
D. Residuals								
D.1	Solids separation	1.05	0.65	10.1	dry t h^{-1} solids	10.1	2.20	2003
D.2	(An)aerobic digestion	1.54	0.6	43	t h^{-1} wastewater	43	1.95	2003

Notes:
SSF: simultaneous saccharification and fermentation.
SSCF: simultaneous saccharification and co-fermentation.
CBP: consolidated bioprocessing.

Table 2.5 Typical Lang factors of various plants for estimating capital investment based on the delivered cost of equipment[9].

Plant	Solid Processing	Solid–Fluid Processing	Fluid Processing
Direct Cost			
Delivered cost of equipment	1.00	1.00	1.00
Installation	0.45	0.39	0.47
Instrumentation and control	0.18	0.26	0.36
Piping	0.16	0.31	0.68
Electrical systems	0.10	0.10	0.11
Buildings (including services)	0.25	0.29	0.18
Yard improvements	0.15	0.12	0.10
Service facilities	0.40	0.55	0.70
Total direct cost, C_D	**2.69**	**3.02**	**3.60**
Indirect Cost			
Engineering and supervision	0.33	0.32	0.33
Construction expenses	0.39	0.34	0.41
Legal expenses	0.04	0.04	0.04
Contractor's fee	0.17	0.19	0.22
Contingency	0.35	0.37	0.44
Total indirect cost, C_{ID}	**1.28**	**1.26**	**1.44**
Working capital	0.7	0.75	0.89
Total capital investment, C_{TCI}	**4.67**	**5.03**	**5.93**

individual equipment are required in Guthrie's method. These factors have taken into account the direct and indirect cost components, but with different assumptions compared to the Lang factor and also working capital is not included. The Lang factors depending upon plant types are shown in Table 2.5.

2.4.2   Operating Cost

The variable operating costs of biorefinery systems include the costs of raw materials (e.g., feedstock, chemical, catalyst, etc.), utilities (e.g., electricity, steam, etc.) and transportation. These are highly dependent on process specification and preferences. Most of the prices of chemicals can be easily obtained from published literature and online resources. Therefore, only the biomass prices (Table 2.6) and its transportation costs (Table 2.7) are shown for year 2007[10].

Table 2.6   *Estimated prices of various biomass feedstocks. (Reproduced with permission from Department of Trade and Industry (2007)[10].)*

Biomass	Central Price ($ GJ^{-1})	Price Range ($ odt^{-1})	($ GJ^{-1})
Forestry woodfuel-chips	5	120	4.0–6.0
Forestry woodfuel-logs	4	80	3.0–5.0
Energy crops			
Short rotation coppice (SRC)	7	140	6.0–8.0
Miscanthus	6	106	5.0–7.0
Arboricultural arisings	5	98	4.0–6.0
Straw	4	70	3.0–5.0
Waste wood (clean)	5	98	4.0–6.0
Waste wood (contaminated)	2	40	1.0–3.0
Pellets to power/industry/commercial from woodfuel	9	180	8.0–10.0
Pellets to power/industry/commercial from SRC	11	220	10.0–12.0
Pellets to power/industry/commercial from miscanthus	10	200	9.0–11.0
Pellets to domestic (including delivery)	14	280	12.0–16.0
Imported biomass (including delivery)	9	180	7.0–11.0

Note: odt = oven dry tonne.

Table 2.7   *Estimated average transportation costs for different biomass feedstocks. (Reproduced with permission from Department of Trade and Industry (2007)[10].)*

Application	Transportation Cost ($ GJ^{-1})		
	Energy Crops	Woodfuel	Straw
Power generation			
1% co-firing, 2000 MW	NA	0.60 (17)	0.60 (17)
5% co-firing, 2000 MW	1.0 (35)	NA	1.60 (52)
10% co-firing, 2000 MW	1.32 (49)	NA	NA
30 MW dedicated	0.72 (24)	0.74 (25)	0.76 (28)
Heat			
0.1–10 MW of heat generation	0.60 (17)	0.60 (17)	NA
CHP			
0.1–10 MW of electricity generation	0.60 (17)	0.60 (17)	NA
> 10 MW of electricity generation	0.72 (24)	0.74 (25)	0.76 (28)

Note: Numbers in parentheses are estimated average transport distance in km. NA = not assessed.

Table 2.8 *Personnel requirement for chemical processing plants. (Reproduced with permission from Seider et al. (2010)[11]. Copyright © 2010, John Wiley & Sons, Ltd.)*

Process	Number of Personnel per Processing Step
Continuous	
i. Fluids processing	1
ii. Solids–fluids processing	2
iii. Solids processing	3
Batch/semi-batch	
i. Fluids processing	2
ii. Solids–fluids processing	3
iii. Solids processing	4

The cost of labor/personnel is difficult to estimate since it is dissimilar from one organization to another. Furthermore, it varies across different chemical sectors (e.g., petroleum and pharmaceutical), different countries and even different locations. Additionally, the personnel position also determines the salary that they earn. For preliminary estimation purposes, the following approach can be used to estimate the cost of labor.

Step 1. Calculate the number of personnel per shift using Equation (2.14)

The operating labor requirement is related to the number of processing steps depending on the process, as shown in Table 2.8. The number of personnel per processing step is also shown:

$$\text{Number of personnel per shift} = \text{Number of processing steps}$$
$$\times \text{Number of personnel per processing step} \quad (2.14)$$

The number of personnel per processing step should be doubled when the size of the continuous process is large, that is, greater than 1000 t d^{-1} of product[11].

Step 2. Calculate the cost of personnel using Equation (2.15)

$$\text{Cost of personnel (\$ y}^{-1}) = \text{Number of personnel per shift} \times 5 \text{ shifts} \times 40 \text{ hours/week} \times 52 \text{ weeks/year}$$
$$\times \text{hourly wages (\$ h}^{-1}) \quad (2.15)$$

Assume 5 shifts per day are required. Each personnel works 40 hours per week and there are 168 hours per week. In addition, consider other factors such as sick leaves, holidays, etc.

The Bureau of Labor Statistics in the US Department of Labor estimated that the mean hourly wages for a chemical engineer in year 2008 is at the rate of $42.67 (€29) per hour.

2.5 Summary

This chapter shows the generic economic analysis concept and fundamental teaching and learning tools for chemical engineering process economics. The time value of money and discounted cash flow are among the important aspects in an economic evaluation. The methods for analyzing capital cost and operating cost have been outlined and translated into easy-to-follow procedures, accompanied by cost estimation correlation and data.

2.6   Exercises

Refer to the ***Online Resource Material, Chapter 2 – Additional Exercises and Examples***, the solutions to all Exercise problems.

1. A company is planning to start a methanol production plant via thermochemical conversion of biomass. The plant encompasses the following main equipment:
    - Dryer
    - Gasifier (direct, oxygen-blown and pressurized)
    - Oxygen plant
    - Tar cracker
    - Water gas shift reactor
    - Methanol synthesis reactor (gas phase)

    Process specifications and assumptions:
    - Biomass feed = 150 dry t h^{-1}
    - Moisture content of biomass = 30 wt%
    - Oxygen requirement in gasifier = 0.45 kg kg^{-1} dry biomass feedstock
    - Yield of product gas from gasification = 100 kmol t^{-1} dry biomass feedstock
    - Standard molar volume of gas = 22.414 m^3 $kmol^{-1}$
    - Mole fraction of components in product gas

    $$CO = 0.15;\ H_2 = 0.20$$

    - Higher heating value (HHV) of

    $$Biomass = 20\,MJ\,kg^{-1}$$

    $$Methanol = 23\,MJ\,kg^{-1}$$

    - Output efficiency = 45% based on HHV
        a. Calculate the purchased cost of each individual piece of equipment and hence the total purchased cost of equipment of the methanol production system.
        b. Calculate the total capital investment of the methanol production plant by applying Guthrie's method.
        c. The cost of equipment is valid in year 2001. Estimate the total cost of equipment in year 2010 using the Chemical Engineering Plant Cost Index.

2. The methanol production plant requires steam and electricity to run the process. It has been estimated that 2.0 kg of steam per kg of methanol produced is needed for the whole system while 235 kW h of electricity per tonne of O_2 are needed for the oxygen plant.
    a. Calculate the utility cost required per annum for this plant using the results calculated from Question 1.
    b. If heat integration is applied on this plant, steam can be generated through heat exchange with process streams. The steam requirement can then be partially satisfied by the steam generated on site. It has been estimated that 40% of the imported steam can be reduced. Calculate the % cost saving from heat integration relative to the steam import scenario.

    Assumptions:
    - 8000 operating hours per year
    - Cost of imported steam = $15 t^{-1}
    - Cost of imported electricity = $0.1 kW h^{-1}

3. The company is at the stage of choosing between two plant options, X and Y. The finance department has provided an estimation of the annual cash flow for a period of 10 years, shown in Table 2.9. Evaluate both options with respect to:
    a. Payback time.
    b. Return on investment.

Table 2.9   *Cash flow for two plant options.*

Year	Cash Flow (million $)	
	Option X	Option Y
−1	−40.0	−40.0
0	−60.0	−60.0
1	30.0	5.0
2	40.0	15.0
3	50.0	25.0
4	65.0	40.0
5	70.0	50.0
6	75.0	65.0
7	85.0	75.0
8	90.0	75.0
9	90.0	80.0
10	90.0	80.0

   c. Discounted cash flow and net present value for each year. Assume an annual discount rate of 10%.

   d. Discounted cash flow rate of return.

   e. Suggest the preferred option based on the results obtained from (a), (b) and (c) and (d).

   f. Illustrate the effect of having different annual discount rates, that is, 5% and 20%, on the overall economic performance of the preferred option in *NPV* versus time plot and draw conclusions based on the trend.

4. By using the results obtained from Questions 1 and 2(b), show the netback calculation to estimate the value of the biomass feedstock. Assume a DCFRR of 15% for the annualized capital cost.

References

1. R.K. Sinnott, *Coulson & Richardson's Chemical Engineering Design Volume 6*, 4th ed., Elsevier: Butterworth-Heinemann, Oxford, 2006.
2. K.S. Ng and J. Sadhukhan, Process integration and economic analysis of bio-oil platform for the production of methanol and combined heat and power, *Biomass and Bioenergy*, **35**(3), 1153–1169 (2011).
3. K.S. Ng and J. Sadhukhan, Techno-economic performance analysis of bio-oil based Fischer–Tropsch and CHP synthesis platform, *Biomass and Bioenergy*, **35**, 3218–3234 (2011).
4. C.N. Hamelinck, A.P.C. Faaij, H. den Uil, H. Boerrigter, Production of FT transportation fuels from biomass; technical options, process analysis and optimisation, and development potential, *Energy*, **29**(11), 1743–1771 (2004).
5. C.N. Hamelinck, G.V. Hooijdonk, A.P.C. Faaij, Ethanol from lignocellulosic biomass: techno-economic performance in short-, middle- and long-term, *Biomass and Bioenergy*, **28**(4), 384–410 (2005).
6. C.N. Hamelinck and A.P.C. Faaij, Future prospects for production of methanol and hydrogen from biomass, *Journal of Power Sources*, **111**(1), 1–22 (2002).
7. S. Jones, C. Valkenburg, C. Walton, D. Elliott, J. Holladay, D. Stevens, C. Kinchin, C. Czernik, *Production of Gasoline and Diesel from Biomass via Fast Pyrolysis, Hydrotreating and Hydrocracking: A Design Case*, Pacific Northwest National Laboratory, Department of Energy, Richland, Washington, 2009.
8. K.M. Guthrie, Data and techniques for preliminary capital cost estimating, *Chemical Engineering*, **76**, 114–142 (1969).
9. H.J. Lang, Simplified approach to preliminary cost estimates, *Chemical Engineering*, **55**(6), 112–113 (1948).
10. *UK Biomass Strategy 2007: Working Paper 1 – Economic Analysis of Biomass Energy*, Energy Technology Unit, Department of Trade and Industry (DTI), 2007.
11. W.D. Seider, J.D. Seader, D.R. Lewin, S. Widagdo, *Product and Process Design Principles: Synthesis, Analysis and Evaluation*, John Wiley & Sons, Ltd, 2010.

3

Heat Integration and Utility System Design

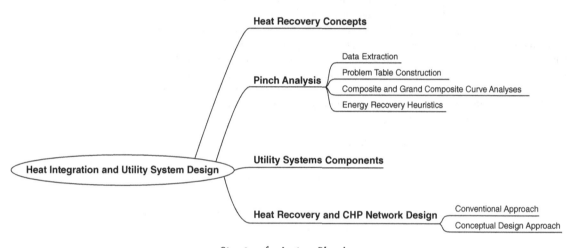

Structure for Lecture Planning

3.1   Introduction

Industrial plants encompass a large number of heat exchanger equipment and involve many heating and cooling duties. Hot and cold utilities such as steam, fuel, cooling water, etc., are hence the essential components within these complexes. Depending on the types and scales of the plant, the associated energy costs can be very high. Therefore, energy reduction and recovery activities are imperative to cut down the demand and cost of energy. A biorefinery complex uses feedstocks with a lower energy content compared to a fossil based refinery. Hence, energy recovery is a very critical aspect within the complex to enhance energy efficiency of the resources.

Heat integration using *pinch analysis* is a well-established methodology and has been demonstrated in many projects in the oil and gas industry throughout the world. The energy saving through the application of pinch analysis is notable and it has become an important practice within the process industry.

Biorefineries and Chemical Processes: Design, Integration and Sustainability Analysis, First Edition.
Jhuma Sadhukhan, Kok Siew Ng and Elias Martinez Hernandez.
© 2014 John Wiley & Sons, Ltd. Published 2014 by John Wiley & Sons, Ltd.
Companion Website: http://www.wiley.com/go/sadhukhan/biorefineries

3.2   Process Integration

The first question that always comes to mind: what is process integration and how is it related to energy efficiency?

Biomass feedstocks are converted into useful products through process units such as reactors, separation units, heat exchangers and other subunits such as dryers, evaporators, mixers, etc., brought together to design an entire biorefinery process plant. Effective utilization of raw materials is desired and so the plant ought to be designed and operated in an energy efficient manner. The traditional method of enhancing the energy efficiency of the processes is through optimization of the operating conditions of individual process units. Does a network of processes achieve the highest efficiency if process units are individually designed? Do the individual effects add up to the overall effect of the plant? The answer is certainly NO.

Changing the condition of one process has an impact on other processes within a plant. Therefore, it is vital to capture the interactions between processes during the analysis. Process integration, in the context of the process industry, is a technique used to analyze interactions between processes within a whole plant and exploiting these interactions to optimize the whole plant for the most efficient use of raw materials, energy and capital. Biorefineries need to adopt a process integration analysis at the outset of design.

Figure 3.1 shows a simple process plant that does not incorporate heat recovery. The feed is converted into products in reactors RX1 and RX2 and further undergoes separation through a distillation column DC1, resulting in Product 1 from the top of DC1 and Product 2 from the bottom of DC1. The feed needs to be preheated in HE1 before entering RX1. After reaction takes place in RX1, the products need to be cooled down in HE2 before entering RX2. Similarly, after reaction takes place in RX2, the products need to be cooled down in HE3 before entering DC1. On top of DC1, a condenser C1 is in place to condense the reflux stream while a reboiler R1 is used to heat the bottom stream of DC1. This system requires external heating and a cooling medium, supplied from the utility system consisting of steam boilers, cooling towers, etc. It can be assumed that the heating and cooling utilities are steam and cooling water.

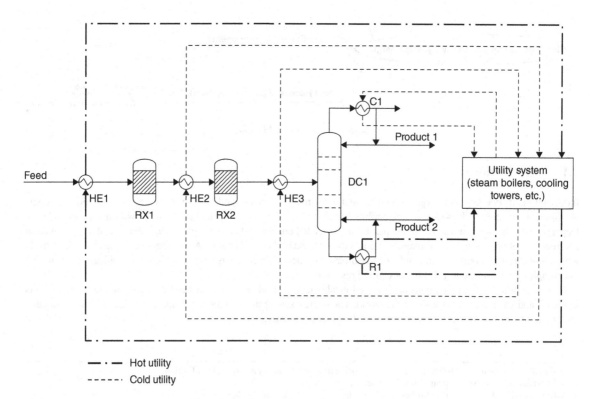

Figure 3.1   *Process plant without heat recovery.*

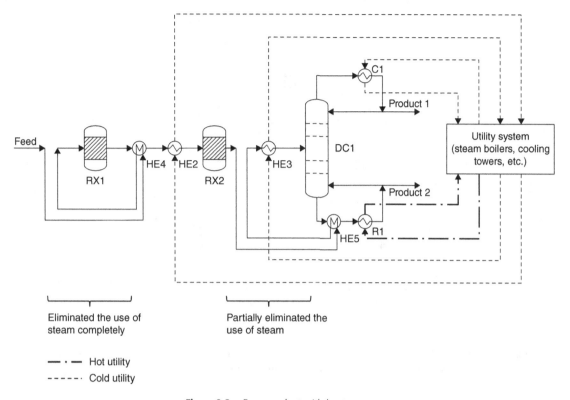

Figure 3.2 *Process plant with heat recovery.*

Figure 3.1 shows a complete system where all the heating and cooling demands within the system are satisfied and the feed is converted into products. Nevertheless, this system is not energy efficient because it uses external utilities. The generation of steam is costly and valuable heat is "dumped" into cooling water.

Figure 3.2 shows a process plant with consideration of heat recovery. The structure of the system is similar to Figure 3.1, apart from the heat exchangers. Figure 3.2 does not include HE1. The outlet stream from RX1 that needs to be cooled down is used to preheat the feed stream, in HE4. This configuration eliminates the need for steam to preheat the feed stream. Also it reduces the heat rejected to the cooling water in HE2. On the other hand, the bottom stream from DC1 requiring heat for the reboiler can first obtain the heat through heat exchange with the outlet stream from RX2. This reduces the heat load required for the reboiler R1 and thus decreases the consumption of steam. Steam is still required in R1 due to the high latent heat of vaporization of the stream, which cannot be satisfied using the heat from the outlet of RX2; hence only partial elimination of steam can be achieved.

The modification of the design from Figure 3.1 to Figure 3.2 shows process integration in terms of heat recovery to enhance the energy efficiency of the system.

Process integration considers a system as a whole wherein the interactions among different processes are taken into account for process improvement. Pinch analysis is a technique used for heat integration.

3.3 Analysis of Heat Exchanger Network Using Pinch Technology

An industrial plant or site has a network of heat exchangers to carry out heating and cooling operations. Heat exchange is carried out using a heating utility or medium such as steam or fuel and a cooling utility or medium such as cooling water and refrigerant after optimum process-to-process heat recovery. It is always desired to recover energy within a process

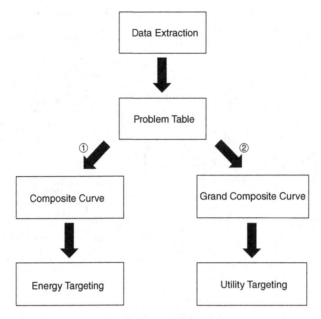

Figure 3.3   *Pinch analysis for energy and utility targeting.*

plant to the greatest possible extent so that the consumption of external fuel (or energy) can be reduced considerably and valuable on-site heat can be utilized efficiently, thus achieving the lowest operating cost scenario.

Pinch analysis consisting of the following activities can be adopted for such purposes.

1. Data extraction. Scope an entire network for important streams by extracting the relevant thermodynamic (heating or cooling requirement) data.
2. Construction of heat profiles. Construct composite curves and/or a grand composite curve to analyze and target the energy recovery. These curves are constructed using the Problem Table algorithm.
3. Graphical analysis. Understand the significance of the profiles. Modification and decision such as placement of the utility and the connection between heat exchangers can be made based on the insights obtained from the graphical approach.

Figure 3.3 shows the pinch technique for energy and utility targeting. Firstly, the extracted thermodynamic data are arranged in Problem Table. Depending on the nature of the problem to be solved, the Problem Table can be transformed into composite curves and a grand composite curve. Composite curves (Route 1) are employed for performing energy targeting, while the grand composite curve (Route 2) is used to carry out utility targeting. In most cases, a combination of the two routes is used to generate an optimum heat integration scenario.

Note that composite curves and Problem Table work in the same way in terms of energy targeting. Composite curves are the graphical representation of data extracted from Problem Table calculations.

3.3.1   Data Extraction

"The first step is always the hardest." In the whole heat integration analysis, data extraction is the most crucial step amongst all. This is because wrong judgment, inappropriate extraction of data and missing data may lead to missed opportunities and thereby result in inefficient heat recovery. Therefore, it is very important to understand the principles and fundamentals of data extraction before carrying out a detailed heat integration analysis. This section shows the data extraction concept which enables what is important and what is not to be learnt when dealing with large number of streams, common in a process plant.

Before beginning the data extraction, let us first learn a few basic terminologies.

Stream – a channel of fluid connecting two process units of which there could be change in the heat load but not the composition.
Hot stream – a stream that contributes heat and acts as a "heat source."
Cold stream – a stream that requires heat and acts as a "heat sink."
Supply temperature (T_S) – the starting temperature of a stream.
Target temperature (T_T) – the terminal temperature of a stream.
Heat capacity flow rate (CP) – the multiplication of mass flow rate and specific heat capacity (C_p) of a stream.
Heat load (ΔH) – the enthalpy change resulting from a heat transfer process.

In most cases for extracting data for heat integration analysis, thermodynamic data comprising T_S, T_T, CP and ΔH are required.

3.3.1.1 *General Guidelines for Data Extraction*

When carrying out heat integration on a process flow diagram (no matter whether it is a new design or an existing design), the million dollar question is: which stream should be extracted for such thermodynamic data sets? If we extract the same number of streams as in the process flow diagram, the consequence is that we would most likely end up with the original design. This is because opportunities for improvement are limited when replicating an existing plant configuration and extracting an excessive data set. Therefore, an effective way is needed to identify the right streams for heat integration in a flowsheet.

Figure 3.4 shows a simple process flow diagram with the reaction, separation and heat exchange without heat integration. Feed is pumped (P1) to the reaction section at 40 °C and preheated in HE1 to 150 °C via a heat exchange with the bottom stream of the fractionation column (FR1). The feed is then further preheated via a fired heater F1 up to 450 °C before entering reactor R1. Saturated low pressure (LP) steam at 5 bar and 150 °C is injected into the reactor. The reactor product exits at 500 °C. The product is cooled down to 400 °C in HE2 and flows into a cyclone C1 to remove ash. The gas coming out from the cyclone undergoes two-stage cooling, from 400 °C to 200 °C in HE3 and further to 100 °C in HE4. An external stream is mixed with the main stream after being heated up from 25 °C to 80 °C in HE5. The resulting temperature after mixing becomes 95 °C. Subsequently, the temperature of the stream is increased to 140 °C in HE6 before entering FR1. This is to partially vaporize the feed entering the column. The top product from FR1 is at 60 °C and

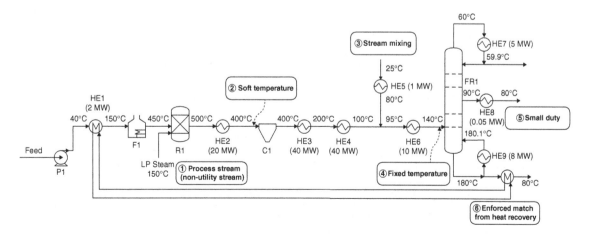

Figure 3.4 *A typical reaction–separation system involving a number of heat exchanger equipments.*

condensed in HE7. A side stream draws part of the product out from FR1 at 90 °C and is cooled down to 80 °C in HE8. The bottom product from FR1 is at 180 °C. Part of the stream is reboiled in HE9, while the remainder is cooled down to 80 °C via heat exchange with HE1.

The process flow diagram in Figure 3.4 is used to discuss the features of data extraction. Here are the useful principles for data extraction:

① Process stream/utility stream

Process streams that experience heating and cooling are normally extracted. Utility streams such as steam and cooling water should not be extracted. However, there are certain designs where these utilities may need to be considered, for example, when they contribute to the mass of products. If a steam is in direct contact with the main process, that is, not used as heating or cooling purposes, then this stream needs to be considered as a process stream. One example is shown in Figure 3.4. LP steam injected into reactor R1 as feed should be extracted as a nonutility stream.

Therefore, the LP steam injected into the reactor should be extracted as a hot stream:

$T_S = 150$ °C; $T_T = 149.9$ °C; $\Delta H = m \times \Delta h_v$, where m is the mass flow rate of steam and Δh_v is the enthalpy change of vaporization of steam (at 5 bar in this case), that is, 2100 kJ kg^{-1}.

② Soft temperature

The temperature of a process may not be fixed at one value. There is flexibility to alter the temperature to improve the process operation. For a stream entering that process, its target temperature is therefore a *soft* number. This is termed "*soft temperature.*" Figure 3.4 shows an example of the soft temperature. The reactor product is sent to cyclone C1 at 400 °C. This temperature is not fixed (i.e., not restricted by the process requirement) and can be varied. The appropriate temperature should be the one that may lead to the highest process performance with the lowest utility consumption. Soft temperatures are variables and can be decided based on economic performance of a plant. A sensitivity analysis can be performed by taking ranges of soft temperatures in a heat exchanger network design exercise.

The stream passing through HE2 should be extracted as a hot stream:

$T_S = 500$ °C; $T_T =$ decided based on minimum utility consumption; $\Delta H = 20$ MW.

③ Stream mixing

It is not recommended to mix several streams with different temperatures because it could result in inefficient heat transfer and in some cases cross-pinch heat transfer. In Figure 3.4, the stream mixing occurred around HE4, HE5 and HE6. It is not advisable to extract the stream nonisothermally in the following way because of possible missed opportunities:

$T_S = 200$ °C; $T_T = 100$ °C; $\Delta H = 40$ MW (HE4)
$T_S = 25$ °C; $T_T = 80$ °C; $\Delta H = 1$ MW (HE5)
$T_S = 95$ °C; $T_T = 140$ °C; $\Delta H = 10$ MW (HE6)

The correct way of extracting the data is to assume that the streams are mixed isothermally, as follows:

$T_S = 200$ °C; $T_T = 140$ °C; $\Delta H =$ calculate using energy balance
$T_S = 25$ °C; $T_I - 140$ °C; $\Delta H =$ calculate using energy balance

④   Fixed temperature

Some streams need particular temperature conditions to be met. For example, the inlet stream to FR1 is needed at 140 °C in Figure 3.4. This is the feed temperature that results in the optimum economic performance of the fractionator. The alteration of feed temperature may lead to increased utility consumption in the condenser and reboiler of FR1 and also may affect the number of stages in the column. Therefore, such a temperature should not be treated as the soft temperature. Streams with fixed conditions and with inflexible heat exchanger design constraints can be omitted from the heat integration analysis.

⑤   Small duty

It is not unusual to have streams with small heating and cooling duties. These streams may be left out from the analysis whenever possible because their energy savings would not be significant compared to the overall improvement. In Figure 3.4, the side draw from FR1 is cooled from 90 °C to 80 °C in HE8, with only 0.05 MW of sensible heat availability. This stream is not extracted.

⑥   Enforced match from heat recovery

In some cases, especially in an existing plant, a stream may already have achieved a proper heat transfer matching with an other stream. In addition, altering the configuration may turn out to be difficult and expensive. Such streams can be omitted from the analysis. Figure 3.4 shows an existing matching between the feed to the reactor and the bottom product stream from FR1. The bottom product stream is cooled from 180 °C to 80 °C and the heat released is used to preheat the feed from 40 °C to 150 °C in HE1.

It is inevitable that assumptions are made while carrying out data extraction. Here are a few suggestions:

1. *Number of streams.* The golden rule is to keep the problem as simple as possible by minimizing the number of streams and hence the data to be extracted. For example, in Figure 3.4, the outlet stream from cyclone C1 undergoes two-stage cooling from 400 °C to 200 °C in HE3 and further cooling to 100 °C in HE4. A proper way to extract the data is to take only one stream from 400 °C to 100 °C with the total heat load of 80 MW. This is because the intermediate temperature of 200 °C is *soft data*. Also, extracting two streams will end up exactly the same as the original design and hence limit the scope for improvement.
2. *Heat capacity.* The heat capacity of a stream is a function of temperature. During heating and cooling, a stream has variable heat capacity across a temperature range. For cases where there is no phase change, it is reasonable to assume a constant heat capacity for a stream, therefore giving a linear line on a temperature–enthalpy diagram. If there is phase change, the heat capacity can be assumed to be constant at different phases and the enthalpy required for phase change is the latent heat at a given saturation pressure and temperature. Thus the temperature–enthalpy profile (or composite curve) is represented by three linear segments. The construction of the temperature–enthalpy diagram is discussed in Section 3.3.2.
3. *Heat losses.* Heat losses from a stream are normally negligible compared to the heating or cooling duty.

3.3.2   Construction of Temperature–Enthalpy Profiles

The following example is to show how the composite curves and grand composite curve are constructed. Consider a four-stream case, shown in Table 3.1. The thermodynamic data for hot and cold streams are provided, including the supply temperature (T_S), target temperature (T_T), heat load (ΔH) and heat capacity flow rate (CP). ΔT_{min} is the minimum temperature difference allowed between hot and cold streams in an exchanger and is assumed to be 10 °C.

Table 3.1   Stream data.

Stream	T_S (°C)	T_T (°C)	ΔH (kW)	CP (kW °C^{-1})
1	250	140	440	4.0
2	180	40	2100	15.0
3	50	180	1560	12.0
4	110	200	1500	16.7

3.3.2.1   Construction of Composite Curves

Step 1: Problem Table.   Hot and cold streams are organized into two separate groups. Each group is separately plotted to generate two respective profiles, called hot and cold composite curves. For the current example, two hot streams are grouped to form a hot composite curve and two cold streams are grouped to form a cold composite curve.

The hot and cold composite curves are plotted based on the Problem Table method shown in Tables 3.2 and 3.3, respectively. The Problem Table method is carried out according to the following procedures:

1. Arrange the streams according to the temperature from high to low.
2. Draw an arrow representing a stream from supply to target temperatures and show its CP value alongside the line.
3. Take a temperature interval and undertake the calculation steps 4 and 5 for each temperature interval.
4. Calculate the total CP of the interval.

Table 3.2   Problem Table for hot streams and hot composite curve.

Interval Temperature (°C)	Stream Population	ΔT (°C)	ΣCP_h (kW °C^{-1})	ΔH interval (kW)	Cumulative heat flow (kW)
250	①				2540
		70	4.0	280	
180	②				2260
		40	4.0 + 15.0 = 19.0	760	
140					1500
		100	15	1500	
40					0

Table 3.3   Problem Table for cold streams and cold composite curve.

Interval Temperature (°C)	Stream Population	ΔT (°C)	ΣCP_c (kW °C^{-1})	ΔH interval (kW)	Cumulative heat flow (kW)
200					3063
		20	16.7	334	
180					2729
		70	12.0 + 16.7 = 28.7	2009	
110	④				720
		60	12.0	720	
50	③				0

5. Calculate the heat (or cold) load of the interval, $\Delta H = \Sigma CP_h \, \Delta T$ or $\Delta H = \Sigma CP_c \, \Delta T$.
6. Assign zero heat flow to the lowest temperature. Sum the heat (or cold) load cumulatively until the highest temperature is reached.

The final value of the cumulative heat load (corresponding to the highest temperature) from hot streams in the Problem Table shows the total heat supply or the demand for cold. Likewise the total heat demand is set by cold streams in the Problem Table. Therefore, the total heat supply and demand are 2540 kW and 3063 kW, respectively.

The problem can be implemented in an Excel spreadsheet. Process simulation software such as Aspen Plus provides tools for generating the heat profile and the corresponding temperature and enthalpy data points.

Step 2: Construction of Composite Curve.   The next step is to plot the composite curves using the data points (temperature and enthalpy) obtained from the Problem Table. The hot composite curve can first be drawn according to the data points shown in Table 3.2. Before drawing the cold composite curve, ΔT_{min}, the minimum temperature difference between hot and cold composite curves, has to be assumed (10 °C in this case). If the cold composite curve is drawn with the true temperatures of the cold streams, there would not be any temperature driving force between the hot and cold streams. Therefore, the cold composite curve should be shifted down by ΔT_{min} to allow the minimum temperature difference between the two composite curves. The final composite curves are shown in Figure 3.5. In an alternative procedure, both composite curves are shifted by $\Delta T_{min}/2$, so that the net minimum temperature difference between the hot and cold composite curves is still ΔT_{min}. Note that without the temperature difference or driving force between the two streams, heat exchange does not occur. ΔT_{min} is to allow at least the minimum temperature difference to occur for a feasible heat exchange and heat exchanger design.

The composite curves provide several important insights into the system, such as follows:

*   Minimum hot utility load, $Q_{h,min} = 1003$ kW.
*   Minimum cold utility load, $Q_{c,min} = 480$ kW.
*   Potential heat recovery, $Q_R = 2060$ kW.

The procedure of obtaining this information is referred to as *energy targeting*. Energy targeting serves as an important procedure in energy integration because it indicates the heat recovery potential of a system prior to any detailed analysis and optimization of the network.

The features of the composite curves are:

1. The gradient of a curve is the inverse of the heat capacity flow rate, that is, $1/CP$. The segments of straight lines imply constant CP in the temperature interval.

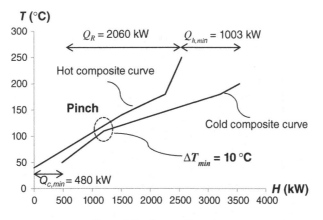

Figure 3.5 *Composite curves.*

2. The hot composite curve is always located above the cold composite curve as the heat is transferred from the hot side to the cold side, according to the second law of thermodynamics.
3. The region enclosed by these two curves signifies the potential for heat recovery from hot streams to cold streams (on the same enthalpy scale). Therefore, the enthalpy change across the overlap region between hot and cold composite curves is the amount of recoverable energy, Q_R.
4. Any excess heat from the heat source (shown on the left hand side of the composite curves) is transferred to the cold utility. This is the minimum cold utility load required, $Q_{c,min}$.
5. Any excess cold from the cold source (shown on the right hand side of the composite curves) is transferred to the hot utility. This is the minimum hot utility load required, $Q_{h,min}$.
6. The two curves are separated from each other by a gap, known as the *pinch* point. The temperature difference at the pinch point is known as the minimum approach temperature, ΔT_{min}. Pinch is a very critical region because it prevents any temperature crossover between hot and cold streams to ensure an effective heat transfer.
7. The relative position of the hot and cold streams can be altered. The energy recovery can be maximized by shifting the curves closer to each other, that is, lower ΔT_{min}.
8. There could be possibilities where no pinch appears on a composite curve. This occurs when either the hot or cold utility is not needed, no matter how the curve is shifted. This type of problem is called the *threshold* problem. Changing the ΔT_{min} does not have any effect on the heat recovery and utility consumption.

Pinch divides the system into the heat source and the heat sink. There are three rules that ought to be followed:

1. No heat transfer is allowed across the pinch.
2. No hot utility below the pinch.
3. No cold utility above the pinch.

ΔT_{min} is an important degree of freedom and there is a trade-off between capital and energy costs. The effect is shown in Figure 3.6.

a. $\Delta T_{min} = 0 \rightarrow$ hot and cold composite curves touch each other. There is no heat transfer driving force implying an infinite heat transfer area and capital requirements.
b. Increase $\Delta T_{min} \rightarrow$ hot and cold composite curves move away from each other. The heat transfer driving force increases and hence the energy cost increases. The heat transfer area decreases and hence the capital cost decreases.
c. Decrease $\Delta T_{min} \rightarrow$ hot and cold composite curves get closer to each other. The heat transfer driving force decreases and hence the energy cost decreases. The heat transfer area increases and hence the capital cost increases.

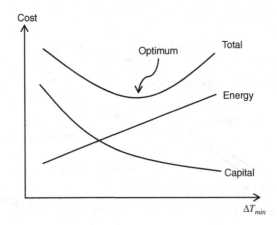

Figure 3.6 ΔT_{min} of the heat exchanger and the economic trade-off.

Therefore, selecting an appropriate ΔT_{min} is an important step to achieve optimum economics. The ΔT_{min} is dependent upon the type of heat exchanger and is normally provided by the manufacturer.

Did you know?

Process integration was initiated in the late 1970s by a group of process engineers in ICI (Imperial Chemicals Industries – formerly a well-known and the largest chemical manufacturer in the Britain). At that time, ICI was planning to expand the capacity of crude distillation unit by 20%. This resulted in an increase in energy demand and because the available space for expansion was limited the debottlenecking of the system was not economically viable.

The first published work was a PhD Thesis "Optimum Networks for Heat Exchange" written by E.C. Hohmann at the University of Southern California, in 1971. T. Umeda and team published a paper "A thermodynamic approach to the synthesis of heat integration systems in chemical processes" in *Computers & Chemical Engineering*, **3**, 273–282, 1979.

A team led by Bodo Linnhoff was also working to minimize the energy consumption. This led to significant savings in capital and energy costs. Since then the pinch technique has received widespread use in other projects in ICI for energy minimization. Apart from heat integration, the technique has also been extended to various applications, including water network, hydrogen network and mass integration.

Exercise 1. Construct composite curves illustrating the effect of varying the ΔT_{min} for the stream data provided in Table 3.1. Consider ΔT_{min} of 20 °C and 30 °C. Calculate the minimum hot utility requirement, minimum cold utility requirement and potential heat recovery.

Solution to Exercise 1. The effect of increasing ΔT_{min} is shown in Figure 3.7. Minimum hot utility and minimum cold utility requirements and potential heat recovery corresponding to ΔT_{min} are shown in Table 3.4.

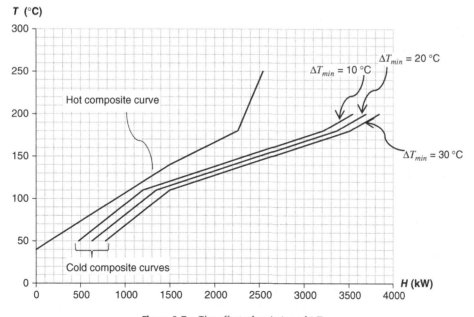

Figure 3.7   *The effect of variation of ΔT_{min}.*

Table 3.4   The effect of variation of ΔT_{min} on the utility loads and potential heat recovery.

ΔT_{min} (°C)	10	20	30
Minimum hot utility load, $Q_{h,min}$ (kW)	1003	1153	1303
Minimum cold utility load, $Q_{c,min}$ (kW)	480	630	780
Potential heat recovery (kW)	2060	1910	1760

Therefore, the lowest ΔT_{min} is favorable because the hot and cold utility requirement is lower while a higher amount of heat can be recovered.

3.3.2.2   Construction of Grand Composite Curves

Step 1: Problem Table.   A Problem Table needs to be established before constructing a grand composite curve, analogous to the methodology for constructing the composite curves. The main differences between the Problem Table used for the grand composite curve compared to the one used for composite curves are as follows:

- All hot and cold streams are embraced into one table, instead of two.
- Shifted temperature (T^*) is used.

Composite curves can be transformed into a grand composite curve, shown in Figure 3.8. Firstly, the temperature of the hot stream is reduced by $\Delta T_{min}/2$ while the temperature of the cold stream is increased by $\Delta T_{min}/2$. The dotted lines on Figure 3.8 represent the shifted composite curves.

Consider the stream data in Table 3.1. All the hot stream temperatures are reduced by $\Delta T_{min}/2$ while the cold stream temperatures are increased by $\Delta T_{min}/2$, shown in Table 3.5.

The Problem Table method is carried out according to the following procedures:

1. Arrange the stream according to the temperature from high to low as before.
2. Draw arrows for streams from the supply to target temperatures and show *CP* values.
3. Calculate the temperature difference of each interval.
4. Calculate the net *CP*, that is, the sum of *CP* of all cold streams minus the sum of *CP* of all hot streams in the interval.
5. Calculate the heat load of the interval, $\Delta H = (\Sigma CP_c - \Sigma CP_h) \Delta T$.
6. For negative ΔH, assign the surplus of heat and for positive ΔH, assign the deficit of heat.
7. Cascade from zero heat flow from the highest temperature and sum the heat load at each interval.
8. Since negative heat flow is not permitted, make the lowest ΔH value (negative number) zero and cascade the heat flow from that point (*pinch point*) to the highest and the lowest temperatures.

From the Problem Table shown in Table 3.6, three pieces of information can be obtained.

- The pinch temperature = 115 °C (occurs at zero heat flow). This is the shifted temperature; hence the hot pinch temperature is 120 °C and the cold pinch temperature is 110 °C (ΔT_{min} = 10 °C).
- Minimum hot utility load = 1003 kW
- Minimum cold utility load = 480 kW.

The grand composite curve is plotted in Figure 3.9 based on the Problem Table method shown in Table 3.6. The significance and application of grand composite curves for designing heat recovery systems is explained in Section 3.3.3.

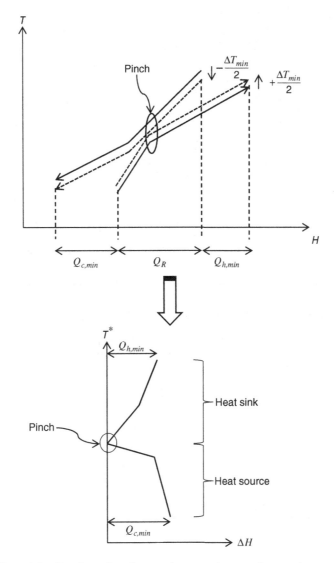

Figure 3.8   *Transformation of composite curves into grand composite curves.*

Table 3.5   *Stream data with shifted temperatures, assuming $\Delta T_{min} = 10\,°C$.*

Stream	T_S^* (°C)	T_T^* (°C)	ΔH (kW)	CP (kW °C^{-1})
1	245	135	440	4.0
2	175	35	2100	15.0
3	55	185	1560	12.0
4	115	205	1500	16.7

Table 3.6 *Problem Table for grand composite curve.*

Interval Temperature (°C)	Stream Population	ΔT (°C)	$\Sigma CP_c - \Sigma CP_h$ (kW °C^{-1})	ΔH Interval (kW)	Surplus/ Deficit	Cumulative Heat Flow 1 (kW)	Cumulative Heat Flow 2 (kW)
245						0	1003
		40	−4.0	−160	Surplus		
205						160	1163
		20	16.7 − 4.0 = 12.7	254	Deficit		
185						−94	909
		10	12.0 + 16.7 − 4.0 = 24.7	247	Deficit		
175						−341	662
		40	12.0 + 16.7 − (4.0 + 15.0) = 9.7	388	Deficit		
135						−729	274
		20	12.0 + 16.7 − 15.0 = 13.7	274	Deficit		
115						−1003	0
		60	12.0 − 15.0 = −3.0	−180	Surplus		
55						−823	180
		20	−15.0	−300	Surplus		
35						−523	480

3.3.3 Application of the Graphical Approach for Energy Recovery

Similar information such as minimum hot and cold utility requirements and potential heat recovery can be obtained from both the composite and grand composite curves. Additionally, the grand composite curve showing the enthalpy gap at each interval is used for the placement of utilities. There are two questions to be addressed:

- Heat sink problem. What are the hot utilities required (and at what temperatures) to result in an optimum economic solution?
- Heat source problem. What is the potential heat recovery strategy?

These problems can be inspected using the following cases.

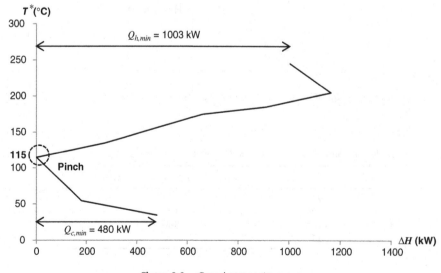

Figure 3.9 *Grand composite curve.*

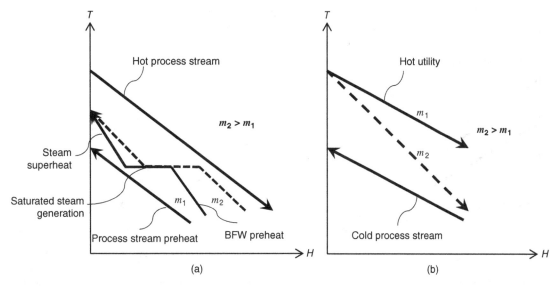

Figure 3.10 *Single process–single utility problem: (a) heat source problem – steam generation/process stream preheat from heat source and (b) heat sink problem – heat supplied to heat sink using a single utility.*

3.3.3.1 Single Process–Single Utility

The heat exchange mechanism between a single process (represented by a single stream), either heating or cooling, and a utility stream is most suitably analyzed using a composite curve. This gives a single segment for each of the hot and cold streams, unless the stream undergoes a phase change.

Figure 3.10 shows two examples for determining the optimum amount of utility to be used. Figure 3.10(a) shows a heat source problem with steam generation using heat from the hot stream. Steam is a cold stream and consists of three segments: boiler feed water (BFW) preheating, latent heat of vaporization generating saturated steam and steam superheat. An initial amount of steam, m_1, can be generated. However, this is not optimum because it is away from the hot stream. The cold composite curve can be moved closer to the hot composite curve, while maintaining a suitable ΔT_{min} so that the maximum amount of steam can be generated. Therefore, m_2 is greater than m_1. This implies that the maximum recovery of heat is attained from the hot process stream into steam generation.

Figure 3.10(b) shows a heat sink problem where a cold process stream is heated by a hot utility, represented by a hot composite curve, with an amount m_1 initially. This is not optimum because the hot composite curve is away from the cold composite curve. The hot composite curve can be shifted closer to the cold composite curve, until it reaches a specific ΔT_{min}. The optimum amount of hot utility, m_2, has a flow rate smaller than m_1. This signifies the minimum use of the hot utility.

Utilizing a hot process stream to generate steam is one of the approaches for heat recovery. Alternatively, it can be used for preheating a process stream as shown in Figure 3.10(a). The decision of whether to preheat streams or generate steam depends on their economic margins.

It should be noted that the composite curve shown here is an alternative representation to Figure 3.5 and has the same physical meaning.

3.3.3.2 Single Process–Multiple Utilities

It is also possible to use a single hot process stream to heat multiple streams, as shown in Figure 3.11(a). Similarly, different hot utilities may be required to heat a cold process stream, shown in Figure 3.11(b). There exists an economic trade-off between using a single utility at the highest temperature level or using multiple utilities at different temperature levels.

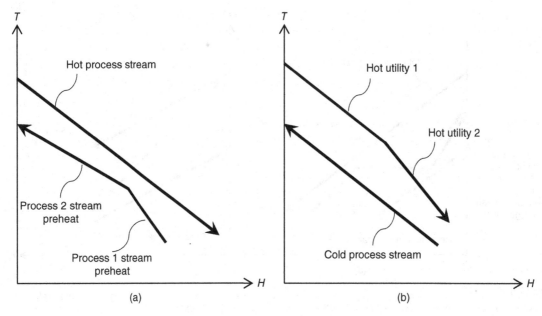

Figure 3.11   Single process–multiple utilities problem: (a) heat source problem – multiple processes preheating using a heat source and (b) heat sink problem – multiple hot utilities for heating the heat sink.

3.3.3.3   Multiple Processes–Single Utility

The earlier problem can be extended to more than one process, if the processes under consideration can recover heat in an economic sense (Figure 3.12(a)) or the processes can share the same utility (Figure 3.12(b)). The resulting composite curve has more than one segment due to multiple processes (not due to a phase change). The heat source and heat sink problems can be dealt with by the same methodology using the composite curve as in the single process–single utility case. The results of optimum steam generation or hot utility consumption may be different from the single process–single utility case due to pinch occurrence.

3.3.3.4   Multiple Processes–Multiple Utilities

A large-scale problem where multiple processes are considered and multiple utilities are required is more convenient to identify matching streams for heat exchange and temperatures of utilities (utility placement) using the grand composite curve. The advantages of using a grand composite curve to address this type of problem are:

- Both heat source and heat sink problems are combined into one graph.
- Process-to-process heat recovery can be seen from the grand composite curve and utility placements for the total site can be systematically organized.

Figure 3.13 shows the grand composite curve with placement of utilities. In Figure 3.13 (left), the heat demands by the heat sinks are satisfied by only one utility, that is, HP steam, while the heat is removed from the heat source into cooling water. The *pockets* show the potential for process-to-process heat exchange and thus no utility is required in these regions. Appropriate placements of multiple utilities can also be realized so that part of the expensive high pressure (HP) steam can be replaced by lower grade steam such as medium pressure (MP) and low pressure (LP) steams. Furthermore, steam can be generated using the low temperature heat source using the entire heat. The placements of multiple utilities and steam generation are shown on the right of Figure 3.13.

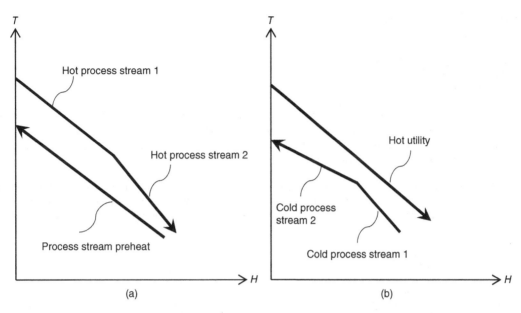

Figure 3.12   *Multiple processes–single utility problem: (a) heat source problem – stream preheating using heat source and (b) heat sink problem – heat supplied to multiple process streams by single hot utility.*

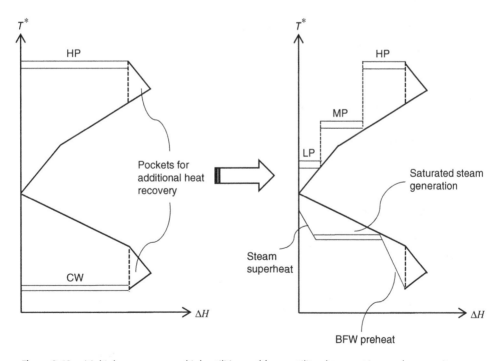

Figure 3.13   *Multiple processes–multiple utilities problem – utility placement in grand composite curve.*

In conclusion, four classes of problems result:

- Single process–single utility
- Single process–multiple utilities
- Multiple processes–single utility
- Multiple processes–multiple utilities

Appropriate utility placements and heat recovery methods are critical in heat integration studies. These can be achieved through either a composite curve or grand composite curve analysis. From the point of view of analytical convenience, the grand composite curve is more suitable to analyze the most complex case, that is, the multiple process–multiple utilities case. Other relatively less complex cases can be examined using the composite curve.

Exercise 2. Figure 3.14(a) and (b) shows two composite curves with different multiple process heat source problems, Case 1 and Case 2, respectively. If there is an option to recover the heat into either steam generation or stream preheating, find which of the heat recovery methods generates more favorable economics. The following data are available.

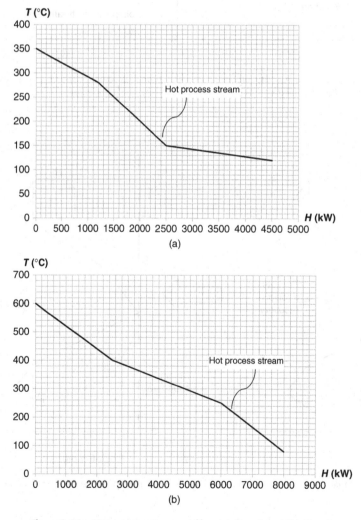

Figure 3.14 *Hot process stream profiles: (a) Case 1 and (b) Case 2.*

For steam generation, saturated MP steam at 12 bar can be considered (at saturation temperature $T_{sat} = 188\ °C$, its latent heat of vaporization is $\Delta h_{vap} = 1985.4\ kJ\ kg^{-1}$). Water is supplied at 25 °C (specific heat capacity of water = 4.2 kJ $kg^{-1}\ K^{-1}$). The water stream is to be preheated from 25 °C to 280 °C, with a total heat load of 4000 kW. Assume ΔT_{min} of 10 °C for both process-to-process and process-to-utility heat exchangers.

Solution to Exercise 2

Case 1. Figure 3.15 shows the heat recovery methods for Case 1, that is, steam generation (Figure 3.15(a)) and stream preheating (Figure 3.15(b)), respectively.

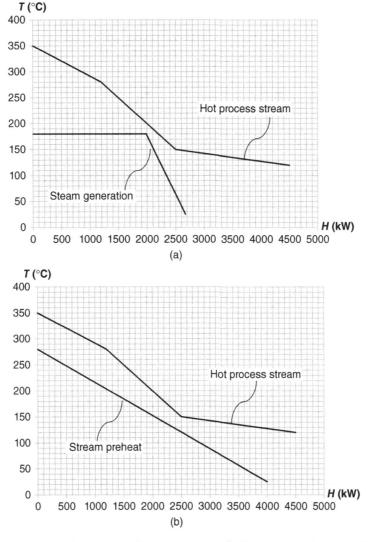

Figure 3.15   *Heat recovery in Case 1: (a) steam generation and (b) stream preheating.*

It can be deduced from Figure 3.15(a) that only 1 kg s^{-1} (3.6 t h^{-1}) of saturated MP steam can be generated. Such an amount is usually negligible in terms of overall plant economics. Preheating the stream is a feasible way of utilizing the heat from the process, as shown in Figure 3.15(b). The stream preheat profile is closely matched with the hot process stream profile, implying that heat utilization is very efficient in this case.

Case 2. Figure 3.16 shows the possible heat recovery methods for Case 2, that is, steam generation (Figure 3.16(a)) and stream preheating (Figure 3.16(b)), respectively.

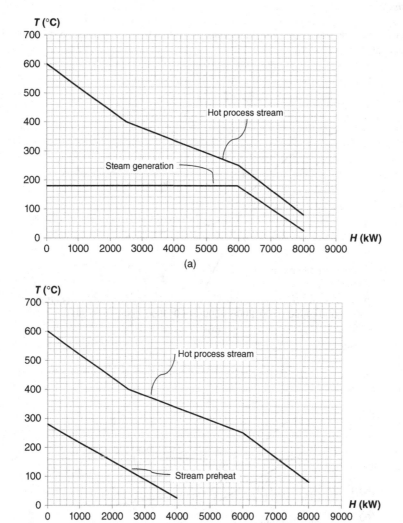

Figure 3.16   *Heat recovery in Case 2: (a) steam generation and (b) stream preheating.*

It can be seen that the steam generated in Case 2 is higher than in Case 1, that is, 3 kg s^{-1} (10.8 t h^{-1}) (Figure 3.16(a)). Such an amount of steam could be useful in the plant for heating or power generation. Nevertheless, utilizing the heat from Case 2 for stream preheating seems to be less favorable since the heat is not appropriately utilized, that is, only 50% of the heat (4000 kW) is used while the rest is wasted (e.g., to cooling water), shown in Figure 3.16(b). The large temperature gap between the hot and cold streams suggests that the heat recovery is less efficient.

3.4 Utility System

A utility system forms an indispensable part of a plant. It is central to a plant comprising all energy flows within the plant: energy generation, consumption, transfer and distribution networks. The interaction between the utility system and processes is very intimate, which means any changes in the process or the utility system would have an impact on the performance of both sides. A highly efficient utility system is thus desirable since it benefits the whole plant in terms of achieving minimum use of energy and minimum energy cost. A utility system consists of a number of components, such as steam boiler, steam turbine, gas turbine, boiler feed water system, cooling water system and furnace. These components are discussed as follows.

3.4.1 Components in Utility System

Steam and power are the main forms of energy streams in a system. An example of a utility system with unit operations for steam and power generation is shown in Figure 3.17. The utility system encompasses a series of processes, starting from the treatment of raw water through boiler feed water (BFW) treatment and deaeration, followed by generation of steam in a steam boiler. The heat recovery steam generator (HRSG) also produces steam. The exhaust gas from the gas turbine after power generation is used in the HRSG for heat recovery into steam generation. Steam at the same pressure level is collected in the steam main and distributed to process sites. The remaining steam is sent to the steam turbine for power generation. Back pressure steam turbines are used for generating power and steam at lower pressure, while

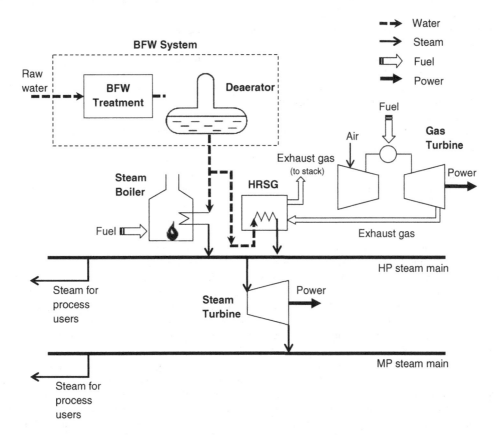

Figure 3.17 *Typical major components in a utility system.*

condensing steam turbines are equipped to generate power and condensate, returned as BFW. This utility system, which embraces the heat and power production in an integrated manner, is commonly termed as the *cogeneration* system.

1. *BFW system*. Raw water (extracted from sea, river or elsewhere) needs to be treated before the production of steam in the steam boiler to remove salts and solids that cause corrosion in the boiler. Treatment methods can be ion exchange or reverse osmosis. A deaerator is also needed to remove the dissolved oxygen, carbon dioxide and miscellaneous gases present in water. This is to prevent corrosion resulting from oxide and acid formation. The treated water is known as BFW. In general, the higher the pressure level of steam to be generated, the more stringent treatment conditions are required to obtain higher quality BFW.
2. *Steam boiler/generator*. A steam boiler is used for generating steam from BFW. Various types of fuel can be used for a steam boiler, such as natural gas, coal and biomass. There are various types of boilers, such as shell type, water-tube, fluidized bed, economizer, preheater, etc. Steam can also be generated by recovering heat from the hot exhaust gas stream. The device is commonly known as a *heat recovery steam generator* (HRSG). The resulting steam can be sent to the steam mains or other processes.
3. *Gas turbine*. The function of a gas turbine in the utility system is to generate power. A gas turbine operates based on the Brayton cycle and consists of three main components: a compressor, combustion chamber and an expander. Air is compressed, mixed with the fuel and heated in the combustion chamber. Pressurized and high temperature exhaust gas resulting from combustion is expanded, by the process of which work is produced. A generator is needed to transform the work into power. Generally, exhaust gas heat is recovered in the HRSG. Some heat may be retained to help the exhaust gas rise through the stack to release to the atmosphere. However, impurities need to be removed before releasing the exhaust gas to the atmosphere. Gas turbine exhaust gas is expected to be free from impurities (except carbon dioxide) because of its stringent fuel composition requirement. There are two types of gas turbines, that is, industrial and aero-derivative.
4. *Steam turbine*. A steam turbine converts thermal energy retained in the steam into mechanical energy. The steam turbine operates based on the Rankine cycle. The whole cycle includes the generation of steam in a boiler, expansion through steam turbines, condensate cooling and BFW recovery. Power is generated through a generator coupled with the shaft that connects to the turbine. Superheated steam is normally used in the steam turbine. Wet saturated steam is not suitable to be used in a steam turbine because the water entrained in the vapor can cause blade erosion in the turbine. There are three main types of steam turbines: back pressure, condensing and extraction.
5. *Cooling water system*. The cooling system plays an important role in taking the heat out from the main system to prevent overheating and provide safety. Many exothermic reactions need cooling so that the reactions can be maintained at a specific temperature to achieve optimal reaction performance. Extra care is taken for the reaction temperature of biological processes involving enzymatic reactions, where precise cooling is essential for the system to perform at the desired level. See Section 11.3.1 for heat transfer strategies of reactors and Section 12.4.4 for heat transfer modeling.

    Cooling can be done via two media: air and water. Air cooling is used when moisture needs to be avoided, if an area where water is scarce or it is expensive to obtain water. Otherwise, cooling water is used. There are essentially three types of cooling water systems, shown in Figure 3.18. The once-through cooling water system is the simplest configuration. Cold water is taken from the source to cool down the system and the warm water is discharged out from the system. This system does not involve recycling of water and is only used once. The once-through system requires vast amounts of water. Thus, this kind of system can normally be used if the plant is built close to the coast and sea water can be supplied to the plant. The closed recirculating cooling water system involves recycling of water after cooling the system. Instead of discharging, the warm water is cooled down and reused within the system. The open recirculating cooling water system, also known as the evaporative system, has a similar configuration to the closed recirculating cooling water system. A distinct feature of this system is that a cooling tower is used instead of a cooler. Warm water (after cooling the system) is sent to a cooling tower and is sprayed inside the tower. Air at normal temperature is supplied to the tower from the bottom to flow countercurrently upward to cool down the water sprayed from the top. Cold water is collected at the bottom of the tower and used in the system. In this system, part of the water is lost through evaporation and drift/windage. Drift/windage refers to the water droplets entrained with air, brought

① Once-through cooling water system

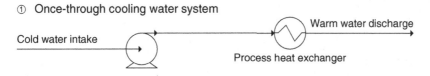

② Closed recirculating cooling water system

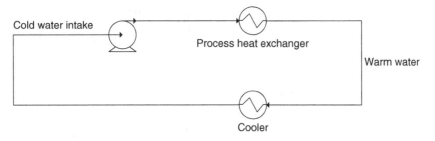

③ Open recirculating cooling water/evaporative system

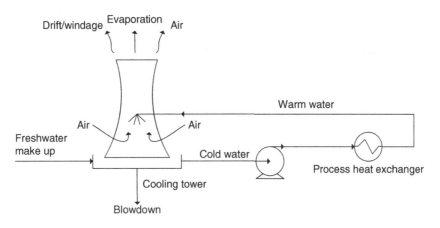

Figure 3.18   *Cooling water systems.*

out from the cooling tower. The loss of water can lead to a buildup of solids and salts and the concentration increases gradually. The solids and salts retained in the cooling tower need to be removed periodically through blowdown. Due to the loss of water through various modes, that is, drift/windage, evaporation and blowdown, freshwater is to be supplied as a makeup to the cooling system. The amount of freshwater makeup is equal to the amount of water loss. In terms of water usage of the three systems, the once-through cooling water system demands the highest (water is not reused), followed by the open recirculating cooling water/evaporative system (moderate water loss) and closed recirculating cooling water system (minor water loss).

Exercise 3. Figure 3.19 shows a utility diagram, consisting of a boiler and a steam turbine. The boiler is fed with BFW (temperature of BFW, $T_{BFW} = 105$ °C and mass flow rate of BFW, $m_{BFW} = 20$ kg s^{-1}) and is fueled with natural gas (density $= 0.8$ kg m^{-3}) to produce 20 kg s^{-1} of superheated steam (m_{steam}) at 50 bar and 500 °C. The efficiency of the boiler, η_{boiler}, is assumed to be 75%; 14 kg s^{-1} of steam is used within the processes, while the remaining 6 kg s^{-1} of steam is used for power generation through the steam turbine. Steam at 50 bar is expanded to 5 bar using a backpressure steam turbine. The steam turbine has an isentropic efficiency, $\eta_{turbine}$, of 60%. Estimate:

a.   The amount of fuel (natural gas) required in the boiler, m_{fuel}.
b.   The power generated from the steam turbine, W.

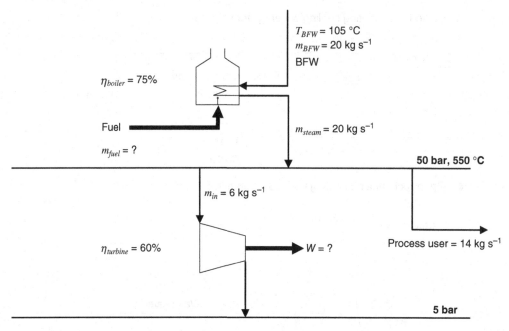

Figure 3.19   *Utility diagram.*

Data:

Heat capacity, C_p, of water $= 4.2$ kJ kg^{-1} K^{-1}
Lower heating value of natural gas $= 45\,000$ kJ kg^{-1}
Saturation temperature of steam, T_{sat}, at 50 bar $= 264$ °C
Enthalpy of vaporization, Δh_{vap}, of water $= 1639.6$ kJ kg^{-1}
Specific enthalpy of saturated steam, h_v, at 50 bar and 264 °C $= 2794.2$ kJ kg^{-1}
Specific enthalpy of superheated steam, h_{sup}, at 50 bar and 500 °C $= 3433.7$ kJ kg^{-1}

Solution to Exercise 3

a.   The heat required for steam generation, $\dot{Q}_{steam}$,

$$= \text{Heat required for heating BFW} + \text{Heat required for generating saturated steam}$$
$$+ \text{Heat required for superheating steam}$$
$$= \dot{Q}_{sensible} + \dot{Q}_{saturate} + \dot{Q}_{superheat}$$
$$= m_{steam} \left[C_p \left(T_{sat} - T_{BFW} \right) + \Delta h_{vap} + \left(h_{sup} - h_v \right) \right]$$
$$= 20 \text{ kg s}^{-1} \left(667.8 \text{ kJ kg}^{-1} + 1639.6 \text{ kJ kg}^{-1} + 639.5 \text{ kJ kg}^{-1} \right)$$
$$= 58\,938 \text{ kW}$$

Note that in this case, $m_{steam} = m_{BFW}$.

$$\eta_{boiler} = \frac{\dot{Q}_{steam}}{\dot{Q}_{fuel}}$$

$$0.75 = \frac{58\ 938\ \text{kW}}{\dot{Q}_{fuel}}$$

$$\dot{Q}_{fuel} = 78\ 584\ \text{kW}$$

Mass flow rate of fuel (natural gas) required $= \dfrac{\dot{Q}_{fuel}}{LHV_{fuel}} = \dfrac{78\ 584\ \text{kW}}{45\ 000\ \text{kJ kg}^{-1}} = \underline{1.75\ \text{kg s}^{-1}}$

Volumetric flow rate of natural gas required $= \dfrac{1.75\ \text{kg s}^{-1}}{0.8\ \text{kg m}^{-3}} = \underline{2.2\ \text{m}^3\ \text{s}^{-1}}$

where

$\dot{Q}_{fuel}$ is the heat supplied from fuel.
LHV_{fuel} is the lower heating value of fuel.

b.   The isentropic efficiency of steam turbine, $\eta_{turbine}$,

$$= \frac{\text{Actual work done, } W_{act}}{\text{Theoretical work done, } W_T}$$

$$= \frac{h_1 - h_{2,act}}{h_1 - h_{2,is}}$$

where

h_1 is the specific enthalpy of the initial state of steam before expansion.
$h_{2,act}$ is the specific enthalpy of the final state of steam after actual expansion.
$h_{2,is}$ is the specific enthalpy of the final state of steam after isentropic expansion.

Figure 3.20 shows the *T–s* diagram illustrating the theoretical and actual expansion of steam in the steam turbine, where

$$h_1 = h_{sup} = 3433.7\ \text{kJ kg}^{-1}$$

From the steam table, the specific entropy of superheated steam at 50 bar and 500 °C, $s_1 = 6.98$ kJ K^{-1}. When $s_{2,is} = s_1$ (isentropic system for theoretical work done), the specific enthalpy of steam at 5 bar, $h_{2,is} = 2817.0$ kJ kg^{-1} (at 182.5 °C),

$$0.60 = \frac{(3433.7 - h_{2,act})\ \text{kJ kg}^{-1}}{(3433.7 - 2817.0)\ \text{kJ kg}^{-1}}$$

$$h_{2,act} = 3063.7\ \text{kJ kg}^{-1}$$

$$W = W_{act} = m_{in}\ (h_1 - h_{2,act})$$

$$= 6\ \text{kg s}^{-1}(3433.7\ \text{kJ kg}^{-1} - 3063.7\ \text{kJ kg}^{-1})$$

$$= 2220\ \text{kW}$$

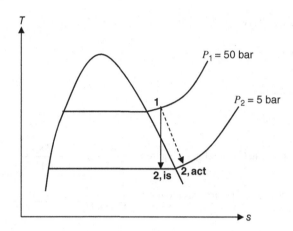

Figure 3.20   *Temperature–entropy (T–s) diagram for expansion of steam through the turbine.*

3.5   Conceptual Design of Heat Recovery System for Cogeneration

Energy efficiency of an overall system can be enhanced by improvement of self-generation or cogeneration of heat and power by energy recovery within the system. Cogeneration not only helps to fulfil the site demand for heat and power but also to export any excess at competitive economics. This concept is known as the *combined heat and power (CHP)* or *cogeneration*. The transfer of heat can be in the form of steam or process-to-process heat exchange. Electricity is the form of power produced.

3.5.1   Conventional Approach

The conventional approach for designing or retrofitting heat exchanger networks and utility systems is to first carry out data extraction, followed by energy targeting and heat exchanger network design using composite and grand composite curves and ultimately optimizing the two systems using a mathematical programming approach. The interaction between the two systems ought to be considered and iterations may be necessary to optimize the whole system. This approach is shown in Figure 3.21. The results from optimization are reliable, subject to fine-tuning of the operating conditions, consideration of other practical constraints and rigorous and accurate models. Hence, the approach is time-consuming and requires certain specialized software such as Aspen Pinch (developed by AspenTech), SuperTarget (developed by KBC), SPRINT and STAR (developed by the Centre for Process Integration, The University of Manchester, UK).

3.5.2   Heuristic Based Approach

Another effective method for designing a cogeneration system without having to use mathematical programming is shown in Figure 3.22. This method is a coupling between process simulation and heuristic based approaches, where a rational decision is made based on some important insights obtained from thermodynamics data[1,2]. *The methodology proves to be most practical for biorefinery systems, heat integration and utility systems design because the flowsheet configurations are new.*

Step 1: Data Extraction.   Important thermodynamic data such as temperature and heat duties across heat exchangers and process units are extracted from process flowsheet simulation. Note that apart from the heat exchanger, the thermo-dynamic data for some of the process units that embrace heat exchange devices such as a heating jacket should also be extracted.

Step 2: Screening and Classifying.   Screening and classification of data are required to ensure appropriate use of heat at different levels. The heat supply and demand of a system are categorized into high and low level heat, respectively,

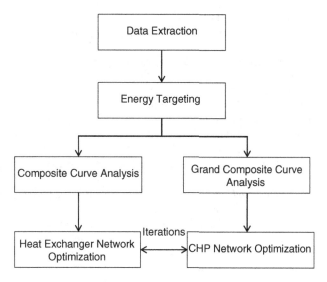

Figure 3.21 *Conventional approach for heat integration and CHP network design and optimization.*

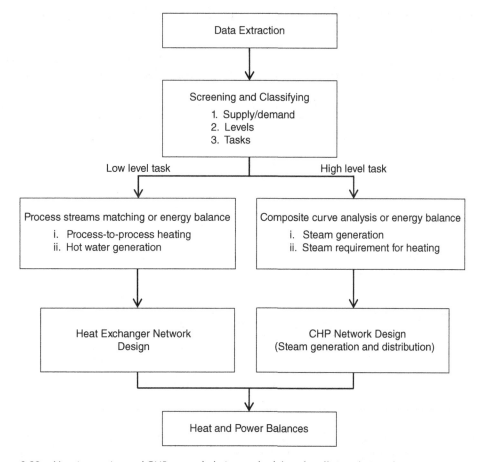

Figure 3.22 *Heat integration and CHP network design methodology for efficient design of a cogeneration system.*

based on the temperature levels and heat duties. In other words, high temperature and/or high heat duties are utilized for high energy intensive tasks, that is, steam generation, whilst low temperature and or low heat duties are utilized for low energy intensive tasks, that is, process-to-process heating, hot water generation, waste heat recovery, etc. This heuristic, which arises from common sense, is analogous to the thermodynamic matching rule, which states that "the hot process and utility streams and the cold process and utility streams are to be matched consecutively in decreasing order of their average stream temperatures" (Liu, 1987)[3]. Another similar heuristic, known as the hottest/highest matching heuristic, was proposed by Ponton and Donaldson (1974)[4] and suggests that "the hot stream having the highest supply temperature should be matched with the cold stream having the highest target temperature." In general, the heat duties and temperature levels should be matched with the same level of task. Violation of this heuristic ends up in a poor steam recovery scheme. A systematic screening and classification method using a Classification Table is shown to examine streams with respect to supply or demand, level of task and nature of task. Exercise 4 demonstrates the concept of the Classification Table.

Step 3: Analysis of High and Low Level Tasks.   Composite curve analysis and energy balance are carried out to estimate the amount of steam that can be generated and the amount of steam needed for heating.

Process stream matching and energy balance are carried out for analyzing low level tasks. The proposed strategy considers a high-to-low level approach, since any excess heat after critical heat recovery can be used to generate hot water.

The final step is to design the structure of the CHP system network (steam generation and distribution) and heat exchanger network (process-to-process heat matching), based on the information obtained from the composite curve and energy balance analyzes.

Step 4: Heat and Power Balances.   The overall heat and power balances are performed to ensure that the energy into/out from the system is well balanced.

Exercise 4.   A list of stream data for a particular system is shown in Table 3.7. Adopt the Classification Table to screen and analyze the characteristics of the streams for their suitability in different levels of heat integration tasks.

Table 3.7   *Example of stream data.*

Stream Number	T_S (°C)	T_T (°C)	ΔH (kW)
1	750	120	12500
2	360	180	1200
3	100	300	1150
4	150	110	2500
5	450	210	8600

Solution to Exercise 4.   By using the Classification Table method, the streams shown in Table 3.7 are categorized by the supply/demand, low/high level task and the nature of the task, shown in Table 3.8.

Table 3.8   *Classification Table.*

Stream Number	T_S (°C)	T_T (°C)	ΔH (kW)	Supply/Demand	Level of Task	Task
1	750	120	12500	Supply	High	Steam generation
2	360	180	1200	Supply	Low	Process-to-process heat exchange
3	100	300	1150	Demand	Low	Process-to-process heat exchange
4	150	110	2500	Supply	Low	Hot water generation
5	450	210	8600	Supply	High	Steam generation

Supply/demand for each stream is designated based on the nature of the stream, that is, hot stream or cold stream. Hot stream is used to supply heat while cold stream requires heat. According to the thermodynamic matching heuristics, the streams with a high level of temperature/heat duty should be assigned to a high level task and vice versa. Intuitively, streams 1 and 5 have higher temperature levels and heat duties and thus are assigned to high level tasks, such as for steam generation. Stream 1 has a higher temperature level than stream 5; hence a higher level (high pressure) steam can be generated from stream 1. Streams 2, 3 and 4 have lower temperature levels and heat duties, compared to streams 1 and 5. There is only one stream that demands heat in the system – stream 3. Stream 3 can obtain heat from streams 1, 2 and 5, but not stream 4. Stream 4 has a target temperature lower than the supply temperature of stream 3, that is, temperature crossover occurs. If streams 1 and 5 are used to heat stream 3, the thermodynamic matching rule is violated; then high temperature level heat duties are used for low level tasks. Such matching may still be feasible, because the remaining heat from stream 1 or stream 5 after exchanging heat with stream 3 can be used for steam generation, but a lesser amount of steam is generated. This also suggests that maximum heat recovery is not yet attained. Therefore, stream 2 is used to satisfy the heat demand of stream 3, because the heat duties and temperatures of both streams are close enough, implying that less heat is wasted and thus heat recovery is maximized. The temperature level of stream 4 is too low for even LP steam generation and hence is used to generate hot water (80–90 °C).

3.6   Summary

This chapter explains the fundamentals and importance of heat integration in the process industry. The heat integration technique using pinch analysis is the focus of this chapter, including data extraction and graphical methods for performing energy targeting to maximize energy recovery. The main aspects of utility system designs are discussed. Conceptual design of heat recovery systems for cogeneration based on the conventional approach as well as a combined heuristic and process simulation based practical approach are covered in this chapter.

Refer to the ***Online Resource Material, Chapter 3 – Additional Exercises and Examples*** for Exercise problems to test your skill.

References

1. K.S. Ng, Y. Lopez, G.M. Campbell, J. Sadhukhan, Heat integration and analysis of decarbonised IGCC sites, *Chemical Engineering Research and Design*, **88**(2), 170–188 (2010).
2. J. Sadhukhan, K.S. Ng, N. Shah, H.J. Simons, Heat integration strategy for economic production of combined heat and power from biomass waste, *Energy & Fuels*, **23**(10), 5106–5120 (2009).
3. Y.A. Liu, Process synthesis: some simple and practical development, in *Recent Developments in Chemical Process and Plant Design*, Y.A. Liu, H.A. McGee, and W.R. Epperly(eds), John Wiley and Sons, Inc., New York, 1987, pp. 147–260.
4. J.W. Ponton and R.A.B. Donaldson, A fast method for the synthesis of optimal heat exchanger networks, *Chemical Engineering Science*, **29**, 2375–2377 (1974).

4

Life Cycle Assessment

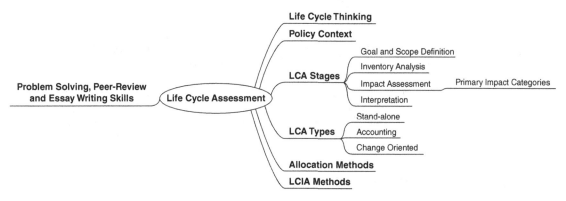

Structure for Lecture Planning

4.1   Life Cycle Thinking

Life cycle thinking (*LCT*) includes analysis of whole system environmental impacts from primary material and energy resource extraction to end of life (*cradle to grave*), much beyond a chemical plant's material and energy efficiency. A system that includes primary resource extraction to manufacturing of products or *upto the plant gate* is called a *cradle to gate* system. If one part of a system has positive impacts, other parts of the system are likely to have negative impacts. System thinking is imperative for environmental sustainability. Environmental sustainability relates to the impact assessments due to emissions to air, water and land and avoidance of emissions by improved process configurations, feedstocks, products and supply chain distributions.

An example is whether a particular nuclear power generation plant can give more energy than it consumes; there is a need to go beyond the generation facilities and consider impacts of other aspects involved such as uranium mining, transportation, associated research and even the marketing services. The systematic process design and development study for one single step can be extended to the formulation of all steps in a product life cycle, raw material acquisition, manufacturing, logistics and, after consumption, reuse, recycling and landfill, etc. Figure 4.1 shows the basic building stages for the production of a product, from raw material extraction through manufacturing to end of life reuse and recycling. However, Figure 4.1 does not include life cycle impacts associated with the construction materials of the

Biorefineries and Chemical Processes: Design, Integration and Sustainability Analysis, First Edition.
Jhuma Sadhukhan, Kok Siew Ng and Elias Martinez Hernandez.
© 2014 John Wiley & Sons, Ltd. Published 2014 by John Wiley & Sons, Ltd.
Companion Website: http://www.wiley.com/go/sadhukhan/biorefineries

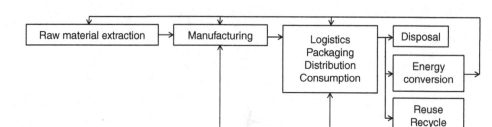

Figure 4.1   *Basic building stages for the production of a product.*

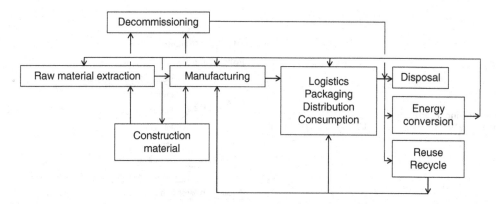

Figure 4.2   *Life cycle stages of a cradle to grave system.*

manufacturing plants. Figure 4.2 shows the complete life cycle stages of a system. Figure 4.2 shows that the emissions from the construction materials life cycles should be included in a life cycle assessment (LCA) study of a manufactured product.

The LCA is a holistic environmental impact assessment tool for cradle to grave systems that systematizes the assessment in a standardized way and format. LCA studies are data intensive. A whole system in Figure 4.2 can be divided into two parts, foreground and background systems, such as in Figure 4.3. This division is primarily applied to minimize the data

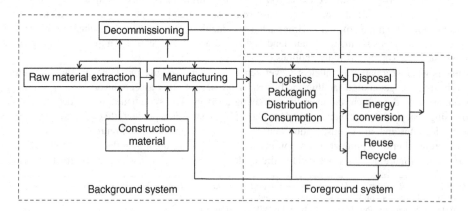

Figure 4.3   *Foreground and background systems.*

need. A decision maker primarily deals with one or some parts of a system, known as *foreground* system. The rest of the cradle to grave system is the *background* system for the decision maker. For the background system the life cycle inventory analysis data may be collected, compared and reconciled for incorporation into the foreground system. The foreground system is of central focus in terms of design, optimization and decision making.

Figure 4.3 shows an example of the foreground system with the rest presented as the background system. Which part of the system is the foreground system depends on the concerned decision maker. In this case, the decision maker is primarily dealing with the product logistics, such as a local authority trying to influence waste recycling as a better policy than landfilling, by taking the LCA approach. Whilst the foreground and the background systems are highly interactive and influenced by each other in terms of environmental impacts, the decision maker may just be concerned with the detailed evaluation of the foreground system, which can include raw material extraction and manufacturing, instead.

Uncertainties in data exist all the way in entire value chains; henceforth, structured approaches are recommended and guidelines are provided for making the most of LCA tools and for the best use of LCA results. Here, the focus is to first get familiar with the LCA approach generically followed by specific practical examples to help in applying LCA to biorefinery problems. The LCT spans not only across the whole lifetime of a process or product but also how the product or process is connected with other networks of products and processes. The networks may span across different geographic regions and hence value or supply chain analysis becomes part of the problem. Thus LCT-inspired problem formulation may take account of interacting products and process network design and decision making across supply chains in different geographic regions as well as time scale. Figure 4.4 shows an example of interacting networks of products and processes[1]. The primary raw material and energy for all kinds of products or processes under consideration come from reserves of coal, crude oil, natural gas and minerals from soil. Soil or land and water are also important resources. These resources contribute to the development of any product shown for cement, iron, electricity and heat, interdependent on each other by the constraints on resource availability. There are also demand side constraints for various products. Policies need to be in place to ensure that supply and demand are constantly balanced without damage to the environment, so as to retain resource reserves and reduce pollutant emissions to land, water and air. In this chapter, the

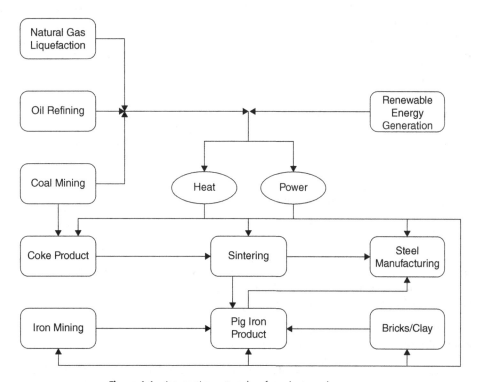

Figure 4.4 *Interacting networks of products and processes.*

focus is to discuss LCA for analyzing interactions between engineering systems and the environment through assessments of resource depletion and pollutant emissions. The learning objectives can be subdivided as follows.

1. Find the likely environmental impacts.
2. Structure pollutants under various impact characterizations.
3. Impact assessments.
4. Compare impacts between materials.

4.2   Policy Context

The policy landscapes are fast evolving by being more concerned about whole system impacts, such as the extended producer policy in which a producer's responsibility for a product is extended to the postconsumer stage of a product's life cycle, defined by OECD (2005)[2]. In this context, the decision maker must also systematically present the whole system impacts; hence the knock-on effects in the background system due to any changes in the foreground system and vice versa must be analyzed for any product or process development.

Modern lifestyle is chemical dependent. Chemicals contribute to household activities (energy appliances, clothes, food, beverages, polymers), pharmaceutical, petrochemical, agrochemical, industrial, etc. Throughout life cycles, chemicals have environmental impacts under various categories. Chemicals may cause global warming, stratospheric ozone layer depletion, acid rain, leaching to soil and water bodies (eutrophication), etc. These various environmental impact categories are commonly known as environmental impact characterizations, which is used in this text. In a broad sense, chemicals contribute to land, atmosphere and water environmental impacts. One of the earliest recognitions of environmental pollutions was by Dangerous Substances Directives, as follows, to regulate discharges to water bodies[3]. Because the earth's water body is connected through rain and soil water absorption, pollutants spread very quickly by water.

The Environmental Protection Act 1990 – for discharges from severe pollutants from industrial processes.
The Water Resources Act 1991 – for discharges to water.
The Water Industry Act 1991 – for discharges to sewer.

These first set of Directives were to limit the concentration and/or the total amount of priority pollutant categories, also known as List I substances, given in Table 4.1.

4.3   Life Cycle Assessment (LCA)

In the 1960s, the life cycle assessment (LCA) technique was applied to energy analysis; however, a group of researchers were commissioned by Coca Cola Amatil around the same period to study the resources and environmental profile of the different packaging materials they used for their products[3]. The impacts of pollutants turned out to be more severe than estimated. The majority of chemicals is related to one or more categories of environmental pollutants. Understanding the impacts of chemicals under various environmental categories or characterizations can help identifying and eliminating them at their sources, if possible at the first instance. If not, chemicals must be treated and regulated before their discharge to the environment.

By the mid-1980s the multicriteria systematic analysis had spread to include many more products. Different terms were used to describe these studies, the term Life Cycle Assessment was proposed and agreed upon at a workshop held by the Society of Environmental Toxicology and Chemistry (SETAC) at Vermont, USA, in 1990. SETAC have since published various guides and advice material on LCA simplification and methods.

The definition of Life Cycle Assessment as given by the *International Organization for Standardization* (ISO) in 1997 is:

Compilation and evaluation of the inputs, outputs and the potential environmental impacts of a product system throughout its life cycle.

According to *ISO standards 14040, 14041 and 14044*, the LCA is carried out in four phases: Goal and Scope definition, Inventory analysis, Impact Assessment and Interpretation[4-6]. All these phases are interdependent as a result of one phase

Table 4.1 *List I of priority pollutant categories.*

1. Organohalogen compounds	For example, CCl_4 used as refrigerant, fire extinguisher, degreasing and dry cleaning agent. Its use declined in recent times due to its effect on stratospheric ozone depletion.
2. Organophosphorous compounds	Phosphates in fertilizer (though insoluble in water) can adhere to soil particles and erode into water bodies.
3. Organotin compounds	Chemicals of hydrocarbons and tin (Sn) generally used as a heat stabilizer in polyvinyl chloride, biocide, wood preservative, antibiofouling agent. Their toxic effects are detrimental to marine lives (1 ng L^{-1})
4. Compounds with carcinogenic properties in aquatic environment	Polycyclic aromatic hydrocarbons have toxic and carcinogenic properties and occur by natural and anthropogenic activities.
5. Mercury and its compounds	Organic mercury is more harmful than inorganic mercury, for human body (brain, nervous system and kidney and more harmful for youngsters). Occurs from rain, rock, soil, thermochemical process of coal and smelting processes.
6. Cadmium and its compounds	Categorized as Category 2 carcinogen by National Occupational Health and Safety Commission. Exposure limit in work place 0.01 mg m^{-3} and 0.002 mg m^{-3} for drinking water. They are present in ores, coal and other fossil fuels and in products, tobacco, fertilizer, PVC, petrol, tires, electronic components, batteries, textile dyes and ceramic glazes.
7. Mineral oils and hydrocarbons of petroleum origin	Fossil hydrocarbons in oil and petrochemical products have environmental impacts in all categories.
8. Synthetic substance in water	Pesticides, solvents and water-borne chemicals are the biggest threat to the fresh water supply and damaging to human reproductive systems.

determining the execution of the next phase. Figure 4.5 shows the different phases involved in an LCA study. Figure 4.6 illustrates the documents published by the ISO and an example of a life cycle management framework developed by 3M, 1997, for the assessments of risks and opportunities throughout the various stages of a product's life cycle.

Life cycle stages include compiling inventories over the complete supply chain, providing a service or product extending from the "cradle" of primary resources – fossil and metal ores, for example – through to the "grave" of recycling or safe disposal; the term "life cycle" also includes the service life of the capital goods needed for a product or process. In the sequence of steps conventionally followed in carrying out an LCA, compiling the material and energy balance is termed

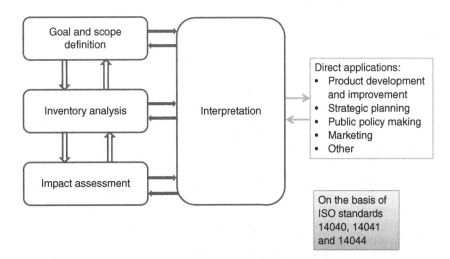

Figure 4.5 *LCA study stages*[4–6].

June 1997	ISO 14040 Life cycle assessment – principles and framework
October 1998	ISO14041 Life cycle assessment – goal and scope definition and inventory analysis
March 2000	ISO 14042 Life cycle assessment – life cycle impact assessment
March 2000	ISO 14043 Life cycle assessment – life cycle interpretation
2002	ISO/TS14048 Life cycle assessment – data documentation format

LCM stage impact	Material acquisition	R&D operations	Manufacturing operations	Customer needs	
				Use	Disposal
Environment					
Energy/Resources					
Health					
Safety					

Figure 4.6   *(above) LCA documents published by the International Organization for Standardization (ISO). (below) 3M, 1997, developed a Life Cycle Management framework for the assessment of risks and opportunities throughout the various stages of a product's life cycle.*

the *Inventory* phase. Apart from the extended system boundary, an inventory analysis differs from a conventional material and energy balance analysis by the need to include trace flows of species whose environmental significance is large, for example, because they have high human or ecotoxicity. There are two terminologies often used in LCA, LCI and LCIA. LCI refers to *Life Cycle Inventory* and LCIA refers to *Life Cycle Impact Assessment*. A study is not an LCA study when the study includes a goal and scope definition and inventory analysis, but not the impact assessment. This is then referred to as an LCI analysis. An LCIA includes an impact assessment, but not necessarily the interpretation. The LCA must include transparent and systematic discussions on all four stages.

It is more common to find chemicals with more profound (or primary) effects on one or more impact characterizations than others (secondary, tertiary, etc.), more commonly known as mid-point and end-point impacts. Pollutants with primary impacts are included in an impact characterization. Several pollutants have mid-point impacts on various characterizations. However, only primary effects of pollutants are considered to avoid double counting in most LCA studies. In Section 4.8, the various life cycle impact assessment (LCIA) methodologies are shown. A study can include a combination of primary, mid- and end-point impacts. Inclusion of mid- or end-point impacts in the LCIA methodology calls for robust reasons for doing so.

The effect of greenhouse gases (GHGs) such as CO_2, CH_4, N_2O, CFC (chlorofluorocarbon), etc., emissions on various levels is explained[3]. These gases absorb solar infrared radiation. Different gases have different abilities to absorb radiation. These gases create a GHG effect in sending the radiation to the earth's surface (Figure 4.7). Earth then absorbs some of this radiation energy and sends back the rest to the atmosphere. This exchange of radiative force between earth and GHGs in the atmosphere is regarded as a primary effect. The change in radiative force is expected to change the global temperature level, known as global warming and regarded as a secondary effect. The global warming effect is not uniform on the earth's surface, causing ice to melt, raised sea levels and severe weather patterns. These effects are known as tertiary effects. Once again, the consequences of tertiary effects on human health and ecological systems vary from place to place. The result may be changes in biodiversity in various ecosystems and impacts on food availability, agricultural patterns, society and economics. Having long chains of cause and effect, describing the environmental effects in a quantitative manner all the way through to the effects on the later stages in geographically segregated regions becomes impossible. As a result, LCIA is mostly done at a primary impact level or mid-point level (see Section 4.8: LCIA Methods).

The global warming potential (GWP) of a gas is determined by the infrared absorption capacity of the gas, which is related to the chemical structure and physicochemical properties of the gas. Similarly, all other impact characterizations depend on physicochemical properties of chemicals. Governed by molecular properties of chemicals, different chemicals have different levels of environmental impact characterizations. Over a 100 year period, methane gas has 25 times more GWP than carbon dioxide gas. Therefore, factors are used to provide an environmental impact characterization in relative

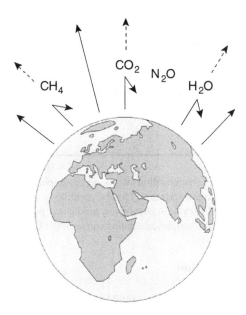

Figure 4.7   *GWP impact of GHGs.*

terms compared to a base chemical with an assumed value of 1 in the environmental impact characterization. Hence, methane gas has a global warming impact characterization factor of 25, over a 100 year period, implying that one unit mass of methane gas has 25 times more intensive global warming impact than the unit mass of carbon dioxide.

Quantities are then associated with each pollutant impacting on every category. These factors are called equivalence factors/equivalents/potentials/category indicators (according to the ISO standard)/characterization factors. Characterization factors are more commonly used in this text. The characterization factors depend on physicochemical properties of pollutants and are provided in relative terms compared to a base pollutant (assumed to have a value of 1 in an impact category). As shown, CO_2 is a base pollutant for the GWP impact category, that is, its GWP impact characterization factor is 1. All other pollutant GWPs are presented relative to CO_2. The GWP of CH_4 is 25 CO_2 equivalent and that of chlorofluorocarbons (CFCs) is 5000 CO_2 equivalent, over a 100 year lifespan.

Their units are represented in terms of: CO_2 equivalent, for example, kg CO_2 equivalent. Various notations used are CO_2 eq., CO_2-eq., etc.

The GHGs have different lifespans and therefore a basis of 25, 100 and 500 years is generally assumed over which the total GWP is predicted.

The volatile organic compounds (VOCs) are carbon containing compounds that take part in photochemical reactions. The VOCs do not include carbon dioxide, carbon monoxide, carbonic acid, metallic carbide, carbonates and ammonium carbonates. VOCs react with NO_x in the presence of sunlight to produce ozone and photochemical pollutants, such as peroxyacetyl nitrate (PAN), formaldehyde and acetic acid in the lower atmosphere. These pollutants and ozone in the lower atmosphere are responsible for urban smog and ground level ozone formation and are classified under photochemical oxidant creation potential.

NO_x (excluding N_2O) has primary effects on acidification and eutrophication potential impact categories. All these environmental impacts discussed are around ecological consequences. Most chemical pollutants also have primary impacts on human health. Depletion of quality and quantity of resources, for example, fossil energy, mineral and ores in earth, land and water, is also a reality and must be considered with due importance, in the context of growing population, energy demands and needs for sustainable lifestyles. Broadly, there are three classifications of environmental impacts of chemicals: resource use, human health and ecological consequences. Table 4.2 shows the most common impact characterizations responsible for environmental consequences that can be assessed. Table 4.3 shows the links between the various types of inventory stressors and their primary and secondary environmental impacts[3].

Table 4.2 *Broad impact characterizations.*

Resource Use	Human Health	Ecological Consequences
1. Energy (can also be subdivided into fossil and renewable) 2. Water 3. Land 4. Abiotic (ores) depletion	1. Toxicological impacts 2. Nontoxicological impacts 3. Work environment impacts	1. Global warming 2. Ozone depletion 3. Acidification 4. Eutrophication 5. Photochemical oxidant creation 6. Aquatic, human, terrestrial ecotoxicity 7. Biodiversity

In evaluating the environmental impact characterization factors, no geographical segregation is considered. Henceforth, assessment of the environmental impact characterization factors, generally restricted up to the primary and mid-point levels of impacts can be applied to analyze life cycle impacts of any products or process systems, irrespective of geographic locations. The following sections explain the various calculation stages of the LCA.

4.4 LCA: Goal and Scope Definition

The goal and scope definition is the first step of LCA. A decision maker sets out objectives of interest for LCA study. The objectives need to be consistent with the intended application. The goal and scope definition involves identification of:

1. Functional unit
2. System definition
3. System boundaries

This phase also includes product definition, coproduct and waste utilization routes, any allocation method considered, LCIA methodology selected, database sources, year and geographic location.

The functional unit is the unit of analysis for the study, and it provides a basis for comparison if more than one alternative is being studied. It should be defined in terms of the **service(s)** provided by any product, process, or activity under analysis.

An example often used to describe a *functional unit* is use of nappies by two babies. Nappies can be used up at various rates. In order to obtain average representative rates of use, the rates can be calculated based on uses over six months or longer. A functional unit can then be presented as the mass of nappies used in six months by each baby.

To compare between two or more washing liquids, the *functional unit* can be the mass of washing done per unit mass of a washing liquid. The functional unit of a system to be studied is associated with the service provided, for comparison

Table 4.3 *Inventory stressors and environmental impacts.*

Inventory Stressor	Initial Impact	Secondary Impact
Acid emission	Acid rain	Acidified lakes
Photochemical oxidants	Smog	Health impairment
Nutrients	Eutrophication	Bogs
Greenhouse gases	Global warming	Sea level rise
Ozone depletors	Ozone depletion	Skin cancer
Toxic chemicals	Toxic effect	Health impairment
Solid waste	Land consumption	Habitat destruction
Chemicals to groundwater	Groundwater impact	Health impairment
Fossil fuel use	Resource depletion	

Table 4.4 *Transferable forms for comparison of LCA results between competing systems.*

Item	Basis: 1	CPE (MJ)	GWP (g CO_2 eq.)
N fertilizer (urea)	kg	49.25	2940
P fertilizer (triple superphosphate)	kg	18.81	1160
K fertilizer	kg	5.6	380
Diesel	MJ	1.114	74.4
Electricity mix	MJ	2.597	173.4
Natural gas (NG)	MJ	1.016	61.2
Liquified petroleum gas (LPG)	MJ	1.06	76.4
Fuel oil	MJ	1.11	94.9
Hexane	kg	37.5	861
Methanol	kg	12.872	2836
Hydrogen	kg	183.2	11888
Soy meal	kg	4.13	726

with other systems. Issues of durability and maintenance should be taken into account for a realistic comparison between the same functional products.

In a goal and scope definition, system boundaries, LCIA methods and purpose must be defined with clear justifications. These will relate to audience and uses of a study. Assumptions, data availability, limitations and the quality of data are important to acknowledge for a system definition. It is important to recognize important impact categories for assessment at an early stage of an LCA. The supply of any component if changed from one region to another can affect impacts in accompanying categories. Thus, geographic regions need to be mentioned. System boundaries can show life cycle stages of productions and any transport involved.

Table 4.4 shows an example of *transferable form* for comparison. To compare between nitrogen fertilizers, a common basis of the mass of nitrogen in a fertilizer rather than the mass of the fertilizer should be selected. If the fertilizer (e.g., urea) is not in the pure form, then 1 kg of nitrogen can be selected as a basis for comparison between N based fertilizers. The unit that is highly dependent on the type of system or functionality to compare can be chosen to reflect the function or service for which a product has been produced. For example, 1 MJ of energy output from each of diesel, electricity, natural gas, LPG and fuel oil products may be compared in terms of the cumulative primary energy (CPE) use and GWP over 100 years. Likewise, if hydrogen rich gases produced from various flowcharts are to be compared for environmental impacts, the amounts of individual gas streams containing 1 kg of hydrogen can be taken as a basis. Thus, 1 kg of hydrogen production is the common basis (functional unit) for comparison of environmental impacts between various gas streams.

Figure 4.8 shows the boundary of an integrated crop and residue based biorefinery system. Assumptions, data availability, limitations and the quality of data are important to acknowledge for a system definition. It is important to recognize important impact categories for assessment in LCA that can be done at any stage and included in the goal and scope definition later. Assumptions and applicability of data over geographic regions and time scales must also be defined. The supply of any component if changed from one region to another may affect impacts in accompanying categories. Generally, 20 years and 100 years of time horizons are chosen for environmental impact assessment. In Figure 4.8, the system boundary shown considers the subsoil preparation outside the boundary, because the subsoil preparation occurs once before a given crop can be grown annually, when a steady state can be assumed, which is further explained in the section on land use.

Various cradle to grave system boundaries, such as for anaerobic digestion of sewage sludge for agricultural and energy generation applications, solar organic photovoltaic (OPV) glass manufacturing system and solar OPV cell manufacturing system, which can be considered within the goal and scope definition of the corresponding systems, are shown in Figures 4.9(a) to (c). The boundaries here are shown for the interactions between systems and the environment in terms of material and energy resource depletion and emissions to the environment. This is the convention followed in process engineering. However, in LCA, it is unnecessary to show such interactions between systems and the environment, because interactions are common: material resource (including land and water), primary energy resource and emissions to atmosphere, water and land.

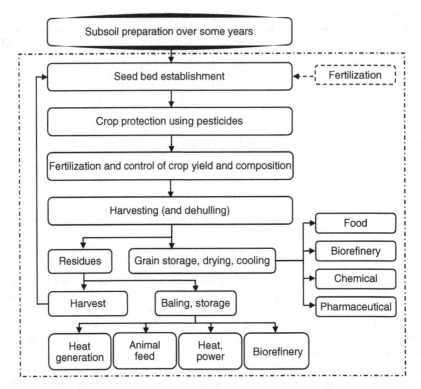

Figure 4.8   *Boundary of an integrated crop and residue based biorefinery system.*

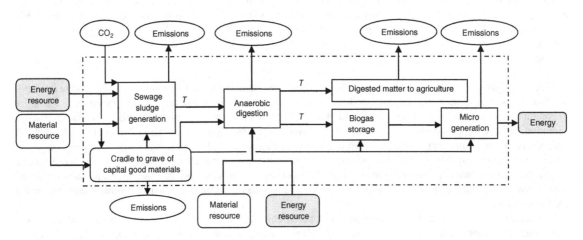

Figure 4.9(a)   *Interactions between anaerobic digestion of sewage sludge system and the environment for LCA. T stands for transport.*

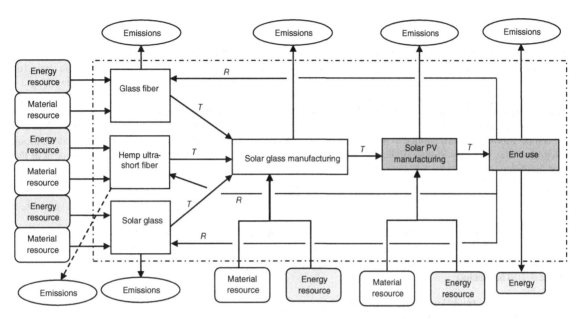

Figure 4.9(b) *Interactions between solar organic photovoltaic (OPV) glass manufacturing system and the environment for LCA. T stands for transport. R stands for recycle. T and R also cause resource depletion and environmental emissions. Only a part of the impacts from the "Solar PV manufacturing" and "End use" blocks that is due to solar glass manufacturing is considered within the system boundary.*

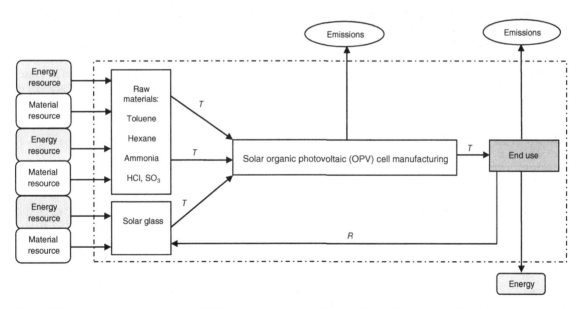

Figure 4.9(c) *Interactions between solar OPV cell manufacturing system and the environment for LCA. T stands for transport. R stands for recycle. T and R also cause resource depletion and environmental emissions.*

4.5 LCA: Inventory Analysis

The inventories are stressors or pollutants causing environmental impacts under various categories. Their analysis is about inlet and outlet mass and energy flow analysis. The inventory analysis is to identify each block of a process life cycle, for example, Figures 4.1 to 4.3, and to quantify the material and energy inputs and outputs for each of these stages.

The inventory analysis includes the following steps, detailed definition of the system, data collection, allocation and quantification of the environmental burdens. Each of the steps in the flowchart in Figures 4.1 to 4.3 can be expanded to look into substeps and to establish the mass and energy flows in and out of the boxes from/to the environment. Further, the compositions of streams emitted to water, land and air must be estimated for the assessment of impacts under various categories. This step may involve simulation/data collection/modeling of each box (e.g., fertilization and control of the crop yield step in Figure 4.8). The environmental burdens (B_j) can be quantified for each step, shown as

$$B_j = \sum bc_{j,i} x_i \tag{4.1}$$

where $bc_{j,i}$ is the burden j from the process i, while x_i is the mass or energy flow associated with that activity. The calculation is further explained by using example problems from biorefinery systems.

Broadly, there are three categories of environmental impacts: resource use, human health and ecological consequences (Table 4.2). They include:

1. Global impacts (global warming, ozone depletion, eutrophication, aquatic ecotoxicity)
2. Regional impacts (acidification)
3. Local impacts (photochemical oxidant creation, biodiversity).

These impact categories are further discussed in the Impact Assessment section.

Various LCI databases exist: Ecoinvent, ILCD/ELCD (ILCD: International reference Life Cycle Data system and ELCD: European reference Life Cycle Database), US-LCI, national ones such as in Australia and older ones include Buwal, IDEMAT, etc. In order to make an LCA study report transparent, the year of data publication, data source and the geographical relevance should be included. The data sets are adaptable and to bridge a gap may exist for process blocks.

Exercise 1. Draw a flowchart schematic with an inventory analysis for a wheat bioethanol plant consisting of processes shown in Table 4.5. The feedstock and raw material consumption rates on a daily basis and production specifications of various processes are shown in Table 4.5[7].

Table 4.5 *Basis for wheat bioethanol process flowsheet mass balance for Exercise 1. (Reproduced with permission from Sadhukhan et al. (2008)[7]. Copyright © 2013, Elsevier.)*

Process Unit	Feedstock	Product	Other Raw Materials	Specification
Milling (hammer)	Wheat: 340 kt	Milled wheat	–	–
Liquefaction	Milled wheat	Liquified grain	Process water	625.6 kt
			Sodium hydroxide	1.70 kt
			Calcium chloride	0.41 kt
			α-Amylase	0.28 kt
Saccharification	Liquified grain	Fermenter feed	Sulfuric acid	0.73 kt
			Glucoamylase	0.37 kt
Fermenter	Fermenter feed	Fermenter product: 88.27% CO_2 rich stream	Yeast	0.07 kt
CO_2 recovery	CO_2 rich stream	CO_2: 90%	–	–
Centrifugation	fermenter product	Ethanol rich stream: 74.68% Waste stream	–	–
Ethanol production	Ethanol rich stream	Ethanol: 17.934%	–	–
Rotary dryer	Waste stream	DDGS: 38.59%	–	–

Solution to Exercise 1. Figure 4.10 shows the flowchart schematic with an operational inventory analysis of a wheat bioethanol plant consisting of processes shown in Table 4.5.

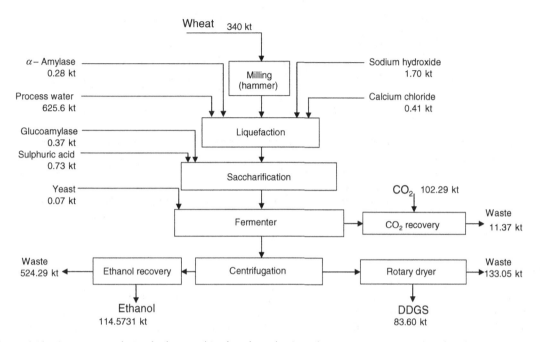

Figure 4.10 *Inventory analysis of wheat to bioethanol production plant operation. (Reproduced with permission from Sadhukhan et al. (2008)[7]. Copyright © 2013, Elsevier.)*

Exercise 2. The mass and energy distribution for a wheat bioethanol and straw combined heat and power (CHP) system in the UK is given in Table 4.6[8].

a. Complete the wheat bioethanol and straw CHP system mass and energy balance for a functional unit of 1 hectare of land use for the cultivation, shown in Figure 4.11(a). Sketch the system with completed operational inventory data.

Table 4.6   *Stream data for Exercise 2. (Reproduced with permission from Martinez-Hernandez et al. (2013)[8]. Copyright © 2013, Elsevier.)*

Subsystem	Product	Yield	LHV (GJ t^{-1})
Wheat cultivation	Wheat	6.96 t DM ha^{-1}	18.6
	Straw[a]	3.49 t ha^{-1}	14.6
Bioethanol plant	Ethanol	0.34 t t^{-1} (DM wheat basis)	26.7
	DDGS	0.25 t t^{-1} (DM wheat basis)	18.2
Straw CHP plant (processes 60% of the total straw cultivated)	Electricity	1060 kW h t^{-1} (straw basis)	
	Heat	567 kW h t^{-1} (straw basis)	

[a]Total amount of straw cultivated before soil retention; 40% of this amount is retained in the soil for enriching nutritional value.
DDGS: dried distillers grains and solubles.
DM: dry matter.
LHV: lower heating value (*c.f.* Section 10.2.2).
1 GJ = 304.2 kW h based on 8000 operating hours per year. This is based on the following calculation: 1 GJ = $\frac{10^6}{3600} \times \frac{365 \times 24}{8000}$ kW h.

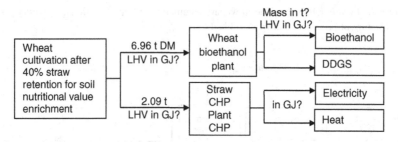

Figure 4.11(a) *Development of wheat bioethanol and straw CHP system mass and energy balance.*

b. Repeat the calculation for a functional unit of 100 t d⁻¹ of bioethanol production. Sketch the mass and energy flow diagram. Calculate the land use for the production of 100 t d⁻¹ of wheat bioethanol in the UK.
c. Calculate the energy efficiency of the straw based CHP plant.
d. Calculate the energy efficiency of the wheat bioethanol plant, if only bioethanol is used as an energy product.
e. In 2010 in the UK, there were 1.2 million hectares of land available for wheat cultivation. The CHP requirement from the biomass in the UK was 48.2 PJ y⁻¹ (1 petajoule = 10¹⁵ joule). Estimate the percentage contribution potential of CHP from the available wheat straw.

Solution to Exercise 2

a. Figure 4.11(b) shows the completed mass and energy balance on the flowchart on the basis of 1 hectare.
b. Hint: the basis a functional unit of 100 t d⁻¹ of ethanol production. Multiply all the mass and energy values in Figure 4.11(a) by 42.26 (obtained from 100/2.37). The UK land requirement for the production of 100 t d⁻¹ of wheat bioethanol is 42.26 hectares.
c. The energy efficiency of the straw based CHP plant = 37% (obtained from the energy balance in Figure 4.11(b): $(7.3 + 3.9)/30.6 \times 100$).
d. The energy efficiency of the wheat bioethanol plant is 49%, if bioethanol is only used as an energy product (obtained from the energy balance in Figure 4.11(b): $63.2/129.5 \times 100$). All the energy inputs and outputs other than the wheat energy input and bioethanol energy output are neglected.
e. The UK straw CHP generation per unit land is 11.2 GJ per hectare (Figure 4.11(b)). Thus, the excess straw CHP energy yield is: $11.2 \times 1.2 = 13.44$ PJ y⁻¹ from 1.2 million hectares of land available for wheat cultivation. The percentage contribution potential of CHP from the available wheat straw towards UK's total CHP requirement of 48.2 PJ y⁻¹ is 28%.

Process simulation can also be used for data extraction for an inventory analysis. An example of the use of Aspen Plus, a process simulator, for the impact assessment of a bio-oil based methanol and a bio-oil based Fischer–Tropsch (FT) liquid synthesis plant[9, 10] is shown as follows in Exercise 3.

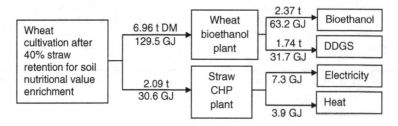

Figure 4.11(b) *A wheat bioethanol and straw CHP system mass and energy balance on the basis of 1 hectare land use.*

Exercise 3. The inventory analysis of a bio-oil based methanol and a bio-oil based Fischer–Tropsch (FT) liquid synthesis plant is shown in Figure 4.12. The details of these process simulations and technoeconomic analyses are available in Ng and Sadhukhan (2011)[9, 10] and Sadhukhan and Ng (2011)[11]. A portion of CO_2, 0.75 kg per kg of bio-oil separated by Sulfinol unit (from the "Heat recovery, Water gas shift and Gas cleaning block" in Figure 4.12) is capture-ready and, therefore, is not accounted for in the plant GWP impact. CO_2 along with water vapor and nitrogen (in air) in the exhaust gas is emitted to the atmosphere after combined heat and power (CHP) generation. Calculate the GHG impact from the exhaust gas for the given plant capacity.

Mass fraction of exhaust gas	Methanol synthesis	FT synthesis	GHG impact kg CO_2 eq.
CO_2	0.16	0.17	1
H_2O	0.15	0.13	0.08218
N_2	0.69	0.7	
Exhaust gas, kt d^{-1}	57	37	
GHG impact, kt CO_2 eq. d^{-1}	9.9	6.7	

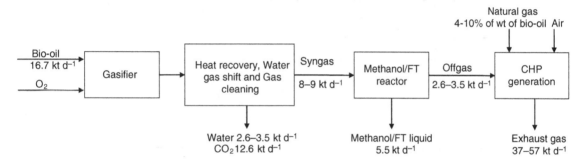

Figure 4.12 *Simulation results of kt CO_2 eq. d^{-1} bio-oil based methanol and a bio-oil based Fischer–Tropsch (FT) liquid synthesis plants for the impact assessment. The GHG impact (in kt CO_2 eq. d^{-1}) from the exhaust gas for the given plant capacity calculated is shown in the inset table.*

Solution to Exercise 3. Figure 4.12 shows the problem data and the inset table shows the solution in terms of the GHG impact from the exhaust gas.

The mass fractions and flow rates of the exhaust gases from the methanol and FT liquid synthesis centralized plants are shown in Figure 4.12. CO_2 and water vapor contribute to the GWP by a total of $(57 \times 0.16 + 57 \times 0.15 \times 0.08218) =$ 9.9 kt CO_2 eq. d^{-1} or $(9.9/16.7) = 0.6$ kg CO_2 eq. kg^{-1} of bio-oil and 6.7 kt CO_2 eq. d^{-1} or 0.4 kg CO_2 eq. kg^{-1} of bio-oil from methanol and FT synthesis centralized plants, respectively.

Exercise 4. A biomass boiler can use straw/wood/RDF (refuse-derived fuel) fuels to produce 50 MW energy output. The ultimate analyses of the fuels are shown in Tables 4.7 to 4.9, respectively. If the fuels are fully combusted in the presence of a stoichiometric amount of oxygen, estimate the composition and the GWP of the exhaust gas resulting from the combustion of the fuels. Assume that nitrogen, sulfur and chlorine present in the fuels do not contribute to the GHG emissions. Assume that ash is removed completely in the solid phase from the biomass boiler.

Solution to Exercise 4. To calculate the emissions resulting from the process operation, mass balance and energy efficiency are used in the first place. The basis assumed in this case is 50 MW of energy output.

Table 4.7 *Proximate and ultimate analyses of straw. (Reproduced with permission from Sadhukhan et al. (2009)[12]. Copyright © 2009, American Chemical Society.)*

Proximate Analysis (wt%)		Ultimate Analysis (wt%)	
Moisture	8.5	C	36.57
Volatile matter	64.98	H	4.91
Fixed C	17.91	O	40.7
Ash	8.61	N	0.57
LHV, MJ kg^{-1}	14.6	S	0.14

Table 4.8 *Proximate and ultimate analyses of wood[12].*

Proximate Analysis (wt%)		Ultimate Analysis (wt%)	
Moisture (after long storage)	25	C	51.8
		H	5.7
Ash (dry basis)	1.1	O	40.9
LHV (dry basis), MJ kg^{-1}	19.3	N	0.1

Table 4.9 *Proximate and ultimate analyses of RDF.*

Proximate Analysis (wt%)		Ultimate Analysis (wt%)	
Moisture	50	C	45.5
Volatile matter	79.6	H	5.8
Fixed C	10	O	37.8
Ash	10.5	N	0.3
		S	0.2
LHV, MJ kg^{-1}	17.73	Cl	0.4

An energy efficiency value should be assumed to determine the biomass feed flow rate from its lower heating value (LHV). The energy efficiency that is calculated using the equation below is found to be 40–45% for combined heat and power (CHP) generation from a biomass[12]:

$$\text{Energy efficiency of CHP generation from biomass} = \frac{\text{Output energy generation}}{\text{Energy input in biomass}} \qquad (4.2)$$

Based on a realistic assumption of an energy efficiency of 0.4 of the biomass CHP plant, the mass flow rates of straw (LHV = 14.6 MJ kg^{-1}), wood (19.3 MJ kg^{-1}) and RDF (17.73 MJ kg^{-1}) biomass feedstocks required are calculated as follows.

Assume 8000 operating hours per year. Then

$$\frac{50}{0.4 \times 14.6} \times 3600 \times \frac{8000}{365} \times \frac{1}{1000} = 675 \text{ t d}^{-1} \text{ straw feedstock}$$

$$\frac{50}{0.4 \times 19.3} \times 3600 \times \frac{8000}{365} \times \frac{1}{1000} = 511 \text{ t d}^{-1} \text{ wood feedstock}$$

$$\frac{50}{0.4 \times 17.73} \times 3600 \times \frac{8000}{365} \times \frac{1}{1000} = 556 \text{ t d}^{-1} \text{ RDF feedstock}$$

However, the energy efficiency is not a flat number for any type of biomass feedstock. For more precise results, a rigorous process simulation must be undertaken to determine the energy efficiency for specific biomass processing plants for a range of given operating conditions.

Table 4.10 Emissions to atmosphere, land and water from the operation of 50 MWe of electricity generation plant for the three types of biomass.

Emissions to:	Pollutant[1] or Component[2]	Straw (t d^{-1})	Wood (t d^{-1})	RDF (t d^{-1})
Atmosphere	[1]CO_2	906	728	923
	[2]N_2	1029	1110	2050
Land	[2]S	0.9	0	1.1
	[1]Ash	58	6	58
Water	[1]Effluent	35.6	32.4	28.9
	Total oxygen (air) requirement	649 (2786)	548 (2350)	719 (3085)

Hence, the following emissions result from the fuel combustion into 50 MW energy generation:
Straw: 906 t CO_2 d^{-1}
Wood: 728 t CO_2 d^{-1}
RDF: 923 t CO_2 d^{-1}

Table 4.10 shows the emissions to atmosphere, land and water from the operation of 50 MWe of an electricity generation plant using mass balance based on the three types of biomass selected. The exhaust gas has primarily carbon dioxide and steam generated from carbon and hydrogen combustion, respectively. Several other assumptions are required.

Nitrogen present in the biomass leaves along with the exhaust gas as gaseous nitrogen. Hence there is no atmospheric emission impact associated with the processing of nitrogen embedded in the biomass body. Though the impact of nitrogen and other GHGs will be significant, if inventories for fertilizer production for biomass growth are considered.

Sulfur present in the biomass is converted into hydrogen sulfide that is generally removed during gas clean-up. Hence, the exhaust gas is free of any sulfur compound. The hydrogen sulfide removed is finally recovered as metallic sulfur using the Claus process. Hence, there is no associated sulfur emission in the exhaust gas from the operation of a biomass boiler.

Land emission from a biomass CHP plant operation results from the ash present in the biomass. A small percentage of the land emissions can be leached into the aqueous body. Metallic components present in the biomass can also be emitted to land and water.

Component Balance

Carbon. The carbon content of the biomass (C in weight %, obtained from the ultimate analysis given in Tables 4.7 to 4.9) is converted into carbon dioxide. The molar mass of carbon dioxide is 44 and that of carbon is 12.

$$CO_2 \text{ mass flow rate in t d}^{-1} = \frac{(C)}{100} \times \frac{44}{12} \times \text{Biomass feedstock mass flow rate in t d}^{-1} \qquad (4.3)$$

Sulfur. Sulfur (S in weight % in Tables 4.7 to 4.9) in a biomass feedstock is recovered as a solid product using the sulfur recovery unit via the production of hydrogen sulfide.

$$S \text{ mass flow rate in t d}^{-1} = \frac{(S)}{100} \times \text{Biomass feedstock mass flow rate in t d}^{-1} \qquad (4.4)$$

Ash. Ash (*Ash* in weight % in Tables 4.7 to 4.9) in a biomass feedstock is disposed to land.

$$Ash \text{ mass flow rate in t d}^{-1} = \frac{(Ash)}{100} \times \text{Biomass feedstock mass flow rate in t d}^{-1} \qquad (4.5)$$

Water. Hydrogen present (H in weight % in Tables 4.7 to 4.9) in a biomass is combusted to generate steam. This steam and moisture present (H_2O in weight % in Tables 4.7 to 4.9) in the body of a biomass give rise to condensate after energy recovery.

$$\text{Effluent purge flow rate in t d}^{-1} = 0.1 \times \left(\frac{(H)}{100} \times \frac{18}{2} + \frac{(H_2O)}{100} \right) \times \text{Biomass feedstock mass flow rate in t d}^{-1} \quad (4.6)$$

Some negligible amount of hydrogen is used up by the sulfur and chlorine present in the biomass. After effluent purge (10% by mass of the condensate), the remaining condensate is recovered as boiler feed water (BFW). An equal amount of fresh BFW is required to make up for the lost amount. Moisture can also emit with the exhaust gas in some process configurations without heat recovery from the exhaust gas. Some industrial systems still adopt this to maintain a required chimney length for releasing exhaust gas to the atmosphere. The downside of such a process is the lost opportunity for heat recovery, resulting in lower energy efficiency. The molar mass of water is 18 and that of hydrogen is 2.

Oxygen. The oxygen required is calculated from the balance of stoichiometric oxygen required for the combustion of carbon and hydrogen present in the biomass and the oxygen available (O in weight % in Tables 4.7 to 4.9) in the biomass.

$$\text{Oxygen required in t d}^{-1} = \left(\frac{\text{(H)}}{100} \times \frac{16}{2} + \frac{\text{(C)}}{100} \times \frac{32}{12} - \frac{\text{(O)}}{100} \right) \times \text{Biomass feedstock mass flow rate in t d}^{-1} \quad (4.7)$$

The source of oxygen is air.

$$\text{Air required in t d}^{-1} = \frac{\text{Oxygen required in t d}^{-1}}{32} \times \frac{(0.79 \times 28 + 0.21 \times 32)}{0.21} \quad (4.8)$$

The atomic mass of oxygen required to combust 1 mole of hydrogen (molar mass = 2) into 1 mole of water is 16. The molar mass of oxygen to combust 1 mole of carbon (molar mass = 12) into 1 mole of carbon dioxide is 32. Assume that air consists of 79 volume or molar % of nitrogen and rest of oxygen. Molar mass of nitrogen is 28 and that of oxygen is 32. A part of the oxygen is produced from the air separation unit (ASU).

For straw:

$$\text{Oxygen produced from the ASU in t d}^{-1} = 0.5 \times \text{Biomass feedstock mass flow rate in t d}^{-1} \quad (4.9)$$

For wood:

$$\text{Oxygen produced from the ASU in t d}^{-1} = 0.67 \times \text{Biomass feedstock mass flow rate in t d}^{-1} \quad (4.10)$$

For RDF:

$$\text{Oxygen produced from the ASU in t d}^{-1} = 0.61 \times \text{Biomass feedstock mass flow rate in t d}^{-1} \quad (4.11)$$

Nitrogen. The source of nitrogen outlet from a biomass based energy generation plant to the atmosphere is that supplied with the air and that present in the biomass feedstock (N in weight % in Tables 4.7 to 4.9), shown in the following equation:

$$\text{Nitrogen outlet flow rate in t d}^{-1}$$
$$= \frac{\left(\text{Oxygen required in t d}^{-1} - \text{Oxygen produced from the ASU in t d}^{-1} \right)}{32} \times \frac{0.79 \times 28}{0.21}$$
$$+ \frac{\text{(N)}}{100} \times \text{Biomass feedstock mass flow rate in t d}^{-1} \quad (4.12)$$

Chlorine in the case of RDF can be removed during process clean-up.

Exercise 5. Biomass gasification processes are commonly integrated to gas turbine for combined heat and power (CHP) generation. Figure 4.13 shows a biomass integrated gasification combined cycle (BIGCC) flowsheet configuration.

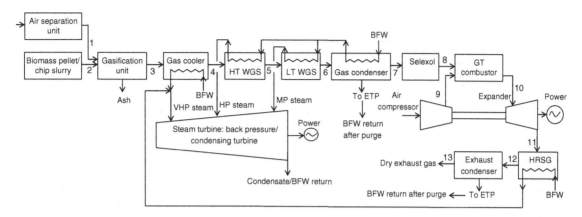

Figure 4.13 *Biomass integrated gasification combined cycle (BIGCC) flowsheet configuration. BFW, boiler feed water; ETP, effluent treatment plant; GT, gas turbine; HRSG, heat recovery steam generator; HP, high pressure; MP, medium pressure; VHP, very high pressure.*

Simulate the BIGCC process flowsheet based on the process models shown in Table 4.11[10]. Table 4.12 shows the correlations for the yields of products: gas, tar and char from biomass primary pyrolysis or devolatilization[12]. Primary pyrolysis or devolatilization is a common occurrence as soon as biomass comes in contact with the hot bed in the gasifier. Biomass breaks down into gas, tar and char products. Gas and tar are steam gasified into a product gas. Char could be combusted in a separate combustor to supply the reforming heat in the gasifier. Alternatively, char could be gasified in the same gasifier. The product gas from the steam gasifier after purification mainly consisting of carbon monoxide and hydrogen is called syngas.

Determine the mass and energy inventories of the BIGCC plant operations for straw, wood chip and refuse-derived fuel (RDF). The proximate and ultimate analyses of straw, wood and RDF are shown in Tables 4.7 to 4.9, respectively.

The solution of this exercise problem is given in the ***Online Resource Material in the Companion Website: Case Study 1.***

4.6 LCA: Impact Assessment

In this phase, the results from the inventory analysis are used to evaluate the potential of different environmental impacts. The inventory data are converted into environmental effects by multiplying a flux of an inventory with an impact characterization factor. The impact characterization factor of an environmental load (or substance or chemical or pollutant) indicates its intensity to an impact category with respect to a reference substance. For example, methane's GWP impact characterization factor of 25 g CO_2 equivalent means that methane has 25 times more GWP compared to carbon dioxide. Various methodologies for LCIA exist, detailed in Section 4.8. Different extents of effects, such as primary, mid- and end-points, normalization, weighting and valuation are used in different LCIA methodologies. Different impact categories are also included in different LCIA methodologies. The assessment is done on the basis of the scope and goal defined for the study. The impact assessment can be carried out using the following steps:

1. Classification
2. Characterization
3. Normalization
4. Valuation.

In classification, inventory data are assigned to different impact categories. Various pollutants resulting from an inventory analysis are classified under various impact categories.

Table 4.11 *BIGCC flowsheet modeling. (Reproduced with permission from Sadhukhan et al. (2009)[12]. Copyright © 2009, American Chemical Society.)*

Unit Names	Modeling Framework	Process Specification
Gasification	Estimate pyrolysis or devolatilization product yield using EXCEL spreadsheet based model shown in Table 4.12. Use RGibbs reactor in Aspen Plus for the gasification of pyrolysis product into product gas.	Temperature = 900–950 °C. Desired product decides the system and the gasification operating pressure. Pressure = near atmospheric when integrating to fuel cell or 25–30 bar for combined cycle integration and fuel and chemical production.
Air compressor	Compressor Isentropic model in Aspen Plus.	Desired product decides the system and compressors' operating pressure. Isentropic efficiency = 75%.
Gas cooler and heat recovery steam generator (HRSG)	Cooler in Aspen Plus.	Temperature is just above the dew point of the gas/flue gas at the system/gasification pressure, such that single-phase gas still leaves the cooler/HRSG without requiring special design of the gas cooler/HRSG.
Gas or exhaust condenser	Flash2, two-phase flash separator in Aspen Plus.	Keep at or lower than the dew point of the gas, so as to dry the gas from the water present in the gas. After purification in the effluent treatment plant (ETP) and 10% purge, the rest of the water is recovered as boiler feed water (BFW).
High temperature water gas shift reactor (HT WGS)	REquil: Rigorous Equilibrium reactor based on stoichiometric approach in Aspen Plus.	Temperature = 450 °C at the system pressure. The water gas shift reaction is as follows: $CO + H_2O \rightleftharpoons CO_2 + H_2$
LT (low temperature) WGS	REquil in Aspen Plus.	Temperature = 350 °C at the system pressure.
Gas clean-up and carbon capture and storage; air separation unit	Sep2: two-outlet component separator based on component purity, flow, etc., for the site flowsheet simulation in Aspen Plus. Individual processes (such as Selexol) can be simulated in detail.	Specify the mole fraction of the component to be separated as 1 in the outlet pure gas.
Gas turbine (GT) combustor	REquil in Aspen Plus.	1300 °C temperature at the system pressure. This temperature restricts the NO_x emission. The combustion reactions of syngas include the following: $CO + 0.5O_2 \rightleftharpoons CO_2$ $H_2 + 0.5O_2 \rightleftharpoons H_2O$ $CH_4 + 2O_2 \rightleftharpoons CO_2 + 2H_2O$
GT expander	Turbine Isentropic model in Aspen Plus.	Exit pressure = Near atmospheric but allowing pressure drop across HRSG. Isentropic efficiency = 75–90%.
Steam gasifier in allothermal gasification	RGibbs reactor in Aspen Plus. Gas and tar yields from the spreadsheet based pyrolysis or devolatilization product yield modeling (Table 4.12) are to be entered as feedstock to the steam gasifier model in Aspen Plus.	Temperature = 900–950 °C. Desired product decides the system and the gasification operating pressure. Pressure = near atmospheric (when integrating to fuel cell) or 25–30 bar (IGCC, fuel and chemical production).
Char combustor cooler	Cooler in Aspen Plus.	Temperature is above the dew point of the flue gas at atmospheric pressure, such that single-phase gas still leaves the cooler without requiring special design of the cooler.
Direct quench	Flash2, two-phase flash separator in Aspen Plus.	Kept at or lower than the dew point of the gas, so as to dry the gas from the water present in the gas. After effluent treatment and 10% purge, the rest of the water is recovered as BFW.
Char combustor in allothermal gasification	RGibbs reactor in Aspen Plus. Char yield from the spreadsheet based pyrolysis or devolatilization product yield model entered as feedstock to the char combustor model in Aspen Plus.	About 50 °C higher temperature than steam gasifier to maintain the temperature gradient and supply exothermic heat of the combustion reaction to the steam gasifier.

Table 4.12 Correlations for the yields of products: gas, tar and char from biomass primary pyrolysis or devolatilization[12].

Component	kg per kg biomass
Total devolatilization	0.96
Total gas	0.48
H_2	0.00
CH_4	0.02
C_2	0.12
CO	0.22
CO_2	0.03
H_2O	0.08
Tar	Total devolatilization – Total gas
Char	1 – Total devolatilization

The characterization factors are assigned to them according to their relative contributions to the environmental impacts. For example, for calculating the GWP, the contribution of methane or any other GHG is given in relation to the impact of CO_2. Some of the impact categories are the GWP over 20, 50 or 100 years, eutrophication potential (EP), acidification potential (AP), ozone depletion, human toxicity and aquatic toxicity. The environmental impact can be calculated using.

$$E_k = \sum_{j=1}^{J} ec_{k,j} B_j \tag{4.13}$$

where $ec_{k,j}$ is the relative contribution of burden B_j (j is the index for pollutant; J is the total number of pollutants) to environmental impact E_k (k is the index for impact category.

In normalizing, impacts are represented with respect to the total emissions in certain areas or over a period of time. This can be useful to assess the potential effects of the activity on a regional or global environment. However the normalization results are not always reliable due to lack of reliable data for many impacts.

Valuation is the final step of impact assessment wherein relative importance of each impact is determined on the basis of a value or weight assigned to it. The environmental impacts (*EI*) can be aggregated into a single environmental impact function, shown as

$$EI = \sum_{k=1}^{K} w_k E_k \tag{4.14}$$

where w_k is the relative importance of E_k and K is the total number of environmental impact strategies.

The commonly occurring pollutants from power plants are carbon dioxide, methane, chlorofluorocarbons (CFCs), VOCs, nitrous oxide, nitrogen oxides (NO_x) and sulfur dioxide. Figure 4.14 shows commonly occurring atmospheric pollutants and their characterization factors, resulting in various environmental impacts. The characterization factors depend on physicochemical properties of pollutants and are provided in relative terms compared to a base pollutant (assumed a value of 1 in an impact category). For example, CO_2 is a base pollutant for the GWP impact category, that is, its GWP characterization factor is 1.

One of the most effective sets of primary impact characterizations and their units, developed by CML, are shown in the following list from the Institute of Environmental Sciences, Leiden University, Netherlands. A spreadsheet with the primary impact characterization factors of all tested chemicals is freely available from their website. Various other

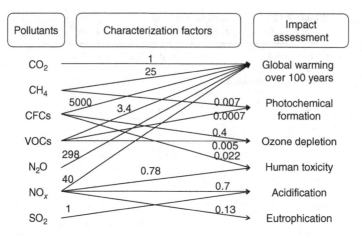

Figure 4.14   *Pollutants with characterization factors resulting in various environmental impacts.*

sources of characterization factors are available[13–20]. 1,4-Dichlorobenzene (DCB) is used as the base chemical for the toxicity categories. The various impact categories for assessment are discussed thereafter.

Resources
1.   Abiotic depletion element (kg Sb equivalent)
2.   Abiotic depletion fossil (MJ)

Emissions
1.   Global warming potential (kg CO_2 equivalent)
2.   Acidification potential (kg SO_2 equivalent)
3.   Eutrophication potential (kg phosphate equivalent)
4.   Ozone layer depletion potential (kg R-11 equivalent; chlorofluorocarbon-11 or CFC-11 or refrigerant-11)
5.   Photochemical oxidant creation potential (kg ethylene equivalent)
6.   Freshwater aquatic ecotoxicity potential (kg DCB equivalent)
7.   Human toxicity potential (kg DCB equivalent)
8.   Marine aquatic ecotoxicity potential (kg DCB equivalent)
9.   Terrestric ecotoxicity potential (kg DCB equivalent)

4.6.1   Global Warming Potential

The global warming potential (GWP) of all pollutants is presented relative to CO_2. The GWP of a pollutant is the ratio between the infrared absorption by the pollutant and the infrared absorption by an equal amount of CO_2. The GWP of CFCs is 5000 over a 100 year lifespan. Their units are also represented in terms of CO_2 equivalent, for example, kg CO_2 equivalent. The GHGs have different lifespans and therefore a basis of 20, 100 and 500 years is generally assumed over which the total GWP is predicted. In evaluating the environmental impact characterization factors, there is no geographical segregation considered. Henceforth, these environmental impact characterization factors, generally restricted up to the primary level of impacts, can be applied to analyze life cycle impacts of any products or process systems, irrespective of geographic locations. The following section provides an assessment of environmental impact characterization factors using example problems. The other important impact categories of global importance for biorefinery systems are explained as follows.

4.6.2   Land Use

Land is an important resource used for renewable energy supply. Biomass uses land for energy and value-added productions. Even though the second generation residues use much less land than the first generation crops, there are land implications during their growth phase. Imagine an agricultural residue, for example, straw, husk, corn stover, sugarcane, etc., is used for energy generation. Without land, these residues cannot be grown. These residues should be used for value-added productions: bulk, fine and specialty chemicals, for which there is no alternative sustainable carbon source. For energy generation, however, land use could be reduced by tidal, hydro, wind, solar, geothermal, etc., alternative to biomass. Municipal wastes have least implication of land use amongst all lignocellulose or waste biomasses.

The type of land differs widely within a geographical boundary. High to low grass lands are used for animal grazing and high fertile lands are used for arable feed crops. Poultry and pigs live on arable crops and therefore rely on fertility of arable lands graded as 3a, while grass lands belong to grade 2. Land use for organically grown systems is more due to building of fertility of the land. From an energy efficiency and environmental impact point of view, the objective is to maximize energy yield, for example, in joules per unit land used, for example, per hectare. Exercise 2 shows the energy yield calculation per unit land use. Sustainable energy crops must use lesser land compared to lignocellulose.

However, fertilizers are needed for growing any kind of crops, arable or energy crops. Fertilizer production and emissions are the hot spots of biomass production systems. The key to successful biorefineries is to mitigate these hot spots and minimize land use for biomass growth and land transformation from it. Both the type of land used and the fertilization application decide the energy or crop yield per unit land used. Thus, within the subject of land use, the steps involved in the agriculture of crops are highlighted from input–output with respect to system boundaries and environmental footprint perspectives. An illustration of the steps involved in the agriculture of arable crops is shown in Figure 4.8. A measure of land use refers to the crop or energy yield per unit land use during a steady-state period of an agricultural system shown within the boundary in Figure 4.8.

Outside this boundary, to prepare a land for agriculture for the first time, the subsoil nutritional value is enriched by nitrogen (N), phosphorus (P) and potassium (K) fertilizer applications. During this time, the input and output of N, P, K around the land are not in balance. Once the land reaches the required maximum nutritional stage for seed bed establishment, a steady state or equilibrium is reached, when N, P, K inputs to the land (fertilizer applications) and outputs from the land (fixation within a plant body and emission to the atmosphere in the case of nitrogen) are in balance. During this time, crops stabilize to steady yields. This is the time for crop rotation. Amongst N, P, K fertilizers, P and K fertilizers remain in solid forms. Hence, these solids form the basis of constitution of higher plants and animals and eventually at the end of their life cycles through various chemicals and species they return to mining extraction, sea bed, etc., via sludge.

However, nitrogen may be emitted as the atmospheric nitrogen gas as well as in eight various forms of NO_x, of which nitrous oxide (N_2O), nitric oxide (NO) and nitrogen dioxide (NO_2) are the more commonly occurring ones. Because of denitrification of the soil, nitrogen fertilizer is applied periodically during a crop rotation period to keep up with the soil nitrogen balance. Once a crop is harvested, dehulled to recover grain, the grain is stored, dried, cooled and transported to various application points. A similar process applies to agricultural residues after dehulling from harvested grains. A part of the residues dehulled from the grain is bedded into the soil to particularly retain the soil carbon balance and other nutritional balances for the next phase of seed bed establishment. Additional fertilizer may be needed at this point to keep the steady balance of N, P, K. Also herbicide is used for crop protection over a rotation.

Figure 4.15 shows the cumulative primary energy and the GWP impacts from a *Jatropha* agricultural system. The inventory refers to the production of 4.21 t of *Jatropha* fruit per hectare of land.

Impacts from fertilizers are the most notoriously known for any agricultural system. The biggest hot spot identified from an agricultural system impact assessment is the application of fertilizer, still primarily sourced from fossil resources.

The field emissions due to NO_x, mainly N_2O, followed by NO and NO_2, and seed conditioning are the second worst impact hot spots. Because of the prominent effect of nitrogen oxides on the environment, many scientists look at the nitrogen footprint, especially for agricultural systems (due to the application of fertilizers), separately from other GHG emissions, but analogous to the carbon footprint to get an idea of the magnitude of the problem.

The GWP, carbon footprint and nitrogen footprint can be estimated from an inventory analysis, shown in the following exercise problem, for two types of fertilizers.

Exercise 6. Calculate the GWP impact and the carbon and nitrogen footprint for two types of fertilizers, ammonium nitrate and urea, for the data shown in Table 4.13.

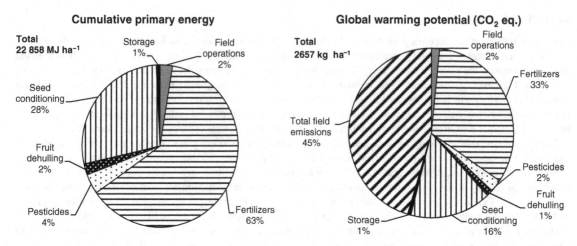

Figure 4.15   *Impact assessment of Jatropha fruit agricultural system for hot spot analysis on the basis of 1 hectare of land use.*

Table 4.13   *Pollutant emission data for ammonium nitrate and urea fertilizers and the characterization factors of the pollutants.*

Impact characterization factor	Pollutants					
	N_2O (kg)	CO_2 (kg)	VOC (kg)	CH_4 (kg)	NO_2 (kg)	CO (kg)
Global warming potential 100 years kg CO_2 eq.	298	1	3.4	25		1.9
Pollutant emissions	N_2O as nitrogen (kg)	CO_2 (kg)	VOC (kg)	CH_4 (kg)	NO_2 (kg)	CO (kg)
Ammonium nitrate as per kg nitrogen	0.005592	2.241140	0.000132	0.007007	0.002030	0.000690
Urea as per kg nitrogen	0.000008	2.713971	0.000123	0.008618	0.002310	0.000810

Solution to Exercise 6

GWP of ammonium nitrate

$$= \left(298 \times 0.005592 \times \frac{28+16}{28}\right) + (1 \times 2.241140) + (3.4 \times 0.000132) + (25 \times 0.007007) + (1.9 \times 0.000690)$$

$$= 5.0 \ CO_2 \text{ eq. per nitrogen, kg kg}^{-1}$$

GWP of urea

$$= \left(298 \times 0.000008 \times \frac{28+16}{28}\right) + (1 \times 2.713971) + (3.4 \times 0.000123) + (25 \times 0.008618) + (1.9 \times 0.000810)$$

$$= 2.9 \ CO_2 \text{ eq. per nitrogen, kg kg}^{-1}$$

Carbon footprint of ammonium nitrate

$$= [(1 \times 2.241140) + (3.4 \times 0.000132) + (25 \times 0.007007) + (1.9 \times 0.000690)] \times \frac{12}{44}$$

$$= 0.66 \text{ carbon per nitrogen, kg kg}^{-1}$$

Carbon footprint of urea

$$= [(1 \times 2.713971) + (3.4 \times 0.000123) + (25 \times 0.008618) + (1.9 \times 0.000810)] \times \frac{12}{44}$$

$$= 0.8 \text{ carbon per nitrogen, kg kg}^{-1}$$

Nitrogen footprint of ammonium nitrate

$$= 0.005592 + \left(0.00203 \times \frac{14}{14 + 32} \right)$$

$$= 0.0062 \text{ nitrogen per nitrogen, kg kg}^{-1}$$

Nitrogen footprint of urea

$$= 0.000008 + \left(0.002310 \times \frac{14}{14 + 32} \right)$$

$$= 0.0007 \text{ nitrogen per nitrogen, kg kg}^{-1}$$

Regarding land use impacts, there are three terms frequently used,

1. *Land use.* This is a measure of the actual use of land per unit yield of biomass, sometimes referred to the dry matter (DM) of the biomass. Its unit is m^2 per kg yield of biomass or m^2 per kg yield of DM of biomass.
2. *Land transformation.* It is a value lower than the value of land use. A fraction of the total land used for a given service (bioenergy production) can be transformed from one type to another, for example, cutting forest into a forest road for transporting timber. The land transformation constantly happens due to production of food, fiber and energy.

   The land transformation can be contributed by a range of activities on arable land, forest, construction, dump sites, industrial, mineral extraction, pasture and meadow, sea and ocean, permanent crop, shrub land, tropical rain forest, rail, road, urban and artificial water bodies and courses. Figure 4.16 shows an overview of the types of land transformation.

   A decision must be made in terms of selecting the feedstock with minimum land use for a required service, for example, heat and power production. The unit of land use or land transformation can be:

   m^2 per m^3 of biomass produced,
   m^2 per kg of biomass produced,
   m^2 per MJ of biomass calorific value,
   m^2 per MJ of net energy production.

3. *Land occupation.* Land transformation data can be converted into land occupational data, by multiplying the land transformation with the number of years for which the land is used after transformation. Hence, the unit for land occupational data is:

   m^2 year per m^3 of biomass produced,
   m^2 year per kg of biomass produced,
   m^2 year per MJ of biomass calorific value,
   m^2 per MJ of net energy production.

On the basis of inventory data of 1 kg of wheat grain production and corresponding straw production, the land occupation is 0.159 m^2 year per kg of wheat grain production; 99% of this land occupation is in the arable nonirrigated land and the remaining 1% is in forest, intensive, normal. Taking account of the yearly rotation of the crop, the land transformation for 1 of wheat grain and corresponding straw production is 0.159 m^2 per kg of wheat grain.

Table 4.14 shows % contributions of various categories in Figure 4.16 in land occupation for wood production. The value of land transformation varies between 0.89 m^2 m^{-3} for soft wood to 1.89 m^2 m^{-3} for hard wood, respectively. If the lifetime of the infrastructure used is assumed as 100 years, the land occupation is 89 m^2 year per m^3 of soft wood and 189 m^2 year per m^3 of hard wood, respectively.

The land use of soft and hard wood is 7.5–13.5 m^2 m^{-3}.

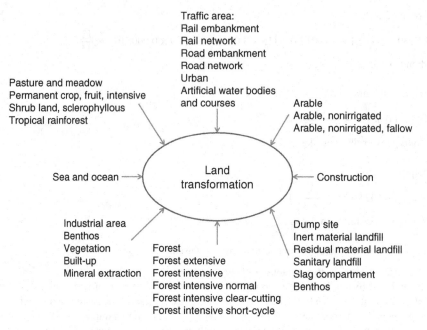

Figure 4.16 *Overview of land transformation types.*

Table 4.14 *The % contributions of various categories in Figure 4.16 in land occupation for wood production.*

Arable	0.00
Arable, nonirrigated	0.06
Arable, nonirrigated, fallow	0.00
Construction site	0.04
Dump site	0.50
Dump site, benthos	0.01
Forest, intensive	3.28
Forest, normal	87.52
Forest, short-cycle	0.00
Industrial area	0.18
Industrial area, benthos	0.00
Industrial area, built-up	0.84
Industrial area, vegetation	0.15
Mineral extraction site	1.49
Permanent crop, fruit, intensive	0.00
Shrub land, sclerophyllous	0.13
Traffic area, rail embankment	0.02
Traffic area, rail network	0.02
Traffic area, road embankment	0.95
Traffic area, road network	0.49
Urban, discontinuously built	0.00
Water bodies, artificial	0.39
Water courses, artificial	0.18

4.6.3 Resource Use

The implication of resource depletion must be understood, mainly that nonrenewables are compared to global reserves when using them for energy commodity productions. Nonrenewable resources including fossil resources, minerals and clays can only be regenerated within human lifetimes. The heaviest to lightest fossil resources are coal (also tar and heavy oils) > medium to light crude oils > natural gas. Unlike renewable resources, solar, wind and tidal, the nonrenewable resources are not continuously regenerated and are depleted over time. Therefore, the depletion of their reserves must be within control all the time and included in environmental impact characterization. The evaluation of the environmental impact characterization factors of nonrenewable resources is dependent on the energy efficiency of their conversion processes. The lower heating value (LHV: an indication of the transferable calorific value) of a nonrenewable resource and the efficiency of conversion processes into useful energy commodities can be multiplied to evaluate the output energy generation. Therefore, if the desired output energy and efficiency of conversion process are known, the nonrenewable resource depletion can be calculated.

Energy resource input (in MJ) × Efficiency of conversion process systems (in fraction) = Output energy generation (in MJ)

Like all other impacts, engineers need to understand the implication of resource depletion, mainly of nonrenewables, compared to global reserves when using them for energy commodity productions.

Exercise 7. Twenty-five million households with an annual average total heating and electricity demands (in the ratio of 2:1) of 172 15 kW h y^{-1} per household are to be supplied by coal and natural gas based energy systems, given in Table 4.15. It is proposed to build city level centralized and household–community level decentralized energy conversion systems. This high level analysis is aimed at assessing whether the resources available are adequate for meeting the household energy demand for two scenarios, the lowest cost scenario and the lowest GHG emission scenario. The LHV of coal and natural gas, the conversion system efficiency into electricity or heat generations, maximum resource supplies, cost ratio of systems compared to the lowest cost technology option and the GHG emission reduction potential are given in Table 4.15. The conversion systems under consideration are electricity generation without carbon capture and storage (CCS) using coal, heat generation using natural gas, electricity generation using natural gas and electricity generation with CCS using coal, from the lowest to the highest cost technology options, respectively. Determine the lowest cost scenario and the lowest GHG emission scenario in terms of the use of resource mix to meet the total household electricity and heat demands (1 PJ = 10^{15} J; 1 MT = 10^6 tonne).

Table 4.15 *Coal and natural gas data to supply household energy.*

	LHV (MJ kg^{-1})	Conversion into	Efficiency (%)	Resource Supply, (MT y^{-1})	Cost Ratio	GHG Emission Reduction (%)
Coal	28	Electricity without CCS	70	<40	1	0
		Electricity with CCS	36	<15	5	90
Natural gas	50	Heat	90	<14	2	40
		Electricity	70	<14	3	30
Electricity		Heat	90			

Solution to Exercise 7. The household electricity and heat demands on a yearly basis are calculated as follows:

Total household electricity and heat demands = $\left(25 \times 10^6\right) \times 17215 \times \frac{1000 \times 3600}{10^{15}} = 1549$ PJ y^{-1}

Total household electricity demand = $1549 \times \frac{1}{3} = 516$ PJ y^{-1}

Total household heat demand = $1549 \times \frac{2}{3} = 1033$ PJ y^{-1}

The maximum energy resource supply constraints from the lowest to the highest cost technology options are as follows:

Coal energy used for electricity generation without CCS = $40 \times 28 = 1120$ PJ y^{-1}
Natural gas energy used for heat generation = $14 \times 50 = 700$ PJ y^{-1}
Natural gas energy used for electricity generation = $14 \times 50 = 700$ PJ y^{-1}
Coal energy used for electricity generation with CCS = $15 \times 28 = 420$ PJ y^{-1}

Lowest Cost Scenario. For the lowest cost energy resource use scenario, electricity generation from coal without CCS is selected first, as the lowest cost option, to meet the entire electricity demand of households of 516 PJ y^{-1}.

The total potential for electricity generation from coal at 70% efficiency is $1120 \times 0.7 = 784$ PJ y^{-1}. After meeting the household electricity demand of 516 PJ y^{-1}, the balance of electricity available from coal without CCS, 268 PJ y^{-1}, is converted into 241 PJ y^{-1} of heat at an efficiency of 90%, assuming that the cost of heat generation from electricity is not considerable.

This consumes all 1120 PJ y^{-1} of coal available.

The balance of heat ($1033 - 241 = 792$ PJ y^{-1}) is supplied by a natural gas based heat generation system at first, followed by heat from a natural gas electrification system. The maximum amount of natural gas energy available for direct heat generation is 700 PJ y^{-1} and can generate 630 PJ y^{-1} of heat, at 90% energy efficiency. Thus, the remaining household heat demand of 162 PJ y^{-1} is to be supplied from the natural gas based electricity system.

Natural gas consumed for meeting the rest of the heat requirements (162 PJ y^{-1}) via electrification is $162/(0.7 \times 0.9) = 257$ PJ y^{-1}.

Natural gas remaining after meeting the heat requirement via electrification is $700 - 257 = 443$ PJ y^{-1}.

The total amount of natural gas resources used is then $700 + 257 = 957$ PJ y^{-1}. The coal to natural gas resource mix used to meet the total household heat and electricity demands is 1.2:1.

Table 4.16 shows the resource supply in terms of energy contents, output heat and electricity generations for households, energy efficiencies and the GHG emission reduction potential compared to the case of the coal energy system without CCS, where all GHGs are emitted to the atmosphere.

The evaluation shows that due to the mix of natural gas by 46% with the rest as coal, as the resources used, the overall GHG emission is reduced by 19% based on the output energy distribution. Overall 75% of energy efficiency is obtained. Note that the actual efficiency would be lower and GHG emission would be higher due to transmission losses, not taken into account in this high level analysis. The cost ratio shown is calculated on the basis of energy resource use as follows:

$$\frac{738 + 382 + 700 \times 2 + 257 \times 3}{2077} = 1.58$$

Lowest GHG Emission Scenario. For electricity generation, the order of preference of technologies for the lowest GHG emission scenario is the electricity generation from coal with CCS and electricity generation from natural gas, respectively.

For the heat supply, the order of preference for the lowest GHG emission scenario is as follows: heat generation from natural gas, heat generation from natural gas based electricity and electricity from coal with CCS systems, respectively.

Table 4.16   *Resource supply, output heat and electricity generations for households, energy efficiency and cost ratio, for the lowest cost resource use scenario.*

	Resource (PJ y^{-1})	Generation (PJ y^{-1})	Efficiency (%)	GHG Emission Reduction (%)	Cost Ratio
Coal without CCS electricity	738	516	70	0	1
Coal without CCS electricity – heat	382	241	63	0	1
Natural gas – heat	700	630	90	40	2
Natural gas electricity – heat	257	162	63	30	3
Total	2077	1549	75	19	1.58

Table 4.17 *Resource supply, output heat and electricity generations for households, energy efficiency and cost ratio, for the lowest cost resource use scenario.*

	Resource (PJ y^{-1})	Generation (PJ y^{-1})	Efficiency (%)	GHG Emission Reduction (%)	Cost Ratio
Coal with CCS electricity	420	151	36	90	5
Natural gas electricity	522	365	70	30	3
Natural gas – heat	700	630	90	40	2
Natural gas electricity – heat	178	112	63	30	3
Coal without CCS electricity – heat	461	291	63	0	1
Total	2281	1549	68	32	2.66

The resource use, output heat and electricity production for the household supply, efficiency, GHG emission reduction potential and cost implications for the lowest GHG emission scenario are shown in Table 4.17. The natural gas use as a resource compared to coal is higher by 1.6 times, for achieving the lowest emission energy scenario. Following the approach for the lowest cost scenario, try working out the solution for the lowest GHG emission scenario.

The GHG emission reduction potential is improved by 32% based on the output energy generation, with a cost implication (ratio) of 2.66 (based on the resource energy used). Only 10% of the output energy can be supplied from a 90% decarbonization technology, coal electricity generation with CCS.

4.6.4 Ozone Layer Depletion

The upper atmosphere called the stratosphere contains ozone that absorbs ultraviolet (UV) rays of the sun, which otherwise would have reached the lower atmosphere and surface of the earth. The ozone layer depletion implies thinning of the stratospheric ozone. UV rays cause skin cancer and endanger polar species. Destruction of ozone creates ozone holes (thinnest ozone layers) at the North and South Poles. The ozone layer depletion is a global problem and hence a global solution is required. NO_x emission is partly responsible for decomposing stratosphere ozone into oxygen:

$$\begin{aligned} O_3 + NO &\rightarrow NO_2 + O_2 \\ 2NO_2 &\rightarrow N_2O_4 \\ 2NO_2 &\rightarrow 2NO + O_2 \end{aligned} \tag{4.15}$$

The stratospheric ozone layer is also destroyed by halocarbons, for example, CCl_4, $CHCl_3$, CFCs that occur naturally by volcanic eruption, marine species and wood fire. This reaction occurs through chlorine, chlorine oxide or halogen radical formations. Catalyzed by these radicals, ozone breaks down into the oxygen molecule, which is not good for the upper atmosphere:

$$\begin{aligned} Cl \cdot + O_3 &\rightarrow ClO \cdot + O_2 \\ ClO \cdot + O &\rightarrow Cl \cdot + O_2 \end{aligned} \tag{4.16}$$

Trichlorofluoromethane, commonly presented as CFC-11, is known to display the highest stratospheric ozone depletion potential. CFC-11 used to be widely applied as a refrigerant due to its phase changing behavior at relatively low pressures compared to other refrigerants. However, due to the presence of chlorine and its associated damaging effect on the environment, its use was banned.

The stratospheric ozone depletion potential is well established for chlorinated and brominated compounds. However, this potential for other halocarbons is yet to be established. The convention is to present the stratospheric ozone depletion potential for any chemical relative to that of the CFC-11 assumed stratospheric ozone depletion potential of 1. If all other CFCs are grouped together, their stratospheric ozone depletion potential is ~0.4 in 100 years, relative to CFC-11. It is seen that CFC-11 has the highest stratospheric ozone depletion potential. Hence, all other chemicals have fractional

Table 4.18   *Inventory analysis in kg of chemicals, for ozone layer depletion on the basis of 1 kg of wheat grain production and corresponding straw production. (Reproduced with permission from GaBi and Ecoinvent database[21].)*

Benzene	4.4311×10^{-14}
Butadiene	4.1979×10^{-14}
Cadmium	2.2211×10^{-17}
Carbon dioxide, fossil	6.9965×10^{-9}
Carbon monoxide, fossil	8.2181×10^{-12}
Chromium	1.1106×10^{-16}
Copper	3.7758×10^{-15}
Dinitrogen monoxide	6.6634×10^{-14}
Ethylene oxide	4.0579×10^{-13}
Formaldehyde	3.4984×10^{-13}
Hydrogen chloride	1.9102×10^{-15}
Lead	4.4423×10^{-17}
Mercury	1.5548×10^{-19}
Methane, fossil	1.1106×10^{-13}
Nickel	1.5548×10^{-16}
Nitrogen oxides	3.1095×10^{-11}
Non-methane volatile organic compounds	1.4903×10^{-12}
Particulates, <2.5 μm	8.4403×10^{-14}
Selenium	2.2211×10^{-17}
Sulfur dioxide	2.2211×10^{-12}
Water	2.7541×10^{-9}
Zinc	2.2211×10^{-15}

stratospheric ozone depletion potentials relative to that of CFC-11 eq. The ozone depletion rate of these substances is considerable over the short term; hence their remediation must also mean controlling the emission of these substances.

Exercise 8. The inventory analysis of wheat cultivation includes the processes of soil cultivation, sowing, weed control, fertilization, pest and pathogen control, harvest and grain drying. Machine infrastructure and a shed for machine sheltering are included. Inputs of fertilizers, pesticides and seed as well as grain transports in the EU regional processing centre (10 km) are considered. The direct emissions on the field are also included.

Table 4.18 shows the cultivation inventory data of various chemicals that have impacts on ozone layer depletion[21]. The basis of inventory data is 1 kg of wheat grain production and a corresponding straw production. The same basis used in Exercise 2, 6.96 t of wheat grains per hectare and corresponding 3.49 t of straw per hectare production, can be assumed.

Figure 4.17 shows the group impact characterization factor of the chemicals in terms of kg CFC-11 eq. over a 5 to 40 year span. For the equation shown, y is the ozone depletion potential impact characterization factor of the chemicals in terms of kg CFC-11 and x is the number of years. The coefficient of determination (defined as $R^2 = 1 -$ SSQ/summation of square (SSQ) of variance) close to 1 shows a very good fit of the empirical equation given.

a. Calculate the total impact potential of the chemicals in kg CFC-11 eq. over a 100 year span generated from 1 kg of wheat grain production and corresponding straw production by extrapolation of the group impact characterization factor.

b. Calculate the impact potential of the chemicals in kg CFC-11 eq. over a 100 year span based on the characterization factor calculated in (a) from 1 kg of straw production. An economic allocation to impact potentials with an allocation factor of 92.5% to wheat grains can be assumed[21].

Solution to Exercise 8

a. Adding all the inventory analysis data of the chemicals, for ozone layer depletion on the basis of 1 kg of wheat grain production and corresponding straw production, shown in Table 4.18, the total inventory obtained is

$$9.79 \times 10^{-9} \text{ kg}$$

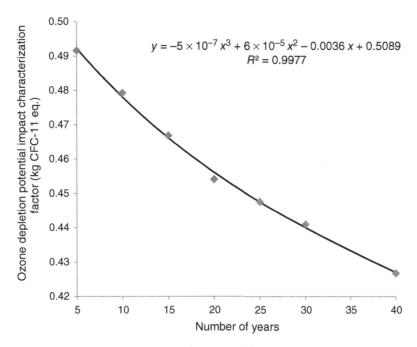

Figure 4.17 *Group impact characterization factor of the chemicals in terms of kg CFC-11 eq. over a 5 to 40 year span.*

By substituting 100 years in the place of x in the equation shown in Figure 4.17, the ozone depletion potential impact characterization factor (y) is calculated as

$$0.25 \text{ kg CFC-11 eq.}$$

The total impact of the chemicals in kg CFC-11 eq. over a 100 year span from 1 kg of wheat grain production and corresponding straw production is

$$9.79 \times 10^{-9} \times 0.25 = 2.44 \times 10^{-9} \text{ kg CFC-11 eq.}$$

Though extrapolation is perfectly feasible for the set of data provided, the reliability of the characterization factor obtained by extrapolation must be examined from physicochemical data and atmospheric science.

b. A kg of straw produced per kg of wheat = $3.49/6.96 = 0.5$ kg.
    Hence, 1.5 kg of wheat grain and straw produce 2.44×10^{-9} kg CFC-11 eq.
    Hence, 1 kg of wheat grain and straw produce 1.63×10^{-9} kg CFC-11 eq.
    Out of 1.63×10^{-9} kg CFC-11 eq. a $(1 - 0.925) = 0.075$ fraction can be allocated to the straw.
    The ozone depletion potential impact of chemicals in kg CFC-11 eq. over a 100 year span, generated from 1 kg straw production is $1.63 \times 10^{-9} \times 0.075 = 1.22 \times 10^{-10}$ kg CFC-11 eq.

4.6.5 Acidification Potential

Sulfur and nitrogen oxides, SO_x and NO_x (except N_2O, which contributes to GWP), are generated from chemical processes and inefficient combustion processes. In addition, carbon oxides emit from the processes. These gaseous emissions dissolve in atmospheric moisture to form sulfurous, sulfuric, nitrous, nitric and carbonic acids, reducing the natural pH from 5–6 to 2–4 (acidic).

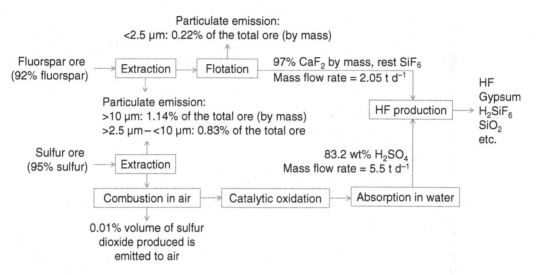

Figure 4.18   *Flow chart from sulfur and fluorspar ore extraction to the production of hydrogen fluoride and other products at the plant gate. The total ore includes both the fluorspar and sulfur ores.*

Acid rain or acidification is a regional problem as acidic clouds can travel. Scandinavia's acid rain is caused by emissions from the UK, Germany and France. The effect is acidification of water that damages water life, water quality, plant life and forests and causes dead lakes and streams. Buildings, particularly of marble and sandstone materials, can also corrode.

The acidifying chemicals commonly generate H^+ ions, responsible for lowering of atmospheric pH. The acid rain potential of a chemical is presented by the maximum number of H^+ ions that can be produced per unit mole of the chemical. In environmental impact potential terms, this potential of a chemical is presented relative to that of sulfur dioxide (SO_2). Sulfur takes part in a number of reactions. It is an important element for a number of chemicals, such as sulfuric acid, sulfites, gypsum, etc. The smelting process that recovers copper from ores can also emit sulfur compounds. When fuels or even desulfurized fuels are processed, sulfur present in the body is emitted mainly in oxidized form, SO_2. Sulfur compounds in gaseous forms, especially sulfur oxides, cause acid rain. Though there is a maximum acid rain potential for every chemical, the actual acid rain effect depends on many factors, such as the buffering capacity of soil and water, sunlight, temperature and moisture content of the atmosphere and air entrapment.

Exercise 9. Hydrogen fluoride is the main component for Teflon (polytetrafluoroethylene) production; 80% of the world's hydrogen fluoride is produced by reaction between extracted fluorspar containing 97% CaF_2 and the rest of SiF_4 by mass and sulfuric acid with a mass concentration of 83%, shown below in Equation (4.17). Gypsum ($CaSO_4.2H_2O$) can be produced as a by-product from the reaction. The SiF_4 present reacts with excess water to produce hexafluoride silica acid and silica, shown below in Equation (4.18). Figure 4.18 shows the flow chart from the extraction of sulfur and fluorspar

Table 4.19   *Molar mass of chemicals.*

Chemical	CaF_2	H_2SO_4	$CaSO_4$	HF	SiF_4	H_2SiF_6	SiO_2	H_2O
Full name	Calcium fluoride	Anhydrous sulfuric acid	Calcium sulfate	Hydrogen fluoride	Silicon tetrafluoride	Hexafluorosilicic acid	Silica	Water
Molar mass	78.07	98	136	20.01	104.08	144.09	60.08	18

Table 4.20   *Molar and mass flow rates of the reactants consumed and products produced by the two reactions in Equations (4.17) and (4.18).*

Reaction	Reactant Consumed				Product Produced			
	kmol d^{-1} t d^{-1}		kmol d^{-1}	t d^{-1}	kmol d^{-1} t d^{-1}		kmol d^{-1}	t d^{-1}
Reaction 1 in Equation (4.17)	CaF_2		$H_2SO_4 . 2H_2O$		Gypsum		HF	
	25.5	1.9885	25.5	3.413	25.5	4.381	51	1.0205
Reaction 2 in Equation (4.18)	SiF_4		H_2O		H_2SiF_6		SiO_2	
	0.591	0.0615	0.394	0.0071	0.394	0.0568	0.197	0.0118

ores to the production of hydrogen fluoride and other products at the plant gate. The required reaction conversion and yield data are also shown. Table 4.19 shows the molar masses of the chemicals involved in the process.

$$CaF_2 + H_2SO_4.2H_2O \rightarrow CaSO_4.2H_2O + 2HF \tag{4.17}$$

$$3SiF_4 + 2H_2O \rightarrow 2H_2SiF_6 + SiO_2 \tag{4.18}$$

a.   Calculate the inlet and outlet mass flow rates of the flowchart in Figure 4.18.
b.   Determine the relevant environmental impact characterizations from the resulting emissions.

Solution to Exercise 9.   Calculate the mass of the fluorspar ore assuming 92% of it is the fluorspar:

$$= 2.05/0.92 = 2.23 \, t \, d^{-1}$$

The ore extraction processes are assumed to give an ideal split of 100%. This assumption will not affect the impact on land as after extraction the residual ore goes back to the land. Hence, the net effect is the depletion of the fluorspar and sulfur from the land. They are amongst the most abundant elements on the earth.

Calculate the mass of sulfur ore using the mass balance, starting from the reactant consumptions in the reactions in Equations (4.17) and (4.18).

Assume complete conversion of CaF_2 and SiF_4 in the two respective reactions. Table 4.20 shows the amounts of reactants consumed and the amounts of products produced by the two reactions.

Assume complete conversion of sulfur dioxide into sulfur trioxide in the catalytic oxidation in the reaction

$$SO_2 + 0.5O_2 \rightarrow SO_3 \tag{4.19}$$

Assume complete sulfur trioxide absorption into water to form sulfuric acid, shown as

$$SO_3 + 3H_2O \rightarrow H_2SO_4.2H_2O \tag{4.20}$$

Doing a backward calculation from the amount of H_2SO_4 solution (at 83.2% in weight basis), the molar flow rate of SO_2 produced from sulfur ore is (1 kmol of H_2SO_4 weighs 98 kg and requires 1 kmol of SO_3, which requires 1 kmol of SO_2, according to Equations (4.19) and (4.20)):

$$\frac{5.5 \times 0.832 \times 1000}{98} = 46.7 \, \text{kmol d}^{-1}$$

The mass flow rate of sulfur ore required (assuming that 0.01% by volume of sulfur dioxide is emitted to air, the ore contains 95% sulfur by mass and a 100% split in the extraction process) is

$$\frac{46.7}{0.9999 \times 1000} \times \frac{32}{0.95} = 1.573 \, t \, d^{-1}$$

(The atomic mass of sulfur is 32.)

Table 4.21 *Product distribution corresponding to unit mass production of HF.*

HF	Gypsum	H_2SiF_6	SiO_2	Anhydrous H_2SO_4
1	4.3	0.056	0.012	2.04

HF is the precursor to many important compounds including pharmaceuticals and polymers such as Teflon. However, upon contact with moisture it forms corrosive hydrofluoric acid. The gas can also cause blindness. Hence, an important aspect of further research will be to find a replacement safer precursor. The results in Table 4.21 will allow comparison with a substitute product and process.

Calculate the total amount of ore:

$$(2.23 + 1.573) = 3.8 \, \text{t d}^{-1}$$

Particulate emissions (calculated based on the total ore including both the fluorspar and sulfur ores):

$$< 2.5 \, \mu\text{m: } 8.4 \, \text{kg d}^{-1}$$
$$> 2.5 \, \mu\text{m} - < 10 \, \mu\text{m: } 31.6 \, \text{kg d}^{-1}$$
$$> 10 \, \mu\text{m: } 43.3 \, \text{kg d}^{-1}$$

Sulfur dioxide emission to air:

$$\frac{46.7 \times 0.0001}{0.9999 \times 1000} \times 64 = 0.3 \, \text{kg d}^{-1}$$

(The molar mass of sulfur dioxide is 64.)
The acidification potential of the flowchart is:

$$0.3 \, \text{kg SO}_2 \text{ equivalent}$$

In addition to the products shown in Table 4.20, there will be unreacted anhydrous sulfuric acid:

$$2.08 \, \text{t d}^{-1}$$

Residual ore (including the particulates present) from fluorspar ore from the extraction process:

$$0.18 \, \text{t d}^{-1}$$

Residual ore (including the particulates present) from sulfur ore from the extraction process:

$$0.08 \, \text{t d}^{-1}$$

On the basis of the unit mass flow rate of HF produced, the product distribution is as shown in Table 4.21.

4.6.6   Photochemical Oxidant Creation Potential

This impact is also known as urban smog, brownish colored air, commonly found in modern cities, especially where air is trapped in a basin, for example, Los Angeles, Mexico City, etc. Photochemical pollutants have harmful effects on living tissue and on buildings. Together, the formation of ozone in the lower atmosphere, the troposphere, adds to detrimental impacts on ecological and human health.

The volatile organic compounds react with NO_x in the presence of sunlight to produce ozone and photochemical pollutants in the lower atmosphere. Photochemical pollutants include peroxyacetyl nitrate (PAN), formaldehyde, acetic acid, etc. These pollutants are named as photochemical pollutants, because their formations are catalyzed by sunlight. These pollutants and ozone, known as photochemical oxidant, in the lower atmosphere are extremely detrimental to health and the environment. The photochemical oxidant creation potential is expressed in terms of the ethylene equivalent.

VOCs are carbon containing compounds that take part in photochemical reactions. VOCs do not include carbon dioxide, carbon monoxide, carbonic acid, metallic carbide, carbonates and ammonium carbonates. VOCs are responsible for urban smog and ground level ozone formation. When there is a mixture of VOCs, the concentration-average photochemical oxidant creation potential of the mixture is estimated.

The various photochemical reaction products they contribute to are also very harmful to health and the environment. High levels of these pollutants cause breathing difficulties and bring on asthma attacks, which can be fatal. Warm weather and still air can exacerbate the problem. Therefore, VOC emissions must be controlled and regulated. The photochemical oxidants are formed when NO_x, VOC and sunlight are present. Therefore, regional effects must be considered in estimating the photochemical oxidant creation potential. See Web Chapter 1 for VOC measurements and mitigation pathways.

4.6.7   Aquatic Ecotoxicity

The effect of manufactured chemicals on aquatic organisms at subcellular, organism, population, community and ecosystem levels is commonly known as aquatic ecotoxicity. There are two main ways of measuring concentration levels of toxic chemicals, at the mark of mortality by 50% of test organisms within a specified time and a specific effect grown by 50% in test organisms within a specified time. Wastewater and sewage treatment discharges are the core causes of aquatic ecotoxicity. PCBs (polychlorinated biphenyls), DDT (dichlorodiphenyltrichloroethane) and dioxins are the main chemical constituents causing aquatic ecotoxicity. As can be noted, some of the toxicities are more human health related than others, with profound impacts on the environment. Thus, human toxicity is differentiated from ecotoxicity, which includes aquatic as well as terrestrial toxicities.

Within aquatic toxicity, both freshwater and marine water toxicities are included. This differentiation is to structure primary impacts of chemicals relevant to different impact characterizations as precisely as possible. The more exhaustive the division the more targeted are the estimations of the environmental impact characterizations of the chemicals. To express toxicity, a critical volume approach is sometimes adopted. The critical volume is the volume of water needed (in the case of aquatic toxicity) to dilute a pollutant to an acceptable or safe limit. At some other times, a toxicity can be measured in terms of the kg of DCB equivalent.

It can be interpreted that a global consensus needed to be agreed to decide on environmentally safe limits for various pollutants to mitigate the toxicity impact potential. The physicochemical properties are also counted to evaluate the primary impacts to aquatic, terrestrial and human bodies. Similar to other characterizations, there is a base chemical, DCB, against which the primary level of toxicity for other chemicals is defined.

4.6.8   Eutrophication Potential

Eutrophication is a result of leaching of soil nutrients into a water body, particularly applicable for biorefinery systems, because of the application of nutrients to agricultural land. It is caused by nutrients, mainly nitrates and phosphates, leaching into a water body. The sources are agricultural, animal and sewage activities. The fertilizers applied in agriculture can be transmitted with rain through soil into the water body. As a result of eutrophication, there is an increase in biomass, for example, algal bloom, on the surface of water bodies, preventing light reaching inside the water bodies. As living creatures, fish lives are endangered. When algae start dying, aerobic bacteria use up dissolved oxygen to decompose algae. Increasing biological oxygen demand (BOD) indicates decreasing dissolved oxygen in water bodies. Nitrates (in fertilizers) are able to enter water bodies easily, because of their high water solubility. Phosphates (also in fertilizers), though water insoluble, can adhere to soil particles eroding into water bodies. All pollutants leaching into a water body have one common characteristic, dissolved oxygen reduction in the water. Thus, the eutrophication potential of a substance can be interpreted in terms of its ability to reduce the dissolved oxygen content in fresh water, with respect to a base chemical, phosphate in this case.

As seen in an agricultural system, nitrogen fertilization is crucial to plant growth, which proportionally increases with increasing nitrogen fertilizer intake. This implies that nitrate leaching into a water body is due to agricultural activity. The main source of phosphate, however, is the sewage discharge. After the treatment of black water (e.g., sewage discharge) and grey water (e.g., water returned from other household activities) the resulting sludge is generally embedded within the sea bed. Given that phosphate is water insoluble, the sea bed discharge of sludge was seen to be a reasonable thing to do. However, soil eroded into water and phosphate content in fresh water increased over the years. One of the ways for reducing phosphate leaching into fresh water is to deposit sludge in deep rocks for mineral formation. It is thus implied that the eutrophication potential is directly linked with human activity, for example, agricultural practices and effluent treatment strategies, etc.

4.6.9   Biodiversity

Amongst all impact categories, biodiversity is the end-point in the cause and effect chain of environmental impacts and directly affects nature. Biodiversity was introduced at the United Nations Conference on Environment and Development in Rio de Janeiro in 1992 (Rio Earth Summit). The International Convention on Biological Diversity was agreed amongst 150 countries and moved to halve the loss of biodiversity, wildlife and habitats. Responsibility at national level includes an action plan and programmes to reduce biodiversity loss of species under their jurisdiction. The UK Biodiversity Action Plan implemented in 1994 focused on the delivery of actions for conservation and protection of priority/targeted species and habitats (1150 species and 65 habitats in 2007–2011). Biodiversity assessment relies on analysis and understanding the whole cause and effect chain of environmental impacts. For example, the aquatic ecotoxicity and marine ecotoxicity can be measured. Their effect can be death to marine life or even species extinction. The latter is the end effect of a cause, ecotoxicity, and a direct effect on nature. An end effect analysis is not straightforward, requiring longer term qualitative as well as quantitative dimensions and multidisciplinary interventions. Often LCA studies are done up to the primary impact characterizations and biodiversity but other end-point impacts are not included because of the evaluation uncertainty.

4.7   LCA: Interpretation

This is the final phase of LCA, where the results of inventory analysis and impact assessment are gathered together with an aim to improve system performance and suggest possible changes. Interpretation of an LCA study includes identification of major burdens, impacts, hot spots; identification of areas with a scope for improvement; sensitivity analysis; robustness of results; evaluation and recommendations. Reliability and applicability of the data are a major issue, for which sensitivity must be analyzed for variability and uncertainty in the data used. Furthermore, there are various issues around the LCA methodology.

Even for primary effect assessments, different characterization factors have been suggested by different research groups, though with little variation. Some works undertake normalization to understand the relative importance and magnitude of the environmental benefits/damages of the technology or product and thereby compare the reliability of such studies, which must be carefully assessed. Characterization and normalization must be drawn on common references and may be related to a given community, country or region over a period of time. After normalization, results are given in the same unit and all the normalized indicator results corresponding to each impact category can be added. A single score for each technology or product is then obtained. It is an effective tool for comparisons of environmental benefits and business generation, if undertaken reliably and responsibly. LCA studies can be of a stand-alone, accounting (consequential) and change oriented (attributional) type. The sets of questions that can be answered using the three different types of LCA follow.

The following chapter discusses the data uncertainty analysis and LCA interpretation.

4.7.1   Stand-Alone LCA

- Product/process/systems/network focused.
- What is the environmental impact of a product/process/system/ …?
- What are the relevant impact characterizations of a product/ … ? What are the key pollutants?
- What are the input parameters that have the maximum impact on the key pollutant emissions and on the relevant impact characterizations?
- Analyze the sensitivity of the significant input parameters on the key pollutant emissions and on the relevant impact characterizations and decide the bounds of their acceptable values.
- What are the limits of the pollutants and what are the optimal design and operating conditions to keep the pollutants within their limits?
- What are the key observations that can be communicated to the policy makers, design and decision makers?
- The approach is stand-alone; hence no comparison is made with an existing product/ …. This implies that the product / … can be functionally different.
- It is a useful attribute of the LCA study that the hot spots and the causes for hot spots are identified and remedial actions are made to mitigate impacts.

4.7.2 Accounting LCA

- This type is about replacement of an existing product/process/systems/network. Hence LCA is to provide a comparison of a decision today with existing ones, retrospectively. This implies that the new product/ ... is intended to replace a similar functionality product/ .... A comparison can be made between a number of different products with similar functionality or providing similar services.
- If a product/ ... is replaced by another type of product/ ..., how can the impact under a given characterization be improved?
- Identify the areas of major improvements with the least cost achievable.
- What would be an appropriate policy measure if a particular product or production route is held responsible for a major impact under a characterization?
- A typical example is the recycling of extracted material in the place of virgin material. What are the additional environmental impacts of using recycled material in place of virgin material? Compare the consequential impacts between landfill disposal and the material being recycled. Take account of the additional material use from construction to decommissioning to process the recycled material in the place of virgin material. Produce the entire flowchart and indicate the impact hot spots.
- The standard set of sensitivity analysis questions can be answered.

4.7.3 Change Oriented LCA

- The change oriented LCA is useful in making comparisons between products/activities prospectively. This is therefore used in making choices of products and processes, while considering growing environmental concerns and policies.
- This LCA is effective in the minimization of resource use, waste management and ways of reusing the materials after the end of product life.
- Thus it allows a comparison amongst a number of options to govern a whole system decision from resource selection through the conversion process decision making to the product end of life, prospectively.
- While the stand-alone LCA is just the beginning of an exercise, the accounting and change oriented LCAs are more appropriate to answer a number of relevant sustainable business development questions.

4.7.4 Allocation Method

The allocation of impact to different products utilizing the same processes and pathways is a difficult decision for an LCA study. The allocation can be done using the by economic value (if market prices are known). If products are functionally the same, allocation by unit functional value or allocation by substitutions/contributions to the service concerned can be useful. In the case of energy production, allocation of impacts by energy values of products is a rational way. Table 4.22 shows the environmental impact allocation to two main products from wheat cultivation, grain and straw. The allocation

Table 4.22 *Allocation of environmental impacts from wheat cultivation. ARU, abiotic resource use.*

Functional Unit	CPE (MJ)	GWP (kg CO_2 eq.)	EP (kg PO_4^{3-} eq.)	AP (kg SO_2 eq.)	ARU (kg Sb eq.)
ha^{-1} grain	18335	3426	16.1	15.8	10.8
ha^{-1} straw	632	77	0.2	0.2	0.30
Total (ha^{-1})	18967	3503	16.3	16.0	11.1
t^{-1} grain	2634	492	2.3	2.3	1.5
t^{-1} straw	181	22	0.1	0.1	0.1
y^{-1} grain	3.16×10^9	5.91×10^8	2.77×10^6	2.72×10^6	1.85×10^6
y^{-1} straw	1.09×10^8	1.33×10^7	4.17×10^4	4.25×10^4	5.26×10^4
Total (y^{-1})	3.27×10^9	6.04×10^8	2.81×10^6	2.76×10^6	1.91×10^6
Land use	Grade 2	Grade 3a	Grade 3b	Grade 4	
ha y^{-1}	151685	172370	186159	193054	
ha t^{-1} grain	0.13	0.14	0.16	0.16	

was done by their relative economic values. Hence, wheat had a bigger burden than straw. Also, note the choices of functional units for comparison with other cultivation systems.

Refer to the ***Online Resource Material in the Companion Website: Case Studies 1 to 4*** for LCA based problem solving approaches for industrial systems.

4.8   LCIA Methods

A number of LCIA methods exist to predict impact under various categories. A method can focus on the primary impact characterizations, such as, by CML. However, the LCIA methods can also include combinations of primary as well as mid- or end-point impacts. A cause–effect chain is created using characterization factors to predict the mid-point and end-point impacts. Needless to say, the latter approaches introduce more uncertainties in LCIA modeling results.

In this book, the CML 2010 method has primarily been used (http://www.cml.leiden.edu/research/industrialecology/researchprojects/finished/new-dutch-lca-guide.html) for comparisons, for example, between biorefinery configurations, feedstock selections, product selections by avoided emissions and focusing on the primary impact characterizations.

The EC-ILCD International Reference Life Cycle Data System and ELCD European Reference Life Cycle Database are available on http://lca.jrc.ec.europa.eu/lcainfohub/index.vm.

ILCD recommendations (http://lct.jrc.ec.europa.eu/pdf-directory/ILCD-Handbook-General-guide-for-LCA-DETAIL-online-12March2010.pdf) include primary as well as mid-point estimations as follows. For some primary impacts, such as the eutrophication potential, the method provides individual impact values in kg N or S equivalent (terrestrial) and in kg P equivalent (for water). For some others, such as the abiotic depletion potential, the combined resource depletion in terms of fossil and mineral reserves is shown.

The USA-TRACI "Tool for the Reduction and Assessment of Chemical and Other Environmental Impacts" method is available on http://www.epa.gov/nrmrl/std/traci/traci.htm and http://www.gabi-software.com/support/gabi/gabi-lcia-documentation/life-cycle-impact-assessment-lcia-methods/traci/. The method provides primary as well as mid-point characterizations. In addition to some of the CML primary categories, but often in different units except GWP, mid-point impacts to human health are estimated.

The various methodologies are in-built and offered by commercially available LCA software, recommended to use. The units used for various impact characterizations included under ILCD recommendations, CML 2001 – November 2010, TRACI and ReCiPe in GaBi, are shown in four parts in Table 4.23[21].

Table 4.23(a)   *Units for various impact characterizations under ILCD recommendations.*

Impact Characterization	Accepted Unit
Climate change	kg CO_2 equivalent
Ozone depletion	kg CFC-11 equivalent
Human toxicity, cancer effects	CTUh (comparative toxic unit for human)
Human toxicity, non-cancer effects	CTUh
Particulate matter, respiratory inorganics	kg PM2.5 equivalent
Ionising radiation, human health	Human exposure efficiency relative to U^{235}
Photochemical ozone formation	kg NMVOC (non-methane volatile organic compounds) equivalent
Acidification	kg N or S equivalent
Eutrophication, terrestrial	kg N or S equivalent
Eutrophication, aquatic	Fraction of nutrients reaching end compartment
Ecotoxicity	CTUe (comparative toxic unit for ecosystem)
Resource depletion, water	Water stress index (WSI)
Resource depletion, mineral, fossil and renewable	kg Sb equivalent

Table 4.23(b) *Units for various impact characterizations under CML 2001 – November 2010.*

Impact Characterization	Accepted Unit
Global warming potential	kg CO_2 equivalent
Acidification depletion potential	kg SO_2 equivalent
Eutrophication potential	kg phosphate equivalent
Ozone layer depletion potential	kg CFC-11 equivalent
Abiotic depletion potential, elements	kg Sb equivalent
Abiotic depletion potential, fossil	MJ
Freshwater aquatic ecotoxicity potential	kg DCB equivalent
Human toxicity potential	kg DCB equivalent
Marine aquatic ecotoxicity potential	kg DCB equivalent
Photochemical ozone creation potential	kg ethylene equivalent
Terrestric ecotoxicity potential	kg DCB equivalent

Table 4.23(c) *Units for various impact characterizations under TRACI.*

Impact Characterization	Accepted Unit
Global warming potential	kg CO_2 equivalent
Acidification depletion potential	kg H^+ equivalent
Eutrophication potential	kg N equivalent
Ozone layer depletion potential	kg CFC-11 equivalent
Ecotoxicity air	PAF (potentially affected fraction) m^3 day kg^{-1}
Ecotoxicity soil	PAF m^3 day kg^{-1}
Ecotoxicity water	PAF m^3 day kg^{-1}
Human health cancer air	Cases
Human health cancer soil	Cases
Human health cancer water	Cases
Human health non-cancer air	Cases
Human health non-cancer soil	Cases
Human health non-cancer water	Cases
Human health criteria air	kg PM10 equivalent
Smog air	kg O_3 equivalent

Table 4.23(d) *Units for various impact characterizations under ReCiPe.*

Impact Characterization	Accepted Unit
Climate change	kg CO_2 equivalent
Terrestrial acidification	kg SO_2 equivalent
Freshwater eutrophication potential	kg P equivalent
Ozone depletion potential	kg CFC-11 equivalent
Fossil depletion	kg oil equivalent
Freshwater ecotoxicity	kg DCB equivalent
Ionizing radiation	kg U235 equivalent
Marine ecotoxicity	kg DCB equivalent
Marine eutrophication	kg N equivalent
Metal depletion	kg Fe equivalent
Natural land transformation	m^2
Particulate matter formation	kg PM10 equivalent
Photochemical oxidant formation	kg NMVOC equivalent
Terrestrial ecotoxicity	kg DCB equivalent
Water depletion	m^3

CML2001 - Dec. 07, Experts IKP (Central Europe)
CML2001 - Dec. 07, Experts IKP (Northern Europe)
CML2001 - Dec. 07, Experts IKP (Southern Europe)
CML2001, Experts IKP (Central Europe)
CML2001, Experts IKP (Northern Europe)
CML2001, Experts IKP (Southern Europe)
CML96, Expert judgement, IKP
CML96, Experts, IKP
CML96, Policy long term, IKP
CML96, Policy short term, IKP
CML96, Policy, IKP
CML96, Population, IKP
CML96, Survey Europe, IKP
CML96, Survey Germany, IKP
CML96, Sustainable development, IKP
EDIP 1997, Env. imp. eval. (PET W, EU 2004)
EDIP 1997, Ress. eval. (PR W 2004)
EDIP 1997, Toxicity eval. (PET EU 2004)
EDIP 1997, Waste eval. (PET DK 2000)
EDIP 2003, Env. imp. eval. (PET W, EU 2004)
EI95
EI95 (incl. ionizing radiation, RF)
EI99, EE (Egalitarian approach)
EI99, HA (Hierarchist approach)
EI99, II (Individualist approach)
PE LCIA Survey 2012 (Europe; CML and ReCiPe)
PE LCIA Survey 2012 (Global; CML, ReCiPe and Traci)
PE LCIA Survey 2012 (North-America; Traci)

Figure 4.19 *A comprehensive list of LCIA methodologies.*

Figure 4.19 shows a comprehensive list of LCIA methodologies and regions applied for, which include CML, EDIP, EI and the PE LCIA survey, offered in GaBi 6, from PE International. The most recent LCIA methodologies are shown in Figure 4.20 (a) to (h). Each of the LCIA methodologies is then elaborated to show the impact characterizations, under the methodology shown in Figure 4.20.

Foundation Concepts Quiz (20 marks in total)

Problem 1. Carries 5 Marks

Consider the emissions from the conversion of two feedstocks, shown on the basis of the weight percentage in Table 4.24.

1. Calculate the global warming potential of the emissions from the two feedstocks. (2.5 marks)
2. Calculate the CO_2 credit gain by environmentally better performing feedstock compared to inferior feedstock. (2.5 marks)

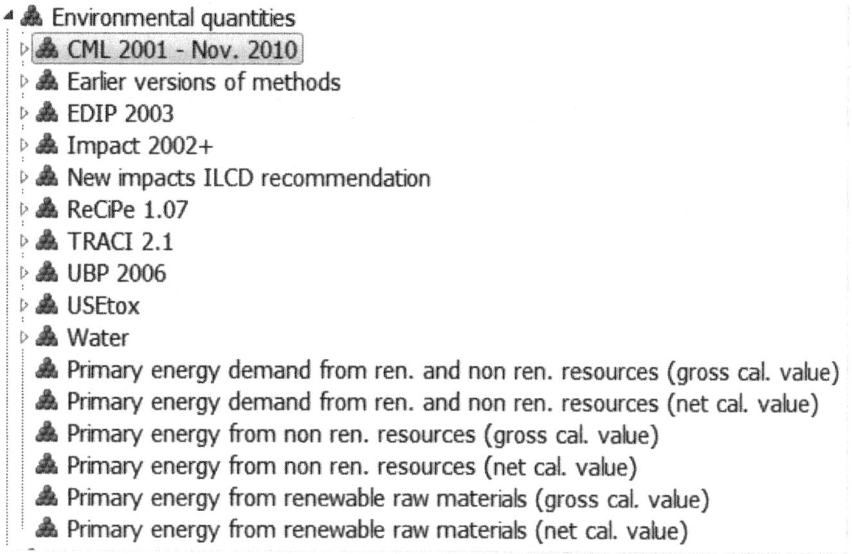

◢ ▲ Environmental quantities
▷ ▲ CML 2001 - Nov. 2010
▷ ▲ Earlier versions of methods
▷ ▲ EDIP 2003
▷ ▲ Impact 2002+
▷ ▲ New impacts ILCD recommendation
▷ ▲ ReCiPe 1.07
▷ ▲ TRACI 2.1
▷ ▲ UBP 2006
▷ ▲ USEtox
▷ ▲ Water
▲ Primary energy demand from ren. and non ren. resources (gross cal. value)
▲ Primary energy demand from ren. and non ren. resources (net cal. value)
▲ Primary energy from non ren. resources (gross cal. value)
▲ Primary energy from non ren. resources (net cal. value)
▲ Primary energy from renewable raw materials (gross cal. value)
▲ Primary energy from renewable raw materials (net cal. value)

Figure 4.20 *LCIA methodologies and their impact characterizations available in LCA software (a to h).* **(a)** *LCIA methodologies available in LCA software.*

▲ CML 2001 - Nov. 2010
▲ CML2001 - Nov. 2010, Abiotic Depletion (ADP elements)
▲ CML2001 - Nov. 2010, Abiotic Depletion (ADP fossil)
▲ CML2001 - Nov. 2010, Acidification Potential (AP)
▲ CML2001 - Nov. 2010, Eutrophication Potential (EP)
▲ CML2001 - Nov. 2010, Freshwater Aquatic Ecotoxicity Pot. (FAETP inf.)
▲ CML2001 - Nov. 2010, Global Warming Potential (GWP 100 years)
▲ CML2001 - Nov. 2010, Global Warming Potential, excl biogenic carbon (GWP 100 years)
▲ CML2001 - Nov. 2010, Human Toxicity Potential (HTP inf.)
▲ CML2001 - Nov. 2010, Marine Aquatic Ecotoxicity Pot. (MAETP inf.)
▲ CML2001 - Nov. 2010, Ozone Layer Depletion Potential (ODP, steady state)
▲ CML2001 - Nov. 2010, Photochem. Ozone Creation Potential (POCP)
▲ CML2001 - Nov. 2010, Terrestric Ecotoxicity Potential (TETP inf.)

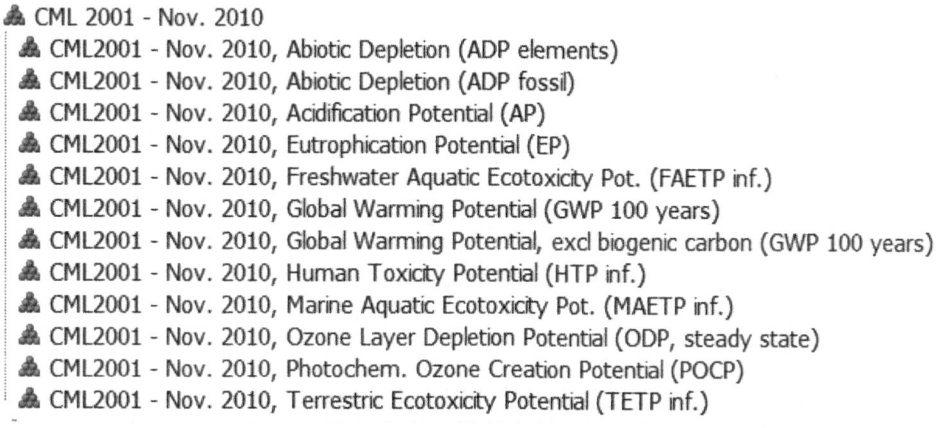

Figure 4.20(b) *Primary impact characterizations included in CML 2001 – November 2010.*

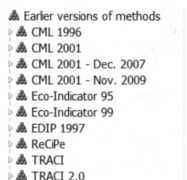

Earlier versions of methods
▷ CML 1996
▷ CML 2001
▷ CML 2001 - Dec. 2007
▷ CML 2001 - Nov. 2009
▷ Eco-Indicator 95
▷ Eco-Indicator 99
▷ EDIP 1997
▷ ReCiPe
▷ TRACI
▷ TRACI 2.0
▷ UBP

Figure 4.20(c)   *Earlier versions of LCIA methods.*

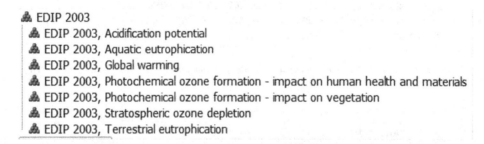

EDIP 2003
EDIP 2003, Acidification potential
EDIP 2003, Aquatic eutrophication
EDIP 2003, Global warming
EDIP 2003, Photochemical ozone formation - impact on human health and materials
EDIP 2003, Photochemical ozone formation - impact on vegetation
EDIP 2003, Stratospheric ozone depletion
EDIP 2003, Terrestrial eutrophication

Figure 4.20(d)   *Primary impact characterizations included in EDIP 2003.*

Impact 2002+
I02+ v2.1 - Aquatic acidification - Midpoint
I02+ v2.1 - Aquatic ecotoxicity - Midpoint
I02+ v2.1 - Aquatic eutrophication - Midpoint
I02+ v2.1 - Carcinogens - Midpoint
I02+ v2.1 - Global warming 500yr - Midpoint
I02+ v2.1 - Ionizing radiation - Midpoint
I02+ v2.1 - Land occupation - Midpoint
I02+ v2.1 - Mineral extraction - Midpoint
I02+ v2.1 - Non-carcinogens - Midpoint
I02+ v2.1 - Non-renewable energy - Midpoint
I02+ v2.1 - Ozone layer depletion - Midpoint
I02+ v2.1 - Photochemical oxidation - Midpoint
I02+ v2.1 - Respiratory effects - Midpoint
I02+ v2.1 - Terrestrial acidification/nutrification - Midpoint
I02+ v2.1 - Terrestrial ecotoxicity - Midpoint

Figure 4.20(e)   *Mid-point impact characterizations included in Impact 2002+.*

🚢 New impacts ILCD recommendation
  🚢 Acidification, accumulated exceedance
  🚢 CML2002 Resource Depletion, fossil and mineral, reserve Based
  🚢 IPCC global warming, excl biogenic carbon
  🚢 IPCC global warming, incl biogenic carbon
  🚢 Particulate matter/Respiratory inorganics, RiskPoll
  🚢 Terrestrial eutrophication, accumulated exceedance
  🚢 Total freshwater consumption, including rainwater (acc. to UBP 2006)

Figure 4.20(f)   *New impacts recommended in ILCD.*

🚢 ReCiPe 1.07
  🚢 ReCiPe 1.07 Endpoint (H) - Agricultural land occupation
  🚢 ReCiPe 1.07 Endpoint (H) - Climate change Ecosystems
  🚢 ReCiPe 1.07 Endpoint (H) - Climate change Human Health
  🚢 ReCiPe 1.07 Endpoint (H) - Fossil depletion
  🚢 ReCiPe 1.07 Endpoint (H) - Freshwater ecotoxicity
  🚢 ReCiPe 1.07 Endpoint (H) - Freshwater eutrophication
  🚢 ReCiPe 1.07 Endpoint (H) - Human toxicity
  🚢 ReCiPe 1.07 Endpoint (H) - Ionising radiation
  🚢 ReCiPe 1.07 Endpoint (H) - Marine ecotoxicity
  🚢 ReCiPe 1.07 Endpoint (H) - Metal depletion
  🚢 ReCiPe 1.07 Endpoint (H) - Natural land transformation
  🚢 ReCiPe 1.07 Endpoint (H) - Ozone depletion
  🚢 ReCiPe 1.07 Endpoint (H) - Particulate matter formation
  🚢 ReCiPe 1.07 Endpoint (H) - Photochemical oxidant formation
  🚢 ReCiPe 1.07 Endpoint (H) - Terrestrial acidification
  🚢 ReCiPe 1.07 Endpoint (H) - Terrestrial ecotoxicity
  🚢 ReCiPe 1.07 Endpoint (H) - Urban land occupation
  🚢 ReCiPe 1.07 Midpoint (H) - Agricultural land occupation
  🚢 ReCiPe 1.07 Midpoint (H) - Climate change
  🚢 ReCiPe 1.07 Midpoint (H) - Fossil depletion
  🚢 ReCiPe 1.07 Midpoint (H) - Freshwater ecotoxicity
  🚢 ReCiPe 1.07 Midpoint (H) - Freshwater eutrophication

Figure 4.20(g)   *End- and mid-point impact characterizations included in ReCiPe.*

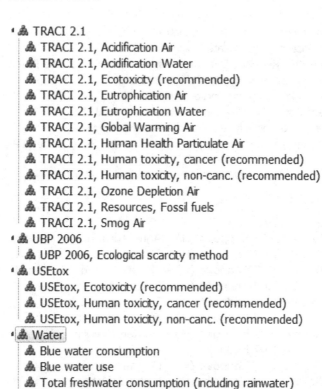

Figure 4.20(h) *Primary impact characterizations included in TRACI, UBP, USEtox and Water.*

Problem 2.   Carries 10 Marks

A comparison of the environmental impact assessment results between anaerobic digestion and natural gas based production systems is shown in Figures 4.21 to 4.28. Neglect the differences due to geographic locations (CH in the figures refers to Switzerland and RNA refers to North America, according to the geographical codes used in GaBi software, http://www.gabi-software.com/support/gabi/geographical-codes/ and <u-so> or <p-agg> indicates the impact potential from that process block only).

1. Calculate the credit gain by environmentally better performing feedstock compared to inferior feedstock in each impact category shown. (2 marks)

Table 4.24 *Emissions from the conversion of two feedstocks on the basis of weight percentage.*

Chemical	Feedstock 1 (weight%)	Feedstock 2 (weight%)	GWP Characterization Factor
CO_2	80	65	1
CH_4	10	30	25
CFC	0.05	0	5000
N_2O	4	5	298
NO	5.95	0	40

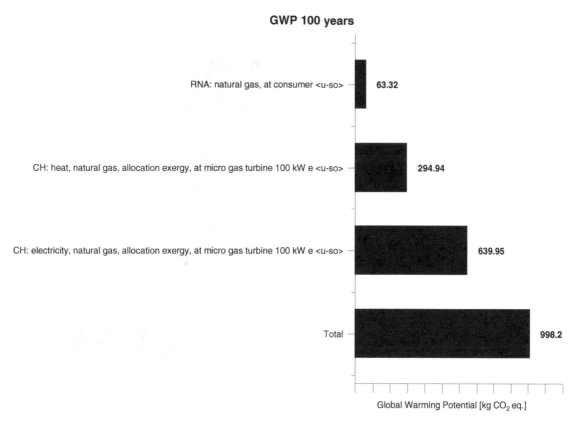

Figure 4.21 *GWP from natural gas based CHP generation system (basis 11 340 MJ energy content in natural gas).*

2. Give the reason for the credit gained by biogas compared to natural gas or otherwise, for each of the categories shown. (3 marks)

3. State the hot spot for each feedstock under each category. (2 marks)

4. Explain how these hot spots may be mitigated. (3 marks)

Problem 3. Carries 5 Marks

A biomass integration gasification combined cycle plant is required to produce 50 MW e output. The biomass feedstock can be wheat straw, waste wood and refuse-derived fuel (RDF), which is produced by shredding and dehydrating municipal solid waste. Their ultimate analyses are shown in Table 4.25.

1. Name the impact categories that can be evaluated to create a hierarchy of biomass feedstocks from most to least beneficial ones. Assume that the toxicity potentials in all categories are comparable between feedstocks. Given that the data generation is a huge constraint in the LCA study, your list must not consist of more than three most important impact categories. (1.5 marks)

2. Give the reasoning for your above selection. (3.5 marks)

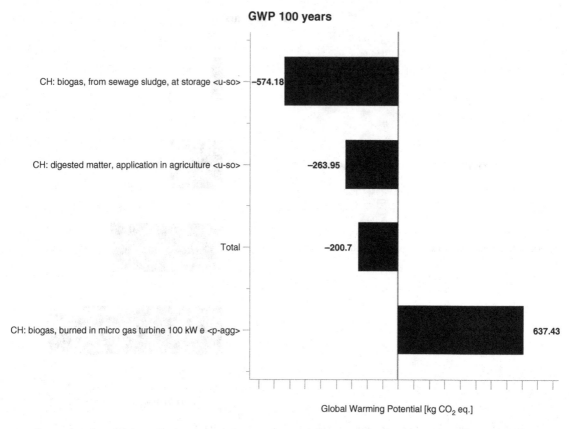

GWP 100 years

CH: biogas, from sewage sludge, at storage <u-so> — **−574.18**

CH: digested matter, application in agriculture <u-so> — **−263.95**

Total — **−200.7**

CH: biogas, burned in micro gas turbine 100 kW e <p-agg> — **637.43**

Global Warming Potential [kg CO_2 eq.]

Figure 4.22 *GWP from anaerobic digestion based production system (basis 11 340 MJ energy content in biogas).*

Test of Peer Review Skill

The LCA study is proving to be effective for LCA studies. Amongst the various environmental impact categories, the most important issues considered are the global warming potential impact saving, land use and water footprint. Various journals provide numerous LCA studies of biorefinery systems. Though not exhaustive, the following list of journals can be searched for biorefinery LCA papers. Select studies under topic areas, such as Bioethanol, Biorefinery, Polygeneration, CHP, Sustainability and LCA and test your review skills.

- Nature journals: http://www.nature.com/siteindex/index.html
- Science: http://www.sciencemag.org/magazine
- Environmental Science and Technology
- Energy and Environmental Science
- Water Research
- Journal of Power Sources
- Applied Energy
- Biomass and Bioenergy
- Atmospheric Environment

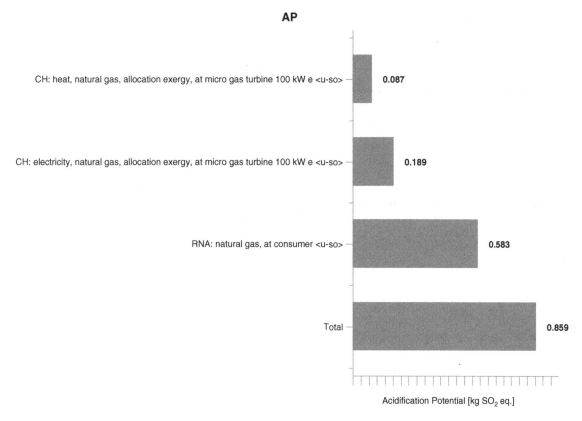

Figure 4.23   *Acidification potential (AP) from natural gas based CHP generation system (basis 11 340 MJ energy content in natural gas).*

- Ecological Economics
- International Journal of Life Cycle Assessment
- Journal of Industrial Ecology
- Energy
- Energy Policy
- Environmental Science and Pollution Research
- Journal of Environmental Management
- Water Resources Management
- Renewable and Sustainable Energy Review
- Solar Energy
- Resources, Conservation and Recycling
- International Journal of Energy Research
- Journal of Cleaner Production
- Technological Forecasting and Social Change
- Energy and Buildings
- Sustainable Development
- Chemical Engineering Research and Design
- Industrial and Engineering Chemistry Research

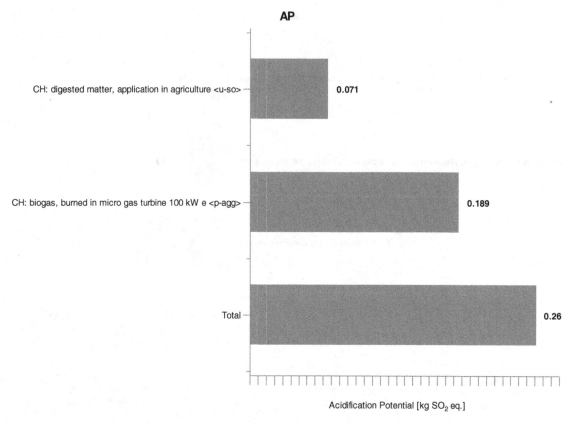

Figure 4.24 *Acidification potential (AP) from anaerobic digestion based production system (basis 11 340 MJ energy content in biogas).*

- Sustainability
- Journal of Applied Microbiology
- Critical Reviews in Environmental Science and Technology

Answer to the Following Questions

Goal and Scope Definition

Are the boundaries defined for the systems appropriate? What other blocks/subsystems could have been added (or otherwise) within the boundaries to make differences in impact assessment values and interpretation?

A biorefinery LCA publication may show the use of various units to present results, such as, per land use, per mass, per energy and per annual bases. Explain the reason for using each unit.

Which other impact categories might have been relevant for the study? Remember "less" is "more" when a study is complete, transparent and coherent. LCA study results should also be in transferable form such that these can be used/adapted and cited in another study (an example is shown in Table 4.4).

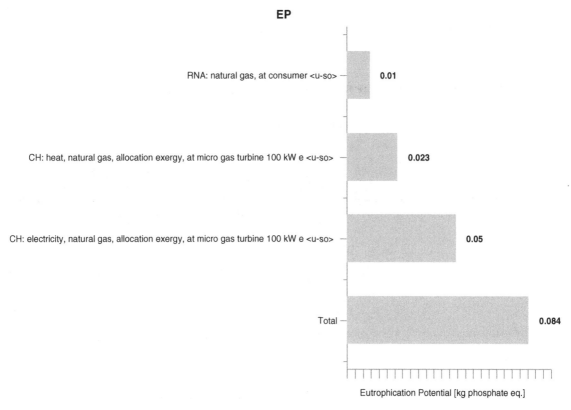

Figure 4.25 *Eutrophication potential (EP) from natural gas based CHP generation system (basis 11 340 MJ energy content in natural gas).*

Inventory Analysis

What are the advantages in showing inventories around various internal boundaries in the system (i.e., cultivation, processing and then for each unit operation)?

What are the advantages and disadvantages of using simulation results, databases from software such as GaBi and primary social science research data?

Interpretation

Are the results clearly shown? Are the expected results clearly stated in the goal and scope definition?

Is this a stand-alone, accounting (consequential) or change oriented (attributional) type of study?

How would you like to improve the presentation of the sensitivity analysis results?

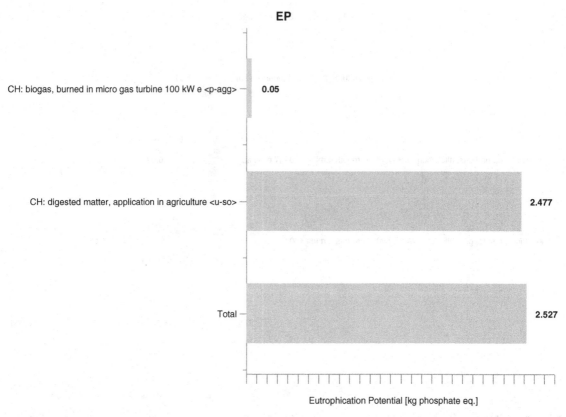

Figure 4.26   *Eutrophication potential (EP) from anaerobic digestion based production system (basis 11 340 MJ energy content in biogas).*

Essay writing (100 marks)

Task

Write a research paper on LCA studies of **your chosen case study or one of the following topics**. It is important that your research refers to a variety of sources, of which a significant proportion should be peer-reviewed journal papers. Answers must be fully referenced using the author–date (Harvard) system. The report may contain solutions for the following issues as appropriate. The list is not exhaustive and only meant to provide a guidance. Please include any other issues as you consider appropriate.

- Identify alternative technology options.
- Select one or two technologies to do LCA in detail.
- Identify system boundary and functional units (suitable for making comparisons with other systems).
- State assumptions and limitations.
- Identify hot spots in alternative technology options.
- Identify key pollutants.
- Identify geographic locations and any associated transport.
- Identify utility requirements (e.g., electricity and heat, etc.) and sources of them.
- Discuss hot spots, types of pollutants and ways to remedy in detail (see the next chapter).
- Make comparison with other equivalent systems for environmental sustainability.

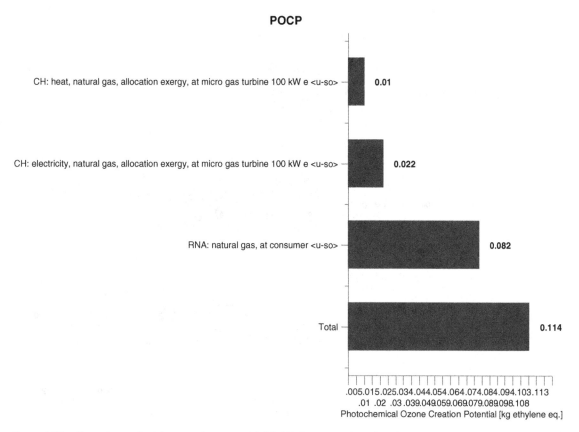

Figure 4.27   *Photochemical oxidant creation potential (POCP) from natural gas based CHP generation system (basis 11 340 MJ energy content in natural gas).*

- Report critical input parameters and sensitivity analysis results.
- Include critical analysis of the quality of results and LCA done by you and others in the field.
- Report results *in transferable form.*
- Include techno-economic and socio-economic analyses as appropriate.
- Discuss policy drivers and barriers; learning points from the case study; action and policy recommendations.
- Discuss key benefits and challenges of the application of LCA.

Option 1. LCA of solar photovoltaic for electricity generation.
Option 2. LCA of energy generation from wastewater sludge.
Option 3. LCA of energy generation from municipal solid wastes.
Option 4. LCA of biofuel production from waste wood.
Option 5. LCA of composite/polymer production from biomass, with applications in aeronautics, automotive industry, construction, etc.

For your own case study:

Describe how LCA can be applied to a specific case. This can be, for example, a specific substance, or a specific product or service or technology.

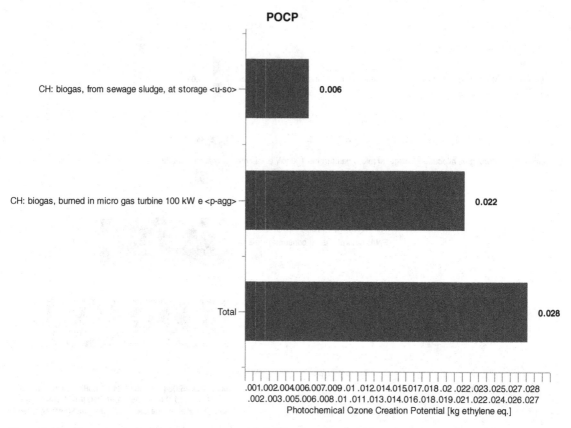

Figure 4.28 *Photochemical oxidant creation potential (POCP) from anaerobic digestion based production system (basis 11 340 MJ energy content in biogas).*

Format

- The content of paper should not exceed 5000 words (this excludes title, abstract, table of contents, reference list and appendix mainly used to show calculations).
- Use standard report format (including abstract, table of contents, headings, subheadings, reference list).

Table 4.25 *Biomass composition in weight %.*

Component in weight %	Straw	Wood	RDF
C	18.66	29.94	35.25
H	4.91	4.28	5.77
O	40.7	30.69	37.59
N	0.57	0.075	0.298
S	0.14	0	0.199
Cl	0	0	0.398
Moisture	8.5	25	0
Fixed C	17.91	8.92	10
Ash	8.61	1.1	10.5
Lower heating value (MJ kg^{-1})	14.6	19.3	17.73

Criteria for Assessment

- Demonstrated understating of LCA and LCT (25%).
- Appreciation of the application of theory to practice in the selected case (25%).
- Evidence of research to develop arguments, and the clarity and comprehensiveness of the arguments (35%).
- Overall structure, quality of the writing and presentation as well as the appropriate use and citing of reference materials (15%).

4.9 Future R&D Needs

Investment for biorefineries should be justified by a more holistic evaluation of social consequences (the agricultural focus may be shifted to energy crops rather than food crops due to subsidies and business cases that would fuel further to the energy demand), economic impacts (food price is likely to increase and affordability is likely to decrease) and long-term environmental implications on land, water and biosphere. It is imperative to compare land to energy yields, life cycle costs and environmental, economic and social sustainability between various energy systems with similar functional products. It is imperative to have an understanding of the value implications across supply chains that are global in some connections, complemented by bottom-up LCA approaches. Crop cultivation mapping on temporal and spatial scales using LCA is also a useful exercise. Research effort in the overall process and energy integration and enhancement of efficiency of integrated biorefinery systems is not apparent. Thus current deployment of process integration and life cycle approaches for the maximization of energy efficiency and minimization of life cycle inventories will be highly relevant for sustainable application of biorefineries.

References

1. A. Banerjee and M. Tierney, Comparison of five exergoenvironmental methods applied to candidate energy systems for rural villages in developing countries, *Energy*, **36**, 2650–2661 (2011).
2. http://www.oecd.org/document/19/0,3746,en_2649_34281_35158227_1_1_1_1,00.html
3. H. Baumann and A.M. Tillman, *The Hitchhiker's Guide to LCA*. Lund, Sweden, Studentlitteratur AB, 2004.
4. ISO 14040, *Environmental Management – Life Cycle Assessment – Principles and Framework*, Geneva, Switzerland, 1997.
5. ISO 14041, *Environmental Management – Life Cycle Assessment – Goal and Scope Definition and Inventory Analysis*, Geneva, Switzerland, 1998.
6. ISO 14044, *Environmental Management – Life Cycle Assessment – Requirements and Guidelines*, Geneva, Switzerland, 2006.
7. J. Sadhukhan, M.M. Mustafa, N. Misailidis, F. Mateos-Salvador, C. Du, G.M. Campbell, Value analysis tool for feasibility studies of biorefineries integrated with value added production, *Chem. Eng. Sci.*, **63**(2), 503–519 (2008).
8. E. Martinez-Hernandez, M.H. Ibrahim, M. Leach, P. Sinclair, G.M. Campbell, J. Sadhukhan, Environmental sustainability analysis of UK whole-wheat bioethanol and CHP systems, *Biomass Bioenergy*, **50**, 52–64 (2013).
9. K.S. Ng and J. Sadhukhan, Process integration and economic analysis of bio-oil platform for the production of methanol and combined heat and power, *Biomass Bioenergy*, **35**(3), 1153–1169 (2011).
10. K.S. Ng and J. Sadhukhan, Techno-economic performance analysis of bio-oil based Fischer–Tropsch and CHP synthesis platform, *Biomass Bioenergy*, **35**(7), 3218–3234 (2011).
11. J. Sadhukhan and K.S. Ng, Economic and European Union environmental sustainability criteria assessment of bio-oil based biofuel systems: refinery integration cases, *Ind. Eng. Chem. Res.*, **50**(11), 6794–6808 (2011).
12. J. Sadhukhan, K.S. Ng, N. Shah, H.J. Simons, Heat integration strategy for economic production of combined heat and power from biomass waste, *Energy Fuels*, **23**, 5106–5120 (2009).
13. A.G. Williams, E. Audsley, D.L. Sandars, *Determining the environmental burdens and resource use in the production of agricultural and horticultural commodities*, In Defra Research Project IS0205, 1–105, Cranfield University, Bedford, and Defra, 2006.
14. W.G. Lattin and V.P. Utgikar, Global warming potential of the sulfur–iodine process using life cycle assessment methodology, *Int. J. Hydrogen Energy*, **34**, 737–744 (2009).
15. J.T. Houghton, G.J. Jenkins, J.J. Ephraums, IPCC. *Climate Change: The Intergovernmental Panel on Climate Change Scientific Assessment*, Cambridge University Press, Cambridge, United Kingdom and New York, USA, 1990.

16. J. Wade, C. Holman, M. Fergusson, Passenger car global warming potential: current and projected levels in the UK, *Energy Policy*, **22**, 509–522 (1994).

17. P. Forster, P. Ramaswamy, P. Artaxo, T. Berntsen, R. Betts, D.W. Fahey, J. Haywood, J. Lean, D.C. Lowe, G. Myhre, J. Nganga, R. Prinn, G. Raga, M. Schulz, R. van Dorland, *Changes in Atmospheric Constituents and in Radiative Forcing, Climate Change, The Physical Science Basis, Contribution of Working Group I to the Fourth Assessment Report of the Intergovernmental Panel on Climate Change*, Edited by S. Solomon, D. Qin, M. Manning, Z. Chen, M. Marquis, K.B. Averyt, M. Tignor and H.L. Miller, Cambridge University Press, Cambridge, United Kingdom and New York, USA, 2007.

18. L. van Oers, A. de Koning, J.B. Guinée, G. Huppes, *Abiotic Resource Depletion in LCA*, Road and Hydraulic Institute, The Netherlands, pp. 41–48, 2002.

19. A. Azapagic, S. Perdan, R. Clift, *Appendix: Life Cycle Thinking and Life Cycle Assessment (LCA), Sustainable Development in Practice: Case Studies For Engineers and Scientists*, John Wiley & Sons Ltd, Chichester, West Sussex, 2004.

20. R.G. Derwent, W.J. Collins, C.E. Johnson, D.S. Stevenson, Transient behaviour of tropospheric ozone precursors in a global 3-D CTM and their indirect greenhouse effects, *Climate Change*, **49**, 463–487 (2001).

21. GaBi and Ecoinvent database: www.gabi-software.com.

5

Data Uncertainty and Multicriteria Analyses

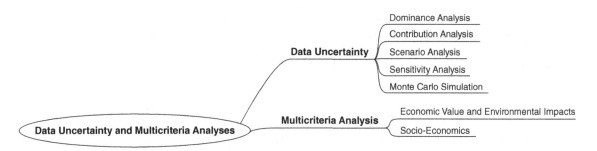

Structure for Lecture Planning

5.1 Data Uncertainty Analysis

LCA (Chapter 4) results have a number of data uncertainties. A transparent approach aims to track the uncertainties in data and reduce the effects of uncertainties in data. Data uncertainty in LCA can result from imprecise measurements of inventories, average or even outdated data using proxies and incomplete data. Various assumptions are made when deducing life cycle inventories (LCI), such as linear correlations and averaged data over time and across regions. During life cycle impact assessments (LCIAs), model approximations can result from average characterization factors deduced from simple biogeochemical models. Missing characterization factors for certain substances and interactions between substances are also approximated. Data uncertainties result from normalization, weighting and valuation stages.

Uncertainties can result due to choices of:

1. Functional units, systems boundaries.
2. LCI.
3. LCIA.
4. Allocation approaches for multioutput and for recycling processes.

Biorefineries and Chemical Processes: Design, Integration and Sustainability Analysis, First Edition.
Jhuma Sadhukhan, Kok Siew Ng and Elias Martinez Hernandez.
© 2014 John Wiley & Sons, Ltd. Published 2014 by John Wiley & Sons, Ltd.
Companion Website: http://www.wiley.com/go/sadhukhan/biorefineries

When all errors are accumulated for data interpretation, the errors can be large. Most LCA results present spatially and temporally averaged data and in this approach resulting errors are also difficult to comprehend. Data uncertainty can be managed by probabilistic or stochastic approaches such as Monte Carlo simulation, and also some deterministic ways such as scenario analysis and sensitivity analysis.

The LCA results include the following aspects:

1. Dominance analysis
2. Contribution analysis
3. Testing robustness of the results
   - Scenario analysis
   - Sensitivity analysis
   - Monte Carlo simulations

5.1.1   Dominance Analysis

A dominance analysis is about hot spot analysis due to environmental activities. The analysis shows what activity has the highest value in an impact category. Dominance analysis results can be shown in various ways. The pie charts in Figure 5.1 show examples of the sequence of activities from the most to the least detrimental activities for the environment. Cumulative primary energy depletion and global warming, acidification and abiotic depletion potential categories are shown[1]. The agricultural system shown is for UK wheat, comprising the FO: Field Operations, F & P: Fertilizers and Pesticides, Dir. & Ind. FE: Direct and Indirect Field Emissions, GC: Grain Conditioning.

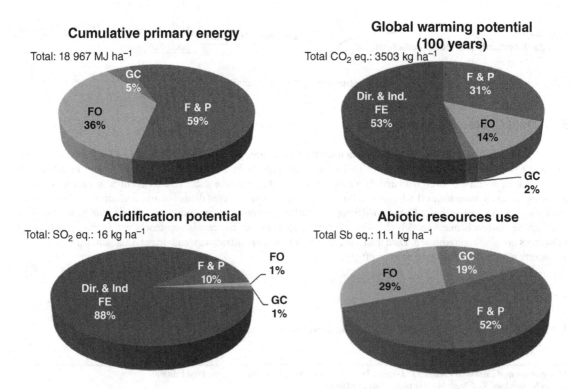

Figure 5.1   *An example of dominance analysis results. (Reproduced with permission from Martinez-Hernandez et al. (2013)[1]. Copyright © 2013, Elsevier.)*

The activities, from the highest to the lowest impacts are as follows:

Cumulative primary energy:

$$F \& P > FO > GC$$

Global warming potential:

$$\text{Dir. \& Ind. FE} > F \& P > FO > GC$$

Acidification potential:

$$\text{Dir. \& Ind. FE} > F \& P > FO/GC$$

Abiotic resource use:

$$F \& P > FO > GC$$

Note that the total values of the impact characterizations are shown on a per hectare basis. This is a transferable form to compare with other equivalent systems, such as first generation crops.

A Sankey diagram is another way to present the hot spot analysis from various activities. In a Sankey diagram, the thickness of a link joining a source to a sink process is proportional to the environmental impact from the source process. A much thicker line from a source process than a line entering to the process shows that the process is environmentally more detrimental. The line thickness increases from upstream to downstream processes as environmental impacts are aggregated. The life cycle global warming potential (GWP) (as CO_2 equivalent) in g MJ^{-1} of two integrated whole *Jatropha* fruit biorefinery systems producing heat and power and (a) biodiesel or (b) green diesel is shown in Figure 5.2. The combustion processes have the highest GWP impacts, followed by the integrated biomass gasification combined cycle (IBGCC) plants producing methanol or hydrogen.

5.1.2   Contribution Analysis

Chemicals causing higher environmental impacts must be identified and replaced with environmentally better performing similar functionality chemicals. This is also a hot spot analysis, but one related to chemicals or loads or polluting substances rather than activities as in the case of dominance analysis. Thus contribution analysis is carried out to track back pollutants to their origins and minimize their emissions from the origin.

Figure 5.3 shows an example of a contribution analysis. The outlet pollutants include carbon dioxide, nitrous oxide and methane in kg per tonne of sewage sludge. These flow rates do not account for the GWP impact characterization factors, such as 298 CO_2 equivalent for nitrous oxide and 25 CO_2 equivalent for methane. You are recommended to use CO_2 equivalent values on the y-axis for direct comparisons, through a contribution analysis. Note that the outlet carbon dioxide quantities do not take account of the inlet carbon dioxide capture during biomass production.

Figure 5.4 shows the contribution analysis of nonrenewable and renewable energy resources for the production of biogas for micro generation (electricity and heat generations) and digested matter for agricultural application (fertilizer production) from sewage sludge anaerobic digestion (AD). Tracking back the GHG emissions, the primary energy resources such as natural gas (by 61%), crude oil (by 5%) and hard coal and lignite (each by 2%), out of a total of 2468.5 MJ of energy input, were found to be responsible. The functional unit is 1000 kg of sewage sludge AD into 315 N m^3 of biogas and 700 kg of digested matter productions. If natural gas is replaced with a renewable energy source, the GHG emissions will be reduced. Note that the organic substance, primary forest and wind and solar energy use is close to zero. Hence, increasing their inputs to the AD system, in particular solar and wind energy, will reduce the environmental emissions. Tracking the routes between primary resources and end uses through conversion processes is essential. The pollutants causing impact hot spots, pollutants with larger impact reduction potentials and the primary resources of pollutants need to be identified. Primary resources causing depletion of energy or material reserves of the earth and emissions can be replaced with better alternatives that reduce these environmental impacts.

A contribution analysis is very useful for identifying resource (elements and fossil) use. The input resources required for the production of polyethylene terephthalate for 1 cm^2 solar organic photovoltaic (OPV) cell fabrication, extracted from Ecoinvent 2.0 and BUWAL databases, are shown in Table 5.1.

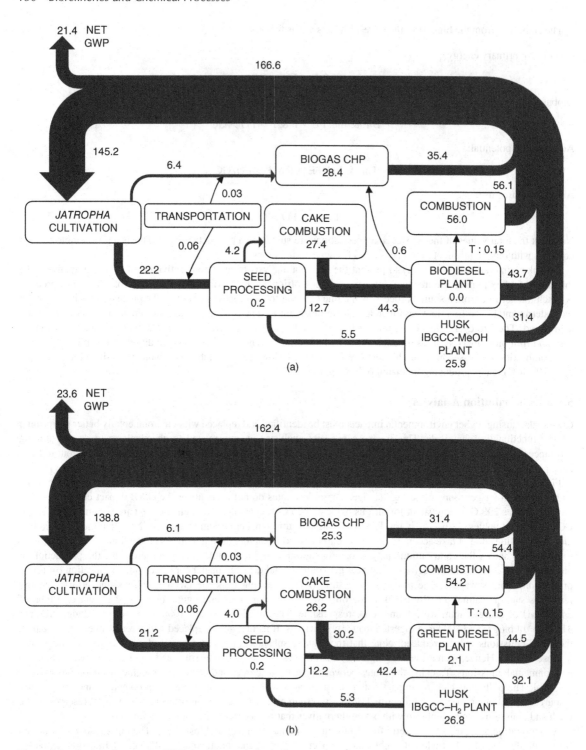

Figure 5.2 *Sankey diagram for GWP (as CO$_2$ equivalent) flows in g MJ^{-1} in two integrated whole Jatropha fruit biorefinery systems producing heat and power and (a) biodiesel or (b) green diesel.*

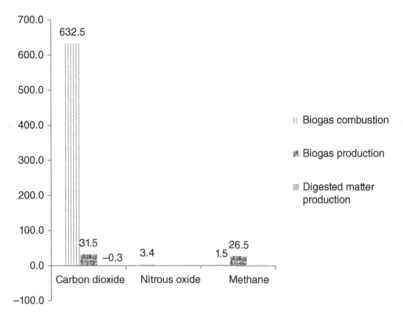

Figure 5.3 *Contribution analysis in kg per tonne of sewage sludge (which environmental load contributes the most).*

5.1.3 Scenario Analysis

At first, independent variables are fixed at certain values for an LCIA. Independent variables can be varied to generate numerous scenarios. One or more independent variables can be varied at the same time. A sensitivity analysis refers to variations in estimated impact potentials due to unit changes in independent variables or due to standard deviations from mean values of independent variables. Independent variables, one or more at a time, can be examined for a sensitivity

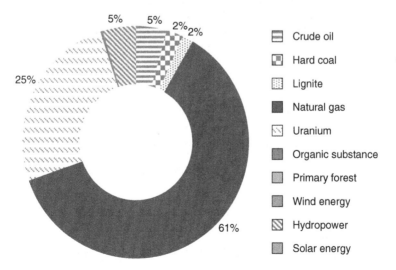

Figure 5.4 *Contribution analysis of nonrenewable and renewable energy resource use for the production of biogas for micro generation (electricity and heat generations) and digested matter for agricultural application.*

Table 5.1   *Resource depletion from the manufacture of a solar OPV cell. Some databases were obtained from BUWAL (Bundesamt für Umwelt, Wald und Landschaft).*

Crude oil free wellhead [crude oil (resource)]	1.07 kg
Raw natural gas (BUWAL) [natural gas (resource)]	0.576 kg
Primary energy from hydropower (BUWAL) (renewable energy resources)	0.5 MJ
Raw hard coal (BUWAL) [hard coal (resource)]	0.13 kg
Raw brown coal (BUWAL) [lignite (resource)]	0.11 kg
Process and cooling water [operating materials]	0.017 kg
Sodium chloride (rock salt) [nonrenewable resources]	0.0049 kg
Iron ore [nonrenewable resources]	0.0005 kg
Bauxite [nonrenewable resources]	0.0003 kg
Limestone (calcium carbonate) [nonrenewable resources]	0.00025 kg
Quartz sand (silica sand; silicon dioxide) [nonrenewable resources]	2.00×10^{-5} kg
Uranium free ore (BUWAL) [uranium (resource)]	1.60×10^{-6} kg

analysis on the environmental impacts. More sensitive independent variables displaying greater chances of variations from their mean values can be selected for further analyses. A number of scenarios can be evaluated using extreme bounds of independent variables and combinations of their values within feasible ranges.

An example of 1 tonne of epoxy resin production from biomass is shown in Table 5.2. Hexane and nitrogen mass flow rates are the two key input independent variables to the LCIA model. Their values are varied within feasible ranges to examine their combined effects on the GWP estimates from the LCIA model. The base value of GWP is 518.46 kg CO_2 equivalent from 1 tonne of epoxy resin production; 3 kg of hexane and 13 kg of nitrogen mass inputs to the system, for example, achieve a GWP reduction by 25%. As can be seen, the hexane mass flow rate can be varied between 3 kg and 99 kg and the nitrogen mass flow rate between 5 kg and 15 kg. For various combinations of their input values within these ranges, impact potentials can be evaluated. This method is called a scenario analysis. Two extreme scenarios can be created: maximum GWP reduction by 25% and minimum GWP reduction by 8% from the base impact value. A number of other scenarios result in 19% and 14% reductions in GWP. One or both flow rates can be varied at the same time.

Table 5.2   *Hexane and nitrogen mass flow rates for various reductions from the base value of GWP impact.*

Hexane Mass Flow Rate (kg)	Nitrogen Mass Flow Rate (kg)	GWP (100 years) Reduction (%)
3	13	
12	12	
31	10	25
60	7	
78	5	
18	15	
27	14	
37	13	19
46	12	
65	10	
93	7	
46	15	
55	14	
65	13	14
74	12	
93	10	
89	14	8
99	13	

5.1.4 Sensitivity Analysis

To do a sensitivity analysis with respect to an independent parameter, a range is specified and variations in LCI and LCIA are estimated for the range. An example is shown in Figure 5.5. A sewage sludge AD system producing biogas for micro generation (electricity and heat generations) and digested matter for fertilizer production is studied for the sensitivity analysis. The independent variable is the biogas volumetric production rate. All other flow rates are dependent on the biogas volumetric production rate. These include digested matter mass flow rate, electricity and heat generation, etc. The biogas volumetric production rate is varied by ±25% standard deviation from the mean value and the LCIA are examined. The variations in the LCIA from their mean values are shown in Figure 5.5. The plot was generated using the graph:stock:high–low–close (this applies as maximum-minimum-mean) option in the Excel spreadsheet. Note that the values are not shown in absolute terms, but in relative terms. Thus, the base value is taken as 100; the mean value is marked at ~100. The minimum and maximum values are thus below and above 100, respectively. The difference between maximum and mean values is the positive standard deviation from the mean value. The difference between minimum and mean values is the negative standard deviation from the mean value. The difference between maximum and minimum values is defined as the range.

Impact categories include the acidification potential (AP) in kg SO_2 equivalent, eutrophication potential (EP) in kg phosphate equivalent, freshwater aquatic ecotoxicity potential (FAETP) in kg 1,4-dichlorobenzene (DCB) equivalent, global warming potential (GWP) in kg CO_2 equivalent, human toxicity potential (HTP) in kg DCB equivalent, marine aquatic ecotoxicity potential (MAETP) in kg DCB equivalent, photochemical oxidant creation potential (POCP) in kg ethylene equivalent and terrestrial ecotoxicity potential (TETP) in kg DCB equivalent.

The maximum range is obtained for the POCP, 38.2%, from 80.92 to 119.12. Its standard deviation is thus ±19.1% from its mean value. The high–low–close stock graphical choice in the Excel spreadsheet was used to create this range: 119.12 (high)–80.92 (low)–100.02 (close). This graph shows the sensitive impact categories with greater ranges. Thus, POCP is the most sensitive and MAETP is the least sensitive impact category. MAETP is the least sensitive impact

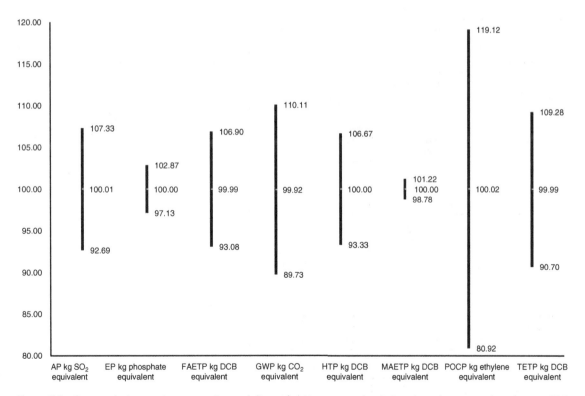

Figure 5.5 *Sewage sludge to micro generation and digested matter: range of variations from the mean values due to ±25% standard deviation in the independent variable (biogas volumetric production rate).*

category due to its very large value and narrow range. The most to the least sensitive life cycle impact categories are as follows:

POCP > GWP > TETP > AP > FAETP > EP > MAETP

5.1.5   Monte Carlo Simulation

Sensitivity analysis can be undertaken for a multiparametric decision making problem using a Monte Carlo simulation combined LCA (MCLCA) approach. With MCLCA important impact characterizations can be selected and optimized to make a choice between various technologies. In Monte Carlo simulation, values of independent variables within their specified standard deviations from their mean values can be randomly selected during a simulation run. All the primary impact characterizations are calculated for the selected set of values of independent variables. At the end of all Monte Carlo simulation runs, the chances or probabilities of occurrence at various values of an impact characterization are counted and plotted (on the *y*-axis) against percentage standard deviations from mean value of the impact characterization (on the *x*-axis). This results in the probability distribution curve of an impact category. Highly sensitive categories show wider probability distributions. Narrow probability distribution curves imply less sensitive categories.

Intergovernmental Panel on Climate Change (IPCC) Guidelines for National Greenhouse Gas Inventories; Geneva, Switzerland, 2006, recommended a Monte Carlo simulation to estimate and mitigate uncertainty in impact assessments[2]. The methodology can be adapted according to the LCA goal and scope definition. Figure 5.6 shows the steps involved in an uncertainty analysis using a Monte Carlo simulation. The MCLCA approach consists of three main steps:

1.   Develop equations or models for the LCI in terms of independent and uncertain input variables to the model.
2.   Select a standard deviation and probability distribution function for each independent variable.

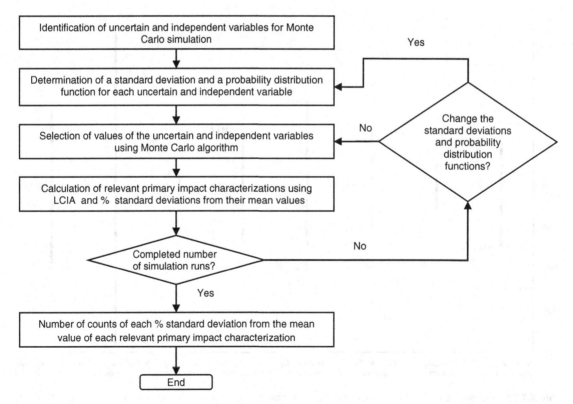

Figure 5.6   *Monte Carlo simulation framework integrated with LCA.*

3. Monte Carlo simulation run. Using the Monte Carlo algorithm (for random number generation), a set of values of independent variables within their given distributions is randomly selected. Determine the LCIA. At the end of a large number of specified simulation runs, count the occurrence of an impact characterization at the estimated values.

4. Repeat the Monte Carlo simulation runs, steps 2 to 3, until enough number of runs is completed for obtaining smooth distribution trends of the LCIA. Calculate the mean value and standard deviations from the mean value for each impact category.

5. Count the chances of occurrence of each model predicted impact characterization by the % standard deviation from their mean values. Ensure that there are enough simulation runs to obtain a smooth and representative probability distribution curve for an impact category.

LCA is data intensive. A dispersed data set makes MCLCA computationally intensive. There lies uncertainty in primary raw material and energy flow data assimilation and in an inventory analysis. A large number of Monte Carlo simulation runs ensures that the approximation can be made more accurately. Monte Carlo simulation runs of ~5000 are recommended in the IPCC Guidelines[2].

The equation below shows the formula for calculating the standard deviation, σ, of n data points: $x_1, x_2, x_3, \ldots x_{n-1}, x_n$, with respect to their average or mean value, $\bar{x}$:

$$\sigma = \sqrt{\frac{\sum_i (x_i - \bar{x})^2}{n}} \tag{5.1}$$

The values of independent variables are generated using their given probability distribution functions. The simplest form of probability distribution function is the uniform probability distribution function. Three other most common forms of probability distribution functions are the normal or Gaussian, lognormal and triangular.

Equations (5.2a) to (5.2c) below show their respective correlations in terms of mean, standard deviation and values of the variable (x_i):

Normal or Gaussian probability distribution function =

$$f(x_i) = \frac{1}{\sqrt{2\pi\sigma^2}} \exp\left(-\frac{(x_i - \bar{x})^2}{2\sigma^2}\right)^\pi \tag{5.2a}$$

Lognormal probability distribution function =

$$f(x_i) = \frac{1}{x_i\sqrt{2\pi\sigma^2}} \exp\left(-\frac{(\ln x_i - \bar{x})^2}{2\sigma^2}\right)^\pi \tag{5.2b}$$

Triangular probability distribution function =

$$f(x_i) = \begin{cases} 0 (x_i \leq a; x_i \geq c) \\ \dfrac{2(x_i - a)}{(b-a)(c-a)} (a < x_i < b) \\ \dfrac{2(c - x_i)}{(c-a)(c-b)} (b \leq x_i < c) \end{cases} \tag{5.2c}$$

where

a = minimum value of x
b = modal value of x
c = maximum value of x

A sewage sludge AD plant can be operated to maximize the output energy generation via biogas production. An AD plant also coproduces digested matter for agricultural application. The biogas and digested matter yields are related by the mass balance for a given sewage sludge mass throughput through an AD plant. As the biogas yield increases, the energy generation increases and digested matter yield decreases, lowering the fertilizer production rate and vice versa. Hence, one

Parameter

Parameter	Formula	/	Value	Minimt Maxim	Standard deviation	Comm
bgf			315	0 %		
dgm	1000-300/315*bgf		700			
ef	3254*bgf/315		3.25E003			
mf	9761.72439914604*bgf/315		9.76E003			
Parameter						

🐾 **LCA**	🛒 **LCC: 0 EUR**	🏦 **LCWE**	📄 **Documentation**

Completeness | No statement   ▾

Inputs

Parameter	Flow	Quantity	Amount	Factor	Unit	Tracked flows /
bgf	➶ biogasflow [Resources]	⚖ Standard volume	315	1	Nm3	X
dgm	➶ digestedmatterflow [Resources]	⚖ Mass	700	1	kg	X
ef	➶ electricityflow [Resources]	⚖ Energy (net calorific value)	3.25E003	1	MJ	X
mf	➶ methaneflow [Resources]	⚖ Energy (net calorific value)	9.76E003	1	MJ	X
	Flow					

Figure 5.7   *Screenshot from GaBi, to define the independent and dependent parameters involved in an AD plant LCA study.*

of the mass yields of biogas and digested matter can be considered as an independent variable and the other can be shown as a function of the independent variable. The following example and data analysis show a systematic decision making about the transfer coefficient of sewage sludge for energy generation an against agricultural application. The problem can be formulated for sensitivity analysis, Monte Carlo simulations, etc., and solved in a spreadsheet environment or any other software supporting such analyses.

Figure 5.7 shows a screenshot from GaBi, LCA software from PE International, to define the independent and dependent variables of an AD plant. Independent variables do not have any formula, while the dependent variables are formulated in terms of independent variables. In the figure, the biogas volumetric flow rate, presented by the symbol bgf, is shown as the independent variable carrying no formula in the formula bar. The dependent variables: digested matter mass flow rate, electricity generation and methane calorific value, using the symbols dgm, ef and mf, respectively, are shown as a function of bgf. The correlations are linear with respect to bgf. A Monte Carlo simulation is then undertaken with:

(a)   5000 runs,
(b)   ±25% standard deviations and normal distributions in bgf.

The probability distribution of the global warming potential impact with respect to its mean value in ±100% standard deviation scale is shown in Figure 5.8. For example, the probability of the global warming potential occurring at the

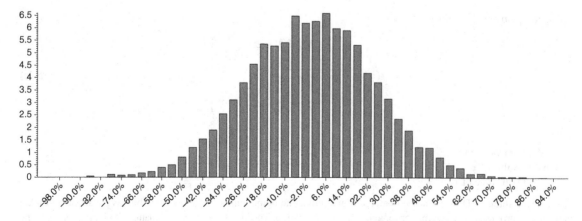

Figure 5.8   *Probability distribution of the global warming potential impact in ±100% standard deviation scale from the mean.*

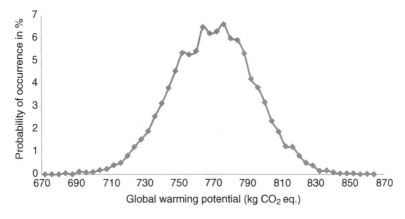

Figure 5.9 *Probability distribution against the global warming potential impact values.*

mean value is ~6.3%. The peak probability, ~6.6%, occurs at the standard deviation of ±6%. The summation of all the probabilities across the ±100% standard deviation scale is 100. The resulting distribution corresponds to the normal distribution.

It is also possible to generate the data points or actual impact values from a probability distribution curve in Figure 5.8, using Equation (5.1). The probability distribution against the actual impact values can then be plotted, such as in Figure 5.9. The mean global warming potential is 768 kg CO_2 equivalent. The number of data points or clusters is 49.

The probability distributions of FAETP, HTP and TETP with respect to standard deviations are shown in Figure 5.10. As TETP displays a wider distribution it has a greater chance of change from the mean value. Figure 5.11 similarly shows that the probability of occurrence of MAETP at the mean value is 98% and hence the probability of reduction (or increase) in the MAETP mean value is only 2%. This is due to the MAETP's very high absolute mean value in the order of 10^4 magnitudes in kg DCB equivalent. EP also shows a narrow distribution, but it is wider than that of MAETP. This analysis thereby helps to screen out the most sensitive set of impact categories for further investigation. For the given example, the GWP showing wider distribution is the most sensitive impact category. The MAETP showing the narrowest

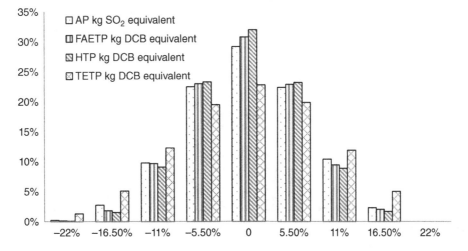

Figure 5.10 *Acidification, freshwater aquatic ecotoxicity, human toxicity and terrestrial ecotoxicity potentials (AP, FAETP, HTP and TETP) impact probability distributions (y axis) with respect to standard deviations (x axis) from their mean values.*

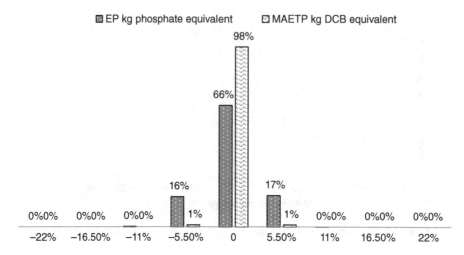

Figure 5.11   *Probability distributions (y axis) of marine aquatic ecotoxicity potential (MAETP) and eutrophication potential (EP) with respect to standard deviations (x axis) from their mean values.*

distribution is the least sensitive impact category. The least sensitive categories also imply that they will not be affected by model uncertainties.

Figure 5.12 shows the probability distribution versus the standard deviation from the mean value of various impact categories using GaBi software.

The recommended specifications for the Monte Carlo simulation are:

(a)   5000 runs
(b)   Normal distributions and standard deviations by ±25% of independent variables
(c)   Capture ±100% standard deviations scale for the impact characterizations.

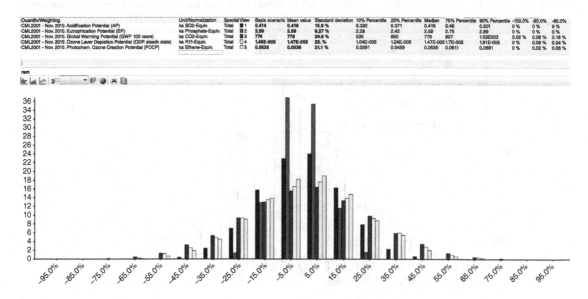

Figure 5.12   *Probability distribution versus standard deviation from the mean value of various impact categories using GaBi software.*

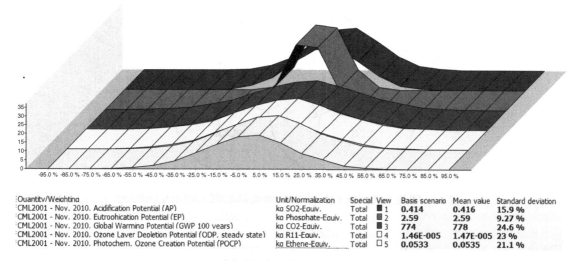

Quantity/Weighting	Unit/Normalization	Special	View	Basis scenario	Mean value	Standard deviation
CML2001 - Nov. 2010. Acidification Potential (AP)	kg SO2-Equiv.	Total	■ 1	0.414	0.416	15.9 %
CML2001 - Nov. 2010. Eutrophication Potential (EP)	kg Phosphate-Equiv.	Total	■ 2	2.59	2.59	9.27 %
CML2001 - Nov. 2010. Global Warming Potential (GWP 100 years)	kg CO2-Equiv.	Total	■ 3	774	778	24.6 %
CML2001 - Nov. 2010. Ozone Layer Depletion Potential (ODP, steady state)	kg R11-Equiv.	Total	☐ 4	1.46E-005	1.47E-005	23 %
CML2001 - Nov. 2010. Photochem. Ozone Creation Potential (POCP)	kg Ethene-Equiv.	Total	☐ 5	0.0533	0.0535	21.1 %

Figure 5.13   *3D plots of probability distributions of impact characterizations given in Figure 5.12.*

Figure 5.13 shows the 3D plots of probability distributions of impact characterizations given in Figure 5.12.

The following list is the summary of MCLCA approach:

- Standard deviations from mean values and nature of distributions are specified for independent variables.
- Values of independent variables can be randomly selected within the given specifications during a simulation run.
- All the primary impact characterizations are calculated for the randomly selected set of values of independent variables.
- Another set of values of independent variables are randomly selected within their specified ranges. This is called a Monte Carlo simulation run. Several runs (~5000) are repeated.
- The total number of runs is specified and the above steps are repeated until all the runs are completed.
- At the end of all Monte Carlo simulation runs, the probability distribution of each impact characterization for various percentage standard deviations from the mean value is counted. The impact characterizations that can be reduced by adjusting values of independent variables show a wider probability distribution and vice versa.

5.2   Multicriteria Analysis

Sustainable development calls for a multicriteria analysis, including social, economic and environmental impact assessments. While LCA is a tool for environmental sustainability analysis, social and economic impacts can also be assessed over life cycles. These are called social LCA (SLCA) and life cycle cost (LCC), respectively. Similar to LCA, SLCA and LCC show corresponding hot spots and ways of mitigation. The hot spots can span across the time scale (life cycle) as well as geographic regions (supply chains). The SLCA and LCC can be applied in the same way as LCA, for accounting (consequential) and change oriented (attributional) systems, discussed in Chapter 4.

Figure 5.14 shows the desirable domain for multicriteria analysis combining LCA, SLCA and LCC tools. Table 5.3 shows the various SLCA categories. The analysis proceeds in the same way, from primary through mid-point to end-point impacts. The analysis is to help decision making about sustainable supply chains by eliminating hot spots and mitigating potential negative or rebound impacts.

SLCA, an evolving tool, is discussed in 2009 UNEP/SETAC *Guidelines for Social Life Cycle Assessment* for SLCA; http://socialhotspot.org/user-portal-2/portal-info also "offers an online database that allows users to browse data on social risks by sector, country, or risk theme. There are choices of 227 countries and 57 economic sectors. The data

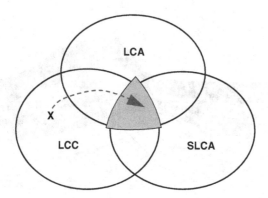

Figure 5.14   *Multicriteria analysis combining LCA, SLCA and LCC tools.*

comprehensively addresses social issues on human rights, working conditions, community impacts and governance issues, via a set of nearly 150 risk indicators grouped within 22 themes. Risks are also expressed, whenever relevant, by country and sector." LCC is implemented using a net present value and discounted cash flow analysis; this is discussed in Chapters 2, 6 and 7.

5.2.1   Economic Value and Environmental Impact Analysis of Biorefinery Systems

The biorefinery system shown in Figure 5.15 has biomass production, product manufacturing, end use and construction materials' life cycle process blocks. The two commonly used system boundaries include cradle to grave and cradle to gate with and without the carbon dioxide sequestration or capture by biomass and biorefinery products' end use blocks, respectively. Each block causes primary resource depletion (input) and emission impacts (output) that are accounted for in the LCA.

The contribution analysis of a biorefinery system shows typical characteristics in the environmental impact against an added cost profile. Raw materials, energy, raw materials for plant installations (capital good raw materials) and emissions are the four main loads interacting between biorefinery systems and the environment. Their environmental impacts decrease in the following sequence:

Emissions > Raw materials > Energy (provided that the energy required by the biorefinery systems can be supplied from renewable sources) > Capital good raw materials

Table 5.3   *SLCA categories.*

Labor Rights	Health and Safety	Human Rights	Governance	Community Infrastructure
1. Child labor	1. Injuries	1. Indigenous rights	1. Legal systems	1. Medical facilities
2. Forced labor	2. Toxics	2. Conflicts	2. Corruption	2. Drinking water
3. Excessive working time	3. Hazards	3. Gender equity		3. Sanitation
4. Wage assessment		4. Human health		4. Children education
5. Poverty				
6. Migrant labor				
7. Freedom of association				
8. Unemployment				
9. Labor laws				

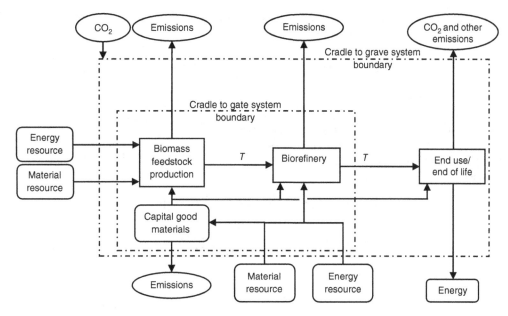

Figure 5.15 *A biorefinery system has biomass production, product manufacturing, end use and construction materials' life cycle process blocks. T shows where the transportation of materials is involved.*

The sequence for annualized costs is different from the sequence for environmental impacts and is shown as follows:

Biomass feedstock > Energy > Raw materials (excluding biomass feedstock) > Capital good raw materials > Emissions

These observations are shown by a generic plot of the environmental impact versus annual added cost of biorefinery systems in Figure 5.16. The cradle to gate biorefinery system has four main loads to analyze for LCA, as follows, shown by numbers 1 to 4 in the figure:

1. Raw materials
2. Energy
3. Capital good raw materials
4. Emissions

Figure 5.16 shows a horizontal line for higher emissions from a fossil based equivalent conversion system that the biorefinery system is designed to replace. The environmental impact from the biorefinery system is shown by the *y* axis value of point 4. The added cost of production is shown by the *x* axis value of point 3 or 4 (assuming that emissions do not add cost to the plant). A biorefinery cradle to gate plant may prove to be unsustainable if the target for emission reduction is higher than its current level of emission reduction and if the market price of its products is lower than their added cost of production.

The following lines are thus shown in Figure 5.16:

- Emission from fossil based equivalent product
- Emission reduction target by policy
- Cost of production
- Market price of product

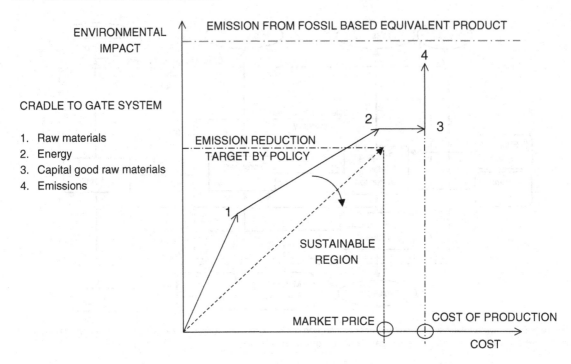

Figure 5.16 *Added cost and environmental impact (contribution analysis) of biorefinery cradle to gate systems.*

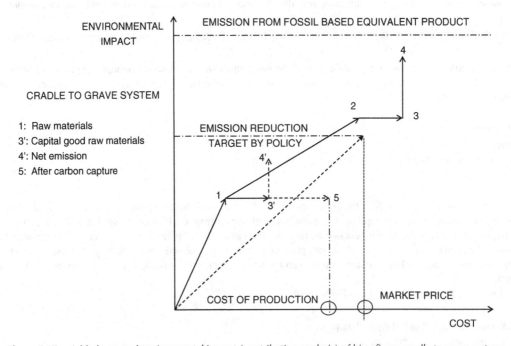

Figure 5.17 *Added cost and environmental impact (contribution analysis) of biorefinery cradle to grave systems.*

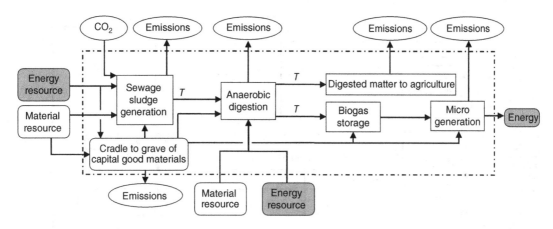

Figure 5.18 *Cradle to grave anaerobic digestion of sewage sludge system for LCA. T stands for transport.*

The target will be to operate the plant below the diagonal and within the lower triangle of the rectangle created by the horizontal line for the "Emission reduction target by policy" and the vertical line for the "Market price" of the product.

A cradle to grave biorefinery system analysis can show a reduction in emissions due to capture of carbon dioxide during biomass production. In addition, the energy balance over the entire cradle to grave system should show net energy production rather than consumption via biomass exploitation. Thus, line 1–2 does not exist in the cradle to grave biorefinery system's added economic value and environmental impact (contribution analysis) profile in Figure 5.17. Line 1–3′ shows the capital good raw material. Line 3′–4′ shows the new net emission after carbon capture (by biomass during production), reduced from line 3–4 for the cradle to gate system. If the entire emission is captured by added investment, line 3′–5 can be created. Therefore, any emission reduction will be accomplished by added costs. The cradle to grave biorefinery system operating at points 4′ and 5, lower than the emission reduction target and product market price, is sustainable.

Further, see the combined economic value and environmental impact analysis calculations in Chapter 7.

Further Challenge Exercise 1. Draw the added cost and environmental impact (contribution analysis) profiles of the following systems in Figures 5.18 to 5.20. Develop hypotheses as necessary.

5.2.2 Socioeconomic Analysis

A socioeconomic analysis is done to estimate the number of job creations and any challenges and barriers associated with an industrial activity. The social performance needs to be compared with other similar systems, in order for an industrial system to operate over the long term. The IChemE provides a metrics to organize the socioeconomic data in the categories of workplace and society:

http://nbis.org/nbisresources/metrics/triple_bottom_line_indicators_process_industries.pdf

These are adaptable to biorefinery systems analysis. Furthermore, the following categories are recommended for biorefinery supply chains.

1. Employment and social well-being: average, highest paid and lowest paid indicative wages and benefit packages
2. Profit as a percentage of payroll expenses
3. External trade
4. Energy security
5. Resource conservation
6. Social acceptability

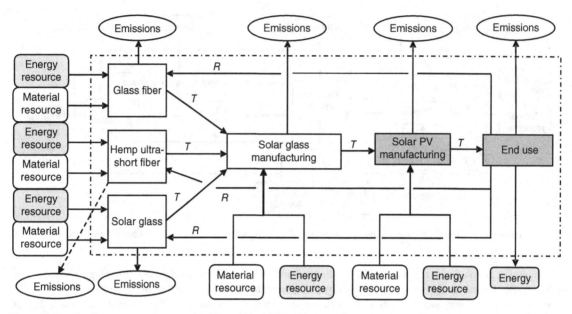

Figure 5.19   *Cradle to grave solar organic photovoltaic (OPV) glass manufacturing system for LCA. T stands for transport. R stands for recycle. T and R also cause resource depletion and environmental emissions. Only a part of "Solar PV manufacturing" and "End use" block impacts due to solar glass manufacturing needs to be considered within the system boundary.*

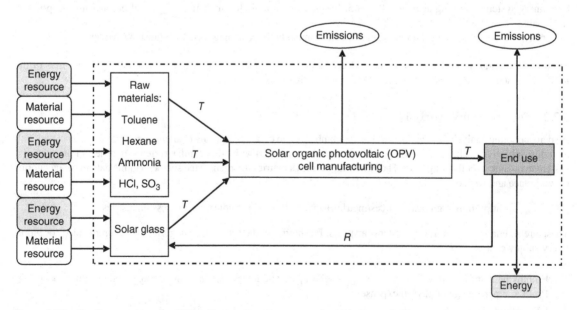

Figure 5.20   *Cradle to grave solar OPV cell manufacturing system for LCA. T stands for transport. R stands for recycle. T and R also cause resource depletion and environmental emissions. Only a part of "End use" block impacts due to solar cell manufacturing needs to be considered within the system boundary.*

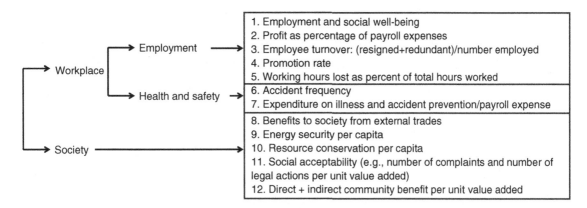

Figure 5.21 *Recommended socioeconomic indicators for biorefinery systems.*

Figure 5.21 shows the recommended structure of the socioeconomic indicators for biorefinery systems.

5.3 Summary

This chapter shows important ways to present LCA results, such that interpretation is accessible. The chapter also outlines multicriteria methods, as well as the incorporation of technoeconomics and socioeconomics in the analysis. Biorefinery sustainability must be assessed using LCA, technoeconomics and socioeconomics. Data uncertainty analysis, sensitivity analysis and Monte Carlo simulations are essential to minimize errors in estimation, to find more important indicators, activity and inventory hot spots. Though these tools are shown to apply for LCA results interpretation, these can also be used for technoeconomic and socioeconomic analyses.

References

1. E. Martinez-Hernandez, M.H. Ibrahim, M. Leach, P. Sinclair, G.M. Campbell, J. Sadhukhan, Environmental sustainability analysis of UK whole-wheat bioethanol and CHP systems, *Biomass Bioenergy*, **50**, 52–64 (2013).
2. *IPCC Guidelines for National Greenhouse Gas Inventories*, Intergovernmental Panel on Climate Change, Geneva, Switzerland, 2006.

6

Value Analysis

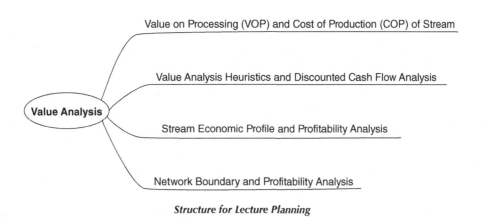

Structure for Lecture Planning

Enhancing flexibility in product portfolios and conversion process configurations depending upon market price and supply–demand variations and environmental regulations should be addressed in the early design phases in conceptual and feasibility studies. Flexible designs of large industrial systems where multiple feedstocks, products and utilities are involved impose large-scale combinatorial problems. The value analysis methodology[1-3] shown in this chapter is used to solve such combinatorial process optimization problems based on economic margins, without mathematical programming. It evaluates economic margins of individual streams and processing pathways in a process network, such that aggregation of their individual economic margins presents the margin of the overall network. Thus, maximizing positive economic margins (profits) and minimizing negative economic margins (losses) of individual pathways ensures maximum profitability of their integrated network.

The value analysis methodology is to answer to very fundamental questions about process network systems. What are the desired products and production routes from a set of feedstocks identified? What is the feedstock mix for maximum value added productions through a given process network? A designer can use the value analysis methodology to decide upon feedstock–process–product combinations and configurations for manufacturing systems based on economic variables, such as market prices, supply and demand, within regulatory constraints.

Biorefineries and Chemical Processes: Design, Integration and Sustainability Analysis, First Edition.
Jhuma Sadhukhan, Kok Siew Ng and Elias Martinez Hernandez.
© 2014 John Wiley & Sons, Ltd. Published 2014 by John Wiley & Sons, Ltd.
Companion Website: http://www.wiley.com/go/sadhukhan/biorefineries

6.1 Value on Processing (VOP) and Cost of Production (COP) of Process Network Streams

Any material stream in a process network can be evaluated in terms of its *value on processing* (VOP) and *cost of production* (COP). The difference between the two, VOP – COP, provides its economic margin.

The evaluation of COP of a stream starts from the known market price of feedstocks, added with the costs incurred from its production and thus proceeds in the forward direction until end product evaluations. The COP of a stream is the summation of all associated cost components (i.e., the costs of feedstocks, utilities, operating and annualized capital costs) that have contributed to the production of that stream up to that point. This must mean inclusion of costs involved with the stream's production.

The VOP evaluation proceeds in the backward direction. From known end product market prices, the VOP of *all* intermediate streams and feedstocks to a process network is evaluated. The VOP of a stream at a point within the process is obtained from the prices of products that will ultimately be produced from it, minus the costs of auxiliary raw materials and utilities and the annualized capital cost of equipment that will contribute to its further processing into these final products.

Thus, the COP of a feedstock to a process flowsheet is its market price and the VOP of an end product from a process flowsheet is its market price. The market dependent economic parameters that are correlated to the VOP and COP of a stream are the known prices of feedstocks, products and utility streams and estimated annualized capital and operating costs associated with its production (included in the COP evaluation) and that associated with its processing (included in the VOP evaluation).

To establish the relationship between the VOP and COP of streams and economic margin of process networks, consider a single process unit in Figure 6.1. The economic margin (Δ) from the unit is the difference between the unit market price (P) of its product multiplied by the product's flow rate (m^P) and the unit market price (F) of the feedstock multiplied by its flow rate (m^F) and the annualized cost (operating plus capital) of the unit (O), given in Equation (6.1).

Assuming an annual basis:

Economic margin = $\sum_{All\,products}$ Unit market price of product × Flow rate of product – $\sum_{All\,feedstocks}$ Unit market price of feedstock × Flow rate of feedstock – Annualized operating costs – Annualized capital cost

The economic margin of a process unit with single feedstock and single product using symbols shown in Figure 6.1 is as follows:

$$\Delta = P \times m^P - (F \times m^F + O) \tag{6.1}$$

The following equation shows a scenario of the COP of a product when $\Delta = 0$:

$$(P \times m^P)_{\Delta=0} = F \times m^F + O \tag{6.2}$$

The economic margin is thus expressed in terms of the COP, $((P)_{COP})$, and the market price of the product (Equation (6.3)) and the COP of the product per unit flow rate is calculated using Equation (6.4):

$$\Delta = P \times m^P - (P \times m^P)_{\Delta=0}$$
$$= \{P - (P)_{COP}\} \times m^P \tag{6.3}$$
$$(P)_{COP} = F + \hat{o} \tag{6.4}$$

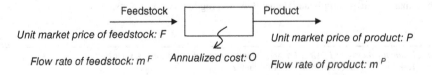

Figure 6.1   *A process unit with single feedstock and single product.*

where

$\hat{o}$ = Annualized cost (operating plus capital) per unit flow rate of the feedstock = O/m^F

Note that the COP of a feedstock is its market price. Hence, the profit of a process unit is the multiplication of the difference between the VOP and COP with the flow rate of its *total or complete* (see Section 6.2) product stream.

Similarly, using the market price of the product, the VOP of the feed, $(F)_{VOP}$, is correlated to the profit of the unit as follows:

$$(F \times m^F)_{\Delta=0} = P \times m^P - O$$

$$\Delta = (F \times m^F)_{\Delta=0} - F \times m^F = \{(F)_{VOP} - F\} \times m^F \tag{6.5}$$

The expression for the VOP of a feedstock per unit flow rate is given by

$$(F)_{VOP} = F_{\Delta=0} = P - \hat{o} \tag{6.6}$$

Analogically, the profit of a process unit is obtained by multiplying the difference between the VOP and COP with the flow rate of its *total or complete* feedstock.

Figure 6.2 shows the VOP, COP and economic margin evaluations of streams in a *Jatropha* biodiesel production process. Note that the annualized capital cost has been taken into account.

The stream exiting the dehusker has a COP of $297.2\ t^{-1}$ based on a market price for the incoming seeds of $296.3\ t^{-1}$, to which is added the cost of the utilities and annualized capital:

$$\frac{296.3 \times 271200 + 248166}{271200} = \$297.2\ t^{-1}$$

Similarly, working backwards from the end products, the VOP of the stream entering the biodiesel distillation is $637.4\ t^{-1}$. This is the value from biodiesel and oily waste streams:

$$\frac{675.0 \times 100000 + (-0.390) \times 5300 - 377773}{105300} = \$637.4\ t^{-1}$$

Returning to the stream exiting the dehusker, its VOP is $462.9\ t^{-1}$, as a result of subtracting the accumulated costs of further processing from the value of the end products. The margin of this stream is $462.9 - 297.2 = \$165.7\ t^{-1}$, indicating that its processing is profitable.

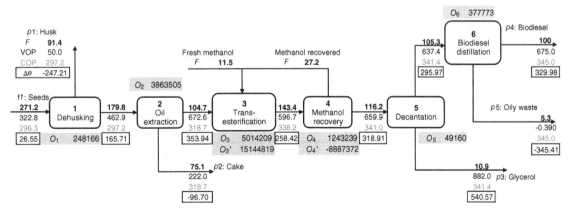

Figure 6.2   *Mass flow rates (F) in kt y^{-1}, VOP, COP and economic margins (Δe) in $ t^{-1} of the streams in a Jatropha based biorefinery. The operating costs O_i are in $ y^{-1}.*

A stream showing a negative margin would imply that it would be better (if possible) to purchase that stream from the market rather than produce it within the process.

$$\text{COP of biodiesel} = \frac{341.4 \times 105300 + 377773}{105300} = \$345.0 \, t^{-1}$$

$$\text{Margin of biodiesel stream} = 675 - 345 = \$330.0 \, t^{-1}$$

The annualized operating and capital costs of the transesterification and methanol recovery units are shown by O_3 and O_4. The calculations in Figure 6.2 are shown when the cost of recycled methanol is added to O_3 (providing O_3') and subtracted from O_4 (providing O_4').

Exercise 1. An oxygen-blown and pressurized gasification unit in Figure 6.3 is used to convert 1 kmol s^{-1} of wood-derived bio-oil, with the composition and higher heating value (HHV) given in Table 6.1, into product gas with the composition provided in Table 6.2. A 2.6 kmol s^{-1} quantity of air with a molar composition of nitrogen:oxygen = 0.79:0.21 is used in the air separation unit to supply the stoichiometric amount of oxygen required for the gasification reactions. Assume complete separation of air. The capital and operating cost data are given in Table 6.3. Assume 8000 operating hours per year and an annualized capital charge of 13%. The capital cost indexes for 1999 and 2009 are 390.6 and 524.2, respectively.

a.   Calculate the cost of production (COP) of the product gas in $ per kmol of bio-oil for its market price of $75 per t.
b.   Estimate the economic margin in $ per kmol of bio-oil from the product gas for its market price of $120 per t.
c.   Show that the economic margins calculated for products and feedstocks individually are equal to the economic margin of the flowsheet.

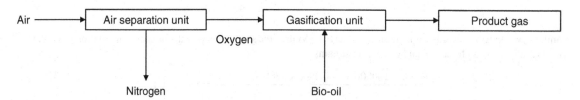

Figure 6.3   *Block flow diagram of bio-oil gasification into product gas.*

Table 6.1   *Bio-oil components, with mole fraction, molar mass and HHV.*

Bio-oil Components	Mole Fraction	Molar Mass (g mol^{-1})
Acetic acid	0.109	60
Acetol	0.109	74
Guaiacol	0.109	124
Water	0.673	18
HHV (MJ kg^{-1})	18	

Table 6.2   *Product gas composition.*

Product Components	Mole Fraction	Molar Mass (g mol^{-1})
Hydrogen	0.305	2
Water	0.253	18
Carbon monoxide	0.320	28
Carbon dioxide	0.122	44

Table 6.3   *Capital and operating costs of the gasification and air separation units.*

Process Unit	Base Cost (million $, 1999)	Scale Factor	Base Scale	Electricity Consumption	High Pressure Steam Consumption
Gasification	30	0.7	400 MW HHV	–	–
Air separation unit	23	0.75	24 tonne oxygen per hour	14696 kW	5.89 kg per second
Utility price				$0.07 per kW h	$16 per t

Solution to Exercise 1

Gasifier

The mass flow rate to be processed

$$= (0.109 \times 60 + 0.109 \times 74 + 0.109 \times 124 + 0.673 \times 18) \times 1 = 40.24 \text{ kg s}^{-1}$$

The scale of the gasifier $= 18 \times 40.24 = 724.25$ MW

The annualized capital charge of the gasifier

$$= \left(\frac{724.25}{400}\right)^{0.7} \times 30 \times 10^6 \times \frac{524.2}{390.6} \times 0.13 \times \frac{1}{8000 \times 3600} = \$0.275 \text{ s}^{-1} = \$0.275 \text{ kmol}^{-1} \text{ of bio-oil}$$

Air separation unit (ASU)

The scale of the ASU = stoichiometric amount of oxygen required

$$= 2.6 \times 0.21 \times \frac{32}{1000} \times 3600 \times \frac{8000}{365 \times 24} = 57.44 \text{ t h}^{-1}$$

$8000/(365 \times 24) = 0.913$ means that 91.3% of the time, the plant is operational.

The annualized capital charge of the ASU

$$= \left(\frac{57.44}{24}\right)^{0.75} \times 23 \times 10^6 \times \frac{524.2}{390.6} \times 0.13 \times \frac{1}{8000 \times 3600} = \$0.27 \text{ s}^{-1} = \$0.27 \text{ kmol}^{-1} \text{ of bio-oil}$$

Operating cost of ASU

Steam cost $= 5.89 \times \dfrac{16}{1000} = \$0.09 \text{ s}^{-1} = \$0.09 \text{ kmol}^{-1}$ of bio-oil

Electricity cost $= 14696 \times 0.07 \times \dfrac{1}{3600} \times \dfrac{365 \times 24}{8000} = \$0.31 \text{ s}^{-1} = \$0.31 \text{ kmol}^{-1}$ of bio-oil

Feedstock and product prices

Bio-oil (at a price of $75 t^{-1}) $= 75 \times \dfrac{40.24}{1000} = \$3.02 \text{ s}^{-1} = \$3.02 \text{ kmol}^{-1}$ of bio-oil

Air and nitrogen: 0 (assuming no end use of nitrogen)

Mass flow rate of the product gas $= 40.24 \times 1 + \dfrac{32 \times 2.6 \times 0.21}{1000} = 0.0577 \text{ t s}^{-1}$

VOP of the product gas (at a price of $120 t^{-1}) $= 120 \times 0.0577 = \$6.9 \text{ s}^{-1} = \$6.9 \text{ kmol}^{-1}$ of bio-oil

Molar mass of the product gas $= 0.305 \times 2 + 0.253 \times 18 + 0.32 \times 28 + 0.122 \times 44 = 19.49 \text{ kg kmol}^{-1}$

The market price of the product gas per kmol of bio-oil $= \$120 \times \dfrac{19.49}{1000} = \$2.34 \text{ kmol}^{-1}$

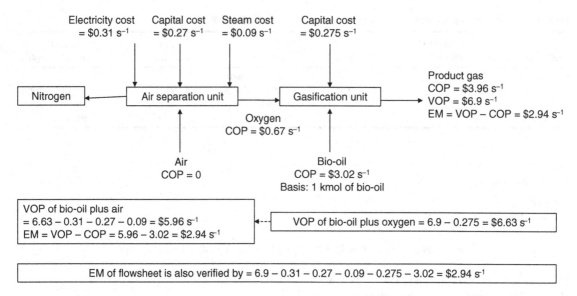

Figure 6.4   *Value analysis and economic margin estimation of the bio-oil gasification flowsheet in Figure 6.3.*

Value analysis in \$ kmol⁻¹ of bio-oil
a.   COP of the product gas $= 0.275 + 0.27 + 0.09 + 0.31 + 3.02 = \$3.96\ s^{-1} = \$3.96\ kmol^{-1}$ of bio-oil
b.   The economic margin (EM) from the product gas $= 6.9 - 3.96 = \$2.94\ kmol^{-1}$ of bio-oil
c.   The product EM, feedstock EM and the flowsheet EM are estimated and equal to $\$2.94\ s^{-1}$ or $\$2.94\ kmol^{-1}$ of bio-oil, as shown in Figure 6.4.

6.2   Value Analysis Heuristics

The following heuristics are developed for the value analysis of a process network. The foundation of process networks is at basic tree and path levels (elements) comprising streams and process units, such that the total marginal contributions of the basic elements in a process network equate to the margin of the overall network. A network example in Figure 6.5 is decomposed into basic elements, paths and trees.

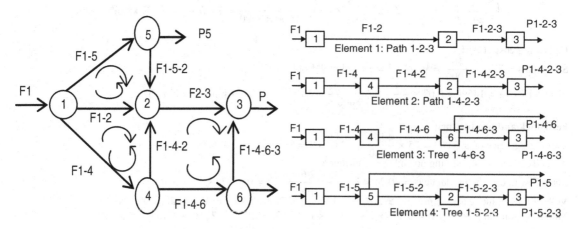

Figure 6.5   *Decomposition of a process network into paths and trees.*

At first, individual streams in basic elements in a process network are evaluated in terms of their *value on processing* (VOP) and *cost of production* (COP) based on a given network configuration, internal and external (market) constraints. The difference between the two (VOP – COP) of a stream provides its marginal contribution.

Next, a *complete* stream in each basic element is identified. The marginal contribution of a *complete* stream in an element is also the marginal contribution of the element.

A *complete* stream in an element, for which no branching stream is available in any other element and the flow rate is equal to the input or output flow rate through its element, is identified. A *complete* stream should satisfy two criteria: (a) it must appear only once in one of the basic elements (paths or trees) in a process network and (b) its flow rate is equal to the input or output flow rate of its element. In the network example in Figure 6.5, although the streams F1 and F1–4 are meeting the second criterion, they are not the complete streams as they appear in more than one element, F1 in all the four elements and F1–4 in the first two paths, respectively. In the trees 1–4–6–3 and 1–5–2–3, although the streams F1–4–6–3 and F1–5–2, F1–5–2–3, respectively, do not appear in any other path, they are not the complete streams as their individual flow rates are not equal to the throughput through these trees.

Thus, the *complete* streams in the first path in Figure 6.5 are F1–2, F1–2–3 and P1–2–3, in the second path F1–4–2, F1–4–2–3 and P1–4–2–3, in the third tree F1–4–6 and in the fourth tree F1–5.

The economic margin of any such complete stream in an element provides the margin of the element in a network. Once the economic margins of the basic elements in a process network are obtained, their summation provides the margin of the overall network.

The cost components involved in the estimation of COP and VOP of a stream also include the capital investments of associated units in the form of tolling fees. Inclusion of the tolling fees (i.e., annualized capital cost per unit flow rate of the stream) in the network value analysis helps to screen the most profitable investment options amongst various processing routes, especially when expansion of an existing network is involved.

6.2.1 Discounted Cash Flow Analysis

The *discounted cash flow* (DCF) economic model for an industrial plant is to find its *net present value* (NPV) from the cumulative differences between the *DCF* in and the *DCF* out of the plant, since the beginning of its construction up to a given year. *NPV* is a function of the plant's *capital investment* (CI), *internal rate of return* (IRR), *annual capital charge* (ACC) and year (yr). The *ACC* is found iteratively for a given value of *IRR*, from the *DCF* model equated to zero $NPV_{yr=Y}$ = 0 (see Equation set (6.7)), where *Y* is the number of years after plant start-up, when all debts are paid off, and the net cash flow is zero (in other words, *Y* is the *payback time* based on the *DCF*). After year *Y*, the net cash flow will be positive.

CI is a product of CI_i (cost of process unit *i* per unit flow rate of the stream fed) and m_i^F (flow rate of the feed stream to process unit *i*). The tolling fee (TF_i) of a process unit *i* is the *ACC* times CI_i. Thus, TF_i can be easily included in the COP and VOP correlation of a stream given in Equations (6.4) and (6.6), respectively. Inclusion of TF_i in the costs takes account of the capital investment in the stream's value analysis.

The *DCF* economic model is as follows

$$NPV_{yr} = \sum_{yr=-Y^c}^{Y} DCF_{yr} = \sum_{yr=-Y^c}^{0} DCF_{yr} + \sum_{yr=0}^{Y} DCF_{yr} = 0 \tag{6.7}$$

For fixed values of *IRR* and *Y*, find the value of ACC that makes Equation (6.7) hold subject to the following constraints:

1. *DCF* in the years during the construction period, $-Y^c \leq yr \leq 0$, assuming year 0 for the plant start-up, when the plant starts operating:

$$DCF_{yr} = -\frac{SCF_{yr} \times CI}{\left(1 + \frac{IRR}{100}\right)^{yr}}$$

where Y^c is the number of years taken to build the plant and SCF_{yr} is the fraction of CI invested in year $= yr$, during the plant construction period ($-Y^c \leq yr \leq 0$). For example, the following parameters can be assumed: $Y^c = 3.25$ years, $SCF_{yr=-3} = 0.1$, $SCF_{yr=-2} = 0.4$, $SCF_{yr=-1} = 0.4$ and $SCF_{yr=0} = 0.1$. Hence, 10%, 40%, 40% and 10% of the total capital investment (CI) are made in the third, second and first years before the plant start-up and the start-up or zeroth year, respectively. Before the plant start-up, the years are counted backwards from its start-up or zeroth year.

2. *DCF* after plant start-up and before NPV is zero, $0 \leq yr \leq Y$ (c.f. Equation (2.4): $C_f =$ the cash flow in a particular year = ACC × CI):

$$DCF_{yr} = \frac{ACC \times CI}{\left(1 + \frac{IRR}{100}\right)^{yr}}$$

3. $CI = \sum_{i \in UNIT} CI_i \times m_i^F$
4. $TF_i = ACC \times CI_i$
5. $0 \leq ACC \leq 1$

ACC is shown as a fraction in the equation shown here, while *IRR* is shown as a percentage.

Figure 6.6 shows an instance of % cash flow with respect to time for given $Y = 10$ years, $IRR = 20\%$ and $ACC = 23.852\%$. Note that the *DCF* curve cuts the positive side of the *x* axis at the tenth year (*payback time* or *Y*). The calculation is iterative with three variables, *ACC*, *IRR* and *Y*. Three payback times, 10 years, 15 years and 25 years, and variations in *IRR* between 10% and 20% are considered in the *DCF* model that results in an *ACC* varying between 11.017% (minimum corresponding to 25 operating years and 10% *IRR*) and 23.852% (maximum corresponding to 10 operating years and 20% *IRR*), shown in the inset in Figure 6.6.

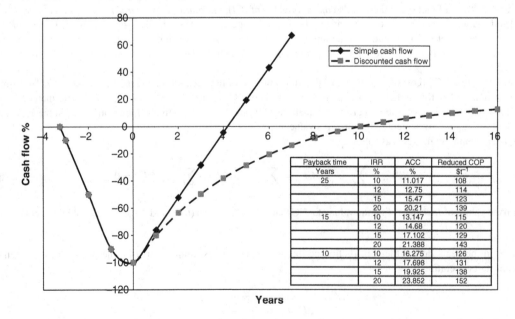

Figure 6.6 *DCF for various investment strategies (see inset); the curve shows an instance of 10 years of payback time.*

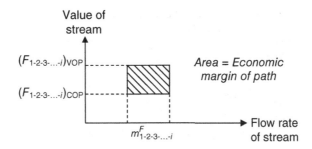

Figure 6.7 *Stream economic profile.*

6.3 Stream Economic Profile

The significance of the difference between the VOP and the COP of a stream is shown in Figure 6.7. The area shaded between VOP and COP of a stream expanded over its flow rate provides the overall margin of its path.

$F_{1-2-3-\cdots-i}$ is the market price of stream F1–2–3–····–i.
$(F_{1-2-3-\cdots-i})_{COP}$ is the COP of stream F1–2–3–····–i.
$(F_{1-2-3-\cdots-i})_{VOP}$ is the VOP of stream F1–2–3–····–i.
$m^F_{1-2-3-\cdots-i}$ is the flow rate of stream F1–2–3–····–i.

6.4 Concept of Boundary and Evaluation of Economic Margin of a Process Network

For a process network, the economic margin of any stream in the network does not represent the total margin of the network. This is because a single stream in a network may not contain the entire mass flow rate through the network. In order to derive the economic margin correlation for a network in terms of the VOP and COP of streams, those streams need to be considered such that summation of their flow rates must be equal to the throughput of the network. In order to do so, boundaries of a process network must be considered. The boundaries with three process units are shown in Figure 6.8.

The boundaries of a network are drawn in such a way that the total inlet flow rate is equal to the total outlet flow rate across a boundary (mass balance) and is individually equal to the total flow rate in or out of the network. In the figure all the boundaries have the inlet stream F1. The intermediate feed F1–2 to unit 2 and the end product P1 from unit 1 are the two outlet streams exiting from boundary 1. Boundary 2 cuts across the two outlet streams P2, F1–2–3, etc. By

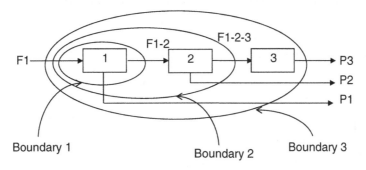

Figure 6.8 *Boundaries shown for a process network.*

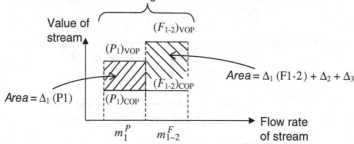

Figure 6.9 *Network economic profile in terms of stream economic profiles across boundary 1 of the process network in Figure 6.8.*

considering different boundaries, the economic margins of different streams can be separately represented. For the outlet streams across boundary 1 (streams F1–2 and P1) the profit function correlations obtained in terms of the VOP of inlet and COP of outlet streams are shown in as follows:

$$\Delta_1(F1\text{--}2) = \left(F_{1-2} \times m_{1-2}^F\right) - \left(F_{1-2} \times m_{1-2}^F\right)_{\Delta_1(F1-2)=0} = \{F_{1-2} - (F_{1-2})_{COP}\} \times m_{1-2}^F \tag{6.8}$$

$$\Delta_1(P1) = P_1 \times m_1^P - \left(P_1 \times m_1^P\right)_{\Delta_1(P1)=0} = \{P_1 - (P_1)_{COP}\} \times m_1^P \tag{6.9}$$

$$\Delta_2 + \Delta_3 = \left(F_{1-2} \times m_{1-2}^F\right)_{\Delta_2+\Delta_3=0} - \left(F_{1-2} \times m_{1-2}^F\right) = \{(F_{1-2})_{VOP} - F_{1-2}\} \times m_{1-2}^F \tag{6.10}$$

where Δ_i is the economic margin of unit i and $\Delta_i(F1 - 2 - 3 - \cdots - i)$ is the economic margin of unit i incurred from stream F1 – 2 – 3 – $\cdots$ – i (denoted as Δe in Figure 6.2 and in Chapter 7).

The summation of the above three equations gives the total margin of the network presented, which is equal to the total economic margins of all the outlet streams across the boundary (given below). This is illustrated in Figure 6.9.

$$\text{Left side: } \Delta_1(F1\text{--}2) + \Delta_1(P1) + \Delta_2 + \Delta_3 = \Delta_1 + \Delta_2 + \Delta_3 \tag{6.11}$$

$$\text{Right side: } \left(F_{1-2} \times m_{1-2}^F\right)_{\Delta_2+\Delta_3=0} - \left(F_{1-2} \times m_{1-2}^F\right)_{\Delta_1(F1-2)=0} + P_1 \times m_1^P - \left(P_1 \times m_1^P\right)_{\Delta_1(P1)=0}$$

$$= \{(F_{1-2})_{\Delta_2+\Delta_3=0} - (F_{1-2})_{\Delta_1(F1-2)=0}\} \times m_{1-2}^F + \{P_1 - (P_1)_{\Delta_1(P1)=0}\} \times m_1^P$$

$$= \{(F_{1-2})_{VOP} - (F_{1-2})_{COP}\} \times m_{1-2}^F + \{(P_1)_{VOP} - (P_1)_{COP}\} \times m_1^P \tag{6.12}$$

The theory that the economic margin of a process network is equal to the total economic margins of all outlet streams across a boundary can be proven across boundary 2 and boundary 3. The boundaries are to check that the value analysis calculations performed on a process network are correct.

6.5 Stream Profitability Analysis

The value analysis tool also helps in planning, scheduling and day-to-day decision making on selling and purchasing strategies for streams. There are 3! arrangements possible among three values of a stream, VOP, COP and market price:

VOP > COP > market price
VOP > market price > COP
Market price > VOP > COP

and so on.

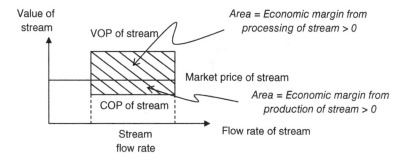

Figure 6.10   *Entirely profitable stream.*

Based on the above scenarios, purchasing and selling strategies can be developed. An ideal case is when the market price of a stream is in between its VOP and COP:

$$(F_{e\text{-}i})_{\text{VOP}} \geq F_{e\text{-}i} \geq (F_{e\text{-}i})_{\text{COP}} \qquad \forall i \in UNIT \tag{6.13a}$$

This means that the production of the stream is profitable. The processing of the stream is also profitable. Hence, the entire element in which the stream belongs to is profitable. This is shown in Figure 6.10.

In Figure 6.11, the market price of the stream is lower than its COP, though overall the stream is still profitable as VOP is more than COP. This means that the production of the stream is no longer profitable under the given market condition:

$$(F_{e\text{-}i})_{\text{VOP}} > (F_{e\text{-}i})_{\text{COP}} > F_{e\text{-}i} \qquad \forall i \in UNIT \tag{6.13b}$$

A further improvement in the economic margin can be achieved if a proper market decision can be taken, so as to reduce the production of the stream and instead increase purchasing of the stream. On purchasing, the new COP of the stream becomes its market price. As the difference between the VOP and the new COP (market price) is increased, the margin incurred from the stream is also increased. The difference between the COP and the market price is the *scope for value improvement* of the stream per unit flow rate. This indicates how much marginal improvement can be achieved by purchasing (instead of producing) a unit flow rate of the stream. Now, to show the total amount of marginal improvement achievable, the amount of stream that can be sold to the market needs to be identified.

For each steam, identify minimum production and minimum requirement levels. Minimum production is determined by minimum capacity of the upstream units to produce the stream. Minimum requirement depends on the downstream units processing the stream. The difference between the current flow rate and minimum production is the *purchasing potential* of the stream. The excess amount of minimum requirement that is processed in the current network is the stream's *selling potential*, shown in Figure 6.12.

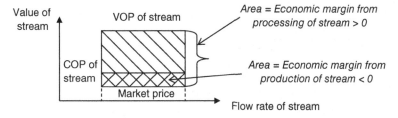

Figure 6.11   *A profitable stream but with nonprofitable production (Case 2).*

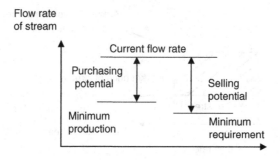

Figure 6.12 *Purchasing and selling potentials of streams.*

When the COP is more than the market price of a profitable stream, the *purchasing potential* and the *scope for value improvement* of the stream are multiplied to find the marginal improvement potential. The current margin and the marginal improvement can be as shown in Figure 6.13. The marginal improvement is given by the area bounded between the COP and market price and the purchasing potential. The direction of arrow for the *scope for value improvement* in the figure shows that the COP of the stream should be brought down to its market price. Thus, the decision is to keep the production rate of the stream to a minimum and buy the balance of its current flow and minimum production rate, known as the purchasing potential.

Decision making for Case 3

In Figure 6.14, the market price of a profitable stream is greater than the VOP:

$$F_{e\text{-}i} > (F_{e\text{-}i})_{\text{VOP}} > (F_{e\text{-}i})_{\text{COP}} \qquad \forall i \in UNIT \tag{6.13c}$$

Thus, processing the stream is no longer profitable. Now, the processing or conversion of the stream needs to be reduced and instead the stream must be sold. By selling the stream the margin per unit flow rate is increased by the difference between its market price and the VOP. Thus, this provides the *scope for value improvement* of the stream per unit flow rate. To estimate the total marginal improvement, the *selling potential* is to be multiplied with the *scope for improvement*. The current margin and marginal improvement are shown in Figure 6.15. The direction for the *scope for value improvement* indicates that the VOP should be increased to the market price at the least. The decisions are to consume a minimum through the current downstream network configuration and sell the amount in excess of the minimum requirement equal to its selling potential.

In all of the above cases, as the VOP has always been greater than the COP, the streams considered are profitable.

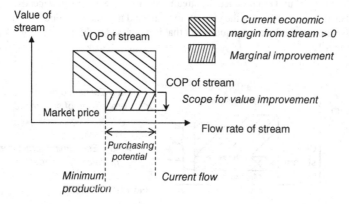

Figure 6.13 *Marginal improvement from purchasing of a profitable stream with nonprofitable production.*

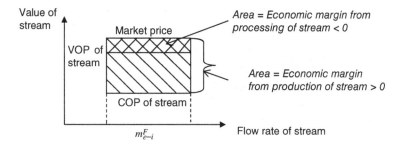

Figure 6.14   *Economic profile of a profitable stream (Case 3).*

The primary criterion for a stream to be nonprofitable is that the COP is greater than its VOP. Nonprofitable streams are very common in process industries when a large network such as a crude oil refinery deals with generation of some waste streams that the industry does not know what to do to them. For the refinery heavy ends, more money is spent to make them than the value gained on processing such a stream. Thus, refiners spend more money to produce these streams and incur losses while converting them into end products. The flow rate of such a stream may be very small but the losses or impacts from the stream can be significant depending on the difference between its COP and VOP. With the value analysis method, the amount of losses these nonprofitable intermediate streams are individually causing to the plant can be determined and graphically analyzed and correct decisions can then be made to make a significant improvement in the economics. The three different scenarios with respect to the position of the market price can occur as in the case of profitable streams.

Decision making for Case 4
Figure 6.16 shows the worst scenario:

$$(F_{e\text{-}i})_{\text{COP}} > F_{e\text{-}i} > (F_{e\text{-}i})_{\text{VOP}} \quad \forall i \in UNIT \tag{6.14a}$$

Both the production and the processing of the stream incur negative profits. Now the industry will have two options either to purchase and reduce production or to sell and reduce processing. The decision will be based upon the position of the market price with respect to the COP and VOP. The greater of the two differences between the COP and market price and between the market price and VOP will decide which route to follow. As both differences are positive, the greater of the two should be avoided for reducing the amount of loss. Thus if producing a stream is more avoidable, keep its production rate to a minimum and buy the rest, equal to its purchasing potential. The economic situation and

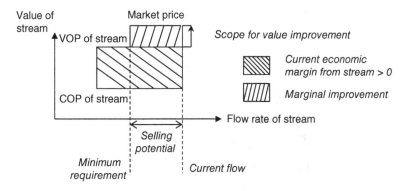

Figure 6.15   *Marginal improvement from selling of profitable stream (Case 3).*

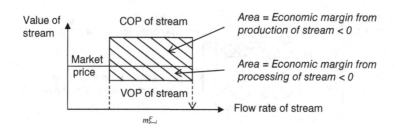

Figure 6.16   *Economic profile of nonprofitable stream (Case 4).*

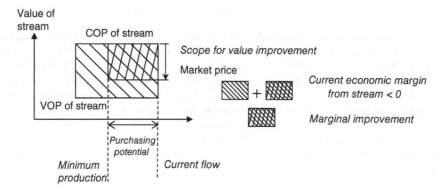

Figure 6.17   *Marginal improvement from purchasing of nonprofitable stream (Case 4).*

improvements are shown in Figure 6.17. The *scope for value improvement* of the stream per unit flow is given by the difference between the COP and the market price. The direction of the *scope for value improvement* indicates that the COP should be brought down to the market price. The marginal improvement is equal to the area bounded between the COP and the market price and the purchasing potential. As can be seen from the figure, the economic margin resulting after improvement is still negative.

If consumption of the stream is more avoidable, consume it to the minimum rate and sell the amount in excess of the minimum requirement equal to its selling potential. The economic scenario is shown in Figure 6.18. The *scope for value improvement* per unit flow of the stream should be towards improving its VOP to its market price, as indicated in the figure. The marginal improvement is equal to the area bounded between the market price and the VOP and the selling potential. The stream remains nonprofitable even after modification, but the loss incurred from it is reduced.

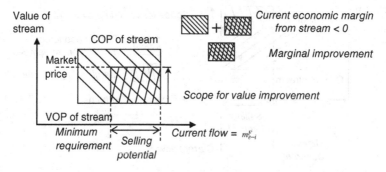

Figure 6.18   *Marginal improvement from selling of nonprofitable stream (Case 4).*

Further Exercise 2. Write decisions for Cases 5 and 6 shown in Equations (6.14b) and (6.14c) respectively, deriving analogical reasoning.

$$(F_{e-i})_{\text{COP}} > (F_{e-i})_{\text{VOP}} > F_{e-i} \qquad (6.14b)$$

$$F_{e-i} > (F_{e-i})_{\text{COP}} > (F_{e-i})_{\text{VOP}} \qquad (6.14c)$$

6.5.1 Value Analysis to Determine Necessary and Sufficient Condition for Streams to be Profitable or Nonprofitable

The margin from an intermediate stream can be determined without knowing its market price using the value analysis approach discussed in this chapter. Thus, a conclusion can be drawn that, independent of the market price of a stream, the marginal contribution of the stream towards the overall margin of the network is always positive as long as its VOP (*value on processing*) is greater than its COP (*cost of production*).

The necessary and sufficient condition for a stream to be profitable is that its value on processing is greater than its cost of production.
Similarly, the necessary and sufficient condition for a stream to be nonprofitable, is when its cost of production is greater than its value on processing.

This explains how the profitability of a stream can be determined using value analysis. Based on profitability of individual streams, the crucial decisions about production versus purchasing and conversion versus selling of streams can be taken. Also, once the streams and elements are evaluated for their individual economic margins, the specific scopes for improvement can be identified. The value analysis method can then be further used to determine the optimum integration opportunities that may or may not involve capital investment.

Exercise 3. It is recognized that, to make bioethanol production economically competitive and commercially feasible, the bioethanol must be produced as one of several coproducts within a biorefinery. Dried distillers grains with solubles (DDGS) is the major coproduct of bioethanol production, but has a low market value. Estimate the VOP, COP and the marginal values of all associated streams (including intermediate streams) in a wheat bioethanol plant coproducing DDGS (see Figure 6.19). Identify the nonprofitable streams and paths. The utility consumptions and recovery of various process units in the wheat bioethanol process flowsheet in Figure 6.19 are provided in Tables 6.4 and 6.5, respectively. Table 6.6 shows the price information for the various raw materials, products and utilities. Tables 6.7 and 6.8 show the wheat and bran compositions, respectively. The data presented are based on 340 kt y^{-1} of wheat processing and 330 days of operation per year. Try the value analysis calculations for arabinoxylans coproduction using the extra data given in this problem and case study discussed elsewhere[1].

Solution to Exercise 3 The unit operating cost of the hammer milling process is calculated as follows:

$$= 1200 \times 0.012 \times \frac{330 \times 24 \times 3600}{1000 \times 1000} \times \frac{1}{340}$$

$$= £1.21 \, t^{-1}$$

The COP of the feed stream to the liquefaction process $= 96 + 1.21 = £97.21 \, t^{-1}$

The unit operating cost of the liquefaction process = cost of calcium chloride + cost of sodium hydroxide + cost of α amylase + cost of process water + cost of electricity + cost of steam

$$= \frac{0.41 \times 130 + 1.7 \times 1600 + 0.28 \times 2000 + 625.6 \times 0.6 + 76.34 \times 0.012 \times \frac{330 \times 24 \times 3600}{1000 \times 1000} + 9.12 \times 7 \times \frac{330 \times 24}{1000}}{340 + 0.41 + 1.7 + 0.28 + 625.6}$$

$$= £4.38 \, t^{-1}$$

The COP of the feed stream to the saccharification $= \dfrac{97.21 \times 340 + 4.38 \times (340 + 0.41 + 1.7 + 0.28 + 625.6)}{340 + 0.41 + 1.7 + 0.28 + 625.6} =$ £38.52 t^{-1}

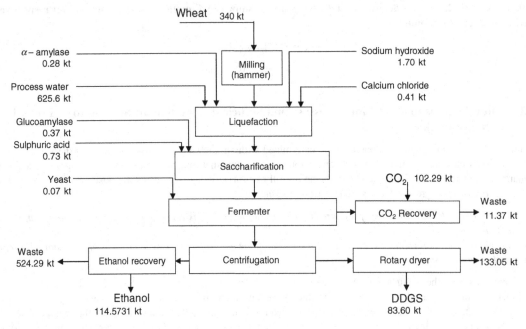

Figure 6.19 *Mass flow rates in kt y^{-1} (in shaded areas) of a wheat bioethanol flowsheet. (Reproduced with permission from Sadhukhan et al. (2008)[1]. Copyright © 2008, Elsevier.)*

The total operating cost of a process unit (e.g., liquefaction) is the multiplication between the unit operating cost of the process unit and the throughput through the unit. The COP of a product from the unit is the total cost accumulated divided by the throughput through the unit.

For saccharification, the unit operating cost = cost of glucoamylase + cost of sulfuric acid + cost of electricity + cost of cooling water

$$= \frac{0.37 \times 2000 + 0.73 \times 40 + 25.56 \times 0.012 \times \frac{330 \times 24 \times 3600}{1000 \times 1000} + 104.59 \times 0.015 \times \frac{330 \times 24}{1000}}{340 + 0.41 + 1.7 + 0.28 + 625.6 + 0.37 + 0.73} = £0.82 \, t^{-1}$$

Table 6.4 *Utility consumptions of various process units in a starch processing facility. (Reproduced with permission from Sadhukhan et al. (2008)[1]. Copyright © 2013, Elsevier.)*

		Utilities				
No.	Equipment	Process Water (v/w)	Electricity (kW)	Steam (t h^{-1})	Natural Gas (t h^{-1})	Cooling Water (t h^{-1})
1	Milling (hammer)	–	1200	–	–	–
2	Liquefaction	1.84	76.34	9.12	–	–
3	Saccharification	–	25.56	–	–	104.59
4	Fermenter	–	136.48	–	–	357.76
5	Centrifugation	–	284.62	–	–	–
6	Ethanol recovery	–	43.75	21.60	–	631.31
7	Rotary dryer	–	284.62	–	0.75	–
8	CO$_2$ recovery	–	1355	–	–	–

Table 6.5 Recovery from various process units in a starch processing facility. (Reproduced with permission from Sadhukhan et al. (2008)[1]. Copyright © 2013, Elsevier.)

Process Units	Recovery
Milling (hammer)	1
Liquefaction	0.98
Saccharification	0.97
Fermenter	0.46
Centrifugation	0.8
Ethanol recovery	0.98
Washing	0.98
Treatment	0.95
Sieving and washing	0.98
Precipitation	0.95
Ultrafiltration	0.9
Centrifugation	0.8
Rotary dryer	0.95
CO_2 recovery	0.9

Table 6.6 Cost of raw materials, products and utilities. (Reproduced with permission from Sadhukhan et al. (2008)[1]. Copyright © 2013, Elsevier.)

Streams		Price in £
Products	Ethanol	590 t^{-1}
	DDGS	65–80 t^{-1}
	CO_2	10.7 t^{-1}
	All waste streams	0 t^{-1}
Raw materials	Wheat	96 t^{-1}
	All enzymes	2000 t^{-1}
	Yeast	6.67 t^{-1}
	$CaCl_2$	130 t^{-1}
	Hydrogen peroxide (2%)	500 t^{-1}
	Sodium hydroxide (25%)	1600 t^{-1}
	Concentrated sulfuric acid	40 t^{-1}
Cost of utilities	Electricity	0.012 MJ^{-1}
	Cooling water (cooling towers)	0.015 t^{-1}
	Steam (from direct fired boilers)	7 t^{-1}
	Refrigeration (0°C)	0.006 MJ^{-1}
	Mains water (process water)	0.6 t^{-1}
	Natural gas	4 per million Btu

Table 6.7 Wheat composition. (Reproduced with permission from Sadhukhan et al. (2008)[1]. Copyright © 2013, Elsevier.)

Components	Composition (weight % dry basis)
Starch	69
Sugar	3
Protein	11.5
Non-starch polysaccharides	11
Lipid	2.5
Ash	2
Lignin	1

Table 6.8   *The composition of wheat bran from various processing options. (Reproduced with permission from Sadhukhan et al. (2008)[1]. Copyright © 2013, Elsevier.)*

Compositions by weight	Bran Produced with a Debranner		Bran Produced with a Roller Mill from Debranned Wheat	
	4%	8%	4%	8%
Protein (%, dry basis)	6.8	11.8	19.3	23.4
Starch (%, dry basis)	17.6	26.9	21.9	21.8
Glucose (%, dry basis)	3.1	3.4	3.7	3.2
Arabinoxylan (%, dry basis)	26.8	18.6	17.2	15.8
Others (%, dry basis)	45.7	39.3	37.9	35.8

The COP of the feed stream to the fermenter

$$= \frac{38.52 \times (340 + 0.41 + 1.7 + 0.28 + 625.6) + 0.82 \times (340 + 0.41 + 1.7 + 0.28 + 625.6 + 0.37 + 0.73)}{340 + 0.41 + 1.7 + 0.28 + 625.6 + 0.37 + 0.73} = £39.3\,t^{-1}$$

The unit operating cost of the fermenter

$$= \frac{0.07 \times 6.67 + 136.48 \times 0.012 \times \frac{330 \times 24 \times 3600}{1000 \times 1000} + 357.76 \times 0.015 \times \frac{330 \times 24}{1000}}{340 + 0.41 + 1.7 + 0.28 + 625.6 + 0.37 + 0.73 + 0.07} = £0.09\,t^{-1}$$

The COP of the feed stream to CO_2 recovery as well as centrifugation

$$= \frac{39.3 \times (340 + 0.41 + 1.7 + 0.28 + 625.6 + 0.37 + 0.73) + 0.09 \times (340 + 0.41 + 1.7 + 0.28 + 625.6 + 0.37 + 0.73 + 0.07)}{340 + 0.41 + 1.7 + 0.28 + 625.6 + 0.37 + 0.73 + 0.07}$$

$$= £39.39\,t^{-1}$$

The COP calculated for the end products, ethanol, DDGS, CO_2, recycle water as well as waste streams, residue (RD) waste and exhaust gas from CO_2 recovery unit thus calculated is shown in Figure 6.20. The VOP of the end streams is their market prices. The VOP of the waste streams and recycle water are assumed to be zero. The market price or the VOP of bioethanol is £590 t^{-1}, for DDGS is £72.5 t^{-1} and for CO_2 is £10.7 t^{-1}, respectively.

The unit operating cost of the ethanol recovery unit should be calculated using the same calculation procedure shown for the unit operating costs of milling, liquefaction, saccharification and fermenter units. Here, the value is shown as £2.02 t^{-1}.

The VOP of the feedstock to the ethanol recovery unit is calculated as follows:

$$= \frac{590 \times 114.57 - 2.02 \times (114.57 + 524.29)}{114.57 + 524.29}$$

$$= £103.8\,t^{-1}$$

The unit operating cost of the rotary dryer can be calculated using the same procedure shown for the unit operating cost calculations for the other process units and its value is £3.88 t^{-1}.

The VOP of the feedstock to the rotary dryer

$$= \frac{72.5 \times 83.6 - 3.88 \times (83.6 + 133.05)}{83.6 + 133.05}$$

$$= £24.09\,t^{-1}$$

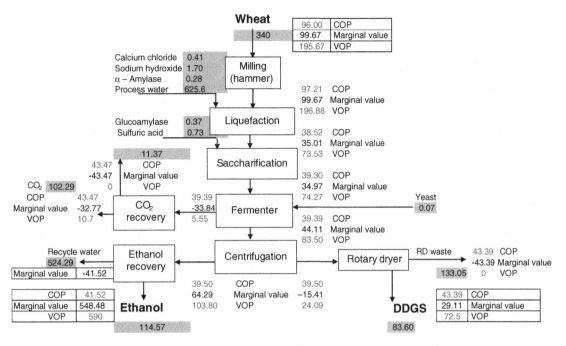

Figure 6.20 *COP, VOP and marginal values of streams in £ t⁻¹ and mass flow rates in kt y⁻¹ (in shaded areas) of wheat bioethanol flowsheet. (Reproduced with permission from Sadhukhan et al. (2008)[1]. Copyright © 2008, Elsevier.)*

The unit operating cost of the centrifugation unit evaluated is £0.11 t^{-1} and that of the CO_2 recovery unit is £4.08 t^{-1}, respectively. The VOP of the feedstock to the centrifugation unit is estimated from those of its two product streams, the feedstocks to the ethanol recovery and rotary dryer units, as follows:

$$= \frac{103.80 \times (114.57 + 524.29) + 24.09 \times (83.6 + 133.05) - 0.11 \times (114.57 + 524.29 + 83.6 + 133.05)}{114.57 + 524.29 + 83.6 + 133.05} = £83.5\,t^{-1}$$

From the unit operating cost of the CO_2 recovery unit and the market price of CO_2, the VOP of the feedstock to the CO_2 recovery unit can be estimated. Its value is

$$= \frac{10.7 \times 102.29 - 4.08 \times (102.29 + 11.37)}{102.29 + 11.37}$$

$$= £5.55\,t^{-1}$$

Now, from the VOP of the two product streams (feedstock to the rotary dryer and feedstock to the CO_2 recovery unit) from the fermenter unit, the VOP of the feedstock to the fermenter unit is predicted:

$$= \frac{83.5 \times (114.57 + 524.29 + 83.6 + 133.05) + 5.55 \times (102.29 + 11.37) - 0.09 \times (114.57 + 524.29 + 83.6 + 133.05 + 102.29 + 11.37)}{114.57 + 524.29 + 83.6 + 133.05 + 102.29 + 11.37}$$

$$= £74.27\,t^{-1}$$

Following the backward calculation procedure, the VOP of all the streams can be calculated, as shown in Figure 6.20. The difference between the VOP and COP of a stream is its marginal contribution.

Thus, the economic margin from the production of bioethanol alone is

$$590 - 41.52 = £548.48\,t^{-1}$$

The economic margin from DDGS is £29.11 t^{-1}.

The losses are incurred from RD waste, CO_2 and the effluent gas from the CO_2 recovery unit. The recycle water is an internal stream and a reasonable value can be assumed to undertake a more correct presentation. However, being an internal stream, its profit margin does not contribute to the overall margin of the flowsheet. In other words, a network margin is independent of any value assumed for a recycle stream and thus its correct value representation is not a requirement for the value analysis.

Further, a network's biggest energy consumer, capital costs or loss of value due to a high flow rate of waste (hot spots in LCA) can be identified, from the huge reduction in the marginal value across a unit, for example, a liquefaction unit or a switch from positive to negative margins across a unit or a reduction in the VOP of a feed stream such that its VOP is lower than its COP, for example, CO_2 recovery and rotary dryer units. The former is due to waste generation, whilst the latter is due to both waste generation and energy consumption. The rotary dryer has been used instead of an air dryer, as the latter could cause odor while the former could make noise. Hence, the social acceptability of technologies must also be assessed.

The overall economic margin of the wheat bioethanol flowsheet is the total marginal value obtained from wheat feedstock:

$$99.67 \times 340/1000 = £33.89 \text{ million per year}$$

It can be verified that the overall economic margin can also be calculated from the economic margins of the end products:

$$(548.48 \times 114.57 + 29.11 \times 83.6 + (-32.77) \times 102.29 + (-41.52) \times 524.29 + (-43.47) \times 11.37 + (-43.39)$$

$$\times 133.05)/1000 = £33.89 \text{ million per year}$$

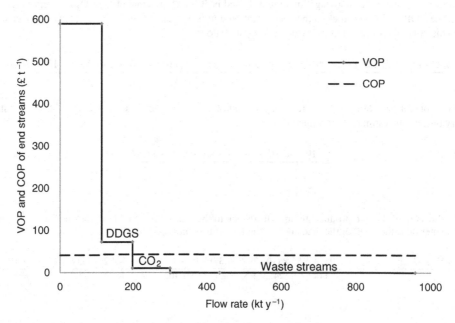

Figure 6.21 *Marginal contributions from end streams of the wheat bioethanol flowsheet in Figures 6.19 and 6.20. (Reproduced with permission from Sadhukhan et al. (2008)[1]. Copyright © 2008, Elsevier.)*

The feedstocks leading to the CO_2 recovery and rotary dryer units and their corresponding processing routes are nonprofitable routes.

Using the value analysis methodology discussed, the overall economic margin for the conventional flowsheet in Figures 6.19 to 6.21 is calculated as £99.67 t^{-1} of wheat processed. The basis of calculation is a wheat based biorefinery processing 340 000 t of wheat per annum with 330 operating days per year producing 99.6% pure ethanol by weight. The economic margin from a typical bioethanol flowsheet in Figure 6.20 can also be shown from the marginal values contributed by wheat or from the net profit from the end streams, ethanol, DDGS, CO_2 and the waste streams, shown in Figure 6.21. Figure 6.21 shows the deduced marginal contributions from individual end streams. The y axis presents the various COP and VOP values of the products while the x axis corresponds to the flow rates of products. The net positive area presented above the cost of production line and between the values on processing of the end streams equates to the economic margin of the overall flowsheet in Figure 6.21. Hence, the economic margins of those streams are negative for which the COP is greater than VOP in the plot.

The total operating cost of the facility is £40.9 million, which corresponds to £120.19 t^{-1} of wheat processed with a 90% contribution from raw materials due to the price of wheat and only 10% due to utilities.

6.6   Summary

The value analysis tool is used for differential economic marginal analysis from process streams to networks. It enables evaluation and graphical presentation of a network margin in terms of the cost of production (COP), value on processing (VOP) and margins of individual streams in a process network. The evaluation of the COP starts from the known market prices of feedstocks and proceeds in the forward direction until end products are reached. The COP of a stream is the summation of all associated cost components (i.e., the costs of feedstocks, utilities and annualized capital costs) that have contributed to the production of that stream up to that point. This must mean the inclusion of costs involved with the stream's production. The VOP evaluation proceeds in the backward direction from the end product market prices until the feedstock in a process network is reached. The VOP of a stream at a point within the process is obtained from the prices of products that will ultimately be produced from it, minus the costs of auxiliary raw materials and utilities and the annualized capital cost of equipment that will contribute to its further processing into these final products. The COP of a feedstock to a process and the VOP of an end product correspond to their respective market prices.

The value analysis methodology ensures the choice of low operating cost and greatest capital saving routes of products with required functionality. Once flowsheet options are identified, the value analysis methodology is applied to select economically favorable designs, if not the optimum. Improvements from one design to another are realized in terms of reduction of waste (marginal loss), capital costs (by process intensification) and energy consumption (by heat recovery and energy efficiency) and enhanced product quality and value. The tasks of design and optimization can become very complex for biorefinery systems, because of the interdependency between process design, operating variables, new component characterization variables and market forces. The variations of market supply–demand and prices of streams and utilities or primary energy sources must be considered. The biorefinery complex must also clearly show an economic incentive over the fossil based reference system.

References

1. J. Sadhukhan, M.M. Mustafa, N. Misailidis, F. Mateos-Salvador, C. Du, G.M. Campbell, Value analysis tool for feasibility studies of biorefineries integrated with value added production, *Chem. Eng. Sci.*, **63**(2), 503–519 (2008).
2. J. Sadhukhan, N. Zhang, X.X. Zhu, Analytical optimisation of industrial systems and applications to refineries, Petrochemicals, *Chem. Eng. Sci.*, **59**(20), 4169–4192 (2004).
3. J. Sadhukhan, N. Zhang, X.X. Zhu, Value analysis of complex systems and industrial application to refineries, *Ind. Eng. Chem. Res.*, **42**(21), 5165–5181 (2003).

7

Combined Economic Value and Environmental Impact (EVEI) Analysis

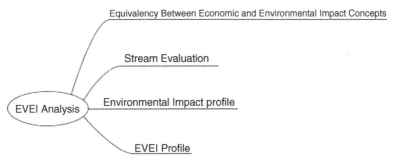

Structure for Lecture Planning

7.1 Introduction

Chemical process designs traditionally involved gate to gate design of a process plant. For example, for an integrated gasification combined cycle (IGCC) plant, the input material stream to the plant gate is the gasification fuel and the output energy stream from the plant gate is the combined heat and power (CHP). The gate to gate design/analysis refers to the IGCC plant from the gasification fuel to the CHP generation and only the design and operation aspects of the plant. The life cycle impacts and costs from the production of fuel to the IGCC plant, maintenance, decommissioning and reuse of the construction material may not be included in the plant design and operation analysis. Life cycle thinking discussed in Chapter 4 has not been traditionally applied.

With governments and industries increasingly looking to replace fossil fuel energy with alternative energy sources, the boundary for chemical process design traditionally considered may not give sustainable solutions. The feedstock to be used for a process plant may already have a significant carbon footprint and other environmental impacts before reaching the plant gate. Increasing a manufacturing plant's energy efficiency with a higher amount of material extraction for equipment construction without quantitative evidence is not a sustainable solution. While energy efficiency of a manufacturing plant needs to be improved, the environmental performance of a whole "system," cradle to grave, must be

Biorefineries and Chemical Processes: Design, Integration and Sustainability Analysis, First Edition.
Jhuma Sadhukhan, Kok Siew Ng and Elias Martinez Hernandez.
© 2014 John Wiley & Sons, Ltd. Published 2014 by John Wiley & Sons, Ltd.
Companion Website: http://www.wiley.com/go/sadhukhan/biorefineries

superior to other competing manufacturing systems. Thus, biorefineries must be designed for environmental sustainability at a very early stage.

Designing a biorefinery according to both economic and environmental objectives is a challenging task due to the wide range of alternatives. In practice, it is essential to be able to identify the most promising process pathways to prioritize for integration into the biorefinery process network. The value analysis approach presented in Chapter 6 is a powerful tool for differential marginal analysis of process networks. It enables evaluation and graphical presentation of the cost of production (COP), value on processing (VOP) and margins of individual components in a network: streams, processes, paths and trees, and systems. Equivalent to the COP and VOP, the environmental impact (EI) cost and credit value can be evaluated to analyze the environmental performance of new products and processes quantitatively.

7.2   Equivalency Between Economic and Environmental Impact Concepts

The *EI cost* of a stream is the embodied environmental impact from its production system per functional unit. In the value analysis, the functional unit for a material of choice is its mass unit. The EI cost can be an aggregated EI under a category or an overall normalized factor per unit mass. For example, the EI cost of a feedstock can refer to the GWP (global warming potential) impact in terms of CO_2 equivalent in kg kg^{-1} of the feedstock. Alternatively, various impact characterizations or categories can be normalized and weighted into one indicator according to the ISO standards 14040 and 14044 (see Section 4.3).

An *EI credit value*, D_p, of a product is the EI cost of an existing product to be displaced or substituted by the biorefinery product multiplied by an equivalency factor β. The *equivalency factor* is the amount of existing product to be displaced or substituted per unit amount of the biorefinery product. For example, β is the ratio between the calorific value of the biofuel and that of the equivalent fossil based transportation fuel. The term I_{peq} is the EI cost of an existing equivalent product to be displaced or substituted:

$$D_p = \beta \times I_{peq} \tag{7.1a}$$

The EI credit value in GWP of a product relates to its carbon credit, which may be certified and traded by industry and, hence, is directly related to business and policy. The two approaches, value analysis and EI analysis, are coupled into an economic value and environmental impact (EVEI) analysis approach[1], as discussed in this chapter.

The EI cost of a biorefinery product should not exceed the EI cost of its counterpart product to achieve an environmentally better performance. Subtracting the EI cost of a counterpart product to be displaced from D_p of a biorefinery product yields the potential EI reduction from the displacement. The resulting EI reduction is the *EI saving margin*, shown below, a concept equivalent to the economic margin:

$$\text{EI saving margin} = D_p - \text{EI cost of biorefinery product} \tag{7.1b}$$

The EI saving margin from any stream in a process is calculated as shown later in Equation (7.9). The EI cost of a stream is determined from the environmental impact assessment of its production system. Methodologies to assess the environmental impact of a system include the environmental footprint analysis[2] and LCA[3,4] (refer to Chapter 4 for the LCA). The adoption of a life cycle approach allows the EI analysis to be carried out in a systematic and holistic way. LCA is an effective methodology that allows evaluation of various EI characterizations, namely, emissions to air: global warming, ozone depletion, photochemical oxidant creation and acid rain potentials; water: eutrophication and aquatic ecotoxicity potentials; land: primary energy and abiotic resource depletion; and human toxicity potentials[5] (see also Section 4.6). Impact categories are selected according to the goals of a given analysis. For consistency with the value analysis methodology, the functional unit for economic value and EI variables is 1 kg for material streams or 1 MJ for energy streams. However, a functional unit is chosen depending on the *service* provided by the product or process (see the definition of a functional unit in Section 4.4).

The EI costs of auxiliary raw materials (I_a) and utilities (I_u) show the embodied environmental impact from their respective productions. The EI costs can be deduced from reported LCA studies or databases, or can be calculated from simulation models. The EI cost should ideally be determined from actual locations and associated supply chains, in spatial (or geographical) and temporal (or time) domains. This exercise can be very data intensive and, hence, must be linked with the objectives to reduce the problem size.

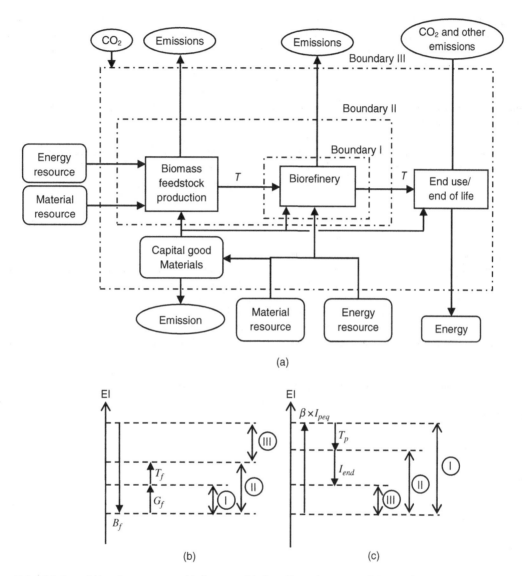

Figure 7.1 *(a) Overall biorefinery system with three possible boundaries for EVEI analysis. T refers to transport. Components of (b) feedstock EI cost (I_f) and (c) product EI credit value (D_p) for various boundaries.*

A differentiation in the EI costs between marketable products and emission and waste streams is carried out in EVEI analysis. The emission and waste streams add an emissions impact cost (I_m) to the system. The emissions impact cost of an emission or waste is calculated from its composition multiplied by its impact characterization factors (shown in Chapter 4). The economic cost for control and treatment of emission and waste streams (C_m) is also considered. I_m and C_m are allocated and distributed to the end products, discussed later.

Boundaries are selected to find the EI cost of feedstock and the EI credit value of products. Figure 7.1(a) (cf. Figure 5.15) shows three possible boundaries for EVEI analysis of biorefinery systems. Boundary I shows a gate to gate system, where process engineering studies have traditionally been applied. The input streams to this system are biomass feedstock, materials used for construction and manufacturing equipment and auxiliary raw materials such as chemicals, reagents, catalysts and utilities. The output streams are the biorefinery products and emissions.

Boundary II covers the biomass feedstock production and conversion or biorefinery plant. Across this boundary, the other input streams are material and energy streams used for biomass production. For agricultural biomasses, fertilizers and pesticides have environmental impacts included in the impacts from the biomass feedstock production. Emissions from the biomass production system are the other output streams to be considered. The biomass feedstock is an internal stream, transported (T) from the production to the biorefinery sites.

Boundary III, which additionally includes "end use" and "capital good materials" blocks, is a *cradle to grave whole biorefinery system* (from primary material and energy resource extraction to end of life; see Section 4.1). Note that CO_2 binding by the biomass, during photosynthesis, is only considered in a cradle to grave system. This is because CO_2 is only released upon the end use of biomass derived products, considered in boundary III.

Depending on the boundary, the EI cost of the biomass feedstock (I_f) or EI credit value of a biorefinery product (D_p) is composed of several factors (Figures 7.1(b) and (c)). Across boundary III, I_f is made up of CO_2 binding by photosynthesis (B_f), EI from transportation (T_f) and EI from production (G_f) of biorefinery feedstock, as shown below:

$$I_f = G_f + T_f - B_f \tag{7.2}$$

Figure 7.1(b) shows Equation (7.2) graphically.

For a biorefinery to be environmentally sustainable, B_f must be greater than the summation of G_f and T_f, resulting in overall negative EI (i.e., a net amount of CO_2 is captured) when boundary III is considered (Figure 7.1(b)). Across boundary II, Equation (7.2) reduces to $G_f + T_f$. Across boundary I, I_f reduces to G_f.

When boundary I is selected (a gate to gate approach), the EI cost of a feedstock corresponds to the EI from its production. Note that B_f would not appear in any impact categories other than the GWP. Carbon dioxide binding is generally reported for the whole crop or unit of land. However, the various crop parts can be utilized in a biorefinery, for example, wheat grain for bioethanol[6] and straw for CHP[7]. In such a case, B_f is allocated to each crop part according to its fractional carbon content in the total biomass.

The net EI credit value of a product is made up of I_{peq}, EI from the "end use/end of life" emissions (I_{end}) and EI from product transportation (T_p). Thus, a more generic expression to determine the EI credit value across boundary III is given below. The I_{end} term is relevant, for example, when a biomass-derived product is combusted in an engine or boiler. Then, the complete equation to calculate D_p is

$$D_p = \beta \times I_{peq} - T_p - I_{end} \tag{7.3}$$

An illustration of Equation (7.3) is shown in Figure 7.1(c). For the EI credit value D_p of biorefinery products to be positive, $\beta \times I_{peq}$ must be greater than the summation of the transport (T_p) and end use emissions (I_{end}). Across boundary II, Equation (7.3) reduces to $D_p = \beta \times I_{peq} - T_p$, excluding end use emissions. Across boundary I, Equation (7.3) only results in $D_p = \beta \times I_{peq}$. Equation (7.3) thus overestimates the EI in the following order: boundary I > boundary II > boundary III.

Because of realistic estimation around boundary III and overestimation within boundary I, a whole system cradle to grave analysis must be performed before investing in any product or process technology.

Boundaries are selected according to the purpose of the analysis, for example, boundary III for identification and mitigation of system hot spots by selection of processing pathways and designs. Data availability can also influence the boundary selection.

Exercise 1. Calculate the net GWP impact of UK wheat grain, with a GWP impact of 0.492 kg CO_2 eq. kg^{-1} from the wheat grain production. The following data are available:

- EI factor from the transportation of wheat grain = 0.0776 kg CO_2 eq. t^{-1} km^{-1}.
- Transportation distance from a field to a biorefinery site: 50 km.
- Whole wheat average CO_2 binding = 10.5 t ha^{-1}.
- Molar mass of CO_2 = 44.01 kg $kmol^{-1}$.
- Straw carbon weight fraction = 0.3675.
- Molar mass of carbon = 12.01 kg $kmol^{-1}$.
- Wheat grain yield = 6.96 t ha^{-1}.

- Straw yield = 3.49 t ha^{-1}.
- Straw harvesting ratio = 0.6.

Solution to Exercise 1. From the data given, the equation for a cradle to grave system boundary is used. The CO_2 binding needs to be converted into a mass basis and split between grain and straw according to the carbon content. Taking the yields and straw harvested fraction into account, CO_2 binding is calculated as

$$B_f = \frac{10.5 - 0.3675 \times (44.01/12.01) \times 3.49 \times 0.6}{6.96} = 1.1 \text{ kg kg}^{-1}$$

Using Equation (7.2), the net GWP impact of UK wheat grain is calculated as

$$I_f = 0.492 + 0.0776 \times 50/1000 - 1.1 = -0.6 \text{ kg kg}^{-1}$$

Exercise 2. Calculate the EI credit value in kg CO_2 eq. kg^{-1} of bioethanol, which has a calorific value of 26.7 MJ kg^{-1} and is delivered to an end use point 50 km from the bioethanol plant. The equivalent product to be displaced is gasoline with a calorific value of 44.5 MJ kg^{-1} and EI cost of 84.6 g CO_2 eq. MJ^{-1}. Assume the system boundary III and EI from transportation given in Exercise 1.

Solution to Exercise 2. Calculate the equivalency factor β from the ratio between the calorific value of bioethanol and that of gasoline. This is the amount of gasoline product that can be displaced or substituted per unit amount of bioethanol product shown as follows:

$$\beta = \frac{26.7}{44.5} = 0.6 \text{ kg kg}^{-1}$$

The EI credit value in kg CO_2 eq. kg^{-1} of bioethanol (D_p) can be calculated across boundary III (Figure 7.1) using Equation (7.3) and substituting the value of β.

Assume that only CO_2 is released as GHG (greenhouse gas) from bioethanol combustion. Hence, 1 mole of C_2H_6O gives rise to 2 moles of CO_2 (molar mass of $C_2H_6O = 46.07$ kg kmol^{-1}, molar mass of $CO_2 = 44.01$ kg kmol^{-1}). Then

$$D_p = 84.6/1000 \times 44.5 \times 0.6 - \frac{2 \times 44.01}{46.07} - 0.0776 \times 50/1000 = 0.34 \text{ kg kg}^{-1}$$

The total economic cost (O_k) associated with a process unit k consists of the costs of auxiliary raw materials, utilities and emissions/wastes and the annualized capital cost. The corresponding EI cost of the process unit k is known as the *total impact cost* (O_{Ik}). The following equation shows combined generic expressions with O_k and O_{Ik} for a process unit k:

$$\begin{bmatrix} O_k \\ O_{Ik} \end{bmatrix} = \begin{bmatrix} \overline{C}_{a,k} \\ \overline{I}_{a,k} \end{bmatrix} \times \overline{A}_k + \begin{bmatrix} \overline{C}_{u,k} \\ \overline{I}_{u,k} \end{bmatrix} \times \overline{U}_k + \begin{bmatrix} \overline{C}_{m,k} \\ \overline{I}_{m,k} \end{bmatrix} \times \overline{M}_k + \begin{bmatrix} CC_k \\ CI_k \end{bmatrix} \tag{7.4}$$

where

$\overline{A}_k$ is a single column vector of flow rates of auxiliary raw materials.
$\overline{U}_k$ is a single column vector of flow rates of utilities.
$\overline{M}_k$ is a single column vector of flow rates of emissions and wastes.
$\overline{C}_{a,k}, \overline{C}_{u,k}$ and $\overline{C}_{m,k}$ represent single row vectors of unit economic costs of auxiliary raw materials, utilities, and emissions and wastes, respectively.
$\overline{I}_{a,k}, \overline{I}_{u,k}$ and $\overline{I}_{m,k}$ are single row vectors containing the EI costs of auxiliary raw materials, utilities, and emissions and wastes, respectively.

The CC_k is the annualized capital cost of the process unit k, whilst CI_k is the annualized impact from the materials of construction up to the system boundary under consideration. CI_k refers to the impact from manufacturing, construction and transportation of commonly used materials for chemical equipment, such as steel, concrete and aluminum. The amounts of the materials required are estimated from equipment sizing and engineering studies. CI_k comes from a linear distribution of total impacts from construction over the biorefinery lifetime:

$$CI_k = \text{Total impact from construction/lifetime} \qquad (7.5)$$

Exercise 3. Determine the vector of total costs for the fermentation unit from a biorefinery processing 100 kt y^{-1} of wheat. The total feed to this unit is 285 000 t y^{-1} (density: 1070 kg m^{-3}) and 19 t y^{-1} of yeast, 833 400 t y^{-1} of cooling water and 1144 GJ y^{-1} of grid electricity are required; 30 000 t y^{-1} of CO_2 is generated. The global warming potential impact (in CO_2 eq.) is to be evaluated.

The following data are available:

- Biorefinery lifetime: 10 years, 330 operating days per year, 10% interest rate.
- Cost of a vessel of 2180 m^3 was \$411 000 in 2006, CEPCI = 499.6.
- Capital cost in 2010, CEPCI = 550.8 using a scaling factor of 0.55.
- Installation factor = 1.10.
- Residence time = 68 h.
- Maximum tank capacity = 2800 m^3.
- Length-to-radius ratio, $L/r = 2.4$.
- Wall thickness $\ell = 4.76$ mm.
- Steel density $\rho_{steel} = 8000$ kg m^{-3}.
- Consider concrete required as 30% of the amount of steel.
- EI cost (CO_2 eq.) of steel = 6.023 kg kg^{-1}.
- EI cost (CO_2 eq.) of concrete = 0.095 kg kg^{-1}.
- EI cost (CO_2 eq.) of CO_2 emissions = 1 kg kg^{-1}.
- Yeast cost = \$5 t^{-1}; EI cost (CO_2 eq.) = 7.7 kg kg^{-1}.
- Cooling water cost = \$0.015 t^{-1}.
- Electricity cost = \$12 GJ^{-1}, EI cost (CO_2 eq.) = 0.186 t GJ^{-1}.

Solution to Exercise 3. The preliminary equipment sizing and costing are as follows.

The fermentor volumetric capacity required is

$$\frac{285\,000}{\dfrac{330 \times 24}{1.07}} \times 68 = 2290 \text{ m}^3 < 2800 \text{ m}^3 \text{ (max. capacity)}$$

The amount of steel required for a cylindrical shape fermentor in terms of its volume capacity (V) and L/r is as follows. Since r and L are unknown, an expression for the amount of steel required in terms of V and (L/r) needs to be developed, shown as follows:

Volume of the cylindrical fermentor: $V = \pi r^2 L$

Total surface area of the cylinder with two circular flat surfaces:

$$A = 2\pi r^2 + 2\pi rL = 2\pi r(r + L)$$

The ratio of volume to area is

$$\frac{V}{A} = \frac{\pi r^2 L}{2\pi r(r+L)} = \frac{rL}{2(r+L)}$$

The surface area can be expressed as

$$A = \frac{V}{\dfrac{rL}{2(r+L)}} = \frac{2V(r+L)}{rL}$$

which when multiplied by $(1/r)/(1/r)$ gives

$$A = \frac{2V(r+L)}{rL} \times \frac{\dfrac{1}{r}}{\dfrac{1}{r}} = \frac{2V\left(1+\dfrac{L}{r}\right)}{r\dfrac{L}{r}} = \frac{2V}{r} \times \frac{1+\dfrac{L}{r}}{\dfrac{L}{r}}$$

Deriving an expression for V/r in terms of V and L/r:

$$\frac{V}{r} = \frac{\pi r^2 L}{r} = \pi r L = \sqrt[3]{\pi^3 r^3 L^3} = \sqrt[3]{(\pi r^2 L)^2 \frac{L}{r} \pi} = \sqrt[3]{V^2 \frac{L}{r} \pi}$$

Substituting in the expression for area gives

$$A = \frac{2 \times \left(1+\dfrac{L}{r}\right) \times \sqrt[3]{V^2 \dfrac{L}{r}\pi}}{L/r}$$

By multiplying the area with the thickness and steel density, the mass of steel is determined:

$$\frac{2\left(1+\dfrac{L}{r}\right)\sqrt[3]{V^2\dfrac{L}{r}\pi}}{L/r} \ell \rho_{\text{steel}} = \frac{2 \times (1+2.4) \times \sqrt[3]{2290^2 \times 2.4 \times 3.1416}}{2.4} \times \frac{4.76}{1000} \times 8000 = 36\,750 \text{ kg}$$

Considering concrete as 30% of the amount of steel, the total EI from construction (as CO_2 eq.) is

$$\frac{36\,750 \times (6.023 + 0.3 \times 0.095)}{1000} = 222.4\,\text{t}$$

The cost of a vessel of 2180 m^3 has been estimated as \$411\,000 in 2006 (CEPCI = 499.6). Thus, the capital cost in 2010 (CEPCI = 550.8) using a scaling factor of 0.55 and installation factor of 1.10 is (refer to Equations (2.1) and (2.2))

$$\$411\,000 \times \left(\frac{2290}{2180}\right)^{0.55} \times \left(\frac{550.8}{499.6}\right) \times 1.10 = \$512\,000$$

Thus, the annualized capital cost and EI from construction materials of the fermentor unit are

$$\begin{bmatrix} CC_{\text{fermentor}} \\ CI_{\text{fermentor}} \end{bmatrix} = \begin{bmatrix} 512\,000 \times \dfrac{0.1 \times (1+0.1)^{10}}{(1+0.1)^{10} - 1} \\[4mm] \dfrac{222.4}{10} \end{bmatrix} = \begin{bmatrix} \$83\,300 \text{ y}^{-1} \\ 22.24 \text{ t y}^{-1} \end{bmatrix}$$

Finally, the total annual unit costs are determined from the earlier result and data provided for utilities, auxiliary raw materials and process emissions using Equation (7.4) as follows:

$$\overline{\mathbf{O}}_{fermentor} = \begin{bmatrix} O_{fermentor} \\ O_{I,fermentor} \end{bmatrix} = \begin{bmatrix} 5 \\ 7.7 \end{bmatrix} \times [19] + \begin{bmatrix} 0.015 & 12 \\ 0 & 0.186 \end{bmatrix} \times \begin{bmatrix} 833\,400 \\ 1144 \end{bmatrix} + \begin{bmatrix} 0 \\ 1 \end{bmatrix} \times 30\,000 + \begin{bmatrix} 83\,300 \\ 22.24 \end{bmatrix}$$

$$\overline{\mathbf{O}}_{fermentor} = \begin{bmatrix} 95 + 26\,229 + 0 + 83\,300 \\ 146.3 + 212.8 + 30\,000 + 22.24 \end{bmatrix} = \begin{bmatrix} \$109\,624 \text{ y}^{-1} \\ 30\,381 \text{ t y}^{-1} \end{bmatrix}$$

7.3 Evaluation of Streams

Equivalent to the cost of production (COP) and the value on processing (VOP) of intermediate streams, the *impact cost of production* (ICP) and the *credit value on processing* (CVP) concepts can be applied to identify intermediate streams that are environmentally unsustainable or hot spots and divert the processing paths towards sustainable production.

For a biorefinery product, CVP = D_p.
For a feedstock, ICP = I_f.

Consider $\overline{\mathbf{V}}$ as a vector containing the "values" (VOP and CVP) of a feed (inlet stream) f to a process unit k. The vector of values is calculated from the known vector of values of the product (outlet) streams p and the total costs of a process unit k, shown as

$$\overline{\mathbf{V}}_f = \left[\sum_{p=1}^{q} \overline{\mathbf{V}}_p P_p - \overline{\mathbf{O}}_k \right] \bigg/ \sum_{f=1}^{g} F_f \tag{7.6}$$

where q is the number of products (excluding emissions/wastes), g is the number of feedstock considered as main material streams (excluding auxiliary raw materials) and P_p and F_f correspond to the mass flow rates of the product and feedstock, respectively.

An allocation by mass or energy can be applied to differentiate between streams generating from a common process path or tree. An impact allocation by economic value allows interactions between the economic and environmental values. If the two trends in the two values can be merged together, such that environmentally sustainable products are also economically profitable products, the economic value can be regarded as a good indicator for impact allocation. The allocation factor (α) of a product stream from a process unit is calculated from the following equation, a direct function of process models. The VOP of intermediate streams can be readily calculated to capture market value variability and process unit costs allocated to the streams:

$$\alpha_p = VOP_p P_p \bigg/ \sum_{p=1}^{q} VOP_p P_p \tag{7.7}$$

Consider now $\overline{\mathbf{C}}$ as a vector containing the costs (COP and ICP) of a product (outlet stream) p from a process unit k. The vector of costs $\overline{\mathbf{C}}$ can be estimated for a product (outlet) stream p from the known vector of costs of the feed (inlet) streams f, the total costs of a process unit k ($\overline{\mathbf{O}}_k$) and the allocation factor α:

$$\overline{\mathbf{C}}_p = \left[\sum_{f=1}^{g} \overline{\mathbf{C}}_f F_f + \overline{\mathbf{O}}_k \right] \alpha_p / P_p \tag{7.8}$$

Calculation of the ICP of a stream proceeds in the forward direction from the resource (cradle) to its production and the calculation of the CVP proceeds in the backward direction from the stream's end use (grave) to the beginning of its processing. The difference between the VOP and COP of a stream p is its economic margin and the difference between the CVP and ICP of the stream gives its impact savings (Δi_p). The relative percentage of impact saving (s_p) from the product by displacement of its counterpart fossil based product is also given:

$$\overline{V}_p - \overline{C}_p = \left[\dfrac{\Delta e_p}{\Delta i_p} \right] \qquad (7.9)$$

$$s_p = \dfrac{\Delta i_p}{(I_{peq} \times \beta)_p} \times 100 \qquad (7.10)$$

Exercise 4. Answer the following questions for a biorefinery producing bioethanol and DDGS (dried distillers grains with solubles) from wheat as shown in Figure 7.2.

a. Calculate the operating costs (i.e., without the fixed costs CC_k and CI_k) for all the process units.
b. Develop the EVEI modeling equations for all streams.
c. Calculate the costs, values and margins for all streams.
d. Estimate the GHG emissions savings from each of the biorefinery products. Calculate the ICP allowed for the 35% saving target from bioethanol and 20% from DDGS.

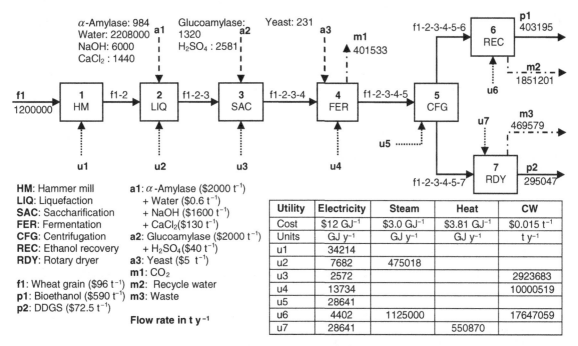

Figure 7.2 *Process network with mass balance and utility requirements for a biorefinery converting wheat grain into bioethanol and DDGS. CW: cooling water.*

Additional information, EI costs of wheat, auxiliary raw materials and natural gas, is given in Table 7.1. Neglect the EI from transportation in calculations. Assume that steam or heat is produced from natural gas at 80% efficiency. For DDGS $\beta = 0.8$ and $I_{peq} = 0.726$ kg kg^{-1}; for bioethanol $\beta = 0.6$ and $I_{peq} = 3.8$ kg kg^{-1} (from Exercise 2).

Table 7.1   *EI cost of the various inputs to the biorefinery of Exercise 4.*

Input	EI Cost (CO_2 eq.)	Units
Wheat	−0.613	kg kg^{-1}
α-Amylase	1	kg kg^{-1}
Glucoamylase	7.7	kg kg^{-1}
Yeast	7.7	kg kg^{-1}
H_2SO_4	4.05	kg kg^{-1}
NaOH	1.12	kg kg^{-1}
$CaCl_2$	1.12	kg kg^{-1}
Natural gas (NG)	0.0612	kg MJ^{-1}
Electricity	0.1857	kg MJ^{-1}

Solution to Exercise 4

a. By substituting the mass and energy flow rates and corresponding costs from Figure 7.2 and Table 7.1 in Equation (7.4), the operating costs for each process unit are calculated and shown in Table 7.2.

Table 7.2   *Operating economic ($ y^{-1}) and EI costs (t CO_2 eq. y^{-1}) for the process units in the biorefinery process network shown in Figure 7.2.*

Unit	O and O_1	Utilities	Auxiliary Raw Materials	Emissions/Wastes	Total
1 HM	O	410 568			410 568
	O_1	6 354			6 354
2 LIQ	O	1 873 500	13 080 000		14 953 500
	O_1	37 765	9 317		47 082
3 SAC	O	74 719	2 743 228		2 817 947
	O_1	478	20 616		21 094
4 FER	O	314 811	1 156		315 967
	O_1	2 551	1 780	401 533	405 864
5 CFG	O	343 692			343 692
	O_1	5 319			5 319
Common subtotal for allocation	O	3 017 290	15 824 384		18 841 674
	O_1	52 467	31 713	401 533	485 713
6 REC	O	4 536 280			4 536 280
	O_1	86 880			86 880
7 RDY	O	2 967 210			2 967 210
	O_1	47 460			47 460
Total	O	10 520 780	15 824 384		26 345 164
	O_1	186 807	31 713	401 533	6 20 053

b. Refer to Chapter 7 – Additional Examples and Exercises in the Companion Website – for the solution strategy for this problem.

c. The ICP of the feedstock corresponds to the EI cost of wheat from Table 7.1.

   The EI credit value of bioethanol is calculated as in Exercise 2 by using Equation (7.3) and neglecting EI from transportation:

$$D_p = 0.347 \text{ kg kg}^{-1}$$

This value becomes the CVP of the bioethanol product.

Similarly, the EI credit value of DDGS is as follows:

$$D_p = 0.8 \times 0.726 = 0.581 \text{ kg kg}^{-1}$$

The results after substitution of the corresponding values in the expressions developed in part b are shown in Table 7.3. The allocation factor for the stream going to the bioethanol recovery is

$$\alpha_{f1\text{-}2\text{-}3\text{-}4\text{-}5\text{-}6} = 0.9268$$

The allocation factor for the stream going to the rotary dryer is

$$\alpha_{f1\text{-}2\text{-}3\text{-}4\text{-}5\text{-}7} = 0.0732$$

Table 7.3 *Stream results from EVEI calculations.*

Stream	Flow Rate (t y^{-1})	VOP ($ t^{-1})	COP ($ t^{-1})	Δe ($ t^{-1})	CVP (kg kg^{-1})	ICP (kg kg^{-1})	Δi (kg kg^{-1})
f1	1 200 000	194.10	96.00	98.10	−0.257	−0.613	0.356
f1-2	1 200 000	194.44	96.34	98.10	−0.252	−0.608	0.356
f1-2-3	3 416 424	72.67	38.22	34.46	−0.075	−0.200	0.125
f1-2-3-4	3 420 325	73.41	39.00	34.42	−0.068	−0.193	0.125
f1-2-3-4-5	3 019 022	83.28	44.29	38.99	0.057	−0.084	0.141
f1-2-3-4-5-6	2 254 395	103.50	55.11	48.40	0.024	−0.103	0.126
f1-2-3-4-5-7	764 627	24.09	12.83	11.27	0.162	−0.024	0.186
p1	403 195	590	319.40	270.60	0.347	−0.358	0.706
p2	295 047	72.50	43.31	29.19	0.581	0.099	0.482

d. From data provided and results in Table 7.3, the EI savings for each product are (Equation (7.10))

$$s_{bioethanol} = \frac{\Delta i_{p1}}{(I_{peq} \times \beta)_{p1}} \times 100 = \frac{0.706}{3.8 \times 0.6} \times 100 = 31\%$$

$$s_{DDGS} = \frac{\Delta i_{p2}}{(I_{peq} \times \beta)_{p2}} \times 100 = \frac{0.482}{0.726 \times 0.8} \times 100 = 83\%$$

From Equation (7.10), the minimum EI saving margin of bioethanol for the target of 35% is

$$\Delta i_{target} = s_{bioethanol} \times (I_{peq} \times \beta)_{p1} = 0.35 \times 3.8 \times 0.6 = 0.798 \text{ kg kg}^{-1}$$

The ICP limit for bioethanol is

$$ICP_{limit} = CVP - \Delta i_{target} = 0.347 - 0.798 = -0.45 \text{ kg kg}^{-1}$$

Similarly, the ICP limit for DDGS for 20% saving target is

$$ICP_{limit} = CVP - \Delta i_{target} = 0.581 - 0.20 \times 0.726 \times 0.8 = 0.46 \text{ kg kg}^{-1}$$

7.4   Environmental Impact Profile

Construction of the EI profile following the same principles of the value analysis profile is shown for analyzing the EI of streams, paths, trees or entire networks. The biorefinery network shown in Figure 7.2 and analyzed in Exercise 4 features two trees with a common path (units 1 to 5). This path has two intermediate product streams forming the two paths 1-2-3-4-5-6 and 1-2-3-4-5-7, leading to two products, bioethanol and DDGS, respectively.

The CVP and ICP of the feedstock and intermediate streams leading to the end products can be plotted sequentially with their EI values on the y axis and cumulative flow rates on the x axis. Such a profile shows the EI costs, values and margins throughout a process network. Figure 7.3 shows the EI profile of the streams in the biorefinery in Figure 7.2.

Emissions or wastes are already accounted for in the CVP and ICP of the feedstock, intermediate and end product streams and are not separately presented in the profiles. This is different from the value analysis profile, where any stream contributing to a network's mass balance is shown. The area bounded between the CVP and ICP of a stream (p) shows the stream's EI credit margin, $(i)_p$, shown in Equation (7.9), multiplied by the flow rate of the stream.

The shaded area bounded between the CVP and ICP of the feedstock or the total area bounded between the CVP and ICP of the products presents the EI credit margin of the biorefinery.

Notably, the ICP remains negative for the tree with the pathway producing bioethanol due to the propagation of the negative EI cost from the biomass feedstock. In the DDGS pathway, a shift in the ICP from a negative to a positive value occurs, after intermediate stream f1-2-3-4-5-7 is processed in unit 7 (rotary dryer). This means that the negative EI cost of the biomass feedstock has been offset by the cumulative EI costs in this pathway, especially in the rotary dryer.

A shift in CVP from positive to negative occurs in the fermentation unit that produces stream f1-2-3-4-5 from stream f1-2-3-4. This means that, at this point, the EI credits gained by the biorefinery products have been offset by the EI costs of the fermentation and downstream units. However, Δi remains positive due to the propagation of the negative EI cost of the feedstock captured in the ICP. The insights obtained from the EI profile are thus into the hot spot analysis in a quantitative and comprehensive way. Note that the condition for a sustainable stream is

$$\text{CVP} > \text{ICP} \quad \text{or} \quad \Delta i > 0 \qquad (7.11)$$

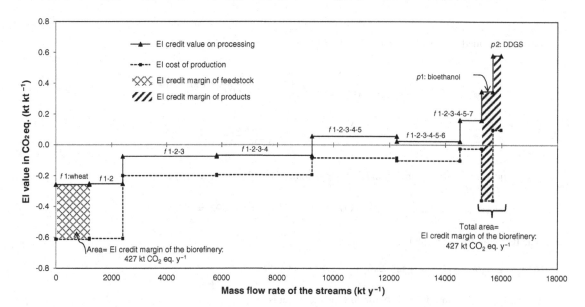

Figure 7.3   *Environmental impact profiles of the streams in a wheat based biorefinery.*

Thus, the CVP line must be above the ICP line in an EI profile for a sustainable biorefinery system. The EI credit margin of a biorefinery system across a boundary is obtained from the EI credit margin of its feedstocks or that of its products across the boundary. To avoid double counting, the EI credit margins of the auxiliary, emission and waste material streams must not be added with those of the biorefinery products.

In this way, Figure 7.3 shows that the EI credit margin of the biorefinery system as obtained from wheat feedstock or bioethanol and DDGS products is 427 kt y^{-1}. This is calculated by multiplying their EI credit margin with their flow rate. For example, using the product or feedstock stream results on Δi and flow rates, the following is obtained:

$$\frac{403\,195 \times 0.706 + 295\,047 \times 0.482}{1000} = 427 \text{ kt y}^{-1}$$

or

$$\frac{1\,200\,000 \times 0.356}{1000} = 427 \text{ kt y}^{-1}$$

7.5   Product Economic Value and Environmental Impact (EVEI) Profile

A product *EVEI profile* is a plot of a product *EI (environmental impact)* on the y axis versus *EV (economic value)* on the x axis. In such a profile, an *EVEI composite curve* shows the total amount of economic cost incurred in each EI interval. The EI cost of a production pathway increases from a feedstock to a product. There are the EI and economic costs from the feedstock, utility, auxiliary raw material and emission streams. To build an EVEI composite curve, the total COP and ICP of the new product is split into four main cost sources: feedstock, utilities, auxiliary raw materials and emissions. The utility cost for the production of a product is the cumulative propagated utility costs right from the beginning to the end of the production chain. The costs of auxiliary raw materials and emissions are calculated in a way similar to the utility cost.

The curve starts at an economic cost equal to zero (COP = 0). Figure 7.4 shows the EVEI composite curve, presenting the total EI cost versus the total COP of the feedstock, utility, auxiliary and emission streams, sequentially.

When evaluating the GHG emissions, the first point of the EI costs is the amount of CO_2 binding by the feedstock as a negative cost. Allocation factors to the product and any intermediate stream between the product and feedstock must be accounted for in propagating this negative amount of CO_2 binding. The final allocation factor (α'_p) is the product of all the allocation factors in a stream's production path.

Hence, the starting point of the EVEI composite curve is

$$(0, -B_f \times F_f \times \alpha'_p) \tag{7.12}$$

where f indicates the feedstock; hence, F_f is the flow rate of the feedstock. B_f is the CO_2 binding by photosynthesis, as noted before.

Calculate the propagated cost of feedstock production using the final allocation factor α'_p, shown below. COP_f and ICP_f refer to the economic and the EI cost of production of the feedstock, respectively:

$$(COP_f \times F_f, ICP_f \times F_f \times \alpha'_p) \tag{7.13}$$

Plot the points using cumulative total costs. The last point must be equal to the coordinates shown below

$$(COP_p \times P_p,\ ICP_p \times P_p) \tag{7.14}$$

where p indicates the alternative new product. The flow rate of the new biorefinery product is P_p; COP_p and ICP_p are the economic and the EI the cost of production of the alternative new product, p.

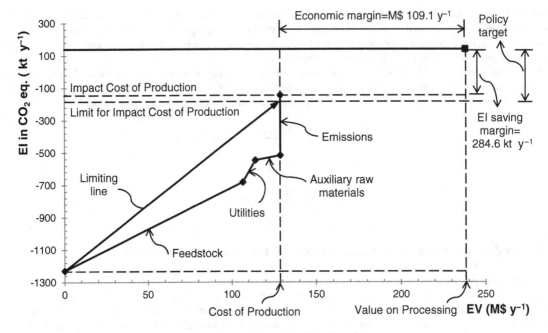

Figure 7.4   *EVEI profile of bioethanol.*

The total EI cost incurred from the existing product is equal to the total EI credit value to the biorefinery product and is shown as a horizontal line at the top of Figure 7.4. The total EI credit value to the biorefinery product can be obtained using

$$\text{Total EI credit value to the biorefinery product} = P_p \times CVP_p \tag{7.15}$$

Another useful measure is the *limiting line*. There are specific EI saving targets for alternative products compared to the EI costs incurred from their counterpart fossil-derived products to be displaced. If ICP_{limit} indicates the maximum limiting EI cost allowed for the new product, then the policy target for an EI saving by the new product due to substitution of an existing product

$$\text{Policy target for EI saving} = P_p \times (CVP_p - ICP_{limit}) \tag{7.16}$$

The *limiting line* is the line between the starting point of the EVEI composite curve $(0, -B_f \times F_f \times \alpha'_p)$ and $(P_p \times COP_p, P_p \times ICP_{limit})$. The slope of the limiting line that sets an attainable policy target indicates the allowable EI per amount of money spent to produce a product. The economic margin $(\Delta e \times P_p)$ and EI saving margin $(\Delta i \times P_p)$ are also shown in Figure 7.4.

Figure 7.4 shows the product EVEI profile for bioethanol from the biorefinery in Figure 7.2. The environmental impact (EI) is presented in kt CO_2 eq. y^{-1} and the economic value (EV) is shown in M$ y^{-1} (M$ = million dollars). The features discussed are shown on the plot. It clearly shows that the bioethanol EVEI composite curve ends above the limiting value of $P_p \times ICP_{limit}$. Hence, the biorefinery in Figure 7.2 cannot achieve the policy target for greenhouse gas emission reduction.

The difference between the EI saving margin $= P_p \times (CVP_p - ICP_p)$ and the policy target for EI saving $= P_p \times (CVP_p - ICP_{limit})$ of a product is

$$P_p \times (ICP_{limit} - ICP_p) \qquad (7.17)$$

Equation (7.17) determines whether the product complies with the policy or not according to the following criteria:

$ICP \le ICP_{limit}$. The product is policy compliant and Equation (7.17) ≥ 0. A result >0 indicates that the product has a surplus EI saving, in relation to the policy target.

$ICP > ICP_{limit}$. The product is not policy compliant and Equation (7.17) < 0. A negative value indicates that the product has a deficit EI saving, in relation to the policy target.

Thus, Equation (7.17) gives useful information for both policy makers and process engineers. A surplus EI can be used to inform policy makers and set a stricter policy target. A deficit EI indicates the need for an emission reduction target for a production to be workable. Emission reduction can be achieved by process integration.

Several options for ICP reduction can be concluded from the EVEI composite curve, as follows. Since most of the EI costs come from the CO_2 emissions from the fermentation unit, a first option could be to capture, partially or completely, that CO_2. It can be seen that although raw materials contribute significantly to the economic costs, their EI costs are minimal. Another improvement option is to reduce the demand for utilities by increasing the thermal efficiency. Although these options are technically feasible, a more detailed analysis must be performed to decide which option or combination of options will be the best. Any modification could incur a capital or operating cost that increases the COP. Thus, the trade-offs between EI and EV must be evaluated for any modification. Figure 7.5 shows how the composite curve is shifted according to different improvement options. The trade-offs can be shown graphically.

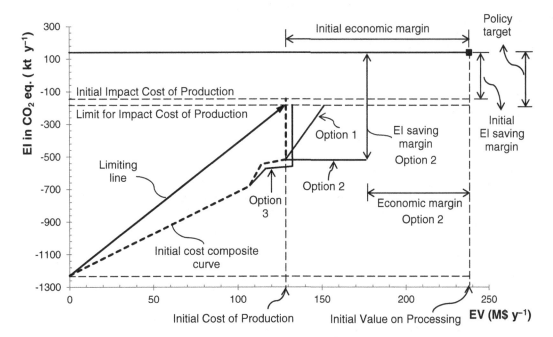

Figure 7.5 *Shifted EVEI composite curves of bioethanol for three different options: 1: partial CO_2 capture for target, 2: total CO_2 capture and 3: efficiency increase in production/usage of utilities to achieve the target savings.*

The shifts in Figure 7.5 show how each option may affect the ICP or COP of the biorefinery product (so and thus its margins). This makes the product EVEI profile a powerful tool to evaluate alternatives to achieve the set targets. Note that the trade-off ratio (difference in EI divided by the difference in EV) is different for each option. The higher the trade-off ratio, the greater the benefits from the biorefinery; this implies that a higher EI saving is possible with lower economic costs.

Option 1 achieves the target but with an increase in economic cost, that is, with a low trade-off ratio from the initial curve.
Option 2 can achieve the target and beyond but at a considerably higher economic cost.
Option 3 seems to be acceptable with almost no increase in COP; however, it may not be enough to reach the target.

Other feasible options can be explored employing the product EVEI profile. The EVEI composite curve can be further differentiated for individual types of raw materials and individual processing steps for detailed differential analyses. For example, utilities can be segregated into electricity, steam, fuels, etc. The option of partially substituting natural gas by DDGS as fuel to produce the steam used in the bioethanol recovery column is explored in Exercise 5.

Exercise 5. Develop the *product EVEI profile* for DDGS from the biorefinery in Example 4. Assume a 35% GHG reduction policy. Use the EVEI composite curves for bioethanol in Figures 7.4 and 7.5 to answer to the following questions:

a.   Calculate the EI saving margin surplus/deficit of DDGS.
b.   Explain a strategy to transfer some of the EI saving margin surpluses between the two products.
c.   Determine the new values and costs of the products from applying the strategy and represent it in the *product EVEI profile*.
d.   Explain the effects of the modifications on the overall biorefinery margins.

Additional information:

• DDGS carbon weight percentage $= 45\%$.
• DDGS heating value $= 21.8$ MJ kg^{-1} and combustion efficiency $= 62\%$.
• Cost of installed biomass boiler[8] $= \$200$ kW^{-1}; assume 5% interest rate.
• CO_2 binding, $B_f = 1.1$ kg kg^{-1} (Exercise 1).
• Wheat cost, $C_f = \$96$ t^{-1} (Exercise 4).
• $I_f(CO_2$ eq.$) = 0.492$ kg kg^{-1} (Exercise 1).

Solution to Exercise 5. Refer to Chapter 7 – Additional Examples and Exercises in the Companion Website – for the solution of this exercise.

These exercises show how important it is to have the right policy target for biorefinery products. In this case, an emission reduction target of 35% used for the bioethanol product from a biorefinery system primarily producing bioethanol can mean that to achieve this target for the bioethanol product there will be a lesser amount of global warming potential reduction from the overall system. However, if assessed differentially for individual products and processing routes, the exact policy targets for emission reductions from individual products can be determined. Thus, the EVEI tool allows a new technology or product assessment and thereby setting the policy targets. This is where the EVEI tool will prove to be powerful and effective.

7.6   Summary

The general steps for the application of the EVEI analysis of biorefinery systems are:

1.   Gather all the economic and EI data and process information required as inputs.
2.   Define the level of detail, EI category, functional units and boundaries of the system to be analyzed.
3.   If the biomass feedstock production is to be included, develop the models to determine its EI cost.

4. Simulate the biorefinery processes for the mass and energy balance calculations.
5. Determine the costs and values of the feedstock and products, according to the boundaries and data availability.
6. Find the cost of utilities, auxiliary raw materials and emissions. Calculate the fixed costs from construction and capital costs. Then calculate the total economic costs of the process units.
7. Perform the EVEI calculations to obtain the economic and environmental costs, values and margins of the streams.
8. Develop the stream economic and EI profiles of the biorefinery streams.
9. Identify the hot spots with high EI costs and positive EI margins in the biorefinery process network.
10. Develop the *product EVEI profiles* for the biorefinery products.
11. Analyze the *product EVEI profiles*, identify hot spots and determine surplus/deficits of product margins.
12. Divide further the cost composite curve if required for a more detailed analysis.
13. Identify alternatives and develop strategies to improve the biorefinery performance.
14. Determine changes in costs and values after each modification.
15. Base the analysis on the greatest global emission potential reduction and set that as the new target.
16. Present the changes as shifted curves in the EVEI profiles and analyze the effects on the overall biorefinery performance.
17. Compare the performance of the alternatives generated.
18. Present the best biorefinery system design and the most stringent policy target that can be achieved by such a design.

References

1. E. Martinez-Hernandez, G.M. Campbell, J. Sadhukhan, Economic value and environmental impact (EVEI) analysis of biorefinery systems, *Chemical Engineering Research and Design*, **91**(8), 1418–1426 (2013).
2. http://www.ccalc.org.uk/.
3. H. Baumann and A.M. Tillman, *The Hitch Hiker's Guide to LCA: An Orientation in Life Cycle*, Studentlitteratur, Lund, Sweden, 2004.
4. A. Azapagic and R. Clift, The application of life cycle assessment to process optimisation, *Computers and Chemical Engineering*, **23**, 1509–1526 (1999).
5. E. Martinez-Hernandez, M.H. Ibrahim, M. Leach, *et al.*, Environmental sustainability analysis of UK whole-wheat bioethanol and CHP systems, *Biomass Bioenergy*, **50**, 52–64 (2013).
6. J. Sadhukhan, M.M. Mustafa, N. Misailidis, *et al.*, Value analysis tool for feasibility studies of biorefineries integrated with value added production, *Chemical Engineering Science*, **63**, 503–519 (2008).
7. J. Sadhukhan, K.S. Ng, N. Shah, H.J. Simons, Heat integration strategy for economic production of CHP from biomass waste, *Energy Fuels*, **23**, 5106–5120 (2009).

8

Optimization

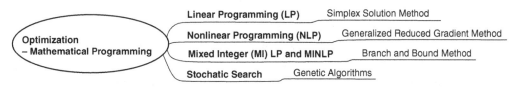

Structure for Lecture Planning

8.1 Introduction

Optimization is a core activity in chemical process design. The focus of this chapter is on mathematical programming based optimization. General purpose software to carry out optimization is readily available, such as GAMS, AMPL, AIMMS, and gPROMS. In addition, an add-in tool (called Solver) is available in the Microsoft Excel spreadsheet for this purpose. Thus, problem formulations and recognition of the algorithm and solution strategy to be used to solve the problem are important. It is also important to understand the philosophy of the search methods, for finding optimal solutions and derivations are a useful way to learn the methods. These aspects are discussed in this chapter.

An optimization problem is concerned with the maximization or minimization of an objective or even simultaneous maximization and minimization of several objectives. An objective function can involve the maximization of economic margin, minimization of environmental footprint, minimization of resource use, etc. Depending on the number of objective functions under consideration, a problem may be defined as a single- or multi-objective optimization problem. In addition to the objective function, there are constraints that must be satisfied by a solution to a problem. Constraints may be related to process capacities, bounds on the variables, production capacity, product demand, raw material availability and others. An optimization problem, consisting of an objective function and related constraints, is formulated in terms of independent and dependent variables and constants. Independent variables are free variables as the name suggests – a solution for an optimization problem is deduced in terms of optimum values of these independent variables. The dependent variables are presented using linear or nonlinear correlations or functions in terms of independent variables. Hence, the key components in an optimization problem are:

- objective function,
- constraints in the form of equalities and/or inequalities,

Biorefineries and Chemical Processes: Design, Integration and Sustainability Analysis, First Edition.
Jhuma Sadhukhan, Kok Siew Ng and Elias Martinez Hernandez.
© 2014 John Wiley & Sons, Ltd. Published 2014 by John Wiley & Sons, Ltd.
Companion Website: http://www.wiley.com/go/sadhukhan/biorefineries

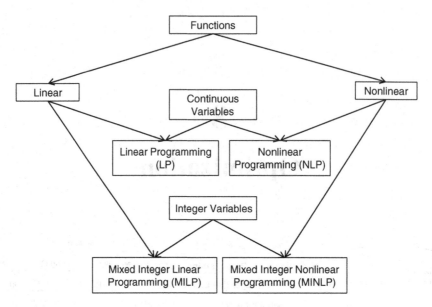

Figure 8.1   *Definitions of LP, NLP, MILP and MINLP optimization problems based on the nature of the variables involved.*

- correlations or functions involving independent and dependent variables,
- constants or parameters.

A generic optimization problem formulation can be shown as follows:

$$
\begin{aligned}
\text{Minimize} \quad & z(x, y) && \text{(objective function)} \\
\text{s.t.} \quad & y = F(x) && \text{(equality constraints)} \\
& a \leq x \leq b && \text{(inequality constraints)} \\
& c \leq y \leq d && \text{(inequality constraints)} \\
& x \in X, y \in Y
\end{aligned}
\tag{8.1}
$$

where x is a vector of independent variables, whilst y is a vector of dependent variables. Their optimal values must be within the bounds a and b, and c and d, respectively. The term y is correlated by functions F with x. The objective function z, a function of x and y, needs to be minimized. Hence, a solution in terms of optimal values of x within the given constraints must be obtained for which the value of z will be a minimum.

If any of the correlations, objective function and constraints is nonlinear, the problem needs to be solved as a nonlinear programming (NLP); otherwise the problem is a linear programming. An LP (linear programming) problem can be solved using gradient based approaches. Special solution techniques may be required to solve an NLP problem. Variables involved can also be integer, such as for the selection of a process unit, thus forming a mixed integer linear programming (MILP) or mixed integer nonlinear programming (MINLP) problem. Figure 8.1 shows the four types of optimization problem, LP, NLP, MILP and MINLP, depending on the nature of the variables.

8.2   Linear Optimization

The field of optimization largely gained prominence through the applications of linear programming (LP) in the planning of petroleum refineries, starting as early as the 1940s[1]. To date, it is common within the petroleum refining industry to refer to the associated tool as the "planning LP." The most popular of such applications was for gasoline or petrol blending. In this problem, LP is used to determine the optimal mix or blend of up to 20 or more different components into four or more grades of gasoline with different qualities, mainly in terms of octane content and other properties, depending on market requirements.

A problem with linear objective functions and constraints can be formulated and solved as an LP model. Since linear functions are convex, a unique global optimum solution can be obtained. A popular solution strategy for an LP is an algorithm called the simplex method, almost synonymous with the LP methodology itself. The development of the simplex method is attributed to the seminal work by George B. Dantzig[2]. In fact, it was really Dantzig's invention of the simplex algorithm that gave birth to LP. The major concepts of simplex are applied in Exercise 1 discussed as follows.

Exercise 1. The primary fossil resource depletion data for five classes of sewer grid system for municipal wastewater treatment with different average per capita equivalents (PCE) and average lengths are shown in Table 8.1. The basis for primary fossil resource depletion data extraction is a sewer grid system of 1000 km length. The cost-effective minimum and maximum acceptable lengths of the five classes of sewer grid system are also shown in Table 8.1. Calculate individual lengths of five classes of sewer grid system for minimum primary fossil resource depletion for a total length of 35 000 km. Calculate the value of minimum primary fossil resource depletion. Assume that the primary fossil resource depletion is proportional to the length of a sewer grid system.

Table 8.1 *Primary fossil resource depletion data for five classes of sewer grid system.*

Class	1	2	3	4	5
PCE	233 225	71 133	24 864	5 321	806
Average length (km)	583	242	109	30	6
Minimum acceptable length (km)	2 000	10 000	15 000	1 000	0
Maximum acceptable length (km)	3 000	15 000	20 000	5 000	1 000
Primary fossil resource depletion (kg)	241 397.8	231 277.8	220 269.9	210 487.2	198 795.9

Solution to Exercise 1. Here i is defined as the index for the class of sewer grid system, while the decision variables x_i denote the optimum length (in km) for i class of the sewer grid system. The objective function for minimization is formulated as follows:

Minimize:

$$241\,397.8 \times \frac{x_1}{1000} + 231\,277.8 \times \frac{x_2}{1000} + 220\,269.9 \times \frac{x_3}{1000} + 210\,487.2 \times \frac{x_4}{1000} + 198\,795.9 \times \frac{x_5}{1000} \qquad (8.2)$$

Subject to constraints:

$$2000 \leq x_1 \leq 3000 \qquad (8.3)$$

$$10\,000 \leq x_2 \leq 15\,000 \qquad (8.4)$$

$$15\,000 \leq x_3 \leq 20\,000 \qquad (8.5)$$

$$1000 \leq x_4 \leq 5000 \qquad (8.6)$$

$$0 \leq x_5 \leq 1000 \qquad (8.7)$$

$$\sum_{i=1}^{5} x_i = 35\,000 \qquad (8.8)$$

As these equations are all linear, LP can be used to solve the problem, or the problem can be solved analytically. By solving the problem using Microsoft Excel's Solver, the optimal length values of individual classes of sewer grid system

Table 8.2   *Optimal length values of individual classes of the sewer grid system and primary fossil resource depletion values.*

Class	Length (km)	Minimum Primary Fossil Resource Depletion (kg)
1	2 000	482 795.6
2	10 000	2 312 778
3	17 000	3 744 588.3
4	5 000	1 052 436
5	1 000	198 795.9
Total	35 000	7 791 393.8

and primary fossil resource depletion values are obtained as shown in Table 8.2. Thus, the minimum objective function value is 7 791 394 kg.

Based on the given numerical data, to avoid the possibility of ill-conditioning of matrices in the solution procedure, scaling of the variables by expressing x_i in terms of multiples of 1000 km can be performed. Thus, the following optimization problem formulation is considered to discuss the solution method:

$$\begin{aligned}
\min \quad & c = 0.241x_1 + 0.231x_2 + 0.220x_3 + 0.210x_4 + 0.199x_5 \\
\text{s.t.} \quad & x_1 \geq 2 \\
& x_1 \leq 3 \\
& x_2 \geq 10 \\
& x_2 \leq 15 \\
& x_3 \geq 15 \\
& x_3 \leq 20 \\
& x_4 \geq 1 \\
& x_4 \leq 5 \\
& x_5 \leq 1 \\
& x_1 + x_2 + x_3 + x_4 + x_5 = 35 \\
& x_1, x_2, x_3, x_4, x_5 \geq 0
\end{aligned} \tag{8.9}$$

This LP problem can be solved in standard form using an algorithm called the simplex method. In the simplex method, all inequality constraints are converted into equality constraints by introducing a nonnegative variable, called the surplus or slack variable, to each inequality constraint, shown as follows.

8.2.1   Step 1: Rewriting in Standard LP Format

The optimization problem is rewritten in the standard form of LP as follows:

$$\min \quad 0.241x_1 + 0.231x_2 + 0.220x_3 + 0.210x_4 + 0.199x_5 \tag{8.10}$$

$$\begin{aligned}
\text{s.t.} \quad & x_1 - s_1 = 2 \\
& x_1 + s_2 = 3 \\
& x_2 - s_3 = 10 \\
& x_2 + s_4 = 15 \\
& x_3 - s_5 = 15 \\
& x_3 + s_6 = 20 \\
& x_4 - s_7 = 1 \\
& x_4 + s_8 = 5 \\
& x_5 - s_9 = 1
\end{aligned} \tag{8.11}$$

$$x_1 + x_2 + x_3 + x_4 + x_5 = 35$$
$$x_1, x_2, x_3, x_4, x_5 \geq 0 \tag{8.12}$$
$$s_1, s_2, s_3, s_4, s_5, s_6, s_7, s_8, s_9 \geq 0$$

Note the introduction of the surplus variables s_1, s_3, s_5, s_7 and s_9, the slack variables s_2, s_4, s_6 and s_8 in order to convert the inequality constraints into equality constraints. No slack or surplus variables are required for equality constraints.

8.2.2 Step 2: Initializing the Simplex Method

The variables $(x_1, x_2, \ldots, x_5)$ are considered as the *nonbasic variables*. The surplus and slack variables $(s_1, s_2, \ldots, s_9)$ are considered as the *basic variables*. This partition of variables allows writing the basic variables in terms of the nonbasic variables shown as follows.

The objective function is written together with the resulting algebraic equations that involve both the basic and nonbasic variables as shown in Equation (8.13), while Equation (8.14) shows the equality constraint in terms of the nonbasic variables only.

$$
\begin{aligned}
s_1 &= -2 & +x_1 \\
s_2 &= 3 & -x_1 \\
s_3 &= -10 & +x_2 \\
s_4 &= 15 & -x_2 \\
s_5 &= -15 & +x_3 \\
s_6 &= 20 & -x_3 \\
s_7 &= -1 & +x_4 \\
s_8 &= 5 & -x_4 \\
s_9 &= -1 & +x_5 \\
35 &= & +x_1 & +x_2 & +x_3 & +x_4 & +x_5 \\
c &= & +0.241x_1 & +0.231x_2 & +0.220x_3 & +0.210x_4 & +0.199x_5
\end{aligned} \tag{8.13}
$$

$$x_1 + x_2 + x_3 + x_4 + x_5 = 35 \tag{8.14}$$

Alternatively, the system is represented in a practical tabulated form known as the *simplex tableau*.

To keep the a record of the iterations and facilitate the algebraic calculations, the problem can be represented in a simplex tableau, shown in Table 8.3. Note that the simplex tableau is another way of representing the standard LP format. The first row in the table contains the objective function, indicated as **c**. The following rows contain the equality

Table 8.3 Simplex tableau 1.

	x_1	x_2	x_3	x_4	x_5	s_1	s_2	s_3	s_4	s_5	s_6	s_7	s_8	s_9	
min c	0.241	0.231	0.220	0.210	0.199	0	0	0	0	0	0	0	0	0	b
A	1	0	0	0	0	−1	0	0	0	0	0	0	0	0	2
	1	0	0	0	0	0	1	0	0	0	0	0	0	0	3
	0	1	0	0	0	0	0	−1	0	0	0	0	0	0	10
	0	1	0	0	0	0	0	0	1	0	0	0	0	0	15
	0	0	1	0	0	0	0	0	0	−1	0	0	0	0	15
	0	0	1	0	0	0	0	0	0	0	1	0	0	0	20
	0	0	0	1	0	0	0	0	0	0	0	−1	0	0	1
	0	0	0	1	0	0	0	0	0	0	0	0	1	0	5
	0	0	0	0	1	0	0	0	0	0	0	0	0	−1	1
	1	1	1	1	1	0	0	0	0	0	0	0	0	0	35

Table 8.4 *Simplex tableau 2.*

	x_1	x_2	x_3	x_4	x_5	s_1	s_2	s_3	s_4	s_5	s_6	s_7	s_8	s_9	
min c	0.241	0.231	0.220	0.210	0.199	0	0	0	0	0	0	0	0	0	**b**
A	1	0	0	0	0	−1	0	0	0	0	0	0	0	0	2
	1	0	0	0	0	0	1	0	0	0	0	0	0	0	3
	0	1	0	0	0	0	0	−1	0	0	0	0	0	0	10
	0	1	0	0	0	0	0	0	1	0	0	0	0	0	15
	0	0	1	0	0	0	0	0	0	−1	0	0	0	0	15
	0	0	1	0	0	0	0	0	0	0	1	0	0	0	20
	0	0	0	1	0	0	0	0	0	0	0	−1	0	0	1
	0	0	0	1	0	0	0	0	0	0	0	0	1	0	5
	0	0	0	0	1	0	0	0	0	0	0	0	0	−1	1
	1	1	1	1	1	0	0	0	0	0	0	0	0	0	35
	N	N	N	N	N	B	B	B	B	B	B	B	B	B	
$\mathbf{x}^{(0)}$	0	0	0	0	0	−2	3	−10	15	−15	20	−1	5	−1	

constraints. Further, in the simplex tableau in Table 8.4, the *nonbasic variables* are indicated with the letter N and the *basic variables* are indicated with the letter B. Note that the nonbasic variables and the objective function are on the left hand side of the table and the basic variables are on the right hand side of the table. The last column in the table shows the values of the constants, denoted as vector **b**. Letter **A** is the matrix of the coefficients of the constraints.

8.2.3 Step 3: Obtaining an Initial Basic Solution

The LP problem is first *relaxed* by dropping out the equality constraint given in Equation (8.14). This will make *all the constraints in the relaxed problem as functions of basic variables.* An initial feasible solution, known as the *basic solution*, is then obtained in order to initialize the simplex search method. An initial basic feasible solution is obtained by setting the nonbasic variables to zero (i.e., $x_1 = x_2 = x_3 = x_4 = x_5 = 0$). The value of the objective function is thus $c = 0$. Then, the initial values of the basic variables obtained from each of the equality constraints given in Equation (8.13) are

$$s_1 = -2,\ s_2 = 3, s_3 = -10, s_4 = 15, s_5 = -15, s_6 = 20, s_7 = -1, s_8 = 5, s_9 = -1 \qquad (8.15)$$

These values form the initial basic solution $x^{(0)}$. This basic solution is recorded into an additional row (the last row) in the simplex tableau, as shown in Table 8.4.

8.2.4 Step 4: Determining Simplex Directions

From the initial solution of the basic variables, the nonbasic variables are moved to the maximum extents without violating the constraints. *The simplex search method is based on the directional movement of the nonbasic variables along the edges or extreme points of a feasible region bounded by the constraint functions.* There are as many possible directions as the number of nonbasic variables. A simplex direction is feasible if the initial constraints ($Ax = b$) are still met when a nonbasic variable x is incremented by Δx (i.e., the constraints $A(x + \Delta x) = b$ must be satisfied). Thus, subtracting $Ax = b$ from $A(x + \Delta x) = b$ gives

$$A\Delta x = 0 \qquad (8.16)$$

Only one nonbasic variable is changed at a time, while leaving the other nonbasic variables unchanged. Equation (8.16) can be presented in matrix form as follows and is solved for all the basic variables. A unit step change is assumed for x_1, that is, $\Delta x_1 = 1$. All other nonbasic variables are kept unchanged, that is, $\Delta x_2 = \Delta x_3 = \Delta x_4 = \Delta x_5 = 0$

$$
\begin{bmatrix}
1 & 0 & 0 & 0 & 0 & -1 & 0 & 0 & 0 & 0 & 0 & 0 & 0 & 0 \\
1 & 0 & 0 & 0 & 0 & 0 & 1 & 0 & 0 & 0 & 0 & 0 & 0 & 0 \\
0 & 1 & 0 & 0 & 0 & 0 & 0 & -1 & 0 & 0 & 0 & 0 & 0 & 0 \\
0 & 1 & 0 & 0 & 0 & 0 & 0 & 0 & 1 & 0 & 0 & 0 & 0 & 0 \\
0 & 0 & 1 & 0 & 0 & 0 & 0 & 0 & 0 & -1 & 0 & 0 & 0 & 0 \\
0 & 0 & 1 & 0 & 0 & 0 & 0 & 0 & 0 & 0 & 1 & 0 & 0 & 0 \\
0 & 0 & 0 & 1 & 0 & 0 & 0 & 0 & 0 & 0 & 0 & -1 & 0 & 0 \\
0 & 0 & 0 & 1 & 0 & 0 & 0 & 0 & 0 & 0 & 0 & 0 & 1 & 0 \\
0 & 0 & 0 & 0 & 1 & 0 & 0 & 0 & 0 & 0 & 0 & 0 & 0 & -1
\end{bmatrix}
\times
\begin{bmatrix}
\Delta x_1 = 1 \\
\Delta x_2 = 0 \\
\Delta x_3 = 0 \\
\Delta x_4 = 0 \\
\Delta x_5 = 0 \\
\Delta s_1 \\
\Delta s_2 \\
\Delta s_3 \\
\Delta s_4 \\
\Delta s_5 \\
\Delta s_6 \\
\Delta s_7 \\
\Delta s_8 \\
\Delta s_9
\end{bmatrix}
=
\begin{bmatrix}
0 \\ 0 \\ 0 \\ 0 \\ 0 \\ 0 \\ 0 \\ 0 \\ 0
\end{bmatrix}
$$

The columns on the left side of the matrix represent $x_1, x_2, \ldots, x_5$ and $s_1, s_2, \ldots, s_9$, respectively. Note that the matrix does not include the equality constraint in Equation (8.14) and the problem defined in this way is a relaxed LP problem.

The same procedure is followed for each of the nonbasic variables (x_1, x_2, x_3, x_4 and x_5) by using a unit step change $\Delta x_j = 1$ of a nonbasic variable x_j, while keeping other nonbasic variables unchanged. The results for each search direction are recorded in the simplex tableau shown in Table 8.5. These results will serve as the basis to determine the point to which the variables are to be moved in the search space, shown in Step 5.

8.2.5   Step 5: Determining the Maximum Step Size by the Minimum Ratio Rule

To continue with the algorithm, select the directions and step sizes for all Δx_j. If all Δx_j are positive, then the solution is improved with any value and the LP problem is unbounded. If there is at least one negative Δx_j, then apply the minimum

Table 8.5   *Simplex tableau 3.*

	x_1	x_2	x_3	x_4	x_5	s_1	s_2	s_3	s_4	s_5	s_6	s_7	s_8	s_9	
min c	0.241	0.231	0.220	0.210	0.199	0	0	0	0	0	0	0	0	0	b
A	1	0	0	0	0	−1	0	0	0	0	0	0	0	0	2
	1	0	0	0	0	0	1	0	0	0	0	0	0	0	3
	0	1	0	0	0	0	0	−1	0	0	0	0	0	0	10
	0	1	0	0	0	0	0	0	1	0	0	0	0	0	15
	0	0	1	0	0	0	0	0	0	−1	0	0	0	0	15
	0	0	1	0	0	0	0	0	0	0	1	0	0	0	20
	0	0	0	1	0	0	0	0	0	0	0	−1	0	0	1
	0	0	0	1	0	0	0	0	0	0	0	0	1	0	5
	0	0	0	0	1	0	0	0	0	0	0	0	0	−1	1
	N	N	N	N	N	B	B	B	B	B	B	B	B	B	
$\mathbf{x}^{(0)}$	0	0	0	0	0	−2	3	−10	15	−15	20	−1	5	−1	
$\Delta \mathbf{x}$ for x_1	1	0	0	0	0	−1	1	0	0	0	0	0	0	0	
$\Delta \mathbf{x}$ for x_2	0	1	0	0	0	0	0	−1	1	0	0	0	0	0	
$\Delta \mathbf{x}$ for x_3	0	0	1	0	0	0	0	0	0	−1	1	0	0	0	
$\Delta \mathbf{x}$ for x_4	0	0	0	1	0	0	0	0	0	0	0	−1	1	0	
$\Delta \mathbf{x}$ for x_5	0	0	0	0	1	0	0	0	0	0	0	0	0	−1	

Table 8.6   *Simplex tableau 4.*

	x_1	x_2	x_3	x_4	x_5	s_1	s_2	s_3	s_4	s_5	s_6	s_7	s_8	s_9
	N	N	N	N	N	B	B	B	B	B	B	B	B	B
$\mathbf{x}^{(0)}$	0	0	0	0	0	−2	3	−10	15	−15	20	−1	5	−1
Δx for x_1	1	0	0	0	0	−1	1	0	0	0	0	0	0	0
	—	—	—	—	—	$\dfrac{-2}{-(-1)}$	—	—	—	—	—	—	—	—

ratio rule given by

$$\lambda = \min\left\{ \frac{x_j^{(t)}}{-\Delta x_j} \right\} \tag{8.17}$$

The simplex search method uses the maximum feasible step size λ as given by the minimum ratio rule. Using *only the negative* Δx_j, find the maximum step size by the *minimum ratio rule* as shown in Equation (8.17). For example, if a vector direction is $\Delta x_j = (1, 0, 1, -2, -3)$ and the variables are $x_j^{(t)} = (0,1,0,2,1)$, the maximum step size is

$$\lambda = \min\left\{ \frac{2}{-(-2)}, \frac{1}{-(-3)} \right\} = \min\left\{ 1, \frac{1}{3} \right\} = \frac{1}{3}$$

Table 8.6 shows the example for Exercise 1 when the search direction for the nonbasic variable x_1 is selected.

Since there is only one negative Δx, that is, for x_1, the ratio $\lambda = -2$ determined for it yields the new solution as shown below:

$$\mathbf{x}^{(1)} = \mathbf{x}^{(0)} + \lambda \Delta \mathbf{x}$$

$$= (0,0,0,0,0,-2,3,-10,15,-15,20,-1,5,-1) + (-2)(1,0,0,0,0,-1,1,0,0,0,0,0,0,0)$$

$$= (-2,0,0,0,0,0,1,-10,15,-15,20,-1,5,-1) \tag{8.18}$$

Thus, the objective function value is updated as follows:

$$0.241(-2) + 0.231(0) + 0.220(0) + 0.210(0) + 0.199(0) = -0.482 \tag{8.19}$$

8.2.6   Step 6: Updating the Basic Variables

Note that in $\mathbf{x}^{(1)}$ (Equation (8.18)), x_1 having a nonzero value becomes a basic variable for the next iteration (iteration 2); s_1, on the other hand, having a zero value becomes a nonbasic variable for the next iteration step.

The simplex tableau is then updated with $\mathbf{x}^{(1)}$ and $\Delta \mathbf{x}$ for x_2.

	x_1	x_2	x_3	x_4	x_5	s_1	s_2	s_3	s_4	s_5	s_6	s_7	s_8	s_9
	B	N	N	N	N	N	B	B	B	B	B	B	B	B
$\mathbf{x}^{(1)}$	−2	0	0	0	0	0	1	−10	15	−15	20	−1	5	−1
Δx for x_2	0	1	0	0	0	0	0	1	1	0	0	0	0	0
	—	—	—	—	—	—	—	$\dfrac{-10}{-(-1)}$	—	—	—	—	—	—

Now follow the same procedure shown in Equations (8.18) and (8.19); for $\lambda = -10$, this yields the new solution as

$$\mathbf{x}^{(2)} = \mathbf{x}^{(1)} + \lambda \Delta \mathbf{x}$$
$$= (-2, 0, 0, 0, 0, 0, 1, -10, 15, -15, 20, -1, 5, -1) + (-10)(0, 1, 0, 0, 0, 0, 0, -1, 1, 0, 0, 0, 0, 0)$$
$$= (-2, -10, 0, 0, 0, 0, 1, 0, 5, -15, 20, -1, 5, -1)$$

In iteration 3, $\mathbf{x}^{(2)}$, x_2 having a nonzero value becomes a basic variable and s_3 having a zero value becomes a nonbasic variable. Thus, the objective function value is updated as follows:

$$0.241(-2) + 0.231(-10) + 0.220(0) + 0.210(0) + 0.199(0) = -2.792$$

The simplex tableau is now updated with $\mathbf{x}^{(2)}$ and $\Delta \mathbf{x}$ for x_3.

	x_1	x_2	x_3	x_4	x_5	s_1	s_2	s_3	s_4	s_5	s_6	s_7	s_8	s_9
	B	B	N	N	N	N	B	N	B	B	B	B	B	B
$\mathbf{x}^{(2)}$	-2	-10	0	0	0	0	1	0	5	-15	20	-1	5	-1
$\Delta \mathbf{x}$ for x_3	0	0	1	0	0	0	0	0	0	-1	1	0	0	0
	—	—	—	—	—		—	—	—	$\frac{-15}{-(-1)}$	—	—	—	—

As $\Delta \mathbf{x}$ has one negative and one positive value, the minimum ratio rule in Step 5 is applied. This suggests that $\lambda = -15$, which yields the new solution as

$$\mathbf{x}^{(3)} = \mathbf{x}^{(2)} + \lambda \Delta \mathbf{x}$$
$$= (-2, -10, 0, 0, 0, 0, 1, 0, 5, -15, 20, -1, 5, -1) + (-15)(0, 0, 1, 0, 0, 0, 0, 0, 0, -1, 1, 0, 0, 0)$$
$$= (-2, -10, -15, 0, 0, 0, 1, 0, 5, 0, 5, -1, 5, -1)$$

In iteration 4, $\mathbf{x}^{(3)}$, x_3 having a nonzero value becomes a basic variable and s_5 having a zero value becomes a nonbasic variable. The objective function value will be updated after the first round simplex tableau is updated for all x variables. The simplex tableau is now updated with $\mathbf{x}^{(3)}$ and $\Delta \mathbf{x}$ for x_4.

	x_1	x_2	x_3	x_4	x_5	s_1	s_2	s_3	s_4	s_5	s_6	s_7	s_8	s_9
	B	B	B	N	N	N	B	N	B	N	B	B	B	B
$\mathbf{x}^{(3)}$	-2	-10	-15	0	0	0	1	0	5	0	5	-1	5	-1
$\Delta \mathbf{x}$ for x_4	0	0	0	1	0	0	0	0	0	0	0	-1	1	0
	—	—	—	—	—		—	—	—	—	—	$\frac{-1}{-(-1)}$	—	—

$$\mathbf{x}^{(4)} = \mathbf{x}^{(3)} + \lambda \Delta \mathbf{x}$$
$$= (-2, -10, -15, 0, 0, 0, 1, 0, 5, 0, 5, -1, 5, -1) + (-1)(0, 0, 0, 1, 0, 0, 0, 0, 0, 0, 0, -1, 1, 0)$$
$$= (-2, -10, -15, -1, 0, 0, 1, 0, 5, 0, 5, 0, 4, -1)$$

In iteration 5, $\mathbf{x}^{(4)}$, x_4 having a nonzero value becomes a basic variable and s_7 having a zero value becomes a nonbasic variable. The simplex tableau is now updated with $\mathbf{x}^{(4)}$ and $\Delta\mathbf{x}$ for x_5.

	x_1	x_2	x_3	x_4	x_5	s_1	s_2	s_3	s_4	s_5	s_6	s_7	s_8	s_9
	B	B	B	B	N	N	B	N	B	N	B	N	B	B
$\mathbf{x}^{(4)}$	−2	−10	−15	−1	0	0	1	0	5	0	5	0	4	−1
$\Delta\mathbf{x}$ for x_5	0	0	0	0	1	0	0	0	0	0	0	0	0	−1
	—	—	—	—	—		—	—	—	—	—	—	—	$\dfrac{-1}{-(-1)}$

$\lambda = -1$

$\mathbf{x}^{(5)} = \mathbf{x}^{(4)} + \lambda\Delta\mathbf{x}$

$\quad = (-2, -10, -15, -1, 0, 0, 1, 0, 5, 0, 5, 0, 4, -1) + (-1)(0, 0, 0, 0, 1, 0, 0, 0, 0, 0, 0, 0, 0, -1)$

$\quad = (-2, -10, -15, -1, -1, 0, 1, 0, 5, 0, 5, 0, 4, 0)$

Thus, the objective function value is updated as follows:

$$0.241(-2) + 0.231(-10) + 0.220(-15) + 0.210(-1) + 0.199(-1) = -6.501$$

Note that the solution obtained corresponds to the minimum values of x, to keep the absolute value of the objective function to the minimum. The corresponding s_2, s_4, s_6, s_8 have the values equal to the ranges between the maximum and minimum limits of x, while s_1, s_3, s_5, s_7, s_9 have a zero value (nonbasic variables). However, this solution is only valid for a relaxed problem without accounting for the constraint in Equation (8.14).

In order to satisfy the constraint in Equation (8.14), some of the x values must be increased to achieve the total of x equal to 35; x_5 has the lowest coefficient in the objective function formulation, having already reached the upper limit. Amongst x_1, x_2, x_3 and x_4, x_4 has the second lowest coefficient and x_3 has the third lowest coefficient in the objective function formulation. Hence, to keep the absolute value of the objective function to the minimum, the decision variable with the current lowest coefficient in the objective function, that is, x_4, is to be first maximized $= 5$, while checking that the summation of *all* x (as in Equation (8.14)) does not exceed the value of 35. Then x_3 is increased, because it has the third lowest coefficient in the objective function after x_4, until the constraint in Equation (8.14) is satisfied. Thus, the optimum values of $\bar{x}^T$ obtained are (2, 10, 17, 5, 1).

The minimum objective function value calculated satisfying all the constraints including Equation (8.14) is as follows:

$$0.241(-2) + 0.231(-10) + 0.220(-17) + 0.210(-5) + 0.199(-1) = -7.781$$

The difference in the optimum objective function values between the spreadsheet based modelling (Table 8.2) and this method is due to rounding-off errors of the coefficients of the decision variables in the objective function formulation in Equation (8.10).

The LP problem of this small size can be easily solved logically, by looking at the coefficients of decision variables in the objective function formulation, as discussed above. Modern commercial LP software packages include CPLEX, XA and XPRESS. For a problem such as the exercise here, such packages will be able to attain an optimal solution in very few iterations (in less than 10 or even just a single iteration, depending on the efficiency of the algorithmic implementation). The following steps now show how Equation (8.18) can be modified to include the equality constraint only involving x

variables given in Equation (8.14). For this reason, s_{10}, another surplus variable, is introduced in to the equation, so as to include it in the iterative search procedure:

$$x_1 + x_2 + x_3 + x_4 + x_5 - s_{10} = 35$$

Equation (8.18) can be rewritten to include the s_{10} variable. The problem is no longer a relaxed problem. However, the same procedure is followed for each of the nonbasic variables (x_1, x_2, x_3, x_4 and x_5) by using a unit step change $\Delta x_j = 1$ of a nonbasic variable x_j, while keeping other nonbasic variables unchanged. Only the evolution of Equation (8.18) is shown, but not the simplex tableau because its structure to store information has already been discussed. The following iteration steps are performed using the simplex method to arrive at the solution, until $s_{10} = 0$. In each iteration step, the minimum ratio rule in Equation (8.17) is applied to decide the step size and Equation (8.18) is applied to decide the direction vector, respectively.

$$\mathbf{x}^{(1)} = \mathbf{x}^{(0)} + \lambda \Delta \mathbf{x}$$
$$= (0,0,0,0,0,-2,3,-10,15,-15,20,-1,5,-1,-35) + (-2)(1,0,0,0,0,-1,1,0,0,0,0,0,0,0,-1)$$
$$= (-2,0,0,0,0,0,1,-10,15,-15,20,-1,5,-1,-33)$$

$$\mathbf{x}^{(2)} = \mathbf{x}^{(1)} + \lambda \Delta \mathbf{x}$$
$$= (-2,0,0,0,0,0,1,-10,15,-15,20,-1,5,-1,-33) + (-10)(0,1,0,0,0,0,0,-1,1,0,0,0,0,0,-1)$$
$$= (-2,-10,0,0,0,0,1,0,5,-15,20,-1,5,-1,-23)$$

$$\mathbf{x}^{(3)} = \mathbf{x}^{(2)} + \lambda \Delta \mathbf{x}$$
$$= (-2,-10,0,0,0,0,1,0,5,-15,20,-1,5,-1,-23) + (-15)(0,0,1,0,0,0,0,0,0,-1,1,0,0,0,-1)$$
$$= (-2,-10,-15,0,0,0,1,0,5,0,5,-1,5,-1,-8)$$

$$\mathbf{x}^{(4)} = \mathbf{x}^{(3)} + \lambda \Delta \mathbf{x}$$
$$= (-2,-10,-15,0,0,0,1,0,5,0,5,-1,5,-1,-8) + (-1)(0,0,0,1,0,0,0,0,0,0,0,-1,1,0,-1)$$
$$= (-2,-10,-15,-1,0,0,1,0,5,0,5,0,4,-1,-7)$$

$$\mathbf{x}^{(5)} = \mathbf{x}^{(4)} + \lambda \Delta \mathbf{x}$$
$$= (-2,-10,-15,-1,0,0,1,0,5,0,5,0,4,-1,-7) + (-1)(0,0,0,0,1,0,0,0,0,0,0,0,0,-1,-1)$$
$$= (-2,-10,-15,-1,-1,0,1,0,5,0,5,0,4,0,-6)$$

$$\mathbf{x}^{(6)} = \mathbf{x}^{(5)} + \lambda \Delta \mathbf{x}$$
$$= (-2,-10,-15,-1,-1,0,1,0,5,0,5,0,4,0,-6) + (-4)(0,0,0,1,0,0,0,0,0,0,-1,1,0,-1)$$
$$= (-2,-10,-15,-5,-1,0,1,0,5,0,5,4,0,0,-2)$$

$$\mathbf{x}^{(7)} = \mathbf{x}^{(6)} + \lambda \Delta \mathbf{x}$$
$$= (-2,-10,-15,-5,-1,0,1,0,5,0,5,4,0,0,-2) + (-2)(0,0,1,0,0,0,0,0,0,-1,1,0,0,0,-1)$$
$$= (-2,-10,-17,-5,-1,0,1,0,5,2,3,4,0,0,0)$$

Thus, the optimum values of $\bar{x}^T$ obtained are $(2, 10, 17, 5, 1)$.

The simplex search method is effective for solving large-scale LP problems with hundreds of variables and a large number of equality and inequality constraints. It gives a structured and systematic way to move the search direction and step sizes along the boundary of the solution space.

Exercise 2. The greenhouse gas emission data for five classes of wastewater treatment plants (WWTPs) for sewage treatment with different average per capita equivalents (PCE) and average annual sewage volumes are shown in Table 8.7. The basis for greenhouse gas emission data extraction is a unit of a wastewater treatment plant. A lifetime of 30 years for a plant has been assumed. The cost-effective minimum and maximum acceptable lengths of the sewer grid system classes are also shown in Table 8.7. Calculate individual lengths of five classes of sewer grid system for minimum primary fossil resource depletion for a total length of 35 000 km. Calculate the value of minimum primary fossil resource depletion. Assume that the primary fossil resource depletion is proportional to the length of a class of sewer grid system.

Table 8.7   *Primary fossil resource depletion data for five classes of sewer grid system.*

	Class				
	1	2	3	4	5
PCE	233 225	71 133	24 864	5 321	806
Average annual sewage volume ($\times 10^6$ m^3 y^{-1})	47.1	14.4	5	1.07	163 000
Minimum acceptable length (km)	2 000	10 000	15 000	1 000	0
Maximum acceptable length (km)	3 000	15 000	20 000	5 000	1 000
Primary fossil resource depletion (kg)	241 397.8	231 277.8	220 269.9	210 487.2	198 795.9

Hint for Exercise 2. The formulation of the objective function and constraints for this problem is largely similar to Exercise 1. Hence, the solution strategy using the simplex method is left as an exercise based on the principles laid out in Exercise 1.

8.3   Nonlinear Optimization

All real-world problems are inherently nonlinear. In the earlier exercise, the relation between resource depletion and the sewer grid system length can likely be described using a certain nonlinear relationship that may be complex. Hence, the formulation of a related optimization problem requires the capability to handle nonlinearity and a suitable strategy to solve such an expectedly more difficult model.

In this regard, nonlinear optimization problems involve one or more nonlinear terms in the objective function and/or constraints. Common examples of nonlinear functions include those involving multiplication such as polynomials like quadratic (x^2) and cubic (x^3) functions for a real number x, as well as those involving division such as fractional terms x/y (for real y). Other nonlinear functions include exponentials $\exp(x)$, logarithmic terms $\ln x$, power terms x^α for real α, β^x for real β, x^y, $|x|$, and trigonometric functions $\sin(x)$, $\cos(x)$, etc.

Solution strategies for nonlinear optimization problems are underpinned by the methods used to determine search directions towards optimality. In this respect, they can be largely categorized into: (i) direct intuitive approaches using function values only and (ii) those using information from derivatives. Figure 8.2 shows the general procedure for the search methods that fall in the first category. Examples of such approaches are random search, grid search, simplex search, conjugate search and Powell's method.

Exercise 3. In a sequence of six numbers, every term after the second term is the sum of the previous two terms. In addition, the last term is five times the first term and the sum of all six terms is 32. What are the first two numbers?

Solution to Exercise 3. The set of algebraic equations can be solved using a suitable NLP optimization algorithm. The Solver tool in Microsoft Excel is employed to carry out the computation.

The problem is formulated as follows. Let x and y be the first two numbers in the sequence. Hence, the first six numbers are given by

$$x, y, x + y, x + 2y, 2x + 3y, 3x + 5y$$

Hence, their sum is $8x + 12y$.

A constraint to the problem stipulates that the value of the last term is five times that of the first term, which yields

$$5x = 3x + 5y \Rightarrow 2x = 5y$$

Minimization of the sum of squared errors is adopted as the objective function:

$$\min [32 - (8x + 12y)]^2$$

The problem is an NLP problem as the objective function is nonlinear. The problem can be solved by trial and error, adjusting the values of x and y such that $8x + 12y = 32$. Alternatively, the problem can be formulated as an optimization problem. By solving this NLP using the generalized reduced gradient (GRG) solution method in Excel Solver, the first two numbers in the Fibonacci sequence are obtained as 1 and 2.5.

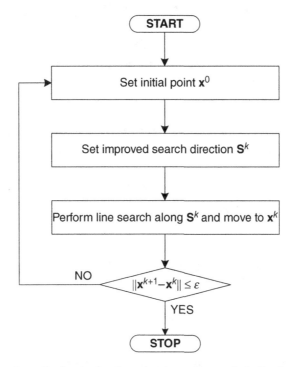

Figure 8.2 *General procedure for methods using function values in nonlinear optimization (note that the parameter ε denotes a small value; index k represents the number of iterations).*

8.3.1 Gradient Based Methods

In contrast to the arbitrary search directions of the foregoing approaches, the second category involves methods that utilize first derivatives (e.g., the gradient method and the conjugate gradient methods) or both first and second derivatives (e.g., Newton's method and quasi-Newton methods). In employing the gradient method for a maximization problem, the search direction is simply the gradient, hence giving rise to what is called a steepest ascent algorithm. Conversely, for a minimization problem, the negative of the gradient provides the search direction for the corresponding steepest descent algorithm (Figure 8.3).

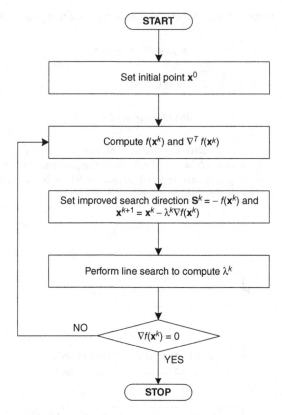

Figure 8.3   *General procedure for the gradient method for nonlinear optimization.*

On the other hand, as indicated, Newton's method uses a second-order quadratic approximation for the transition from a current point x^k to a new point x^{k+1} as given by the following expression:

$$f(\mathbf{x}) \approx f(\mathbf{x}^k) + \nabla^T f(\mathbf{x}^k) \Delta \mathbf{x}^k + \frac{1}{2} \left(\Delta \mathbf{x}^k \right)^T \mathbf{H}(\mathbf{x}^k) \Delta \mathbf{x}^k \tag{8.20}$$

where $\mathbf{H}(\mathbf{x}^k)$ is the Hessian matrix (of second partial derivatives with respect to $\mathbf{x}$ as evaluated at point $\mathbf{x}^k$) and $\Delta \mathbf{x}^k = \mathbf{x}^{k+1} - \mathbf{x}^k$.

For further details on these algorithms, extensive discussion is available in standard references on nonlinear optimization that include Nocedal and Wright[3], Bazaraa, Sherali and Shetty[4] and Edgar, Himmelblau and Lasdon[5].

Figure 8.4 shows the main methods that can be employed to solve nonlinear optimization problems. As one of the methods most commonly applied to NLP optimization, the generalized reduced gradient (GRG) algorithm is discussed further.

NLP can be constrained or unconstrained. As most of the problems found in chemical engineering are of the constrained type, this section focuses on nonlinear optimization of constrained problems of the form shown below. However, the essential difference in the solution strategies between the two types of NLP problems has been discussed.

$$\min f(\mathbf{x}), \mathbf{x} \in R^N \tag{8.21}$$

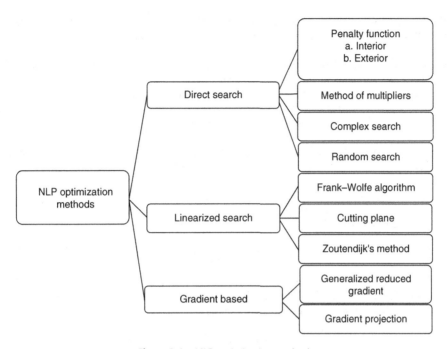

Figure 8.4 *NLP optimization methods.*

subject to

$$g_m(\mathbf{x}) = 0, m = 1, \ldots, M_e$$

$$g_m(\mathbf{x}) \geq 0, m = M_e + 1, \ldots, M$$

$$l_i \leq x_i \leq u_i, i = 1, \ldots, N$$

where

$\mathbf{x}$ is an N-dimensional vector of decision variables.
$f(\mathbf{x})$ is an objective function that is continuously differentiable in the space of real numbers R^N.
$g_m(\mathbf{x})$ are the functions for the equality and inequality constraints.
M_e is the number of equality constraints and M is the total number of constraints.
l_i and u_i are the lower and upper bounds ($u_i > l_i$) of the variable x_i belonging in vector $\mathbf{x}$.

The formulation can be modified to express inequality constraints as equality constraints or to consider the variable bounds within the constraints. To express an inequality constraint $g_m(\mathbf{x}) \geq 0$ in the form of an equality constraint a vector of *slack variable* $\mathbf{x}_{N+s}$ is added to the constraint function below, as shown earlier in Equation (8.11) in the simplex method:

$$g_m(\mathbf{x}) - \mathbf{x}_{N+s} = 0 \tag{8.22}$$

If there is a point at which $\mathbf{x}_{N+s} = 0$, then that point lies on the boundary of the solution space of a given problem. This manipulation of the inequality constraints is known as the slack variable strategy.

In the GRG algorithm, a solution of a constrained NLP problem is found by solving reduced formulations of the original problem. Problem reduction is made by the following two steps: (i) and (ii).

i.   After formulation of all constraints as equalities $g_m(\mathbf{x}) = 0, i = 1 \ldots, M$; the equalities are used to express M of the variables, called *basic variables*, in terms of the remaining $N - M$ variables, called *nonbasic variables*. As a result, M number of variables is eliminated and the NLP problem is reduced to functions of nonbasic variables only:

$$\text{Number of nonbasic variables} = N - M$$

$\mathbf{x}$ is partitioned into M number of basic variables $\hat{\mathbf{x}} \in x_1, x_2, \ldots, x_M$ and $N - M$ number of nonbasic variables $\bar{\mathbf{x}} \in x_{M+1}, x_{M+2}, \ldots, x_N$. The basic variables $\hat{\mathbf{x}}$ are expressed as functions of nonbasic variables $\bar{\mathbf{x}}$, as shown in below.

$$\hat{\mathbf{x}} = F(\bar{\mathbf{x}}) \tag{8.23}$$

The constraints are now of the form:

$$g_m(\mathbf{x}) = g_m(\hat{\mathbf{x}}, \bar{\mathbf{x}}) = \bar{g}_m(\bar{\mathbf{x}}) = 0, m = 1, \ldots, M$$

These constraints may be linearized using Taylor's expansion, shown later, to help the solution method. Note that the nonbasic variables are independent variables, while the basic variables can be seen as the dependent variables. The objective function can also be expressed in terms of the nonbasic variables, as shown below:

$$\text{i.e. min} f(\mathbf{x}) \quad or \quad f(\hat{\mathbf{x}}, \bar{\mathbf{x}}) \quad or \quad \bar{f}(\bar{\mathbf{x}}) \tag{8.24}$$

subject to

$$l_i \leq x_i \leq u_i, \quad i = M + 1, \ldots, N$$

where l_i and u_i are the lower and upper bounds ($u_i > l_i$) of the nonbasic variables x_i belonging in vector $\mathbf{x}$.

ii. Thus, the reduced formulation of the NLP problem in Equations 8.23 and 8.24 is obtained. Such a reduced problem can be solved by modifications or adaptations of unconstrained minimization methods such as the simplex method discussed in Section 8.2.

Next, the gradient or the first-order derivative $\nabla f(\mathbf{x})$ of an objective function $f(\mathbf{x})$ is used to verify that the optimality conditions are met and thus to find the solution for the optimization problem. *The necessary optimality condition for unconstrained optimization problems is that the first-order derivative of the objective function is equal to zero, that is,* $\nabla f(\mathbf{x}) = 0$:

$$\nabla f(\mathbf{x}) = \left(\frac{\partial}{\partial x_1} f(\mathbf{x}), \ldots, \frac{\partial}{\partial x_N} f(\mathbf{x}) \right) = 0 \tag{8.25}$$

For a necessary and sufficient condition for an optimality, the second derivate must be evaluated and checked whether this is lesser than zero to obtain a maximum or greater than zero to obtain a minimum.

The second-order derivatives of a multivariable function are represented in the Hessian matrix $\mathbf{H}(\mathbf{x})$. Thus, if $\mathbf{H}(\mathbf{x}) > 0$, there is a minimum; if $\mathbf{H}(\mathbf{x}) < 0$, there is a maximum.

As the problem is a reduced problem now, the first-order derivative of the objective function is only expressed in terms of the nonbasic variables (Equation (8.24)) *and is differentiated with respect to nonbasic variables and equated to zero.*

The *necessary condition for a feasible solution of a reduced problem* is shown as

$$\nabla f(\mathbf{x}) = \left(\frac{\partial}{\partial x_{M+1}} f(\mathbf{x}), \ldots, \frac{\partial}{\partial x_N} f(\mathbf{x}) \right) = 0 \tag{8.26}$$

This means that the values of $x_i, i = M + 1, \ldots, N$ are displaced by $dx_i, i = M + 1, \ldots, N$, such that at the new point, $x_i + dx_i, i = M + 1, \ldots, N$, the constraints are still satisfied:

$$g_m(\bar{\mathbf{x}} + d\bar{\mathbf{x}}) = 0, m = 1, \ldots, M$$

Noting that $dx_i, i = M + 1, \ldots, N$ are sufficiently small, such that they still satisfy the constraints, gives

$$\nabla g_m(\mathbf{x}) = \left(\frac{\partial}{\partial x_{M+1}} g_m(\mathbf{x}), \ldots, \frac{\partial}{\partial x_N} g_m(\mathbf{x})\right) = 0 \tag{8.27}$$

For a generic nonlinear constrained optimization problem defined in Equation (8.21), a *Jacobian matrix* is thus defined as shown below. This equation shows the *necessary condition for an optimal solution of a generic nonlinear constrained optimization problem*:

$$J_i\left(\frac{f, g_1, g_2, \ldots, g_M}{x_{M+1}, x_{M+2}, \ldots, x_N}\right) = \begin{vmatrix} \partial f/\partial x_{M+1} & \partial f/\partial x_{M+2} & \cdots\cdots & \partial f/\partial x_N \\ \partial g_1/\partial x_{M+1} & \partial g_1/\partial x_{M+2} & \cdots\cdots & \partial g_1/\partial x_N \\ \partial g_2/\partial x_{M+1} & & \cdots\cdots\cdots & \partial g_2/\partial x_N \\ \vdots & & \ddots & \vdots \\ \partial g_M/\partial x_{M+1} & & \cdots & \partial g_M/\partial x_N \end{vmatrix} = 0 \tag{8.28}$$

J_i is the *Jacobian matrix* with:

$(M + 1)$ number of rows (objective function + number of constraints) and
$(N - M)$ number of columns, that is, number of non-basic variables.

The difference between Equation (8.25) (unconstrained NLP) and Equation (8.28) (constrained NLP) is the presence of constraints in the latter equation.

Equation (8.28) implies that the determinant of each of $(N - M)$ number of Jacobian functions is to be set equal to zero and solved, in order to find an optimal solution for a given nonlinear constrained optimization problem. However, this is not only a necessary condition but can ensure that local solutions exist at the feasible points that satisfy the constraints.

For a *necessary and sufficient condition for an optimality, the second derivate must be evaluated and checked as to whether this is less than zero to obtain a maximum or greater than zero to obtain a minimum.*

The method of Karush–Kuhn–Tucker (KKT) establishes the *necessary* and *sufficient* conditions for a solution to be optimal. The KKT approach shown in Equations (8.29) to (8.33) below is an analytical method to solve NLP problems. First, all constraints expressed as equalities are linearized according to Taylor's series expansion at the point $\mathbf{x} = \mathbf{x}^k$, where k is the iteration number:

$$g_m(\mathbf{x}^k) + \nabla g_m(\mathbf{x}^k)(\mathbf{x} - \mathbf{x}^k) = 0, \quad \forall m \in 1, 2, \ldots, M \tag{8.29}$$

At $\mathbf{x} = \mathbf{x}^k$, the original constraint functions are noted as $g_m(\mathbf{x}^k) = 0$. Equation (8.29) reduces to

$$\nabla g_m(\mathbf{x}^k)(\mathbf{x} - \mathbf{x}^k) = 0, \quad \forall m \in 1, 2, \ldots, M \tag{8.30}$$

The gradients in Equation (8.30) are also partitioned into the matrix J of dimensions $M \times M$ and the matrix C of dimensions $M \times (N - M)$, represented as

$$J = \begin{bmatrix} \nabla \hat{g}_1 \\ \vdots \\ \nabla \hat{g}_M \end{bmatrix}, C = \begin{bmatrix} \nabla \bar{g}_1 \\ \vdots \\ \nabla \bar{g}_M \end{bmatrix} \tag{8.31}$$

Each row in matrix J (each $\nabla \hat{g}_m$) contains partial derivatives of the constraint with respect to individual basic variables, such as

$$\nabla \hat{g}_m = \frac{\partial g_m}{\partial x_1}, \frac{\partial g_m}{\partial x_2}, \ldots, \frac{\partial g_m}{\partial x_M}$$

Each row in matrix C (each $\nabla \bar{g}_m$) contains partial derivatives of the constraint with respect to individual nonbasic variables, such as

$$\nabla \bar{g}_m = \frac{\partial g_m}{\partial x_{M+1}}, \frac{\partial g_m}{\partial x_{M+2}}, \ldots, \frac{\partial g_m}{\partial x_N}$$

Then, applying the addition properties of matrices, the following equations are obtained:

$$J\left(\hat{\mathbf{x}} - \hat{\mathbf{x}}^k\right) + C(\bar{\mathbf{x}} - \bar{\mathbf{x}}^k) = 0$$

$$\hat{\mathbf{x}} = \hat{\mathbf{x}}^k - J^{-1}C(\bar{\mathbf{x}} - \bar{\mathbf{x}}^k) \tag{8.32}$$

Equation 8.32 is the generic form to evaluate the basic variables from the non-basic variables.

Linearization of the objective function by Taylor's series expansion and substitution of variables result in the reduced gradient of the linearized objective function:

$$\nabla \tilde{f}\left(\mathbf{x}^k\right) = \nabla \bar{f}\left(\mathbf{x}^k\right) - \nabla \hat{f}\left(\mathbf{x}^k\right) J^{-1}C \tag{8.33}$$

where

$\nabla \bar{f}\left(\mathbf{x}^k\right)$ is the reduced gradient vector of the first-order partial derivatives of the objective function with respect to nonbasic variables.

$\nabla \hat{f}\left(\mathbf{x}^k\right)$ is the reduced gradient vector of the first-order partial derivatives of the objective function with respect to basic variables.

Hence, $\nabla \tilde{f}\left(\mathbf{x}^k\right)$ is the reduced gradient vector of the first-order partial derivatives of the objective function with respect to all basic and nonbasic variables.

J^{-1} is the inverse of matrix J

The following KKT conditions are verified for the necessary and sufficient condition of optimality:

a. When the value of $\nabla \tilde{f}\left(\mathbf{x}^k\right)$ is negative, the optimal value corresponds to a maximum, or when the value of $\nabla \tilde{f}\left(\mathbf{x}^k\right)$ is positive, the optimal value corresponds to a minimum.
b. Additionally, if the value of $\nabla \tilde{f}\left(\mathbf{x}^k\right)$ does not change over a sufficient number of iterations (this is shown in equation), a reliable optimal solution can be obtained:

$$\left|\nabla \tilde{f}\left(\mathbf{x}^{k+1}\right) - \nabla \tilde{f}\left(\mathbf{x}^k\right)\right| \ll \varepsilon_1 \tag{8.34}$$

where ε_1 is a very small positive number, the convergence factor for the objective function.

There are two more convergence criteria defined for satisfying constraints and basic variables: ε_2 is the convergence factor for satisfying the constraints:

$$\left|g_m(\mathbf{x}^{k+1})\right| \leq \varepsilon_2 \tag{8.35}$$

ε_3 is the convergence factor for the basic variables:

$$\hat{\mathbf{x}}^{k+1} - \hat{\mathbf{x}}^k \leq \varepsilon_3 \tag{8.36}$$

If all three convergence criteria are met, the KKT condition for the reduced gradient of the linearized objective function and for the constraints and basic variables (Equations (8.34) to (8.36)), the solution obtained is an optimal solution.

The search procedure has some limitations as the solution is forced to be on the boundary of the feasible region of the solution of a constrained NLP problem. As discussed earlier in the simplex method, a new set of nonbasic variables may be needed in an iteration k. The feasibility criteria for the $\mathbf{x}^{k+1}$ solution must be checked by verifying whether the constraints still hold. If a solution is not feasible, the variables need to be changed in a systematic way by first obtaining a feasible solution that satisfies all the constraints and then setting the magnitude of the step change such as using the minimum ratio rule in Equation (8.17) and the direction of changes of the variables shown in Equation (8.18).

The Newton–Raphson method in the following equation is applied to ensure that a local solution is feasible (all constraints are satisfied). j is the iteration number of the subproblem to obtain a feasible solution, starting from an infeasible point $j = 0$ to a feasible local solution that satisfies all constraints in a given iteration number k of the main reduced problem (Equations (8.29) to (8.33)):

$$\mathbf{x}^{j+1} = \mathbf{x}^j - \left(\nabla g_m\left(\mathbf{x}^j\right)\right)^{-1} g_m(\mathbf{x}^j) \qquad \forall m \in 1, 2, \ldots, M \tag{8.37}$$

Exercise 4. The concentration of a highly toxic compound C, formed from a reaction between A and B, has been correlated to the concentration of the reactants as

$$C_C = (C_A - 5)^2 + (C_B - 3)^2$$

The mass balance constraints are

$$2C_A - C_B^2 - 1 \geq 0$$
$$8 - C_A^2 - 2C_B \geq 0$$

It is desirable to minimize the concentration of the dangerous product C to improve health and safety working conditions in the production plant. Formulate the NLP problem in a form that can be solved using the GRG algorithm. Use the initial values of $C_A = C_B = 1$. To avoid corrosion, the concentration of the reactants can be up to 4 mol L^{-1}.

Approach to solution to Exercise 4. Let $C_A = x_1$ and $C_B = x_2$ and X be the vector containing the decision variables, x_1 and x_2. As the amount of the dangerous product must be minimized by varying the amounts of the reactants, the original NLP problem can be written as

$$\min f(X) = (x_1 - 5)^2 + (x_2 - 3)^2$$

subject to

$$g_1(X) = 2x_1 - x_2^2 - 1 \geq 0$$
$$g_2(X) = 8 - x_1^2 - 2x_2 \geq 0$$
$$0 \leq X \leq 4$$

The constraints $g_1(X)$ and $g_2(X)$ represent the mass balance constraints. The reactant concentration values have to be positive and the upper bound has been given as 4. These bounds are expressed in the last constraint.

As the GRG method requires the inequality constraints $g_1(X)$ and $g_2(X)$ in the form of equality constraints, the constraints $g_1(X)$ and $g_2(X)$ are transformed by adding slack variables x_3 and x_4 as

$$g_1(x_1, x_2) - x_3 = 0$$

$$g_2(x_1, x_2) - x_4 = 0$$

The initial values for x_3 and x_4 are found from the original constraint equations as

$$x_3 = 2x_1 - x_2^2 - 1 = 2(1) - (1)^2 - 1 = 0$$

$$x_4 = 8 - (1)^2 - 2(1) = 5$$

Now, a vector **x** is defined to include all x_1, x_2, x_3 and x_4 variables. Therefore, the initial set of values is expanded to $\mathbf{x}^0 = (1, 1, 0, 5)$. The number of nonbasic variables is equal to the total number of variables N minus the number of constraints M, that is,

$$\text{Number of nonbasic variables} = N - M$$

For this exercise, the required number of nonbasic variables is $4 - 2 = 2$.

8.3.2 Generalized Reduced Gradient (GRG) Algorithm

The GRG method discussed in Section 8.3.1 is shown in the form of an algorithm in Figure 8.5 and is used in Exercise 5. While this exercise shows the calculation steps for understanding the method, Exercises 6 and 7 show how an NLP problem can be formulated and solved using the commercial software GAMS and Excel, respectively.

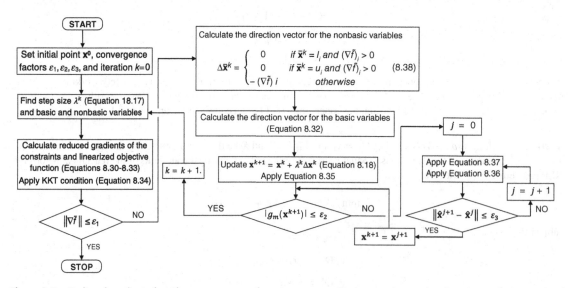

Figure 8.5 *Reduced gradient algorithm: $\varepsilon_1, \varepsilon_2, \varepsilon_3$ are the convergence criteria; γ and α set the direction and magnitude of changes in decision variable values.*

Exercise 5. Consider the problem of Exercise 4 and use the GRG algorithm to find the optimum values of X that minimize the generation of the toxic compound C.

Solution to Exercise 5. The GRG algorithm shown in Figure 8.5 can be divided into four general steps shown as follows:

Step 1. From Exercise 4, the optimization problem is defined as follows:

$$\min f(X) = (x_1 - 5)^2 + (x_2 - 3)^2$$

subject to

$$g_1(\mathbf{x}) = 2x_1 - x_2^2 - 1 - x_3 = 0$$

$$g_2(\mathbf{x}) = 8 - x_1^2 - 2x_2 - x_4 = 0$$

$$l_i = 0 \le x_i \le u_i = 4$$

Set the initial point $\mathbf{x}^0 = (1, 1, 0, 5)$, convergence factor for the objective function $\varepsilon_1 = 0.0001$ and start the iteration $k = 0$.

Step 2. To select the basic variables, apply the minimum ratio rule to each x_i^k variable. The simplex search method uses the maximum feasible step size λ as given by the minimum ratio rule (Equation (8.17)):

$$\lambda_i^k = \frac{\min\{(x_i^k - l_i), (u_i - x_i^k)\}}{u_i - l_i}$$

For the variable x_1:

$$\lambda_1^0 = \frac{\min\{(x_1^0 - l_1), (u_1 - x_1^0)\}}{u_i - l_i} = \frac{\min\{(1-0), (4-1)\}}{4-0} = 1/4$$

Doing similar calculations for each x_i variable:

$$\lambda_2^0 = \frac{1}{4}, \lambda_3^0 = 0, \lambda_4^0 = -\frac{1}{4}$$

The number of basic variables required is 2 ($M = 2$) equal to the number of constraints. Select the two basic variables with the highest value of λ_i^k. Thus the two basic variables are x_1 and x_2. The variables are partitioned into two basic and two nonbasic variables, at the initial point:

$$\hat{x} = (x_1^0, x_2^0), \quad \bar{x} = (x_3^0, x_4^0)$$

Now, proceed to calculate the matrices J for $\hat{x}$ and C for $\bar{x}$ and the reduced gradients of the constraints and the linearized objective function, using Equations (8.31) to (8.33). The gradients for the first constraint are calculated at $\mathbf{x}^0$ as

$$\nabla g_1(\mathbf{x}^0) = \left(\frac{\partial g_1(\mathbf{x}^0)}{\partial x_1}, \frac{\partial g_1(\mathbf{x}^0)}{\partial x_2}, \frac{\partial g_1(\mathbf{x}^0)}{\partial x_3}, \frac{\partial g_1(\mathbf{x}^0)}{\partial x_4} \right)$$

$$\nabla g_1(\mathbf{x}^0) = (2, \ -2x_2^0, \ -1, \ 0) = (2, \ -2(1), \ -1, \ 0) = (2, -2, -1, 0)$$

Similarly,

$$\nabla g_2(\mathbf{x}^0) = (-2x_1^0, \ -2, \ 0, \ -1) = (-2, -2, 0, -1)$$

and the gradient of the objective function:

$$\nabla f\left(\mathbf{x}^0\right) = \left(2\left(x_1^0 - 5\right),\ 2\left(x_2^0 - 3\right)\ 0\ 0\right) = (-8, -4, 0, 0)$$

Using Equation (8.31) and taking the corresponding parts $\nabla \hat{g}_m$ and $\nabla \bar{g}_m$ from each $\nabla g_m\left(\mathbf{x}^0\right)$ gives

$$J = \begin{bmatrix} \nabla \hat{g}_1 \\ \nabla \hat{g}_2 \end{bmatrix} = \begin{bmatrix} 2 & -2 \\ -2 & -2 \end{bmatrix}, C = \begin{bmatrix} \nabla \bar{g}_1 \\ \nabla \bar{g}_2 \end{bmatrix} = \begin{bmatrix} -1 & 0 \\ 0 & -1 \end{bmatrix}$$

The inverse of the matrix J is

$$J^{-1} = \frac{1}{(2 \times (-2) - (-2) \times (-2))} \begin{bmatrix} -2 & 2 \\ 2 & 2 \end{bmatrix} = \frac{1}{8} \begin{bmatrix} 2 & -2 \\ -2 & -2 \end{bmatrix}$$

The reduced gradient is then calculated using Equation (8.33):

$$\nabla \tilde{f}\left(\mathbf{x}^0\right) = \nabla \hat{f}\left(x_3^0, x_4^0\right) - \nabla \hat{f}\left(x_1^0, x_2^0\right) J^{-1} C$$

$$\nabla \tilde{f}\left(\mathbf{x}^0\right) = (0, 0) - \frac{(-8, -4)}{8} \begin{bmatrix} 2 & -2 \\ -2 & -2 \end{bmatrix} \begin{bmatrix} -1 & 0 \\ 0 & -1 \end{bmatrix}$$

$$\nabla \tilde{f}\left(\mathbf{x}^0\right) = \frac{1}{8}(-8, 24) = (-1, 3)$$

Step 3. Examine the KKT conditions:

$$\nabla \tilde{f} = \sqrt{(-1)^2 + 3^2} \gg 0$$

Hence, a minimum exists. Therefore, the search strategy must continue. The direction vector for the nonbasic and the basic variables can be set using the following equations, respectively:

$$\Delta \bar{\mathbf{x}} = \begin{cases} 0 & \text{if } \bar{\mathbf{x}} = l_i \text{ and } (\nabla \tilde{f})_i > 0 \\ 0 & \text{if } \bar{\mathbf{x}} = u_i \text{ and } (\nabla \tilde{f})_i < 0 \\ -(\nabla \tilde{f})_i & \text{otherwise} \end{cases} \tag{8.38}$$

Equation (8.39) is derived from Equation (8.32) to deduce the basic variables from the nonbasic variables:

$$\hat{\mathbf{x}} = \hat{\mathbf{x}}^k - J^{-1} C \left(\bar{\mathbf{x}} - \bar{\mathbf{x}}^k\right)$$
$$\Delta \hat{\mathbf{x}} = -J^{-1} C \Delta \bar{\mathbf{x}} \tag{8.39}$$

In this example, $\nabla \tilde{f}_1 < 0$ but x_1 is not at the upper bound. Therefore,

$$\Delta x_1 = -\left(\nabla \tilde{f}\right)_i = -\frac{-8}{8} = 1$$

Similarly, $\nabla \tilde{f}_2 > 0$ but x_2 is not at the lower bound (i.e., $\bar{x}_i \neq l_i$). Therefore,

$$\Delta x_2 = -\left(\nabla \tilde{f}\right)_i = -\frac{24}{8} = -3$$

$$\Delta \bar{\mathbf{x}} = \begin{bmatrix} \Delta x_1 \\ \Delta x_2 \end{bmatrix} = \begin{bmatrix} 1 \\ -3 \end{bmatrix}$$

Using Equation (8.39):

$$\Delta \hat{\mathbf{x}} = -\frac{1}{8} \begin{bmatrix} 2 & -2 \\ -2 & -2 \end{bmatrix} \begin{bmatrix} -1 & 0 \\ 0 & -1 \end{bmatrix} \begin{bmatrix} 1 \\ -3 \end{bmatrix} = \begin{bmatrix} 1 \\ 0.5 \end{bmatrix}$$

The complete direction vector $\Delta \mathbf{x}$ is obtained using

$$\Delta \mathbf{x} = \begin{bmatrix} \Delta \hat{\mathbf{x}} \\ \Delta \bar{\mathbf{x}} \end{bmatrix} = \begin{bmatrix} x_1 \\ x_2 \\ x_3 \\ x_4 \end{bmatrix} \tag{8.40}$$

$$\Delta \mathbf{x} = \begin{bmatrix} 1 \\ 0.5 \\ 1 \\ -3 \end{bmatrix}$$

Step 4. For the given problem, the convergence factors for the constraints and basic variables are $\varepsilon_2 = \varepsilon_3 = 0.0001$ (Equations (8.35) and (8.36)). Assume a step size of 2 in iteration $k = 0$.

4a. Equation (8.18) as shown in the simplex method is now applied to update the x_i^k variable for the next iteration.

$$\mathbf{x}^{k+1} = \mathbf{x}^k + 2\Delta \mathbf{x}^k \tag{8.18}$$

$$\mathbf{x}^1 = \begin{bmatrix} 1 \\ 1 \\ 0 \\ 5 \end{bmatrix} + 2 \begin{bmatrix} 1 \\ 0.5 \\ 1 \\ -3 \end{bmatrix} = \begin{bmatrix} 3 \\ 2 \\ 2 \\ -1 \end{bmatrix}$$

Substituting $\mathbf{x}^1$ in the constraint functions gives

$$g_1 \left(\mathbf{x}^1 \right) = -1; g_2 \left(\mathbf{x}^1 \right) = -4$$

Since $\left| g_m(\mathbf{x}^1) \right| > \varepsilon_2$, the criterion in Equation (8.35) is not satisfied; continue the search and go to Step 4b.

4b. To obtain the values of the basic variables that satisfy the constraints, Equation (8.37) (Newton–Raphson method) is applied as follows. A subiteration j is introduced for the iteration steps in the Newton–Raphson method and repeated until Equation (8.36) is satisfied. Substitute $\hat{x}^0 = \hat{x}^k$ when $j = 0$ in Equation (8.37) and begin the iteration in the Newton–Raphson method:

$$\hat{\mathbf{x}}^{j+1} = \hat{\mathbf{x}}^j - \left(J \left(\hat{\mathbf{x}}^j \right) \right)^{-1} g(\hat{\mathbf{x}}^j) \tag{8.37}$$

$$\hat{\mathbf{x}}^1 = \begin{bmatrix} 3 \\ 2 \end{bmatrix} - \frac{1}{8} \begin{bmatrix} 2 & -2 \\ -2 & -2 \end{bmatrix} \begin{bmatrix} -1 & 0 \\ 0 & -1 \end{bmatrix} \begin{bmatrix} -1 \\ -4 \end{bmatrix} = \begin{bmatrix} 3.75 \\ 3.25 \end{bmatrix}$$

Check the convergence criterion for the basic variables (Equation (8.36)):

$$\| \hat{\mathbf{x}}^1 - \hat{\mathbf{x}}^0 \| = \sqrt{(3.75 - 3)^2 + (3.25 - 2)^2} > \varepsilon_3$$

As Equation (8.36) is not satisfied, set $j = j + 1$, apply Equation (8.37) and check the convergence criterion in Equation (8.36). This step is continued until Equation (8.37) is satisfied, in which case $\hat{\mathbf{x}}^{k+1}$ values in the current iteration $k + 1 = 1$ are updated with the converged $\hat{\mathbf{x}}^{j+1}$ values, $\hat{\mathbf{x}}^{k+1} = \hat{\mathbf{x}}^{j+1}$. Go to Step 4c.

4c. If $\left| g_m(\mathbf{x}^{k+1}) \right| \leq \varepsilon_2$ or Equation (8.35) is satisfied, set $k = k + 1$ and go to Step 2. Note that until now the step size of Step 2 did not change and the values of the nonbasic variables $\bar{\mathbf{x}}$ remained the same. If Equation (8.35) is not satisfied, go to Step 4b.

4d.   Steps 2 to 4 are continued until all three convergence criteria, KKT conditions for the reduced gradient of the objective function and Equations (8.34) to (8.36) are satisfied. This solution method should be carried out using the computer based GRG algorithm. The optimum values X^* are obtained as follows:

$$X^* = \begin{bmatrix} 2.15445 \\ 1.67916 \\ 0.48932 \\ 0 \end{bmatrix}$$

Then the objective function has a value of

$$f(X^*) = (2.15445 - 5)^2 + (1.67916 - 3)^2 = 9.84$$

and the constraints are satisfied:

$$g_1(X^*) = 2(2.15445) - (1.67916)^2 - 1 - 0.48932 = 0$$

$$g_2(X^*) = 8 - (2.15445)^2 - 2(1.67916) - 0 = 0$$

Figure 8.6 shows the pathway from the initial point to the first iteration of step 4 and the final solution X^*. Note that the final solution lies on the boundaries of the feasible region and at the point where the original constraints $g_1(X)$ and $g_2(X)$ intersect.

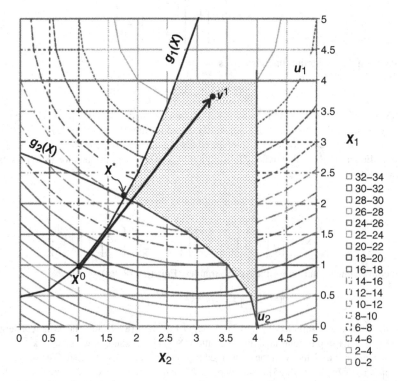

Figure 8.6   *Contour plot of the objective function, constraint functions. The first iteration point and the optimum solution of the NLP of Exercise 5 are shown. The enclosed dotted area represents the feasible region.*

Exercise 6. Water network synthesis is a design problem concerned with the simultaneous consideration of both water-using units and wastewater treatment operations typically found in an industrial process plant. Water-using units encompass water sources and sinks including freshwater sources. Water treatment operations act as intermediate regeneration processes, where necessary, before water sources can be made available for reuse or recycle in the sinks. The goal is to synthesize a network integrating these water-using and water regeneration operations in order to meet a certain objective, typically economic based, by determining the stream piping interconnections and their water flow rates and contaminant concentrations. Find the optimal flowsheet for a minimum cost operation. A nonlinear programming (NLP) formulation is to be considered for a problem based on the paper by Khor, Chachuat and Shar[6]. Background information on water network synthesis problems can be found in a paper by Khor *et al.*[7]

Tables 8.8 and 8.9 show the data on the flow rates and concentrations W of the chemical oxygen demand (COD) for a set of water sources and sinks, respectively. The term x_W denotes the freshwater supply required while x_D is the wastewater discharged to the environment. The respective unit costs are given in Table 8.10. The removal ratio R is a measure of the efficiency for removing contaminants by a certain regenerator technology. For this example, a single regenerator such as a reverse osmosis unit is considered. Table 8.11 gives the variable costs of operating piping interconnections between any two nodes. Table 8.12 gives the capital costs of piping interconnections between two nodes.

Table 8.8 Data on water sources.

Source Node	Flow Rate x (L h^{-1})	Concentration w (mg L^{-1})
1	23	40
2	3.5	37
3	1.8	35
4	x_W	10

Table 8.9 Data on water sinks.

Sink Node	Flow Rate x (L h^{-1})	Maximum Inlet Concentration (w_{max}) (mg L^{-1})
6	25.6	30
7	115	10
8	28.4	20
9	x_D	25

Table 8.10 Data and parameters.

Parameter	Value
Unit cost for freshwater (c_W)	$1.00 L^{-1}
Unit cost for effluent treatment (c_D)	$1.00 L^{-1}
Removal ratio (R)	0.84

Table 8.11 Operating cost in $ L^{-1} for piping interconnections between two nodes.

Source	Sink				5 (Regenerator)
	6	7	8	9	
1	5	6	5	50	5
2	7	5	5	50	7
3	6	7	5	50	6
4	10	10	11	50	20
5 (Regenerator)	6	5	6	50	–

Table 8.12   *Capital cost in $ h^{-1}$ for piping interconnections between two nodes.*

Source	Sink				5 (Regenerator)
	6	7	8	9	
1	10	12	10	100	10
2	14	10	10	100	14
3	12	14	10	100	12
4	10	10	11	100	20
5 (Regenerator)	12	10	12	100	–

Solution to Exercise 6.   Based on the data provided, a superstructure representation of the problem is first constructed that involves a network of all possible piping interconnections of the nodes representing the water sources, regenerators and sinks, as shown in Figure 8.7.

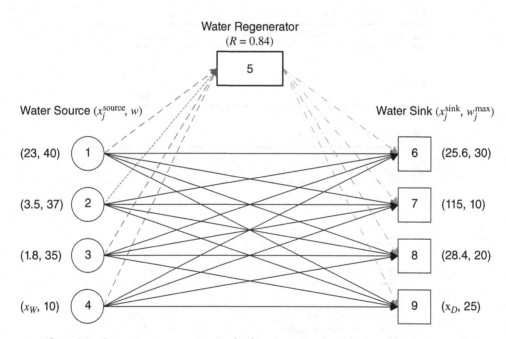

Figure 8.7   *Superstructure representation for the water network synthesis problem in Exercise 6.*

The water network synthesis problem can be formulated as the following optimization model based on the superstructure in Figure 8.7. The decision variables $x_{i,j}$ are the water flows in each of the piping interconnections from node i to node j. A total cost minimization objective function can be formulated as follows:

$$\min f = 5x_{1,5} + 5x_{1,6} + 6x_{1,7} + 5x_{1,8} + 50x_{1,9} + 7x_{2,5} + 7x_{2,6} + 5x_{2,7} + 5x_{2,8} + 50x_{2,9}$$
$$+ 6x_{3,5} + 6x_{3,6} + 7x_{3,7} + 5x_{3,8} + 50x_{3,9} + 20x_{4,5} + 10x_{4,6} + 10x_{4,7} + 11x_{4,8} + 50x_{4,9}$$
$$+ 6x_{5,6} + 5x_{5,7} + 6x_{5,8} + 50x_{5,9} + c_W x_W + c_D x_D$$

To write the formulation in a compact representation, indices or subscripts for the two main dimensions of the water network synthesis problem are employed, namely an origin node i, $i \in I = \{1, \dots, l\}$, and a destination node j, $j \in J = \{1, \dots, m\}$. Thus, a compact representation of the objective function is given by

$$\min f = \sum_{i \in I} \sum_{j \in J} c_{i,j} x_{i,j}$$

subject to the following constraints (a to f):

a. Mass balances around the sources:

$$x_{1,5} + x_{1,6} + x_{1,7} + x_{1,8} + x_{1,9} = 23 \quad \text{(node 1)} \quad \text{(NLP1)}$$
$$x_{2,5} + x_{2,6} + x_{2,7} + x_{2,8} + x_{2,9} = 3.5 \quad \text{(node 2)} \quad \text{(NLP2)}$$
$$x_{3,5} + x_{3,6} + x_{3,7} + x_{3,8} + x_{3,9} = 1.8 \quad \text{(node 3)} \quad \text{(NLP3)}$$
$$x_{4,5} + x_{4,6} + x_{4,7} + x_{4,8} + x_{4,9} = x_W \quad \text{(node 4)} \quad \text{(NLP4)}$$

or written in compact form:

$$\sum_{j \in J} x_{i,j} = x_i^{\text{source}}, \quad \text{for all } i \in I$$

where x_i^{source} is the fixed outlet flow from a source i.
b. Mass balances around the regenerator:

$$x_{1,5} + x_{2,5} + x_{3,5} + x_{4,5} = x_{5,6} + x_{5,7} + x_{5,8} + x_{5,9} \quad \text{(node 5)} \quad \text{(NLP5)}$$

c. Mass balances around the sinks:

$$x_{1,6} + x_{2,6} + x_{3,6} + x_{4,6} + x_{5,6} = 25.6 \quad \text{(node 6)} \quad \text{(NLP6)}$$
$$x_{1,7} + x_{2,7} + x_{3,7} + x_{4,7} + x_{5,7} = 115 \quad \text{(node 7)} \quad \text{(NLP7)}$$
$$x_{1,8} + x_{2,8} + x_{3,8} + x_{4,8} + x_{5,8} = 28.4 \quad \text{(node 8)} \quad \text{(NLP8)}$$
$$x_{1,9} + x_{2,9} + x_{3,9} + x_{4,9} + x_{5,9} = x_D \quad \text{(node 9)} \quad \text{(NLP9)}$$

or written in compact form:

$$\sum_{i \in I} x_{i,j} = x_j^{\text{sink}}, \quad \text{for all } j \in J$$

where x_j^{sink} is the fixed inlet flow to a sink j.
Note that the concentration balances around the sources are redundant to the mass balances (since the concentration terms remain constant throughout, hence they are not required).
d. Concentration balance with respect to COD levels around the regenerator (node 5):

$$\left(1 - R_R\right)\left(40x_{1,5} + 37x_{2,5} + 35x_{3,5} + 10x_{4,5}\right) = w_R \left(x_{5,6} + x_{5,7} + x_{5,8} + x_{5,9}\right)$$

or written in compact form:

$$\left(1 - R_k\right) \sum_{i \in I} w_i x_{i,k} = w_k \sum_{j \in J} x_{k,j}, \quad \text{for all } k \in K$$

with the removal ratio $R_k = R_R = 0.84$ (Table 8.10), which yields

$$6.4x_{1,5} + 5.92x_{2,5} + 5.6x_{3,5} + 1.6x_{4,5} - w_R \left(x_{5,6} + x_{5,7} + x_{5,8} + x_{5,9}\right) = 0 \quad \text{(NLP10)}$$

where w_R is the outlet concentration from the regenerator. Note that this is the only nonlinear equality constraint in the problem.

e. Quality requirements for each of the sinks as dictated by the respective maximum allowable inlet concentration limits:

$$40x_{1,6} + 37x_{2,6} + 35x_{3,6} + 10x_{4,6} + w_R x_{5,6} \le 25.6 \times 30 = 768 \quad \text{(node 6)} \quad \text{(NLP11)}$$

$$40x_{1,7} + 37x_{2,7} + 35x_{3,7} + 10x_{4,7} + w_R x_{5,7} \le 115 \times 10 = 1150 \quad \text{(node 7)} \quad \text{(NLP12)}$$

$$40x_{1,8} + 37x_{2,8} + 35x_{3,8} + 10x_{4,8} + w_R x_{5,8} \le 28.4 \times 20 = 568 \quad \text{(node 8)} \quad \text{(NLP13)}$$

$$40x_{1,9} + 37x_{2,9} + 35x_{3,9} + 10x_{4,9} + w_R x_{5,9} \le 25x_D \quad \text{(node 9)} \quad \text{(NLP14)}$$

which can be generalized in compact representation as

$$\sum_{i \in I} w_i x_{i,j} \le w_j^{\max} x_j^{\text{sink}}, \quad \text{for all } j \in J$$

f. Variable lower and upper bounds:

$$\text{all } x_{i,j} \ge 0$$
$$\text{all } x_{i,j} \le 300$$
$$x_W \le 300$$
$$x_D \le 300$$
$$w_R \le 100.$$

The formulation gives rise to a nonlinear programming (NLP) due to the regenerator concentration balance as a result of contaminant mixing. Specifically, the nonlinearity is due to a bilinear term from the multiplication of the unknowns of the outlet flow and outlet concentration from the regenerator.

The NLP in this exercise can be solved by using an optimization based modeling system such as GAMS with access to an NLP solver such as CONOPT that employs the GRG algorithm. Alternatively, the NEOS optimization server (http://www.neos-server.org/) provides a free service for solving optimization problems using CONOPT and a range of other state-of-the-art optimization solvers.

The GRG technique is particularly suitable for large-scale NLPs involving sparse nonlinear constraints and a number of linear constraints, such as in this example. The idea behind the reduced gradient methods is to solve a sequence of subproblems with linearized constraints, where the subproblems are solved by variable elimination. This is similar to a two-level optimization procedure. The outer problem takes the Newton step in the reduced space (Equations (8.20) to (8.33)), and the inner subproblem is the linearly constrained optimization problem that can be solved using the simplex method discussed to solve LP problems.

The gradient of the objective function is given by

$$\Delta f(x) = \left(\frac{\partial f(x)}{\partial x_{1,5}}, \frac{\partial f(x)}{\partial x_{1,6}}, \cdots, \frac{\partial f(x)}{\partial w_R} \right)^T$$

$$\Delta f(x) = (5,5,6,5,50,7,7,5,5,50,6,6,7,5,50,20,10,10,11,50,6,5,6,50,1,1,0)^T$$

Gradients of the constraints, in accordance with the order of constraints numbered as NLP1 to NLP14, are

$$\nabla g_n = \left(\frac{\partial g_{\mathrm{NLP}_n}}{\partial x_{1,5}}, \frac{\partial g_{\mathrm{NLP}_n}}{\partial x_{1,6}}, \frac{\partial g_{\mathrm{NLP}_n}}{\partial x_{1,7}}, \frac{\partial g_{\mathrm{NLP}_n}}{\partial x_{1,8}}, \frac{\partial g_{\mathrm{NLP}_n}}{\partial x_{1,9}}, \frac{\partial g_{\mathrm{NLP}_n}}{\partial x_{2,5}}, \frac{\partial g_{\mathrm{NLP}_n}}{\partial x_{2,6}}, \frac{\partial g_{\mathrm{NLP}_n}}{\partial x_{2,7}}, \frac{\partial g_{\mathrm{NLP}_n}}{\partial x_{2,8}}, \right.$$
$$\frac{\partial g_{\mathrm{NLP}_n}}{\partial x_{2,9}}, \frac{\partial g_{\mathrm{NLP}_n}}{\partial x_{3,5}}, \frac{\partial g_{\mathrm{NLP}_n}}{\partial x_{3,6}}, \frac{\partial g_{\mathrm{NLP}_n}}{\partial x_{3,7}}, \frac{\partial g_{\mathrm{NLP}_n}}{\partial x_{3,8}}, \frac{\partial g_{\mathrm{NLP}_n}}{\partial x_{3,9}}, \frac{\partial g_{\mathrm{NLP}_n}}{\partial x_{4,5}}, \frac{\partial g_{\mathrm{NLP}_n}}{\partial x_{4,6}}, \frac{\partial g_{\mathrm{NLP}_n}}{\partial x_{4,7}}, \frac{\partial g_{\mathrm{NLP}_n}}{\partial x_{4,5}},$$
$$\left. \frac{\partial g_{\mathrm{NLP}_n}}{\partial x_{4,9}}, \frac{\partial g_{\mathrm{NLP}_n}}{\partial x_{5,6}}, \frac{\partial g_{\mathrm{NLP}_n}}{\partial x_{5,7}}, \frac{\partial g_{\mathrm{NLP}_n}}{\partial x_{5,8}}, \frac{\partial g_{\mathrm{NLP}_n}}{\partial x_{5,9}}, \frac{\partial g_{\mathrm{NLP}_n}}{\partial x_W}, \frac{\partial g_{\mathrm{NLP}_n}}{\partial x_D}, \frac{\partial g_{\mathrm{NLP}_n}}{\partial x_R} \right)$$

where

$$\nabla g_1 = (1,1,1,1,1,0,0,0,0,0,0,0,0,0,0,0,0,0,0,0,0,0,0,0,0,0,0)^T$$
$$\nabla g_2 = (0,0,0,0,0,1,1,1,1,1,0,0,0,0,0,0,0,0,0,0,0,0,0,0,0,0,0)^T$$
$$\nabla g_3 = (0,0,0,0,0,0,0,0,0,0,1,1,1,1,1,0,0,0,0,0,0,0,0,0,0,0,0)^T$$
$$\nabla g_4 = (0,0,0,0,0,0,0,0,0,0,0,0,0,0,0,1,1,1,1,1,0,0,0,0,0,0,0)^T$$
$$\nabla g_5 = (1,0,0,0,0,1,0,0,0,0,1,0,0,0,0,1,0,0,0,0,-1,-1,-1,-1,0,0,0)^T$$
$$\nabla g_6 = (0,1,0,0,0,0,1,0,0,0,0,1,0,0,0,0,1,0,0,0,1,0,0,0,0,0,0)^T$$
$$\nabla g_7 = (0,0,1,0,0,0,0,1,0,0,0,0,1,0,0,0,0,1,0,0,0,1,0,0,0,0,0)^T$$
$$\nabla g_8 = (0,0,0,1,0,0,0,0,1,0,0,0,0,1,0,0,0,0,1,0,0,0,1,0,0,0,0)^T$$
$$\nabla g_9 = (0,0,0,0,1,0,0,0,0,1,0,0,0,0,1,0,0,0,0,1,0,0,0,1,0,-1,0)^T$$
$$\nabla g_{10} = (6.4,0,0,0,0,5.92,0,0,0,0,5.6,0,0,0,0,1.6,0,0,0,0,-w_R,-w_R,-w_R,-w_R,0,0,-(x_{5,6}+x_{5,7}+x_{5,8}+x_{5,9}))^T$$
$$\nabla g_{11} = (0,40,0,0,0,0,37,0,0,0,0,35,0,0,0,0,10,0,0,0,0,w_R,0,0,0,0,0,x_{5,6})^T$$
$$\nabla g_{12} = (0,0,40,0,0,0,0,37,0,0,0,0,35,0,0,0,0,10,0,0,0,0,w_R,0,0,0,0,x_{5,7})^T$$
$$\nabla g_{13} = (0,0,0,40,0,0,0,0,37,0,0,0,0,35,0,0,0,0,10,0,0,0,0,w_R,0,0,0,x_{5,8})^T$$
$$\nabla g_{14} = (0,0,0,0,40,0,0,0,0,37,0,0,0,0,35,0,0,0,0,10,0,0,0,0,w_R,0,-25,x_{5,9})^T$$

Each gradient (∇g_n) is multiplied by a corresponding vector v_n. The summation of these results is equal to the gradient of the objective function. This corresponds to the gradient equation of the KKT conditions:

$$
\begin{aligned}
\sum \nabla g_n v_n =\ & (1,1,1,1,1,0,0,0,0,0,0,0,0,0,0,0,0,0,0,0,0,0,0,0,0,0,0,0,0,0)^T v_1 \\
& + (0,0,0,0,0,1,1,1,1,1,0,0,0,0,0,0,0,0,0,0,0,0,0,0,0,0,0,0,0,0)^T v_2 \\
& + (0,0,0,0,0,0,0,0,0,0,1,1,1,1,1,0,0,0,0,0,0,0,0,0,0,0,0,0,0,0)^T v_3 \\
& + (0,0,0,0,0,0,0,0,0,0,0,0,0,0,0,1,1,1,1,1,0,0,0,0,0,0,0,0,0,0)^T v_4 \\
& + (1,0,0,0,0,1,0,0,0,0,1,0,0,0,0,1,0,0,0,0,-1,-1,-1,-1,0,0,0)^T v_5 \\
& + (0,1,0,0,0,0,1,0,0,0,0,1,0,0,0,0,1,0,0,0,1,0,0,0,0,0,0)^T v_6 \\
& + (0,0,1,0,0,0,0,1,0,0,0,0,1,0,0,0,0,1,0,0,0,1,0,0,0,0,0,0)^T v_7 \\
& + (0,0,0,1,0,0,0,0,1,0,0,0,0,1,0,0,0,0,1,0,0,0,1,0,0,0,0,0)^T v_8 \\
& + (0,0,0,0,1,0,0,0,0,1,0,0,0,0,1,0,0,0,0,1,0,0,0,1,0,-1,0)^T v_9 \\
& + (6.4,0,0,0,0,5.92,0,0,0,0,5.6,0,0,0,0,1.6,0,0,0,0,-w_R,-w_R,-w_R,-w_R,0,0, \\
& \quad - (x_{5,6}+x_{5,7}+x_{5,8}+x_{5,9}))^T v_{10} \\
& + (0,40,0,0,0,0,37,0,0,0,0,35,0,0,0,0,10,0,0,0,0,w_R,0,0,0,0,0,x_{5,6})^T v_{11} \\
& + (0,0,40,0,0,0,0,37,0,0,0,0,35,0,0,0,0,10,0,0,0,0,w_R,0,0,0,0,x_{5,7})^T v_{12} \\
& + (0,0,0,40,0,0,0,0,37,0,0,0,0,35,0,0,0,0,10,0,0,0,0,w_R,0,0,0,x_{5,8})^T v_{13} \\
& + (0,0,0,0,40,0,0,0,0,37,0,0,0,0,35,0,0,0,0,10,0,0,0,0,w_R,0,-25,x_{5,9})^T v_{14} \\
=\ & (5,5,6,5,50,7,7,5,5,50,6,6,7,5,50,20,10,10,11,50,6,5,6,50,1,1)^T
\end{aligned}
$$

Complementary slackness for the inequality constraints are as follows:

$$
v_{11}\left(768 - 40x_{1,6} - 37x_{2,6} - 35x_{3,6} - 10x_{4,6} - w_R x_{5,6}\right) = 0
$$

$$
v_{12}\left(1150 - 40x_{1,7} - 37x_{2,7} - 35x_{3,7} - 10x_{4,7} - w_R x_{5,7}\right) = 0
$$

$$
v_{13}\left(568 - 40x_{1,8} - 37x_{2,8} - 35x_{3,8} - 10x_{4,8} - w_R x_{5,8}\right) = 0
$$

$$
v_{14}\left(-40x_{1,9} - 37x_{2,9} - 35x_{3,9} - 10x_{4,9} - w_R x_{5,9} + 25x_D\right) = 0
$$

Sign restrictions on the KKT multipliers are

$$
v_{11}, v_{12}, v_{13}, v_{14} \leq 0
$$

Finally, the primal constraints are as given by NLP1–NLP14. The KKT necessary and sufficient conditions of optimality for the problem can be stated as follows. Any solution $x_{i,j}$ for which there exists a corresponding vector $v = \{v_1, \ldots, v_{14}\}$ satisfying these conditions is a local minimum for the NLP problem.

A suitable numerical method can be used to solve this system of nonlinear equations such as a gradient direction based method offered by the CONOPT solver (specifically the version CONOPT3). An optimal solution to this NLP problem is shown in Figure 8.8, with a total cost of $1712 (the operating cost from the connections is $1571.7 and the cost of fresh water is $140.7). Interestingly, the network gives rise to a zero liquid discharge network since all water flows from the sources undergo reuse/recycle (with or without regeneration).

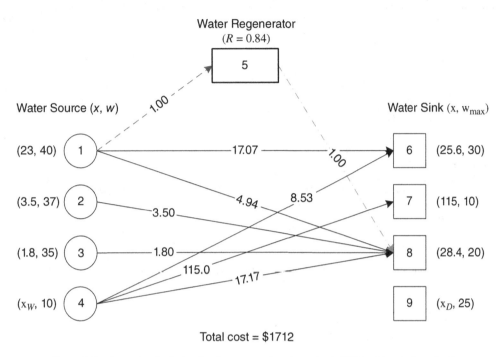

Figure 8.8 *An optimal solution for the water network synthesis NLP problem in Exercise 6.*

Exercise 7. Bioethanol is produced by pretreatment, saccharification and fermentation reactions from lignocellulosic biomass. The lignocellulosic biomass is composed of cellulose, hemicellulose, galactan, mannan and arabinan, with mass compositions and chemical formulae as shown in Table 8.13. The reaction steps and the fractional molar conversions of the reactants involved in the pretreatment, saccharification and fermentation reactions are shown in Table 8.14.

Table 8.13 *Mass compositions and chemical formulae of components in lignocellulosic biomass feedstock (on dry basis).*

Component Name	Component Formula	Mass Composition
Cellulose	$C_6H_{10}O_5$	0.45
Hemicellulose	$C_5H_8O_4$	0.28
Galactan	$C_6H_{10}O_5$	0.20
Mannan	$C_6H_{10}O_5$	0.07
Arabinan	$C_5H_8O_4$	0.01

Calculate the maximum amount of bioethanol that can be produced per unit mass of the biomass feedstock. The decision variables to be deduced are the fractional molar conversions of the various reaction steps. Their base values obtained from process simulation are shown in Table 8.14.

Table 8.14   *Reactions steps and fractional molar conversions of the reactants in pretreatment, saccharification and fermentation reactions.*

Reaction	Equation	Reactant	Fractional Molar Conversion
Pretreatment reaction (PR)			Base Values
PR1	$C_6H_{10}O_5 + H_2O \rightarrow C_6H_{12}O_6$	Cellulose	0.065
PR2	$C_6H_{10}O_5 + 0.5H_2O \rightarrow 0.5C_{12}H_{22}O_{11}$	Cellulose	0.007
PR3	$C_5H_8O_4 + H_2O \rightarrow C_5H_{10}O_5$	Hemicellulose	0.75
PR4	$C_5H_8O_4 \rightarrow C_4H_3OCHO + 2H_2O$	Hemicellulose	0.1
PR5	$C_6H_{10}O_5 + H_2O \rightarrow C_6H_{12}O_6$	Galactan	0.75
PR6	$C_6H_{10}O_5 \rightarrow C_6H_6O_3 + 2H_2O$	Galactan	0.15
PR7	$C_6H_{10}O_5 + H_2O \rightarrow C_6H_{12}O_6$	Mannan	0.75
PR8	$C_6H_{10}O_5 \rightarrow C_6H_6O_3 + 2H_2O$	Mannan	0.15
PR9	$C_5H_8O_4 + H_2O \rightarrow C_5H_{10}O_5$	Arabinan	0.75
PR10	$C_5H_8O_4 \rightarrow C_4H_3OCHO + 2H_2O$	Arabinan	0.1
Saccharification reaction (SR)			
SR1	$C_6H_{10}O_5 + 0.5H_2O \rightarrow 0.5C_{12}H_{22}O_{11}$	Cellulose	0.012
SR2	$C_6H_{10}O_5 + H_2O \rightarrow C_6H_{12}O_6$	Cellulose	0.8
SR3	$C_{12}H_{22}O_{11} + H_2O \rightarrow 2C_6H_{12}O_6$	Cellobiose	1
Fermentation reaction (FR)			
FR1	$C_6H_{12}O_6 \rightarrow 2CH_3CH_2OH + 2CO_2$	Glucose	0.92
FR2	$C_6H_{12}O_6 + 1.2NH_3 \rightarrow 6CH_{1.8}O_{0.5}N_{0.2} + 2.4H_2O + 0.3O_2$	Glucose	0.027
FR3	$C_6H_{12}O_6 + 2H_2O \rightarrow 2C_3H_8O_3 + O_2$	Glucose	0.002
FR4	$C_6H_{12}O_6 + 2CO_2 \rightarrow 2HOOCCH_2CH_2COOH + O_2$	Glucose	0.008
FR5	$C_6H_{12}O_6 \rightarrow 3CH_3COOH$	Glucose	0.022
FR6	$C_6H_{12}O_6 \rightarrow 2CH_3CHOHCOOH$	Glucose	0.013
FR7	$3C_5H_{10}O_5 \rightarrow 5CH_3CH_2OH + 5CO_2$	Xylose	0.85
FR8	$C_5H_{10}O_5 + NH_3 \rightarrow 5CH_{1.8}O_{0.5}N_{0.2} + 2H_2O + 0.25O_2$	Xylose	0.029
FR9	$3C_5H_{10}O_5 + 5H_2O \rightarrow 5C_3H_8O_3 + 2.5O_2$	Xylose	0.002
FR10	$C_5H_{10}O_5 + H_2O \rightarrow C_5H_{12}O_5 + 0.5O_2$	Xylose	0.006
FR11	$3C_5H_{10}O_5 + 5CO_2 \rightarrow 5HOOCCH_2CH_2COOH + 2.5O_2$	Xylose	0.009
FR12	$2C_5H_{10}O_5 \rightarrow 5CH_3COOH$	Xylose	0.024
FR13	$3C_5H_{10}O_5 \rightarrow 5CH_3CHOHCOOH$	Xylose	0.014

Complete conversion of the biomass feedstock on dry basis should be achieved by the process. Deduce the range of by-product productions with chemical formulae shown in Table 8.15, per unit mass of the biomass feedstock, for various reductions in their base production values on the basis of molar conversions shown in Table 8.14. Intermediate reactants with chemical formulae are shown in Table 8.16.

Table 8.15   *By-product names and chemical formulae.*

By-product	Component Formula
Furfural	C_4H_3OCHO
Hydroxyquinol	$C_6H_6O_3$
Glycerol	$C_3H_8O_3$
Succinic acid	$HOOCCH_2CH_2COOH$
Acetic acid	CH_3COOH
Lactic acid	$CH_3CHOHCOOH$
Xylitol	$C_5H_{12}O_5$

Table 8.16 *Intermediate reactant names and chemical formulae.*

Component Name	Component Formula
Glucose	$C_6H_{12}O_6$
Cellobiose	$C_{12}H_{22}O_{11}$
Xylose	$C_5H_{10}O_5$
Carbon dioxide	CO_2
Ammonia	NH_3
Microorganism/fermentation bacteria, e.g., *Acetobacter aceti*	$CH_{1.8}O_{0.5}N_{0.2}$
Oxygen	O_2
Glycerol	$C_3H_8O_3$
Succinic acid	$HOOCCH_2CH_2COOH$
Acetic acid	CH_3COOH
Lactic acid	$CH_3CHOHCOOH$
Xylitol	$C_5H_{12}O_5$

The molar ratio of microorganism/fermentation bacteria, for example, *Acetobacter aceti* (chemical formula: $CH_{1.8}O_{0.5}N_{0.2}$), to the bioethanol product must be maintained at 0.1. The atomic masses of carbon, hydrogen and oxygen are 12, 1 and 16, respectively.

Solution to Exercise 7. The solution is shown in Chapter 8 – Additional Examples and Exercises.

8.4 Mixed Integer Linear or Nonlinear Optimization

In Exercise 6, another important decision, besides determining the interconnection flows and concentrations, is the capital cost for the piping interconnections that is incurred in their purchasing and installation. To accomplish this, an explicit variable for the selection of the interconnections is needed. Such a formulation gives rise to a mixed integer linear optimization technique, which provides a useful modeling technique for incorporating such a discrete type of decision. The resulting mixed integer linear programming (MILP) can handle both discrete or integer variables, for example, the selection of streams and units or facilities, on top of the more typical continuous variables, as shown later in Exercise 9.

In addition to the complexities that arise from constraints and nonlinearity of the functions, many optimization problems have the additional feature of noncontinuous functions to perform the selection or activation of options (e.g., chemical route, process technology, product, node interconnection, etc.) by using discrete or integer variables. In this section, problems combining continuous variables related by linear relationships (LP) and integer variables are discussed. These types of problems are known as mixed integer linear problems (MILP).

As the name suggests, mixed integer nonlinear programming (MINLP) involves both continuous and discrete or integer type of decision variables, with nonlinearity occurring in at least one or more of the constraints and the objective function. In terms of historical development, although MINLP gained acceptance later than the foregoing other optimization formulation techniques, it has found wide-ranging applications in the last 10 years, particularly in the chemical engineering community. Examples of these applications in the process industry include the optimal synthesis and design of plants, optimization of planning and scheduling activities and optimization of supply chain design and management. A recent review of MINLP applications in this field is provided by Grossmann and Guillén-Gosálbez[8].

Consider the cost function having the form

$$\min f = \sum_i \sum_j (x_{i,j} c_j + C_j)$$

where c_j is the operating cost of a process j per unit of mass flow rate of feedstock i and C_j is the capital cost required for equipment in the process j.

Assume that the cost function is used to select, from a range of technologies, those processes that minimize the total costs of a biorefinery. The formulation must account for the case when certain technology is not selected.

For the hypothetical case when there is one feedstock $i = 1$ and three technology options $j = \{1,2,3\}$ the expanded objective function is

$$\min f = \left(x_{1,1}c_1 + C_1\right) + \left(x_{1,2}c_2 + C_2\right) + \left(x_{1,3}c_3 + C_3\right)$$

Now assume that only technology $j = 1$ and $j = 3$ are selected at a certain iteration of the solution search. The solver then sets the flow rate to the technology $j = 2$ equal to $x_{1,2} = 0$. This automatically removes the operating cost from that technology; however, the capital cost term C_2 remains active in the function, adding a cost that is not being incurred. This cost has to be made equal to zero in this iteration.

Those terms in a function that are not directly related to the decision variables but that need to be activated or deactivated when a certain selection is made require the use of an integer variable. For this case, a variable y is introduced that can have the value of $y = 0$ when a technology is not selected or $y = 1$ when the technology is selected. This is implemented in the objective function by multiplying the capital cost of each technology by a corresponding integer or binary variable y_j as

$$\min f = \left(x_{1,1}c_1 + C_1y_1\right) + \left(x_{1,2}c_2 + C_2y_2\right) + \left(x_{1,3}c_3 + C_3y_3\right)$$

Thus, for the case when the technology $j = 2$ is not selected, $x_{1,2} = 0$ and $y_2 = 0$ and now the objective function only adds the costs incurred by the technologies selected and the corresponding integer variables are set as $y_1 = y_3 = 1$. Integer variables may be needed in the formulation of constraints and variable bounds.

There are two main methods to handle mixed integer optimization problems, the penalty function method and the branch and bound method.

8.4.1   Branch and Bound Method

In the branch and bound method, the integer variables are first relaxed. This relaxed problem leads to an NLP or LP problem, depending upon the nature of the formulation. The resulting noninteger problems are then solved by NLP or LP optimization methods. The results of the integer variables are used to guide the branching of the NLP problem into complimentary subproblems.

Nodes and *branches* are considered from the initial node to partition and organize the solution space. This step is known as *branching*. The structure consisting of nodes and branches is called a *search tree*. Figure 8.9 shows that the

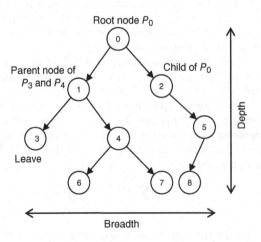

Figure 8.9   *Search tree for an MILP problem.*

root node (P_0) is the starting node of the tree and contains the solution of the relaxed problem ($f^{(0)}$ and $X^{(0)}$). The internal nodes are *child* nodes of a parent node. The outermost nodes are called *leaves* (P_3, P_6, P_7, P_8).

The main feature of the branch and bound method is the reduction of the search space or feasibility region by eliminating infeasible search trees under each node. This is done by calculating the bounds of the objective function within the reduced search space (*subregion*) and retaining the search tree with a minimum objective function value (in the case of a minimization problem). This second procedure is known as *bounding*.

The main guideline at the bounding step is that if the lower bound of the *active* node is greater than the global upper bound found from other iterations, then the active node is discarded from the search. This step is called *pruning*. A global variable is implemented to record the minimum upper bound amongst all the bounds resulting from the nodes previously evaluated. This is the strategy to follow for the minimization problems.

The search tree is developed dynamically during the solution process and consists initially of only the root node. The current best solution, called the *incumbent*, is retained and propagated to the next step, until it is replaced by a better solution. The order of the branching and bounding steps depends on the strategy followed.

In the *eager strategy*, branching from a parent node is performed first and constraints are added to the subproblems. In this way, the search space is divided into smaller subregions. For each of these nodes or subregions, the bound (objective function value) is calculated. In the case were the node corresponds to a feasible solution, the solution is compared to the incumbent (the current best solution). If the value is better (less than the incumbent in the case of a minimization problem), the new solution replaces the old one as the new incumbent. If the value is no better than the incumbent, the subproblem is discarded, since no feasible solution of the subproblem can be better than the incumbent. In the case where no feasible solutions to the subproblem exist the subproblem is eliminated (i.e., the node and its children are deactivated). Otherwise, the possibility of a better solution remains at the node in question and is thus kept as an active node (with its bound stored).

In the *lazy strategy*, the bound of the selected node is calculated first and then branching on the node is carried out if necessary. The active children nodes are stored with the bound of their parent node as part of the information. In this strategy, active nodes of maximal depth in a search tree are evaluated (i.e., the node at or closest to the bottom of the tree), before moving to another node.

Figure 8.10 shows strategies used for the selection of the next node in a tree to be evaluated. These strategies are:

- In the best first search (BeFS), subproblems of the node with the lowest bound are retained until the global solution is found (Figure 8.10(a)). The strategy can be computationally expensive when various critical subproblems are found (that cannot be discarded) as more memory is required to store their information.

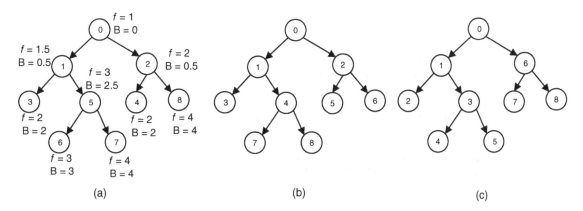

Figure 8.10 *Search strategies in a tree: (a) best first, (b) breadth first and (c) depth first. The numbers in each node correspond to the sequence in which the nodes are processed. B means bound.*

- In the breadth first search (BFS), all nodes at the same depth level of the search tree are evaluated for bounds before proceeding to the next depth level of nodes (Figure 8.10(b)). The number of nodes at each depth level increases exponentially as the search proceeds to the outermost leaves. This makes the breadth search computationally expensive for problems with numerous nodes with larger ranges.
- In the depth first search (DFS), active nodes in a search tree at all depth levels are selected for exploration first (Figure 8.10(c)), before moving to another search tree from the route node. The problem size is usually manageable. The DFS can be used with both lazy and eager strategies. The DFS is advantageous for programming, since the use of recursive search enables the recording of information about the current node in an incremental fashion and constraints are only added during the branching of a node.

The search terminates when there are no unexplored nodes and subregions of the feasible solution search space. The optimal solution corresponds to the "current best" solution. For a maximization problem, the objective function is multiplied by −1 in order to treat the problem as a minimization problem.

Exercise 8. A biodiesel production company is planning to build a capacity of 100 000 t y^{-1} of biodiesel. The company is keen to choose locations to grow biomass and produce biodiesel so that the environmental impact (EI) in terms of the global warming potential in CO_2 equivalent from the cradle-to-plant gate is least. Using the information in Table 8.17, and in the following text, determine the capacity and location of the biorefineries and their feedstock supplies to help the company achieve their goal.[9]

Table 8.17   *Data for the feedstock supply network planning problem in Exercise 8.*

Source	Plantation Land (ha)	Yield (t ha^{-1})	EI Seed Production (t CO_2-eq t^{-1})	Distance to B_1 (km)	Distance to B_2 (km)	Distance to B_3 (km)	Distance to B_4 (km)
S_1	25 000	4.2	0.626	436	1136	315	1068
S_2	20 000	3.8	0.414	915	325	1140	275
S_3	10 000	4.0	0.451	1542	1180	1770	902
S_4	15 000	4.1	0.523	887	1708	675	1520

There are four biomass supply locations (S_1 to S_4) and four choices for biorefinery location (B_1 to B_4). The distances between the various biomass sources and biorefinery locations are also given in Table 8.17.

Minimum biorefinery capacity: 30 000 t y^{-1}
EI from the plant operation in t CO_2 equivalent y^{-1} is related to the capacity according to the following equation:

$$EI_{op} = 2.54 \times Capacity + 350.6 \quad 30000 < Capacity\ (t\ y^{-1}) < 60\,000$$

EI from transportation of biomass: 0.00008 t CO_2 eq. t^{-1} km^{-1}
Yield factor for conversion of biomass into biodiesel: 0.37 kg biodiesel per kg biomass

Figure 8.11 shows a schematic of the network for the instance of the problem considered here.

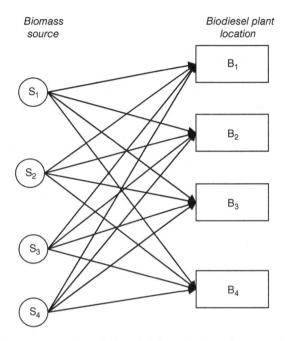

Biomass source

Biodiesel plant location

Figure 8.11   *Schematic network for the bioethanol supply chain planning problem in Exercise 8.*

Exercise 9. Consider the water network synthesis problem in Exercise 6. Formulate the problem employing an explicit approach as to whether a piping interconnection should be built between any two nodes in terms of a yes (valued at 1) or no (valued at 0) decision, along with the imposition of an associated capital cost, shown in Table 8.18, using a fixed-charge cost model.

Table 8.18   *Capital cost in $ h^{-1} for piping interconnections between two nodes.*

Source	Sink				
	6	7	8	9	5 (Regenerator)
1	10	12	10	100	10
2	14	10	10	100	14
3	12	14	10	100	12
4	10	10	11	100	20
5 (Regenerator)	12	10	12	100	—

Solution to Exercise 9. The problem formulation is discussed in Chapter 8 – Additional Examples and Exercises.

8.5   Stochastic Method

In real-world problems, it is more often the case that the model parameters are uncertain. This implies that the parameters are unknown, or at least not well known, and thus may assume many possible different values. Uncertainties due to unknown parameters are identified as uncertain model parameters and variable process parameters from the observability point of view[10]. In the context of a design or planning problem, the exact values of uncertain model parameters are never known, although the expected values and confidence regions may be known. These include model parameters determined from offline experimental studies such as kinetic parameters of reactions or unmeasured and unobservable disturbances such as the influence of wind and sunshine on renewable energy generation. On the other hand, variable process parameters, although unknown at the design or planning stage, can be specified deterministically or measured accurately at later operating stages. Examples of these are internal unmeasured disturbances such as feed flow rates, product demands and process conditions and inputs, for example, temperatures and pressures, and also external unmeasured uncertainties such as ambient conditions where an operation of interest takes place. A stochastic programming technique such as the genetic algorithm discussed here or simulated annealing is applied to an optimization problem with uncertain model parameters.

Exercise 10.   Consider the water network synthesis in Exercise 9. Formulate and solve the problem using stochastic optimization approaches[11].

Solution to Exercise 10.   The approach to the solution is discussed in Chapter 8 – Additional Examples and Exercises.

8.5.1   Genetic Algorithm (GA)

Genetic algorithms (GAs) are adaptive heuristic search algorithms premised on the evolutionary ideas of natural selection and genetics. Various GA based methods are available, outlined here. References are given for more detailed discussion. GA starts with a set of initial assumptions for decision variables, represented by chromosomes, called a *population*. Solutions from one population are taken and used to form a new population (*offspring*). Solutions are selected to form new solutions (*offspring*) according to their fitness (values of the objective function) – the more suitable they are the more chances they have to reproduce. When applying a GA to problems with constraints, it is common to add a penalty term to the objective function for fitness evaluation. This technique transforms the constrained problem into an unconstrained problem by penalizing infeasible solutions, in which a penalty term is added to the objective function for any violation in the constraints. While optimizing a function, one therefore also needs to minimize the constraint violation.

A general genetic algorithm requires the determination of five fundamental issues: (i) creation of the initial population, (ii) fitness evaluation, (iii) selection function, (iv) reproduction function (genetic operators) and (v) termination criteria.

Figure 8.12 shows an outline of a genetic algorithm. The main steps of the algorithm can be shown as follows.

1. An initial population $(P_i(N))$ of N chromosomes is generated randomly in iteration $i = 1$. Each chromosome is a representation of probable solutions in terms of decision variables that are solved for optimal values using iterative steps as follows.
2. Evaluate the fitness of each chromosome in $P_i(N)$ based on the objective function value. If the problem is constrained, a suitable penalty function to the objective function is added for evaluation.
3. A new population $(P_i'(N))$ of N chromosomes is generated by a probabilistic selection function. The function is performed based on individual fitness evaluated in step 2, such that the better individuals have an increased chance of being selected.
4. Genetic operators, represented by a reproduction function, are used to create a new population $P_i''(N)$ of N chromosomes from the population $P_i'(N)$. GA employs two important operators, crossover and mutation, which are responsible for creating the chromosomes for a new population. The mechanisms of mutation and crossover are shown in Figure 8.13.
5. The generation to generation steps 3 and 4 are iterative. These steps are followed until termination conditions are satisfied. The most frequently used stopping criterion is the maximum number of generations or the convergence factor of objective function. If the criterion is not met, a new iteration starts (step 3). Iteration $i = i + 1$ is set, and population $P_i(N)$ is substituted with $P_{i-1}''(N)$.

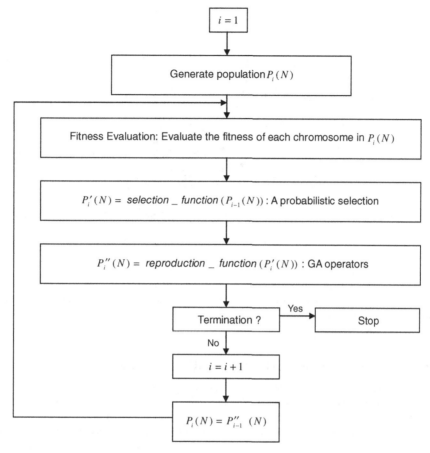

Figure 8.12 *Genetic algorithm overview. (Reproduced with permission from Xu et al. (2009)[12]. Copyright © 2009, Elsevier.)*

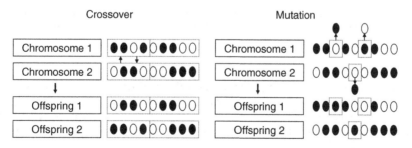

Figure 8.13 *GA operators of crossover and mutation. (Reproduced with permission from Xu et al. (2009)[12]. Copyright © 2009, Elsevier.)*

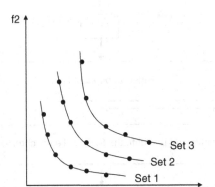

Figure 8.14   *Non-dominated sorting sets (Pareto sets). (Reproduced with permission from Xu et al. (2009)[12]. Copyright © 2009, Elsevier.)*

However, most real-world problems involve simultaneous optimization of several objective functions, which can be noncommensurate. For those cases, multiobjective multiobjective optimization problem formulation is developed to find the optimal trade-offs in an overall sense.

8.5.2   Non-dominated Sorting Genetic Algorithm (NSGA) Optimization

Since there are multiple objective functions to be optimized simultaneously, a whole set of possible solutions of equivalent quality, known as a Pareto set, can be generated. The nondominated sorting genetic algorithm (NSGA) computes successive generations on a population of solutions partitioned into nondominated fronts (Figure 8.14). Thus, several nondominated sets (Pareto sets) are identified and each of them constitutes its own nondominated fronts. This process is called nondominated sorting. Once the partition is complete, an optimal Pareto set of several equally good solutions is provided. These can provide useful insights into the decision maker to decide upon the preferred solution[13].

The GA algorithm shown here is called NSGA-II-JP. The difference between NSGA-I and NSGA-II is that the latter applies the concept of elitism, borrowed from nature evolution. In NSGA-I, chromosomes in new generations are always formed from the crossover or mutation of their parents. However, in NSGA-II, better parents from the old generation are given a chance to join the new generation with their daughters (those after crossover or mutation) as best members. This modification obviously speeds up the processing, but also decreases the diversity. Therefore, another adaptation, inspired by the concept of jumping genes (JGs, or transposons) in biology is developed, which not only exploits the benefits of elitism but also maintains the genetic diversity. The NSGA-II-JP framework is shown in Figure 8.15 and the steps involved are discussed as follows[13-15]:

1. Generate an initial population $(P_i(N))$ of N chromosomes randomly in iteration $i = 1$. These chromosomes are given a sequence of genes represented by binary digits (0 and 1).
2. Classify these chromosomes in population $P_i(N)$ into fronts based on nondomination. Make N copies randomly (duplication permissible) of the best chromosomes based on $P_i(N)$ after classification. Put them in a new generation $P_i'(N)$ of N chromosomes.
3. Do crossover and mutation to chromosomes in population $P_i'(N)$ (Figure 8.13). Put the results into a new daughter generation $P_i''(N)$ of N chromosomes.
4. Do jumping gene (JG) alteration to chromosomes in population $P_i''(N)$.

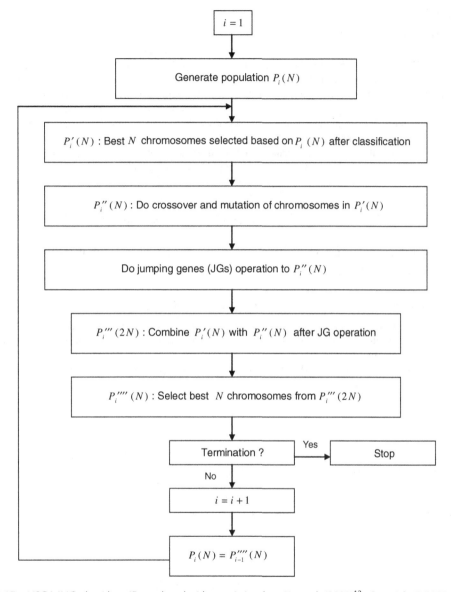

Figure 8.15   *NSGA-II-JG algorithm. (Reproduced with permission from Xu et al. (2009)[12]. Copyright © 2009, Elsevier.)*

5. Elitism: two steps are involved for elitism:
   a. Copy all the N chromosomes from best parent population $P_i'(N)$ and all the N chromosomes from daughter population $P_i''(N)$ after JG alteration into a new generation $P_i'''(N)$ of $2N$ chromosomes.
   b. Reclassify these $2N$ chromosomes of $P_i'''(N)$ into fronts using nondomination (as described in step 2). Select the best N chromosomes and put them in a new generation $P_i''''(N)$.
6. Determine whether the termination criterion is met or not. If so, stop the programme. Otherwise, set iteration $i = i + 1$ and substitute population $P_i(N)$ with $P_{i-1}''''(N)$.

8.5.3   GA in MATLAB

Since the performance of genetic algorithms is significantly affected by constraints, it is important to design a suitable penalty function (with coefficients having very large bounds) to penalize infeasible points and favor the convergence (the convergence factor is a very small number, as shown in Exercise 5). Although the NSGA-II-JG technique is generic and can be applied to both equality and inequality constraints, more specialized methods can be designed to handle particular types of constraints. Too severe penalties make it difficult to drive the population toward the optimum. Conversely, if the penalty function is not severe enough, a large region is searched and much of the search time will be used to explore regions far from feasibility. Most complex engineering problems include a large number of linear equality and inequality constraints in addition to nonlinear equality and inequality constraints. The speed of convergence of NSGA-II-JG depends on how efficiently such penalty functions are manipulated and defined. Based on the application of penalty functions, a constrained problem is transformed into a nonconstrained one, shown as follows[16]:

$$F(x) = \begin{cases} f(x) & x \in \text{feasible region} \\ f(x) + \text{penalty}(x) & x \notin \text{feasible region} \end{cases} \tag{8.41}$$

where x, $f(x)$, penalty(x) and $F(x)$ indicate the decision variables, objective function, penalty function and constrained optimization problem, respectively.

8.5.3.1   *Optimization of Metabolic Productivity Using a Genetic Algorithm*

Three optimization problem formulations, a two-step optimization by GAMS, a nondominated sorting genetic algorithm and a MATLAB based combined genetic algorithm and linear programming tool are shown using a cellular system metabolic productivity and thermodynamic performance optimization as an example elsewhere[12,17]. The objective was to deduce optimum flux distributions through a unique set of enzyme-catalyzed pathways for maximizing the production of a given metabolic product of importance.

8.6   Summary

This chapter discusses various optimization approaches and problem formulations in a systematic way. The LP, NLP, MILP and MINLP problems and stochastic search methodologies are discussed. To a certain extent, a suitable formulation type largely depends on the nature of the problem and the major decisions of interest. By extension, the latter consideration is associated with the type of decision variables, which are (mainly) either continuous or discrete. The main optimization search methodologies used to arrive at solutions are discussed with example problems.

References

1. C.E. Bodington and T.E. Baker, A history of mathematical-programming in the petroleum-industry, *Interfaces*, **20**, 117–127 (1990).
2. G.B. Dantzig, *Linear Programming and Extensions*, Princeton University Press, Princeton, 1963.
3. J. Nocedal and S.J. Wright, *Numerical Optimization*, Springer, New York, 1999.
4. M.S. Bazaraa, H.D. Sherali, C.M. Shetty, *Nonlinear Programming: Theory and Algorithms*, John Wiley & Sons, Inc., New York, 1993.
5. T.F. Edgar, D.M. Himmelblau, L.S. Lasdon, *Optimization of Chemical Processes*, McGraw-Hill, New York, 2001.
6. C.S. Khor, B. Chachuat, N. Shah, A superstructure optimisation approach for water network synthesis with membrane separation-based regenerators, *Computers and Chemical Engineering*, **42**, 48–63 (2012).
7. C.S. Khor, S. Mahadzir, A. Elkamel, N. Shah, An environmentally-conscious retrofit design of sustainable integrated refinery water management network systems incorporating reuse, regeneration, and recycle strategies, *Canadian Journal of Chemical Engineering*, **90**, 137–143 (2012).
8. I.E. Grossmann and G. Guillén-Gosálbez, Scope for the application of mathematical programming techniques in the synthesis and planning of sustainable processes, *Computers and Chemical Engineering*, **34**, 1365–1376 (2010).

9. O. Akgul, N. Shah, L.G. Papageorgiou, Economic optimisation of a UK advanced biofuel supply chain, *Biomass Bioenergy*, **41**, 57–72 (2012).

10. W.C. Rooney and L.T. Biegler, Optimal process design with model parameter uncertainty and process variability, *AIChE Journal*, **49**, 438–449 (2003).

11. C.S. Khor, A. Elkamel, K. Ponnambalam, P.L. Douglas, Two-stage stochastic programming with fixed recourse via scenario planning with economic and operational risk management for petroleum refinery planning under uncertainty, *Chemical Engineering and Processing: Process Intensification*, **47**, 1744–1764 (2008).

12. M. Xu, S.A. Bhat, R. Smith *et al.*, Multi-objective optimisation of metabolic productivity and thermodynamic performance, *Computers and Chemical Engineering*, **33**(9), 1438–1450 (2009).

13. K. Deb, *Optimisation for Engineering Design: Algorithms and Examples*, Prentice Hall, New Delhi, 1995.

14. K. Deb *Multi-objective Optimisation Using Evolutionary Algorithms*, John Wiley & Sons, Ltd, Chichester, 2001.

15. M. Gen and R. Cheng, *Genetic Algorithms and Engineering Design*, John Wiley & Sons, Inc., New York, 1997.

16. A.E. Smith and D.W. Coit, Penalty function (Section C 5.2), in *Handbook of Evolutionary Computation* (eds T. Back, D.B. Fogel and Z. Michalewicz), Oxford University Press and Institute of Physics Publishing, Bristol, 1997.

17. M. Xu, R. Smith, J. Sadhukhan, Optimization of productivity and thermodynamic performance of metabolic pathways, *Industrial & Engineering Chemistry Research*, **47**(15), 5669–5679 (2008).

Part III

Process Synthesis and Design

9

Generic Reactors: Thermochemical Processing of Biomass

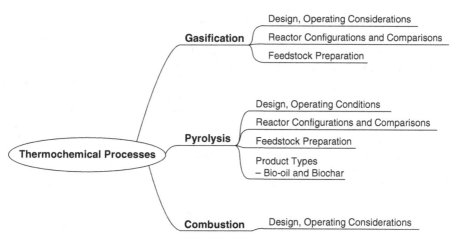

Structure for Lecture Planning

9.1 Introduction

Thermochemical processing is one of the main biorefinery platforms, where biomass is converted into high value intermediate products. Gasification is the most effective of all thermochemical technologies, generating a product gas that upon clean-up is known as *syngas*, which consists of a mixture of carbon monoxide and hydrogen. Syngas is versatile in terms of a wide range of applications, such as for production of fuels, chemicals and power. The thermochemical processing route has several advantages:

- high conversion of feed into products;
- feedstock flexibility with acceptance of a wide range of biomass;

Biorefineries and Chemical Processes: Design, Integration and Sustainability Analysis, First Edition.
Jhuma Sadhukhan, Kok Siew Ng and Elias Martinez Hernandez.
© 2014 John Wiley & Sons, Ltd. Published 2014 by John Wiley & Sons, Ltd.
Companion Website: http://www.wiley.com/go/sadhukhan/biorefineries

- compatibility with existing facilities using fossil fuel with minimal modification and upgrade;
- production yield is stable and large-scale production can be achieved.

Combustion and pyrolysis are the two other technologies that constitute a thermochemical biorefinery. These technologies have long been practiced in industry; however, the application of these technologies in a biorefinery context is relatively new compared to fossil fuels. The integration of these technologies with downstream processes such as combined cycle, Fischer–Tropsch and methanol synthesis is techno-economically feasible.

This chapter gives a comprehensive overview of the thermochemical processing technologies relevant to biorefineries. The learning outcomes from this chapter are:

- distinguishing between combustion, gasification and pyrolysis;
- understanding the process operating principles of gasification and pyrolysis;
- gaining insights into the design strategies for gasification and pyrolysis.

9.2   General Features of Thermochemical Conversion Processes

Carbonaceous feedstocks such as coal and biomass contain a valuable element – carbon, a very important source of energy as well as constituent for chemical production. Combustion, gasification and pyrolysis are the major thermochemical processing technologies. These processes extract energy from the carbonaceous materials, transforming carbon, hydrogen and oxygen atoms in the materials into other forms of energy. The forms of energy can be heat, syngas and oil depending on the processing route and operating condition. Figure 9.1 shows the energy conversion pathways from carbonaceous material to intermediate products via thermochemical processes. The intermediate products are subsequently converted into fuel, electricity and chemicals. Thermochemical processes are highly favorable platforms for conversion of any lignocellulosic material into energy products. However, notably, without any predrying, biomass with a high moisture content (>50%) is not efficient for thermochemical processes. It is then more suitable for use in biological processes.

Combustion, gasification and pyrolysis already have a long history in the industry for fossil fuel processing. These technologies have received considerable attention for biomass processing in recent decades. It is possible to coprocess biomass feedstock in existing fossil fuel based thermochemical conversion facilities. However, the characteristics of biomass are fairly different from a fossil fuel such as coal and thus such facilities have to be modified when biomass is used as feedstock. From the facet of energy content, biomass has a lower energy value than fossil fuels. The configuration of certain machinery and operating conditions need to be altered to account for this. Cofiring with other fossil fuel feedstocks such as natural gas and coal can also be one of the approaches to dealing with the shortcomings related to the energy content of biomass. In addition, different types of biomass feedstock also result in variation and inconsistency with respect to the heating value as well as the composition of the products. From the environmental standpoint, biomass is cleaner than fossil fuel. Carbon dioxide released from biomass is captured during photosynthesis, thus achieving carbon neutrality or even carbon negativity (with carbon capture and storage), when considering the whole carbon cycle. Furthermore, biomass has a negligible sulfur content and hence has a low emission of SO_x, which poses no threat to the environment.

Did you know?

The history of gasification can be dated back to the late eighteenth and early ninteenth centuries. At that time, Phillipe Lebon from France and William Murdoch, who worked for Boulton & Watt in England, invented the method independently for producing gas from wood and coal (Figure 9.2). The process was later commercialized and used widely throughout Europe for producing "town gas," serving as lighting, heating and cooking purposes. The first utility company supplying "town gas" was the Gas Light and Coke Company based in London, established in 1812. In 1839, the first wood gasifier was built by Bischof. During World War II, many gasification plants were constructed in Germany to produce synthetic liquid fuels through Bergius and Fischer–Tropsch processes to address the shortage of petroleum. After World War II, the technology was exported to the South African oil company Sasol.

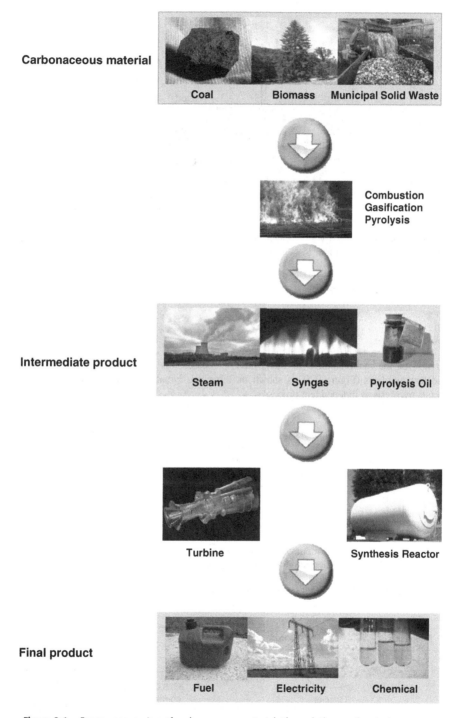

Figure 9.1 *Energy conversion of carbonaceous materials through thermochemical processes.*

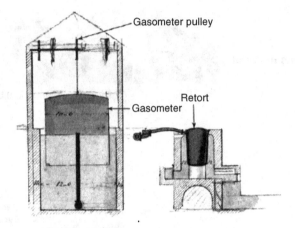

Figure 9.2 *A gas plant built by Boulton & Watt. Original images by wood, Daintry & Wood, 28 July 1805, BWA MS 3147/3/478 #46; and Murdoch to B&W, 1 January 1806, BWA MS 3147/3/289 #18*[1].

The three main thermochemical processes are:

1. Combustion
2. Gasification
3. Fast pyrolysis.

A holistic overview of the main features of these three processes, that is, combustion (Figure 9.3), gasification (Figure 9.4) and fast pyrolysis (Figure 9.5), is shown in Table 9.1. Notably, the comparison considers combustion, gasification and fast pyrolysis as independent processes.

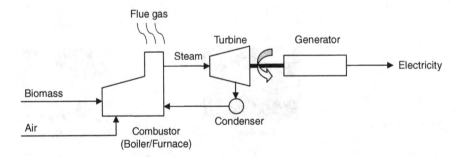

Figure 9.3 *Combustion process – transforming biomass into heat and electricity.*

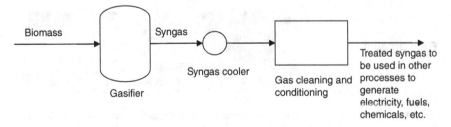

Figure 9.4 *Gasification process – transforming biomass into syngas.*

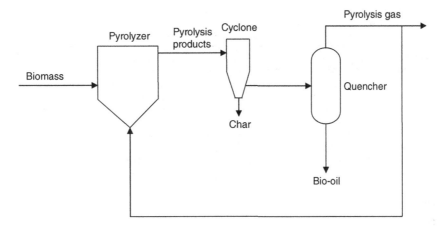

Figure 9.5 *Fast pyrolysis process – transforming biomass into char, bio-oil and gas.*

9.3 Combustion

Combustion is an exothermic chemical reaction, where a substance (which often refers to a fuel, carbonaceous material or hydrocarbon) reacts with oxygen to produce mainly heat along with other by-products (known as exhaust) such as carbon dioxide and water. There are essentially three conditions to initiate a combustion reaction:

1. Combustible material. This can be any substance containing carbon.
2. Oxygen. This is an important element during combustion. Combustion will not take place without oxygen.
3. Ignition temperature. For a reaction to take place, the amount of energy supplied to the reactants has to be sufficient to overcome the activation energy of the reaction. The ignition temperature is the point or the minimum temperature at which the chemical reaction can occur.

Hence, a combustion reaction can proceed when:

1. The correct fuel-to-air ratio is satisfied. The amount of oxygen in air has to be in excess to complete the combustion. If the reaction takes place under a lean oxygen environment, the combustion process will not be efficient.
2. The operating temperature is greater than the ignition temperature. This makes sure that the process is operated in an efficient manner.

A good combustion should fulfil the following '3T' rule:

1. Temperature. The temperature must be sufficiently high to overcome the activation energy of reaction.
2. Turbulent. Gives good mixing between fuel and air.
3. Time. Must be long enough to reach complete combustion.

The main reactions involved in a combustion reaction are described by

$$C(s) + O_2(g) \rightarrow CO_2(g) \qquad \Delta H = -393.5 \ \text{kJ mol}^{-1} \qquad (9.1)$$

$$H_2(g) + \tfrac{1}{2}O_2(g) \rightarrow H_2O(l) \qquad \Delta H = -241.8 \ \text{kJ mol}^{-1} \qquad (9.2)$$

$$CH_4(g) + 2\,O_2(g) \rightarrow CO_2(g) + 2\,H_2O(l) \qquad \Delta H = -802.8 \ \text{kJ mol}^{-1} \qquad (9.3)$$

Table 9.1 *Comparison of the major thermochemical conversion processes.*

Specification	Combustion	Gasification	Fast Pyrolysis
Process Fundamentals			
Definition (based on NREL)	Thermal conversion of organic matter with excess oxidant (normally oxygen) to produce primarily carbon dioxide and water	Thermal conversion of organic materials through partial oxidation/indirect heating at elevated temperature and reducing conditions to produce primarily permanent gases, with char, water and condensables as minor products	Thermal conversion (destruction) of organics into liquid in the absence of oxygen
Reaction mechanism	Full oxidation	Partial oxidation	Thermal degradation
Process Input–Output			
Feed	Any	Any	Any
Primary product	Heat (steam)	Combustible gas (syngas)	Liquid (bio-oil)
Main components in product	CO_2 and H_2O	CO and H_2	Various chemical constituents (aldehydes, carboxylic acids, ketones, phenols, etc.)
By-products component	NO, NO_2, SO_2, SO_3,	HCN, NH_3, N_2, H_2S, COS	Water and char
Process Operating Requirements			
Operating temperature (°C)	High (700–1400)	High (500–1300)	Moderate (~500)
Operating pressure (bar)	Low to moderate (1–60)	Low to moderate (1–60)	Low (1–5)
Oxygen requirement	Oxygen-rich environment	Oxygen-lean environment	No oxygen
Air requirement	Excess	Partial	No
Steam requirement (NOT for feed drying purposes)	No	Yes (steam may not be necessary when the moisture content of feed is high)	No
Process Performance			
Efficiency (%)	90 and higher	60–70	80
Yield of primary product (%)		85	75
Residence time	Long	Long	Short (<2 s)
Further processing of products	Steam produced can be used to generate electricity by expanding through turbine and generator	Gas cleaning and conditioning	Removal of oxygen content in oil for stability reasons

9.4 Gasification

In a combustion process, carbonaceous fuel undergoes complete reaction with oxygen to generate heat. Gasification, on the other hand, provides a cleaner and higher efficiency platform for partially oxidizing carbonaceous fuels to a gas product that upon clean-up and purification contains mainly CO and H_2, known as *syngas* (synthesis gas). The valuable energy is retained in the syngas. It is a versatile way to produce chemicals, fuels, heat and power.

9.4.1 The Process

The chemical reactions that take place inside the gasifier are complex. The main reactions can be categorized into four zones (Figure 9.6). Figure 9.6 shows an updraft gasifier, where the gasifying medium (air/steam) enters from the bottom

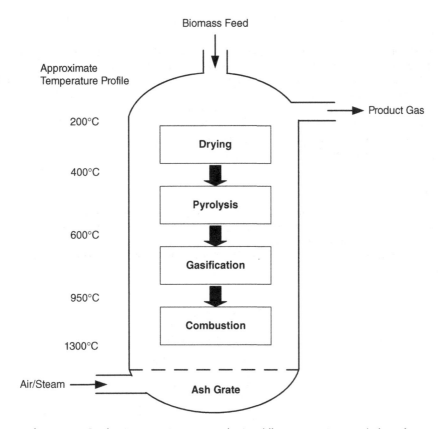

Figure 9.6 *Gasification process represented using different zones in an updraft gasifier.*

of the gasifier, while biomass is fed from the top of the gasifier. The gasifier is operated in a countercurrent mode and the product gas comes out from the top of the gasifier. More discussion on this type of gasifier is included in Section 9.4.2.

A typical biomass gasification process involves four main reaction steps: drying, pyrolysis, gasification and combustion, discussed as follows.

9.4.1.1 Drying

The moisture content in biomass is normally high, typically 30–60 wt%. Without the drying process, more energy is needed to vaporize the water; approximately 2260 kJ of energy from the gasifier is used for drying 1 kg of biomass (note that this is the heat of vaporization of water at 1 atm and 100 °C = 2260 kJ kg^{-1}). Biomass feed is dried using the heat released from the downstream zones. The process takes places at around 100 to 200 °C. The moisture content in biomass has a close relationship with the heating value of the corresponding biomass. A lower moisture content of around 10–15% is desirable, so that the heating value of the biomass can be maintained at a higher level.

9.4.1.2 Pyrolysis/Devolatilization

This process does not involve interaction with air or oxygen and occurs at around 500 °C. This is a thermal degradation process, where large molecules are broken down into smaller fragments, primarily producing a *liquid*, known as *pyrolysis oil* or *bio-oil* that is eventually gasified in a gasifier. Primary pyrolysis or devolatilization occurs as soon as solid biomass comes into contact with the hot bed of gasifier before the main gasification reactions (discussed next) take place. Some *tar* may be formed. Tar has hundreds of types of chemical constituents and is a notorious material produced from pyrolysis. Tar can cause severe operational issues such as clogging in equipment. The product gas from an updraft gasifier has a considerable amount of tar. A downdraft gasifier produces less tar because the products from pyrolysis undergo further gasification and combustion, which breaks down the tar. The reaction time, temperature and oxygen feed flow rate are the controlling factors. High temperature and slower reactions produce more gas (gasification with partial oxidation at one extreme); pyrolysis at a lower temperature and with little or no oxygen provides more liquid. Pyrolysis can be of two types, fast and slow. A fast pyrolysis process has a larger liquid yield. As the temperature is lowered and reaction rates are slowed and properly controlled, a higher proportion of solid biochar is obtained (in another extreme), detailed discussion of pyrolysis is given in Section 9.5.

9.4.1.3   Gasification

Solid biomass and residual biomass, for example, char and tar, undergo gasification in the presence of a gasifying medium such as air, oxygen and steam, at around 800–1000 °C. Typical gasification reactions are shown below in the following equations. Exothermic partial oxidation (Equation (9.4)) and endothermic steam methane reforming (Equation (9.9)) reactions are the two main reactions. The others are side reactions. Thus, the product gas from the gasifier has two main constituents, CO and H_2.

Heterogeneous reactions:

$$C + \tfrac{1}{2}O_2 \rightarrow CO \qquad \Delta H = -110 \text{ kJ mol}^{-1} \text{ (partial oxidation reaction)} \tag{9.4}$$

$$C + H_2O \rightleftharpoons CO + H_2 \qquad \Delta H = +131 \text{ kJ mol}^{-1} \text{ (water gas reaction)} \tag{9.5}$$

$$C + CO_2 \rightleftharpoons 2CO \qquad \Delta H = +172 \text{ kJ mol}^{-1} \text{ (Boudouard reaction)} \tag{9.6}$$

$$C + 2H_2 \rightleftharpoons CH_4 \qquad \Delta H = -75 \text{ kJ mol}^{-1} \text{ (methanation reaction)} \tag{9.7}$$

Homogeneous reactions:

$$CO + H_2O \rightleftharpoons CO_2 + H_2 \qquad \Delta H = -41 \text{ kJ mol}^{-1} \text{ (water gas shift reaction)} \tag{9.8}$$

$$CH_4 + H_2O \rightleftharpoons CO + 3H_2 \qquad \Delta H = +206 \text{ kJ mol}^{-1} \text{ (steam methane reforming)} \tag{9.9}$$

9.4.1.4   Combustion/Oxidation

This is a highly exothermic reaction step that supplies the majority of the heat to other reaction zones, mostly endothermic reactions. The main products are carbon dioxide and water. The reaction occurs at approximately 1300 °C. The following equations show the main oxidation reactions occurring in the combustion zone.

Oxidation reactions:

$$C + O_2 \rightarrow CO_2 \qquad \Delta H = -393 \text{ kJ mol}^{-1} \tag{9.10}$$

$$H_2 + \tfrac{1}{2}O_2 \rightarrow H_2O \qquad \Delta H = -242 \text{ kJ mol}^{-1} \tag{9.11}$$

9.4.2   Types of Gasifier

Figure 9.7 shows several types of gasifier. They are compared in Table 9.2.

9.4.3   Design Considerations

The specification of the gasification is very important since it affects the efficiency of utilizing the feedstock, the quality of products (composition, heating value, etc.) and also the downstream processes. Therefore, design considerations have to be emphasized and are shown in Table 9.3.

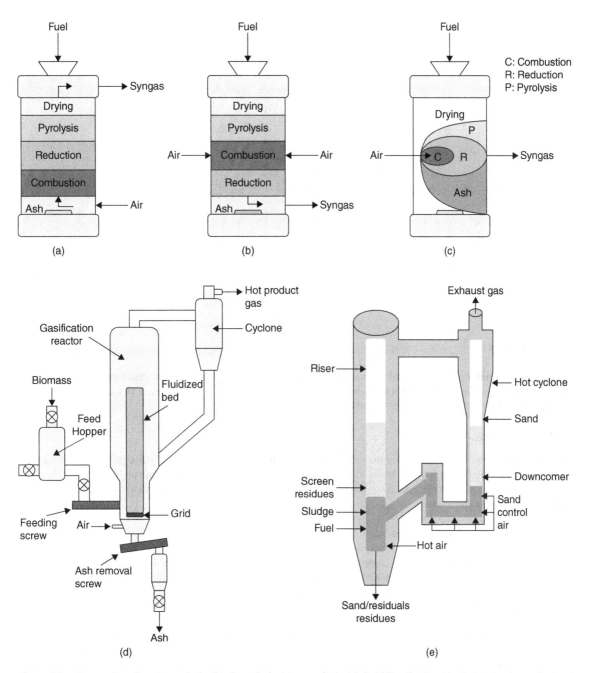

Figure 9.7 *Types of gasifier: (a) updraft; (b) downdraft; (c) crossdraft; (d) bubbling fluidized bed; (e) circulating fluidized bed; (f) entrained bed.*

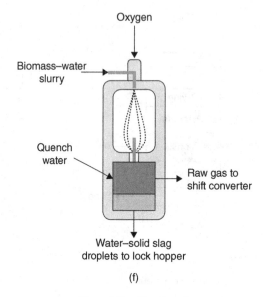

Oxygen

Biomass–water
slurry

Quench
water

Raw gas to
shift converter

Water–solid slag
droplets to lock hopper

(f)

Figure 9.7   *(Continued)*

9.5   Pyrolysis

Transportation, handling and storage of biomass are amongst the major challenges in the development of large-scale biomass based processing. To solve this issue, biomass can be processed into more convenient, cleaner (without tar, char and ash) and transportable forms such as liquid bio-oil.

9.5.1   What is Bio-Oil?

Bio-oil (also known as *bio-crude* or *pyrolysis oil* as the main product of pyrolysis and liquefaction processes, discussed in Section 9.5.2) is essentially a complex liquid mixture made up from different classes of compounds, ketones, aldehydes, carboxylic acids and other oxygenated compounds. The bio-oil composition varies depending on the biomass feedstock as well as the process conditions.

1. *Physical appearance and characteristics*
    *   Dark brown in color and has a smoky smell.
    *   Immiscible with petroleum and cannot be introduced into a conventional refinery directly.
    *   The viscosity of bio-oil, typically within the range of 40 to 100 cP at 40 °C with 25% water, varies depending on the feedstock and process conditions, and increases with time. This induces a storage problem for bio-oil and it can only be kept for weeks to a few months.
2. *Chemical properties*
    *   Has high proportions of oxygen (45–50 wt%) and water (15–35 wt%), which reduce the heating value of biomass. Typically, bio-oil has a lower higher heating value (HHV) of 17 MJ kg^{-1} (with approximately 25 wt% water), compared to diesel or fuel oil, which has HHV of 45 MJ kg^{-1}. Before processing a biomass feedstock, the water content should be reduced to less than 10 wt%, since water is also produced during the pyrolysis reaction.
    *   It is chemically unstable due to the existence of significant amounts of reactive components such as aldehydes, which polymerize at higher temperature. This results in difficulties for further processing of the bio-oil and application in turbines and boilers.
    *   Highly acidic (pH 2–3) due to the presence of large amount of carboxylic acids, which lead to corrosion in equipment such as turbines. Upgrading of bio-oil is necessary to eliminate or decrease the adverse effects, that is, incompatibility with a conventional petroleum refinery, chemical instability and increasing viscosity.

Table 9.2 *Comparison of gasifier technologies.*

	Type of Gasifier		
Parameter	Fixed Bed/Moving Bed (Figure 9.7(a)–(c))	Fluidized Bed (Figure 9.7(d),(e))	Entrained Bed (Figure 9.7(f))
Configuration	Updraft, downdraft, crossdraft	Bubbling, circulating	Top-fed, side-fed
Principle of operation	Fuel is fed from the top of the gasifier and flows downward. Air/oxygen is supplied from the bottom (updraft) or a certain height from the bottom (downdraft/ crossdraft). Depending on the configuration, the syngas product leaves the gasifier either from the top (updraft), bottom (downdraft) or from the side (crossdraft). The flow of fuel and air can be countercurrent (updraft) or cocurrent (downdraft and crossdraft). Ash is collected at the bottom.	Fuel particles float on a stream of air and sand at a velocity that is enough to keep suspension of the solid. Char particles leaving with the product gas at the top of the gasifier are recovered via a cyclone and recycled back to the gasifier. The operating temperature is kept below the ash fusion temperature.	Fuel particles and oxidant are gasified under high temperature (higher than the slag fusion temperature) in a very short residence time to achieve high carbon conversion. Oxygen is generally used as the oxidant.
Oxidant requirement	Low	Moderate	High
Steam requirement	High	Moderate	Low
Outlet gas temperature (°C)	Low (450–650)	Moderate (800–1000)	High (1200–1600)
Feed particles size	6–50 mm	6–10 mm	<100 μm
Capacity (the suffix 'th' in the power unit indicates thermal output)	Small (10 kW_{th}–10 MW_{th})	Moderate (5 MW_{th}–100 MW_{th})	Large (>50 MW_{th})
Cold gas efficiency (indicative) (%)	80	89	80
Developer(s)	• Lurgi (dry ash gasifier) • BGL	• KBR (transport gasifier) • Winkler	• Conoco Phillips • Future Energy • GE Energy • Shell • Siemens
Advantages	• Need less oxidants	• Uniform temperature distribution • Good mixing • Lower risk of agglomeration • Produce less tar and oil	• All types of fuel rank can be gasified • Elimination of tar
Disadvantages	• High maintenance cost • Poor mixing and heat transfer • Produce considerable amount of tar and oil • Agglomeration	• Extensive solid (char) recycling	• Require a large amount of oxidant • High temperature slagging operation • Require very fine fuel particles as feedstock

Table 9.3 *Design consideration of gasifier.*

Specification	Options
Pressure	
Atmospheric	Suitable and more economical for small-scale application. However, high cost would be incurred in the downstream processes, such as syngas cleaning and conditioning, synthesis processes and electricity generation, which demand higher pressure. Thus, higher compression power is required for the downstream processes.
Pressurized	High pressure is favorable for most of the downstream processes. This would result in a more economic design, since less compression power is needed. However, extra caution has to be taken for the safety of the gasifier.
Gasifying Medium	
Air	Air is cheaper than oxygen. However, air has a significant amount of nitrogen ($\sim$79 mol%), which has a dilution effect on the product gas stream. This would lower the heating value of the product gas. The dilution effect of nitrogen leads to inefficiency in CO_2 removal and also the downstream production. Furthermore, larger unit operations are needed for the entire system.
Oxygen	Highly pure oxygen is expensive compared to air since it requires high capital investment and operating cost for an air separation unit or an oxygen plant. However, it prevents product gas from dilution and retains the heating value of product gas. In addition, the unit operations are smaller for the whole system.
Syngas Cooling	
Quench	Heat is not recovered and is downgraded to process heat. A lower capital cost is needed and thus it is suitable for the low value feedstock. If water gas shift is used downstream, then the quench design provides enough water for the reaction.
Heat recovery	Allows heat recovery and produces steam, leading to a more energy efficient system.

3. *Applications.* Bio-oil has a wide range of applications, such as electricity generation via turbines, engines or the boiler as well as fuel for transportation and chemical production including food flavoring.
4. *Pros and cons.* Regardless of the disadvantages, bio-oil is still a more attractive choice compared to solid biomass, mainly because it is cleaner, with a negligible amount of sulfur, nitrogen, ash and other impurities. Moreover, liquid bio-oil has a higher bulk energy density than solid biomass, which mitigates logistics difficulties and thereby minimizes the cost for transportation and requires less space for storage. Bio-oil has an advantage over crude oil during transportation, due to its inert and nontoxic properties and separation into heavy organic fractions that sink, instead of spreading over the water surface. Therefore, it does not cause severe environmental pollution during a spill or pipeline leakage unlike the Gulf of Mexico oil spill in 2010.

9.5.2   How Is Bio-Oil Obtained from Biomass?

There are two primary processes for liquefying biomass into bio-oil, that is, pyrolysis and liquefaction. Pyrolysis is a thermal decomposition process, which occurs in the absence of or little oxygen environment. Fast pyrolysis signifies that the reaction occurs within a few seconds or even less with modest temperature conditions ($\sim$500 °C). Fast pyrolysis can produce a high yield of liquid bio-oil (up to 75 wt% on a dry feed basis) with up to seven times the energy density of biomass, helpful for transport and storage. Liquefaction also produces bio-oil; however, its operating conditions differ from pyrolysis. Comparisons between pyrolysis and liquefaction of biomass in terms of operating conditions (temperature and pressure), product quality (oxygen and moisture content and heating value) and impact on capital costs are summarized in Table 9.4. The composition of bio-oil varies according to the process adopted. The pyrolysis process is more widely used and the technology is commercially available.

Table 9.4 *Comparison of pyrolysis and liquefaction of biomass processes.*

Specification	Pyrolysis	Liquefaction
Temperature (°C)	High temperature (450–550)	Low temperature (250–450)
Pressure (atm)	Low pressure (1–5)	High pressure (50–200)
Oxygen content in bio-oil	Higher	Lower
Moisture content in bio-oil	Higher	Lower
Heating value in bio-oil (MJ kg^{-1})	Lower (~17)	Higher (~34)
Capital cost	Lower	Higher

9.5.3 How Fast Pyrolysis Works

The operating conditions of biomass fast pyrolysis are fairly stringent, particularly the temperature applied and the vapor residence time. This is because the gaseous product predominates at a higher temperature with a longer vapor residence time, whilst the formation of char is favored by a lower temperature and a longer vapor residence time. A moderate temperature condition and short vapor residence time are about right for producing bio-oil. Beyond just the temperature and vapor residence time, the heat transfer rate is another crucial parameter for fast pyrolysis. This is to avoid the formation of char, considering that the char sticks to the surface of the particle and hence the heat transfer pathway to the centre of particles is blocked. In addition, char is undesirable, since it causes secondary vapor cracking. Therefore, char has to be removed in the vapor phase before condensing into liquid. Nevertheless, char has been found to have a considerable effect on increasing the viscosity of bio-oil. The presence of alkali metals such as potassium has also been identified to have caused the formation of char. A fast heat transfer rate is desired and can be achieved by using very small particles of biomass, ranging from 200 μm to 6 mm, depending on the type of reactor, to produce a higher yield of bio-oil.

Figure 9.8 shows the fast pyrolysis processing steps, from biomass feed preparation that involves comminution (size reduction) and drying to the fast pyrolysis reaction generating final products, that is, char, gas and oil. The product distribution from slow pyrolysis is given for comparison with the product distribution from fast pyrolysis.

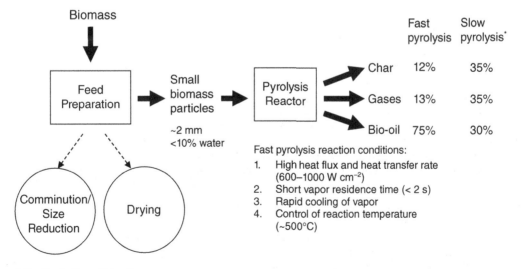

Figure 9.8 *Production of bio-oil.*
Slow pyrolysis (also known as carbonization) has a longer vapor residence time and the reaction occurs at a lower temperature compared to fast pyrolysis, leading to a different distribution of products.

9.5.3.1 Feed Preparation

Biomass needs to be ground into fine particles before the pyrolysis reaction. The reasons are:

1. to achieve high surface area per unit volume for reaction;
2. to avoid build-up of char on the surface of biomass; char hinders heat transfer to the center of particles.

In addition, the water content in biomass has a negative impact on the heating value and hence the quality of products formed. Therefore, the water content should be reduced to less than 10 wt%.

9.5.3.2 Fast Pyrolysis Reaction

A fast pyrolysis reaction has to be controlled carefully to attain the optimum yield of products. This includes controlling the heating rates, vapor residence time, quenching of vapor and reaction temperature. These conditions are essential because oxygen in the biomass causes thermal instability and polymerization of the biomass components.

Did you know?

Biochar is an organic solid material produced from biomass pyrolysis reaction as a by-product. The conditions that favor biochar production are a low temperature (below 700 °C) and long vapor residence time – hence the yield of biochar is greater in slow pyrolysis (35%) compared to fast pyrolysis (12%). Although biochar is a by-product, it is potentially useful in various applications:

- Soil amendment by improving the soil properties (e.g., pH adjustment, nutrient-retention properties) and thus improving the crop yield. It was discovered by pre-Columbian Indians in the Amazon Basin (500–2500 BC)!
- Carbon sequestration and reducing greenhouse gas emissions (biochar is a stable form of carbon and can be retained in soil for a long period of time).

There are various types of reactor configurations for the fast pyrolysis reaction; the prevalent technologies are shown in Figure 9.9. A comparison of design configurations and process operations between various fast pyrolysis reactor technologies is given in Table 9.5.

Apart from the notable pyrolysis reactors shown in Figure 9.9 and Table 9.5, there are also other types of reactors such as fixed bed and entrained flow pyrolysis reactors. These reactors are less common and are not commercialized due to inherent disadvantages such as maintenance problems and poor heat transfer.

Amongst the technologies mentioned above, the fluidized bed reactor (bubbling/circulating) has gained considerable attention and its performance has been well proven in commercial applications. Apart from fulfilling the essential requirements of biomass pyrolysis as aforementioned (e.g., high heat transfer rate), a fluidized bed reactor is also capable of large throughput applications, such as large-scale production of bio-oil.

Several challenges for the large-scale production of bio-oil have been identified:

- Expensive: the cost of bio-oil is 10–100% higher than fossil fuel.
- Availability: limited supplies for testing.
- Quality: specification of bio-oil has not been standardized and has inconsistent quality.
- Compatibility: incompatible with fossil fuels.
- Familiarization: many users are not familiar with bio-oil.
- Environmental health and safety issues.

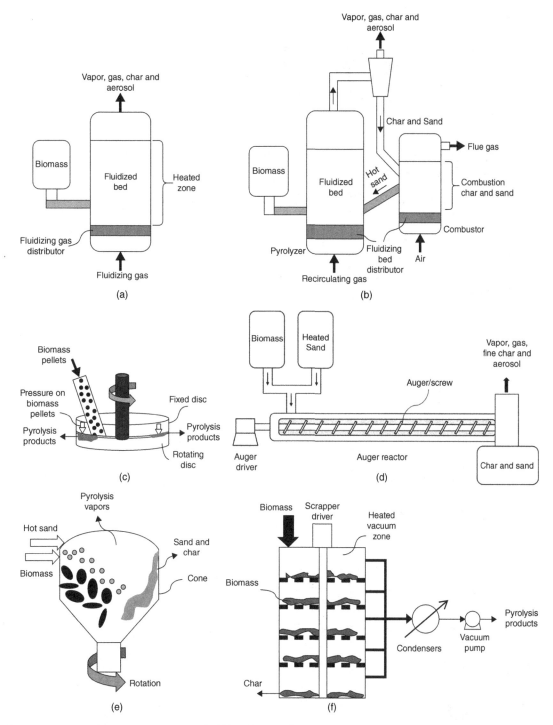

Figure 9.9 *Types of pyrolysis reactors: (a) bubbling fluidized bed pyrolysis reactor; (b) circulating fluidized bed pyrolysis reactor; (c) ablative pyrolysis reactor; (d) auger/screw pyrolysis reactor; (e) rotating cone pyrolysis reactor; (f) vacuum pyrolysis reactor.*

Table 9.5 Comparison of pyrolysis reactor technologies.

Type of Pyrolyzer	Bubbling Fluidized Bed (Figure 9.9(a))	Circulating Fluidized Bed (Figure 9.9(b))	Ablative (Figure 9.9(c))	Auger Screw (Figure 9.9(d))	Rotating Cone (Figure 9.9(e))	Vacuum (Figure 9.9(f))
Principle of operation	A stream of gas (normally recycled pyrolysis gas) is used to fluidize biomass particles together with hot sand. Heat is supplied to the bed using either the process heat through the preheated fluidized gas (adiabatic) or sand that is indirectly heated by fired tubes. The former method normally allows a smaller throughput for the reactor design while the latter gives a better performance.	The principle is similar to fluidized bed except that it involves solid transport between two vessels, i.e., pyrolyzer and combustor. Sand is transported together with char to combustor and recirculated back to pyrolyzer.	Biomass particle is melted under pressure on the hot surface of reactor (similar concept to melting a slice of butter over a frying pan).	Sand and biomass are fed to screw and transported to reactor. Mixing and heating are carried out using rotational screw and transported to the reactor. The concept originated from Lurgi coal gasifier.	Centrifugal force drives the biomass and sand from the bottom of the cone to the top. High speed rotation provides good mixing and heat transfer. Vapor is directed to a condenser. Sand is reheated in a combustor and sent back to the base of the cone.	The boiling point of biomass is lowered due to reduced pressure condition (~0.15 bar). After decomposition through heating, the small fragment is immediately removed using a vacuum pump and recovered as liquid product through condensation. This results in a shorter vapor residence time for reaction and, hence, secondary reactions can be minimized.
Vapor residence time (s)	0.5–2.0	0.5–1.0	50–100 ms	0.5–2.0	0.3	0.5–2.0
Operating temperature (°C)	500–550	~500	450–600	400–500	550	~450
Feed particle size (mm)	<2–3	1–2	Up to 20	1–3	0.05–3	2–5 cm
Inert carrier gas requirement	Yes	Yes	No	No	No	No
Capacity (kg h^{-1})	200–400	20–1000	350	200	2000	3000
Bio-oil yield (indicative) (%)	75	75	65	30–65	75	30–45

Table 9.5 (Continued)

Type of Pyrolyzer	Bubbling Fluidized Bed (Figure 9.9(a))	Circulating Fluidized Bed (Figure 9.9(b))	Ablative (Figure 9.9(c))	Auger Screw (Figure 9.9(d))	Rotating Cone (Figure 9.9(e))	Vacuum (Figure 9.9(f))
Developer(s)	• Union Fenosa (Spain) • Dynamotive/RTI (Canada) • University of Waterloo (Canada) • Wellman (UK)	• Ensyn/Red Arrow RTP technology (US) • CRES (Greece) • ENEL (Italy) • VTT (Finland)	• NREL, formerly SERI (US) • Aston University (UK)	• ROI • Mississippi State University • BIOGREEN	• University of Twente (The Netherlands) • BTG (The Netherlands)	• Université Laval and Pyrovac (Canada)
Advantages	• Well established with long operating history • High heat transfer rate • Good temperature control • Simple design	• High heat transfer rate • Large throughput is possible • Good temperature control	• Large feed particles can be used	• Low operating temperature	• Carrier gas is not needed, recovery of bio-oil product is easier. • Less aggressive transport dynamics of sand and biomass, hence wear problems can be minimized.	• Produces clean oil with little char • Large particles size of biomass can be used • Carrier gas is not needed, minimizing aerosol formation
Disadvantages	• Normally small throughput (heat supply is a major consideration for scaling up the reactor) • Feed has to be carefully sized (narrow particle size distribution)	• High energy need for circulating sand • Wear problem in the reactor • High gas velocity leading to higher char abrasion and hence higher char content • Ash build-up in the circulating solid causing further cracking and thus reducing the bio-oil yield	• Wear problem in the solid recycle loop • Scale-up is difficult • High gas velocity causing erosion	• Heat transfer at large scale can be a problem	• Integration is complex • Scale-up is uncertain	• Slow pyrolysis process due to poor heat and mass transfer rate • Low bio-oil yield • Generates significant amount of water • High cost for maintenance

9.6   Summary

This chapter features various aspects of the thermochemical conversion of biomass, including combustion, gasification and pyrolysis. The operating conditions, reactions, comparison between reactor technologies and design considerations have been discussed.

Exercises

1. Explain how the operating conditions of a gasifier (e.g., temperature and pressure) play an important role for gas turbine power generation in a biomass integrated gasification combined cycle plant.

    Hint: The answer can be found in References 2 and 3. The sensitivity study of gasifier operating conditions on the plant performance can be examined.

2. Air or oxygen is used as the gasifying medium in gasification processes. A study on biomass integrated gasification combined cycle[2] has used air as the gasifying medium, while oxygen was used in the gasifier in a study on bio-oil integrated gasification and methanol synthesis[4]. Comment on the impact of using different gasifying media (air/oxygen) on the downstream process.

3. Biomass is regarded as a sustainable fuel that causes less pollution to the environment compared to fossil fuel. However, combustion of either biomass or fossil fuels results in CO_2 emission. Explain why biomass is still a preferred source compared to fossil fuel from the environmental perspective.

4. Biomass can be used directly as the feedstock in thermochemical conversion units for generating electricity, transportation fuels, chemicals, etc. Explain why fast pyrolysis as an alternative technology is useful to convert biomass into bio-oil as an intermediate platform, for processing into products.

5. Discuss the process variables that decide the optimum yield of bio-oil produced from a fast pyrolysis process.

6. Feed preparation is a very important step before the fast pyrolysis process. These include size reduction of biomass particles and drying. Explain the importance of the feed preparation step.

References

1. L. Tomory, Gaslight, distillation, and the industrial revolution, *History of Science*, **49**, 395–424 (2011).
2. J. Sadhukhan, K.S. Ng, N. Shah, H.J. Simons, Heat integration strategy for economic production of combined heat and power from biomass waste, *Energy Fuels*, **23**, 5106–5120 (2009).
3. K.S. Ng and J. Sadhukhan, Techno-economic performance analysis of bio-oil based Fischer–Tropsch and CHP synthesis platform, *Biomass Bioenergy*, **35**, 3218–3234 (2011).
4. K.S. Ng and J. Sadhukhan, Process integration and economic analysis of bio-oil platform for the production of methanol and combined heat and power, *Biomass Bioenergy*, **35**, 1153–1169 (2011).

10

Reaction Thermodynamics

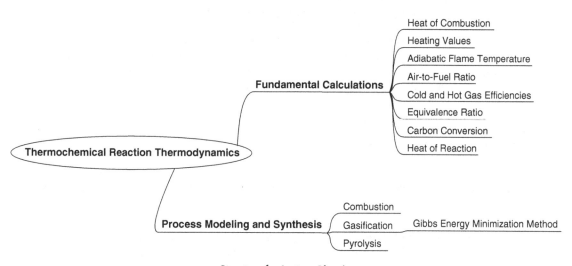

Structure for Lecture Planning

10.1 Introduction

Designing the thermochemical conversion processes of biomass involves the assessment of process performance with respect to material utilization and energy efficiency. Modeling of these processes requires the application of fundamental concepts of reaction thermodynamics, and thereby identification of key operating parameters and performance indicators depending upon feed definition and process yields. The first part of this chapter shows various fundamental design calculations for combustion, gasification and pyrolysis, providing important tools to assess the process performance. The second part of this chapter shows the techniques for modeling the combustion, gasification and pyrolysis processes.

   This chapter gives the reader a solid understanding of commonly occurring problems as well as exposure to practical techniques to solve these problems in relation to thermochemical process modeling.

Biorefineries and Chemical Processes: Design, Integration and Sustainability Analysis, First Edition.
Jhuma Sadhukhan, Kok Siew Ng and Elias Martinez Hernandez.
© 2014 John Wiley & Sons, Ltd. Published 2014 by John Wiley & Sons, Ltd.
Companion Website: http://www.wiley.com/go/sadhukhan/biorefineries

10.2   Fundamentals of Design Calculation

Part 1: Combustion

10.2.1   Heat of Combustion

The heat of combustion can be determined experimentally using a bomb calorimeter. The heat of combustion of most of the common chemical compounds has been determined experimentally at standard conditions, that is, 1 atm and 298 K. This information can be found in most of the thermodynamic handbooks (available in Knovel database, why.knovel.com) such as *Yaws' Handbook of Thermodynamic and Physical Properties of Chemical Compounds*[1] and *Chemical Properties Handbook*[2] and databases such as *NIST Chemistry WebBook* (webbook.nist.gov/chemistry/). However, it is not always convenient to carry out experiments to determine the heat of combustion at various temperatures. Thermodynamic handbooks as above-mentioned contain heat of combustion of various compounds at different temperatures, presented in the form of plots and correlations for a range of temperatures. The most common issue arising in engineering calculations is the limited data availability to obtain results at various operating conditions. A more straightforward calculation approach is vital to estimate the heat of combustion at different temperatures.

Hess's law is currently the most favorable method in predicting the heat of combustion of a chemical compound at a particular temperature. Hess's law states that the enthalpy change of a reaction is independent of the path between the initial stage and final stage, by which the reactant(s) is/are converted into the product(s). This is a very useful method for determining the enthalpy change of a reaction, when direct measurement is not available. Hess's law follows the first law of thermodynamics, the law of conservation of energy.

To determine the heat of combustion of a compound at a desired temperature using Hess's law, the minimum data requirements are:

- Initial temperature and final temperature of the reaction
- Standard heat of formation of reactant(s) and product(s)
- Heat capacities of reactant(s) and product(s).

Assuming that there is no information on the standard heat of combustion, the following methods applying Hess's law should be adopted.

Step 1: Determine the Standard Heat of Combustion   The standard heat of a reaction (in this case the reaction refers to combustion) can be determined using Hess's law, as shown in Figure 10.1. The information required is:

- Standard heat of formation of the reactant(s) to form the intermediate product(s)
- Standard heat of formation of the intermediate product(s) to form the product(s).

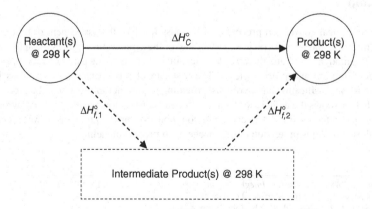

Figure 10.1   *Calculation of the standard heat of combustion of a reaction using an alternative pathway.*

This information is used in Equation (10.1) for calculating the standard heat of combustion. Note that:

1. All the enthalpy change should be calculated based on standard conditions, that is, 1 atm and 298 K.
2. The alternative pathway shown in Figure 10.1 is the simplest possible route. Other pathways that involve several intermediate products are also viable, as long as the initial and final stages remain the same.
3. The directions of the arrows shown in Figure 10.1 are just an example. It can be different depending on the nature of a specific case. The change of sign in the enthalpy change calculation may be expected.

Therefore,

$$\Delta H_C^\circ = \Delta H_{f,1}^\circ + \Delta H_{f,2}^\circ \tag{10.1}$$

where

ΔH_C° is the standard heat of combustion. Note the symbol $^\circ$ that indicates the standard condition, for example ΔH_C° is the change in enthalpy or heat of combustion at the standard condition, while ΔH_C is the heat of combustion at any temperature and pressure condition.

$\Delta H_{f,1}^\circ$ and $\Delta H_{f,2}^\circ$ are the standard heats of formation of reaction 1 (Reactant(s) $\rightarrow$ Intermediate Product(s)) and reaction 2 (Intermediate Product(s) $\rightarrow$ Product(s)) respectively.

Equation (10.2) shows an alternative form of Equation (10.1) for calculating the heat of combustion. In this case, the overall reaction form is *reactant* $\rightarrow$ *product*:

$$\Delta H_C^\circ = \sum \left(n \Delta H_{f,product}^\circ \right) - \sum \left(n \Delta H_{f,reactant}^\circ \right) \tag{10.2}$$

where n is the stoichiometric number of moles and $\Delta H_{f,product}^\circ$ and $\Delta H_{f,reactant}^\circ$ are the standard heats of formation of the product and reactant, respectively.

Step 2: Determine the Heat of Combustion at the Desired Initial and Final Temperatures The heat of combustion of a reaction, ΔH_C, is determined by calculating the enthalpy change involved along an alternative pathway, shown in Figure 10.2. This is because enthalpy changes along this pathway can be calculated from known thermal properties, heat capacities of reactants and products and heats of reaction at the standard condition. The pathway for calculating an overall enthalpy change follows this sequence:

1. Cooling of reactant(s) from an initial temperature, T_1, to the standard temperature of 298 K, to calculate the enthalpy change due to cooling of the reactant(s), denoted by ΔH_R.
2. Heat of reaction converting reactant(s) into product(s) at the standard temperature of 298 K, where ΔH_C° is noted.
3. Heating of the product(s) from the standard temperature of 298 K to the final temperature, T_2, to calculate the enthalpy change due to heating of the product(s), denoted by ΔH_P.

This method as shown in Figure 10.2 is in accordance with Hess's law. For calculating the heat of combustion of a reaction, the following information is required:

*   Initial temperature and final temperature of the combustion reaction.
*   Standard heat of combustion, obtained in Step 1.
*   Heat capacities of reactant(s) and product(s).

The heat of combustion of a reaction, ΔH_C, can be expressed using the following equations:

$$\Delta H_C = \Delta H_R + \Delta H_C^\circ + \Delta H_P \tag{10.3}$$

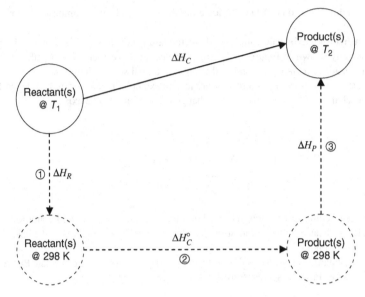

① Cooling of reactant(s) from T_1 to the standard temperature of 298 K.

② Combustion of reaction at the standard temperature of 298 K.

③ Heating of product(s) from the standard temperature of 298 K to T_2.

Figure 10.2 *Calculation of heat of combustion of a reaction using an alternative pathway.*

where

$$\Delta H_R = \int_{T_1}^{298} C_{p,reactant}\, dT \tag{10.4}$$

and $C_{p,reactant}$ is the specific heat capacity of reactant(s). ΔH_C° is the standard heat of combustion of reaction at 298 K. Examples of standard heat of combustion of some common components are shown in Table 10.1. Also

$$\Delta H_P = \int_{298}^{T_2} C_{p,product}\, dT \tag{10.5}$$

where $C_{p,product}$ is the specific heat capacity of product(s).

Table 10.1 *Examples of the standard heat of combustion at 298 K[1].*

Reaction	Standard Heat of Combustion @ 298 K, ΔH_C° (kJ·mol^{-1})
$C\,(s) + O_2\,(g) \rightarrow CO_2\,(g)$	−393.5
$CO\,(g) + \frac{1}{2}O_2\,(g) \rightarrow CO_2\,(g)$	−283.0
$H_2\,(g) + \frac{1}{2}O_2\,(g) \rightarrow H_2O\,(g)$	−241.8
$CH_4\,(g) + 2O_2\,(g) \rightarrow CO_2\,(g) + 2H_2O\,(g)$	−802.7
$S\,(s) + O_2\,(g) \rightarrow SO_2\,(g)$	−296.8

Exercise 1. Calculate the standard heat of combustion of methane. If the combustion reaction is carried out at 1000 °C, estimate the enthalpy change of combustion of methane assuming that the final temperature is at 1200 °C. The standard heats of formation and the coefficients of correlation for estimating the heat capacities of the relevant components are provided in Tables 10.2 and 10.3.

Table 10.2 *Standard heat of formation of components involved in methane combustion*[3].

Component	Standard Heat of Formation @ 298 K, ΔH_f° (kJ mol^{-1})
CH_4	−74.5
CO_2	−393.5
H_2O	−241.8

Note. The standard heat of formation for pure elements such as O_2 is zero.

Data:

$$C_p = A + BT + CT^2 + DT^3 + ET^4 + ET^5 + FT^6 + GT^7 \qquad (10.6)$$

where

C_p is the heat capacity, J mol^{-1}.
A, B, C, D, E, F and G are the coefficients of correlation for estimating the heat capacities of components.
T is the temperature, K.

Table 10.3 *Coefficient of correlation (Equation (10.6)) for estimating heat capacities of components involved in methane combustion*[3].

Component	A	B	C	D	E	F	G
CH_4	44.35658	−0.14623	0.0006	-8.7×10^{-7}	6.78×10^{-10}	-2.8×10^{-13}	4.58×10^{-17}
CO_2	23.5061	0.038066	7.4×10^{-5}	-2.2×10^{-7}	2.34×10^{-10}	-1.1×10^{-13}	2.17×10^{-17}
H_2O	33.17438	−0.00325	1.74×10^{-5}	-6×10^{-9}	0	0	0
O_2	29.79024	−0.00949	2.86×10^{-5}	9.87×10^{-9}	-5.7×10^{-11}	4.3×10^{-14}	-1.0×10^{-17}

Note. The coefficient of correlation for estimating heat capacity is valid within the temperature range of 150 K to 1500 K.

Solution to Exercise 1

Standard heat of combustion of methane. The combustion reaction of methane can be shown as follows:

$$CH_4 \;+\; 2O_2 \xrightarrow{\;\Delta H_C^\circ\;} CO_2 \;+\; 2H_2O \qquad (10.7)$$

$$\uparrow \Delta H_{f,1}^\circ \quad \uparrow \Delta H_{f,2}^\circ \qquad \uparrow \Delta H_{f,3}^\circ \quad \uparrow \Delta H_{f,4}^\circ$$

By applying Hess's law, the following equations can be derived:

$$\Delta H_{f,1}^\circ + \Delta H_{f,2}^\circ + \Delta H_C^\circ = \Delta H_{f,3}^\circ + \Delta H_{f,4}^\circ \qquad (10.8)$$

$$\Delta H_C^\circ = \left(\Delta H_{f,3}^\circ + \Delta H_{f,4}^\circ \right) - \left(\Delta H_{f,1}^\circ + \Delta H_{f,2}^\circ \right) \qquad (10.9)$$

Hence, Equation (10.9) is analogous to Equation (10.2). Using the standard heat of formation of the related components from Table 10.2,

$$\Delta H_C^\circ = \sum \left(n \Delta H_{f,product}^\circ \right) - \sum \left(n \Delta H_{f,reactant}^\circ \right) \qquad (10.2)$$

$$= (1 \times (-393.5)) + (2 \times (-241.8)) - (1 \times (-74.5))$$

$$= -802.6 \text{ kJ mol}^{-1}$$

Heat of combustion of methane from 1000 °C to 1200 °C. The heat of combustion of methane from 1273 K to 1473 K can be predicted by applying Equations (10.3) to (10.5), the standard heat of combustion of methane and the heat capacity data provided in Table 10.3 (refer to Figure 10.2 for the concept):

$$\Delta H_R = \int_{1273}^{298} C_{p,reactant} \, dT \qquad (10.4)$$

$$= -92.2 \text{ kJ mol}^{-1} \text{ (enthalpy change of cooling } CH_4 \text{ and } O_2 \text{ from 1273 K to 298 K)}$$

$$\Delta H_P = \int_{298}^{1473} C_{p,product} \, dT \qquad (10.5)$$

$$= 107.1 \text{ kJ mol}^{-1} \text{ (enthalpy change of heating } CO_2 \text{ and } H_2O \text{ from 298 K to 1473 K)}$$

$$\Delta H_C = \Delta H_R + \Delta H_C^\circ + \Delta H_P \qquad (10.3)$$

$$= -92.2 + (-802.6) + 107.1$$

$$= -787.7 \text{ kJ mol}^{-1}$$

Note that the integration technique of C_p (as a function of temperature in Equation (10.6)) is prerequisite for solving this question. Detailed calculus steps are not shown here.

10.2.2   Higher and Lower Heating Values

Heating values are used to compare the quality among different fuels and also to deduce the thermodynamic efficiency of the process, in which the fuel is used as the feedstock. In general, the heating value is the quantity of heat produced from complete combustion of a unit quantity of fuel. There are two types of heating values commonly reported: the higher heating value (gross calorific value) and the lower heating value (net calorific value). The distinction between these two values is the state of water. The higher heating value considers water to be in the liquid phase and thus the heat of vaporization of water is included. The lower heating value refers to water in vapor form. The relationship between the higher and lower heating values is described as

$$HHV = LHV + m_{COND} h_{vap} \qquad (10.10)$$

where

HHV is the higher heating value of fuel.
LHV is the lower heating value of fuel.
m_{COND} is the mass of water formed.
h_{vap} is the latent heat of vaporization of water at the standard condition, that is, 25 °C.

In the case of biomass, LHV can be related to HHV using[4]

$$LHV_{dry} = HHV_{dry} - \frac{21.822\ H}{100} \tag{10.11}$$

where

LHV_{dry} and HHV_{dry} are the lower and higher heating values on dry basis, MJ kg^{-1}.
H is the weight percent of hydrogen, determined in the ultimate analysis.

The heating value of a fuel is measured using a bomb calorimeter. It is not always possible to obtain the heating values through experiments all the time; hence empirical methods should be exploited. There are generally three ways of estimating the heating value of a fuel, based on proximate analysis, ultimate analysis and chemical composition. The following presents the correlations for estimating the higher heating value of biomass based on proximate (Equations (10.12) and (10.13)) and ultimate analyses (Equation (10.14)). *Note that the correlations should not be used for other types of fuel because the coefficients may be different.* The estimation of higher heating value of biomass based on chemical compositions of hemicellulose, lignin and other extractives is less popular and does not give a satisfactory accuracy. Sheng and Azevedo (2005) developed the correlation for estimating HHV with high accuracy, shown by the following equations[5]. Other existing correlations developed by other authors can also be found in their paper or the ECN website (http://www.ecn.nl/phyllis/defs.asp)[4].

1. *Estimation of HHV based on proximate analysis*

$$HHV = -3.0368 + 0.2218\ VM + 0.2601\ FC \tag{10.12}$$

$$HHV = 19.914 - 0.2324\ Ash \tag{10.13}$$

2. *Estimation of HHV based on ultimate analysis*

$$HHV = -1.3675 + 0.3137\ C + 0.7009\ H + 0.0318\ O \tag{10.14}$$

where *HHV* is measured in MJ kg^{-1} and *VM, FC, C, H* and *O* represent volatile matters, fixed carbon, carbon, hydrogen and oxygen, respectively. These are in weight percent based on dry biomass. The *O* element in Equation (10.14) embraces all other elements such as S, N, Cl, etc., that is, $O = 100 - C - H - Ash$.

A comparison of heating values between different types of biomass is shown in Table 10.4. Note that ECN provides a detailed database of various biomass resources.

Table 10.4   *Proximate and ultimate analyses of biomass. (Reproduced with permission from Phyllis2, database for biomass and waste, ECN[6].)*

Fuel Type	Proximate Analysis (dry wt%)			Ultimate Analysis (dry wt%)					HHV, Measured (MJ kg^{-1})
	Fixed Carbon	Volatile Matters	Ash	C	H	O	N	S	
Redwood	19.92	79.72	0.36	50.64	5.98	42.88	0.05	0.03	20.72
Poplar	28.10	71.40	0.50	48.06	5.77	45.57	0.10	0.00	18.53
Rice straw	16.56	66.62	16.82	39.75	4.93	37.38	0.94	0.19	14.70
Wheat straw	17.00	77.80	5.20	47.12	5.78	40.20	0.47	0.19	17.33

Exercise 2. Estimate the HHV values of redwood based on proximate and ultimate analyses shown in Table 10.4 and compare them against the measured HHV. Deduce the LHV of redwood.

Solution to Exercise 2. HHV of redwood can be estimated using the correlations shown in Equations (10.12) to (10.14), by substituting the proximate and ultimate values provided in Table 10.4. This can be compared with the measured HHV value of redwood, i.e. 20.72 MJ kg^{-1}. The results are shown in Table 10.5.

Table 10.5 *Comparison of estimated and measured HHV of redwood.*

Correlation	HHV (MJ kg^{-1}), Estimated Using Correlations	Discrepancy Compared to the Measured HHV Value
Equation (10.12)	19.83	4.3% less
Equation (10.13)	19.83	4.3% less
Equation (10.14)	20.07	3.14% less

The correlation is a useful tool to evaluate the HHV of biomass, if direct measurement is not readily available. LHV of redwood is estimated to be **19.4 MJ kg^{-1}**, using Equation (10.11) and the data provided in Table 10.4. LHV is preferred as the basis for evaluating efficiency of a biomass conversion process, in particular the combustion process (e.g., in a boiler), where water remains in the vapor form.

10.2.3 Adiabatic Flame Temperature

Adiabatic flame temperature is the maximum temperature that can be achieved in a combustion process, when there is no heat and work transferred to or from the system. This is a crucial condition because combustion is a fast reaction and the safety precaution needs to be considered. The reactor temperature must be prevented from rising too fast.

Equation (10.15) can be applied to estimate the adiabatic flame temperature. $\Delta H_C = 0$ for adiabatic combustion reaction (cf Equation (10.3)):

$$0 = \Delta H_R + \Delta H_C^\circ + \Delta H_P \tag{10.15}$$

Equations (10.4) and (10.5) express ΔH_R and ΔH_P in terms of heat capacity, C_p. Heat capacity is a function of temperature. It is also worth noting that variation of heat capacity with temperature of a combustion product in the gaseous phase is significant. Therefore, when solving Equation (10.15), the problem is associated with determining the C_p values. C_p values can be estimated using a polynomial expression in terms of temperatures, as shown in Equation (10.16). Equation (10.17) is the integrated form of Equation (10.16). The data of the coefficients can be found in Perry's handbook or in the Knovel library (why.knovel.com), such as Yaws' handbook. A spreadsheet solver is recommended for solving this kind of problem.

$$C_p = A + BT + CT^2 + DT^3 \tag{10.16}$$

$$\int_{T_1}^{T_2} C_p \, dT = \int_{T_1}^{T_2} A + BT + CT^2 + DT^3 \, dT = \left[A\left(T_2 - T_1\right) + \frac{B}{2}\left(T_2^2 - T_1^2\right) + \frac{C}{3}\left(T_2^3 - T_1^3\right) + \frac{D}{4}\left(T_2^4 - T_1^4\right) \right] \tag{10.17}$$

where A, B, C and D are coefficients. It should be noted that different reference sources give different units for heat capacity and temperature and also different degrees of polynomials (cf Equation (10.6)). The units should be checked carefully when applying the above equation.

An alternative method, which would result in a lower degree of accuracy, is to estimate the C_p using mean heat capacity, $C_{p,mean}$, as described by

$$C_{p,mean} = \frac{\displaystyle\int_{T_1}^{T_2} C_p \, dT}{T_2 - T_1} \tag{10.18}$$

This method of using the mean heat capacity is more commonly used, when the spreadsheet solver is not available.

Exercise 3. Calculate the theoretical flame temperature of a fuel gas, assuming 100% methane, when combusted with a stoichiometric amount of air at 25 °C.

Solution to Exercise 3. A generic equation for combustion of hydrocarbon is given by

$$C_mH_n + \left(m + \frac{n}{4}\right)O_2 \rightarrow m\,CO_2 + \left(\frac{n}{2}\right)H_2O \tag{10.19}$$

$$CH_4 + 2O_2 \rightarrow CO_2 + 2H_2O \tag{10.7}$$

Since the reaction is carried out at the standard condition at 25 °C, $\Delta H_R = 0$ and hence Equation (10.15) can be simplified to

$$\Delta H_P = -\Delta H_C^\circ \tag{10.20}$$

Assume the adiabatic flame temperature, $T_{AFT} = 2000$ °C. Values of $C_{p,mean}$ at the current assumed temperature shown in Table 10.6 can be obtained from general thermodynamic reference handbooks.

Table 10.6 *Calculation of enthalpy change of products.*

Component	Number of Moles, N (kmol)	Mean Heat Capacity, $C_{p,mean}$ (kJ kmol^{-1} K^{-1})	$N \times C_{p,mean}$ (kJ kmol^{-1})
N_2	7.52	33.16	249.36
CO_2	1	55.14	55.14
H_2O	2	43.44	86.88
			391.38

Note. 1 kmol of air contains 79% N_2 and 21% O_2. Therefore 1 kmol of O_2 is accompanied by (1 × 0.79/0.21) = 3.76 kmol of N_2.

The enthalpy change of products, ΔH_P is calculated as follows:

$$\begin{aligned} \Delta H_P &= NC_{p,mean}\left(T_{AFT} - 25\right) \\ &= 391.38\left(T_{AFT} - 25\right) \end{aligned} \tag{10.21}$$

The standard heat of combustion of methane is -802.7×10^3 kJ kmol^{-1}, as given in Table 10.1. Therefore, by combining Equations (10.20) and (10.21),

$$391.38\left(T_{AFT} - 25\right) = -(-802.7 \times 10^3)$$

$$T_{AFT} = \mathbf{2076°C}$$

The calculated T_{AFT} is close enough with the initial estimation, that is, approximately 4% discrepancy. This is acceptable since the calculation using the mean heat capacity is a crude way of estimation. If the result is significantly different from the initial guess, then iteration has to be carried out by using another guess and the same steps of calculation as above need to be performed.

10.2.4 Theoretical Air-to-Fuel Ratio

The air-to-fuel ratio (R_{AF}) is the ratio of mass of air (m_{air}) to the mass of fuel (m_{fuel}) in a mixture, in the context of a combustion process, defined as

$$R_{AF} = \frac{m_{air}}{m_{fuel}} \tag{10.22}$$

If a fuel is completely combusted in an environment with a strictly sufficient amount of air, that is, without the presence of any excess of air, this is known as *stoichiometric* or *theoretical combustion*. Therefore, the minimum amount of air required for a complete combustion is known as *stoichiometric* or *theoretical air*.

Another common parameter used as the indicator for the combustion process – the stoichiometric ratio, λ – is the actual air-to-fuel ratio to the stoichiometric air-to-fuel ratio, defined as

$$\lambda = \frac{(R_{AF})_{actual}}{(R_{AF})_{stoic}} = \frac{(m_{air}/m_{fuel})_{actual}}{(m_{air}/m_{fuel})_{stoic}} \qquad (10.23)$$

Therefore,

$\lambda = 1.0 \rightarrow$ stoichiometric condition
$\lambda > 1.0 \rightarrow$ fuel-lean condition
$\lambda < 1.0 \rightarrow$ fuel-rich condition

Excess air is often needed for complete combustion. Typically, biomass with a high moisture content requires approximately 70% excess air, equivalent to 10 tonne of air for 1 tonne of biomass[7]. The excess air requirement is in the order of solid fuel > liquid fuel > gaseous fuel.

Part 2: Gasification

10.2.5   Cold Gas Efficiency

The product gas of gasification at high temperature is cooled down to a lower temperature in order to recover heat for satisfying the demand by the downstream processes. The cold gas efficiency is calculated based on the products after being cooled; in other words, the sensible heat is assumed not to be recoverable. Cold gas efficiency, η_{CG}, is shown in the following equation. A higher or lower heating value can be used and must be mentioned. The cold gas efficiency often falls within the range of 60–80% in the modern commercial gasification unit.

$$\eta_{CG}(\%) = \frac{\text{Heating value in cooled product gas}}{\text{Heating value in fuel}} \times 100 \qquad (10.24)$$

Exercise 4. Biomass at a mass flow rate of 17 t h^{-1} is fed to a gasifier. The gasifier generates a product gas stream of 16 t h^{-1}. The mass loss is due to the formation of char. The product gas of gasifier has the following composition:

Component	Composition (mol%)
CO	35.5
H$_2$	24.0
CO$_2$	5.4
N$_2$	30.6
CH$_4$	4.5

Assuming the biomass feedstock has an LHV of 14 MJ kg^{-1}, calculate the cold gas efficiency:

LHV of CO = 10.1 MJ kg^{-1}; H$_2$ = 120 MJ kg^{-1} and CH$_4$ = 50 MJ kg^{-1}

Solution to Exercise 4. The first step is to calculate the LHV of the product gas. Since the flow rate of the feed and LHV of the gaseous components are provided in terms of mass, it makes sense to first convert the molar composition of the product gas into mass composition, shown in Table 10.7.

Table 10.7 Product gas composition conversion from molar composition to mass composition.

Component	Molar Composition (mol%)	Molar Flow Rate (kmol s^{-1}), Assume a total 100 kmol s^{-1}	Molar Mass (kg kmol^{-1})	Mass Flow Rate (kg s^{-1})	Mass Composition (wt%)
CO	35.5	35.5	28	994	45.0
H$_2$	24.0	24.0	2	48	2.2
CO$_2$	5.4	5.4	44	237.6	10.8
N$_2$	30.6	30.6	28	856.8	38.8
CH$_4$	4.5	4.5	16	72	3.3
Total	100	100		2208.4	100

The next step is to calculate the LHV of the product gas based on the LHV of the combustible components, that is, CO, H$_2$ and CH$_4$. The mass flow rate of the product gas is 16 t h^{-1} (4.44 kg s^{-1}). This is shown in Table 10.8.

Table 10.8 Determination of the LHV of the product gas.

Component	Mass Flow Rate (kg s^{-1})	LHV (MJ kg^{-1})	Mass Flow Rate × LHV (MW)
CO	2.00	10.1	20.2
H$_2$	0.098	120	11.76
CH$_4$	0.147	50	7.35
Total			**39.3**

The feed flow rate is 17 t h^{-1} (4.72 kg s^{-1}) and the LHV is 14 MJ kg^{-1}. The total LHV of the feed is thus 66.1 MW. Therefore, the cold gas efficiency of the gasifier is

$$\eta_{CG}(\%) = \frac{\text{Heating value in cooled product gas}}{\text{Heating value in fuel}} \times 100 \tag{10.24}$$

$$= \frac{39.3}{66.1} \times 100$$

$$= 59.5\%$$

10.2.6 Hot Gas Efficiency

Sometimes, the high temperature product gas from gasification is utilized directly in other equipment such as a boiler and thus cooling of the gas may not be needed. The calculation of the hot gas efficiency assumes that the product gas remains at a high temperature so that the sensible heat of the hot gas is also utilized. Hot gas efficiency, η_{HG}, is shown in the following equation. The higher or lower heating value can be used and, as for cold gas efficiency, this has to be clearly indicated. It is obvious that the hot gas efficiency is greater than the cold gas efficiency:

$$\eta_{HG}(\%) = \frac{\text{Heating value in hot product gas} + \text{Sensible heat of the hot gas}}{\text{Heating value in fuel}} \times 100 \tag{10.25}$$

10.2.7 Equivalence Ratio

The equivalence ratio, ϕ, defined below is a strong parameter that determines the performance of the gasifier as well as the quality of the product gas. The equivalence ratio is defined as the ratio of actual (*actual* in Equation (10.26)) fuel-to-air ratio to the stoichiometric (*stoic* in Equation (10.26)) fuel-to-air ratio in a gasification or combustion process:

$$\phi = \frac{(m_{fuel}/m_{air})_{actual}}{(m_{fuel}/m_{air})_{stoic}} \tag{10.26}$$

where m_{fuel} is the mass flow rate of fuel and m_{air} is the mass flow rate of air.

- For gasification, $\phi > 1$ since the reaction is carried out in an air-deficient environment, in contrast to the combustion reaction, where excess air is used ($\phi < 1$).
- $\phi = 1$ is when stoichiometric air is used with respect to the fuel.
- For gasification of biomass, $3 < \phi < 5$. $\phi = 4$ is normally a good initial guess for modeling gasification of biomass[8].
- Adjustment of ϕ is needed, because it dictates the quality of the products formed.
- High ϕ results in a low gas yield while producing a high proportion of char and tar, that is, the pyrolysis reaction dominates the gasification reaction.
- Low ϕ results in more combustion products formation, that is, the combustion reaction dominates the gasification reaction.
- Both scenarios (low and high ϕ) would lead to a low heating value of the product gas. This is because at high ϕ, char and tar are produced rather than the product gas; at low ϕ, more CO_2 and H_2O are generated instead of CO and H_2.

10.2.8   Carbon Conversion

Carbon conversion, X_C, defined below shows the efficiency of utilization of the carbon element in the fuel:

$$X_C(\%) = \frac{\text{Carbon converted in gasification}}{\text{Carbon in the feed}} \times 100 \qquad (10.27)$$

10.2.9   Heat of Reaction

The heat of reaction for gasification can be determined in the same way as for the combustion reaction, using Hess's law concept, discussed earlier. However, there are numerous reactions involved in a gasification process, and thus it is not straightforward to determine the heat of reaction by specifying all the reactions. It is recommended that the main reactions such as partial oxidation and steam reforming reactions are taken into account in the calculation. From a practical point of view, such an estimation of the heat of reaction still lacks accuracy and the values obtained are only indicative of the order of magnitude.

10.3   Process Design: Synthesis and Modeling

10.3.1   Combustion Model

The combustion reaction can be defined as the reaction of fuel with excess oxygen, either pure oxygen or air, resulting in a mixture of gas containing mainly CO_2 and H_2O. Excess air or oxygen is used to ensure complete combustion. Some of the impurities such as ash and nitrogen in the fuel act as inerts, while other impurities may participate in the reaction forming oxides, such as SO_2.

The combustion reaction can be modeled using Equation (10.19), shown earlier:

$$C_mH_n + \left(m + \frac{n}{4}\right)O_2 \rightarrow mCO_2 + \left(\frac{n}{2}\right)H_2O \qquad (10.19)$$

If the reactions of other impurities with oxygen are taken into account, Equation (10.19) can be extended to

$$CH_aO_bN_cS_d + y\left(O_2 + 3.76N_2\right) \rightarrow CO_2 + \left(\frac{a}{2}\right)H_2O + \left(\frac{c}{2} + 3.76y\right)N_2 + d\,SO_2 \qquad (10.28)$$

where the number of moles of air, y, can be defined as

$$y = 1 + \frac{a}{4} + d - \frac{b}{2} \qquad (10.29)$$

Table 10.9 Deducing molar composition of an element in the fuel based on ultimate analysis.

Element	wt% (dry basis)	Mass (g), Assume Total = 100 g	Atomic Mass	Atom	Normalized Atom Based on C
C	p	p	12	$p/12$	1
H	q	q	1	q	$\dfrac{q}{(p/12)} = \dfrac{12q}{p}$
O	r	r	16	$r/16$	$\dfrac{(r/16)}{(p/12)} = \dfrac{3r}{4p}$
N	s	s	14	$s/14$	$\dfrac{(s/14)}{(p/12)} = \dfrac{6s}{7p}$
S	t	t	32	$t/32$	$\dfrac{(t/32)}{(p/12)} = \dfrac{3t}{8p}$
Ash		$1 - (p + q + r + s + t)$			

Equation (10.29) is formulated based on the balance of oxygen on the left (feed) and right-hand sides (products) of Equation (10.28).

The chemical formula $CH_aO_bN_cS_d$ is deduced from the ultimate analysis of fuel, shown in Table 10.9. Hence, the coefficients a, b, c and d in the chemical formula $CH_aO_bN_cS_d$ can be represented by

$$a = \frac{12q}{p}; \ b = \frac{3r}{4p}; \ c = \frac{6s}{7p}; d = \frac{3t}{8p}$$

The number of moles as well as the mole fraction (composition) of products can be determined through the stoichiometric method, as shown in Equation (10.28) and also in Table 10.10.

10.3.2 Gasification Model

There are two gasification models described here, that is, the stoichiometric equilibrium and the Gibbs free energy minimization models.

Table 10.10 Calculation of number of moles and mole fraction of products from combustion.

Components in Product	Number of Moles	Mole Fraction
CO_2	1	$\dfrac{1}{n_T}$
H_2O	$\left(\dfrac{a}{2}\right)$	$\dfrac{a}{2n_T}$
N_2	$\left(\dfrac{c}{2} + 3.76y\right)$	$\dfrac{\left(\dfrac{c}{2} + 3.76y\right)}{n_T}$
SO_2	d	$\dfrac{d}{n_T}$
Total	$n_T = 1 + \left(\dfrac{a}{2}\right) + \left(\dfrac{c}{2} + 3.76y\right) + d$	1

Note. n_T is the total number of moles.

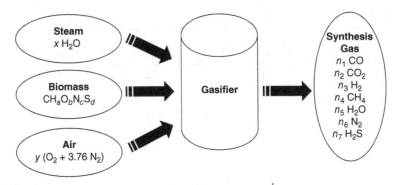

Figure 10.3   *Stoichiometric equilibrium gasification model.*

10.3.2.1   *Stoichiometric Equilibrium Model*

The principle of this model is based on the conservation of mass, where the reactants are converted stoichiometrically into products and an equilibrium state is reached. This model is thermodynamically driven, while eliminating other considerations such as reactor geometric and hydraulic parameters[8]. The yield of products is predicted by balancing the molar proportion of the reactants and this also represents the maximum achievable amount generated under a particular reaction condition.

Figure 10.3 shows a schematic representation of the gasification model based on the stoichiometric equilibrium method. Biomass with the chemical formula $CH_aO_bN_cS_d$ is gasified in the presence of x moles of steam and y moles of air. The product of gasification is a mixture of components, simplified to CO, CO_2, H_2, CH_4, H_2O, N_2 and H_2S. The gasification reaction can be represented using

$$CH_aO_bN_cS_d + x\,H_2O + y(O_2 + 3.76N_2) \rightarrow n_1CO + n_2CO_2 + n_3H_2 + n_4CH_4 + n_5H_2O + n_6N_2 + n_7H_2S \quad (10.30)$$

Equation (10.30) can be modified to include more components or to eliminate certain components depending on the desired modeling requirement, as necessary. For example:

- If pure oxygen instead of air is used as the gasifying medium, then $n_6 = c/2$ and $y(O_2 + 3.76N_2)$ becomes yO_2.
- For gasification of treated bio-oil containing negligible amounts of sulfur, S_d in biomass and H_2S in the product, on the left and right hand sides of Equation (10.30) can be eliminated.
- For some gasifier type such as entrained-flow gasifier, or when a high-moisture/liquid feedstock is used, steam may not be needed as the gasifying medium. Hence, H_2O on the left hand side of Equation (10.30) can be eliminated. However, the term H_2O on the right hand side of Equation (10.30) should remain, because the water gas shift reaction may occur in equilibrium (i.e., only partial oxidation is carried out).

Taking Equation (10.30) as the current example, atomic balance can be performed on the elements C, H, O, N and S. The atomic balances are shown in the following equations:
Atomic balance:

$$C: \quad 1 = n_1 + n_2 + n_4 \quad (10.31)$$

$$H: \quad a + 2x = 2n_3 + 4n_4 + 2n_7 \quad (10.32)$$

$$O: \quad b + x + 2y = n_1 + 2n_2 + n_5 \quad (10.33)$$

$$N: \quad c + 7.52y = 2n_6 \quad (10.34)$$

$$S: \quad d = n_7 \quad (10.35)$$

There are seven unknown variables (n_1, n_2, n_3, n_4, n_5, n_6 and n_7), while only five equations are presented. Two additional equations are thus required to solve all seven unknowns. These two equations can be obtained by working out the equilibrium constants of gasification equations such as the water gas reaction and the Boudouard reaction as follows (see Chapter 9):

$$C + H_2O \rightleftharpoons CO + H_2$$

$$C + CO_2 \rightleftharpoons 2CO$$

The equilibrium constant of a reaction at constant pressure, K_p, can be expressed by the ratio of the multiplication of the partial pressure of products to the multiplication of the partial pressure of reactants, shown below. Dalton's law of partial pressure (i.e., $p_i = y_i\, P$) is applied, where the ideal gas law is assumed.

$$K_p = \frac{\displaystyle\prod_{i=1}^{i=n_P} p_i}{\displaystyle\prod_{j=1}^{j=n_R} p_j} = \frac{\displaystyle\prod_{i=1}^{i=n_P} y_i\, P}{\displaystyle\prod_{j=1}^{j=n_R} y_j\, P} \tag{10.36}$$

where

p_i is the partial pressure of product component i.
p_j is the partial pressure of reactant component j.
y_i is the mole fraction of product component i in gaseous phase.
y_j is the mole fraction of reactant component j in gaseous phase.
n_P is the total number of products.
n_R is the total number of reactants.
P is the total pressure of the system.

The equilibrium constants for these two reactions are shown in the following equations, respectively. Note that only gaseous components are involved in the equilibrium constant calculation; the solid component is not included.

$$K_{p,1} = \frac{y_{CO}\, y_{H_2}\, P}{y_{H_2O}} = \frac{\dfrac{n_1}{n_T}\dfrac{n_3}{n_T}P}{\dfrac{n_5}{n_T}} \tag{10.37}$$

$$K_{p,2} = \frac{y_{CO}^2\, P}{y_{CO_2}} = \frac{\left(\dfrac{n_1}{n_T}\right)^2 P}{\dfrac{n_2}{n_T}} \tag{10.38}$$

The equilibrium constant, K_{eq}, can be related to the Gibbs free energy, $\Delta G°$, shown by

$$K_{eq} = \exp\left(-\frac{\Delta G°}{RT}\right) \tag{10.39}$$

Hence, the following equations can be obtained by equating Equation (10.37) with Equation (10.39) and Equation (10.38) with Equation (10.39):

$$\frac{n_1 n_3 P}{n_5 n_T} = \exp\left(-\frac{\Delta G°_1}{RT}\right) \tag{10.40}$$

$$\frac{n_1^2 P}{n_2 n_T} = \exp\left(-\frac{\Delta G°_2}{RT}\right) \tag{10.41}$$

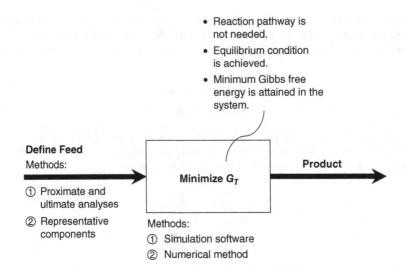

Figure 10.4   *Overview of Gibbs energy minimization method. G_T is the total Gibbs free energy of a system.*

The seven variables (n_1, n_2, n_3, n_4, n_5, n_6 and n_7) can now be determined by solving the seven Equations (10.31) to (10.35) and (10.40) and (10.41).

10.3.2.2   Gibbs Free Energy Minimization Model

When dealing with complex reactions such as gasification, it is tedious to define all the chemical reactions or to select appropriate reaction pathways. Another gasification modeling approach using the minimization of Gibbs free energy (Figure 10.4) can eliminate such hassles, as discussed in the equilibrium modeling of gasification using an equilibrium constant. The information required in the Gibbs free energy minimization approach is the reactant and product component specifications, while the detailed reactions are not needed. This simplifies the overall problem solution and minimizes uncertainties and inaccuracy in defining reaction equations and the equilibrium constants. In most circumstances, only a hypothetical set of chemical reactions can be defined for gasification. This is because only the main reactions are well understood, while most of the side reactions cannot be well defined. Let us now look at how to define a feedstock.

Defining Feed   The most straightforward method of defining a feed is by elemental composition. In most cases, this information can be obtained from the proximate and ultimate analyses of a feedstock (see Table 10.4). The question now is which analysis should be adopted in the modeling? The feedstock contains a certain level of moisture and ash. Moisture refers to the water content in the feedstock, while ash refers to the solid residue composed of metallic compounds. These should be separated from the elemental composition during gasification modeling. Fixed carbon is the nonvolatile carbon that is left after devolatilization. This should also be separated from the carbon in the volatile matter that is involved in gasification. The fixed carbon and ash content can be ignored in some circumstances, when a simplified gasification modeling is desirable and especially the syngas composition is the main focus. This is because the fixed carbon mostly ends up as char upon gasification, while ash is separated almost completely from the syngas; thus the subsequent process is not affected.

The feed analysis, which consists of normalized H_2O, elemental composition of C, H, O, N and S, and ash, is normally required, shown in Figure 10.5.

Using the elemental composition of feedstocks, such as coal in a Gibbs free energy minimization based gasification reaction model, gives reliable results. However, in some cases, using the elemental composition to model the gasification process may not yield a realistic syngas composition due to the inherent complexity of the feedstock's chemical

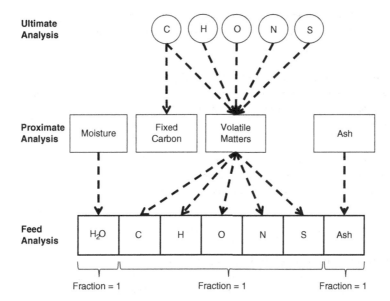

Figure 10.5   *Normalization of feed analysis for gasification modeling.*

constituents. This issue sometimes arises when modeling biomass feedstock. It is not recommended to use complex compounds such as cellulose, hemicellulose and lignin, which are not readily available in most of the component databanks in simulation software. This is tedious, since it may require user-defined parameters for physical property information, thus leading to uncertain results. Therefore, using a reasonable number of *representative components* to model the feedstock is a more effective method. For example, bio-oil can be modeled using three representative components: acetic acid, acetol and guaiacol, as shown in the bio-oil gasification cases[9, 10]. It should be noted that the selection of representative components and the number of components are not straightforward and requires rigorous validation and statistical analysis to establish a reliable modeling framework. The following rules of thumb can be applied when selecting representative components:

- For the first attempt, one should choose the dominant chemical components that are present in the feedstock.
- If the feedstock consists of more than one phase, such as bio-oil, where there are aqueous and lignin phases, the selection of representative chemical compounds should be considered for each phase.
- Avoid using complex compounds. Common chemical compounds, for which the physical properties can be established from calculations or from component databank within simulation software, should be considered.
- Normally one representative chemical compound is not sufficient to model the feedstock. Selecting a number of representative chemical compounds is preferred but requires a trial-and-error approach. A minimum number of representative chemical compounds is desirable.

Even though the representative chemical compounds are also composed of C, H, O, N and S elements, the resulting syngas composition estimated using the Gibbs free energy minimization method is different from using the feed defined solely by the elemental composition. This is because the transformation of the elemental composition into syngas components only involves the Gibbs energy of formation, while transformation of a fixed compound in the feedstock into syngas components involves both the energy needed to break down the bonds between the atoms and also the formation of new bonds to generate the products. The choice of adopting either the representative chemical compositions or the elemental compositions of the feedstock relies on which procedure comes closer to experimental or plant results in terms of syngas composition, obtained by the Gibbs free energy minimization method. Therefore, validation of the results is crucial in this case.

Gibbs Free Energy Minimization Method   There are two ways of performing the calculation: the *numerical method and simulation software.* The concept of the Gibbs free energy minimization is first discussed, before proceeding with the calculation.

What is Gibbs free energy and why it has to be minimized?

Gibbs free energy, G, is the maximum amount of nonexpansion work that can be extracted from a closed system in a reversible process. The change in Gibbs free energy, ΔG, is a thermodynamic quantity used for measuring the spontaneity of a reaction, as follows:

- If $\Delta G > 0$, then the reaction is not spontaneous and not favorable.
- If $\Delta G < 0$, then the reaction is spontaneous and favorable.
- If $\Delta G = 0$, the reaction is in the equilibrium condition.

In short, the ΔG tends to decrease in order to favor a reaction.

Derivation of Gibbs free energy minimization equation

The total Gibbs free energy of a system, G_T, is related to the chemical potential using the following equation. The derivation of Equation (10.42) can be found in the ***Online Resource Material, Chapter 10 – Additional Exercises and Examples***.

$$G_T = \sum_{i=1}^{N} n_i \mu_i \tag{10.42}$$

where n_i is the number of moles of component i, μ_i is the chemical potential of component i and N is the number of components.

Assuming the mixture of gases is an ideal gas at 1 atm, the chemical potential of component i, μ_i, can be expressed as a function of mole fraction of the component i, y_i, shown in the following equation. The derivation of Equation (10.43) can be found in the ***Online Resource Material, Chapter 10 – Additional Exercises and Examples***.

$$\mu_i = \Delta \overline{G}_{f,i}^{\circ} + RT \ln (y_i) \tag{10.43}$$

where $\Delta \overline{G}_{f,i}^{\circ}$ is the standard Gibbs energy of formation of component i, in kJ kmol^{-1}, R is the universal gas constant, 8.314 kJ kmol^{-1} K^{-1}, and T is the temperature, K.

Equation (10.44) can be obtained by substituting Equation (10.43) into Equation (10.42):

$$G_T = \sum_{i=1}^{N} n_i \Delta \overline{G}_{f,i}^{\circ} + \sum_{i=1}^{N} n_i RT \ln \left(\frac{n_i}{n_T} \right) \tag{10.44}$$

Equation (10.44) represents the total Gibbs free energy of the gasification system by considering all the products involved in it. G_T should be minimized in order to deduce the number of moles of component n_i.

Equation (10.44) can be solved using either the numerical method or the simulation method, as follows:

1. *Numerical method.* If Equation (10.44) is to be solved numerically, the Lagrange multiplier method can be adopted. The constraint to this problem is the mass balance of elements of the feed compared to the product, which has to be

the same. In other words, the number of atoms of a particular element (C, H, N, O, S) entering the reactor is equal to the total number of atoms in all components in the product gas. This statement can be represented by

$$\underbrace{\sum_{i=1}^{N} a_{ij}\, n_i}_{\text{product}} = \underbrace{A_j}_{\text{feed}} \tag{10.45}$$

where a_{ij} is the number of atoms of the element j in component i, A_j is the total number of atoms of element j in the feed, $j = 1, 2, \ldots, k$ and k is the total number of element j.

The Lagrange function, L, is derived by multiplying the constraint in Equation (10.45) by the Lagrange multipliers, λ_j, and subtracting it from G_T, hence defined as

$$L = G_T - \sum_{j=1}^{k} \lambda_j \left(\sum_{i=1}^{N} a_{ij} n_i - A_j \right) \tag{10.46}$$

The partial derivative of L with respect to n_i should be equal to zero in order to deduce the extreme point, shown by

$$\frac{1}{RT} \sum_{i=1}^{N} \Delta \overline{G}_{f,i}^{\circ} + \sum_{i=1}^{N} \ln \left(\frac{n_i}{n_T} \right) + \frac{1}{RT} \sum_{j=1}^{k} \lambda_j \left(\sum_{i=1}^{N} a_{ij} \right) = 0 \tag{10.47}$$

Solving Equation (10.47) for the Gibbs free energy minimization gives the product composition n_i from gasification. MATLAB and Excel based calculations are recommended for solving Equation (10.47) numerically.

2. *Simulation method.* Alternatively, the simulation method is a quicker way compared to the numerical method to solve a gasification model based on the Gibbs free energy minimization method. Process simulators such as Aspen Plus, Aspen HYSYS and PRO/II have an in-built Gibbs reactor model for such a purpose. Also, thermodynamic property packages are available to take account of the nonideality of the system concerned. The concept of the Gibbs reactor model is as described earlier.

10.3.3 Pyrolysis Model

The thermal decomposition of biomass via the pyrolysis reaction is widely modeled using the Waterloo concept[11], shown in Figure 10.6. The model assumes that the pyrolysis reactions proceed in a two-stage mechanism. The primary reactions involve the formation of gas, oil and char. The secondary reactions involve the conversion of oil into further products in the forms of gas, oil and char. The rate of the secondary reactions is much lower than the primary reactions. The secondary conversion of bio-oil into char is negligible and thus can be omitted.

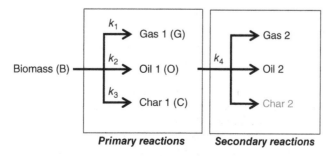

Figure 10.6 *Waterloo concept for modeling the biomass pyrolysis reactions*[11].

The rate equations of the biomass pyrolysis reactions as proposed in the Waterloo concept can be expressed by the following equations:

$$\frac{dm_B(t)}{dt} = -(k_1 + k_2 + k_3)\, m_B(t) = -k\, m_B(t) \tag{10.48}$$

$$\frac{dm_G(t)}{dt} = k_1 m_B(t) + k_4\, m_O(t) \tag{10.49}$$

$$\frac{dm_O(t)}{dt} = k_2 m_B(t) - k_4 m_O(t) \tag{10.50}$$

$$\frac{dm_C(t)}{dt} = k_3\, m_B(t) \tag{10.51}$$

where m_B, m_G, m_O and m_C denote the mass fraction of biomass, gas, oil and char, respectively, and k_1, k_2, k_3 and k_4 are the rate constants. For first-order reaction rate constants, the unit is s^{-1}:

$$k = k_1 + k_2 + k_3 \tag{10.52}$$

and t is any time during the reaction duration.

At $t = 0$, $m_B = 1$, $m_G = 0$, $m_O = 0$ and $m_C = 0$. These boundary conditions are applied during the integration of the set of first-order differential equations, of which the results after integration are shown below[11]

$$m_B(t) = \exp(-kt) \tag{10.53}$$

$$m_G(t) = -\frac{k - k_4}{k}\left[kk_1 \exp(-kt) - k_1 k_4 \exp(-kt) - k_2 k_4 \exp(-kt) + kk_2 \exp(-k_4 t) - kk_1 + k_1 k_4 - kk_2 + k_2 k_4\right] \tag{10.54}$$

$$m_O(t) = -\frac{k_2}{k - k_4} \exp(-k_4 t)\left[\exp(-t(k - k_4)) - 1\right] \tag{10.55}$$

$$m_C(t) = \frac{k_3}{k}\left[1 - \exp(-kt)\right] \tag{10.56}$$

Did you know?

Instead of doing the integration of the first-order differential Equations (10.48) to (10.51) by hand, which is tedious and time consuming, a quicker method using MATLAB is recommended. Try using the "dsolve" function in MATLAB, as shown below:

```
[Mb,Mc,Mg,Mo]
=dsolve('DMb=k*Mb','DMc=k3*Mb','DMg=k1*Mb+k4*Mo','DMo=k2
*Mb-k4*Mo','Mb(0)=1','Mc(0)=0','Mg(0)=0','Mo(0)=0')
```

The mass fractions of biomass, gas, oil and char are determined by solving Equations (10.53) to (10.56), amongst which the yield of bio-oil is of particular interest (Equation (10.55)). In order to solve these equations, the rate constants (k, k_1, k_2, k_3 and k_4) and the residence time are needed.

The global rate constant k can be determined through experiments such as thermogravimetric (TGA) analysis, k_1 and k_4 can be estimated using Equations (10.57) and (10.59), while k_2 and k_3 can be calculated simultaneously using Equations (10.52) and (10.58). Equation (10.58) is derived from Equation (10.56) at $t = \infty$ based on the observation from experiments that the amount of char is constant after the completion of pyrolysis reaction.

$$k_1 = 14\,300 \, \exp\left(\frac{-106\,500}{RT}\right) \tag{10.57}$$

$$k_3 = \frac{m_{C,\infty}}{1 - m_{C,\infty}}(k_1 + k_2) \tag{10.58}$$

$$k_4 = 7900 \exp\left(\frac{-81\,000}{RT}\right) \tag{10.59}$$

where R is the universal gas constant, 8.314 kJ kmol^{-1} K^{-1}, and T is the pyrolysis reaction temperature, K.

10.4   Summary

This chapter details various design calculations needed for combustion and gasification reaction processes. Furthermore, the techniques for modeling combustion, gasification and pyrolysis, adequate for predicting the yields of product, are also discussed.

Exercises

Refer to *Online Resource Material, Chapter 10 – Additional Exercises and Examples*, for solutions to the Exercise Problems.

1. A boiler using biomass as the feedstock is in place to generate steam. Assuming that the biomass can be represented by glucose ($C_6H_{12}O_6$) and all reactions are taking place at the gaseous phase with stoichiometric amounts of pure oxygen, calculate:
    a.   The standard heat of combustion. [The heat of formation of glucose $= -1274$ kJ mol^{-1}; $CO_2 = -393.5$ kJ mol^{-1} and $H_2O = -241.8$ kJ mol^{-1}]
    b.   The final temperature of combustion. The combustion reaction of biomass takes place at an initial temperature of 900 °C and the heat of combustion is -2800 kJ mol^{-1}. The coefficients of correlation for calculating the heat capacities of components are given in Table 10.11. Note that the coefficients of correlation for estimating heat capacities are valid within the temperature range of 150 K to 1500 K.

$$C_p = A + BT + CT^2 + DT^3 + ET^4 + ET^5 + FT^6 + GT^7 \tag{10.6}$$

    where C_p is the heat capacity, J mol^{-1}, A, B, C, D, E, F and G are the coefficients of correlation for estimating heat capacity and T is the temperature, K.
    c.   The mass flow rate of the biomass required if 500 kW of steam is generated at the conditions given in 1b.

Table 10.11   Coefficient of correlation for estimating heat capacities of components involved in glucose combustion[3].

Component	A	B	C	D	E	F	G
$C_6H_{12}O_6$	313.8621	−2.12485	0.010449	-2×10^{-5}	1.86×10^{-8}	-8.6×10^{-12}	1.58×10^{-15}
CO_2	23.5061	0.038066	7.4×10^{-5}	-2.2×10^{-7}	2.34×10^{-10}	-1.1×10^{-13}	2.17×10^{-17}
H_2O	33.17438	−0.00325	1.74×10^{-5}	-6×10^{-9}	0	0	0
O_2	29.79024	−0.00949	2.86×10^{-5}	9.87×10^{-9}	-5.7×10^{-11}	4.3×10^{-14}	-1.0×10^{-17}

Table 10.12   Feed and product specifications for bio-oil gasification.

Specification	Parameter
Feed to Gasifier	
Bio-oil	1 kmol s^{-1}
Oil	29.6 mol%
Water / Moisture	70.4 mol%
Oil analyses (ultimate, moisture and ash free, wt%)	
C	56.1
H	36.9
O	7
Oxygen	0.57 kmol s^{-1}
Product from Gasifier	
Mass flow rate of syngas	60.4 kg s^{-1}
Syngas elementary distribution by mass flow rate	
C	16.55 kg s^{-1}
H	3.47 kg s^{-1}
O	40.38 kg s^{-1}

2. Bio-oil is transported from some distributed pyrolysis sites to a centralized gasification site to produce syngas. The quality of syngas in terms of composition and yield is vital in deciding the performance of the downstream processes. Considering the difficulties in identifying the chemical components in bio-oil, the bio-oil is modeled using representative chemicals, that is, acetic acid, acetol and guaiacol. The bio-oil undergoes gasification under conditions of 1300 °C and 30 bar. Information on the input and output of the gasifier is shown in Table 10.12:

   a. Using the information given in Table 10.12, determine the composition (mass fraction and mole fraction) of bio-oil with respect to the representative components, that is, acetic acid, acetol and guaiacol.

   b. Given the operating conditions as above-mentioned, predict the syngas composition (CO, CO_2, H_2, H_2O and CH_4) in terms of mole and mass fractions, using the Gibbs free energy minimization method. The use of the simulation approach is highly recommended. Confirm the simulation results by comparing with the given mass flow rates of the output products.

   c. If the bio-oil has an LHV of 23.3 MJ kg^{-1} (moisture and ash free), estimate the cold gas efficiency of the gasification process based on the LHV of the feed. [LHV of CO = 10.1 MJ kg^{-1}, H_2 = 120 MJ kg^{-1}, CH_4 = 50 MJ kg^{-1}]

3. A biorefinery company is planning to set up a fast pyrolysis plant using biomass as feedstock to produce bio-oil. Spruce has been chosen as the desired feedstock due to its availability in the local region. Use the Waterloo concept to predict the yield of bio-oil. Assume a pyrolysis temperature of 500 °C with a residence time of 2.5 s and heating rate of 100 K min^{-1}. The yield of char at $t = \infty$ is assumed to be 26.2 wt%. The kinetic data for spruce[11] are activation energy E_a = 68 400 kJ kmol^{-1} and pre-exponential factor A = 34 700 s^{-1}.

References

1. C.L. Yaws, *Yaws' Handbook of Thermodynamic and Physical Properties of Chemical Compounds*, Knovel, New York, 2003.
2. C.L. Yaws, *Chemical Properties Handbook*, Knovel, New York, 1999.
3. C.L. Yaws, *Yaws' Handbook of Thermodynamic Properties for Hydrocarbons and Chemicals*, Knovel, New York, 2009.
4. ECN. Available from: http://www.ecn.nl/phyllis/defs.asp.
5. C. Sheng and J.L.T. Azevedo, Estimating the higher heating value of biomass fuels from basic analysis data, *Biomass Bioenergy*, **28**, 499–507 (2005).
6. ECN. Phyllis 2 Database, Database for biomass and waste.

7. R.P. Overend, Direct combustion of biomass, in *Renewable Energy Sources Charged with Energy from the Sun and Originated from Earth–Moon Interaction* – Volume 1, Encyclopedia of Life Support Systems, E.E. Shpilrain (ed.), EOLSS, USA, 2009.
8. P. Basu, *Biomass Gasification and Pyrolysis: Practical Design and Theory*, Academic Press, Burlington, USA, 2010.
9. K.S. Ng and J. Sadhukhan, Techno-economic performance analysis of bio-oil based Fischer–Tropsch and CHP synthesis platform, *Biomass Bioenergy*, **35**, 3218–3234 (2011).
10. K.S. Ng and J. Sadhukhan, Process integration and economic analysis of bio-oil platform for the production of methanol and combined heat and power, *Biomass Bioenergy*, **35**, 1153–1169 (2011).
11. M. Van de Velden, J. Baeyens, I. Boukis, Modeling CFB biomass pyrolysis reactors, *Biomass Bioenergy*, **32**, 128–139 (2008).

11

Reaction and Separation Process Synthesis: Chemical Production from Biomass

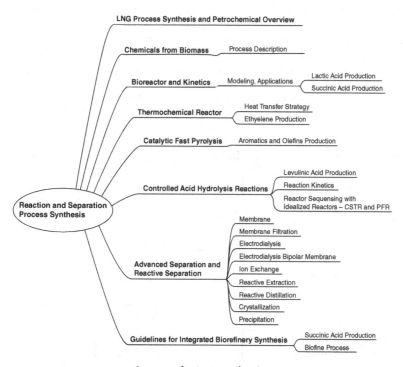

Structure for Lecture Planning

Biomass is the single source of functionalized organic chemicals and materials. Biomass is essentially made up of the same chemical elements (carbon, hydrogen and oxygen) as crude oil in varying proportions. This opens the possibility of producing biomass based products that could directly replace chemically identical crude oil derivatives (e.g., ethylene

Biorefineries and Chemical Processes: Design, Integration and Sustainability Analysis, First Edition.
Jhuma Sadhukhan, Kok Siew Ng and Elias Martinez Hernandez.
© 2014 John Wiley & Sons, Ltd. Published 2014 by John Wiley & Sons, Ltd.
Companion Website: http://www.wiley.com/go/sadhukhan/biorefineries

from bioethanol can replace ethylene from natural gas). It is also possible to substitute chemically different products but having similar functionality (e.g., poly(lactic acid) can substitute poly(ethylene terephthalate) for plastic bottles). In this chapter, reaction and separation process synthesis and process integration approaches are shown for biomass based products identified as potential building blocks for chemical synthesis. State-of-the-art and emerging process technologies for the production of such chemicals and their subsequent derivatives are shown alongside their design, modeling and simulation frameworks.

Refer to the ***Online Resource Material, Chapter 11 – Additional Exercises and Examples*** for a petrochemicals overview and description of natural gas liquids fractionation and naphtha steam cracking. Problems on distillation sequencing heuristics, distillation column design and estimation of energy requirements and greenhouse gas emissions are shown in the online material for a natural gas processing plant.

11.1 Chemicals from Biomass: An Overview

Figure 11.1 shows the process routes to major intermediate petrochemicals using biomass feedstocks. Ethanol and higher alcohols (propanol, butanol) can be used as precursors to the corresponding olefins (ethylene, propylene and butylenes) via dehydration, creating a bridge between the biorefinery and the current infrastructure of the petrochemical industry.

Figure 11.2 shows two additional routes for ethylene production from biomass based methanol. Starting from organic residues, (a) ethylene is obtained in a four-step route. First, organic residues are processed via anaerobic digestion to produce methane-rich biogas. Biogas is then processed by steam reforming to produce synthesis gas (CO + H_2). Synthesis gas is used for methanol production. In the final step, a methanol-to-olefins (MTO) process converts methanol into ethylene product. The lignocellulosic residues route (b) is a three-step pathway comprising biomass gasification, syngas fermentation and methanol dehydration.

The choice of a particular route depends on various factors, amongst which the following are important: (a) the feedstock: chemical composition, moisture content, physical properties, transport properties; (b) the reaction technologies involved and the information available from experiments or pilot plant trials; (c) the operating and capital costs involved; (d) the

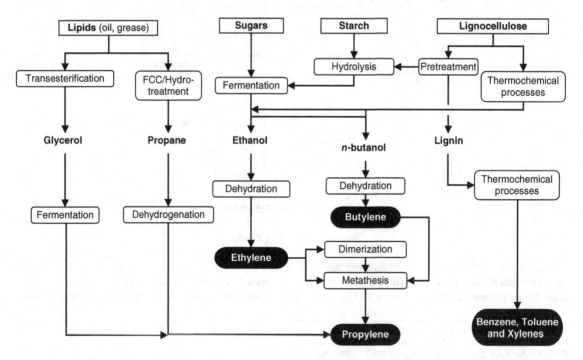

Figure 11.1 *Pathways to major petrochemicals from biomass. FCC, fluid catalytic cracking.*

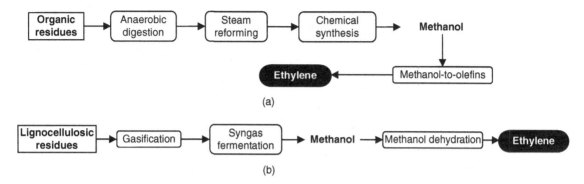

Figure 11.2   *Two alternative routes to obtain biomass based ethylene from (a) organic and (b) lignocellulosic residues.*

environmental impact of the process, energy and raw materials required; (e) the potential market for the products. As a general strategy, the coproduction of a low volume and high value specialty product needs to be produced alongside a large volume and low value product to enhance economic and environmental sustainability. The identification of the potential products and their pathways for their synthesis is crucial.

Refer to the ***Online Resource Material, Chapter 11 – Additional Exercises and Examples*** for chemical specific synthesis routes. Chemicals included are: ethanol, ethylene, glycerol and its derivatives, propylene glycol, lactic acid and its derivatives, succinic acid and its derivatives, levulinic acid and its derivatives, diphenolic acid, furan, furfural, hydroxymethylfurfural, sorbitol, xylitol and its derivatives, hydrocarbon and biohydrocarbon fuels.

11.2   Bioreactor and Kinetics

Figure 11.3 shows the most common bioreactor or fermentor types: batch, fed-batch and continuous reactors. High product concentrations are obtained in batch and fed-batch reactors. High productivity, but lower concentrations are achieved in continuous reactors. Dilution makes separation more difficult and may require more downstream processing steps. The trade-offs between productivity and purity of the product must be evaluated.

In a *batch fermentation* process, the reactor is charged with sterilized substrate and nutrients and inoculated with the bacteria. The bacteria are allowed to grow until substrate is depleted or product formation stops. After this time, the fermentation broth is discharged from the reactor and is sent to downstream processing for product recovery.

In a *continuous fermentation* process, fresh nutrients and substrates are continuously supplied to the reactor. Fermentation broth is also continuously withdrawn. The system reaches a steady state at which the cell, product and substrate concentrations remain constant. *Continuous stirred tank* and *plug flow reactors* are the two main *continuous ideal reactors.*

In a *fed-batch* process, the feed is continuously or semi-continuously fed while fermentation broth is removed after a certain period of time, when substrate is depleted or product formation stops. Fed-batch reactors are usually used to overcome substrate inhibition.

In continuous systems with biomass recycle, the biomass is externally separated from the outlet stream and recycled back to the bioreactor while the product-containing stream goes to downstream processing, discussed later.

Kinetic models allow the prediction of fermentation performance under varying conditions and serve as the basis for bioreactor modeling and design. Kinetic models can be developed from small-scale batch reactor experimentations and used to design the scaled-up processes. The bacterial growth dynamics in the batch culture can be idealized by the curve shown in Figure 11.4. The dynamic profile consists of a lag phase, an exponential phase, a stationary phase and a death phase. Different products may be predominant at different stages. It is possible to stop the fermentation at the phase with the maximum production of the main product. In general, such a phase is the stationary phase.

Elementary metabolic pathways can be considered to optimize fermentation conditions. Alternatively, an overall reaction model as a function of substrate concentrations can be used.

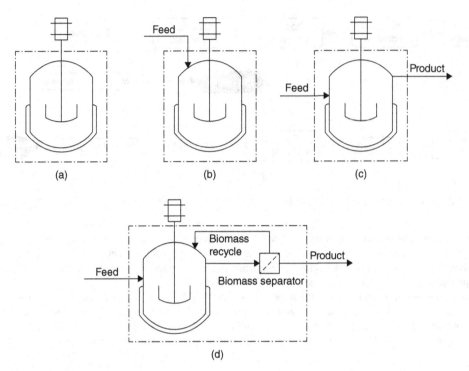

Figure 11.3 *Common bioreactor types: (a) batch (b) fed-batch, (c) simple continuous stirred tank reactor (CSTR) and (d) CSTR with biomass separation and recycle.*

In the overall reaction model, growth rates are expressed as the biomass specific growth rate μ. The *Monod kinetic equation* is used to describe the growth of a single microbial culture limited by substrate concentration[1]. The model is expressed as

$$\mu = \frac{\mu_{max}C_S}{K_S + C_S} \qquad (11.1)$$

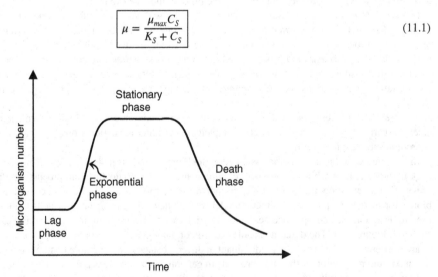

Figure 11.4 *Bacterial growth curve.*

where μ is the specific growth rate, h^{-1}, μ_{max} is the maximum achievable growth rate, h^{-1}, C_S is the substrate concentration, $g\ L^{-1}$ and K_S is the saturation constant corresponding to the concentration of the rate-limiting substrate when $\mu = 0.5$ μ_{max}, $g\ L^{-1}$. The parameters μ_{max} and K_S can be estimated from the specific growth rate data as a function of substrate concentrations during the exponential phase of growth. This simplified model can be extended to apply to the mixture of substrates and microbial cultures. See Equations (17.22) to (17.28) for application of Monod kinetics for algae growth.

Exercise 1. Determining the parameters of the Monod equation. Determine the parameters of the Monod equation for the fermentation of a low cost feedstock into lactic acid using data given in Table 11.1.

Table 11.1   *Data for lactic acid production from low cost feedstock fermentation.*

C_S (g L^{-1})	μ (h^{-1})
100	0.2671
80	0.2665
63	0.2662
55	0.2653
40	0.2641
19	0.2607
10.2	0.2509

Solution to Exercise 1. The parameters are estimated by using graphical methods. To manipulate the data, the Monod model in Equation (11.1) should be linearized as shown below. This equation is also known as the Lineweaver–Burk equation[2]:

$$\frac{1}{\mu} = \frac{K_S}{\mu_{max}}\frac{1}{C_S} + \frac{1}{\mu_{max}} \tag{11.2}$$

By plotting the reciprocal of the specific growth rate ($1/\mu$) against the reciprocal of the substrate concentration ($1/C_S$), the plot in Figure 11.5 is obtained. A straight line can be approximated so that the y axis intercept corresponds to $1/\mu_{max}$ while the gradient corresponds to K_S/μ_{max} of the Lineweaver–Burk equation.

   From the equation for the linear correlation in Figure 11.5, the value of the specific maximum growth rate is

$$\mu_{max} = 1/3.716 = 0.27\ h^{-1}$$

The value of the saturation constant is

$$K_S = 2.6617\ \mu_{max} = 2.6617 \times 0.27 = 0.72\ g\ L^{-1}$$

11.2.1   An Example of Lactic Acid Production

The kinetics of lactic acid bacteria growth can be described by a model that takes into account the effect of biomass and lactic acid concentrations. In this case biomass refers to the microbial biomass produced during fermentation and not to

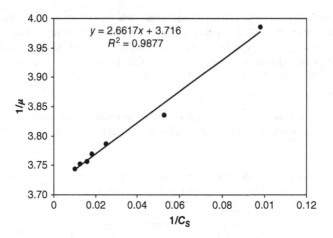

Figure 11.5   *Lineweaver–Burk plot of data given in Table 11.1.*

the biomass feedstock. The expression for the biomass production rate is

$$\frac{dC_F}{dt} = \mu \left(1 - \frac{C_F}{C_F^{max}}\right)^{f'} \left(1 - \frac{C_P}{C_P^{max}}\right)^{p'} \tag{11.3}$$

where C_F is the bacterial biomass concentration, g L^{-1}, C_F^{max} is the maximum attainable biomass concentration, g L^{-1}, C_p is the product concentration, g L^{-1}, C_p^{max} is the maximum product concentration above which bacteria do not grow, g L^{-1}, f' and p' are dimensionless parameters related to the inhibitory effect of biomass and product, respectively, and t is the time.

The product formation rate is described by a linear dependency on the biomass concentration and growth rate[3]:

$$\frac{dC_p}{dt} = \alpha \frac{dC_F}{dt} + \beta C_F \tag{11.4}$$

where α is the growth-associated product formation coefficient, g lactic acid g^{-1} biomass, and β is the nongrowth-associated product formation coefficient, h^{-1}.

The substrate utilization can be expressed as considering the substrate conversion into the product and the substrate consumption to maintain the microbial growth[4]:

$$\frac{dC_S}{dt} = -\frac{1}{Y_p} \frac{dC_p}{dt} - m_S C_F \tag{11.5}$$

where m_S is the substrate maintenance coefficient, h^{-1}, and Y_p is the dimensionless product yield coefficient.

Exercise 2. Modeling and optimization of lactic acid production. Whey is a major by-product of the dairy industry. It contains lactose, protein, fat and mineral salts. Due to the high lactose content, whey degradation has a high chemical oxygen demand. This poses a disposal and pollution problem for the dairy industry. However, whey lactose is a low cost feedstock with a high substrate concentration and thus is effective and favorable for the production of lactic acid. A cheese maker company is looking for a process to use its whey waste and avoid the treatment and disposal problems.

Experimental results using *Lactobacillus casei* to determine kinetic parameters have been reported as shown in Table 11.2. Using this information:

a. Determine the biomass, substrate and lactic acid profiles in a batch reactor for a feed with an initial lactose concentration of 21.4 g L^{-1} and an initial biomass concentration of 1.0 g L^{-1}.
b. Determine the lactose concentration that produces the maximum amount of lactic acid.

Table 11.2 *Kinetic parameters for whey powder fermentation into lactic acid. (Reproduced with permission from Altiok, Tokatli and Harsa (2006)[5]. Copyright © 2006, Society of Chemical Industry.)*

Kinetic Parameter	Value
μ_{max} (h^{-1})	0.265
K_S (g L^{-1})	0.72
α (g lactic acid g^{-1} biomass)	$0.029C_{S0} + 2.686$
	C_{S0} = initial substrate concentration
β (h^{-1})	0.06
f'	0.5
p'	0.5
Y_P (g lactic acid g^{-1} lactose)	0.682
m_S (h^{-1})	0.03
C_F^{max} (g L^{-1})	8

Solution to Exercise 2

a. The profiles for the biomass, substrate and lactic acid can be determined from solving the differential equations, Equation (11.3) to Equation (11.5), and using the Monod kinetic model in Equation (11.1) to express the specific growth rate. Note that the growth-related coefficient for the product formation rate is a function of the initial substrate concentration. In this case, the substrate is lactose. The three differential Equations (11.3) to (11.5) are solved using MATLAB command ode45 which uses a fourth-order Runge–Kutta numerical method. The concentration profiles are shown in Figure 11.6. It can be observed that lactose consumption is complete in about 8.5 hours when the lactose initial concentration is 21.4 g L^{-1}.

```
% This is the function file including the differential equation system for
% modeling the mass balance of batch whey fermentation into lactic acid
function dy=wheyferm_func(t,y,Mmax,Ks,a,b,Yp,CFmax,CPmax,ms,f,p)
dy=zeros(3,1);
if (y(3)<0) y(3)=0; end
M=Mmax*y(3)/(Ks+y(3));   % Equation for specific growth
dy(1)=M*(1-y(1)/CFmax)^f*(1-y(2)/CPmax)^p*y(1);   % Biomass production rate
dy(2)=a*dy(1)+b*y(1);   % Product formation rate
dy(3)=-1/Yp*dy(2)-ms*y(1);   % Substrate consumption rate


% This is another MATLAB file which solves the system of differential equations
% previously written based on the declared constants and initial conditions
% Declaration of constants
Mmax=0.265; Ks=0.72; b=0.06;
f=0.5; p=0.5; Yp=0.682;
```

```
ms=0.03; CFmax=8; CPmax=90;
CF0=1; CS0=21.4; CP0=0;   % Initial conditions (concentrations)
Ci=[CF0 CP0 CS0];
a=0.029*CS0+2.686;   % Growth-associated coefficient in function of CS0
Ti=0; Tf=9;
tspan=linspace(0,Tf, Tf*2);   %Time span


% ODE Solution
[t C]=ode45(@(t,y) wheyferm_func(t,y,Mmax,Ks,a,b,Yp,CFmax,CPmax,ms,f,p),tspan, Ci);

plot(t,C(:,1), '-or', t,C(:,2),'-sk',t,C(:,3),'-db')
% C(:,1)=CF, C(:,2)=CP, C(:,3)=CS
title ('Concentration profiles')
xlabel('Time (h)', 'fontsize', 12, 'fontweight','b')
ylabel('Concentration (g/L)', 'fontsize', 12, 'fontweight','b')
h=legend('Cell biomass','Lactic acid','Substrate');
set(h, 'fontsize', 8);
figure
```

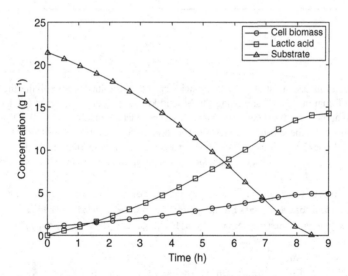

Figure 11.6   *Concentration profiles of whey lactose fermentation into lactic acid.*

b.   To find out the optimum lactose concentration, a sensitivity analysis is performed by changing the values of the initial substrate concentration in the MATLAB code and calculating the productivity in g L^{-1} h^{-1}, defined as the difference between the *final and initial product concentrations divided by batch time*:

$$Productivity = \frac{C_p - C_{po}}{t - t_o}$$

(11.6)

where t_0 is the initial time and t is the fermentation time, C_{p0} is the initial product concentration and C_p is the final product concentration when all substrates have been consumed. The time and product concentration results are shown in Table 11.3.

Table 11.3   *Total whey fermentation time and final lactic acid concentration results at total substrate consumption predicted by the kinetic model.*

C_{S0} (g L^{-1})	t (h)	C_p (g L^{-1})
10	6	6.55
20	8.3	13.1
35	11	23.1
40	12	26.35
45	13	29.5
50	14.7	32.2
60	21.2	38.5
75	30	47.3

The productivities calculated using Equation (11.6) are plotted against initial lactose concentrations as shown in Figure 11.7. The maximum productivity is found for a lactose concentration of about 45 g L^{-1}.

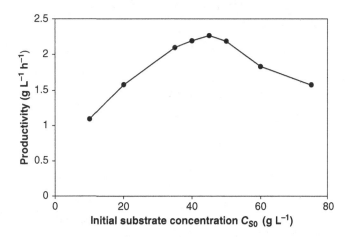

Figure 11.7   *Lactic acid productivities at different initial substrate concentrations.*

Current industrial fermentation gives about 90% yield of calcium lactate on a glucose basis. The process carried out in batch mode takes 4 to 6 days to complete. Batch fermentation is affected by substrate and lactate concentrations. These factors can affect the productivity due to inhibition of cell growth and product formation. By-product formation can be another problem. Other types of reactor and integrated reaction and separation systems can help to alleviate these complications. Table 11.4 shows examples of fermentation–separation systems discussed in Section 11.5.

Table 11.4   *Examples of fermentation–separation systems.*

Fed-batch bioreactor with solvent extraction product removal
Batch bioreactor with electrodialysis product removal
Continuous bioreactor with electrodialysis product removal
Continuous bioreactor with cell–recycle via membrane
Continuous bioreactor with ion-exchange product removal

Refer to the ***Online Resource Material, Chapter 11 – Additional Exercises and Examples*** for lactic acid process and derivatives description.

11.2.2   An Example of Succinic Acid Production

The biotechnological route for succinic acid production comprises microbial transformation of sugars or glycerol into succinic acid production using fermentation. Succinic acid is a metabolite participating in the tricarboxylic acid cycle (TCA) and also results as a metabolism product of various fungi and bacteria. Figure 11.8 shows a simplified scheme of metabolic pathways leading to succinate production. Note the CO_2 fixation during the metabolism. This is a promising CO_2 capture and reuse route to produce succinic acid and thereby reduce greenhouse gas emissions from an integrated

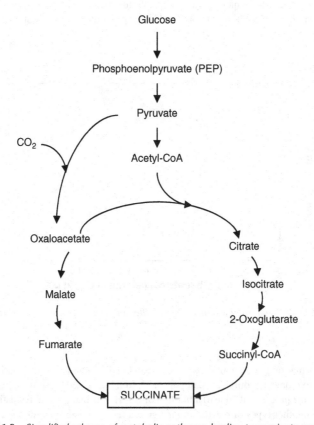

Figure 11.8   *Simplified scheme of metabolic pathways leading to succinate production.*

process plant (e.g., carbon dioxide released from glucose fermentation into the bioethanol production process can be used for on-site succinic acid production). The succinate salt rather than the free acid is normally present in the fermentation broth.

Depending on the microorganism, fermentation technology and operating conditions, other metabolites such as ethanol, lactic acid, acetic acid and formic acid are formed as by-products. The viability for a commercial technology resides in applying metabolic engineering to favor the fermentative pathways that lead to no or least by-product formation.

The stoichiometric reaction for succinic acid production from glucose is shown by

$$C_6H_{12}O_6 + CO_2 \longrightarrow C_4H_6O_4 + CH_3COOH + HCOOH \tag{11.7}$$

Succinic acid yield can be improved with strains that consume H_2 or HCOOH, as shown by

$$C_6H_{12}O_6 + 2CO_2 + 2H_2 \longrightarrow 2C_4H_6O_4 + 2H_2O \tag{11.8}$$

$$C_6H_{12}O_6 + 2HCOOH \longrightarrow 2C_4H_6O_4 + 2H_2O \tag{11.9}$$

A maximum theoretical yield of two moles of succinic acid per mole of glucose can be achieved when using a mixture of CO_2 and H_2 (Equation (11.8)). Carbon feedstock, pH and carbon dioxide and hydrogen reactants are critical for the production of succinate. A proper combination of these parameters must be selected for each microorganism, as they use different pathways for succinate production and tolerate different levels of CO_2, pH and H_2. CO_2 is an electron acceptor that diverts metabolism to pyruvate and lactate/ethanol when present at low levels but to succinate when present at high levels. CO_2 can be supplied from a treated flue gas stream, a purified CO_2 stream, carbonates (e.g., $MgCO_3$, $NaCO_3$, $NaHCO_3$ or $CaCO_3$) or from a combination of these sources.

Modeling of CO_2 dissolution from a gas stream in the fermentation broth is helpful to determine the amount of CO_2 needed for the microorganisms. The equilibrium concentration of CO_2 (C_{CO_2} in mol L^{-1}) dissolved in the fermentation broth from the gas can be estimated from the following modeling equations[6].

Henry's law for solubility of a gas into a liquid is shown by

$$C_{CO_2} = \frac{p_{CO_2}}{H} \tag{11.10}$$

where p_{CO_2} is the partial pressure of CO_2, kPa, and H is the Henry's constant in fermentation broth, kPa L mol^{-1}. The Henry's constant in the fermentation broth is determined from the relation between gas solubility in pure water and in fermentation broth, given (for the succinic acid case) by

$$\log\left(\frac{H}{H'}\right) = 0.0275 \tag{11.11}$$

where H' is the Henry's constant in pure water $= 4320$ kPa L mol^{-1}.

More CO_2 can be dissolved when carbonate or bicarbonate salts are also added to the fermentation broth due to the equilibrium shift driven by a change in pH. The total dissolved CO_2 is influenced by the equilibria between water, carbonic acid and carbonate and bicarbonate species. Thus, the total dissolved CO_2 concentration can be determined from

$$C_{CO_2} = \frac{\frac{p_{CO_2}}{H} + C_{CO_2,carb}}{1 + \frac{K_{1,H_2CO_3}}{[H^+]} + \frac{K_{1,H_2CO_3}K_{2,H_2CO_3}}{[H^+]^2}} \tag{11.12}$$

where

$C_{CO_2,carb}$ is the CO_2 concentration due to dissolution of carbonate salts, mol L^{-1}.

K_{1,H_2CO_3} is the first dissociation constant for carbonic acid, including the equilibrium between CO_2 and $H_2CO_3 = 5.3502 \times 10^{-7}$ mol L^{-1}.

K_{2,H_2CO_3} is the second dissociation constant for carbonic acid, corresponding to the dissociation of $HCO_3^- = 6.1245 \times 10^{-11}$ mol L^{-1}.

$[H^+]$ is the concentration of H^+ ion, mol L^{-1}.

The concentration of H^+ can be determined from the definition of pH given by

$$pH = -\log[H^+] \tag{11.13}$$

Refer to the **Online Resource Material, Chapter 11 – Additional Exercises and Examples** for discussion on succinic acid production from various bacteria species and substrates.

Exercise 3. Calculation of batch bioreactor volume and amount of CO_2 for succinic acid production. The effect of dissolved CO_2 concentration on the growth of *Mannheimia succiniciproducens* and succinic acid production has been studied at pH = 6.5 and 39 °C[6]. A succinate concentration of 10.51 g L^{-1} has been obtained after 6 hours of fermentation from 19.0 g L^{-1} of glucose. The dissolved CO_2 concentration of 0.052 mol L^{-1} was supplied from a gas stream at $p_{CO_2} = 101.325$ kPa and $NaHCO_3$. Using the information provided, answer the following questions:

a. Calculate the bioreactor volume for succinic acid (as succinate) production of 250 kg h^{-1}.
b. Determine the amounts of CO_2 and $NaHCO_3$ required.

Solution to Exercise 3

a. The bioreactor volume V can be determined from the final succinic acid concentration C_{SA}, the batch time t and the production rate desired Y_{SA}, as shown by

$$V = \frac{Y_{SA} \times t}{C_{SA}} \tag{11.14}$$

The batch time is the total time resulting from the fermentation time and the discharge and cleaning time for the next batch. Assuming a discharge and cleaning time of 30 min, the total time for the batch is

$$t = 6 + 0.5 = 6.5 \text{ h}$$

From Equation (11.14), the bioreactor volume is

$$V = \frac{250 \times 6.5}{10.51} = 155 \text{ m}^3$$

b. To determine the amount of CO_2 and $NaHCO_3$, the total concentration is determined from Equation (11.12). First, the concentration of H^+ is calculated from the definition of pH (Equation (11.13)):

$$[H^*] = 10^{-pH} = 10^{-6.5} = 3.16 \times 10^{-7} \text{ mol } L^{-1}$$

Using Equation (11.12) and the given dissolved concentration, the total CO_2 concentration is obtained:

$$= 0.052 \times \left[1 + \frac{5.3502 \times 10^{-7}}{3.16 \times 10^{-7}} + \frac{\left(5.3502 \times 10^{-7}\right) \times \left(6.1245 \times 10^{-11}\right)}{(3.16 \times 10^{-7})^2} \right] = 0.140 \text{ mol L}^{-1}$$

The contribution from the gas stream is determined as follows. First, the Henry's constant for the solubility of CO_2 in the fermentation broth is calculated using Equation (11.11):

$$H = H' \times 10^{0.0275} = 4320 \times 10^{0.0275} = 4602 \text{ kPa L mol}^{-1}$$

Then, the dissolved concentration from the gas stream at the given partial pressure is calculated using Equation (11.10):

$$101.325/4602 = 0.022 \text{ mol L}^{-1}$$

The required concentration of $NaHCO_3$ is the difference between the total CO_2 concentration and CO_2 dissolution from the gas:

$$0.140 - 0.022 = 0.118 \text{ mol L}^{-1}$$

The required mass flow rate of CO_2 or $NaHCO_3$ is the product of concentration, bioreactor volume and corresponding molar mass divided by the batch time:

$$F_{CO_2} = \frac{0.022 \times 155 \times 44.01}{6.5} = 23 \text{ kg h}^{-1}$$

$$F_{NaHCO3} = \frac{0.118 \times 155 \times 84.01}{6.5} = 236 \text{ kg h}^{-1}$$

This is about 0.1 kg of CO_2 and 0.93 kg of $NaHCO_3$ per kg of succinic acid.

Further Questions Calculate the CO_2 pressure required to provide the total CO_2 concentration required by the fermentation broth from a gas stream only. Refer to the *Online Resource Material, Chapter 11 – Additional Exercises and Examples* for the succinic acid process and derivative products' description.

As in any fermentation process, the pH is a key parameter. $MgCO_3$ is commonly used for a pH buffer as other salts present inhibitory, precipitation or flocculation problems. The pH is also a key factor in the availability of CO_2 for microorganisms because it affects CO_2 solubility in the medium. In pure water, the solubility of carbon dioxide is low at a pH value of 1 to 5. The dissolved CO_2 forms carbonic acid, H_2CO_3. The acid–base equilibrium is favored towards the dissociation of the acid at higher (basic) pH values. This drives the equilibrium towards the formation of the species HCO_3^- and CO_3^{2-}. As dissociation occurs, more H_2CO_3 needs to be formed in the aqueous solution. This consumes the CO_2 in the solution and as a result more CO_2 is dissolved. Addition of alkali is required for neutralization of succinic acid to favor CO_2 dissolution and to avoid inhibition of bacterial growth. However, engineered yeast can be grown at low pH $\sim$3.5 requiring no neutralization with an alkali and thus simplifying the downstream processing.

Fermentation technology still presents some issues such as by-product formation, costly separation and utilization of feedstocks that interfere with the food supply (e.g., corn or wheat). The sustainability of the biorefinery and its products may be undermined by the large amounts of land and water used to grow the feedstock and the energy required for processing them into products. Agricultural residues such as corn stover, rice husks and straws, wheat straws, sugarcane bagasse, etc., could be more sustainable fermentation feedstocks than first generation crops, because land and resources are used anyway to grow the crops. The microorganisms, operating conditions and such alternative feedstocks need to be selected

Table 11.5 *Reactor technologies for ethanol dehydration to ethylene[7–9]. LHSV: liquid hourly space velocity.*

Technology	Conventional	Lummus	Chematur	Petrobras
Reactor type	Fixed bed	Fluidized bed	Fixed bed	Fixed bed
Operation mode	Isothermal	Adiabatic	Adiabatic	Adiabatic
Temperature (°C)	330–380	399	315–425	355–390
Catalyst	Alumina or silica	Various	Syndol (MgO—Al$_2$O$_3$/SiO$_2$)	Various
Typical LHSV (h^{-1})	0.2–0.4	Residence time 2.7 s	0.15–0.5	0.15–0.5
Regeneration cycle (months)	1–2	–	8–12	6–12
Ethylene selectivity (%)	95–99	99.6	97–99	97–99
Ethylene yield (%)	93–98	99.5	>96	99.2

by techno economic and environmental sustainability evaluations. The implications to downstream separation must be considered in conjunction to identify opportunities for integration and reduction in costs and environmental impacts.

11.2.2.1 Thermochemical Reactor

Earlier two chapters discuss the gasification and pyrolysis reactors used for syngas or bio-oil production. Here the focus is on the chemical production using thermochemical reactors.

Table 11.5 shows the various reactor technologies, operating temperatures, catalyst types, regeneration cycles, ethylene product selectivity and yield using an example of ethanol dehydration to produce ethylene. The conventional technology is an isothermal fixed bed reactor, where the reactor bed is randomly packed with catalyst. An isothermal reactor has a multitubular fixed bed arrangement circulating a fluid that supplies heat for the reaction and for maintaining the temperature. Common heating fluids are limited to temperatures below 370 °C due to thermal degradation. This constraint limits the operation to a temperature region where the conversion of reactants and selectivity of products are reduced. Also, the area requirements for large-scale production can increase the capital cost of the reactor and make temperature control difficult. Strategies for temperature control are outlined in this section followed by a discussion on fixed bed and fluidized bed reactor technologies.

11.2.3 Heat Transfer Strategies for Reactors

As mentioned before, in reactors designed for isothermal operation, the control of temperature is an important component. Adiabatic operation is another operating mode for reactors. In adiabatic operation there is no heat transfer with the surroundings and the heat released by exothermic reaction increases the reaction medium temperature. On the other hand, an endothermic reaction decreases the reaction temperature. The operating temperature decides the performance and safety of the reactor. Temperature control is required to avoid an unacceptable temperature increment for exothermic reactions or an unacceptable temperature decrement for endothermic reactions. Strategies used to transfer heat from or to the reaction mixture for temperature control in common types of reactors (e.g., batch, fixed bed, plug flow reactor, continuous stirred tank reactor) include:

- Direct contact heat transfer by injecting a cold or hot shot of fresh feed into the reactor. This also allows variation in concentration of the reactants to control the temperature by controlling the reaction rate and reaction heat release or consumption.
- Indirect heating or cooling through the reactor surface using a vessel shell or a coil inserted in the reaction mixture. A fluid circulates through the shell or the immersed coil to absorb or provide heat. For highly exothermic reactions at high temperature, steam can be generated from the heat released by the reaction. Another alternative is to take a stream from the reactor to an external heat transfer device. After heat exchange, the stream is returned to the reactor. Intercooling or interheating can be used in reactor configurations, forming a cascade of reaction stages. Cooling or heating is carried out in intermediate sections and between the reaction stages.
- Introduction of an inert material with the reactor feed. This helps to reduce the temperature change by transferring part of the heat to or from the inert material. When possible, the existing process fluids should be used as heat carriers.

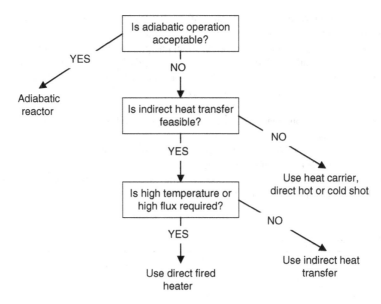

Figure 11.9   *Decision tree to select a heat transfer strategy for temperature control in a reactor.*

Product or by-product recycling can be used for heat control, if this does not decrease selectivity of the reaction or reactor yield.
- Manipulation of catalyst profiles by changing the distribution of active material in the catalyst bed or using a different catalyst in different sections of the reactor. Alternatively, a mixture of catalyst and inert solid can be used. By these strategies, the rate of reaction can be easily controlled in the various sections of a catalyst bed.

Overall, in all of these strategies, the flow rate of the most influential stream is increased to dissipate heat or cold so as to maintain the reactor temperature at a desired level. Figure 11.9 shows a decision tree to select the heat transfer strategy for temperature control of a reactor.

Sometimes the reactor effluent may need to be cooled rapidly for practical reasons. This rapid cooling is called *quenching* and can be carried out by indirect heat transfer (using heat exchangers) or by direct heat transfer (mixing the effluent with another fluid). Quenching by direct heat transfer is commonly used to stop a fast reaction to minimize by-product formation. Direct heat transfer is preferred when the reactor product can cause fouling. When special construction materials or designs are required due to high temperatures or corrosiveness of the reaction products, the direct heat transfer can also be used. For a gas stream, quenching can be carried out with a liquid stream. The liquid should be readily separable and must not contaminate the product or cause environmental problems.

11.2.4   An Example of Ethylene Production

A fixed bed reactor technology for ethanol dehydration to produce ethylene is licensed by Chematur Engineering AB[7]. Petrobras has developed a similar technology while Lummus has developed both fixed bed and fluidized bed reactor technologies[8,9]. These reactors' operations are adiabatic. In adiabatic operation, the heat of reaction is supplied by steam as an inert energy carrier. Steam-to-ethanol weight ratios of 2:1 to 3:1 improve catalyst life and product yield while decreasing coke formation and the regeneration cycle time. Three or four fixed bed reactors in series are typically used, with additional ethanol makeup between stages and intermediate heating. The capital cost of adiabatic reactors is relatively lower than the capital cost of isothermal reactors. The fluidized bed reactor technology allows better control of process conditions and minimizes coke and by-product formation. Table 11.6 shows results obtained from the Chematur and Lummus technologies.

Exercise 4. Feedstock, reaction system and economics of ethylene production. Ethanol conversion up to 99.9% and high selectivity towards ethylene (94.5–99%) are required from ethanol dehydration by proper combination of reactor technology, reactor conditions and catalyst. Furthermore, fossil energy use of bioethylene can be up to 60% lower than ethylene from naphtha cracking while GHG emissions can be up to 40% lower. Advise a renewable chemical company planning to invest in a plant with a capacity of 250 kt y^{-1} of polymer grade bioethylene. The data required are shown in Tables 11.6 to 11.9. The company has the following questions:

a. Recommend the most suitable combination of reactor type, reactor conditions and catalyst to produce polymer grade bioethylene (Tables 11.6 and 11.7).
b. Recommend the biomass feedstock for ethanol production that gives the lowest fossil energy consumption and GHG emissions amongst all feedstocks shown in Table 11.8.
c. Provide an order-of-magnitude estimate of the required capital investment (Table 11.9). CEPCI (Chemical Engineering's Plant Cost Index, Preliminary October 2012) = 575.4.

Table 11.6 *Results from two ethanol-to-ethylene reactor technologies.*

Chematur Engineering AB Fixed Bed Reactor[7]	Lummus Fluidized Bed Reactor[9]
Ethanol purity required: 95% by volume	Pressure: 1.7 bar
Ethylene selectivity at 99% ethanol conversion:	Effluent composition (% mol):
Ethylene 96.8	Water 50.02
Ethane 0.5	Ethylene 49.75
Propylene 0.06	Acetaldehyde 0.04
Butylenes 2.4	Ethanol 0.19
Acetaldehyde 0.2	

Table 11.7 *Ethylene content in reactor effluent using different catalyst for ethanol dehydration in fixed bed reactors (LHSV = 3 h^{-1} and ethanol partial pressure = 0.7 atm). (Reproduced with permission from Zhang et al. (2008)[10]. Copyright © 2008, Elsevier.)*

Catalyst	γ-Al_2O_3	HZSM-5	SAPO-34[a]	NiAPSO-34[b]
Temperature	450	300	350	350
Ethanol (%)	14.4	4.9	8.8	6.1
Ethylene (%)	78.7	93.7	86	92.3

[a]SAPO-34: Silicoaluminophosphate.
[b]NiAPSO-34: Ni-substituted SAPO-34.

Table 11.8 *GHG emissions and fossil energy consumption from cradle to gate bioethylene production and estimated costs of producing bioethylene from various feedstocks.*

Feedstock	Corn	Sugar cane	Lignocellulosic	Naphtha
Fossil energy use (GJ t^{-1})	50	−20	3	80
GHG emissions (t CO_2 eq. t^{-1})	3.1	−0.6	1.5	2.0
Production cost of bioethylene (US$ t^{-1})	2000 (US)	1200 (Brazil), 2600 (sugar beet, EU)	2000 (US)	1100

Table 11.9 Capital investments (M$ = million $) reported for bioethanol-to-ethylene plants.

Company	Capacity (t y^{-1})	Investment M$	Year Reported	CEPCI
Solvay Indupa[11]	60 000	135	2007	525.4
Braskem[12, 13]	200 000	278	2010	550.8
Dow-Mitsui[14]	350 000	400	2011	585.7

Solution to Exercise 4

a. Although the fluidized bed reactor shows great benefits in terms of yield and selectivity (Table 11.6), it has not been demonstrated at a large scale. Therefore a fixed bed reactor is recommended. Due to the large-scale production rate, adiabatic reactors are recommended for lower capital cost and ease of control.

   According to the experimental data in Table 11.7, the order of preference of catalysts from the highest to lowest yield of ethylene is: HZSM-5 > NiAPSO-34 > SAPO- 34 > γ-Al$_2$O$_3$. NiAPSO-34 and SAPO-34 are currently proved only at the laboratory scale. HZSM-5 is recommended as high conversion and selectivity of ethylene and lower effluent are achieved at a lower operating temperature. γ-Al$_2$O$_3$ can also be used at the proper temperature. Reactors in series are recommended to increase the overall yield.

b. According to the data in Table 11.8, the production from sugar cane gives the highest savings on fossil energy and GHG emissions with reference to the production from naphtha cracking. Production costs are comparable. The best place to install the plant could be in Brazil due to economic, energetic and environmental advantages. If the plant is to be based in another part of the world, the options between imported bioethanol from Brazil or indigenous lignocellulosic ethanol production must be analyzed. Although the production cost of bioethylene is currently between 1.1 and 2.3 times the average cost of fossil based ethylene, lignocellulosic ethanol is expected to decrease this ratio. The production cost from sugar beet is the highest in the EU and therefore it may not be the best feedstock in terms of economics.

c. Using the data in Table 11.9, the order-of-magnitude estimate can be performed by plotting the cost updated to a common date against the plant capacities and applying a potential regression. The costs reported for each plant are updated using the following equation and CEPCI values provided, where year A is the present year and year B is the year of the original cost (refer to Equation (2.1)):

$$\text{Cost in year A} = \text{Cost in year B} \times \left(\frac{\text{CEPCI year A}}{\text{CEPCI year B}} \right) \tag{11.15}$$

Figure 11.10 shows the resulting plot of capital investment versus bioethylene capacity. The regression equation shows a power of 0.5553, which reflects the economy of scale. Using such an equation, the order-of-magnitude estimate is

$$\text{Capital investment estimate} = 0.3289 \times (250\,000)^{0.5553} = \$327 \text{ million}$$

Refer to the ***Online Resource Material, Chapter 11 – Additional Exercises and Examples***, for ethanol and ethylene process synthesis.

11.2.5 An Example of Catalytic Fast Pyrolysis

Catalytic fast pyrolysis (CFP) is a biorefining technology for the production of aromatics (benzene, toluene and xylenes: BTX) and olefins (ethylene and propylene) from lignocellulosic feedstocks. In the CFP process, cellulose and hemicellulose in biomass solids are instantaneously decomposed by fast pyrolysis into anhydrous sugars that undergo dehydration to furans. The furans are then contacted with a catalyst bed in the same reactor for conversion into aromatics, olefins, CO, CO$_2$ and H$_2$O. Coke is produced from the lignin fraction of the biomass.

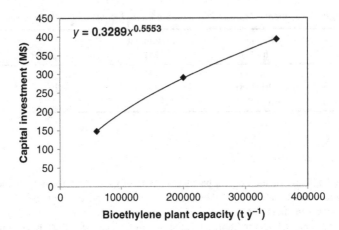

Figure 11.10   *Correlation between capital investment and bioethylene capacity.*

Different reactor types can be used for CFP of wood and furan including the pyroprobe reactor, fixed bed reactor and fluidized bed reactor (FBR). The pyroprobe reactor is a semibatch reactor, where small samples of biomass and catalyst are admixed together and heated to the reaction temperature. The fixed bed and fluidized bed reactors operate in a continuous regime. Figure 11.11 shows the experimental results of the CFP of wood and furan using different reactors[15]. Figure 11.11(a) shows the overall yield of the major fractions. Figure 11.11(b) shows the selectivity towards different aromatic compounds.

Note that the yield is shown as the carbon percentage as defined in Equation (11.16). Aromatic selectivity is defined in Equation (11.17).

$$\text{Yield (\% carbon)} = \frac{\text{Carbon content in product}}{\text{Carbon content in feedstock}} \times 100 \qquad (11.16)$$

$$\text{Selectivity (\% carbon)} = \frac{\text{Carbon content in product}}{\text{Total carbon in aromatic fraction}} \times 100 \qquad (11.17)$$

Exercise 5. Feedstock and reactor choice for catalytic flash pyrolysis. According to the reactor performances shown in Figure 11.11, evaluate the following:

a. Which is the reactor of choice for the CFP of pine wood sawdust?
b. Which of the following residues is better feedstock for CFP?
   i.   Nut shells (cellulose 25%, hemicellulose 30%, lignin 37%, others 8%)
   ii.   Switchgrass (cellulose 45%, hemicellulose 31%, lignin 12%, others 12%)
   iii. Softwood (cellulose 45%, hemicellulose 25%, lignin 26%, others 4%)
   iv.   Wheat straw (cellulose 36%, hemicellulose 39%, lignin 15%, others 10%).

Solution to Exercise 5

a. From Figure 11.11(a), the pyroprobe reactor gives a higher aromatic yield than the fluidized bed reactor and does not produce olefins. However, the fluidized bed reactor produces less coke. Figure 11.11(b) shows that the pyroprobe reactor produces more naphthalene, which is not as valuable as BTX aromatics. Furthermore, it is not economical to scale a pyroprobe reactor to a large reactor. Thus, the fluidized bed reactor is recommended for CFP of pine wood sawdust.
b. It can be seen in Figure 11.11(a) that furan results in higher aromatic and olefin yields and has lower coke formation than wood sawdust in the pyroprobe reactor. Furan is a model compound for the intermediate pyrolysis product of cellulose and hemicellulose. This suggests that the reactions are sensitive to the lignin present in the pine wood. Thus,

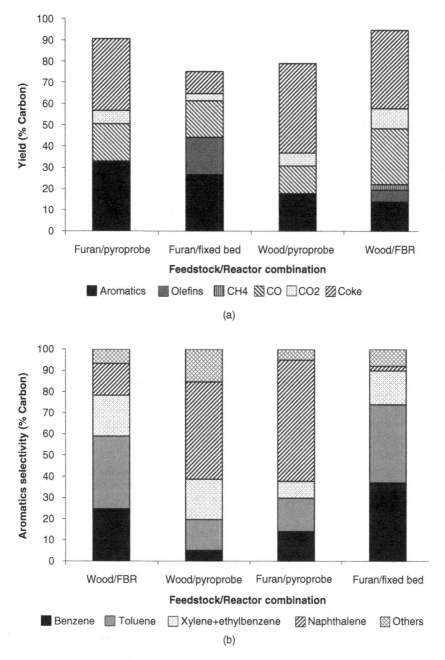

Figure 11.11 *(a) Yield of the various fractions produced from CFP of pine wood sawdust and furan in different reactor types and (b) the selectivity towards aromatic compounds.*

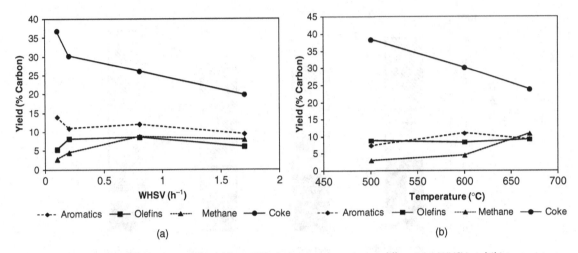

Figure 11.12 *Yield of the fractions produced from CFP of pine wood sawdust at different (a) WHSV and (b) temperature.*

the sequence of feedstocks of choice governed by lignin content is: switchgrass>wheat straw>softwood>nut shells (feedstock with least lignin content is preferred). However, moisture content has an impact on energy efficiency and product yields and must be considered for feedstock choice.

Figures 11.12 and 11.13 show the yield results from CFP of pine wood saw dust in a fluidized bed reactor[15]. The *weight hourly space velocity* (*WHSV*) and the temperature of the reactor affect the yields and selectivities of aromatics and olefins. The study of these variables is important for optimization of reaction conditions and reactor design. *WHSV* is defined by

$$WHSV = \frac{\text{Mass flow rate of feed}}{\text{Mass of catalyst in reactor}} \qquad (11.18)$$

Figure 11.12(a) shows the effect of *WHSV* on the yields of aromatics, olefins, methane and coke. The aromatics and coke yields decrease with increasing *WHSV* from 0.1 to 1.7 h^{-1}. The highest aromatic yield of 14% carbon is obtained at *WHSV* = 0.1 h^{-1}. The methane yield increases with increasing *WHSV*, while the olefin yield increases at the beginning

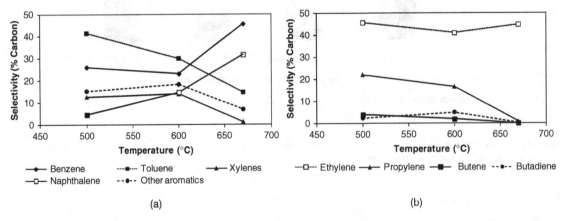

Figure 11.13 *Selectivity of the (a) aromatics and (b) olefins produced from CFP of pine wood sawdust at different temperatures.*

and decreases after $WHSV = 0.2$ h^{-1}. Lowering the catalyst charge to a catalytic reactor is expected to lower the desired product yields.

Figure 11.12(b) shows the yields from CFP of pine wood sawdust at three different temperatures. The maximum aromatic yield can be observed at 600 °C. The amount of methane increases at higher temperatures. The amount of coke decreases with increasing temperature.

Figure 11.13 shows that benzene is favored at higher temperatures. However, the amount of naphthalene is also increased at the cost of the more valuable aromatics. Overall, the BTX aromatics are favored at lower temperatures. Lighter olefins are expected at higher temperatures. The influence of temperature is considerable on olefin and aromatic selectivities. These variables, along with feedstock composition, can be adjusted to produce the desired product. For example, when aromatics are to be sold as gasoline additives, toluene and xylene are the best compounds to produce as they have a higher octane number than benzene and naphthalene. Hence, a lower operating temperature is preferred.

A way to improve the CFP process is by designing more effective catalysts. Improved aromatic carbon yields from CFP of pine wood sawdust are achieved using a Ga-promoted zeolite based catalyst (Ga-ZSM-5)[16]. In an experimental set up, the fluidized bed reactor is operated at 550 °C and WHSV of 0.35 h^{-1}. The solid particles are removed from the reaction products by a cyclone separator. The clean stream then passes through a series of condensers wherein the liquids are extracted using ethanol. Further condensation is achieved in four condensers using acetone as refrigerant at −55 °C. The gaseous stream is collected in air bags. The spent catalyst and coke are sent to a regenerator where the coke is burned to provide process heat. CO is converted into CO_2 during regeneration[15, 16].

Exercise 6. CFP of pine wood sawdust. Table 11.10 shows the product carbon yields from CFP of pinewood sawdust using the Ga-ZSM-5 catalyst.

a. Determine the amounts of the various products from a feed flow rate of 1000 kg h^{-1}. Assume that toluene (91.3% carbon mass content) and propylene (85.7% carbon mass content) are the major components in the aromatic and olefin fractions, respectively. The carbon content in pine wood is 51% by mass.

b. Determine the energy ratio of the process as: energy content in aromatics and olefins/energy in feedstock. Assume that the average higher heating value (HHV) of aromatic and olefin products is 42 MJ kg^{-1}. HHV of pine wood is 18.9 MJ kg^{-1}.

c. Discuss the advantages of this process technology in terms of complexity, product types and energy conversion efficiency over (i) gasification combined with Fischer–Tropsch synthesis for diesel production and (ii) cellulose hydrolysis combined with fermentation for ethanol production. In these processes, around 20% and 49% of the energy in biomass feed is transferred to diesel and ethanol, respectively.

d. Determine the annual economic potential (revenue from products – cost of feedstock) of the process if the price of sawdust chips is $50 t^{-1} and the price of raw naphtha from crude oil is $980 t^{-1}. Assume that the mixture of aromatics and olefins could be sold at the same price as naphtha and that the plant operates 330 days per year.

e. Write the potential applications of the gas stream within the biorefinery.

Table 11.10 *Product carbon yields from CFP of pine wood sawdust. (Reproduced with permission from Cheng et al. (2012)[16]. Copyright © 2012, Wiley-VCH Verlag GmbH & Co. KGaA, Weinheim.)*

Fraction	Product Carbon Yield (%)
Aromatics	23.2
Olefins	8.9
Methane	1.5
CO_2	5.4
CO	17.1
Coke	33.3
Other compounds	10.6

Solution to Exercise 6

a. The product carbon yield is defined as shown in Equation (11.16). The yields of aromatics in Table 11.10, assuming toluene as the main component, can be converted from a carbon basis to a total weight basis as (Equations (11.16) and (11.17))

$$\text{Aromatics yield on mass basis} = \text{C content in feedstock} \times \left(\frac{\text{Aromatics carbon yield}}{\text{C content in toluene}} \right)$$

$$= 0.51 \times \left(\frac{0.232}{0.913} \right) = 0.1296 \text{ kg aromatics kg}^{-1} \text{ feedstock}$$

Thus, the amount of aromatics produced is: $1000 \times 0.1296 = 129.6 \text{ kg h}^{-1}$
The amount of olefins produced is:

$$1000 \times 0.51 \times \left(\frac{0.089}{0.857} \right) = 53.0 \text{ kg h}^{-1}$$

Aromatics + olefins $= 182.6 \text{ kg h}^{-1}$
Similar calculations give 10.2 kg h^{-1} of CH_4, 100.9 kg h^{-1} of CO_2, 203.4 kg h^{-1} of CO and 169.8 kg h^{-1} of coke. Carbon contents in mass fraction in these molecules are $\frac{12}{16}, \frac{12}{44}, \frac{12}{28}$ and 1, respectively. The balance corresponds to other carbon components and water:

$$= 1000 - 129.6 - 53.0 - 10.2 - 100.9 - 203.4 - 169.8 = 333.1 \text{ kg h}^{-1}$$

b. Considering aromatics and olefins as the marketable products, the energy produced is:

$$= 182.6 \times 42 = 7667 \text{ MJ h}^{-1}$$

The amount of energy in the pinewood sawdust is:

$$= 1000 \times 18.9 = 18900 \text{ MJ h}^{-1}$$

Thus, the energy ratio is:

$$= \frac{7667}{18900} \times 100 = 40.6\%$$

c. Gasification plus Fischer–Tropsch synthesis for production of synthetic diesel requires various process steps (gasification, gas clean-up, chemical reaction, etc.) and more severe conditions than CFP. Ethanol production requires large reaction times and reactor volume for hydrolysis and fermentation. These complexities can lead to capital costs similar to or higher than CFP. The energy ratio is higher in the CFP process than in Fischer–Tropsch fuel production and is similar to the energy ratio of ethanol production. Furthermore, aromatics are more valuable than diesel fuel or ethanol. Unlike ethanol, the aromatics and olefins from biomass CFP are renewable feedstocks compatible with existing infrastructure for petrochemicals.

d. The total cost of the feedstock is:

$$= 1000 \times \frac{50}{1000} = \$50 \text{ h}^{-1}$$

The revenue from selling the aromatics and olefins mixture as a naphtha substitute is:

$$= 182.6 \times \frac{980}{1000} = \$178.9 \text{ h}^{-1}$$

Thus, the economic potential is:

$$= ((178.9 - 50) \times 330 \times 24)/10^6 = \$1.02 \text{ million y}^{-1}$$

The co-feed of propylene streams have shown increased olefin yields. This suggests that recycling the olefin stream could be favorable. According to data in Table 11.10, if the entire olefin stream could be converted into aromatics production, the theoretical yield of aromatics from pine wood sawdust can be up to 32.1% carbon.

e. The gaseous stream contains methane and could be used as fuel to generate heat required by the process. This will increase the energy efficiency of the overall process as less energy input from external utilities will be required.

Exercise 7. Process flowsheet for a CFP based biorefinery. RENPET is a renewable petrochemicals company interested in producing and commercializing the renewable aromatics produced from CFP of pine wood sawdust. Draw a scalable process from the experiments discussed in Section 11.3. Some modifications may be required as the equipment or process conditions used in the experiments may not be practical on an industrial scale. Draw a conceptual process flowsheet and provide a brief process description. Identify process integration opportunities.

Solution to Exercise 7. The proposed flowsheet is shown in Figure 11.14 and described as follows.

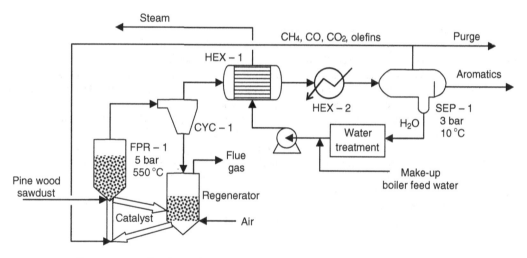

Figure 11.14 *Flowsheet of the CFP of pine wood sawdust for aromatics production.*

Main process

The pine wood sawdust is fed to the fluidized bed reactor FPR-1 wherein CFP occurs at 550 °C temperature and 5 bar pressure. FPR-1 is connected to a regenerator where ZSM-5 catalyst is regenerated and recycled into the fluidized bed reactor. The gaseous stream from FPR-1 is passed to the cyclone CYC-1 for solids removal. The cleaned gas is cooled down in the heat exchanger HEX-1 to 50–60 °C so that cooling water can be used. As discussed in the experiments, further cooling is required to improve the separation between the aqueous and organic phases. However, cooling down to −50 °C could be cost-prohibitive due to the high cost of refrigeration. Therefore, moderate refrigeration to 10 °C is considered for the heat exchanger HEX-2. This temperature can be adjusted according to the desired purity and recovery of the aromatics, while also considering the economic trade-off. Water condenses and forms an aqueous phase and is separated from the organic and gas phases. Then, the multiphase stream enters the three-phase separator SEP-1. Water, aromatics and noncondensable gases are recovered. The aromatics can be sent for further refining in distillation columns if required. Part of the gaseous stream from SEP-1, which contains the olefins produced, is recycled to the reactor FPR-1 for further conversion into aromatics. A purge is necessary to avoid accumulation of CO, CO_2 and CH_4.

Process integration opportunities

The coke along with the solids removed in CYC-1 can be burned to provide process heat for the regeneration and the pyrolysis reactor FPR-1.

The gaseous reactor effluent is available at a high temperature (550 °C). This stream can be used to generate medium pressure (MP) steam in HEX-1. Furthermore, the water recovered from the separator SEP-1 can be sent to a treatment plant so that it can be used for MP steam generation in HEX-1. Depending on the hydrocarbon content of the purge, this stream can also be used as fuel.

CFP is a promising thermochemical reactor technology for renewable petrochemical production. The main advantages of CFP are[15]:

- Conversion occurs in one single reactor.
- No process water is required.
- Flexibility on lignocellulosic feedstock.
- Simple biomass preprocessing (drying and grinding).
- Fluidized bed reactor is a reactor technology widely proven in petroleum refineries.
- Products are fully compatible with existing petrochemical infrastructure.

11.3   Controlled Acid Hydrolysis Reactions

Acid hydrolysis is commonly used to extract C6 sugars from cellulose in lignocellulosic feedstocks and convert these C6 sugars into specialty products such as levulinic acid under controlled temperature and pressure. Understanding kinetic mechanisms and models is essential for reactor design, optimization and control and thereby to select the conditions that improve conversion efficiency and economics of the process. The reaction rate of hydrolysis is affected by the cellulose, hemicellulose and lignin contents and the proportions of crystalline and amorphous cellulose in the lignocellulose. Low acid concentrations require higher temperatures and pressures and longer reaction times for better productivity. However, working with high concentrated acid requires more special and expensive materials for process equipment to prevent corrosion. The ash content of feedstocks needs to be least, because ash can neutralize the acid catalyst and reduce catalytic activity. This means more acid may be required at higher ash contents. Measuring the alkalinity of feedstocks helps to have a better estimation of the acid requirements.

Consider the acid hydrolysis reaction scheme in Figure 11.15 producing levulinic acid to develop a kinetic model[17]. The first step is the break down of cellulose by the acid catalyst to yield glucose. Glucose then produces

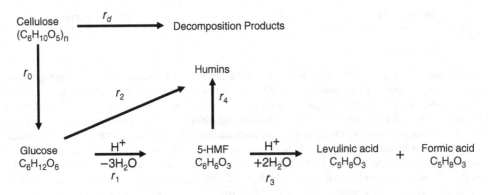

Figure 11.15   *Reaction scheme for the acid hydrolysis of cellulose to levulinic acid.*

5-hydroxymethylfurfural (5-HMF) as an intermediate product, which is later converted into levulinic acid and formic acid. Both glucose and 5-HMF can decompose into humins and form tar. Cellulose decomposition is also considered in the reaction scheme. Other possible hydrolysis by-products such as levoglucosan and other anhydrosugars are generally not taken into account.

Reaction rates (r) of the reaction scheme in Figure 11.15 can be shown as power law equations in Equation (11.19) to (11.24). For each reaction rate expression, a reaction constant (k) is multiplied by the concentration of the reactant (C_C: cellulose, C_G: glucose, C_{HMF}: 5-hydroxymethylfurfural). The concentration is powered to the order of reaction indicated by the exponents a to f:

$$r_0 = k_0 C_C^a \tag{11.19}$$

$$r_d = k_d C_C^b \tag{11.20}$$

$$r_1 = k_1 C_G^c \tag{11.21}$$

$$r_2 = k_2 C_G^d \tag{11.22}$$

$$r_3 = k_3 C_{HMF}^e \tag{11.23}$$

$$r_4 = k_4 C_{HMF}^f \tag{11.24}$$

The effect of temperature and the acid concentration can be included in a general expression for the reaction constant as the modified Arrhenius equation[18] shown by

$$k = (C_{H+})^g k_o \exp\left[\frac{E_a}{R}\left(\frac{T - T_o}{T_o T}\right)\right] \tag{11.25}$$

In this expression, k_o is the reaction constant at the reference temperature T_o (in K) and g is the power of the catalytic activity of the acid. The catalyst activity is expressed in terms of concentration of hydrogen cations (C_{H+}). E_a is the activation energy and R is the universal gas constant. For the case of sulfuric acid, a diprotic acid having two protons to donate per molecule, the second dissociation donating one proton can be taken into account for the equilibrium ($HSO_4^- \rightarrow H^+ + SO_4^{2-}$) using

$$C_{H+} = C_{H_2SO_4} + \frac{1}{2}\left(\sqrt{(K_{a2}^2 + 4C_{H_2SO_4}K_{a2})} - K_{a2}\right) \tag{11.26}$$

where $C_{H_2SO_4}$ is the concentration of the acid used as catalyst (H_2SO_4) and K_{a2} is the second dissociation constant of H_2SO_4 corresponding to the monoprotic species (e.g., HSO_4^-) at the reaction temperature.

Equations (11.19) to (11.26) can be developed using batch experiments. Kinetic modeling allows the generation of useful insights into reactor selection and design. Further understanding of the reaction networks and their mechanisms during the acid hydrolysis of the whole lignocellulosic complex is required. The following exercise shows the usefulness of the kinetic models for the selection and conceptual design of a network of idealized reactors.

Exercise 8. Reactor sequencing and design for controlled acid hydrolysis. Table 11.11 shows kinetic parameters of the models shown in Equations (11.19) to (11.24) for acid hydrolysis of cellulose to levulinic acid. Use the information provided to perform the following tasks:

a. Recommend the reactor type or reactor sequence that favors continuous levulinic acid production.
b. Calculate the required volumes of the reactors for 99% conversion of cellulose. The reactors are to be designed to process 100 000 kg h^{-1} of feed containing 10% mass of cellulose (0.617 mol L^{-1}) at 200 °C and H_2SO_4 concentration of 2% on mass basis (0.208 mol L^{-1}).

Table 11.11   *Kinetic parameters of the modeling equations for acid hydrolysis of cellulose into levulinic acid production. (Reproduced with permission from Girisuta, Janssen and Heeres (2007)[17]. Copyright © 2007, American Chemical Society.)*

Equation/reaction	Parameter	Value	Units
1. Cellulose to glucose	a	0.98	–
	k_o	0.41	$L^{0.94}$ min^{-1} $mol^{-0.94}$
	T_o	175	°C
	E_a	151.5	kJ mol^{-1}
	g	0.96	–
2. Cellulose decomposition	b	1.01	–
	k_o	0.065	$L^{0.95}$ min^{-1} $mol^{-0.95}$
	T_o	175	°C
	E_a	174.7	kJ mol^{-1}
	g	0.94	–
3. Glucose to 5-HMF	c	1.09	–
	k_o	0.013	$L^{1.22}$ min^{-1} $mol^{-1.22}$
	T_o	140	°C
	E_a	152.2	kJ mol^{-1}
	g	1.13	–
4. Glucose to humins	d	1.3	–
	k_o	0.013	$L^{1.42}$ min^{-1} $mol^{-1.42}$
	T_o	140	°C
	E_a	164.7	kJ mol^{-1}
	g	1.13	–
5. 5-HMF to levulinic acid	e	0.88	–
	k_o	0.34	$L^{1.26}$ min^{-1} $mol^{-1.26}$
	T_o	140	°C
	E_a	110.5	kJ mol^{-1}
	g	1.38	–
6. 5-HMF to humins	f	1.23	–
	k_o	0.117	$L^{1.30}$ min^{-1} $mol^{-1.30}$
	T_o	140	°C
	E_a	111	kJ mol^{-1}
	g	1.07	–
7. Equilibrium constant (at 200 °C)	K_{a2}	3.16×10^5	mol L^{-1}

Solution to Exercise 8

a.   A network of idealized continuous reactors, for example, a continuous stirred tank reactor (CSTR) and a plug flow reactor (PFR), can be created to achieve desired production goals. The CSTR has uniform concentration in the reactor, equal to the outlet concentration. Because the *outlet concentration* $(C_{out,i})$ of reactant species i is lower than the inlet concentration $(C_{in,i})$, the reaction rate is constant and low in an idealized CSTR. In the PFR, the concentration of reactant reduces as its conversion increases with the length of the PFR. Hence, the reaction rate is the highest at the inlet and lowest at the outlet of the PFR. The presence of an intermediate in the reaction scheme suggests that a two-stage reactor system should be used. The first stage must aim at maximum intermediate production while the second stage must aim at maximum levulinic production. In both stages, the cellulose decomposition reaction or the degradation of glucose and 5-HMF to humins must be minimized. With this in mind and considering the order of the reactions involved, the type of reactor can be recommended.

For parallel reaction systems of the form shown below:

$$\text{Feed} \longrightarrow \text{Product} \qquad r_1 = k_1 (C_{Feed})^{\alpha 1} \qquad\qquad (11.27)$$

$$\text{Feed} \longrightarrow \text{By-product} \qquad r_2 = k_2 (C_{Feed})^{\alpha 2} \qquad\qquad (11.28)$$

The ratio to minimize is shown by

$$\frac{r_2}{r_1} = \frac{k_2}{k_1} \left(C_{Feed} \right)^{\alpha2-\alpha1} \tag{11.29}$$

When the undesired reaction has a higher order than the desired one, that is, $\alpha_2 > \alpha_1$, a CSTR is preferred to keep the reaction rate. This is shown in Figure 11.16(a). When the desired reaction is faster than the undesired reaction, that is, $\alpha_2 < \alpha_1$, the reactor of choice is a PFR. As shown in Figure 11.16(b), the reaction rate reduces along the length of the PFR. Furthermore, the reactor volume required to achieve a given conversion is lower for a PFR than a CSTR. This is shown by the dotted areas in Figure 11.16.

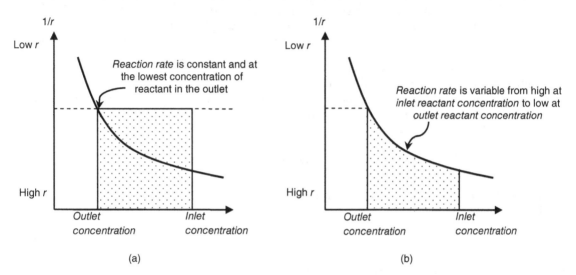

Figure 11.16 *Inverse of reaction rate versus reactant concentration in (a) CSTR and (b) PFR.*

The cellulose decomposition reaction (an undesired reaction) has a higher order than the cellulose conversion reaction into glucose production (a desired reaction): $\alpha_2 = 1.01 > \alpha_1 = 0.98$. In the case of glucose reactions, $\alpha_2 = 1.3 > \alpha_1 = 1.09$. This suggests using a CSTR for the first reaction stage to keep the reaction rates low.

If the catalyst is considered as a second reactant, the parallel reaction system is of the form shown below:

$$\text{Feed 1} + \text{Feed 2} \longrightarrow \text{Product} \qquad r_1 = k_1 (C_{Feed1})^{\alpha1} (C_{Feed2})^{\beta1} \tag{11.30}$$

$$\text{Feed 1} + \text{Feed 2} \longrightarrow \text{By-product} \qquad r_2 = k_2 (C_{Feed1})^{\alpha2} (C_{Feed2})^{\beta2} \tag{11.31}$$

The ratio to minimize is shown by

$$\frac{r_2}{r_1} = \frac{k_2}{k_1} \left(C_{Feed1} \right)^{\alpha2-\alpha1} \left(C_{Feed2} \right)^{\beta2-\beta1} \tag{11.32}$$

When the undesired reaction is of a higher order than the desired ones with respect to reactant 1, that is, $\alpha_2 > \alpha_1$, then:

if $\beta_2 > \beta_1$ a CSTR is preferred;
if $\beta_2 < \beta_1$ a semi-batch or semi-PFR is preferred.

When the undesired reaction is of a lower order than the desired ones with respect to reactant 1, that is, $\alpha_2 < \alpha_1$, then:

if $\beta_2 > \beta_1$ a semi-batch or semi-PFR is preferred;
if $\beta_2 < \beta_1$ a batch or PFR is preferred.

For the glucose reactions, $\alpha_2 > \alpha_1$ and $\beta_2 = \beta_1$. Thus, with regard to glucose, the choice remains as before, a CSTR. In the case of cellulose reactions, $\alpha_2 > \alpha_1$ and $\beta_2 < \beta_1$; this changes the choice to a semi-batch reactor or a semi-PFR. A clear option cannot be suggested until now. Looking at the formation of the intermediate 5-HMF, it can be seen that it consists of two sequential reactions. A reaction pathway consisting of a series of reactions is always favored in a PFR or batch than in a CSTR. As a continuous process is preferred for large-scale production, the choice is a PFR for the first reaction stage to favor production of glucose and 5-HMF.

The higher order for the undesired reactions of cellulose and glucose suggests that the concentrations of these components must be kept low to prevent high humins formation. This can be achieved by feeding the cellulose feedstock at a low concentration.

In the second stage, for the 5-HMF reactions: $\alpha_2 > \alpha_1$. Therefore, a CSTR is the reactor of choice for this stage. Finally, a sequence of a PFR followed by a CSTR is suggested.

b. *Design of a PFR for the first stage*, To determine the volume of the reactors, the design equations derived from mass balances for a reactor volume V are used. For a PFR, the mass balance equation for a reactant i and a feed flow rate F is shown by

$$FdC_i = -r_i dV \qquad (11.33)$$

As this is a continuous reactor, there is no accumulation term $(dC_i/dt = 0)$. Changing Equation (11.33) to the *space–time* domain τ results in

$$dC_i = -r_i d\tau \qquad (11.34)$$

where τ is the time required to process one reactor volume of feed, defined as

$$\tau = \frac{V}{F} \qquad (11.35)$$

Then, for the main species in the first reactor stage, the set of ordinary differential equations (ODEs) shown in the following equations and also Equation (11.41) must be solved simultaneously (refer to Figure 11.15 for the symbols used):

$$\frac{dC_C}{d\tau} = -k_o C_C^a - k_d C_C^b \qquad (11.36)$$

$$\frac{dC_G}{d\tau} = k_o C_C^a - k_1 C_G^c - k_2 C_G^d \qquad (11.37)$$

$$\frac{dC_{HMF}}{d\tau} = k_1 C_G^c - k_3 C_{HMF}^e - k_4 C_{HMF}^f \qquad (11.38)$$

$$\frac{dC_{LA}}{d\tau} = k_3 C_{HMF}^e \qquad (11.39)$$

The concentration of formic acid C_{FA} can be obtained from reaction stoichiometry and is found to be equal to the levulinic acid concentration C_{LA}, as shown by

$$C_{FA} = C_{LA} \qquad (11.40)$$

The concentration of humins C_h is obtained from the mass balance given as

$$\frac{dC_h}{d\tau} = k_d C_C^b + k_2 C_G^d + k_4 C_{HMF}^f \tag{11.41}$$

Equations (11.36) to (11.39) and (11.41) can be solved in MATLAB following a code similar to that in Exercise 2. The resulting concentration profiles for space–time from 0 to 5 min are shown in Figure 11.17. The cellulose profile clearly shows the occurrence of a fast hydrolysis reaction. Although complete conversion can be achieved at a space–time of just $\tau = 5$ min, most of this is converted into humins, the undesired product. Therefore, the residence time must be chosen to favor one of the intermediate products. In this case, the main intermediate product is glucose. The highest glucose concentration of 0.176 mol L^{-1} is achieved at a space–time of $\tau = 1$ min. This means that a small reactor will be required. Note that the corresponding final cellulose concentration is about 0.215 mol L^{-1}.

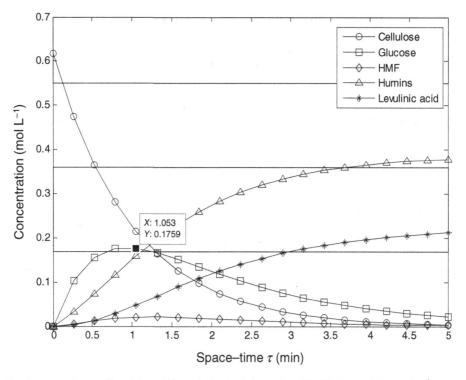

Figure 11.17 *Concentration profiles of the acid hydrolysis of cellulose at T = 200 °C, $C_{in,C}$ = 0.617 mol L^{-1} (10% mass basis) and C_{H2SO4} = 0.208 mol L^{-1}.*

The volume of a PFR at the reaction conditions given is thus calculated from the space–time relation (Equation (11.35)) as shown by

$$\boxed{V_{PFR} = F \times \tau} \tag{11.42}$$

If the feed contains 10% cellulose by mass, then the amount of water in 100 000 kg h^{-1} is 90 000 kg h^{-1}. By assuming a density of 1000 kg m^{-3}, the volumetric flow rate is 90 m^3 h^{-1}. Therefore, the estimated reactor volume for $\tau = 1$ min is

$$V_{PFR} = 90 \times \frac{1}{60} \times 1 = 1.5 \text{ m}^3$$

Assuming a tube length of 5.8 m and inner diameter of 0.04 m, a quick calculation of the number of tubes gives

$$\frac{1.5}{\pi \times \left(\frac{0.04}{2}\right)^2 \times 5.8} = 206 \text{ tubes}$$

The conversion X_i is calculated from the inlet concentration $C_{in,i}$ and the outlet concentration $C_{out,i}$ of component i, as shown by

$$X_i = \frac{C_{in,i} - C_{out,i}}{C_{in,i}} \tag{11.43}$$

Thus, the conversion of cellulose (X_C) in the PFR reactor is calculated at $\tau = 1$ min as follows. From Figure 11.17, $C_{in,c} = 0.617$ and $C_{out,c} = 0.215$. Thus

$$X_C = \frac{0.617 - 0.215}{0.617} = 0.651$$

The selectivity towards a product is expressed as the ratio between the rate of the desired reaction and the rate of undesired reaction. Thus, the selectivity towards glucose (S_G) at the beginning of the reaction can be expressed from the reaction rates of cellulose as shown below:

$$S_G = \frac{r_0}{r_d} = \frac{k_{o,0}}{k_{o,d}} \exp\left[\frac{E_{a0} - E_{ad}}{R}\left(\frac{T - T_o}{T_o T}\right)\right] (C_C)^{a-b} (C_{H+})^{g0-gd} \tag{11.44}$$

From the values in Table 11.11, the activation energies of the desired reaction $(E_{a0} = 151.5 \text{ kJ mol}^{-1})$ is lower than that of the side reaction $(E_{ad} = 174.7 \text{ kJ mol}^{-1})$. This means that the selectivity of the reaction is favored at low temperatures. However, the low temperature reduces the cellulose conversion.

The reaction order with respect to acid of the main reaction of cellulose to glucose $(g_0 = 0.96)$ is higher than that of the humins formation $(g_d = 0.94)$, which means that high acid concentrations will have a positive effect on the selectivity of the reaction. In conclusion, glucose production can be increased by low cellulose concentration, high acid concentration and moderate temperature.

Design of CSTR for second stage, The outlet concentrations from the first reactor are shown in Table 11.12. To determine the volume of the CSTR reactor, the generic design equation for a reactant i is

$$F(C_{in,i} - C_{out,i}) = -r_i V \tag{11.45}$$

Table 11.12 *Outlet concentrations from the PFR used for the first reaction stage.*

Component	Concentration (mol L^{-1})
Glucose	0.176
Cellulose	0.215
5-HMF	0.020
Humins	0.158
Levulinic acid	0.048
Formic acid	0.048

In the space–time domain Equation (11.45) can be rearranged to

$$\tau = \frac{C_{in,i} - C_{out,i}}{-r_i} \tag{11.46}$$

Note that now the reaction rates in the CSTR remain constant because of well-mixed reactions. The reaction rate now is dictated by the outlet concentration, that is, $C_i = C_{out,i}$. The equation for cellulose ($i = C$) results in

$$\tau = \frac{C_{in,C} - C_C}{k_o C_C^a + k_d C_C^b} \tag{11.47}$$

Equation (11.47) is expressed in terms of conversion (Equation (11.43)) as

$$\tau = \frac{X_C C_{in,C}}{k_o C_C^a + k_d C_C^b} \tag{11.48}$$

First, the final outlet concentration from the CSTR is calculated from the overall conversion desired, which is the concentration that dictates the rate of reaction:

$$C_C = C_{out,C} = 0.617 \times (1 - 0.99) = 0.0062$$

The rate of reaction of cellulose conversion in the second reactor can be calculated using data in Tables 11.11 and 11.12 and, assuming the same acid concentration as in the first reactor:

$$k_o C_C^a + k_d C_C^b = 0.7884 \times 0.0062^{0.98} + 0.1792 \times 0.0062^{1.01} = 0.0064 \text{ mol L}^{-1} \text{ min}^{-1}$$

To achieve an overall 99% conversion of cellulose, the conversion in the CSTR must be

$$X_C = \frac{0.215 - 0.617 \times (1 - 0.99)}{0.215} = 0.9713$$

Thus, substituting the earlier values in Equation (11.48), the space–time required is

$$\tau = \frac{0.9713 \times 0.215}{0.0064} = 33 \text{ min}$$

Figure 11.18 shows the variation of the overall cellulose conversion, that is, after the PFR reactor followed by the CSTR reactor, with space–time. To achieve conversions beyond 99%, the volume required increases significantly while the increase in conversion is small. For example, to achieve 99.99% conversion, the space–time should be more than three times the value required for 99%. This means that three times more reactor volume is required. Therefore, the conversion currently achieved for the calculated time of 33 min is justified.

The required volume is calculated as

$$V_{CSTR} = F \times \tau = 90 \times \frac{1}{60} \times 33 = 50 \text{ m}^3$$

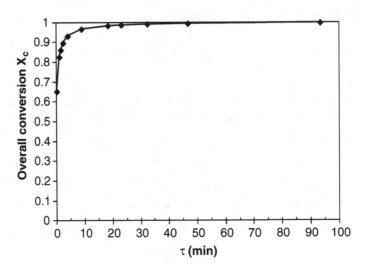

Figure 11.18   *Variation of overall conversion of cellulose with the space–time.*

The concentration of any intermediate product and the levulinic acid can be determined from the mass balance. For example, the balance equation for glucose combined with Equation (11.48) results in

$$\tau = \frac{C_{in,G} - C_G}{-k_o C_C^a + k_1 C_G^c + k_2 C_G^d} = \frac{X_C C_{in,C}}{k_o C_C^a + k_d C_C^b} \tag{11.49}$$

This equation allows solving for the concentration of glucose C_G as shown by

$$C_G = C_{in,G} - \frac{X_C C_{in,C}(-k_o C_C^a + k_1 C_G^c + k_2 C_G^d)}{k_o C_C^a + k_d C_C^b} \tag{11.50}$$

Similarly, the following equations show the expressions to calculate the concentration of HMF and levulinic acid (LA), respectively:

$$C_{HMF} = C_{in,HMF} - \frac{X_C C_{in,C}(-k_1 C_G^c + k_3 C_{HMF}^e + k_4 C_{HMF}^f)}{k_o C_C^a + k_d C_C^b} \tag{11.51}$$

$$C_{LA} = C_{in,LA} + \frac{X_C C_{in,C}(k_3 C_{HMF}^e)}{k_o C_C^a + k_d C_C^b} \tag{11.52}$$

The system of nonlinear Equations (11.50) to (11.52) are solved simultaneously using a numerical method (e.g., Newton's method). Results of molar concentrations and yields are shown in Table 11.13. The formation of humins has been significant. Thus, the process requires further analysis to minimize production of humins.

Further Questions   In Exercise 8, the mass yield of levulinic acid from cellulose is just 27% compared to 60% of humins. Carry out a sensitivity analysis to improve the results reported in Table 11.13 in order to maximize levulinic acid production. Explore other possible reactor configurations such as two CSTRs, two PFRs or CSTR followed by a PFR.

Table 11.13   *Outlet concentrations from the CSTR.*

Component	Concentration (mol L^{-1})	Yield (% mol)
Cellulose	0.006	Base component
Glucose	0.015	2.4
5-HMF	0.001	0.2
Levulinic acid	0.257	41.7
Formic acid	0.257	41.7
Humins	0.158	54.8

11.4   Advanced Separation and Reactive Separation

Advanced separation and purification technologies such as membrane filtration, electrodialysis, ion exchange, reactive extraction, crystallization and precipitation can reduce the pollution and costs of bio based chemical production. These technologies will be prominent operations in biorefineries, especially in those using fermentation, and are discussed as follows.

11.4.1   Membrane Based Separations

Membranes are semipermeable barriers used to perform separations by controlling the transport of species across them. The separation can involve two vapor or gas phases, two liquid fractions or phases, or a vapor and a liquid phase. The main principle for the operation of membranes is difference in transport properties between substances such as permeability, size of molecules or particles, solubility, diffusivity, volatility, etc. Such a difference or driving force may be caused, for example, by the process conditions, solvent, electrical field or the properties of the membrane itself. Membrane based processes can be:

a. Driven by a pressure gradient: microfiltration, ultrafiltration, nanofiltration, reverse osmosis, gas separation, pervaporation
b. Driven by a concentration gradient: dialysis, osmosis, forward osmosis
c. Driven by an electrical potential gradient: electrodialysis, membrane electrolysis, electrophoresis
d. Driven by a temperature gradient: membrane distillation.

For all the membrane processes the condition of the feed has a significant influence on the performance of the unit. Feed pretreatment is often necessary to minimize fouling and the degradation of the membrane.

Figure 11.19 shows a simple membrane separation process. The feed is separated into two streams: *retentate* and *permeate*. The retentate is the fraction containing the species from the feed that does not pass through the membrane. The

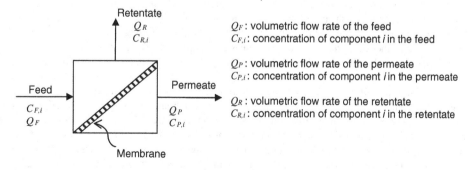

Figure 11.19   *Simple membrane separation.*

permeate is the fraction containing the species that passes through the membrane. For example, in a separation of lactic acid from water, membrane retains the lactic acid product in the retentate while most of the water permeates through the membrane. In this example, the main fraction of interest is the retentate.

A membrane must have a high *permeance* ratio between the species being separated to achieve an effective separation. The *permeance* of a component i ($\bar{P}_{M,i}$) is defined as the *permeability* of the component ($P_{M,i}$) divided by the membrane thickness (δ_M):

$$\bar{P}_{M,i} = \frac{P_{M,i}}{\delta_M} \tag{11.53}$$

The flux or rate of transport of a component across a membrane (J_i) is expressed as

$$J_i = \frac{P_{M,i}}{\delta_M} \times \text{(Driving force)} \tag{11.54}$$

Permeability values are estimated from experimental studies. Its unit depends on the driving force of the process concerned. For a molar concentration gradient, the unit is [length]^2 [time]$^{-1}$ (e.g., cm^2 s^{-1}). Note that this is the same unit as the diffusivity. Example of units for the molar flux is mol s^{-1} cm^{-2}. Flux can also be expressed in terms of volumetric or mass flow rates. For the process in Figure 11.19, the molar flux of a component in Equation (11.54) can be expressed as

$$\boxed{\frac{Q_P C_{P,i}}{A_M} = \frac{P_{M,i}}{\delta_M} \times (C_{R,i} - C_{P,i})} \tag{11.55}$$

where Q_P is the volumetric flow of the permeate stream, $C_{P,i}$ is the concentration of the component i in the permeate stream, $C_{R,i}$ is the concentration of the component i in the retentate stream and A_M is the membrane area.

Equation (11.55) allows the calculation of membrane area required for a desired separation. The ratio between the permeabilities of two components (i and j) is the separation factor or selectivity $\alpha_{i,j}$, as shown by

$$\alpha_{i,j} = P_{M,i}/P_{M,j} \tag{11.56}$$

Another important parameter in membrane based separation is the fraction of feed permeated (cut fraction) θ defined as

$$\theta = Q_P/Q_F \tag{11.57}$$

where θ is the fraction of feed permeated and Q_F is the volumetric flow rate of the feed.

Membranes can be microporous or dense solid materials. Microporous membranes are made of materials constituted by interconnected pores. A membrane with a certain pore size is chosen according to the size or molar mass of the molecules of the substance to be separated. Molecules of size larger than the pores will not pass through the membrane and remain in the retentate. Although the permeability of microporous membranes is high, giving rise to high productivity or yield of products, all molecules smaller than the pore size would diffuse through the membrane. This could result in low selectivity. On the other hand, dense solid membranes are nonporous, giving rise to high selectivity or purity of products. Hence, there is a trade-off between the productivity (or yield) and selectivity (or purity) of the products through a membrane.

The mechanism of transport through a membrane is as follows. First, the components are diffused from the solution to the surface of the membrane. Then they are dissolved into the membrane material and diffused through the solid. Finally, the components are desorbed and diffused in the solution at the other side of the membrane. The permeability depends on both the diffusivity and solubility of solution components in the membrane. Although diffusion through nonporous membranes can be slow and the permeance can be low, high selectivity can be achieved for the separation of small molecules. The permeance of a component can be high if the thickness of the membrane is very small according to Equation (11.53).

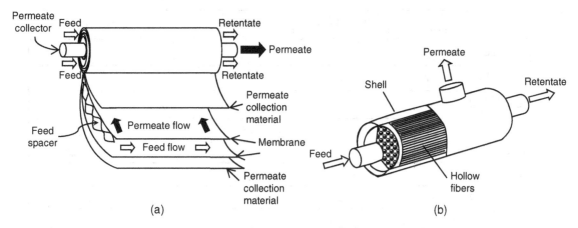

Figure 11.20 *Scheme of (a) spiral wound membrane module and (b) hollow fiber membrane module.*

Most membranes are manufactured from synthetic polymers. In general, these materials can only operate at temperatures below 100 °C and cannot be used with reactive species. These factors produce degradation of the membrane material. When operation at high temperatures is required or the species are reactive, ceramic microporous membranes or metallic nonporous membranes can be used.

Various flow patterns (well-mixed, countercurrent, cross-flow and co-current flow) occur in membrane equipment and can be idealized for modeling and simulation. Under the same conditions, the well-mixed pattern gives the lowest separation efficiency, while the countercurrent flow pattern gives the highest efficiency. Cross-flow and co-current flow patterns are in between. The countercurrent flow requires the lowest membrane area among all the patterns mentioned. The flow pattern depends on the permeation rate and the geometry of the membrane arrangement.

Figure 11.20 shows the two most common arrangements in membrane modules. The spiral wound arrangement in Figure 11.20(a) consists of wound flat membrane sheets forming a spiral that is inserted into a vessel. The two layers of membranes are separated by flow spacers. Two additional layers of impermeable material are used to collect and direct the flow of permeate to a perforated collection pipe in the centre of the arrangement. The cross-flow pattern in the scheme can also be noted. The hollow fiber membrane module shown in Figure 11.20(b) consists of cylindrical membranes arranged as tubes in a shell-and-tube heat exchanger. Various arrangements are possible with the feed entering either the shell side or inside the hollow membrane. For example, a gas stream such as combustion flue gas enters the hollow membranes, through which the CO_2 is diffused and separated as permeate. Nitrogen and other components are recovered as the retentate, which flows through the channels of the hollow membranes. In other applications such as hemodialysis, a dialysate stream is fed to the shell side and the blood stream is fed to the hollow fibers in the countercurrent flow. In this case, the shell has both inlet and outlet ports.

In practical applications, more than one membrane module may be required and can be arranged in parallel or in series. The number of modules depends on the membrane area required to achieve the level of separation desired. Depending on the stream to be treated, the membrane can suffer from fouling and mechanical or chemical degradation and must be cleaned. Membrane lifetime is an important parameter when selecting membrane separation processes and depends on the type of material and application.

Wastewater treatment and gas separation are the two most important applications of membrane based separations. Membrane technology is also used to develop cost-effective systems for clarification, filtration, concentration and purification of biorefinery streams. It is useful for the recovery of chemicals produced in diluted concentrations (e.g., in fermentation) and to perform separations difficult to achieve by other methods such as distillation (e.g., in the production of anhydrous bioethanol). Membrane filtration (micro- and ultrafiltration) and electrodialysis are playing an important role in the development of biorefinery processes. These types of membrane based separations are further discussed.

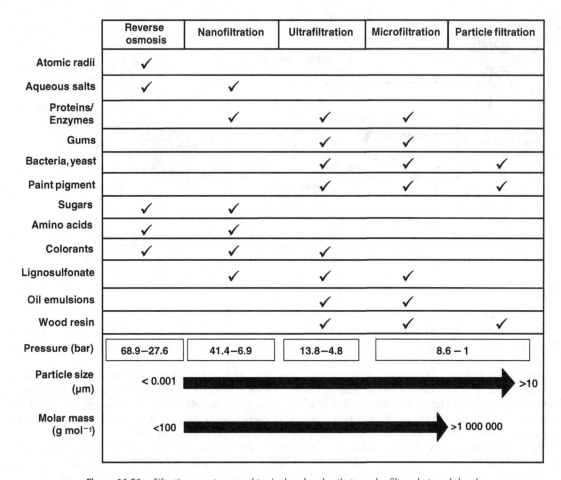

	Reverse osmosis	Nanofiltration	Ultrafiltration	Microfiltration	Particle filtration
Atomic radii	✓				
Aqueous salts	✓	✓			
Proteins/ Enzymes		✓	✓	✓	
Gums			✓	✓	
Bacteria, yeast			✓	✓	✓
Paint pigment			✓	✓	✓
Sugars	✓	✓			
Amino acids	✓	✓			
Colorants	✓	✓	✓		
Lignosulfonate		✓	✓	✓	
Oil emulsions			✓	✓	
Wood resin			✓	✓	✓
Pressure (bar)	68.9–27.6	41.4–6.9	13.8–4.8	8.6 – 1	
Particle size (μm)	< 0.001				>10
Molar mass (g mol⁻¹)	<100			>1 000 000	

Figure 11.21   *Filtration spectrum and typical molecules that can be filtered at each level.*

11.4.2   Membrane Filtration

The various levels of filtration required to separate particles or molecules form a filtration spectrum, shown in Figure 11.21. From right to left, the spectrum goes from separation of coarse particles to separation of atomic particles. Filtration of coarse particles is most easily achieved by using screens to separate, for example, sand, wheat flour, activated carbon, etc. The next level is microfiltration, which can be used to separate suspended particles and can be used for clarification or pretreatment. Moving further, the ultrafiltration level is reached at which macromolecules like proteins, enzymes and fat are rejected from the mixture or solution. Micro- and ultrafiltration are used in continuous fermentation processes with cell recycle. The lactate salt product and components of a similar molecular size pass through the membrane along with water. The yeast or bacterial cells and other big components are retained and recycled back to the fermentation tank. The next levels, nanofiltration and reverse osmosis, are used for more selective separations. Basically only small molecules (e.g., water) or atoms are allowed to pass through the membrane. Nanofiltration and reverse osmosis help to concentrate components such as salts, sugars, organic acids, etc. For example, reverse osmosis could be used as a concentrating step for the production of lactic acid. As the spectrum goes from right to left, the pore size of the membranes used for filtration is reduced. Figure 11.21 gives a sense of the particle size and molar mass of the molecules that can be separated along with the required operating pressures at each level of the spectrum.

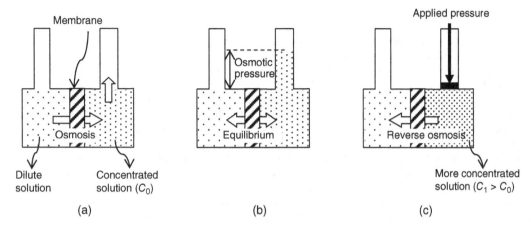

Figure 11.22 (a) Osmosis, (b) equilibrium and (c) reverse osmosis.

In membrane filtration processes, the solutes are diffused through the membrane under the influence of a pressure gradient. A pressure difference is generated across the membrane (ΔP). The required pressure difference depends on the membrane application and the size of the molecules to be separated. Figure 11.21 shows increases in the operating pressure range with decreasing size of the particles (e.g., <10 bar for microfiltration and >30 bar in reverse osmosis).

Pressure difference is also needed between the osmotic pressure of the feed solution and that of the permeate solution ($\Delta \pi$). *Osmosis* is the natural tendency of water to migrate from a dilute solution to a concentrated solution through the membrane separating the solutions, as shown in Figure 11.22(a). Water flows, until equilibrium is reached. At this state the flow in both directions is equal, as shown in Figure 11.22(b). The pressure difference generated between the two sides of the membrane is the *osmotic pressure* π. The osmotic pressure (in bar) can be approximated for dilute solutions by the van't Hoff correlation, shown by

$$\pi = nRT\frac{N_s}{V} \tag{11.58}$$

where n is the number of ions formed when the solute molecule dissociates, for example, $n = 2$ for NaCl, R is the universal gas constant, 0.083145 bar m^3 kmol^{-1} K^{-1}, T is the absolute temperature, K, N_s is the number of moles of solute, kmol, and V is the volume of pure solvent, m^3.

An operating pressure higher than the osmotic pressure is required, because osmotic pressure acts in the opposite direction to the pressure applied for a desired separation. Therefore, the factor driving the process is the difference between ΔP and $\Delta \pi$. From Equation (11.54), the general expression for the flux through a filtration membrane is

$$J_i = \frac{P_{M,i}}{\delta_M} \times (\Delta P - \Delta \pi) \tag{11.59}$$

According to Equation (11.59), the higher this pressure difference, the higher is the flux of the component i. In general, the micro- and ultrafiltration processes are not affected by osmotic pressure and the driving force is just ΔP. The osmotic pressure is especially important in reverse osmosis applications.

As a result of the osmosis, the solvent mixes with the concentrated solution in the opposite side of the membrane. However, this mixing is not helpful for the purpose of further concentrating the solute. In Figure 11.22(c), *reverse osmosis* is produced by applying a pressure higher than the osmotic pressure to the concentrated solution. Now the solvent is forced to go from the concentrated solution to the dilute solution through the semipermeable membrane. Because molecules of solute are, in general, larger than water molecules, the solute does not pass through the membrane. Since the pressure applied is higher than the osmotic pressure, the number of solvent molecules transferred to the dilute solution is higher than the number of solvent molecules that crossed the membrane by natural osmosis. This produces an increased solute concentration in the concentrated solution.

In reverse osmosis, dense membranes that only allow the solvent to pass are used. This creates concentration polarization on the concentrated side of the membrane and can limit the permeation process. Concentration polarization refers to the difference between the concentration in the liquid adjacent to the surface of the membrane and the concentration in the bulk liquid. The concentration of the solute in the liquid adjacent to the membrane surface is higher than that in the bulk due to the loss of solvent that passes to the other side of the membrane. Concentration polarization is measured as the ratio between the concentration at the membrane surface and that in the bulk. Concentration polarization has the effect of increasing the osmotic pressure due to the increase in concentration on the membrane surface. This means that, for a constant applied pressure ΔP, the overall driving force ($\Delta P - \Delta \pi$) decreases and the flux of the solvent also decreases. As a result, a higher operating pressure or higher membrane area is required. Energy is required for the reverse osmosis process. This energy will depend on the overall driving force required to achieve a desired separation.

Exercise 9. Concentration of lactic acid by reverse osmosis. A stream of lactic acid in aqueous solution needs to be concentrated by reverse osmosis (RO). The following data are available for an RO process using a phenolic polyamide membrane:

Feed concentration $= 0.372$ kmol m^{-3}
Pressure difference $\Delta P = 30$ bar
Permeance $\bar{P}_M = 0.001$ m^3 bar^{-1} m^{-2} h^{-1}
Cut fraction $\theta = 0.67$
Operating temperature $= 25\ °C$ (298 K)

Determine the membrane area required for 10.5 m^3 h^{-1} of a pretreated lactic acid stream to achieve a final concentration of 1.11 kmol m^{-3} (100 g L^{-1}). Assume that Equation (11.59) is valid to determine the osmotic pressure and neglect the effect of concentration polarization.

Solution to Exercise 9. First, the concentration of permeate is found from mass balances to determine the osmotic pressure. A balance for the volumetric flow is given by the following expression:

$$Q_F = Q_P + Q_R$$

As the cut fraction and feed flow rate are given, it is possible to calculate the permeate flow rate from Equation (11.57), as follows:

$$Q_P = Q_F \times \theta = 10.5 \times 0.67 = 7.0\ \text{m}^3\ \text{h}^{-1}$$

Then, the volumetric flow rate of retentate is as follows:

$$Q_R = Q_F - Q_P = 10.5 - 7.0 = 3.5\ \text{m}^3\ \text{h}^{-1}$$

The mass balance for lactic acid thus has the following form:

$$Q_F C_F = Q_P C_P + Q_R C_R$$

The concentration of permeate can be calculated from the concentration of lactic acid required in the retentate, shown as follows:

$$C_P = \frac{Q_F C_F - Q_R C_R}{Q_P} = \frac{10.5 \times 0.372 - 3.5 \times 1.11}{7.0} = 0.003\ \text{kmol m}^{-3}$$

Now the osmotic pressure of the feed and permeate solutions can be determined. Assuming that the lactic acid molecule dissociates into one lactate anion and one hydrogen cation, $n = 2$.

For the feed solution (using Equation (11.58)), the osmotic pressure is

$$\pi = 2 \times 0.083145 \times 298 \times 0.372 = 18.43 \text{ bar}$$

For the permeate solution, the osmotic pressure is

$$\pi = 2 \times 0.083145 \times 298 \times 0.003 = 0.149 \text{ bar}$$

The osmotic pressure difference is calculated as follows:

$$\Delta\pi = 18.43 - 0.149 = 18.28 \text{ bar}$$

Then, the permeate flux can be calculated from Equation (11.59) using the given permeance value ($\bar{P}_M = P_M/\delta_M$):

$$J = 0.001 \times (30 - 18.28) = 0.012 \text{ m}^3 \text{ h}^{-1} \text{ m}^{-2}$$

Finally, the area can be calculated from a relation similar to Equation (11.55), shown as follows:

$$\frac{Q_P}{A_M} = \bar{P}_M \times (\Delta P - \Delta\pi) = J$$

The area required for the given specifications is

$$A_M = \frac{Q_P}{J} = \frac{7.0}{0.012} = 583 \text{ m}^2$$

As with lactic acid, membrane separation processes can be used for recovery and purification of succinic acid. The filtrate is generally concentrated under vacuum and then crystallized. However, membrane technology can be more expensive than the other separation technologies as a membrane needs to be replaced due to degradation or fouling.

Succinic acid fermentation broth is treated by microfiltration, ultrafiltration and nanofiltration[19]. Subsequent concentration and crystallization gives a purity of succinic acid greater than 99.4%. Other techniques for succinic acid recovery include electrodialysis, which achieves a lower recovery yield (60%) due to succinic acid losses. The equipment and operating costs for electrodialysis can be higher than other processes.

11.4.3 Electrodialysis

Dialysis refers to the process of separating molecules or ions based on the difference in size, as shown in Figure 11.23(a). It is mainly used to separate solutes with a large difference in their rates of diffusion through a semipermeable membrane. The dialysis process is driven by a concentration gradient across the membrane. The flux in this process is in general lower than in other membrane processes. To enhance the separation process, the dialysis can be performed under the influence of an electrical field. This process is called *electrodialysis* and is useful to separate charged particles such as ions. Electrodialysis separates an aqueous solution into a dilute solution (called the *diluate*) and a concentrated solution (called the *concentrate*).

Electrodialysis is based on the principle that positively (cations) or negatively (anions) charged particles migrate towards the electrodes with opposite charges, cathode and anode, respectively. A simple electrodialysis cell formed by two membranes and two electrodes is shown in Figure 11.23(b). When an electrical current is applied between the electrodes, the anions migrate towards the positively charged electrode (anode). The anion-selective membrane allows only anions to pass. The cations move in the opposite direction towards the negatively charged electrode (cathode). The cation-selective membrane allows only cations to pass.

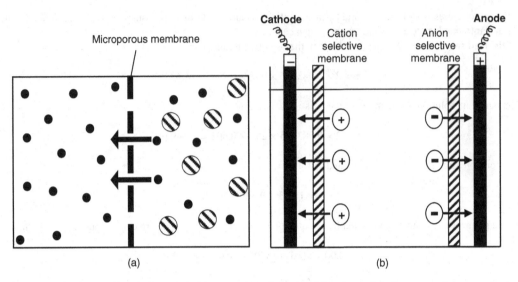

Figure 11.23   *Schematics of (a) dialysis and (b) electrodialysis.*

In practical applications, multicompartment electrodialysis stacks are used. As shown in Figure 11.24(a), feed enters the various compartments formed by the anion- and cation-selective membranes in alternate arrangements. The anions from a compartment move towards the anode through the anion-selective membrane and enter the adjacent compartment. However, they are not able to pass further by the next membrane, which is cation-selective. As a result, the anions are trapped between the two membranes of the compartment. A similar process occurs for the cations, which move in the opposite direction and are also trapped in an adjacent compartment. Cations neutralize any excess charge due to accumulation of anions. As a result, the solution in the original compartment becomes diluted while the solution in the adjacent compartment becomes concentrated. By using this process a highly concentrated solution of the salt (e.g., lactate) can be obtained in the concentrate stream.

Another possible system is the electrodialysis with bipolar membranes (EDBM), as shown in Figure 11.24(b). This system combines the conventional electrodialysis with EDBM for the conversion of a salt into its corresponding acid and base. Bipolar membranes induce the splitting of water into hydrogen protons (H^+) and hydroxide ions (OH^-). Because of this, the process is also called water-splitting electrodialysis. Bipolar membranes consist of a cation exchange (i.e., cation selective) membrane, an anion exchange (i.e., anion-selective) membrane and a catalytic intermediate layer to promote the splitting of the water. The anions are neutralized by the protons generated at the surface of the anion exchange membrane. The cations are neutralized by the hydroxide ions generated at the surface of the cation exchange membrane. By using this process an organic salt can be converted into the free acid form. The base used for pH adjustment may also be recovered.

EDBM requires a feed previously treated to remove large molecules. Clarification of the fermentation broth is first performed by microfiltration. An important issue for EDBM operation is the presence of multivalent cations such as Ca^{2+}, Mg^{2+} and Al^{3+}. These cations form insoluble hydroxides at the interface of the bipolar membrane where the ions separate. Thus, their concentrations must be reduced to about 1 ppm. Bipolar membranes require integration with other separation processes. The organic acid (e.g., lactic acid) obtained after EDBM can be further purified by ion exchange resins.

11.4.4   Ion Exchange

The removal of ionic impurities from a stream containing the main product (e.g., the fermentation broth) is necessary as a polishing step or as a pretreatment step to obtain good performance in downstream processes (e.g., reverse osmosis). This

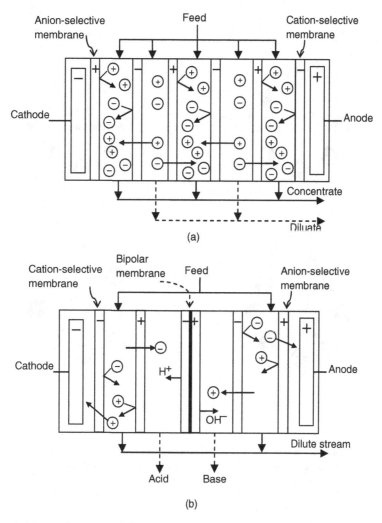

Figure 11.24 *(a) Conventional electrodialysis and (b) electrodialysis with biopolar membrane.*

purpose is generally achieved by *ion exchange* between a solution and an exchanger complex. Typical *ion exchangers* are resins (functionalized porous or gel polymer), zeolites, clay, etc. Ion exchangers can be selective for certain ions or classes of ions, depending on their chemical structure and electrical charge. *Cation exchangers* are able to exchange positively charged ions whereas *anion exchangers* are able to exchange negatively charged ions.

Ion exchange is a form of sorption driven by electrostatic interactions between the exchanger and ionic species to be separated. Figure 11.25 shows the ion exchange in a cation exchanger material and an anion exchange material. The exchanger complex is formed by a base structure with a fixed electrostatic charge. This charge is negative for cation exchangers and positive for anion exchangers. The exchanger complex base is neutralized with opposite ions from the solution to be treated. When the exchanger particles are flushed with the ionic solution, ion exchange occurs on the surface of the particles. For example, if a solution containing Ca^{2+} ions flushes a bed of cation exchanger particles neutralized with H^+ ions, the Ca^{2+} ions replace the H^+ ions. In this way, the Ca^{2+} ions are bound to the exchanger base while the H^+ ions go into the solution. The anions will not attach to the exchanger base as this also has a negative charge. Another

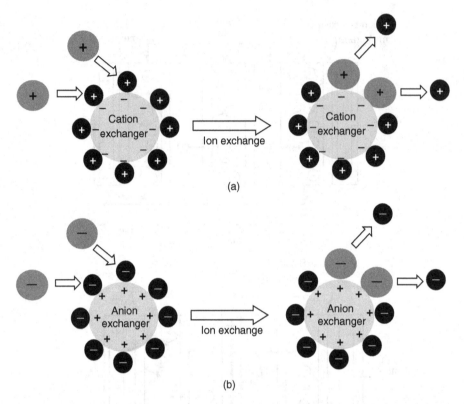

Figure 11.25   *Ion exchange in (a) a cation exchanger and (b) an anion exchanger.*

example is the demineralization of water, where calcium and magnesium ions in water are exchanged by hydrogen or sodium ions initially bound to the exchanger particles. Anions like chloride and sulfate stay in the solution.

For the exchange to be effective, the affinity of the exchanger particles for the ions in the solution to be treated must be greater than the affinity of the ions originally bound to the exchanger particles. The affinity depends on the size and electrical charge of the ions. Larger ions with a higher charge are more easily bound, as shown in the following list for sulfonic acid resins (a cation exchanger):

$$Fe^{3+} > Al^{3+} > Pb^{2+} > Sr^{2+} > Ca^{2+} > Co^{2+} > Ni^{2+} > Cu^{2+} > Zn^{2+} > Mg^{2+} > Mn^{2+} > Ag^+ > Cs^+ > Cd^{2+}$$

$$> K^+ \approx NH^{4+} > Na^+ > H^+ > Li^+$$

For exchangers based on quaternary amines, the order of affinity for anions is

$$Citrate > PO_4^{3-} > CrO_4^{2-} > SO_4^{2-} > HSO_4^- > NO_3^- > Br^- > Cl^- > HCO_3^- > CH3COO^- > OH^- \approx F^-$$

Ion exchange is a reversible process. Consider a resin R with ion X bond and with greater affinity for ion Z. If a solution containing the ion Z is put in contact with the resin, ion exchange is produced according to

$$RX + Z^{\pm} \rightleftarrows RZ + X^{\pm} \tag{11.60}$$

where R is an anion or cation exchanger depending on the charge of X and Z (+ or −). After a certain period of time, the resin will be mostly loaded with ion Z. The ion exchange may continue until equilibrium is reached. An equilibrium quotient Q can be expressed from the concentrations of the species (represented in square brackets), as shown by

$$Q = \frac{[RZ]\,[X^{\pm}]}{[RX]\,[Z^{\pm}]} \tag{11.61}$$

where Q is a constant specific for the pair of ions and type of resin involved in the process. Once equilibrium is reached, the resin becomes saturated and needs to be cleaned and regenerated. If a concentrated solution containing ion X is now passed through the saturated resin particles, the resin will regenerate into the RX form, ready for reuse. The ion Z is washed out by the cleaning or regenerating fluid. The equilibrium conditions decide the selection of an ion exchange resin. Ion exchangers are characterized by a retention capacity for a particular ion or solute (in g per g of ion exchanger) obtained from isothermal equilibrium studies. The pH and feed flow rate can affect the performance of the ion exchanger.

In an ion exchange process, the exchanger particles are loaded into a bed contained in a vessel. The whole set is known as an *ion exchange column*. An ion exchange module is generally made up of two ion exchange columns. One of those columns is used as an operational column, while the other column is in the cleaning and regeneration mode. Figure 11.26 shows the modes of operation of the ion exchange columns.

At start-up, only the process lines passing through the left column operate. When the bed of ion exchanger particles is saturated, the regenerated column starts operation for the main process. The saturated column is then cleaned by a fluid in reverse flow. This step is also known as backwash. More than one backwash passes may be needed. After cleaning,

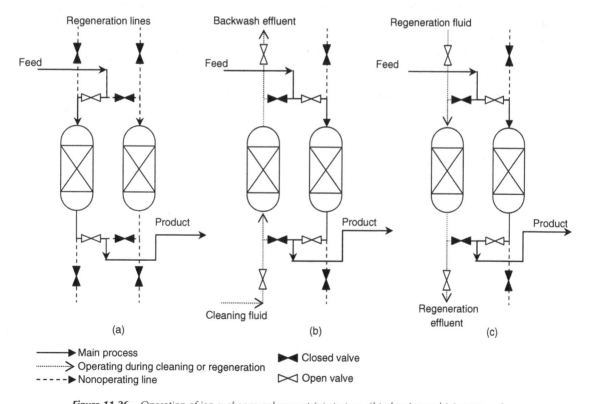

Figure 11.26 *Operation of ion exchange columns at (a) start-up, (b) cleaning and (c) regeneration.*

Table 11.14 *Examples of commercial ion exchange resins for lactate retention.*

Resin	Functional Group	Type
Amberlite IRA 900	Quaternary ammonium	Strongly basic
Amberlite IRA 400	Quaternary ammonium	Strongly basic
Amberlite IRA 96	Polyamine	Weakly basic
Amberlite IRA 67	Polyamine	Weakly basic

the particles are reactivated or regenerated by flushing a chemical containing the original ion attached to the particles. Another cleaning step may be required before the column is ready to be used in another cycle.

Ion exchange is widely used for softening water for steam generation and in wastewater treatment processes. Ion exchange resins offer more process flexibility than zeolites and are mainly based on cross-linked polystyrene molecules. Resins have functional groups that confer different degrees of acidity and basicity. Thus, resins can be (a) strongly acidic (sulfonic acid groups); (b) strongly basic (quaternary amino groups); (c) weakly acidic (carboxylic acid groups) or weakly basic (amino groups, e.g., polyethylene amine). Table 11.14 shows examples of commercial anion exchange resins for lactate solutions. These resins can be used for the recovery of the lactate ions that can be converted into lactic acid form during the regeneration step by passing an acid solution (e.g., HCl).

Ion exchange resins can be effective in biorefinery processes such as biodiesel filtration and the removal of inorganic salts from fermentation broths. Ion exchange is also used to convert the lactate salt into lactic acid. The separation of by-product organic salts is possible by selecting the proper ion exchange resin.

The fermentation broth contains organic and inorganic cations and anions. Table 11.15 shows an example of the ionic composition of a fermentation broth after being clarified by ultrafiltration. A two-stage ion exchange process can be used to remove the undesired ions. In the first stage, cation exchange columns separate the cations. In a second stage, anion exchange columns are used to separate the anions. A mixed bed of both cation and anion exchangers in the same ion exchange column is also possible.

Anion exchange resins can be used to recover succinic acid from fermentation broth. The succinic acid solution is passed through ion exchange columns. The succinate is adsorbed in the resin and converted into the acid form during regeneration of the ion exchange resin. Regeneration of the resin involves water washing to remove unbounded material followed by washing with an acid solution to recover succinic acid. The process is selective towards succinic acid and allows separation from other organic acids (e.g., acetic acid and lactic acid). The succinic acid is purified by crystallization to produce succinic acid with purity higher than 99%.

11.4.5 Integrated Processes

Integration between various separation technologies helps to overcome major barriers in downstream processing of fermentation and other biorefinery products. Advances in electrodialysis (ED) have allowed the development of an

Table 11.15 *Composition of a fermentation broth from lactic acid production.*

Species	Concentration g L^{-1}
Sodium	10
Potassium	2
Magnesium	0.2
Calcium	0.5
Lactate	50
Chloride	3
Phosphate	2

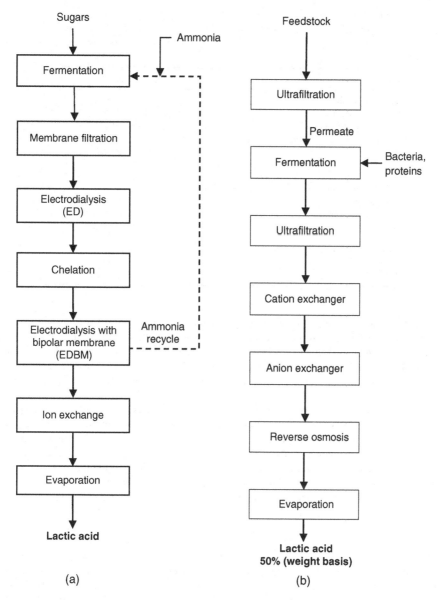

Figure 11.27 *Block diagrams for (a) the double ED process and (b) a process using membrane and ion-exchange processes for the production of lactic acid.*

integrated separation process called the *double ED* process. The process combines ED followed by EDBM (also called water-splitting ED) as shown in Figure 11.27(a) for lactic acid production. The fermentation broth is first clarified by microfiltration to remove bacterial cells, proteins, etc. Ultrafiltration can also be used to allow cell recycling. Then the ED step removes the inorganic salts.

Despite this, an intermediate chelation step is required for efficient operation of EDBM. Chelation refers to the separation of ions by forming coordinated complex arrangements with organic compounds wherein the separated ions are the central atoms. After chelation, the EDBM step allows the conversion of lactate into lactic acid and its concentration. Then,

Table 11.16 *Process specifications for lactic acid production from starch using electrodialysis and ion-exchange units in downstream processing. (Reproduced with permission from Datta et al. (1995)[20]. Copyright © 2006, John Wiley and Sons, Ltd.)*

Fermentation Yield	0.95 g Lactate g^{-1} Starch
Electrodialysis	
Lactate recovery	95%
Desalting efficiency	90%
Rejection of divalent ions (e.g., Ca^{2+})	98–99%
Power consumption	0.33 kW h kg^{-1} lactate
Membrane life	1 year
EDBM (Electrodialysis with Bipolar Membrane)	
Lactate recovery as lactic acid	99%
Power consumption	0.55 kW h kg^{-1} lactic acid
Membrane life	2 years
Ion-exchange refining	
Lactic acid recovery	99%
Resin life	2 years

polishing is carried out in ion-exchange columns. Finally, evaporation is used for further concentration and purification to produce technical grade lactic acid. To produce high purity lactic acid, the esterification–hydrolysis strategy used in the conventional process can be used. In the process of Figure 11.27(a), ammonium salt is used for neutralization in the fermentation step. EDBM allows the recovery and recycling of this salt. Alternatively, the ammonium salt can be sold as a low cost fertilizer. Table 11.16 shows some parameters obtained from experimental and pilot plant trials by the process developers.

Figure 11.27(b) shows a process for production of food grade lactic acid using membrane based separations and ion exchange. The feedstock first passes to an ultrafiltration module. Permeate contains the substrate sugars (lactose mainly) fed to fermentation. The bacteria and proteins clarified by ultrafiltration are recycled to the fermentation. The clarified broth goes to a cation-exchange column followed by an anion-exchange column for removal of undesired salts. The conversion of lactate into lactic acid is achieved in the cation-exchange column. Most of the impurities are removed after the anion exchange column. But the lactic acid remains in a dilute aqueous solution. Reverse osmosis is then used as a concentrating step. Lactic acid solution is further concentrated to a 50% content by weight using evaporation. Parameters for the main process steps are shown in Table 11.17.

More advanced, integrated reaction–separation methods can be developed to alleviate the product inhibitory effects and the product recovery hurdles. These methods aim at the removal of desired product from the fermentation broth in situ to avoid product inhibition. At the same time, the downstream separation is simplified.

In situ separation combining succinic acid fermentation and ion exchange is more effective to achieve high purity and productivity with energy and capital savings. In terms of downstream processing, the low pH yeast technology is advantageous over bacteria based conversion technologies. The Reverdia process (using yeast) at low pH allows the direct conversion of the feedstock into succinic acid, with less by-product formation[21]. Thus, the additional chemical processing, equipment and energy required to convert intermediate salts into succinic acid are avoided. Figure 11.28 shows a process using bacteria based fermentation and a process using yeast based fermentation. Simpler unit operations and less downstream processing steps can be observed in yeast based fermentation.

Electrodialysis fermentation (EDF) is designed for in situ product removal. The system allows continuous removal of product by electrochemically controlling the pH of the fermentation mixture. The process eliminates the need for the addition of an alkali and could potentially simplify the downstream processing. However, EDF suffers from the issue of membrane fouling by microorganism cells. Some strategies to overcome fouling issues include immobilization of the microorganisms or using yeast. This may open the possibility of converting lignocellulosic sugars into a high value chemical.

Table 11.17 *Process specifications for lactic acid production from a dairy waste stream using membrane based separations and ion exchange.*

Feedstock ultrafiltration	
Cut fraction θ	0.2
Lactose recovery	20%
Fermentation	
Lactose conversion	>99%
Temperature	40 °C
pH	5.8
Batch time	2–3 days
Broth ultrafiltration	
Membrane	Ceramic membrane
Temperature	40 °C
Pressure	3 bar
Cut fraction θ	0.8
Rejection	
Microorganisms and proteins	100%
Lactate	1.5%
Cation exchange	
Selectivity for cations	>99%
Temperature	40 °C
Anion exchange	
Selectivity for anions	>99.5%
Temperature	40 °C
Reverse osmosis	
Cut fraction θ	0.67
Temperature	25 °C
Pressure	30 bar
Final concentration / initial concentration	3.0

11.4.6 Reactive Extraction

Extraction is a separation process based on the distribution of a substance in immiscible solvents. To perform the separation, the solution containing the substance of interest is put in contact with the extraction solvent. For the extraction to be effective, the solubility of the substance of interest must be higher in the extraction solvent than in the original solvent. In a liquid–liquid extraction process, the solvents involved are generally water and an organic compound. Thus, an aqueous phase and an organic phase are present in the process. Fermentation products such as lactic acid are difficult to extract from aqueous solutions due to their hydrophilic nature (i.e., affinity for water). However, an effective separation can be achieved by *reactive extraction*.

In reactive extraction, the compound of interest is extracted from the original solution by reaction with an *extractant* dissolved in a second solvent or *diluent*. Fermentation processes produce the substance of interest in an aqueous phase. The mixture containing the extractant dissolved in an organic diluent forms the organic phase. The extractant reacts with the compound of interest forming a reversible chemical complex. The chemical complex is immediately extracted into the organic phase. Ideally, such a complex is soluble in the organic phase but not in the aqueous phase. This allows an efficient separation. The extractant and diluent have low solubility in the aqueous phase.

The starting point for the development of a reactive extraction process is the selection of the extractant–diluent combination. The distribution coefficient gives an indication of the extent of separation that can be achieved by a certain solvent. The distribution coefficient K_d is defined as the ratio of the concentration of the solute in the organic phase

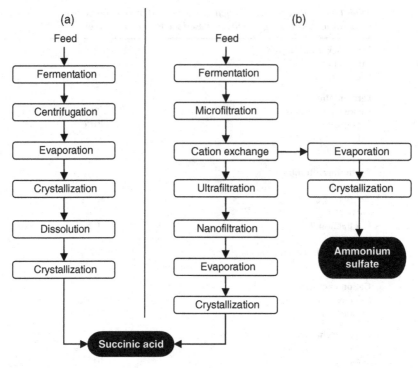

Figure 11.28   *Comparison between succinic acid production using (a) yeast based fermentation and (b) using bacteria based fermentation.*

(y) to the concentration of the solute in the aqueous phase (x), as shown in the following equation, where x and y are concentrations at equilibrium:

$$K_d = \frac{y}{x}$$

(11.62)

For an efficient and economical recovery, a high distribution coefficient is desirable. On the other hand, the distribution coefficient for the impurities (e.g., residual sugars and other organic acids) must be low. Thus, a high selectivity of desired solute is also required. The distribution coefficient varies with the extractant–diluent system. Table 11.18 shows the distribution coefficients of lactic acid in various extractant–diluent mixtures. For the same diluent, K_d is increased by increasing the composition of the extractant. For the methyl isobutyl ketone (MIBK) and alamine 336 system, a 10% increase in the composition of MIBK increases the value of K_d by 270%. However, after a certain value of extractant concentration, K_d remains constant. Formation of a third phase can occur and complicate the extraction process at high extractant concentrations; 30% alamine 336 in octanol provides the highest value of K_d among the systems listed in the table. However, toxicity is an important factor to consider when selecting the extractant–diluent mixture for extractive fermentation. The distribution coefficient is also known as the partition coefficient.

The reactive extraction of lactic acid from the fermentation broth is another promising method for in situ product removal. The process is also called *extractive fermentation*. Figure 11.29 shows a simplified block diagram of an extractive fermentation system. The fermentation broth is first filtered to remove the microorganism cells, which are recycled to the fermentor. The clarified broth then passes through an extraction unit. A stream containing the lactic acid, extraction reactant and the solvent is generated along with waste broth. The waste broth can also be recycled to the fermentor. Then, lactic acid is separated from the extraction reactant and solvent in a regeneration unit. In the regeneration

Table 11.18 *Distribution coefficients K_d for lactic acid in various extraction mixtures. (Reproduced with permission from Wasewar et al. (2004)[22]. Copyright © 2004, American Chemical Society.)*

Diluent	Extractant	Composition	K_d
Oleyl alcohol	Alamine 336	15% (volume basis)	3
Oleyl alcohol	Alamine 336	30% (volume basis)	4.5
Oleyl alcohol	Alamine 336	50% (volume basis)	6.5
MIBK	Alamine 336	20% (volume basis)	0.72
MIBK	Alamine 336	30% (volume basis)	2.68
MIBK	Alamine 336	40% (volume basis)	4.24
Decanol	Alamine 336	20% (volume basis)	12.57
Decanol	Alamine 336	30% (volume basis)	16.44
Decanol	Alamine 336	40% (volume basis)	23.37
Octanol	Alamine 336	10% (volume basis)	15.35
Octanol	Alamine 336	20% (volume basis)	19.69
Octanol	Alamine 336	30% (volume basis)	25.95
Oleyl alcohol	Alamine 336	50% (volume basis)	2.62
Oleyl alcohol	Di-*n*-octylamine	50% (volume basis)	11.1
Oleyl alcohol	Di-*n*-decylamine	50% (volume basis)	7.76
Oleyl alcohol	Tri-*n*-hexylamine	50% (volume basis)	2.94

step, the extractant and solvent are separated and recycled to the extraction unit. The extractant and its solvent must not be toxic to the microorganisms used in fermentation. Various extraction systems use amines dissolved in organic solvents. Alamine 336 in octanol or oleyl alcohol is the most efficient system as it gives a high distribution coefficient, low toxicity and scope for regeneration or back extraction. The concepts of extraction processes are shown as follows.

Equilibrium and kinetic models are used for the modeling of reactive extraction. Both physical liquid–liquid equilibrium and chemical equilibrium are relevant. Figure 11.30 shows a scheme for the reactive extraction of organic acids from aqueous solution. Physical liquid–liquid equilibrium of the nondissociated acid exists between the two phases ($HA_{(aq)}$ and $HA_{(org)}$). Another physical equilibrium exists for the dissociated ions between the aqueous phase ($H^+_{(aq)}$, $A^-_{(aq)}$) and the interface ($H^+_{(int)}$, $A^-_{(int)}$). The extractant B forms a complex with the acid called *B-HA*; they also form an equilibrium between the organic phase ($B_{(org)}$, *B-HA*$_{(org)}$) and the interfacial region ($B_{(int)}$, *B-HA*$_{(int)}$). Chemical equilibrium exists for the dissociation of the acid in the aqueous phase and for the reaction between the acid and extractant at the interface.

Models are developed for a specific system considering the characteristics of the species involved. The models can be simplified to describe only the phenomena that limit the mass transfer. Apart from the reactions in Figure 11.30, other possible reactions are included in the modeling: for example, the dimerization of acid in the organic phase ($2HA_{(org)} \leftrightarrows (HA)_{2\,(org)}$) and the direct reaction of the extractant with the nondissociated acid ($HA_{(int)} + B_{(int)} \leftrightarrows$ *B-HA*$_{(int)}$). The reactions governing the transfer of the solute from the aqueous phase to the organic phase depends on the extraction system used. Consider a simple system using only a hydrocarbon solvent (e.g., hexane) for extraction. Due to the nonpolar

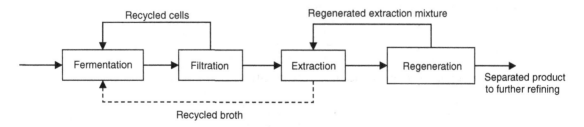

Figure 11.29 *Block diagram of extractive fermentation.*

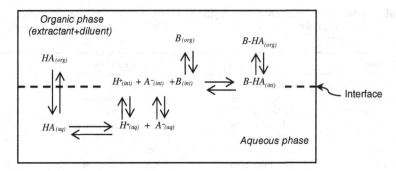

Figure 11.30   *Equilibrium and reaction scheme for the reactive extraction of an organic acid (HA).*

nature of hydrocarbons, it can be assumed that there is no interaction between the solvent and water. The following three basic equations can be used to describe the equilibrium of the solute between the aqueous and organic phases[23]:

1. Ionization of the acid in water (chemical equilibrium), shown by

$$HA_{(aq)} \leftrightarrows H^+_{(aq)} + A^-_{(aq)} \tag{11.63}$$

for which the equilibrium constant corresponds to the ionization constant of the acid K_a shown below, where [i] denotes the concentration of species i:

$$K_a = \frac{\left[H^+\right]_{aq} [A^-]_{aq}}{[HA]_{aq}} \tag{11.64}$$

2. Physical equilibrium of the solute between the two phases, shown by

$$HA_{(aq)} \leftrightarrows HA_{(org)} \tag{11.65}$$

The equilibrium constant in the following equation corresponds to the distribution coefficient in Equation (11.62):

$$K_d = \frac{[HA]_{org}}{[HA]_{aq}} \tag{11.66}$$

3. The dimerization of the acid in the organic phase is shown by

$$2HA_{(org)} \leftrightarrows (HA)_{2(org)} \tag{11.67}$$

The equilibrium constant of dimerization reactions is expressed as

$$K_{dim} = \frac{\left[(HA)_2\right]_{org}}{[HA]^2_{org}} \tag{11.68}$$

An overall distribution coefficient can be formulated accounting for all the species in which the acid is involved. With $C_{HA(org)}$ and $C_{HA(aq)}$ being the total concentrations in the organic and the aqueous phases, respectively, the overall distribution coefficient is given by

$$K_{d,overall} = \frac{C_{HA(org)}}{C_{HA(aq)}} = \frac{[HA]_{org} + 2\left[(HA)_2\right]_{org}}{[HA]_{aq} + [A^-]_{aq}} \tag{11.69}$$

Using expressions for $[A^-]_{aq}$, $[HA]_{org}$ and $\left[(HA)_2\right]_{org}$ from Equations (11.64), (11.66) and (11.68) and substituting in Equation (11.69) gives

$$K_{d,overall} = \frac{[HA]_{aq}K_d + 2K_{dim}[HA]_{org}^2}{[HA]_{aq} + K_a[HA]_{aq}/[H^+]_{aq}}$$

$$K_{d,overall} = \frac{[HA]_{aq}K_d + 2K_{dim}K_d^2[HA]_{aq}^2}{[HA]_{aq}\left(1 + K_a/[H^+]_{aq}\right)}$$

$$K_{d,overall} = \frac{K_d + 2K_{dim}K_d^2[HA]_{aq}}{1 + K_a/[H^+]_{aq}} \tag{11.70}$$

The overall coefficient is determined by experiment. In the above expressions concentrations are in mol L^{-1} or kmol m^{-3}. For nonideal system thermodynamic modeling, the activity of species, shown by [Species name], must be determined (cf. UNIQUAC activity model in Chapter 18) and used in the equations. Mutual miscibility between solvents may also need to be considered.

For reactive extraction, the chemical equilibrium between the extractant and the solute must be considered. For the case of aliphatic amine extractants, an acid–base type of reaction occurs between the extractant B and the acid HA to form the complex BHA. This complex, formed according to the following reaction[23], can be regarded as an ammonium salt of the extracted acid:

$$HA_{(aq)} + B_{(org)} \leftrightarrows BHA_{(org)} \tag{11.71}$$

An equilibrium constant for this reaction can be written as

$$K_c = \frac{[BHA]_{org}}{[B]_{org}[HA]_{aq}} \tag{11.72}$$

However, the simple stoichiometric reaction may not represent the process. The formation of a chemical complex occurs with n molecules of solute[22,23]. This reaction can be represented as

$$nHA_{(aq)} + B_{(org)} \leftrightarrows B(HA)_{n_{(org)}} \tag{11.73}$$

The equilibrium constant for this reaction is then expressed as

$$K_{c,n} = \frac{\left[B(HA)_n\right]_{org}}{[B]_{org}[HA]_{aq}^n} \tag{11.74}$$

Exercise 10. Overall distribution coefficient and effect of pH on reactive extraction of lactic acid. The following data for the reactive extraction of lactic acid using the system Alamine 336 ($C_{24}H_{51}N$) in MIBK has been reported at 25 °C[24,25]:

Concentration of Alamine 336 = 0.29 kmol m^{-3}
Molar mass of lactic acid = 90.08
$K_{c,1}$ = 10.15 m^3 kmol^{-1}
K_d = 0.11
K_{dim} = 0.56 m^3 kmol^{-1}
$pK_a = -\log K_a = 3.86$

a. Determine the overall distribution coefficient of a solution containing 20 g L^{-1} of lactic acid. For such a dilute solution, consider that only one molecule of lactic acid forms a complex with one molecule of Alamine 336.
b. Plot the effects of pH on the overall distribution coefficient, assuming that the expression is valid for a pH ranging from 2 to 6.

Solution to Exercise 10

a. An expression for the overall distribution coefficient needs to be developed. As $n = 1$, the reaction in Equation (11.73) reduces to the reaction in Equation (11.71). The equilibrium constant is as given in Equation (11.72). The additional species to be accounted is $[BHA]_{org}$. Considering this, an expression for $K_{d,overall}$ is developed as follows:

$$K_{d,overall} = \frac{C_{HA(org)}}{C_{HA(aq)}} = \frac{[HA]_{org} + 2\left[(HA)_2\right]_{org} + [BHA]_{org}}{[HA]_{aq} + [A^-]_{aq}}$$

From Equation (11.72):

$$[BHA]_{org} = K_{c,1} [B]_{org} [HA]_{aq}$$

Substituting this expression and the expressions for concentrations of the other species (as in Equation (11.70)):

$$K_{d,overall} = \frac{[HA]_{aq} K_d + 2K_{dim}K_d^2[HA]_{aq}^2 + K_{c,1} [B]_{org} [HA]_{aq}}{[HA]_{aq} \left(1 + K_a / [H^+]_{aq}\right)}$$

$$K_{d,overall} = \frac{K_d + 2K_{dim}K_d^2[HA]_{aq} + K_{c,1} [B]_{org}}{1 + K_a / [H^+]_{aq}}$$

From mass balance for the extractant B in the organic phase:

$$[B]_{org} = C_{B(org)} - [BHA]_{org}$$

where $C_{B(org)}$ is the total concentration of the extractant (free form + with lactic acid complex). Substituting the expression for $[BHA]_{org}$:

$$[B]_{org} = C_{B(org)} - K_{c,1}[B]_{org}[HA]_{aq}$$

Solving for $[B]_{org}$:

$$[B]_{org} = \frac{C_{B(org)}}{1 + K_{c,1}[HA]_{aq}}$$

Substituting $[B]_{org}$ in the expression for $K_{d,overall}$:

$$K_{d,overall} = \frac{K_d + 2K_{dim}K_d^2[HA]_{aq} + K_{c,1}\left(\frac{C_{B(org)}}{1+K_{c,1}[HA]_{aq}}\right)}{1 + Ka/[H^+]_{aq}}$$

The concentrations of H^+ and HA in the aqueous phase can be found using the K_a of the acid according to the following mass balance table, where x is the dissociated concentration and $[HA]_{0,aq}$ is the initial concentration of lactic acid in the aqueous solution:

Stage	$HA_{(aq)}$	$\leftrightarrows$	$H^+_{(aq)}$	$+$	$A^-_{(aq)}$
Initial	$[HA]_{0,aq}$				
Reaction	$-x$		x		x
Equilibrium	$[HA]_{0,aq} - x$		x		x

Substituting concentrations at equilibrium in Equation (11.64):

$$K_a = \frac{x^2}{[HA]_{0,aq} - x}$$

Rearranging:

$$x^2 + K_a x - K_a[HA]_{0,aq} = 0$$

Solving this equation for the positive and real value of x:

$$x = \frac{-b + \sqrt{b^2 - 4ac}}{2a}$$

where $a = 1$, $b = K_a$ and $c = -K_a[HA]_{0,aq}$.
   The initial concentration in the aqueous solution is given as

$$[HA]_{0,aq} = \frac{20}{90.08} = 0.22 \text{ kmol m}^{-3}$$

The value of the dissociation constant is found from the value and definition of pK_a (provided in the problem statement), as

$$K_a = 10^{-pK_a} = 10^{-3.86} = 0.00014 \text{ kmol m}^{-3}$$

Thus, the value of $[HA]_{aq} = [H^+]_{aq}$ at equilibrium is:

$$x = \frac{-0.00014 + \sqrt{0.00014^2 - 4(-0.00014 \times 0.22)}}{2} = 0.0054 \text{ kmol m}^{-3}$$

Now, substituting all the given values, the overall distribution coefficient can be calculated as

$$K_{d,overall} = \frac{0.11 + 2 \times 0.56 \times (0.11)^2 \times 0.0054 + 10.15 \left(\dfrac{0.29}{1 + 10.15 \times 0.0054} \right)}{1 + 0.00014/0.0054}$$

$$K_{d,overall} = 2.8$$

b. Assuming that the expression developed for $K_{d,overall}$ is valid for the range of pH given, the overall distribution coefficient as a function of pH is shown in Figure 11.31. It can be observed that the extraction is enhanced at low pH values. In general, pH values lower than the pK_a favors the extraction. A maximum value for $K_{d,overall}$ can be achieved at a pH of about 2.5. However, the practicality of operation at this very low pH must be evaluated because pH affects the performance of the fermentation. The optimum value for the fermentation process to produce lactic acid is between 5.0 and 5.5. Thus, an optimal reaction–separation system must be determined. Extractant–diluent systems that could achieve a high distribution coefficient at pH values closer to the operating pH of fermentation are desirable. Alternatively, engineered microorganisms that tolerate low pH values could facilitate highly efficient extractant–diluent systems.

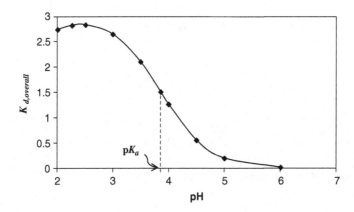

Figure 11.31   *Effect of pH on the overall distribution factor using an equilibrium model for a diluted solution.*

Figure 11.32 shows a simple black box scheme for a one-stage extraction unit. The flow rate of the solvents (water or diluent) are used rather than total stream flow rates. Mass fractions rather than molar concentrations are used and are specified on a solvent-only basis. These conventions enable simpler mathematical manipulation of mass balances. The

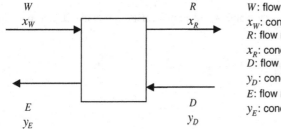

W: flow rate of aqueous phase feed (water only)
x_W: concentration in aqueous feed, kg solute kg^{-1} water
R: flow rate of raffinate (water only)
x_R: concentration in raffinate, kg solute kg^{-1} water
D: flow rate of organic phase feed, (diluent only)
y_D: concentration of solute in organic feed, kg solute kg^{-1} diluent
E: flow rate of extract (diluent only)
y_E: concentration of solute in the extract, kg solute kg^{-1} diluent

Figure 11.32   *Black box scheme for a single-stage reactive extraction process.*

depleted aqueous stream is called the *raffinate* (R) while the enriched organic stream is called the *extract* (E). Using the nomenclature in Figure 11.32, the component mass balance for the solute is expressed as

$$x_w W + y_D D = x_R R + y_E E \tag{11.75}$$

Assuming that the organic feed does not contain any solute dissolved, $y_D = 0$. If the mutual miscibility of the solvents can be neglected, the output flow rates of the solvents are equal to their inlet flow rates, that is, $R = W$, and $E = D$. After these assumptions, the mass balance is simplified to

$$\boxed{x_w W = x_R W + y_E D} \tag{11.76}$$

The concentration in the extractant y_E is the total amount of solute extracted from the aqueous phase and dissolved in the diluent (either as free solute or in the complex formed with the extractant). Using the overall distribution factor in terms of mass fractions, y_E can be found from

$$y_E = x_R K_{d,overall} \tag{11.77}$$

This expression is valid for the equilibrium model for one extraction stage only. As shown earlier, $K_{d,overall}$ varies with changes in concentrations, pH and other process conditions. In multistage extraction, the stages have different concentration values and therefore different values of $K_{d,overall}$. After this observation, the mass balance can be rewritten as

$$x_w W = x_R W + x_R K_{d,overall} D \tag{11.78}$$

Rearranging Equation (11.78) gives

$$x_w = x_R \left(1 + K_{d,overall} \frac{D}{W}\right) \tag{11.79}$$

The product $K_{d,overall} D/W$ is called the extraction factor. The ratio of solvent or diluent to treated feed D/W is a design specification that is subject to optimization. The composition of the solute in the raffinate can be calculated by rearranging Equation (11.79) into

$$x_R = \frac{x_W}{\left(1 + K_{d,overall} \frac{D}{W}\right)} \tag{11.80}$$

The recovery or extraction percentage of the solute can be determined from

$$e\,(\%) = 100 \times \left(1 - \frac{x_R}{x_W}\right) \tag{11.81}$$

In earlier modeling, it is assumed that the extractant is not dissolved in the aqueous phase and its total flow rate remains unchanged.

Exercise 11. Reactive extraction process for lactic acid recovery. Evaluate the recovery of lactic acid from a single-stage reactive extraction process using 1000 kg h^{-1} of a feed stream containing 2% lactic acid (total mass basis). The Alamine 336/MIBK system is used. The following data are available from Exercise 10:

Concentration of Alamine 336 $= 0.29$ kmol m^{-3} ≈ 0.128 kg Alamine 336 kg^{-1} MIBK
Temperature $= 25\,°C$
$K_{d,overall} = 2.8$

Use the results from the analysis to find a feasible recovery percentage in the extract and the corresponding flow rate of the diluent and extractant.

Solution to Exercise 11. To determine a feasible recovery percentage, the analysis must start by evaluating this parameter for various values of the diluent-to-feed ratio D/W. First, the flow of water and the lactic acid concentration in a solvent-only basis are determined.

The mass flow rate of lactic acid in the feed is: $1000 \times 0.02 = 20$ kg h^{-1}.
The mass flow rate of water is then: $1000 - 20 = 980$ kg h^{-1}.

The feed concentration of lactic acid on a water-only basis is then:

$$x_W = 20/980 = 0.0204 \text{ kg kg}^{-1}$$

Equation (11.80) is used to determine the outlet concentration in the raffinate x_R for a range of values of D/W. Then Equation (11.81) is used to determine the extraction percentage $e\,(\%)$, which is plotted for a range of D/W ratios in Figure 11.33(a). It can be observed that the gradient of the curve is high at low ratios. This means that for low ratios, a greater increase in the extraction percentage can be achieved from a small increment in D/W. However, the gradient of the curve is lower after an extraction percentage of about 85%. This implies that to increase the extraction percentage by one unit, the D/W ratio must be increased by several times. Thus, a higher flow rate of diluent is required, increasing the cost of the extraction process.

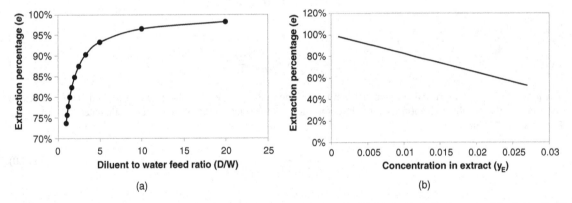

Figure 11.33   *Plot showing (a) the variation of extraction percentage with the diluent-to-water feed ratio and (b) the trade-off between recovery and concentration in the extract.*

Figure 11.33(b) shows the trade-off between the extraction percentage and the concentration of lactic acid in the extract. Higher extraction percentages require higher diluent flow rates. Therefore, the extracted lactic acid becomes more and more diluted. This has an important effect on the separation of the lactic acid in the regeneration unit. A more diluted solution may mean a more difficult separation and higher separation costs. An economic study is required to select an appropriate D/W. From this preliminary analysis, a single-stage extraction unit can be suggested to operate at an extraction percentage of 80% using a D/W ratio of 1.43. The corresponding concentration of lactic acid in the extract is 0.011 kg kg^{-1}. For this design, the following flow rates are calculated.

The diluent flow rate is: $980 \times 1.43 = 1401$ kg h^{-1}.
The amount of extractant required is: $1401 \times 0.128 = 179$ kg h^{-1}.
The amount of lactic acid recovered in the organic phase is: $20 \times 0.8 = 16$ kg h^{-1}.

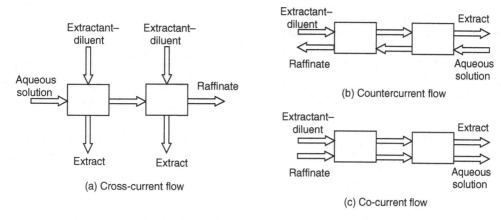

Figure 11.34   *Flow patterns in extraction processes.*

As shown in this exercise, a single-stage extraction unit may not be enough to achieve a high recovery of the solute. In practice, various stages with certain flow patterns are used to achieve a better performance. Figure 11.34 shows the various flow patterns in the extraction process. In batch extraction processes requiring few stages, cross-current flow is used. However, this option uses more solvent flow rates and achieves a lower extraction efficiency than the rest of the patterns. In large-scale processes, a counter-current or co-current flow pattern is used for efficient solvent utilization. To obtain this flow pattern, mixer-settlers or columns are used. *Mixer-settlers* are a series of mixers with intermediate settling stages. *Counter-current column contactors* are predominant in the chemical industry and can be static or agitated. *Centrifugal extractors* are also commercially used. This type of extractor requires a low residence time to separate the two liquid phases formed. The properties of the substances involved, the number of stages and the flow pattern are important factors for selection of extraction equipment. Table 11.19 shows some characteristics of the extraction equipment.

For the recovery of the solute, the equilibrium is shifted by changing the conditions of the extract solution. The shift is known as the *swing* process and can be carried out by a temperature swing, pH swing or concentration swing. In the temperature swing, the fact that the equilibrium is favored at low temperatures helps to break the chemical complex between the solute and the extractant. Thus, the extraction is carried out at a low temperature, whilst regeneration is carried out at a higher temperature. As shown in an earlier exercise, high pH values for extraction are not favored. This means that the equilibrium can be shifted to the left-hand reaction by increasing the pH during the solvent regeneration stage. A concentration swing is carried out by removing one of the components or by adding a second diluent. The solute is then separated due to a change in composition that destabilizes the solute–extractant complex. This method is also called a *diluent swing* or *back extraction*.

Due to the differences in optimum operating pH values between extraction and fermentation, external extraction of the acid is used. A typical flowsheet constitutes an integrated process of an extraction column followed by a solvent regeneration unit. Figure 11.35 shows a process producing high purity lactic acid. Reactive extraction and back extraction is used for the recovery of the lactic acid. The transesterification–hydrolysis strategy is used for the purification of the product as in the conventional lactic acid process.

Table 11.19   *Main characteristics of the extraction equipment.*

Characteristic	Mixer-Settler	Static Column	Agitated Column	Centrifugal Extractor
Number of stages	Low	Moderate	High	Low
Flow rate	High	Moderate	Moderate	Low
Residence time	Very high	Moderate	Moderate	Very low
Floor space	High	Low	Low	Moderate

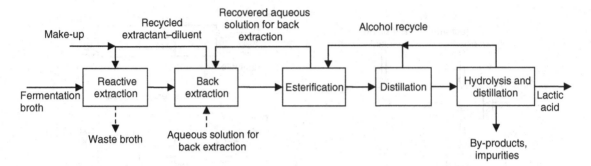

Figure 11.35   *A process using reactive extraction for recovery of lactic acid from fermentation broth.*

The analysis of reactive extraction processes discussed so far has considered equilibrium models. Although this type of model is useful to gain insights into the chemistry involved, their validity is constrained to conditions at which the process is equilibrium limited. For a detailed engineering design, mass transfer rate models combined with equilibrium models are used. Mass transfer rate models consider a kinetically limited system and also the hydrodynamics of the process. Mass transfer rate models are mathematically complex and require computer based simulations. However, these models allow optimizing, for example, the required solvent-to-feed ratio, integration of the number of extraction stages, etc. An example of the expression for the rate of mass transfer is that formulated for the extraction of lactic acid with Alamine 336/MIBK[25]. The system is characterized by a regime where extraction is accompanied by a fast chemical reaction in the interfacial diffusion film. The reaction rate equation is of zero order with respect to Alamine 336 and first order with respect to lactic acid, finally reducing to

$$r = C_{HA,(org)} \sqrt{D_{HA} k_1} \tag{11.82}$$

where $C_{HA,(org)}$ is the concentration of lactic acid at equilibrium in a diluent-only basis, kmol m^{-3}, D_{HA} is the diffusivity of lactic acid in the diluent $= 1.16 \times 10^{-9}$ m^2 s^{-1} and k_1 is the first order rate constant $= 1.38$ s^{-1}.

Reactive extraction is a promising technology for improving the overall process economics of biorefinery products from fermentation. In particular, fermentation coupled with reactive extraction can achieve a high product yield and removal of by-products. The developments of solvents that are less toxic to the fermentation microorganisms are required for successful extractive fermentation. Also, pH constraints must be overcome for a high separation efficiency.

In case of lactic acid, aliphatic amines have been found to be effective for separation of succinic acid. About 95% of succinic acid can be extracted from an aqueous solution (0.423 mol L^{-1} at pH $= 2.0$) using trihexylamine in 1-octanol[26]. Selective separation of succinic acid from organic acid by-products (e.g., acetic, formic, and lactic acids) is possible by controlling the pH of the extraction. Using tri-n-octylamine in 1-octanol, succinic acid remains in the raffinate while other organic acids are extracted. Vacuum distillation and crystallization can then be used to obtain pure succinic acid.

11.4.7   Reactive Distillation

Reactive distillation allows both reaction and distillation of a reaction mixture to be performed in a single column. This is useful for equilibrium-limited reactions, to break azeotropes between components and to reduce undesired side reactions. To achieve a highly efficient system, the operating conditions must favor both the reaction and the separation of components. For example, the product to be separated to shift the equilibrium must have higher relative volatility than other components. This means that the difference in boiling points must be high. In the case of azeotropes, the extraction process can be carried out within the same column. As an example consider the production of methyl acetate by the reaction:

Methanol (MeOH) + Acetic acid (HOAc) $\rightleftharpoons$ Methyl acetate (MeOAc) + water (H$_2$O)

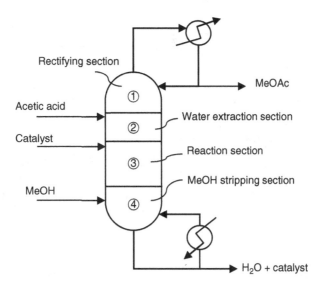

Figure 11.36 *Reactive distillation column for the production of methyl acetate.*

The components of the reaction mixture complicate their separation due to the possibility of the formation of azeotropes MeOAc/H$_2$O and MeOAc/MeOH and the near azeotrope HOAc/H$_2$O. The conventional flowsheet uses 8 to 10 distillation columns. However, by using reactive distillation the process can be simplified to the single multifunctional column shown in Figure 11.36. The various column sections perform the following functions:

1. Rectifying section. Separation of MeOAc from HOAc taking advantage of no azeotrope formation.
2. Extractive distillation section. Use HOAc feed as entrainer to break MeOAc / H$_2$O azeotrope
3. Reaction section. MeOAc is vaporized, removing heat of reaction and shifting equilibrium
4. Stripping section. Unreacted HOAc and MeOH are stripped out of the water byproduct

The advantages of reactive distillation are:

- Reaction equilibrium is shifted to higher (or even complete) conversion.
- Considerable reductions in plant costs.
- Exothermic reaction heat can drive separation and reduce operating costs.
- Reduction or elimination of additional separation system.
- Suppression of side reactions and enhanced selectivity.

Figure 11.37 shows that a process of separation of succinic acid comprises the simultaneous acidification and esterification of succinate salts[27]. Diethyl succinate rather than the free succinic acid is produced in this process. The broth is first clarified by microfiltration. The clarified broth then passes to a water evaporator. Some salts are precipitated and separated. Then, acidification using H$_2$SO$_4$ is carried out to recover free succinic acid in the presence of ethanol. Esterification takes place simultaneously, converting some acid into mono- and diethyl succinate. The mixture is completely esterified via reactive distillation. A recovery of succinate greater than 95% can be achieved. The inorganic sulfate salt is separated from the ethanol solution and could be sold as a by-product. Diethyl succinate can be used as the chemical feedstock for the synthesis of other chemicals (e.g. tetrahydrofuran, 1,4-butanediol and γ-butyrolactone). Excess ethanol is recovered using distillation columns and recycled to the acidification and esterification steps. Alternatively, a hydrolysis step can be used to return the ester to the acid form.

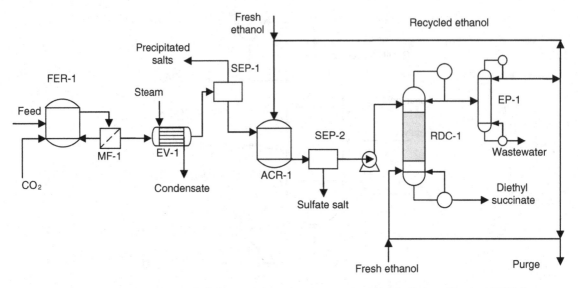

Figure 11.37   *Succinic acid recovery as diethyl succinate using reactive extraction and distillation with ethanol. FER: fermentor, MF: microfiltration, EV: multistage evaporator, SEP-1: salts separator, SEP-2: sulfate salt separator, ACR: acidification reactor, RDC: reactive distillation column, EP: ethanol purification.*

11.4.8   Crystallization

Crystallization involves the separation of a compound as a solid from a homogeneous liquid solution. The solution can be made up of the compound of interest or a mixture of substances with varying solubility in the solvent. Crystallization occurs depending on the solubility of a substance. The solubility s of a substance is the total amount dissolved at the equilibrium state and at a given temperature. The resulting concentration is also called concentration of saturation $C^* = s$. In this case, concentration is generally expressed as the mass of solute per unit mass of solvent. A plot of concentration at saturation against temperature is called a solubility curve. Generic solubility curves are shown in Figure 11.38.

Solutions generally allow more solute to be dissolved than the amount dissolved at saturation conditions. Such a condition is called supersaturation. Supersaturated solutions present a region thermodynamically metastable between the *equilibrium solubility curve* and a *supersolubility curve*, shown in Figure 11.38(a). Because of this metastable region, crystals are not immediately formed when a saturated solution is cooled down.

Nucleus formation is necessary for crystals to begin to grow. The supersolubility curve represents the points at which the nucleation rate is rapid enough for crystallization to start. A mechanism called *primary nucleation* occurs when molecules of a solute associate to form clusters in an ordered arrangement, until enough molecules are present to form the crystals. Primary nucleation must first occur, when no solid particles are present in the solution. As the solution becomes more supersaturated, nuclei formation is increased. Thus, a driving force for crystallization is the difference between a concentration of supersaturation in the unstable region and the corresponding concentration at saturation or equilibrium. The more supersaturated a solution is, the more is the nucleation and the formation of crystals. In industrial practice, a condition of supersaturation can be created by cooling, evaporation, precipitation or by adding a substance as a secondary solvent, precipitant or pH modifier. These methods are represented by the two general processes shown in Figure 11.38(a). Cooling is shown by the line between points 1 and 4, starting with an unsaturated solution point 1. Lowering the temperature produces a saturated solution in the equilibrium solubility curve at point 2. As cooling proceeds, the solution starts to become supersaturated and crosses the metastable region until the supersolubility curve at point 3 is reached, where crystallization starts and continues as the solution is further cooled until point 4 in the unstable

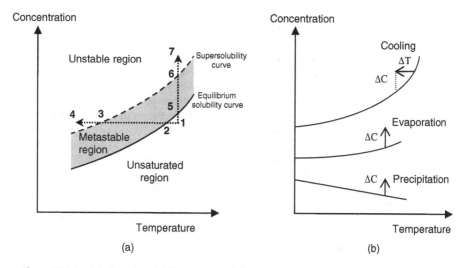

Figure 11.38   *(a) Generic solubility curves and (b) supersaturation in crystallization processes.*

solution region is reached. On the other hand, solvent evaporation, precipitation or addition of a secondary substance is represented by the line between points 1 and 7. Starting with an unsaturated solution at point 1, any of these methods occurs at constant temperature. The removal of solvent from the solution increases the concentration until the equilibrium solubility is reached at point 5. As the process proceeds, the metastable region of supersaturation is crossed until the supersolubility curve at point 6 is reached. Crystallization starts and continues until a point 7 in the region above the supersaturation curve is reached.

The selection of any of the methods can be guided by the solubility and the ratio $\Delta C/\Delta T$, as represented in Figure 11.38(b). Table 11.20 shows the guidelines according to these criteria. In cooling crystallization, the driving force is the difference in concentration created by decreasing the temperature. In precipitation, a difference in concentration is created by removing the solvent from the solution. In evaporative crystallization, the solution becomes supersaturated by removing the solvent from the solution and creates the driving force required. Addition of *a secondary solvent, also*

Table 11.20   *Guidelines for selection of methods to create supersaturation.*

Criteria	Method Preferred
Very soluble substances $s > 0.2$ g g^{-1} $\Delta C/\Delta T > 0.005$ g g^{-1}°C^{-1}	Cooling crystallization
Soluble substances 0.01 g g$^{-1} < s < 0.2$ g g^{-1}	Cooling or evaporative crystallization
Slightly soluble substances $(s < 0.01$ g g$^{-1})$	Precipitation induced by chemical reaction
$\Delta C/\Delta T < 0.005$ g g^{-1}°C^{-1}	Evaporative crystallization
All cases	Adding a secondary solvent, also called nonsolvent or antisolvent

called nonsolvent or antisolvent, can be used in any case for the criteria shown in Table 11.20. The nonsolvent must be miscible with the solvent, but of different polarity, and must change the solubility of the solute. Although fouling of equipment surfaces is reduced, further separation is required to recover and recycle the secondary solvent.

Secondary nucleation (as opposed to the *primary nucleation mechanism*) can occur when a crystalline substance is present in the solution. This crystalline substance can be the crystals formed from primary nucleation that induce nucleation and further crystal growth by interacting with other crystals or with a solid wall. Such interaction is favored by agitation. Agitation moves the supersolubility curve towards the equilibrium solubility curve, which remains unchanged. This decreases the barrier created by the metastable region and facilitates the crystallization. Secondary nucleation can also be generated by adding seed crystals to the supersaturated solution.

The design of a crystallization process and equipment is complex due to difficulty to determine nucleation and growth rates. The process is a dynamic process and hence a time-derivative equation of crystal growth has to be simulated using suitable boundary conditions. The complexities involved (mass and energy balances, multiple solutes or phases, varying crystal size, etc.) are described in more detail in books specializing in the crystallization subject[28–30]. In this book, mass and energy balances and shortcut calculations are shown for conceptual designs.

The degree of supersaturation is an important decision variable for the design of crystallizer. Detailed design is required to define this with more certainty. Although it is possible to model a supersolubility curve, this involves dynamic state calculations. A crystallizer can initially be designed to operate under supersaturated conditions in the metastable region. The basic design requires the purity or yield desired, property data for solubility, heat capacity, etc., as well as the selection of the supersaturation method, operation mode and type of crystallizer[31]. Batch operation is used for small-scale processes and offers the flexibility to use the same crystallizer in various production lines. Batch crystallization is also recommended for substances with low crystal growth rates. Continuous crystallization is used for large-scale production and offers better consistency in product quality. Multiple effect crystallizers create scope to save energy by heat integration and to provide products with a narrow crystal size distribution. However, this option implies higher capital investment compared with the other operation modes.

The *preliminary sizing* of crystallization equipment makes use of steady-state mass and energy balances to determine the product flow rates, heat duties and evaporation or cooling rates. The size of a continuous crystallizer can be determined from *residence time* (τ). The mean value of τ can be estimated using

$$\tau = \frac{L_m}{4G_m} \tag{11.83}$$

where L_m is the mean crystal size and G_m is the mean crystal growth rate. For estimation, L_m values between 0.2 and 1 mm and G_m between 10^{-8} and 10^{-7} m s^{-1} can be assumed for highly soluble substances[32]. The product of residence time and volumetric flow rate of the liquid gives the crystallizer volume. For estimation of the heat exchanger area, a mean heat transfer coefficient in the range of 800 to 1100 W m^{-2} K^{-1} can be assumed[32].

The following exercise shows an example of the basic design of a crystallization process.

Exercise 12. Crystallization process for succinic acid purification. B-Succinum, Inc. is aiming to set up a succinic acid purification unit in its biorefinery. Succinic acid is to be refined from a liquid stream previously treated to remove most of the impurities. Your task is to select between cooling and evaporative crystallization and perform the basic design of a crystallization process producing succinic acid with a yield higher than 80%. The feed liquid stream has been saturated at 60 °C. Cooling below 20 °C should be avoided if possible. Assume that succinic acid is the only solute in water.

The solubility of succinic acid (g of succinic acid g^{-1} of water) as a function of temperature (°C) is shown by

$$C^* = s = 7.003 \times 10^{-5}T^2 - 2.05 \times 10^{-4}T + 0.0388 \tag{11.84}$$

Note that h_v or the latent heat of vaporization of water is 2.35 MJ kg^{-1}.

Solution to Exercise 12

Crystallization by cooling

The cooling method is shown first. No data are available on the supersolubility of the system. The resulting liquid, also called the mother liquid, is assumed to be at saturated conditions. During cooling, the volume of water can be considered as constant. Therefore, the yield of the process can be determined from

$$\text{Yield} = \frac{C_{in} - C_{out}}{C_{in}} \times 100 \tag{11.85}$$

The yield can be calculated at various temperatures to determine the required temperature and then determine the cooling duty. The inlet and outlet concentrations at equilibrium solubility can be determined using Equation (11.84).

The yield results at various temperatures are shown in Figure 11.39. The yield increases with decreasing temperature. However, cooling beyond 10 °C is not worthwhile as the gain in yield is marginal and requires refrigeration, implying a higher cost of production. Cooling at 17 °C gives a yield of 80%. A yield higher than of 80% can be obtained at 10 °C. As noted earlier, *this would require refrigeration below the constraint of 20 °C.* Evaporative crystallization is discussed as an alternative to see if the cooling process can be avoided to obtain similar or higher yields.

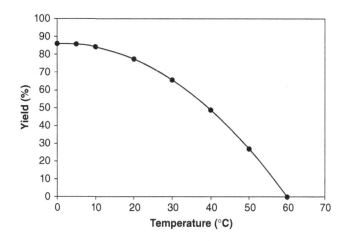

Figure 11.39   *Yield as a function of temperature in the crystallization of succinic acid by cooling.*

Evaporative crystallization

The process can be represented as shown in Figure 11.40.

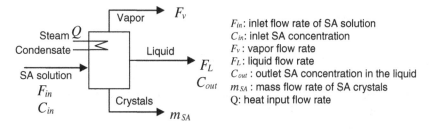

F_{in}: inlet flow rate of SA solution
C_{in}: inlet SA concentration
F_v: vapor flow rate
F_L: liquid flow rate
C_{out} : outlet SA concentration in the liquid
m_{SA} : mass flow rate of SA crystals
Q: heat input flow rate

Figure 11.40   *Process scheme of the evaporative crystallization process for succinic acid (SA) purification.*

Mass balance on the solvent, water in this case, is given by

$$F_{in}\left(1 - C_{in}\right) = F_L\left(1 - C_{out}\right) + F_v \tag{11.86}$$

Mass balance on succinic acid (SA) is shown by

$$C_{in}F_{in} = C_{out}F_L + m_{SA} \tag{11.87}$$

The yield of SA crystals can be expressed as

$$\text{Yield} = \frac{m_{SA}}{C_{in}F_{in}} \times 100 \tag{11.88}$$

Combining Equation (11.87) with Equation (11.88) gives

$$\text{Yield} = \frac{C_{in}F_{in} - C_{out}F_L}{C_{in}F_{in}} \times 100 \tag{11.89}$$

Considering the mass balance for water and solving for F_L gives

$$F_L = \frac{F_{in}1 - C_{in} - F_v}{\left(1 - C_{out}\right)} \tag{11.90}$$

Substitution of Equation (11.90) in Equation (11.89) gives

$$\frac{\text{Yield}}{100} = 1 - \frac{C_{out}}{C_{in}F_{in}\left(1 - C_{out}\right)}\left(F_{in}\left(1 - C_{in}\right) - F_v\right) \tag{11.91}$$

As no information on the supersaturation conditions is given, it is difficult to determine concentration in the liquid stream C_{out}. Assuming that the liquid stream is at saturated equilibrium conditions at the operating temperature, then $C_{out} = C_{in}$ and Equation (11.91) becomes

$$\frac{\text{Yield}}{100} = 1 - \frac{F_{in}\left(1 - C_{in}\right) - F_v}{F_{in}\left(1 - C_{in}\right)} = \frac{F_v}{F_{in}\left(1 - C_{in}\right)} \tag{11.92}$$

According to Equation (11.92), the yield increases as the rate of evaporation (F_v) or the concentration increases. F_v depends on the heat input (Q) to provide the latent heat of vaporization, h_v. From the energy balance, F_v can be expressed as

$$F_v = \frac{Q}{h_v} \tag{11.93}$$

Thus, the yield can be expressed as a function of heat input as

$$\text{Yield} = \frac{Q}{F_{in}h_v\left(1 - C_{in}\right)} \times 100 \tag{11.94}$$

Taking a total feed flow rate of 100 kg h^{-1} as the basis, the amount of heat required for 85% yield is determined. First, C_{in} is calculated as the saturated concentration at 60 °C using Equation (11.84):

$$C_{in} = 7.003 \times 10^{-5}(60)^2 - 2.05 \times 10^{-4}(60) + 0.0388 = 0.2786 \text{ g SA g}^{-1} \text{ H}_2\text{O}$$

The feed flow rate is the sum of SA and water. Thus, this concentration is converted into a total solution weight basis. The amount of SA and water in 100 kg h^{-1} of solution is

$$\text{Water} = \frac{100}{C_{in} + 1} = 78.2 \text{ kg h}^{-1}$$

$$\text{SA} = 100 - 78.2 = 21.8 \text{ kg h}^{-1}$$

The concentration of SA in total weight basis is then 0.218 kg kg^{-1}.

From Equation (11.94) and the given latent heat of vaporization of water, the required amount of heat for 85% yield is

$$Q = 0.85 \times 100 \times 2.35 \times (1 - 0.218) = 156.23 \text{ MJ h}^{-1}$$

The amount of SA crystals produced is

$$m_{SA} = 21.8 \times 0.85 = 17.4 \text{ kg SA h}^{-1}$$

This means that for 85% yield, about 9 MJ of heat per kg of SA recovered is needed. Evaporative crystallization could be an effective option for succinic acid purification at a yield higher than 80%. Note that the calculations neglect the sensible heat to bring water from 60 °C to the boiling temperature (e.g., 100 °C).

Further Questions Determine the yield that can be achieved if the mother liquor, F_L, is recycled to the crystallizer in the evaporative process. Repeat all calculations for an operating temperature of 80 °C and describe the effect on yield and energy requirements of the evaporative process.

Crystallization is based on the differences in solubility of the nondissociated carboxylic acids at different pH values. Using crystallization, succinic acid can be selectively crystallized, while other organic acid by-products remain in the solution. Crystallization from clarified fermentation broth gives a succinic acid yield purity over 90%. To obtain a higher succinic acid purity (>99%), further purification is needed. The direct crystallization process can simplify downstream processing if optimized for higher yields and purities.

Figure 11.41 shows another process using acidification and purification by electrodialysis with bipolar membrane and evaporative crystallization[33]. When magnesium or calcium hydroxides are used for broth neutralization, chemical chelation of the divalent ions is required before bipolar membrane electrodialysis, such as for lactic acid.

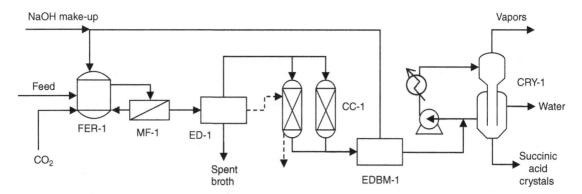

Figure 11.41 *Succinic acid production process using simultaneous acidification and purification by electrodialysis and crystallization. FER: fermentor, MF: microfiltration, ED: desalting electrodialysis, EDBM: electrodialysis with bipolar membrane, CC: chelation columns, CRY: crystallizer. Dashed lines operate only during regeneration of the ion exchanger.*

Crystallization plays an important role as a purification step in biorefineries. Crystallization and precipitation represent the main biorefining techniques to separate solid materials. Crystallization yields a solid product with high purity at relatively low temperature and energy consumption. Many organic chemicals derived from biomass can be purified by crystallization.

11.4.9 Precipitation

Precipitation is readily adaptable in the lactic acid, citric acid and succinic acid industries. Separation of succinic acid from the fermentation broth can be achieved by precipitation with $Ca(OH)_2$ or CaO. The resulting calcium succinate is then filtered. Succinate is converted into succinic acid by reaction with H_2SO_4. Ion exchange is used to remove ionic impurities. Water is evaporated to concentrate and crystallize the succinic acid product. However, the auxiliary chemicals cannot be recovered and recycled. The calcium sulfate product a has very low value and requires further treatment to be sold as a coproduct. Other technologies include using ammonia as a recyclable precipitation agent. However, equipment corrosion can be an issue due to the low pH requirement to regenerate ammonia. Additional energy is required for ammonia recovery.

11.5 Guidelines for Integrated Biorefinery Design

The following heuristics are shown for integrated biorefinery design:

1. Carbon dioxide reuse. Coproducts needing CO_2 as reactant can be integrated with a CO_2 source process. Synthesizing a CO_2 sink process or product is an effective strategy for reusing CO_2 within a sustainable biorefinery.
2. Conversion of a coproduct to higher value products. A biorefinery coproduct should be used as a substrate for another added value production, provided positive economic margin and avoided emissions are achieved through the conversion. Coproduction shares economic and environmental costs.
3. Metabolite optimized for higher value products and market adaptation. Microorganisms can be engineered to favor certain metabolic pathways towards higher added value productions and inhibit others. If there is potential for production of more than one product, the growing conditions can be manipulated to favor the product with higher market adaptability. It is best to target a desired product and minimize by-product formations at the source.
4. Process sequencing and intensification and recycling for optimization of coproduct proportions. Process simulation has to be carried out to arrive at the optimal process sequence, recycle streams, operating conditions, intermediate and end product proportions. This means that having a robust process flowsheet *superstructure* is important as the first place to analyze options. Wherever synergies in reaction and separation process integration exist and separation processes can enhance productivity and selectivity, in situ separations or reactive separation must be investigated. After conceptual synthesis of such a process, dynamic simulation must be undertaken (with mass and energy transfer equations; also see Chapter 12) for detailed design and operating conditions. Reactive separations are an excellent way to intensify processes to enhance capital and energy saving, productivity and purity.
5. Petrochemical replacement by biohydrocarbons. Production of existing hydrocarbons from biomass can favor the successful deployment of biorefineries due to well-established markets. The cost of production of hydrocarbon products from biorefineries must be made competitive against existing petrochemicals.
6. Waste-to-energy generation and at least one main chemical production. Waste streams generating within a biorefinery must be converted into energy products for on-site use. Any surplus energy can be exported. This reduces dependency on fossil resources, environmental impacts and costs and improves the sustainability of the biorefinery. A biorefinery can be developed around one major product, preferably a chemical with high market demand, while the rest of the whole ton of biomass feedstocks can be used to produce numerous coproducts with better market adaptability (preferably platform, fine and specialty chemicals) or for in-process utilization. Value generation from each production route can be ensured using a value analysis approach (Chapter 6).
7. Decide most suitable upstream conversion process depending upon feedstock constituents. Moving to wastes as feedstocks implies more by-products with respect to the main biorefinery product (e.g., more char and ash with respect to the biofuel, chemical or polymer). Cost-effective technologies must be evaluated according to feedstock composition. Anaerobic digestion is preferred for high moisture feedstocks such as wastewater, slurries, organic juices, etc. Fermentation is also generally made in an aqueous medium but is preferred to produce added value products from sugar-rich feedstocks. Low lignin and ash content are preferred if the feedstock is converted via fermentation.

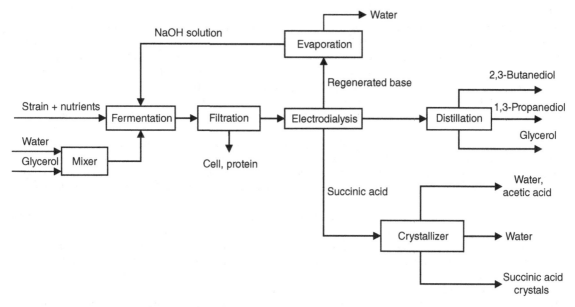

Figure 11.42 *Process to produce 1,3-propanediol and succinate from glycerol.*

Feedstock moisture and ash content becomes more important as the technologies move towards hydrothermolysis, pyrolysis, gasification and combustion. In these cases, low moisture and ash contents are preferred to avoid thermal inefficiencies. Electrodialysis, ion exchange and electrochemical processes will be increasingly important for metal recovery from wastes.

Example of heuristic 1 (carbon dioxide reuse). Due to the chemical versatility of succinic acid and its derivatives, there is potential for process integration and coproduction in a biorefinery. Succinate production can be integrated with a CO_2 sourcing process such as fermentation, combustion or gasification. An interesting integration pathway emerges when both ethanol and succinate production are combined. While two moles of CO_2 are emitted during fermentation of one mole of glucose into ethanol, the production of one mole of succinate consumes at least one mole of CO_2. The integration allows CO_2 emission reduction from the bioethanol and succinic acid integrated production plant. Furthermore, the succinate and ethanol can be combined to form diethyl succinate. Diethyl succinate can be formed and recovered from fermentation broth by reactive distillation. Thus, integration opportunities exist for coproduction of ethanol, succinic acid and diethyl succinate.

Glycerol is available as a substrate for succinic acid production from biodiesel plants. This represents another integration pathway by coupling biodiesel production with glycerol conversion into succinic acid production.

Example of heuristic 2 (conversion of a coproduct to higher value products). A process to produce 1,3-propanediol and succinate from glycerol is shown in Figure 11.42. An integrated strategy for the simultaneous separation of the products and NaOH recycling comprises electrodialysis with bipolar membrane and crystallization. This is an example of a promising opportunity for making products economically competitive and sustainable by sharing production costs and environmental impacts.

Example of heuristic 3 (metabolite manipulation for higher value products and market adaptation). Escherichia coli KJ073 can be used for the coproduction of succinate and maleate when succinic acid is more desirable. The ratio of succinate to maleate production can be up to 5:1 using batch fermentation of glucose. On the other hand, *E. coli* KJ071 produces about two times more maleate than succinate. Thus, the production rate of the two products can be adjusted according to market opportunities by selecting one or another microorganism.

Another interesting opportunity is generated by the engineered *E. coli* QZ1112 for simultaneous production of succinate and polyhydroxybutyrate. *Escherichia coli* KNSP1 can also produce both succinate and polyhydroxybutyraldehydes (PHAs) from glycerol.

Example of heuristic 4 (process sequencing and intensification and recycling for optimization of coproduct proportions). In another integrated process, gas phase hydrogenation of succinic acid can be used to produce γ-butyrolactone, tetrahydrofuran (THF) and 1,4-butanediol (BDO) in a two-stage reaction system. Catalytic hydrogenation is used first to produce γ-butyrolactone (GBL) and THF as intermediates. Succinic acid anhydride is a reaction by-product separated by partial condensation. The gaseous stream containing THF, GBL and water is fed to another hydrogenation step. Recycle of this stream may be examined for better control over the product composition and reactor conversion. The products of the second stage are THF, GBL, BDO and water, which are separated by distillation. γ-Butyrolactone can be recycled to the second step. Product distribution can be adapted to market conditions by manipulating hydrogenation temperatures and recycle rates.

Integration opportunities can be devised with the aim to reduce downstream processing steps and its related costs for a more economically viable biorefinery. Traditional separation techniques should be improved and coupled in synergy with other process and biotechnological developments. For example, metabolic engineering to eliminate pathways leading to the formation of organic acid salts similar to succinate reduces the amounts of by-products and facilitates downstream separation.

Example of heuristic 5 (petrochemical replacement by biohydrocarbons). Numerous processes are available for the development of integrated biorefineries for hydrocarbon production. The advanced technologies offer the required linking between biomass and petroleum refining. The various alternatives and their integration must be evaluated and optimized to make biorefining competitive and a sustainable alternative. By producing established hydrocarbons sufficiently cheaply in a biorefinery, the products can be introduced and accepted in the market. An overview of the integrated biorefinery pathways is shown in Figure 11.43. Although in this particular example a biodiesel blend is shown, other fuel blends are possible by modifying the process conditions of the corresponding process units. Basically, the major petroleum refinery products (transportation fuels and basic petrochemicals) can be made available by the integration of the biohydrocarbon pathways in a biorefinery.

As shown in Figure 11.43, all the processes can be strategically integrated. Interestingly, the role played by hydrogen is as important as in petroleum refining for producing clean and efficient fuels. Apart from catalytic flash pyrolysis, hydrogen is required in all pathways. Self-sufficiency can be achieved by the integrated utilization of coproducts such as propane, char and lignin. Available technologies such as steam reforming and gasification can be integrated as hydrogen sources. This will enable the integration of hydrogen supply networks within the biorefinery, as in petroleum refineries. In addition, heat and power could be produced to meet energetic demands of the processes.

The utilization of the whole ton of biomass is possible by the synergistic combination of pathways, feedstock and product portfolio. A whole crop can be divided into the basic fractions of oil, sugars and lignocellulose. If such fractionation is feasible, then the full range of options in Figure 11.43 can be realizable. Furthermore, the creation of any residual stream can also be processed via anaerobic digestion to produce methane, which can serve as another source of hydrogen or energy.

The production of products conventionally derived from crude oil (e.g., hydrocarbons, petrochemicals, polymers) is a key to successful development of biorefineries. Recognizing this as a strategic point, novel processes for the production of biohydrocarbons are under development. These processes would enable a direct bridge between the emerging biorefining industry and the well-established crude oil refining and petrochemical industries. Thermochemical processes adapted from crude oil processing would benefit from proven technologies for achieving these goals.

Example of heuristics 6 (waste-to-energy generation and one main chemical production) and 7 (decide most suitable upstream conversion process depending upon feedstock constituents). See CFP and Biofine examples in Sections 11.3 and 11.6, respectively.

An integrated approach consists of the combination of fermentation and in situ product removal, which also reduces product inhibition during microbial growth. In the following exercise, the conceptual integration for the Novomer chemical process in a biorefinery is shown. In the Novomer process, ethylene oxide and CO are reacted in the presence of a catalyst to produce succinic anhydride. Succinic anhydride is then hydrated to succinic acid.

Exercise 13. Integrated pathways for succinic acid production via chemical processes. Succinic acid can be synthesized entirely from biomass using the Novomer process. Draw an integrated biorefinery block flow diagram with an analysis of the process integration heuristics.

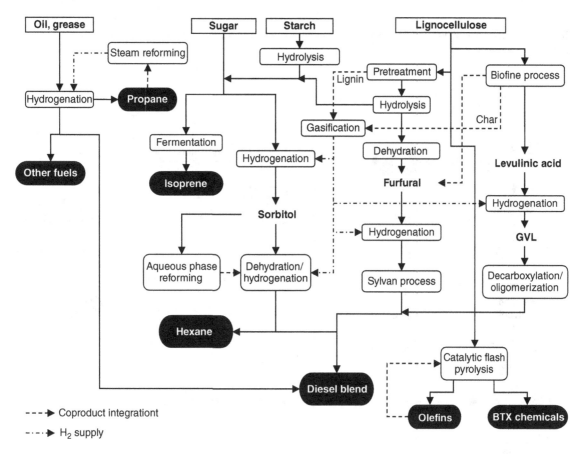

Figure 11.43 *Integrated advanced biorefinery pathways for biohydrocarbon production GVL, γ-valerolactone.*

Solution to Exercise 13. The raw materials used in Novomer's process are ethylene oxide and CO. Ethylene oxide can be obtained from biomass based ethylene glycol via ethylene chlorohydrin (2-chloroethanol). A more direct route avoiding the toxic 2-chloroethanol is the direct oxidation of ethylene. The reactions for these routes are shown in Figure 11.44. CO can be produced from glycerol or biomass gasification.

Figure 11.44 *Alternative chemical routes for ethylene oxide production.*

One integration opportunity exists in a biorefinery, wherein glycerol is converted into ethylene glycol and syngas. Ethylene glycol can be converted into ethylene oxide. CO_2 and H_2 can be separated from syngas to have a CO rich stream by using absorption processes. Then, the raw materials for succinic acid can be produced using the pathways shown in Figure 11.45(a). Another option is to integrate the ethylene production in a biorefinery producing ethanol from lignocellulosic feedstock. Ethylene can then be converted into ethylene oxide via direct oxidation. The lignin or part of the feedstock can be gasified to produce CO.

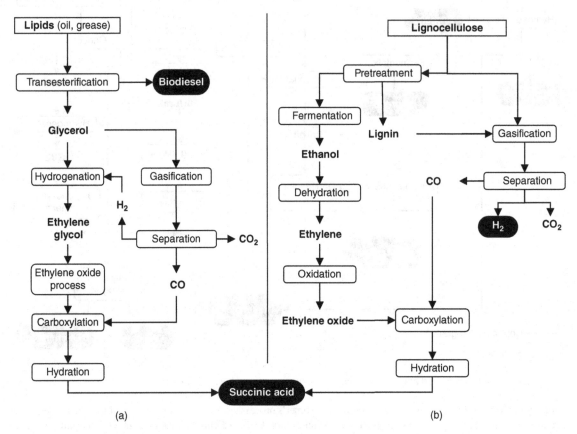

(a) (b)

Figure 11.45 *Pathways integrating the Novomer chemical process for succinic acid production from biomass based raw materials.*

Further Questions Develop a similar integrated process scheme for the production of fermentation based succinic acid in a biorefinery using lignocellulosic feedstock. Identify other integration opportunities for fermentation based succinic acid in the biorefinery using oily crops as feedstock. Name the coproducts that can be produced.

In terms of sustainability, one advantage of the biochemical production under anaerobic conditions is that CO_2 is consumed rather than produced. As a commodity chemical, succinic acid could replace many other commodity petrochemicals, resulting in a reduction in the environmental impact.

Exercise 14. COP and GHG emissions from diethyl succinate production. Estimate the direct operating cost of production (COP) and GHG emissions from the production of 1 kg of diethyl succinate (DES) within a biorefinery already producing bioethanol from wheat. DES is produced from fermentation broth after acidification with sulfuric acid and ethanol.

Reactive distillation with ethanol is then carried out to complete the esterification of succinic acid. A block diagram flowsheet is shown in Figure 11.46 including the main inventory data. Assume that CO_2 is available from ethanol fermentation and no internal cost is charged. Credit from CO_2 emission reduction can be accounted for in the final result.

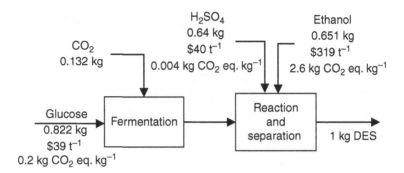

Figure 11.46 *Simplified flowsheet for diethyl succinate production.*

The electricity input is 0.139 MJ kg^{-1} with a cost of \$0.02 MJ^{-1} and GHG emission factor of 0.186 kg MJ^{-1}. The natural gas input for heat production is 1.13 MJ kg^{-1} with a cost of \$5.6 GJ^{-1} and GHG emission factor of 0.061 kg MJ^{-1}.

Solution to Exercise 14. Direct operating costs:

$$\frac{0.822 \times 39 + 0.64 \times 40 + 0.651 \times 319 + 1.13 \times 5.6}{1000} + 0.139 \times 0.02 = \$0.27 \text{ kg}^{-1}$$

GHG emissions:

$$0.822 \times 0.2 + 0.64 \times 0.004 + 0.651 \times 2.6 + 1.13 \times 0.061 + 0.139 \times 0.186 - 0.132 \times 1 = 1.82 \text{ kg } CO_2 \text{ eq. kg}^{-1}$$

Thus, there is potential to produce DES at a competitive price in an integrated biorefinery with low GHG emissions. A more detailed study including capital costs is required to determine the actual cost of production. Detailed allocation in the integrated biorefinery system could also modify the results.

Further Question When the CO_2 capture during biomass growth is accounted, the GHG emissions from glucose and ethanol are -0.72 kg kg^{-1} and -0.35 kg kg^{-1}, respectively. Determine the GHG emissions as kg CO_2-eq per kg of DES based on this information.

11.5.1 An Example of Levulinic Acid Production: The Biofine Process

The Biofine process (discussed under the controlled acid hydrolysis reaction) is an example of a process exploiting integration and coproduction opportunities. The Biofine process consists of the cracking of lignocellulose into cellulose, hemicellulose and lignin using sulfuric acid as the catalyst at moderate temperature and pressure[34–36]. In the Biofine process, the cellulose fraction is converted into levulinic acid with formic acid as the coproduct. The hemicellulose fraction is decomposed into furfural, which can be sold as a product or upgraded to levulinic acid. Lignin, along with

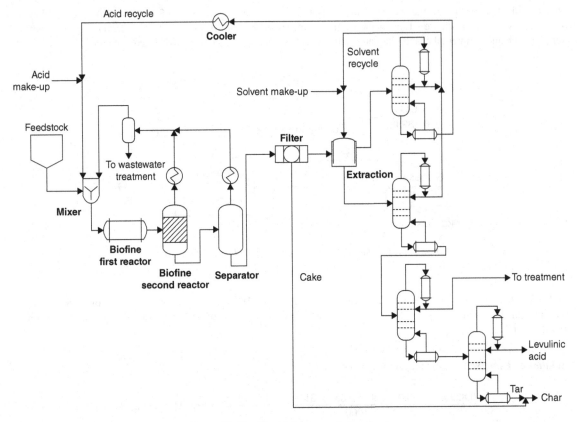

Figure 11.47 *The Biofine process flowsheet.*

some degraded cellulose and hemicellulose and ash produces char, which can be used as fuel to provide heat and power for the process. Coproduction shares the economic and environmental impacts and can make the biorefinery economically competitive (see Chapters 6 and 7 for the assessment tools).

A Biofine process flowsheet is shown in Figure 11.47. The Biofine process requires biomass shredding to reduce the particle size. The particles are then conveyed by a high-pressure steam injection system to a mixing tank. The feedstock is mixed with dilute sulfuric acid (concentration of 1.5–3%) and pumped into the Biofine reactors.

The reactor system consists of two stages with operating conditions optimized to give the greatest yield and least product degradation and tar formation. The first stage is targeted towards acid hydrolysis of cellulose and hemicellulose to generate sugars of six and five carbons (C6 and C5, respectively). The C5 sugars are mainly converted into furfural. The C6 sugars form the intermediate compound 5-hydroxymethylfurfural (5-HMF). This reaction is carried out at 210–230 °C and 25 bar in a plug flow reactor[34].

The second stage uses a continuous stirred tank reactor. This reactor favors the first-order reaction of 5-HMF conversion into levulinic acid and formic acid. The second reactor operates at the same acid concentration as in the first reactor but at a lower temperature of 195–215 °C and pressure of 14 bar[34]. The reactions occurring here are not as fast as the ones in the first reactor and need a longer residence time.

Furfural, formic acid and other light products volatilize from the reaction mixture at the operating conditions. The vapor stream is condensed and these products can be recovered by distillation. The levulinic acid and heavy residues in the liquid stream pass to a centrifuge or filter press. The solid residue or cake forms part of the char product used as fuel. The liquid stream then passes through a series of treatments for recovery and purification of levulinic acid.

Levulinic acid is first separated by solvent extraction from the liquid acidic solution. Then, the solvent is stripped from the stream containing the levulinic acid using distillation. The recovered solvent is recycled to the extraction unit. Levulinic acid is recovered from the bottom stream of the stripping column and is further refined by distillation. The raw levulinic acid obtained is purified up to 98% purity by vacuum distillation. The tars from the refining step are incorporated with the cake resulting from the centrifuge or filter press and form the char product. Sulfuric acid is recovered from the effluent of the extraction step in the acid stripping column and recycled back to the reactants mixing tank.

The maximum theoretical yield of levulinic acid from C6 sugars is 71.6% on a weight basis and the rest goes to formic acid. With the use of an efficient reactor system and polymerization inhibitors to reduce tar formation, the Biofine process achieves levulinic acid yields of up to 50–57% from C6 sugars[34]. The lignin and ash components of the feedstock form part of the char.

The overall yields (kg of product per kg of input feedstock) of the Biofine process depend on the operating conditions and the initial composition of the feedstock. Typical product yields generated from the cellulose, hemicellulose and lignin fractions are as follows: (a) 46% levulinic acid, 18% formic acid and 36% char is obtained from the initial cellulose mass; (b) 40% furfural, 35% char and 25% water are obtained from the initial hemicellulose mass; and (c) 100% of initial lignin and char goes to the char. Although the process offers great flexibility with regard to water content, feedstocks with high cellulose content but low ash and lignin contents are preferred.

The advantages of the Biofine process can be summarized as follows[34]:

- Any biomass with sufficient cellulose and without excess ash is a potential feedstock (wood, forest and agricultural residues, food wastes, recycled paper, organic fraction of municipal solid waste, etc.).
- Feedstock with water content up to 50% can be processed without a significant effect on the economic performance.
- Lignin content does not show an inhibition effect on the Biofine process as in the enzymatic hydrolysis technology, which requires feedstock pretreatment.
- Avoids the need for large fermentation tanks, resulting in lower capital costs and floor space.
- No microorganisms are needed.

Each of the products of the Biofine process (levulinic acid, furfural, formic acid and char) can be platform chemicals for synthesis of useful derivatives. This gives potential process integration and flexibility so that production rates can be adapted to market conditions. Figure 11.48 shows possible integration pathways in a biorefinery processing lignocellulosic feedstock using the Biofine process.

The formic acid coproduct is a commodity chemical that can be purified by distillation. This product has application in decalcification, leather tanning and the textile industry. As a chemical platform, formic acid is a raw material for the synthesis of formaldehyde, pharmaceuticals, dyes, insecticides, refrigerants and road salt. Formic acid is also useful for the preparation of catalysts and regeneration of desulfurization catalysts. Methyl and ethyl formate can find application as fuel additives and chemical building blocks. Alternatively, formic acid can be used in the biorefinery to provide energy via gasification or anaerobic digestion.

Furfural is mainly used to produce furan resins, lubricating oils and textiles. Furfural can be used to generate furfuryl alcohol, tetrahydrofuran (THF) and levulinic acid. THF is obtained by conversion of furfural into furan followed by catalytic hydrogenation. The possibility of furfural conversion into levulinic acid is interesting for process flexibility and market adaptability. In this case, furfural is first converted into furfuryl alcohol by hydrogenation. Then, furfuryl alcohol is reacted with ethyl methyl ketone in the presence of HCl. This conversion gives a levulinic acid yield of up to 90–93%[34]. Alternatively, furfuryl alcohol can be sold to the market.

The char energy content ranges between 17.4 and 25.6 MJ kg^{-1}. The energy in the char can be used to provide the steam and electricity for the Biofine process. Any excess electricity can be exported. Char is also a potential soil conditioner and a lignin source for carbon fiber. Another possible pathway for the utilization of char within a biorefinery is steam gasification to produce syngas ($CO + H_2$). Syngas can then be used in alcohol synthesis and combined heat and power generation. Hydrogen can be separated as a product using pressure swing adsorption. Further investigation is required for the feasibility of char gasification.

The following exercise shows how process integration methods such as heat and water pinch analysis could be used to reduce production costs of a biorefinery.

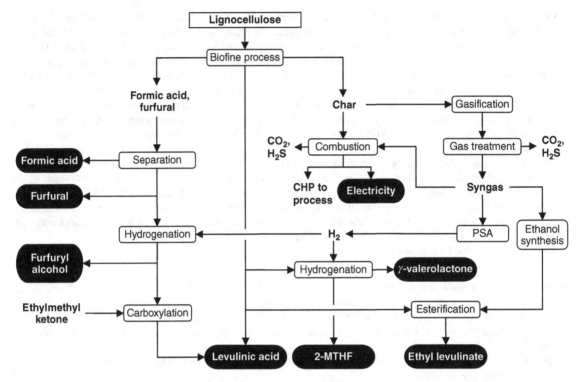

Figure 11.48 *Integration pathways between the Biofine process for the production of levulinic acid and the further conversion processes of its coproducts. PSA, pressure swing adsorption.*

Exercise 15. Integration of the Biofine process. Figure 11.49 shows a simulation flowsheet for a Biofine process using 17 857 kg h^{-1} of corn cob as feedstock (30% moisture content, mass basis). Some input and output mass flow rates are presented along with energy requirements. Answer the following questions.

a.   Calculate the overall levulinic acid yield.
b.   Calculate the solvent and acid requirements after recycling the amounts recovered in downstream processing.
c.   Extract the data in the form of sink-source tables to identify heat and water integration opportunities.

Solution to Exercise 15. Refer to the ***Online Resource Material, Chapter 11 – Additional Exercises and Examples*** for the solution to Exercise 15.

11.6   Summary

This chapter shows examples of biomass processing routes into chemicals. Process integration makes chemical production more efficient, economically viable and sustainable. Combinations of various chemical pathways, products and coproducts must be taken into account when setting up new or expanded configurations to exploit the biomass potential in biorefineries fully. Such integrations, combined with the co-evolutionary developments of high throughput process technologies, need to be investigated. Integrative reaction–separation systems must be investigated to increase efficiencies of downstream processing. Development of models for the understanding of unit operations predominant in biorefineries such as bioreactors, extraction columns, ion exchange columns, membranes and crystallizers is also urgent. Much experimental

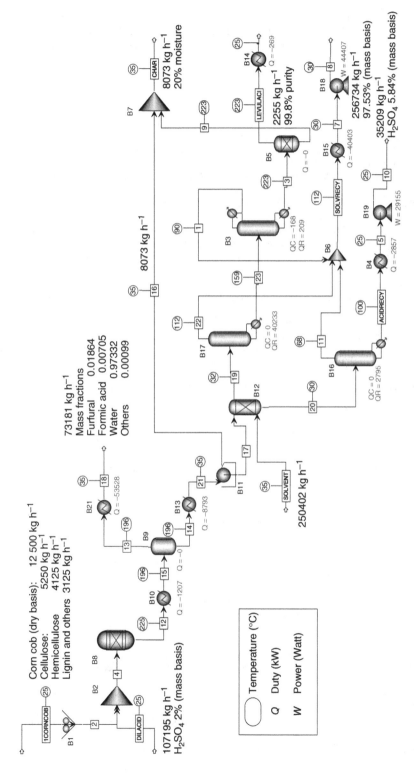

Figure 11.49 *Simulation flowsheet of the Biofine process. QC, condenser cooling duty; QR, reboiler heat duty.*

work is involved, especially for the determination of physicochemical properties of organic compounds present in biorefinery processes. Such knowledge will be helpful for more accurate simulation models. With the help of tools, such as those shown in the various exercises in this book, a well-informed process design and decision making is possible.

References

1. J. Monod, *Research of the Growth of Bacterial Cultures*, Hermann, Paris, France, 1942.
2. H. Lineweaver and D. Burk, The determination of enzyme dissociation constants, *J. Amer. Chem. Soc.*, **56**, 658–666 (1934).
3. R. Luedeking and E.L. Piret, A kinetic study of the lactic acid fermentation. Batch process at controlled pH, *J. Biochem. Microbiol. Technol. Eng.*, **4**, 393–412 (1959).
4. J. Biazar, M. Tango, E. Babolian, R. Islam, Solution of the kinetic modelling of lactic acid fermentation using Adomian decomposition method, *Appl. Math. Comput.*, **144**, 433–439 (2003).
5. D. Altiok, F. Tokatli, S. Harsa, Kinetic modelling of lactic acid production from whey by *Lactobacillus casei* (NRRL B-441), *J. Chem. Technol. Biotechnol.*, **81**, 1190–1197 (2006).
6. H.H. Song, J,W, Lee, S. Choi, J.K. You, W.H. Hong, S.Y. Lee, Effects of dissolved CO_2 levels on the growth of mannheimia succiniciproducens and succinic acid production, *Biotechnol. Bioeng.*, **98**, 1296–1304 (2007).
7. Chematur Engineering AB, *Ethylene from ethanol brochure*, http://www.chematur.se/sok/download/Ethylene_rev_0904.pdf.
8. V. Barraca, B. Joao, R. Coutinho, *Process for preparing ethane*, US Patent 4232179, 1978.
9. U. Tsao and B. Howard, *Production of ethylene from ethanol*, US Patent 4134926, 1979.
10. X. Zhang, R. Wang, X. Yang, F. Zhang, Comparison of four catalysts in the catalytic dehydration of ethanol to ethylene, *Microporous Mesoporous Mater.*, **116**, 210–215 (2008).
11. R. Schill, *Braskem starts up ethanol-to-ethylene plant*, Ethanol Producer Magazine, 2010, http://ethanolproducer.com/articles/7022/braskem-starts-up-ethanol-to-ethylene-plant/.
12. Solvay, *Press release: Solvay Indupa will produce bioethanol-based vinyl in Brazil and considers state-of-the-art power generation in Argentina. Polyvinyl chloride (PVC) derived from sugar cane and salt*, 2007, http://www.solvay.com/EN/NewsPress/Documents/2007/20071214_ColomboII_EN.pdf.
13. Chemicals Technology, *Braskem ethanol-to-ethylene plant, Brazil*, http://www.chemicals-technology.com/projects/braskem-ethanol/.
14. Mitsui & Co. Ltd., *News release: Participation in production of sugarcane-derived chemicals in Brazil with The Dow Chemical Company*, 2011, http://www.mitsui.com/jp/en/release/2011/1194650_1803.html.
15. T.R. Carlson, Y.T. Cheng, J. Jae, G. W. Huber, Production of green aromatics and olefins by catalytic fast pyrolysis of wood sawdust, *Energy Environ. Sci.*, **4**, 145–161 (2011).
16. Y.T. Cheng, J. Jae, J. Shi, W. Fan, G.W. Huber, Production of renewable aromatic compounds by catalytic fast pyrolysis of lignocellulosic biomass with bifunctional Ga/ZSM-5 catalysts, *Angew. Chem*, **124**, 1416–1419 (2012).
17. B. Girisuta, L.P.B.M. Janssen, H.J. Heeres, Kinetic study on the acid-catalyzed hydrolysis of cellulose to levulinic acid, *Ind. Eng. Chem. Res.*, **46**, 1969–1708 (2007).
18. J. Horvat, B. Klaic, B. Metelko, V. Sunjic, Mechanism of levulinic acid formation, *Tetrahedron Lett.*, **26**, 2111–2114 (1985).
19. H. Wu, M. Jiang, P. Wei, D. Lei, Z. Yao, P. Zuo, *Nanofiltration method for separation of succinic acid from its fermented broth*, CN Patent 200910025531.5, 2011.
20. R. Datta, S.P. Tsai, P. Bonsignore, S.H. Moon, J.R. Frank, Technological and economic potential of poly(lactic acid) and lactic acid derivatives, *FEMS Microbiol. Rev.*, **16**, 221–231 (1995).
21. Reverdia, http://www.reverdia.com/technology/.
22. K.L. Wasewar, A.A. Yawalkar, J.A. Moulijn, V.G. Pangarkar, Fermentation of glucose to lactic acid coupled with reactive extraction: A Review, *Ind. Eng. Chem. Res.*, **43**, 5969–5982 (2004).
23. A.S. Kertes and C.J. King, Extraction chemistry of fermentation product carboxylic acids, *Biotechnol. Bioeng.*, **28**, 269–282 (1986).
24. J.A. Tamada and C.J. King, Extraction of carboxylic acids with amine extractants. 3. Effect of temperature, water coextraction and process consideration, *Ind. Eng. Chem. Res.*, **29**, 1333–1338 (1990).
25. K.L. Wasewar, A.B.M. Heesink, G.F. Versteeg, V.G. Pangarkar, Reactive extraction of lactic acid using Alamine 336 in MIBK: equilibria and kinetics, *J. Biotechnol.*, **97**, 59–68 (2002).
26. T. Kurzrock and D. Weuster-Botz, New reactive extraction systems for separation of bio-succinic acid, *Bioprocess. Biosys. Eng.*, **34**, 779–787 (2011).
27. A. Orjuela, A.J. Yanez, L. Peereboom, C.T. Lira, D.J. Miller, A novel process for recovery of fermentation-derived succinic acid, *Sep. Purif. Technol.*, **83**, 31–37 (2011).
28. J. Nývlt, *Design of Crystallizers*, CRC Press, Boca Raton, 1992.

29. N. Tavare, *Industrial Crystallization: Process Simulation Analysis and Design*, Plenum Press, New York, 1995.
30. A. Mersmann, *Crystallization Technology Handbook*, Marcel Dekker, New York, 1994.
31. W.L. McCabe, J.C. Smith, P. Harriot, *Unit Operations of Chemical Engineering*, 5th ed., McGraw-Hill, New York, 1993.
32. M.M. Seckler, *Tecnologia da Cristalização*, Instituto de Pesquisas Tecnológicas, São Paulo, 2000.
33. G.J. Zeikus, M.K. Jain, P. Elankovan, Biotechnology of succinic acid production and markets for derived industrial products, *Appl. Microbiol. Biotechnol.*, **51**, 545–552 (1999).
34. D.J. Hayes, S.W. Fitzpatrick, M.H.B. Hayes, J.R.H. Ross, The Biofine Process – Production of levulinic acid, furfural, and formic acid from lignocellulosic feedstocks, in *Biorefineries – Industrial Processes and Products*, B. Kamm, P. R. Gruber and M. Kamm (eds), Wiley, Weinheim, Germany, 139–164, 2005.
35. S.W. Fitzpatrick, *Lignocellulose degradation to furfural and levulinic acid*, U.S. Patent 4897497, 1990.
36. S.W. Fitzpatrick, *Production of levulinic acid from carbohydrate-containing materials*, U.S. Patent 5608105, 1997.

12

Polymer Processes

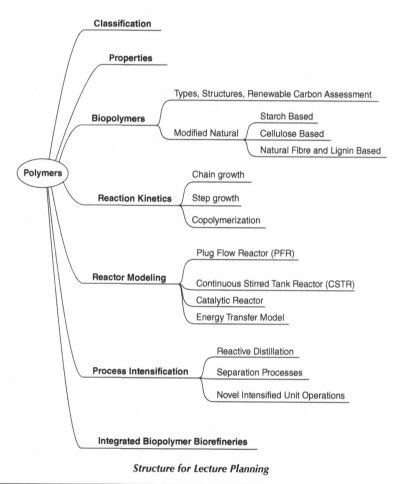

Structure for Lecture Planning

Biorefineries and Chemical Processes: Design, Integration and Sustainability Analysis, First Edition.
Jhuma Sadhukhan, Kok Siew Ng and Elias Martinez Hernandez.
© 2014 John Wiley & Sons, Ltd. Published 2014 by John Wiley & Sons, Ltd.
Companion Website: http://www.wiley.com/go/sadhukhan/biorefineries

Polymers are present in almost every aspect of human daily life, in textiles, automobile, computer parts, insulation, food protection and packaging, etc. Polymers have contributed to energy saving and environmental impact reduction by more efficient energy use in lightweight automobiles, insulation of houses, refrigeration, etc. However, the majority of the polymers used today are produced using petrochemicals derived from fossil resources. Diminishing fossil resource and stronger market competition for production of environmentally more benign products are making companies look at innovative synthesis routes for polymer materials. This scenario, along with consumer awareness of environmental impacts of polymers, is playing an important role in the introduction of *biopolymers* to the market. Low cost petrochemicals like ethylene, propylene and butadiene allowed the development and innovation of fossil based polymers. Likewise, the availability of cheap biomass based building blocks or monomers and the development of polymers with more functionality and enhanced performance are crucial for successful market uptake of biopolymers.

If biopolymers are going to be part of the product portfolio of a biorefinery, understanding the structures and properties of polymer and monomer and polymerization reaction kinetics is important to design reactors and integrated process flowsheets. These basic concepts are discussed in this chapter.

12.1   Polymer Concepts

The literal meaning of the term polymer is *many (poly) parts or units (mer)*. It describes the multiplicity of simple chemical units repeated in the structure of a macromolecule. The *repeat unit* is the basic group of atoms forming a monomer molecule. Repeat units in a polymer molecule are successively linked by covalent bonds. Obtained from a chemical precursor, a chemical formula repeatedly occurring in a polymer molecule is called the *monomer*. Structures of repeat units of the common polymers are shown in Table 12.1. The number *n* is a large number indicating how many repeat units form a polymer molecule. This number can range from a few dozens to many thousands. This number is called the polymer *chain length*. Polymers with different chain lengths and molar masses are generally formed during a polymerization reaction.

Since various polymer molecules are formed during polymerization, the resulting polymer products have different molar masses. Macromolecules with a different *degree of polymerization* (DP) are produced. As a consequence, there is a distribution of molar masses characterizing a polymer product. The molar mass of a polymer is calculated as an average from that distribution. The most common methods used to find an average molar mass of a polymer are the *number average molar mass* (M_n) and the *weight average molar mass* (M_w), shown by the following equations, respectively:

$$M_n = \sum_i M_i N_i / \sum_i N_i \qquad (12.1)$$

$$M_w = \sum_i M_i^2 N_i / \sum_i M_i N_i \qquad (12.2)$$

where N_i is the number of moles of polymer species i with molar mass M_i. Properties of a polymer can be correlated to the chain length and thus to the number average molar mass. Molar mass is also referred to as the molecular or molar weight.

The ratio of M_w to M_n is called *polydispersity*. Polydispersity depends on the molar mass distribution of a polymer. An ideal polymer has only chains of identical molar mass and is *monodisperse*, that is, $M_w/M_n = 1$. However, all the polymers are *polydisperse*. Polydispersity shows how wide or how narrow is the distribution of molar masses and whether a range of chain sizes predominates in the polymer sample (narrow distribution) or not (wide distribution). The molar mass range is generally optimized during polymerization to obtain properties appropriate for a specific polymer application. The value of polydispersity is determined by the reaction mechanism of the polymerization and by the reaction conditions.

Polymers can be crystalline or amorphous. Crystalline polymeric materials have better mechanical and thermal properties compared to their amorphous counterparts. Amorphous polymers, such as polystyrene, are formed when the repeat unit has a bulky functional group such as benzene. Other common classifications of polymers are discussed as follows.

Table 12.1 *Monomer and structure of common polymers.*

Polymer Name	Monomer	Structure (repeat unit)$_n$
Polyethylene (PE)	Ethylene	
Polypropylene (PP)	Propylene	
Polystyrene (PS)	Styrene	
Polyvinyl chloride (PVC)	Vinyl chloride	
Polycaprolactone (PCL)	ε-Caprolactone	
Polycaprolactam (PC or Polyamide 6)	Caprolactam	
Polycarbonate (diphenyl carbonate)	Bisphenol A	
Poly(ethylene terephthalate) (PET)	Terephthalic acid and ethylene glycol	
Polyurethane	Methylene diphenyl diisocyanate (MDI) and ethylene glycol	
Polyvinyl alcohol	Vinyl acetate	

12.1.1 Polymer Classification

Polymers can be classified according to various factors that describe their origin, formation mechanism, structure and thermal behavior, as shown in Figure 12.1. A polymer can be obtained from various reaction mechanisms. These mechanisms involve step growth or chain growth, discussed as follows, producing *step-growth polymers* or *chain-growth polymers*, respectively.

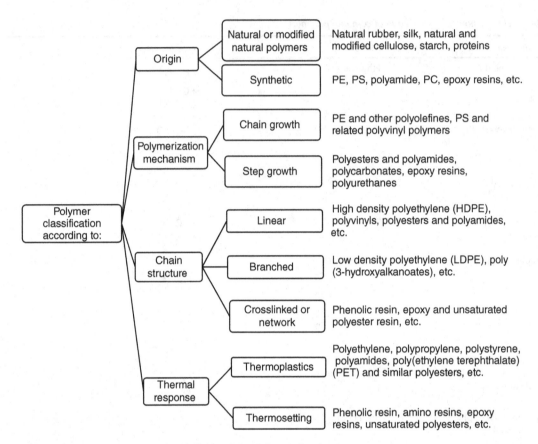

Figure 12.1   *Polymer classification.*

Polymer classification according to thermal behavior is as follows. *Thermoplastic* polymers are soft solid materials that become fluid when heated. They can be moulded and then cooled to solidify as a finished product. No chemical changes occur during the process. Reshaping is possible by heating. *Thermoset* polymers undergo crosslinking reactions on heating. They cannot be reshaped by applying heat and they generally decompose at high temperatures.

Classification based on molecular structure is important, as structure plays a key role in the final properties of a polymer. Linear-chain polymers are generally thermoplastic materials. Linear-chain polymers result from monomers having two functional groups such as dicarboxylic acids or diols or a double bond. Branched-chain polymers are formed by monomers with more than two functional groups. This allows the polymer chains to grow in various directions. An example of a multifunctional monomer is glycerol. Network polymers form three-dimensional crosslinked structures and are called gels.

12.1.2 Polymer Properties

Chemical structure and molar mass distribution are the main characteristics of a polymer. The main properties of polymers are shown as follows.

Degree of crystallinity. Polymers forming crystals can have crystalline regions or fragments with the rest amorphous. Polymers are characterized by their degree of crystallinity. The degree of crystallinity (in %) is the proportion of the mass of crystalline material to the mass of a polymer. A high degree of crystallinity gives higher density, stiffness, strength, toughness and heat resistance.

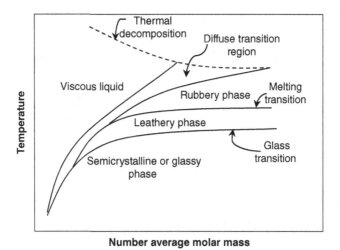

Figure 12.2 *Phase diagram of semicrystalline polymers relating temperature to the number average molar mass.*

Tacticity. This property is described as the degree of different configurations or arrangements of the atoms or groups of atoms in a molecule. It refers to the stereochemistry of the polymer. Polymers can be:
a.   Isotactic: all substituents of the main chain are in the same side.
b.   Atactic: has no pattern on the arrangement of substituents in the polymer chain.
c.   Syndiotactic: alternating arrangement of the substituents in the main polymer chain.

Glass transition temperature (T_g). The temperature at which an amorphous polymer undergoes phase change from a rubbery, viscous liquid to a glassy amorphous solid is called the glass transition temperature. T_g can be adjusted by changing the structure of the polymer chains or using plasticizers. T_g can be estimated from the number average molar mass (M_n) using

$$T_g = T_g^\infty - \frac{K}{M_n} \tag{12.3}$$

where T_g^∞ is the glass transition temperature of a hypothetical polymer having an infinite molar mass and K is the constant for a polymer.

Melting point (T_m). At this temperature, semicrystalline polymers melt. As the molar mass of the polymer increases, both T_g and T_m increase. Amorphous polymers do not have a melting point. Figure 12.2 shows a generic phase diagram for semicrystalline polymers. The upper dashed line indicates the thermal decomposition temperatures of polymer molecules of different number average molar masses. At the glass transition temperature, transition from the semicrystalline to the leathery phase occurs. The leathery phase does not appear in a diagram for amorphous polymers. Transition from the leathery to the rubbery phase occurs along the line for the melting transition temperature. There is also a diffuse transition region to go from any of the other phases to a viscous liquid phase.

Tensile strength. This property refers to the amount of stress or pressure that a polymer can support without suffering permanent deformation. The higher the degree of polymerization and crosslinking of the polymer chains, the higher is the tensile strength. Figure 12.3(a) shows that the tensile strength increases with the number average molar mass.

Elasticity. This property is a measure of the stiffness of elastic or rubbery materials. Polymers with high elasticity are called *elastomers*. Elasticity is measured by Young's modulus, defined as the ratio of the tensile stress and the deformation or tensile strain produced. Young's modulus has units of pressure. A generic diagram for the variation of Young's modulus with temperature is shown in Figure 12.3(b). Semicrystalline and amorphous polymers are the extremes with high and low modulus, respectively. In between these two types, crosslinked polymers appear.

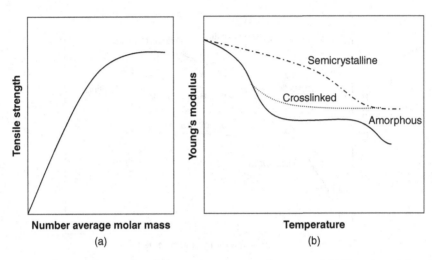

Figure 12.3   *Variation of (a) tensile strength with the number average molar mass and (b) Young's modulus with temperature for amorphous, crosslinked and semicrystalline polymers.*

Copolymers generally have a melting point and tensile strength lower than that of the respective single polymers (also called *homopolymers*). The glass transition temperature of copolymers has an intermediate value between the glass transition temperatures of the single polymers (T_{g1} and T_{g2}) and can be determined using

$$\frac{1}{T_g} = \frac{x_1}{T_{g1}} + \frac{1 - x_1}{T_{g2}} \qquad (12.4)$$

where x_1 is the weight fraction of the single polymer with glass transition temperature T_{g1}. Block copolymers may show various T_g values, each one corresponding to each block.

Other more common physical properties of materials, such as density, viscosity, thermal conductivity, deformation or elongation, etc., are important during production of polymers in a reactor and processing into finished articles by extrusion, spinning, blowing or other methods. Consider a polymer finished into fibers using melt-spinning. The viscosity of the melt solution has an effect on its degree of polymerization and polymer formation by melt-spinning. Estimation of the average molar mass of a polymer by a direct method is difficult. Indirect methods such as the Mark–Houwink equation are commonly used. This method relates the molar mass of a polymer (M_w) to the intrinsic viscosity (Λ). Intrinsic viscosity is a measure of the contribution of a solute to the viscosity of its solution. The Mark–Houwink equation is expressed as

$$\Lambda = \Theta \times M_w{}^u$$

where Θ and u are empirical parameters.

Exercise 1.   Determination of the average molar mass of a polymer from its intrinsic viscosity. Poly(trimethylene terephthalate) or PTT is made by the transesterification reaction between 1,3-propanediol and dimethyl terephthalate. The parameters for the Mark–Houwink equation for PTT are as follows[1]:

$$\Theta = 0.082 \text{ mL g}^{-1}$$

$$u = 0.63$$

For good spinning and fiber properties, a weight average molar mass of 50 000–60 000 g mol^{-1} of the polymer is desirable. The intrinsic viscosity of a solution from a PTT production process has been measured as 80 mL g^{-1}. Find whether the fiber produced achieves the desired weight average molar mass of the polymer.

Solution to Exercise 1.   The weight average molar mass obtained from the Mark–Houwink equation is as follows:

$$M_w = \left(\frac{\Lambda}{\Theta} \right)^{1/u} = \left(\frac{80}{0.082} \right)^{1/0.63} = 55\,580 \text{ g mol}^{-1}$$

Therefore, the polymer is suitable to be finished as fibers by spinning.

12.1.3   From Petrochemical Based Polymers to Biopolymers

Polymers made of at least one component derived from biomass are called *biopolymers*. Biopolymers can potentially contribute to a more sustainable polymer production by:

a.   Reducing environmental impact by using plant materials as feedstocks.
b.   Improving environmental performance of products.
c.   Substitution of conventional polymers with enhanced biodegradability and ease of recovery from wastes.

An example of a successful biopolymer development is poly(lactic acid) or PLA. PLA is obtained by polymerization of lactic acid, a chemical building block produced in a biorefinery by fermentation of sugars. This is a biodegradable polymer featuring enhanced properties and more functionality than some fossil based polymers. PLA can replace polyesters in fibers, films, food, beverage containers, textiles, coated papers and many other applications. Other examples of biomass based building blocks for biopolymers include 3-hydroxypropionic acid (Cargill), succinic acid (BASF) and biopropanediol (used by DuPont for the Sorona™ polymer).

Biopolymers can be generated from biomass by the following three general pathways:

1.   From naturally occurring polymers such as starch, cellulose, hemicellulose, lignin and proteins. When modifications are required to produce end products, the resulting polymers are called modified natural biopolymers.
2.   From primary conversion of biomass into polymers such as sugar fermentation into polyhydroxyalkanoate (PHA) production. These are called primary biomass based polymers.
3.   From polymerization of a chemical building block or precursor produced in a biorefinery. An example includes PLA production from lactic acid as the chemical building block. Biomass sugars are first separated and fermented to lactic acid; then lactic acid is polymerized to PLA. Another example is polyethylene production from polymerization of ethylene as the chemical building block. Ethylene is a product of bioethanol dehydration. These polymers are called secondary biomass based polymers.

Figure 12.4 shows an overview of the biopolymers discussed in this chapter. These biopolymers belong to one of the following types of polymers: polysaccharides and other natural polymeric materials (fibers and lignin), polyesters, polyurethanes, polyamides and polycarbonates.

Refer to *Online Resource Material, Chapter 12 – Additional Exercises and Examples* for a detailed process description of each biopolymer, poly(lactic acid) or PLA, polyhydroxyalkanoates (PHAs), poly(trimethylene terephthalate) or PTT and other polyesters, poly(ethylene terephthalate) or PET, polyethylene furanoate or PEF, polyurethanes (PURs), polyamides, polycaprolactam (PC) and polycarbonates. Table 12.2 shows the structures of the biopolymers shown in this chapter (including *Companion Website*) and challenges for their production.

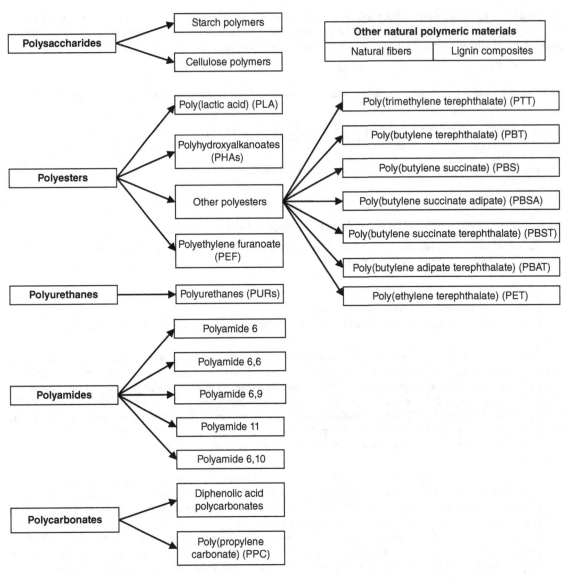

Figure 12.4   *Overview of biopolymers.*

Biopolymers have only a fraction of carbon from biomass. This fraction of carbon is captured (or renewed) by the biomass during production. The following exercise shows how to determine the fraction of biomass based carbon present in biopolymers.

Exercise 2.   Calculation of the renewable carbon content of a polymer. Calculate the percentage of renewable mass content of PTT with the general formula $[COOC_6H_4CO_2(CH_2)_3]_n$. Calculate the percentage of biogenic carbon that can be considered as balanced if the monomer 1,3-propanediol (PDO) is derived from renewable biomass feedstock. When the polymer is incinerated, the biogenic carbon dioxide generated is captured during biomass growth.

Table 12.2 *Structures of biopolymers and production challenges.*

Starch Polymers

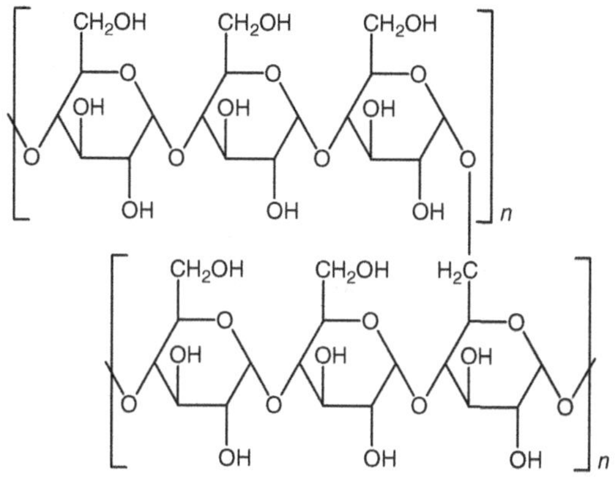

Amylopectin

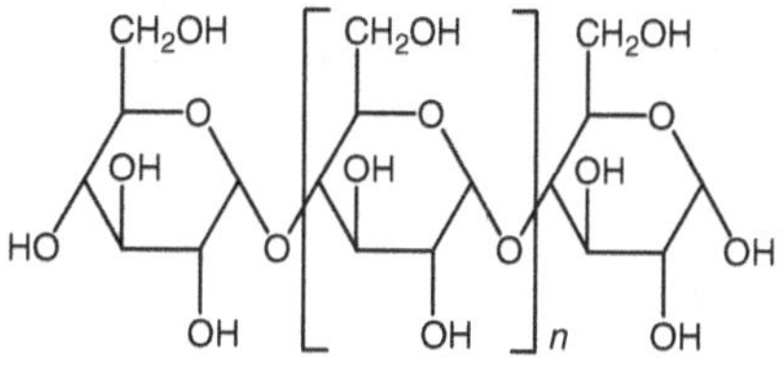

Amylose

Reduce raw material need for manufacturing thermoplastic starch (TPS) polymer films.
Enable starch films for wider applications such as membrane separations.
Improve environmental performance, especially in terms of eutrophication potential and fossil based energy use.

Cellulose Polymers

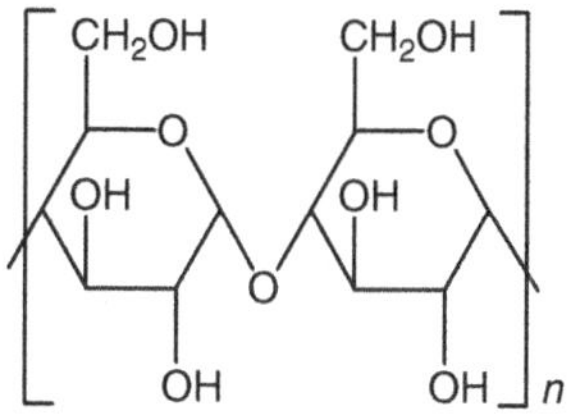

Develop or improve processes for cellulose extraction and modification using solvents or reagents that are nontoxic, easy to remove and recyclable. Also, see the Introduction chapter.

Natural Fiber and Lignin Composites
Improve permeability, reinforcement properties and resistance to chemical and biological degradation.

Poly(lactic acid)

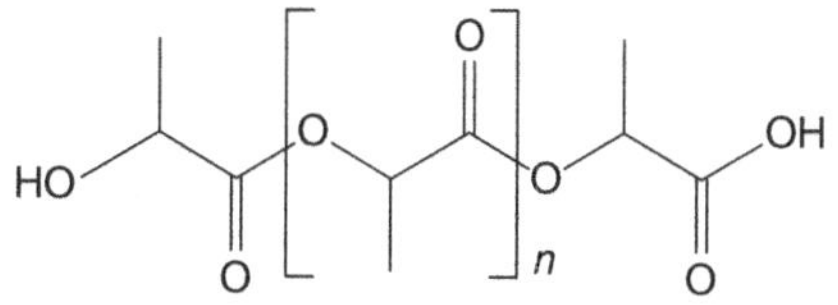

Ease its separation and recycling.
Enable direct polymerization of lactic acid to obtain high molar mass polymers.
Optimize indirect polymerization route.
Integrate with lactic acid production to reduce overall costs.

(Continued)

Table 12.2   (*Continued*)

Polyhydroxyalkanoates (PHAs)

Polyhydroxybutyrate (PHB)   *Polyhydroxyvalerate (PHV)*

Reduce production costs and environmental impact by:
Using abundant and low cost feedstocks such as lignocellulosic residues, whey, sugar cane molasses, etc.
Developing bacteria strains producing high yields.
Scaling the fermentation process and optimization of product separation and recovery.

Other polyesters
Poly(trimethylene terephthalate) or PTT

Poly(butylene terephthalate) or PBT

Poly(butylene succinate) or PBS

Poly(butylene succinate adipate) or PBSA

Table 12.2 *(Continued)*

Poly(butylene succinate terephthalate) or PBST

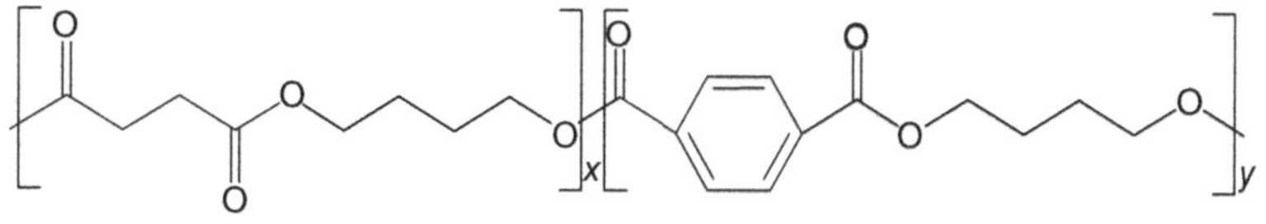

Poly(butylene adipate terephthalate) or PBAT

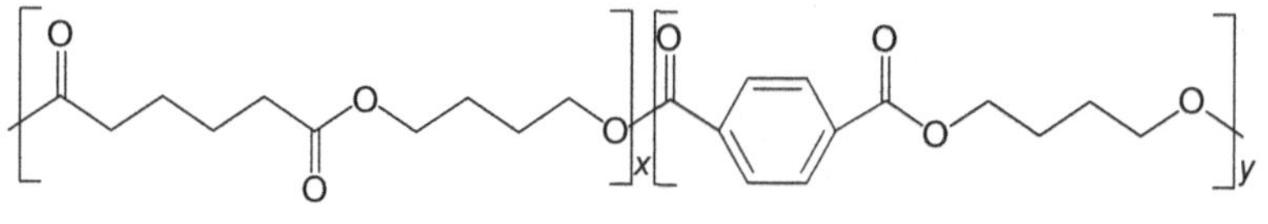

Poly(ethylene terephthalate) or PET

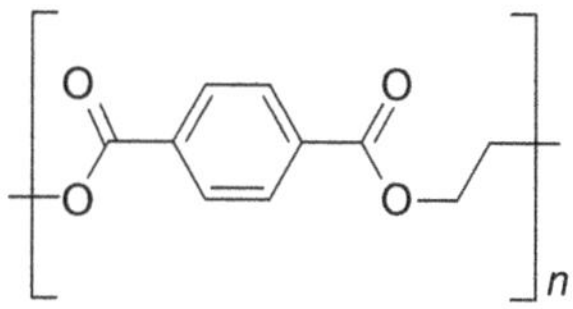

Poly(ethylene furanoate) or PEF

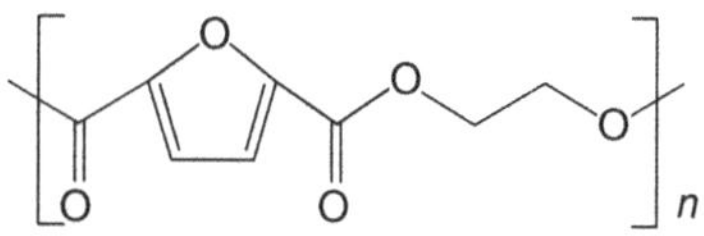

Reduction of costs of production including those of main or coproducts and expansion of the market for bio based polyesters.
Improve polymer functionality and performance.
New polymer synthesis pathways to produce high molar mass polyesters with appropriate properties for a wider range of applications.
Develop the pathway for the production of terephthalic acid and adipic acid from biomass.
Develop scalable process for production of PEF.

Polyurethanes (PURs)

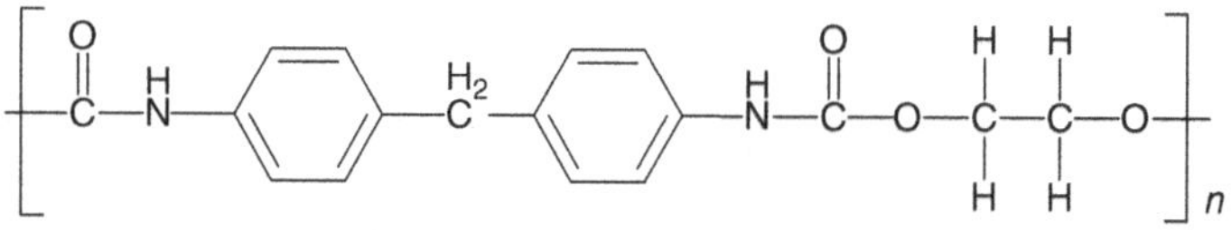

Quality consistency of monomers from the biorefinery.
Produce biorefinery chemicals used for PURs production at low cost.

Polyamides
Polyamide 6,6

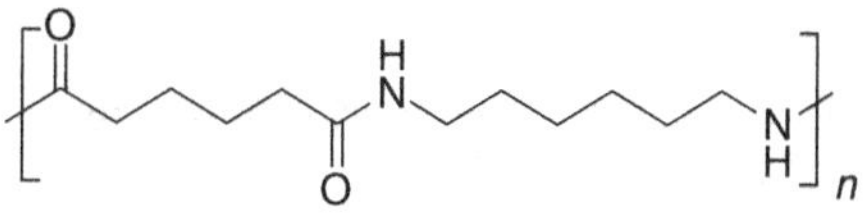

Table 12.2   (*Continued*)

Polyamide 6

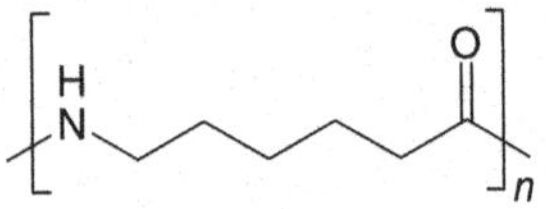

Polyamide 6,9

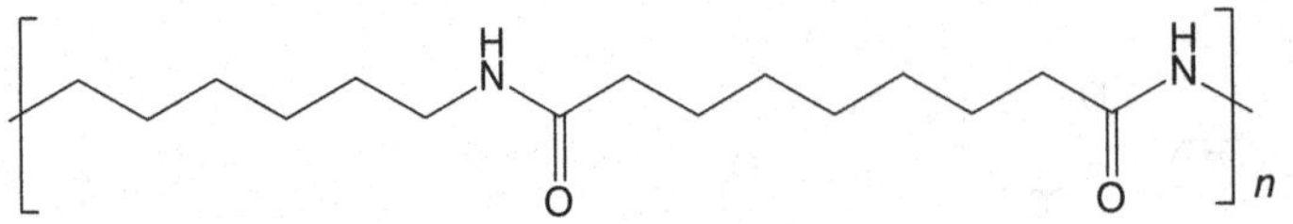

Polyamide 11

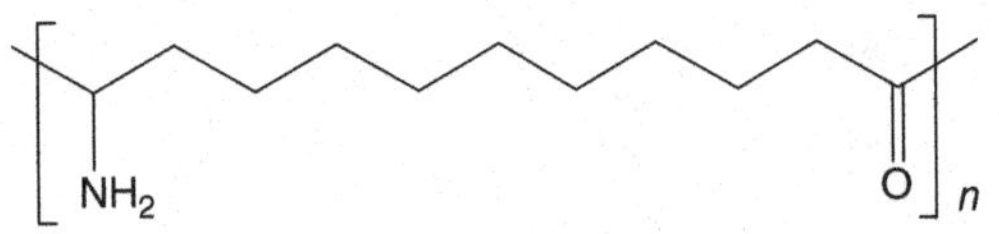

Polyamide 6,10

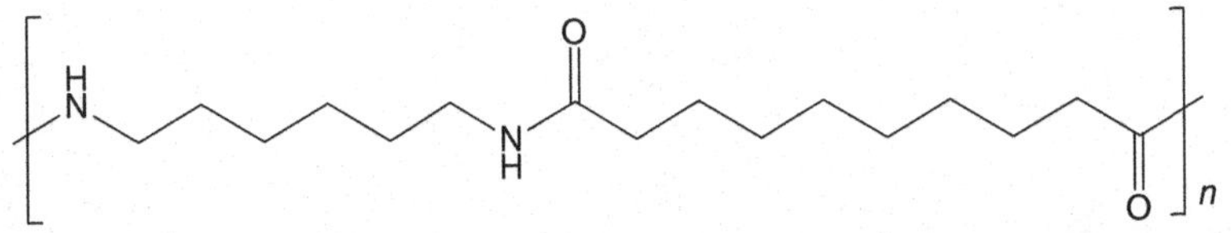

Find a route to amines or substitutes from biomass.
Reduce energy requirements by production processes.
Develop commercial process for production of caprolactam (used in the production of polyamide 6) from biomass via lysine.
Develop shorter process pathways with higher yields.

Polycarbonates
Diphenolic acid polycarbonate
Poly(propylene carbonate) or PCC

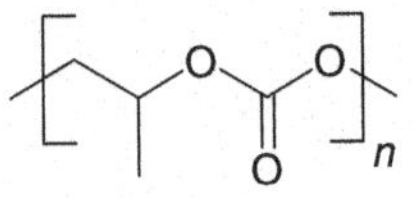

Find an economically practical polycarbonate route from CO_2.
New catalytic systems.
Improve polymer performance.

Solution to Exercise 2.   The formula for the repeat unit is $C_{11}H_{10}O_4$. The molar mass is thus

$$11 \times 12 + 10 \times 1 + 4 \times 16 = 206 \text{ g mol}^{-1}$$

From Table 12.2, the PTT repeat unit has three carbons from renewable PDO and the corresponding hydrogen atoms, that is, $(CH_2)_3$. The molar mass of the polymer fragment $(CH_2)_3$ is

$$3 \times (1 \times 12 + 2 \times 1) = 42 \text{ g mol}^{-1}$$

Therefore, the PTT polymer has a renewable embedded mass content of

$$\frac{42}{206} \times 100 = 20\%$$

The biogenic carbon content that can be accounted as balanced is

$$\frac{12 \times 3}{206} \times 100 = 17\%$$

12.2   Modified Natural Biopolymers

As noted earlier, biomass itself consists of polymers such as cellulose and hemicellulose having carbohydrates as their main building block. These biopolymers belong to the group of polysaccharides. They can be used as they are, with conditioning, or as the framework for deriving polymers and materials. Polymer blends, composites and nanocomposites with enhanced properties can also be formed by mixing natural polymers with other natural and synthetic polymers. The rest of lignocellulosic biomass is lignin formed by polymeric phenolic units and mainly used in composites.

12.2.1   Starch Polymers

Starch is the major carbohydrate (polysaccharide) stored in plants such as potato, wheat, corn, rice and cassava. Starch is a mixture of two main polysaccharides: amylose and amylopectin. Amylose is a linear chain polymer, whilst amylopectin is a branched polymer. A starch chain has between 500 and 2000 glucose units. Starch polymers are currently an economically competitive alternative to fossil based polymers.

The main sources of starch are crops such as potatoes and corn or starchy waste streams from the food industry. For most applications, the flow behavior of the natural starch is unsuitable and modification is required to improve its properties. Figure 12.5 shows the methods used for the modification of natural starch into useful polymeric materials. Starch for polymer production can be generated as:

i.   Pure starch produced from a typical wet milling process.
ii.   Partial fermentation starch by pretreatment of starchy waste streams.
iii.   Chemically modified starch by esterification or etherification to substitute some of the hydroxyl groups in the starch chains. Starch derivatives resulting from chemical modification include esters, ethers, phosphates, acetates

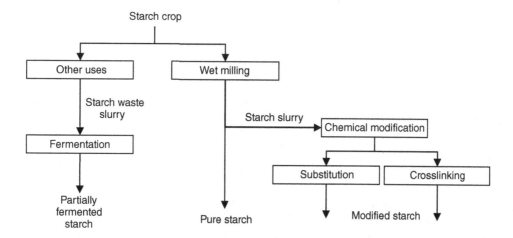

Figure 12.5   *General methods used for the modification of natural starch.*

and others. Another method for chemical modification is based on the crosslinking of starch structures with other polymers.

Extrusion, reactive extrusion and blending of conditioned starch with additives (e.g., plasticizers, bleaching and decolorizing agents) are used to obtain products with tailored properties. The products are mainly finished as pellets or nanoparticle fillers. The pellets can be further processed into films, fibers, sheets and other forms. The main starch polymer products include:

(A) *Thermoplastic starch (TPS)* polymers. Starch can be made as a thermoplastic polymer using plasticizers (e.g., glycerol). TPS may be blended with biodegradable polyesters to increase flexibility and resistance to moisture. Polyesters such as PHA, PLA and PCL can be used.

(B) *Starch based films* made by copolymerization with thermoplastic polyesters form biodegradable and compostable products such as composting bags, packaging, hygienic and agriculture products. The properties of the resulting films are similar to those made of low density polyethylene (LDPE). Starch based films have also started to substitute for polystyrene films. Biodegradable starch films can reduce the environmental impact from biomass production when used for the controlled release of fertilizers and by returning their carbon content back to the soil. This reduction then propagates towards end products, including biopolymer production itself.

(C) *Nonplasticized starch films* are formed by blending starch with poly(ethylene-co-acrylic acid) (PEAA). They have a potential application in membrane separation, where resistance to water and other solvents is required. Starch-PEAA blends can be processed into extrusion-blown films with semipermeable properties. Starch-PEAA films also show oxygen permeability much lower than LDPE films and can be used in food packaging. Their capability for the separation of organic and inorganic solutes from aqueous solutions offers great promise for the application of starch based membranes in a biorefinery.

(D) *Starch nanoparticle fillers* are being used in tyres, resulting in lower tyre weight and reduced rolling resistance.

The environmental impact of polymers needs to be assessed and compared against that of fossil based polymers. See Chapters 4 and 5 for LCA and a multicriteria analysis. The impact from the end of life stage depends on whether polymers go to recycling, incineration with or without energy recovery, disposal in landfills or composting. Depending on the difficulty of recycling and energy requirements for recycling, the end of life stage can add more impact (or otherwise) to the life cycle of the polymer. Composting of the polymer product can reduce the environmental impact when fast degradation occurs and the composting product can be directly used as fertilizer. In contrast, unfavorable landfill conditions can degrade polymers to methane, H_2S and chemicals harmful for the environment. Methane is a gas having a global warming potential impact 25 times higher than CO_2. H_2S has the potential for acidification and human toxicity.

Comparisons of life cycle impacts by functionality between bio based and fossil based polymers are necessary. This is because most of the bio based polymers cannot perform at the same levels as fossil based polymers. Often, blends or composites rather than pure bio based polymers need to be used. Thus, careful considerations of allocation factors for LCA are required. The following exercise shows an example of LCA for blends of TPS polymer with a petrochemical based polymer.

Exercise 3. LCA of TPS blends with poly(ε-caprolactone). BioStarch Film is producing biodegradable polymers by blending thermoplastic starch (TPS) polymers and aliphatic polyesters such as polycaprolactone (PCL) polymer. The company wants to manufacture a biodegradable polymer film with a tensile strength of 5 MPa (at the least) and, in doing so, wants to find out ways to mitigate environmental trade-offs.

The feedstocks for TPS production are starch, glycerol and water, for which the weight % contributing to the production of two types of TPS polymers is shown in Table 12.3. Table 12.3 also shows the weight % of TPS and PCL in various blends.

Table 12.4 shows the environmental impact assessments of various feedstocks to produce TPS and the environmental impact assessments to produce PCL. The environmental impact categories evaluated are cumulative fossil primary energy (CPE), global warming potential (GWP) 100-year horizon, photochemical ozone creation potential (POCP), acidification potential (AP) and eutrophication potential (EP).

Table 12.3 Composition (mass basis) and tensile strength of two TPSs, PCL and their blends.

TPS Polymer	Starch(%)	Glycerol(%)	Water(%)	TPS(%)	PCL(%)	Tensile Strength (MPa)
1	70	18	12	100	0	3
	70	18	12	75	25	6
2	74	10	16	100	0	21
	74	10	16	75	25	10
	74	10	16	60	40	9
	–	–	–	0	100	14

Table 12.4 Environmental impact factors for the production of 1 kg of materials.

Impact Category	Starch	Glycerol	Water	PCL	Units
CPE	49.1	26.6	0.0106	83	MJ
GHG	1.43	1.4	0.0007	3.1	kg CO_2 eq.
POCP	0.0012	0.00093	Negligible	0.0061	kg C_2H_4 eq.
AP	0.0104	0.0319	Negligible	0.0055	kg SO_2 eq.
EP	0.005	0.0242	0.0001	0.0005	kg PO_4^{3-}

Carry out environmental impact assessments to produce feasible blends using information in Tables 12.3 and 12.4. Report results in a format to help BioStarch Film make a decision.

Approach to Solution to Exercise 3

1. *Goal and scope:*
   a. To compare the aggregated environmental impact (EI) between polymer blends of TPS and PCL.
   b. To identify environmental trade-offs when blending different proportions of two TPS polymers and PCL and ways to mitigate the trade-offs.
   c. Boundary of the system. From the information available, the boundary of the system is from cradle to gate, producing the polymer blends from primary resource extraction through conversion processes. The functional unit is 1 kg of a polymer blend production. According to tensile strength data, 100% TPS-1 does not meet the minimum tensile strength of 5 MPa. Thus, the blends to be evaluated are the last five listed in Table 12.3. The categories to be evaluated are those for which characterization values are shown in Table 12.4.
2. *Life cycle inventory.* The inventories of resources, emissions and wastes from the production systems of starch, glycerol and water (for the production of TPS) and PCL have already been accounted for, in the impact assessments of the systems. The weighted average of the impact assessments of individual materials (Table 12.5) producing polymer blends is to be calculated according to the compositions shown in Table 12.3.
3. *Impact assessment.* The environmental impact values are going to vary according to polymer blend composition. As an example, the calculations are shown for TPS-1 in Table 12.5. Then, for 1 kg of the polymer blend with 75% TPS-1 and 15% CPL, the calculations of the environmental impact are shown in Table 12.6.

Table 12.5 Calculations of the environmental impact factors for the production of 1 kg of the TPS-1 polymer.

Impact Category	Starch	Glycerol	Water	Sum	Units
CPE	49.1 × 0.7	26.6 × 0.18	0.0106 × 0.12	39.2	MJ
GHG	1.43 × 0.7	1.4 × 0.18	0.0007 × 0.12	1.3	kg CO_2 eq.
POCP	0.0012 × 0.7	0.00093 × 0.18	Negligible	0.0010	kg C_2H_4 eq.
AP	0.0104 × 0.7	0.0319 × 0.18	Negligible	0.0130	kg SO_2 eq.
EP	0.005 × 0.7	0.0242 × 0.18	0.0001 × 0.12	0.0079	kg PO_4^{3-} eq.

Similar calculations are followed for the rest of the options. A generic equation for the calculation of each environmental impact (EI) is

$$EI\ blend = [(EI \times x)_s + (EI \times x)_g + (EI \times x)_w] \times Composition\ of\ TPS\ in\ blend$$

$$+EI_{CPL} \times Composition\ of\ CPL\ in\ blend$$

where EI is the impact factor for a particular environmental impact category, x is the mass fraction of starch (s), glycerol (g) and water (w) to produce TPS. Resulting comparisons are shown in Figure 12.6.

Table 12.6 *Calculations of the environmental impact factors for the production of 1 kg of the polymer blend with 75% TPS-1 and 25% PCL.*

Impact Category	TPS-1	PCL	Sum	Units
CPE	39.2×0.75	83×0.25	50.1	MJ
GHG	1.3×0.75	3.1×0.25	1.75	kg CO_2 eq.
POCP	0.0010×0.75	0.0061×0.25	0.0023	kg C_2H_4 eq.
AP	0.0130×0.75	0.0055×0.25	0.0111	kg SO_2 eq.
EP	0.0079×0.75	0.0005×0.25	0.0060	kg PO_4^{3-} eq.

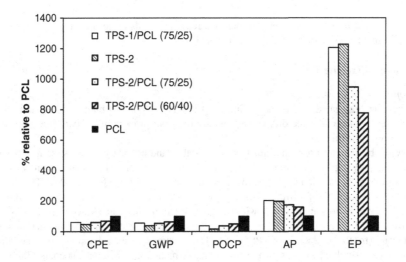

Figure 12.6 *Comparison of environmental impacts between TPS, PCL and their blends.*

4. *Interpretation.* Thermoplastic starch TPS-2 and the blends of TPS-1 and TPS-2 with PCL show better environmental performance than PCL in categories: cumulative fossil primary energy depletion, global warming potential and photochemical ozone creation potential. However, for acidification and eutrophication potential, PCL performs far better than the TPS based polymers. TPS-2 and 75% TPS-1 and 25% PCL blends have double the acidification potential of PCL. TPS polymer blends have up to 12 times higher eutrophication potential than PCL. Thus, an environmental impact trade-off exists between CPE-GWP-POCP and AP-EP. Such trade-offs are common for bio based products. Bio based products have higher nitrogen emissions and footprint due to application of fertilizers during biomass growth. Bio based products have a lower GWP and urban smog and save primary fossil energy resources. Comparing between TPS-1 and TPS-2 blends with the same compositions (75% TPS and 25 % PCL), it can be observed that TPS-1 has more impact with respect to EP and AP and almost an equal impact in the rest of the categories. The higher impact is due to the higher content of glycerol in TPS-1 (Table 12.3). For the lowest

environmental trade-off and with an acceptable tensile strength value of 9 MPa, a blend of 60% TPS and 40% PCL is acceptable. Pure PCL providing a tensile strength value of 14 MPa can also be chosen. The environmental impact potentials can be reduced by sourcing starch and glycerol from renewable feedstocks for TPS production.

12.2.2   Cellulose Polymers

Cellulose is found in nature as the primary cell wall component of plants and algae and is produced as biofilms by some bacteria. Unlike starch, cellulose is a straight chain polysaccharide polymer. These bio based polymers have lost market presence due to their replacement with petrochemical based polymers. However, this scenario is being reversed by significant advances in the development of new cellulose polymers for films, fibers and composites. Cellulose polymers are mainly finished as fibers or films, used in textiles, food (e.g., for sausages) and nonplastic products (e.g., varnishes). Rayon, cellophane and lyocell are examples of commercial cellulose products.

Wood pulp or cotton linters are the main sources of cellulose. The technologies for cellulose extraction into useful fibers and films are well known. The main processes are shown in Table 12.7. Refer to ***Online Resource Material, Chapter 12 – Additional Exercises and Examples*** for process flowsheets and a description of cellulose materials.

12.2.3   Natural Fiber and Lignin Composites

(A) *Natural fibers* are lignocellulosic complexes, in which cellulose is embedded in the lignin and hemicellulose matrix. Examples of natural fibers are those extracted from henequen or agave leaves, fruit peels, cotton seeds, bamboo, grass, etc. These natural materials are renewable, biodegradable, strong and abundant and have low densities and cost. Despite these advantages, their applications are limited because of degradation by microorganisms, UV light or high temperatures and because of their hydrophilic nature. The properties of natural fibers are improved by forming composites with other materials. To allow strong adhesion between natural fibers and the composite matrix, chemical pretreatment with stearic acid, mineral oil or maleated ethylene is required.

Natural fibers are used in textiles, in construction materials to provide insulation properties and in plastic composites for automotive parts and other applications. Other examples of natural fibers (e.g., flax fiber, hemp fiber, china reed fiber) in composite materials include underfloor panels, car interior side panels and fiberglass pallets. To allow strong adhesion between natural fibers and the composite matrix, chemical pretreatment with stearic acid, mineral oil or maleated ethylene is required.

(B) *Lignin* can form composites with epoxy resins. Epoxy resins are commonly produced from the reaction between epichlorohydrin and bisphenol A. Bisphenol A can be subsituted by less harmful diphenolic acid (see Chapter 11). Epoxy resins have a phenolic structure as shown in Figure 12.7. The phenolic structure of lignin makes it suitable as a filler in composites with epoxy resins, substituting a fraction of the petroleum based material. Lignin composites with epoxy resins have been used in microelectronics for the manufacture of printed wiring boards (PWBs). Lignin composites improve the tractability and recyclability of the epoxy fiberglass used in the PWBs. Lignin is also used to reinforce rubber. One example on epoxy resin manufacturing from soya or algal oil applying LCA based decision making is shown in ***Case Study 2 in the Companion Website***.

Lignin is a by-product from paper kraft pulping mills. High purity lignin can be produced in a biorefinery using the organosolv process. Refer to ***Online Resource Material, Chapter 12 – Additional Exercises and Examples*** for a description and process flowsheet of the organosolv process.

Exercise 4. Trade-off between glucan purity and lignin recovery in the Organosolv process. A pilot scale investigation of acid hydrolysis of woody biomass in Organosolv shows that the yield of polysaccharides (in % w/w dry biomass basis) is inversely proportional to the lignin recovery (lignin in product/lignin in feedstock in % w/w) according to the following equation:

$$\text{Yield of polysaccharides} = \frac{310}{(\text{Lignin recovery})^{0.375}}$$

Table 12.7 *Process technologies for extraction of cellulose and derivatives.*

Technology	Viscose Process	Lyocell Process	Celsol Process	Carbamate Process	Other Processes
Type of treatment	Chemical derivative (Xanthate)	Solvent extraction	Biochemical (enzymatic)	Chemical derivative (Cellulose carbamate)	Chemical or solvent extraction
Reagents or solvents used	NaOH CS_2 H_2SO_4	N-methylmorpholine N-oxide (NMMO) Stabilizers (propyl gallate)	Cellulases NaOH	Urea NH_3 NaOH Catalyst	(a) $(CH_2O)_x$/DMSO (b) N_2O_4/DMSO (c) Ionic liquids (e.g., 1-butyl-3-methylimidazolium chloride) See Introduction chapter. DMSO: dimethyl sulfoxide
Product	Regenerated cellulose (e.g., rayon fiber or cellophane films)	lyocell	Biotransformed cellulose	Nitrogenated cellulose	Methylol cellulose (a) Nitrogenated cellulose (b) Extracted cellulose (c)
Advantages	Advanced developments possible	Cellulose is not chemically modified Solvent is nontoxic and can be regenerated and recycled	More soluble and functionalized cellulose Avoids toxic solvents	More soluble and functionalized cellulose Avoids toxic reagents	(a) and (b) Produce useful derivatives (c) Direct extraction of cellulose
Challenges	Control of toxic CS_2 and H_2S emissions Recovery and recycle of reagents	Degradation by-products must be minimized by optimizing process conditions and using cellulose stabilizers	Pretreatment to improve enzymatic effectiveness and efficiency Process scale-up	Minimize reaction by-products Scale-up of the process	Separation and regeneration of solvents or reactants is difficult Process scale-up

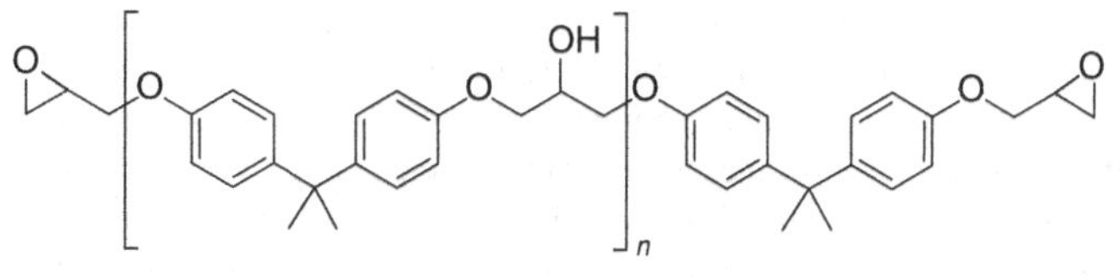

Figure 12.7 *Epoxy resin structure.*

With increasing yield of polysaccharides, purity of glucan (a glucose polysaccharide) recovered reduces according to the following correlation.

$$\text{Purity of glucan} = \frac{16281}{e^{0.088 \times \text{yield of polysaccharides}}}$$

Purity is in % w/w polysaccharides basis. Using the two correlations given, answer to the following questions:

a. Find the maximum level of lignin recovery achievable for a maximum glucan purity of 90%.
b. Find the purity of glucan for a maximum polysaccharide yield of 66%. Calculate the corresponding minimum yield of lignin.
c. Plot the graphs of yield of lignin (*y* axis) versus yield of polysaccharides (*x* axis) and purity of glucan (*y* axis) versus yield of polysaccharides (*x* axis). Discuss the main observations and trade-offs. Discuss how optimal trade-offs can be achieved. Determine the optimal value of the polysaccharide yield, if any.

Solution to Exercise 4. Refer to *Online Resource Material, Chapter 12 – Additional Exercises and Examples* for the solution to Exercise 4.

Exercise 5. Economic aspects of the Organosolv process. A company developing an Organosolv proprietary technology is interested to understand the critical cost parameters that would influence the product yields and purity. The Organosolv process can be operated over a range of operating conditions with negligible cost implications. The operating condition can lead to the following options: (1) high lignin recovery and low polysaccharides yield, but with high purity of glucan or (2) low lignin recovery and high polysaccharides yield, but with low purity of glucan. However, (2) requires downstream separation of polysaccharides. Depending on the cost indications, the company would decide whether to invest for downstream separation processes towards a fully developed biorefinery or to sell the Organosolv products of (1). Give correlations for the critical cost parameters using equations for polysaccharides yields and glucan purity shown in Exercise 4. The following assumptions can be made:

i. The price of lignin does not depend on the purity of lignin. Lignin product from Organosolv has a market potential, though small, to be sold as a substrate for other end products.
ii. Polysaccharides stream with lower than 90% by weight glucan purity has a very low market value.
iii. The capital and operating costs of the Organosolv process are not much influenced by the operating conditions leading to various product grades.

Solution to Exercise 5. Refer to *Online Resource Material, Chapter 12 – Additional Exercises and Examples* for the solution to Exercise 5.

12.3 Modeling of Polymerization Reaction Kinetics

As discussed earlier, primary biopolymers are derived from direct conversion of biomass and secondary biopolymers are produced by the reaction between biomass based monomers. To understand and manipulate the production process of these types of biopolymers, the polymerization kinetics needs to be modeled. Kinetic models of polymerization reactions are needed to find the concentration profiles of chemical constituents in the reaction mixture. Kinetic models are also needed for polymer reactor design. One of the main characteristics of a polymer, upon which other properties depend, is the molar mass. The kinetic model must follow the evolution of the degree of polymerization and molar mass distribution according to the various reaction mechanisms, discussed in this section. The modeling of bacterial growth kinetics, relevant to produce primary biopolymers, was shown in an earlier chapter. The modeling of polymerization reactions for the production of secondary biopolymers is shown as follows.

12.3.1   Chain-Growth or Addition Polymerization

In chain-growth polymerization, polymers are formed by sequential addition of a monomer (M) to the growing polymer chain (P_n) without the elimination of any part of the monomer molecule. The final product from this type of polymerization is called a chain-growth or addition polymer (Figure 12.1). Chain-growth polymerization can be shown as

$$P_n + M \rightarrow P_{n+1} \tag{12.5}$$

In general, monomers containing a double bond form *addition polymers*. The polymerization is initiated by decomposition of an unstable compound. Depending on the type of *initiator* or catalyst, a free radical, anion or cation of the monomer is formed. Most monomers can be polymerized by the free radical mechanism. Monomers are selective to the type of ionic mechanism. The presence of an electron-donating group (e.g., $-OH$, $-CH_2CH_3$, $-HC=CH_2$, $-C_6H_5$) in a monomer generates a partial negative charge and polymerization is favored by cationic initiators. A partial charge refers to a noninteger charge value. The presence of an electron-withdrawing group (e.g., $-CHO$, $-CH_2COCH_3$, $-CH_2COOH$, $-CH_2COOCH_3$) in a monomer generates a partial positive charge and polymerization is favored by anionic initiators. However, alkenyl and phenyl substituent polymers can undergo both anionic and cationic polymerization. Another mechanism is the coordination polymerization. As free radical polymerization is suitable to most of the monomers, this mechanism is discussed as follows.

The free radical polymerization mechanism has three major stages: *initiation*, *propagation* and *termination*. The initiation step can be induced by peroxides, azo compounds, redox systems, or irradiation. In the *initiation step*, free radicals (R^*) are generated from the initiator (I) with a rate constant k_d. Depending on the initiator, two radical species produced could be identical or different or a radical and a neutral molecule. The initiation reaction assuming that two identical radicals are formed is shown as

$$I \overset{k_d}{\rightarrow} 2R^* \tag{12.6}$$

Then a radical attaches to a monomer molecule to form the chain-initiating radical M^* at an initiation rate constant k_i, as shown by

$$R^* + M \overset{k_i}{\rightarrow} M^* \tag{12.7}$$

The formation of the monomer radical is much faster due to the high reactivity of the initiator radical. Therefore, the rate of initiation r_{is} is expressed only in terms of the dissociation of the initiator (as the rate limiting step):

$$r_{is} \cong r_d = \frac{d[M^*]}{dt} = 2fk_d[I] \tag{12.8}$$

where f is the initiator efficiency defined as the fraction of the radicals successfully initiating a polymer chain and $[I]$ is the activity (for a nonideal system) or concentration (for an ideal system) of the initiator. The stoichiometric factor of 2 results from the assumption that an initiator splits into two identical radicals, as shown in Equation (12.6).

In the *propagation* step, the addition of more monomer molecules to the free radicals M^* leads to radicals of increasing chain length (i.e., from n to $n+1$). This *propagation* reaction is shown by

$$M_n^* + M \xrightarrow{k_p} M_{n+1}^* \qquad (12.9)$$

Assuming that the rates of reaction for all free radicals of chain length n are the same, the rate of propagation with rate constant k_p can be expressed as

$$r_p = -\frac{dM}{dt} = k_p[M][M^*] \qquad (12.10)$$

The *termination step* involves two reactions: *combination* and *disproportionation* between two radicals. Two growing chain radicals M_n^* and M_m^* combine to form one polymer molecule of chain size $n+m$ with a termination rate constant k_{tc}, as shown by

$$M_n^* + M_m^* \xrightarrow{k_{tc}} M_{n+m} \qquad (12.11)$$

In the termination by the *disproportionation* reaction, with a rate constant k_{td}, a hydrogen atom (or another functional group) is transferred from one radical to another, forming two polymer molecules. A double bond is formed in the molecule donating the hydrogen atom. This can be shown by

$$M_n^* + M_m^* \xrightarrow{k_{td}} M_n + M_m \qquad (12.12)$$

Both types of termination can take place at the same time and at different proportions depending on the monomer, the initiator and the polymerization conditions. This affects the degree of polymerization and the average molar mass of the polymer.

The rate of termination r_t with the overall rate constant k_t for two growing chain radicals M^* having the same reactivity can be expressed as

$$r_t = -\frac{dM^*}{dt} = 2k_t[M^*]^2 \qquad (12.13)$$

The model equations shown in Equations 12.10 to 12.13 assume that k_p and k_t are independent of the length chain of the polymeric radical. This simplification allows a generic kinetic model to be formulated for the overall polymerization reaction, shown as follows.

The monomer molecules are consumed by the initiation and propagation reactions according to Equations (12.7) and (12.9). Then the rate of consumption of the monomer during polymerization can be expressed as follows. Equation (12.14) shows the *rate of polymerization*:

$$-\frac{dM}{dt} = r_{is} + r_p \qquad (12.14)$$

For a reaction producing a long-chain polymer the number of monomer molecules reacting in the initiation step can be assumed to be less than the number in the propagation step, that is: $r_p \gg r_{is}$. Then the rate of polymerization r_{pol} can be expressed as

$$r_{pol} = -\frac{dM}{dt} = r_p = k_p[M][M^*] \qquad (12.15)$$

$$nCH_2CHCl \rightarrow (CH_2CHCl)_n \qquad nCH_2OCH_2 \rightarrow (CH_2CH_2O)_n$$

(a) (b)

Figure 12.8 *(a) Polyvinyl chloride and (b) poly(ethylene oxide) are formed by chain-growth polymerization.*

A steady state can also be assumed when the rate of formation and the rate of consumption of radicals M^* becomes equal. Therefore, the concentration of M^* remains constant, that is, $dM^*/dt = 0$ and $r_{is} = r_t$. The steady-state assumption then leads to the following equation for the concentration of the radicals M^* (from Equations (12.8) and (12.13)):

$$[M^*] = \left(\frac{fk_d[I]}{k_t}\right)^{1/2} \tag{12.16}$$

Substituting Equation (12.16) into Equation (12.15), the rate of polymerization can be approximated as

$$r_{pol} = k_p \left(\frac{fk_d[I]}{k_t}\right)^{1/2} [M] \tag{12.17}$$

Since the rate of initiation can have various forms according to the type of initiator, the rate of polymerization can be written as follows in terms of the rate of initiation from Equation (12.8):

$$r_{pol} = k_p[M]\left(\frac{r_{is}}{2k_t}\right)^{1/2} \tag{12.18}$$

The reaction constants are determined for each initiator. The reaction constants depend also on the reaction medium when a solvent is used. Figure 12.8 shows two examples of polymers formed by chain-growth polymerization. The formation of poly(ethylene oxide) takes place by a ring-opening reaction and follows a chain-growth or addition mechanism. A ring-opening polymerization reaction involves a molecule having a cyclic structure. The ring is opened to form polymer chains.

The *number average degree of polymerization (DP)* is the average number of monomer molecules contained in a polymer molecule. The DP can be determined from the kinetic chain length. The kinetic chain length l is the average number of monomer molecules consumed per radical that initiates a polymer chain. The kinetic chain length is the ratio of the rate of formation of propagating chains and the rate of formation of the active radicals in the initiation step:

$$l = \left(\frac{r_p}{r_{is}}\right) \tag{12.19}$$

At the steady state, $r_{is} = r_t$. Thus, the following equation is deduced by substituting Equations (12.10) and (12.13) into Equation (12.19):

$$l = \left(\frac{r_p}{r_t}\right) = \frac{k_p[M][M^*]}{2k_t[M^*]^2} = \frac{k_p[M]}{2k_t[M^*]} \tag{12.20}$$

Combining Equation (12.20) with Equation (12.16) for a steady-state assumption gives

$$l = \frac{k_p[M]}{2\left(k_t fk_d[I]\right)^{1/2}} \tag{12.21}$$

If two propagating radicals *combine* to form a final or dead polymer molecule in the termination step, the relation between the DP and the kinetic chain length is shown by

$$DP = 2l \tag{12.22}$$

Equation (12.23) below is valid when the dead polymer molecule is formed by *disproportionation* of the growing chain:

$$DP = l \tag{12.23}$$

However, both types of termination can occur. Thus, the DP can be expressed as shown by

$$DP = \frac{2l}{(2-a)} \tag{12.24}$$

where a is the fraction of propagating chains terminated by *combination*.

The number average molar mass of a polymer to the degree of polymerization and the molar mass of the repeat unit (M_0) is given by

$$M_n = M_0 \times DP \tag{12.25}$$

In general, polymers with a high M_n are desirable. At high initiator concentrations, according to Equation (12.21), the kinetic chain length decreases. This means that the DP decreases, leading to a polymer with low M_n. However, according to Equation (12.17), high initiator concentrations increase the rate of polymerization and, hence, the polymer formation. This trade-off can be used to control the molar mass of a polymer. It can also be observed that the DP decreases as conversion decreases. This is due to the lower availability of monomer molecules.

Exercise 6. Number average molar mass of a chain-growth polymer. Find the number average molar mass of a polymer P after 10% conversion of the monomer in a continuous process. Assume termination by combination only and neglect the change in concentration of the initiator. Available data are as follows:

$$fk_d = 3 \times 10^{-6} \text{ s}^{-1}$$

$$k_p = 176 \text{ L mol}^{-1}\text{s}^{-1}$$

$$k_t = 72 \times 10^6 \text{ L mol}^{-1}\text{s}^{-1}$$

$$[M]_{in} = 8 \text{ mol L}^{-1}$$

$$[I]_{in} = 0.02 \text{ mol L}^{-1}$$

Molar mass of the repeat unit $= 104 \text{ g mol}^{-1}$

Solution to Exercise 6. The final concentration of the monomer is

$$[M] = 8 \times (1 - 0.1) = 7.2 \text{ mol L}^{-1}$$

In a continuous process, the steady state can be assumed and then $r_p = r_{pol}$. Thus, using Equations (12.21) and (12.22) with $[I] = [I]_{in}$:

$$DP = \frac{k_p[M]}{(k_t fk_d[I])^{1/2}} = \frac{176 \times 7.2}{(72 \times 10^6 \times 3 \times 10^{-6} \times 0.02)^{1/2}} = 610$$

Thus, the number average molar mass is

$$M_n = 610 \times 104 = 63400 \text{ g mol}^{-1}$$

12.3.2   Step-Growth Polymerization

Step-growth polymerization occurs when monomers with more than one functional group react to form polymer molecules with or without the elimination of a compound such as H_2O. A special case of step-growth polymerization with product elimination is called condensation polymerization. Polymers formed from this type of step-growth polymerization are condensation polymers (Figure 12.1).

In the step-growth mechanism, monomer molecules react fast to form low molar mass molecules. These molecules continue to react with each other producing continually growing chains. The polymerization reaction between any two growing polymer molecules (e.g., P_m and P_n) is shown by

$$P_m + P_n \leftrightarrows P_{m+n} + E \tag{12.26}$$

where P_{m+n} is the resulting polymer molecule and E is the elimination product, if any. Note that step-growth polymerization is reversible. Formation of polyesters (reaction between carboxylic acid and polyol) and polyamides (reaction between amine and carboxylic acid) are examples of step-growth condensation polymerization.

Condensation polymerization is the reaction between two different functional groups of a monomer (e.g., hydroxycarboxylic acids) or two different monomers with identical functional groups (e.g., diols and dicarboxylic acids). For the modeling of condensation polymerization, a generic equilibrium reaction is considered:

$$M1 + M2 \leftrightarrows P + E \tag{12.27}$$

The constants k_1 and k_{-1} are the forward and backward reaction rate constants. $M1$ and $M2$ are the two monomers, P is the polymer and E is the elimination product. At equilibrium, the reaction rates of the forward and backward reactions are equal. This is shown by

$$k_1 [M1] [M2] = k_{-1} [P] [E] \tag{12.28}$$

Therefore, the equilibrium constant (K_{eq}) is expressed as

$$K_{eq} = \frac{[P] [E]}{[M1] [M2]} = \frac{k_1}{k_{-1}} \tag{12.29}$$

At the beginning of polymerization, the forward reaction predominates and the backward reaction can be neglected. As the reaction proceeds, equilibrium may be reached. Equilibrium limits the conversion of the monomers into high molar mass polymers. Formation of the product or forward reaction can be facilitated by removing a product from the reaction mixture.

The kinetics of condensation polymerization between diol and dicarboxylic acid is taken as an example[2]. This is a condensation or step-growth reaction producing polyester with elimination of water. The first step is the reaction between the diol and diacid monomers, also known as esterification. Representing a diol as $HO-R-OH$ and a dicarboxylic acid as $HOOC-R'-COOH$, the reaction can be expressed as

$$HO-R-OH + HOOC-R'-COOH \xrightarrow{\text{catalyst}} HO-R-OCO-R'-COOH + H_2O \tag{12.30}$$

This product can react with another diol monomer to give a polymer made up of three monomers (trimer): two diols and one dicarboxylic acid. Another trimer can be formed if the product reacts with the dicarboxylic acid. The product in Equation (12.30) can react with itself to form a tetramer. These short-chain polymeric molecules are known as *oligomers*. Then all these species generated react between themselves in a stepwise fashion forming polymer molecules of growing size. As noted earlier, the formations of dimer, trimer and tetramer molecules quickly consume the monomer at the beginning of polymerization. The condensation polymerization reactions are catalyzed by acids.

The generation of multiple species with varying sizes complicates the modeling as every polymer molecule of size n has its own reactions. To simplify the problem, the reaction is shown in terms of the reactivity of the functional groups of the monomers, rather than the reactivity of each dimer, trimer, etc. It is assumed that the reactivity of a functional group is independent of its molar mass. Such reactivity can then be expressed as concentration (or activity for a nonideal system) of the functional group. In the following equation, the concentration or activity of carboxyl groups (–COOH) is shown as $[A]$ and that of hydroxyl groups (–OH) is shown as $[B]$. The rate of consumption of the carboxyl groups is thus

$$\frac{d[A]}{dt} = -k'[A][B] \tag{12.31}$$

where k' is the overall reaction constant including the effect of the catalyst. For equal concentrations of the functional groups, Equation (12.31) becomes

$$\frac{d[A]}{dt} = -k'[A]^2 \tag{12.32}$$

This assumption is known as the concept of *equal reactivity of functional groups*. Integration of Equation (12.32) gives the concentration of carboxyl groups as a function of time. This allows tracking the conversion of the monomers into polymer production. For an initial condition of $[A] = [A]_0$ at $t = 0$ and integrating over a time t gives

$$\left[\frac{1}{[A]} - \frac{1}{[A]_0} \right] = -k't \tag{12.33}$$

By rearranging to a linear equation ($y = mx + c$), the following equation is formed:

$$\boxed{\left[\frac{1}{[A]} \right] = \frac{1}{[A]_0} - k't} \tag{12.34}$$

Thus, a plot of the inverse of concentration against time gives a line with slope equal to $-k'$ and the intercept is $1/[A]_0$.

A correlation between the degree of polymerization (DP) and conversion is needed to follow the course of the reaction and the molar mass of the polymer product. During the reaction, the total number of repeat units remains unchanged. This number is equal to the initial number of monomer molecules N_0. The repeat units from the monomer are now part of the various polymeric molecules formed during the reaction. The number of repeat units in the polymeric molecules coming from the initial monomer molecules is N. The degree of polymerization is expressed as

$$DP = \frac{N_0}{N} \tag{12.35}$$

The DP in terms of the concentrations of the functional groups involved is shown as

$$DP = \frac{[A]_0}{[A]} = \frac{[B]_0}{[B]} \tag{12.36}$$

Conversion (X) can be defined as

$$X = 1 - \frac{[A]}{[A]_0} \tag{12.37}$$

Combining Equation (12.37) with Equation (12.36) gives

$$DP = \frac{1}{1 - X} \tag{12.38}$$

The number average molar mass can be obtained using Equation (12.25). Note that when full conversion of the monomer containing the functional group A is approached, the DP tends to be an infinite number. This is different to the radical polymerization, wherein the higher conversion leads to a lower DP. However, the model shown may be oversimplified for some systems and thus limited according to the various assumptions stated earlier (e.g., equal reactivity of functional groups). The equations developed are relevant for batch processes and can serve as a basis for the development of other reactor configurations.

12.3.3   Copolymerization

Copolymers are produced by polymerizing two or more monomers. These polymers have repeat units coming from the monomers involved. Most polymer products used in daily life are copolymers. Copolymerization is used to adjust the properties of polymers for a particular application by variation of the composition of the repeat units. The degree of modification of the properties depends on the copolymer structure. The common copolymer structures are:

Random polymers are formed when the reactivity of the monomers involved is similar and polymer chains are generated without any particular pattern. For a copolymer of A and B, an example of the structure formed by this type of polymer can be shown as: –BBAABBBABABBBB–. An overall property of a copolymer is the weighted average of properties of monomer repeat units, according to their molar compositions in the copolymer.

Statistical polymer has monomeric units arranged according to a known statistical probability.

Alternating structure results when radicals of monomer A react only with the radicals of monomer B. A monomer radical never reacts with itself. Alternating polymers (e.g., –BABABA–) have the combined properties of the monomers involved. Polymers with a highly regular alternating pattern are generally crystalline.

Block structure is generated by performing the polymerization in multiple stages. Examples of structures are –BBBBAAAA–, –BBBAAAABBB–, etc. The blocks of molecules are repeated in the polymer chain.

Graft and branched structures are formed due to an intramolecular rearrangement of the macromolecules. In general, the repeat unit in the branches is different to that in the main chain.

The complexity of a copolymerization system makes its modeling more difficult than for a simple polymer. Copolymerization can be simplified considering the following four reactions for a free radical mechanism[2]:

1.   The reaction of the first monomer with a radical (also of the first monomer) to give a homopolymer radical:

$$M_1^* + M_1 \rightarrow M_1M_1^* \tag{12.39}$$

for which the reaction rate equation is given by

$$r_{11} = k_{11}[M_1^*][M_1] \tag{12.40}$$

2.   The copolymerization between a radical of the first monomer with a neutral molecule of the second monomer gives rise to

$$M_1^* + M_2 \rightarrow M_1M_2^* \tag{12.41}$$

for which the reaction rate equation is given by

$$r_{12} = k_{12}[M_1^*][M_2] \tag{12.42}$$

3.   The reaction of the second monomer with a radical (also of the second monomer) to give the corresponding homopolymer radical:

$$M_2^* + M_2 \rightarrow M_2M_2^* \tag{12.43}$$

for which the reaction rate equation is given by

$$r_{22} = k_{22}[M_2^*][M_2] \tag{12.44}$$

4. The copolymerization between a radical of the second monomer with a neutral molecule of the first monomer:

$$M_2^* + M_1 \rightarrow M_2M_1^* \tag{12.45}$$

for which the reaction rate equation is given in Equation 12.46.

$$r_{21} = k_{21}[M_2^*][M_1] \tag{12.46}$$

To simplify the modeling, the steady state is assumed. Then, the concentrations of the radical species are constant and the reaction rates r_{12} and r_{21} are equal, as shown by

$$r_{12} = r_{21} \Rightarrow k_{12}[M_1^*][M_2] = k_{21}[M_2^*][M_1] \tag{12.47}$$

The rates of polymerization of the monomers can be expressed from their rates of consumption by the following equations:

$$\frac{d[M_1]}{dt} = -k_{11}[M_1^*][M_1] - k_{21}[M_2^*][M_1] \tag{12.48}$$

$$\frac{d[M_2]}{dt} = -k_{12}[M_1^*][M_2] - k_{22}[M_2^*][M_2] \tag{12.49}$$

Dividing Equation (12.48) by Equation (12.49), the ratio of monomers present in the polymer molecule is given by

$$\frac{d[M_1]}{d[M_2]} = \frac{k_{11}[M_1^*][M_1] + k_{21}[M_2^*][M_1]}{k_{12}[M_1^*][M_2] + k_{22}[M_2^*][M_2]} \tag{12.50}$$

Mathematical manipulation allows having a measurable ratio of concentrations $[M_1]$ and $[M_2]$. Multiplying Equation (12.50) with $(1/[M_1^*])/(1/[M_1^*])$ results in

$$\frac{d[M_1]}{d[M_2]} = \frac{k_{11}[M_1] + k_{21}[M_1]\dfrac{[M_2^*]}{[M_1^*]}}{k_{12}[M_2] + k_{22}[M_2]\dfrac{[M_2^*]}{[M_1^*]}} \tag{12.51}$$

The following equation is obtained from Equation (12.47) for the ratio of radical concentrations:

$$\frac{[M_2^*]}{[M_1^*]} = \frac{k_{12}}{k_{21}}\frac{[M_2]}{[M_1]} \tag{12.52}$$

Substituting Equation (12.52) into Equation (12.51) and multiplying with the unity $(1/k_{12})/(1/k_{12})$ gives

$$\frac{d[M_1]}{d[M_2]} = \frac{\dfrac{k_{11}}{k_{12}}[M_1] + [M_2]}{[M_2] + \dfrac{k_{22}}{k_{21}}\dfrac{[M_2]^2}{[M_1]}} \tag{12.53}$$

According to Equation (12.53), the ratio of monomers present in the copolymer molecules can be varied by manipulating the concentrations of the monomers. The ratio of reaction constants in Equation (12.53) is the selectivity of a given monomer radical to react with a molecule of the same monomer or with a molecule of the comonomer. Such ratios are given in the following equations:

For the monomer M_1:

$$S_{M1} = \frac{k_{11}}{k_{12}} \tag{12.54}$$

For the monomer M_2:

$$S_{M2} = \frac{k_{22}}{k_{21}} \tag{12.55}$$

Depending on the ratio given in Equations (12.54) and (12.55), copolymers can be ideal, random, alternating, block, graft and branched copolymers. Ideal copolymers are formed if $S_{M1}S_{M2} = 1$. Random copolymers are formed if $S_{M1} \approx S_{M2}$. Alternating copolymers are formed if $S_{M1}S_{M2} = 0$; perfect alternation occurs when both ratios are equal to zero. For $S_{M1} > S_{M2}$, block copolymers are formed.

Exercise 7. Kinetic modeling of ring-opening polymerization. Consider the homogeneous ring-opening polymerization of lactide to produce PLA. The ring opening polymerization (ROP) can be shown by a mechanism similar to chain-growth polymerization (in reality a combination of step-growth and chain-growth polymerizations may occur)[3]. Thus, the reaction scheme consists of initiation, propagation and termination reactions. Develop a kinetic model with the following assumptions:

- Concentrations are independent of spatial position within the reactor.
- Termination occurs by transfer of the polymerization activity from a growing chain to a monomer molecule and is irreversible.
- Propagation rate constants are functions of the growing chain length or degree of polymerization. Thus, the simplification for the same reaction constant for all the chain lengths is not applicable.
- Only monomer units are added to the growing chain during propagation.

Solution to Exercise 7. Refer to ***Online Resource Material, Chapter 12 – Additional Exercises and Examples*** for the solution to Exercise 7.

12.4   Reactor Design for Biomass Based Monomers and Biopolymers

The aim of polymer reactor design is to optimize molar mass distribution for desired polymer properties. Design parameters involved are reaction conditions such as temperature, reactant concentrations and by-product removal strategies. Production of primary biopolymers (i.e., directly produced from biomass via fermentation) is affected by the pH, substrate and product inhibition in the fermenter. The following sections show the reactor design for the production of monomers and biopolymers via fermentation or chemical synthesis using kinetic models.

12.4.1   Plug Flow Reactor (PFR) Design for Reaction in Gaseous Phase

Some monomers or intermediate precursors to monomers or polymers are synthesized in the gaseous phase. Polyethylene is a polymer produced by gaseous phase reaction. Reactions in gaseous or more generally in the fluid phase are carried out in a plug flow reactor (PFR). In this section, the design of a PFR for reactions in the gaseous phase is shown.

The volume (V) of a PFR can be estimated from the mass balance of reactant species i across a control volume (see Equation (11.33)), as shown by

$$\boxed{r_i dV = FC_{in,i} dX}$$ (12.56)

The reactor design equation can be shown in terms of molar flow, X is the reaction conversion and $FC_{in,i}$ is the inlet molar flow rate ($n_{in,i}$) of reactant i to the reactor. Thus, the mass balance equation takes the form of

$$r_i dV = n_{in,i} dX$$ (12.57)

The equation of rate of reaction (r_i) of order a and rate constant k can be expressed in terms of the partial pressure of the reaction species (p_i):

$$r_i = kp_i^a$$ (12.58)

The partial pressure needs to be expressed in terms of the difference in stoichiometric coefficients (v_i) between the products and reactants as shown below. This is because the total number of moles vary with the PFR length, as discussed in Solution to Exercise 8 in Chapter 11.

$$\Delta n = \sum v_{i_{products}} - \sum v_{i_{reactants}}$$ (12.59)

Note that Δn can be positive or negative depending on the stoichiometry of the reaction.

Assuming that the reactor is isothermal and the pressure loss in the reactor is negligible, *Dalton's law* can be used to relate the number of moles to partial pressure of a component, as shown by

$$\boxed{p_i = y_i P}$$ (12.60)

where P is the total gas pressure and y_i is the molar fraction of component i in the gas phase. Changes in the total number of moles vary with the conversion according to

$$n_T = n_{in,T} + n_{in,T} y_{in,i} X \Delta n$$ (12.61)

where $y_{in,i}$ is the molar fraction of reactant i in the inlet with the total inlet molar flow rate of $n_{in,T}$. The number of moles of reactant i at a certain length in the rector is its inlet moles minus moles converted by the reaction at that point, as shown by

$$n_i = n_{in,i}(1 - X)$$ (12.62)

Combining Equations (12.61) and (12.62), the molar fraction can be expressed as

$$y_i = \frac{n_{in,i}(1 - X)}{n_{in,T} + n_{in,T} y_{in,i} X \Delta n}$$ (12.63)

Then, substitution of Equation (12.63) into Equation (12.60) yields, for the partial pressure of i,

$$p_i = \frac{n_{in,i}(1 - X)}{n_{in,T} + n_{in,T} \, y_{in,i} X \Delta n} P$$ (12.64)

Equation (12.64) can be substituted into Equation (12.57) to obtain the following equation 12.65 for the mass balance equation of a PFR with a first-order reaction:

$$\frac{dV}{n_{in,i}} = \frac{dX}{k \frac{n_{in,i}(1-X)P}{n_{in,T} + n_{in,T} \, y_{in,i} X \Delta n}}$$ (12.65)

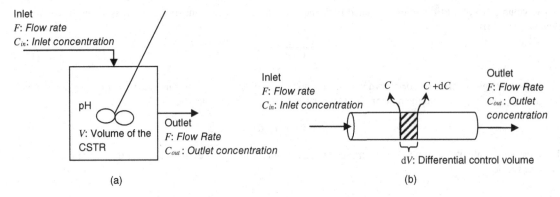

Figure 12.9   *Schematic representation of (a) a CSTR and (b) a PFR.*

Integrating Equation (12.65) from $V = 0$ to V for conversion from $X = 0$ to X_{out} (from the inlet to the outlet, respectively), the reactor design equation for an isothermal PFR with a first-order reaction in the gas phase at constant pressure is shown as

$$V = \int_0^{X_{out}} \frac{dX}{k \dfrac{(1-X)P}{n_{in,T}(1+y_{in,i}X\Delta n)}} \tag{12.66}$$

The PFR can be used for synthesis of methyl 10-undecylenate ($C_{12}H_{22}O_2$). This compound is an intermediate in the synthesis pathway of monomer 11-aminoundecanoic acid ($NH_2(CH_2)_{10}COOH$)[4]. The polymer of this monomer is polyamide PA 11. An example of the production of the monomer in a PFR reactor is shown as a further exercise in the ***Online Resource Material, Chapter 12 – Additional Exercises and Examples***.

12.4.2   Bioreactor Design for Biopolymer Production – An Example of Polyhydroxyalkanoates

The production of primary biopolymers involves fermentation in bioreactors that can be batch reactors, continuous stirred tank reactors (CSTR) or PFR. Consider polyhydroxyalkanoates (PHAs) as an example of primary biopolymer production. PHAs are polyesters of hydroxyalkanoates synthesized by numerous bacteria as carbon and energy storage compounds (the equivalent to fat for humans). The reactor volume to carry out the fermentation process to produce PHAs can be estimated from kinetic modeling. This is shown as follows for CSTR and PFR reactors.

Figure 12.9 shows a schematic of the two reactors. Mass balances are formulated accordingly.

1.   CSTR. The mass balance for the production of PHA is

Mass flow rate of PHA in the inlet + Mass of PHA generated = Mass flow rate of PHA in the outlet

For a constant density system, the volumetric flow rate input and output are the same and the mass balance (see Equation (11.45)) for the CSTR is given by

$$FC_{in} + r_i V = FC_{out} \tag{12.67}$$

where F is the volumetric flow rate, m³ h⁻¹, C_{in} is the concentration of PHA in the inlet stream, kg m⁻³, r_i is the production rate of PHA, kg m⁻³ h⁻¹, C_{out} is the concentration of PHA in the outlet stream, kg m⁻³, and V is the reactor volume, m^3. Rearranging Equation (12.67), the reactor design equation for a CSTR is as shown below (cf. Equation (11.45)). Note that for a product, its concentration in the outlet is greater than its concentration in the inlet:

$$V_{CSTR} = \frac{F(C_{out} - C_{in})}{r_i} \qquad (12.68)$$

2. PFR. In a PFR, the general mass balance equation across an element of volume dV can be formulated as

Mass flow rate of PHA in the inlet to dV + mass of PHA generated in dV = Mass flow rate of PHA in the outlet from dV

For a constant density system, the volumetric flow rate input and output are the same. Then the mass balance in the PFR is as shown in the following equation (cf. Equation (11.33); the only difference is that there is no minus sign with r because a product's rate formation is already a positive term):

$$FC + r_i dV = F(C + dC)$$

$$r_i dV = FdC \qquad (12.69)$$

Integrating Equation (12.69) over the entire length of the PFR and hence between the PHA concentrations at the inlet and the outlet, C_{in} and C_{out}, respectively, the volume of the PFR is obtained:

$$V_{PFR} = F \int_{C_{in}}^{C_{out}} \frac{dC}{r_i} \qquad (12.70)$$

In both cases the volume of the reactors is correlated with the rate of production.

Exercise 8. Calculation of reactor volume for PHA production in CSTR and PFR. 1 kg m^{-3} of PHA is produced using microorganism Y in a substrate containing glucose. The rate of production of PHA is dependent on pH and the initial substrate concentration $C_{in,S}$ (kg m^{-3}) following the kinetic equation (kg m^{-3} h^{-1}) for a pH value in the range 6–9:

$$r_i = 0.005[pH]^{1.8} C_{in,S}$$

Deduce the correlation between the volume of a reactor and the initial substrate concentration if the reactor is (1) an ideal continuous stirred tank reactor and (2) a plug flow reactor. The pH value is uniform at 6.8 in the CSTR and is going to change in the axial direction in a plug flow reactor. The glucose feed stream (10 m^3 h^{-1}) contains 0.01 kg m^{-3} PHA at the inlet of the reactor.

Solution to Exercise 8. Refer to **Online Resource Material, Chapter 12 – Additional Exercises and Examples** for the solution to Exercise 8.

12.4.3   Catalytic Reactor Design

As with conventional chemical processes, the conversion efficiency of most of the biorefinery chemical processes depends on the catalyst. Catalysts can be (a) homogeneous if they are in the same phase as the reactants (e.g., H_2SO_4 catalyst is in the same aqueous solution for hydrolysis of sugars) or (b) heterogeneous if the catalyst and reaction medium are in separate phases (e.g., Cu based catalysts for glycerol hydrogenolysis is solid while the reaction medium is aqueous solution or gas).

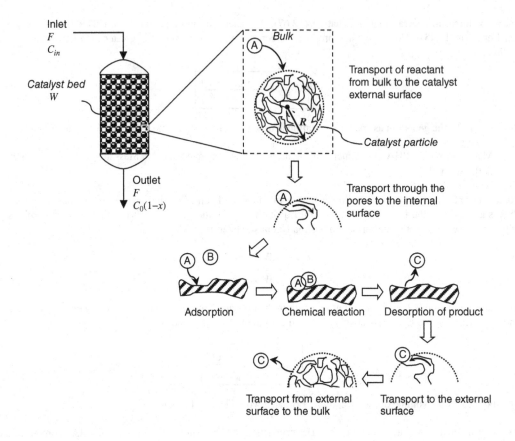

Figure 12.10   *A catalytic fixed bed reactor with catalytic process steps.*

As discussed in Chapter 11, catalytic reactors can have a fixed or a fluidized catalytic bed. Consider the fixed bed reactor in Figure 12.10. For the reaction to occur, the steps shown in the diagram are needed:

1. Transport of reactants from the reaction bulk mixture to the external catalyst surface.
2. Transport of reactants to the internal surface through the catalyst pores.
3. Adsorption of reagents, chemical reaction and desorption of products at the catalytic sites.
4. Transport of products from the catalyst interior to the external catalyst surface.
5. Transport of products into the reaction bulk mixture.

It is important to know the steps limiting the overall rate of reaction. To identify the limiting steps, the rates at which each step occurs are compared. When the first step is slow and the rest of the steps are fast, the reaction is controlled by the external transport. When step 3 is the slowest, the process is reaction limited. When step 2 is the slowest, the catalytic process is limited by the intraparticle transport or diffusion of the species. More quantitative details are needed to determine the *rate controlling step* (also see Chapter 18 for the rate limiting steps in elementary reactions).

Differences in transport and reaction rates generate concentration gradients inside the catalyst particle. Such differences depend on the characteristics of the catalyst particles. As can be observed in the expanded view of Figure 12.10, adsorption of the reactants on to the catalyst active sites is a prerequisite for a reaction. Thus, the availability of such active sites is dependent on the internal surface of the catalyst pores. Therefore, the catalyst characteristics such as pore size, pore volume, area and tortuosity are critical factors in a catalytic process. The pore size can be from the order of magnitude

Table 12.8   *Characteristic lengths of common catalyst particle shapes.*

Catalyst Particle Shape	Characteristic Length of the Catalyst Particle
Sphere	$R \longrightarrow$ $L = \dfrac{R}{3}$
Cylinder	$R \longrightarrow$ $L = \dfrac{R}{2}$
Slab	$L = L$, L

of a molecule radius (nm) to the order of millimetres. The total pore area is generally determined by physical adsorption experiments with N_2. The multiscale aspects of solid catalytic processes are discussed in Section 18.3. For the concerns of this chapter, the macroscopic characteristics are shown to study the effect of a catalyst on the reaction rate, the amount of catalyst required and the reactor design. Such characteristics are the characteristic length and bulk density of the catalyst particles. Table 12.8 shows the characteristic lengths (volume-to-area ratio of the active surfaces) of commonly occurring catalytic particle shapes.

As shown in the expanded view of a spherical particle in Figure 12.10, the pores are not straight, their sectional area can vary along the pore path and they can be interconnected. To account for this, the concept of *effective diffusivity* D_{eff} is introduced (see Section 18.2 for the modeling of effective diffusivity). The effective diffusivity is a measure of the average diffusivities of reaction species within the pores accounting for the effect of pore structure.

Now consider a catalyst particle of spherical shape with radius R shown in Figure 12.11. The mass balance of A through a section of the sphere between radius r and $r + \Delta r$ is

> Moles of A entering the spherical section by diffusion − Moles of A consumed by the reaction within the spherical section = Moles of A leaving the spherical section by diffusion

Similar to the mass transfer equation, the energy transfer equation can also be developed for adiabatic reactors or reactors operating under variable temperatures. See the following section and Web Chapter 2 for energy transfer modeling.

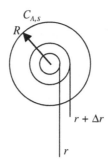

Figure 12.11   *Scheme of the section of a spherical catalyst particle used for mass balance.*

The mass transfer equation is the balance of molar flux N_A (e.g., in mol m^{-2} s^{-1}) of a reactant A in and out of a cross-sectional area at r and $r + \Delta r$, as shown in terms of its rate of reaction r_A:

$$N_A 4\pi r^2 \big|_{r=r} - N_A 4\pi r^2 \big|_{r=r+\Delta r} - r_A 4\pi r^2 \Delta r = 0 \tag{12.71}$$

Dividing by $4\pi\Delta r$ and taking the limit when $\Delta r \to 0$, the following equation is obtained in differential form:

$$-\frac{dN_A}{dr}r^2 - r_A r^2 = 0 \tag{12.72}$$

The flux by diffusion can be expressed using the effective diffusivity (D_{eff}) and *Fick's law* in spherical coordinates, defined as

$$N_A = -D_{eff}\frac{dC_A}{dr} \tag{12.73}$$

Thus, the mass transfer is derived by substituting Equation (12.73) into Equation (12.72):

$$\frac{d}{dr}\left(D_{eff}\frac{dC_A}{dr}r^2\right) - r_A r^2 = 0 \tag{12.74}$$

Dividing by $D_{eff}r^2$, Equation (12.74) becomes

$$\frac{1}{r^2}\frac{d}{dr}\left(\frac{dC_A}{dr}r^2\right) - \frac{r_A}{D_{eff}} = 0 \tag{12.75}$$

Differentiation of Equation (12.75) results in

$$\boxed{\frac{d^2C_A}{dr^2} + \frac{2}{r}\frac{dC_A}{dr} - \frac{r_A}{D_{eff}} = 0} \tag{12.76}$$

To solve the second-order differential equation, the following boundary conditions are used:

1. Concentration at radius $r = R$, the concentration is equal to the concentration at the particle external surface $C_{A,s}$:

$$C_A\big|_{r=R} = C_{A,s} \tag{12.77}$$

2. At the centre of the sphere $r = 0$, the concentration is symmetric and the slope of the concentration with respect to the radius is zero:

$$\frac{dC_A}{dr}\bigg|_{r=0} = 0 \tag{12.78}$$

For a reaction of order n the general reaction rate equation is given by

$$r_A = kC_A^n \tag{12.79}$$

Then Equation (12.76) becomes

$$\frac{d^2C_A}{dr^2} + \frac{2}{r}\frac{dC_A}{dr} - \frac{kC_A^n}{D_{eff}} = 0 \tag{12.80}$$

The equation can be transformed to a nondimensional form using the following equations:

$$\bar{r} = \frac{r}{L} \tag{12.81}$$

$$\bar{C} = \frac{C_A}{C_{A,s}} \tag{12.82}$$

where L is the characteristic length resulting from the particle volume-to-surface ratio (shown in Table 12.8). With dimensionless variables, Equation (12.80) is transformed into

$$\frac{C_{A,s}}{L^2}\frac{d^2\bar{C}}{d\bar{r}^2} + \frac{C_{A,s}}{L^2}\frac{2}{\bar{r}}\frac{d\bar{C}}{d\bar{r}} - \frac{k\bar{C}^n C_{A,s}^n}{D_{eff}} = 0 \tag{12.83}$$

Dividing Equation (12.83) by $C_{A,s}$ and multiplying by L^2 gives

$$\frac{d^2\bar{C}}{d\bar{r}^2} + \frac{2}{\bar{r}}\frac{d\bar{C}}{d\bar{r}} - \frac{kC_{A,s}^{n-1}L^2}{D_{eff}}\bar{C}^n = 0 \tag{12.84}$$

Rearranging the third member in Equation (12.84) gives

$$\frac{kC_{A,s}^{n-1}L^2}{D_{eff}} = \frac{L^2/D_{eff}}{1/kC_{A,s}^{n-1}} = \frac{\text{Characteristic diffusion time}}{\text{Characteristic reaction time}} \tag{12.85}$$

Equation (12.85) is the ratio of the characteristic diffusion time over the characteristic reaction time and gives a sense of the slowest step. The square root of Equation (12.85) defines the dimensionless *Thiele modulus* ϕ as

$$\phi = \sqrt{\frac{kL^2 C_{A,s}^{n-1}}{D_{eff}}} \tag{12.86}$$

and then:

If $\phi \gg 1$, the catalytic process is diffusion limited.

If $\phi \ll 1$, the catalytic process is kinetically or reaction rate limited.

For a first-order reaction, Equation (12.86) has the form of

$$\boxed{\phi = \sqrt{\frac{kL^2}{D_{eff}}}} \tag{12.87}$$

Note that the constant k in Equation (12.87) is the reaction constant with unit of time^{-1} (e.g., s^{-1}) for a first-order reaction. The corresponding differential equation for the mass transfer in a catalyst particle is shown by

$$\boxed{\frac{d^2\bar{C}}{d\bar{r}^2} + \frac{2}{\bar{r}}\frac{d\bar{C}}{d\bar{r}} - \phi^2\bar{C} = 0} \tag{12.88}$$

To facilitate the solution of Equation (12.88), the transformation variable v is introduced, as shown by

$$\bar{C}(\bar{r}) = \frac{v(r)}{\bar{r}} \tag{12.89}$$

Then Equation (12.88) in terms of v is simplified to

$$\frac{d^2v}{dv^2} - \phi^2 v = 0 \tag{12.90}$$

The boundary conditions for the new variables are (according to Equation (12.89)):

1. $\bar{C} = 1$ at $\bar{r} = R/L = R/(R/3) = 3$, for the spherical shape. Then $v = 3$.
2. $\bar{C}$ is symmetric everywhere and $v = 0$ at $\bar{r} = 0$

The solution for $\bar{C}$ and for the given boundary conditions (see also derivations of Equations (18.46) and (18.53)) is given by

$$\boxed{\bar{C}(r) = \frac{C_A}{C_{A,s}} = \frac{3}{\bar{r}}\left(\frac{\sinh\phi\bar{r}}{\sinh 3\phi}\right)} \tag{12.91}$$

The overall consumption rate of A, $r_{A,cat}$, in the catalyst particle is found by integrating the reaction rate over the volume of the catalyst particle, as shown by

$$r_{A,cat} = -\frac{1}{V_{cat}}\int_0^R r_{A,cat}(r)\,4\pi r^2 dr \tag{12.92}$$

Equation (12.92) can be written in terms of the transformation variables (Equations (12.81), (12.82) and (12.89)), as shown by

$$r_{A,cat} = -\frac{kC_{A,s}}{9}\int_0^3 \bar{C}(\bar{r})\,\bar{r}^2 d\bar{r} \tag{12.93}$$

Using the solution for $\bar{C}(r)$ and in terms of the Thiele modulus with $y = \phi r$ gives

$$r_{A,cat} = -\frac{kC_{A,s}}{3\phi^2 \sinh 3\phi}\int_0^{3\phi} y \sinh y\, dy \tag{12.94}$$

Integration of Equation (12.94) gives the solution for the reaction rate in the catalyst particle for a first-order reaction in terms of the Thiele modulus. This solution is given by

$$\boxed{r_{A,cat} = -\frac{kC_{A,s}}{\phi}\left[\frac{1}{\tanh 3\phi} - \frac{1}{3\phi}\right]} \tag{12.95}$$

where $kC_{A,s}$ is the rate of reaction with fast diffusion. Then the concentration inside the particle is equal to the surface concentration. For a fast diffusion, D_{eff} is high and the Thiele modulus tends to have a limiting value $\phi = 0$. This particular situation helps to define the *effectiveness factor* of a catalytic particle as the ratio of the actual reaction rate to the reaction rate in the absence of diffusion. Thus, for the first-order reaction, the effectiveness factor η is given by

$$\boxed{\eta = \frac{r_{A,cat}}{kC_{A,s}} = \frac{1}{\phi}\left[\frac{1}{\tanh 3\phi} - \frac{1}{3\phi}\right]} \tag{12.96}$$

Exercise 9. Analysis of the Thiele modulus and effectiveness factor for a first-order reaction. For a first-order reaction:

a. Plot the concentration in a spherical catalyst particle versus the nondimensional radius $0 < \bar{r} < 3$ for Thiele modulus values of 0.1, 1 and 2. Analyze your observations.
b. Plot the effectiveness factor versus the Thiele modulus values between 0.001 and 20. Analyze your observations.

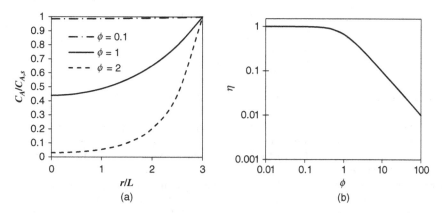

Figure 12.12 (a) Concentration profile in the spherical catalytic particles and (b) effectiveness factor at various values of the Thiele modulus.

Solution to Exercise 9. Figure 12.12(a) shows that for small values of the Thiele modulus (e.g., $\phi = 0.1$), the concentration inside the particle remains unchanged and is almost the same as the concentration in the external surface. Thus, the diffusion is fast and the reaction is slow. This is also reflected in Figure 12.12(b). When the diffusion rate is much larger than the reaction rate, the Thiele modulus is close to unity and the effectiveness factor approximates to unity. At these conditions, the catalytic process is *reaction-limited*. In comparison, when the reaction rate is much faster than diffusion, the reactant consumption is almost complete, as shown for $\phi = 2$. Thus, at values of the Thiele modulus $\phi > 1$, the efficiency is less than unity and the catalytic process is *diffusion-limited*.

Now that the modeling of the effect of the catalyst has been discussed, the design equation for a plug flow reactor with a fixed catalytic bed can be formulated. As shown in Figure 11.16, a plug flow reactor requires lesser volume than a CSTR, for a given reaction conversion. Consider the catalytic fixed bed reactor in Figure 12.10. The mass balance for the reactor can be expressed in a similar way to a plug flow reactor for an element of volume dV as

Moles of A entering dV − Moles of A consumed by the reaction in dV = Moles of A leaving dV

For a constant molar flow rate F, the mass balance is obtained as

$$FC_A - r_A dV = F(C_A + dC_A) \tag{12.97}$$

Rearranging Equation (12.97) gives (cf. Equation (11.33))

$$- r_A dV = F dC_A \tag{12.98}$$

In terms of conversion X (cf. Equation (12.56)), Equation (12.98) becomes

$$r_A dV = FC_{in,A} dX \tag{12.99}$$

Relating catalyst weight W_c to catalyst bed density ρ_b (not catalyst particle density) gives

$$W_c = \rho_b V \tag{12.100}$$

Substituting Equation (12.100) into Equation (12.99) and rearranging gives

$$\boxed{\frac{dX}{dW} = \frac{r_A}{FC_{in,A}\rho_b}}$$

(12.101)

This equation finds the amount of catalyst required to achieve a conversion.

Exercise 10. Using the effectiveness factor in catalytic reactor design for the production of the monomer 2,5-furandicarboxylic acid. A biorefinery plant produces 600 L min^{-1} of 5-hydroxymethylfurfural (HMF) as a coproduct of the extraction of xylitol from corn cobs. The company has identified a potential market for 2,5-furandicarboxylic acid (FDCA), a derivative of HMF and a monomer for the biopolymer poly(ethylene furanoate) or PEF. PEF is a potential substitute for polyethylene terephthalate in plastic bottles. The company is planning to invest for a reactor for direct oxidation of HMF into FDCA production. The oxidation reaction is of first order with respect to the concentration of HMF when oxygen is supplied in excess and in the presence of NaOH. The oxidation of HMF ($C_6H_6O_3$) produces FDCA ($C_6H_4O_5$) and 5-hydroxymethyl-2-furancarboxylic acid ($C_6H_6O_4$) as a by-product. The overall reaction can be written as

$$2C_6H_6O_3 + 2O_2 \rightarrow C_6H_4O_5 + C_6H_6O_4 + H_2O$$

The following data using a platinum based catalyst at 25 °C for the oxidation is available from pilot plant trials:

Catalyst shape: spherical
Catalyst particle diameter $d_p = 0.5$ cm
Density of the particle $\rho_p = 1909.8$ g L^{-1}
Void volume of catalyst bed $\varepsilon_b = 0.5$

$$k = 0.374 \text{ s}^{-1}$$
$$D_{eff} = 0.0005 \text{ cm}^2 \text{ s}^{-1}$$

HMF concentration in feed $C_{in,A} = 0.5$ mol L^{-1}

Using this information, determine:

a.   The Thiele modulus and effectiveness of the catalyst. Identify the dominant regime in the catalytic process.
b.   The catalyst mass and reactor volume for a conversion of 95%. At this conversion the selectivity of FDCA is 31%.
c.   If conversion is increased to 99.9%, the selectivity increases to 80%. Find the new catalyst mass requirement and reactor volume.

Refer to **Online Resource Material, Chapter 12 – Additional Exercises and Examples** for the solution to Exercise 10.

So far the models have been developed for a spherical catalytic particle shape and first-order reaction. For a reaction order n, the Thiele modulus can be approximated as

$$\phi = \sqrt{\frac{n+1}{2} \frac{kC_{A,s}^{n-1}L^2}{D_{eff}}}$$

(12.102)

The effectiveness factor for a cylinder is approximated as

$$\eta = \frac{1}{\phi} \frac{I_1(2\phi)}{I_0(2\phi)}$$

(12.103)

where I_1 and I_0 are the Bessel functions of order 1 and order 0, respectively. For the slab shape, the effectiveness factor is of a simpler form, as shown by

$$\eta = \frac{\tanh \phi}{\phi} \tag{12.104}$$

A reference specialized in transport phenomena or catalytic reaction engineering can be consulted for the derivations of these expressions and the definition of the Bessel functions[5-7].

Equations (12.71) to (12.96) considered fast diffusion between the bulk fluid and particle external surface and no reaction in the bulk fluid. Thus, the concentration in the external particle surface $C_{A,s}$ has been considered equal to the concentration in the bulk fluid $C_{A,f}$, that is, $C_{A,s} = C_{A,f}$. When the external mass transport is slow, the flux of reactants from the bulk fluid to the surface of a spherical catalyst particle needs to be considered:

$$\frac{dC_{A,s}}{dr} = \frac{k_m}{D_{eff}}(C_{A,f} - C_{A,s}) \tag{12.105}$$

where k_m is the mass transfer coefficient and $C_{A,f}$ is the concentration in the bulk fluid. Mass transfer resistance in the external phase makes the catalyst less efficient. To measure to what extent the external mass transport rate affects a catalytic process, the *Biot number* is introduced. The Biot number is a dimensionless quantity defined as

$$\text{Biot number} = \frac{k_m L}{D_{eff}} = \frac{D_{A,fluid}}{\delta}\frac{L}{D_{eff}} = \frac{L/D_{eff}}{\delta/D_{A,fluid}} = \frac{\text{Internal resistance}}{\text{External resistance}} \tag{12.106}$$

where δ is the thickness of the fluid film around the catalyst particle and $D_{A,fluid}$ is the diffusivity of reactant A in the bulk fluid. At low values of the Biot number, the external transport is limiting the catalytic process. At large values of the Biot number, the external mass transport is faster than the internal transport.

The effectiveness factor in Equation (12.96) can now be modified as the ratio of the reaction rate within a spherical catalyst particle in the presence of external transport limitation to the reaction rate without such limitation. The following equation can be derived by replacing the boundary condition at the catalyst particle surface in Equation (12.105) and solving the resulting equation by introducing the Biot number, following the procedure shown in Equations (12.90) to (12.96):

$$\eta = \frac{1}{\phi}\left[\frac{\dfrac{1}{\tanh 3\phi} - \dfrac{1}{3\phi}}{1 + \dfrac{\phi\left(\dfrac{1}{\tanh 3\phi} - \dfrac{1}{3\phi}\right)}{\text{Biot number}}}\right] \tag{12.107}$$

Finally, the regimes governing the overall catalytic process can be distinguished as follows:

1. For *Biot number* <1:
   a. If $\phi <$ *Biot number*, the process is governed by the reaction.
   b. If *Biot number* $< \phi < 1$, the process is governed by external transport.
   c. If $\phi > 1$, the process is governed by both internal and external diffusion.
2. For *Biot number* >1:
   a. If $\phi < 1$, the process is governed by the reaction.
   b. If $1 < \phi <$ *Biot number*, the process is governed by internal transport.
   c. $\phi >$ *Biot number*, the process is governed by both internal and external diffusion.

Similar expressions can be derived for other shapes, orders and mechanisms of reaction.

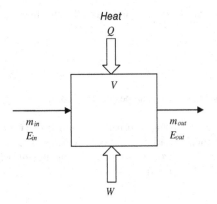

Heat
Q

V

m_{in}
E_{in}

m_{out}
E_{out}

W

m_{in}: inlet mass flow rate in units of [mass] [time] $^{-1}$
E_{in}: inlet energy in units of [energy] [mass] $^{-1}$
m_{out}: outlet mass flow rate in units of [mass] [time] $^{-1}$
E_{out}: outlet energy in units of [energy] [mass] $^{-1}$
Q: heat flow in units of [energy] [time]$^{-1}$
W: work flow in units of [energy] [time]$^{-1}$
V: volume in units of [volume]

Figure 12.13   *Energy balance around a reactor.*

12.4.4   Energy Transfer Models of Reactors

Energy balances need to be performed to minimize energy consumption and control the temperature, operability and productivity of a process.

The energy flows required for the energy balance across a reactor are shown in Figure 12.13. The generic energy balance equation of a reactor with respect to time (t) is as

$$\frac{dE}{dt} = m_{in}E_{in} - m_{out}E_{out} + Q + W \qquad (12.108)$$

The total energy term (E) is made up of internal energy (U), kinetic energy (K) and potential energy (PE), as shown by

$$E = U + K + PE \qquad (12.109)$$

The total work (W) is made up of shaft work applied to a system (W_s, e.g., by stirrers), work done by the flow of the streams ($W_f = m \times P/\rho$) and work due to the change in volume ($W_v = -P\, dV/dt$), as shown by

$$W = W_s + m_{in}\frac{P_{in}}{\rho_{in}} - m_{out}\frac{P_{out}}{\rho_{out}} - P\frac{dV}{dt} \qquad (12.110)$$

where P is the pressure, ρ is the density and V is the reactor volume. The subscripts *in* and *out* refer to the inlet and outlet energy or mass streams.

These terms complicate the analysis of energy balance of a reactor but can be simplified making assumptions as follows:

- The differences in kinetic (due to change in velocity in the reactor) and potential energy (due to change in height) between inlet and outlet streams are relatively small compared to the difference in internal energy. Thus, E can be reduced to $E = U$.
- For low viscosity reaction mixtures, the shaft work by stirrers can be negligible compared to Q or the change in U.

From thermodynamics, the enthalpy (H) of a fluid can be defined as

$$H = U + PV = U + \frac{P}{\rho} \qquad (12.111)$$

Using the earlier definitions and assumptions, the energy balance can be rewritten as

$$\frac{dU}{dt} = m_{in}\left(H_{in} - \frac{P_{in}}{\rho_{in}}\right) - m_{out}\left(H_{out} - \frac{P_{out}}{\rho_{out}}\right) + m_{in}\frac{P_{in}}{\rho_{in}} - m_{out}\frac{P_{out}}{\rho_{out}} - P\frac{dV}{dt} + Q \qquad (12.112)$$

which reduces to

$$\frac{dU}{dt} = m_{in}H_{in} - m_{out}H_{out} - P\frac{dV}{dt} + Q \qquad (12.113)$$

Taking the differential of the definition of enthalpy, the following equation is obtained for enthalpy change:

$$dH = dU + PdV + VdP \qquad (12.114)$$

and substitution of expression for dU in the energy balance equation gives

$$\frac{dH}{dt} - P\frac{dV}{dt} - V\frac{dP}{dt} = m_{in}H_{in} - m_{out}H_{out} - P\frac{dV}{dt} + Q \qquad (12.115)$$

This reduces, for a constant reactor volume, to

$$\frac{dH}{dt} - V\frac{dP}{dt} = m_{in}H_{in} - m_{out}H_{out} + Q \qquad (12.116)$$

Since the enthalpy is a function of temperature (T), pressure (P) and the number of moles (n_i), the differential dH can be expressed in terms of partial derivatives, as shown by

$$dH = \left(\frac{\partial H}{\partial T}\right)_{P,n_i} dT + \left(\frac{\partial H}{\partial P}\right)_{T,n_i} dP + \sum_i \left(\frac{\partial H}{\partial n_i}\right)_{P,T,n_{j\neq i}} dn_i \qquad (12.117)$$

Each of the partial derivatives can be related to a measurable quantity of the system.

The equation below shows the definition of the heat capacity at constant pressure and can be related to the specific heat capacity (C_p) and density of the reaction mixture (ρ) as

$$\left(\frac{\partial H}{\partial T}\right)_{P,n_i} = V\rho C_p \qquad (12.118)$$

The partial derivative of enthalpy with respect to pressure related to the coefficient of expansion of the mixture (α) is given as

$$\left(\frac{\partial H}{\partial P}\right)_{T,n_i} = V - T\left(\frac{\partial V}{\partial T}\right)_{P,n_i} = V(1 - \alpha T) \qquad (12.119)$$

The last derivative in Equation (12.117) is the summation of the partial molar enthalpies ($\bar{H}_i$). Then, dH can be expressed as

$$dH = V\rho C_p dT + V(1 - \alpha T)dP + \sum_i \bar{H}_i dn_i \qquad (12.120)$$

Equation (12.120) is substituted into Equation (12.116) to give rise to the generic energy balance for reactors shown by

$$\boxed{V\rho C_p \frac{dT}{dt} - \alpha T V \frac{dP}{dt} + \sum_i \bar{H}_i \frac{dn_i}{dt} = m_{in}H_{in} - m_{out}H_{out} + Q} \qquad (12.121)$$

The term dn_i/dt corresponds to the mass balance of the reactor and can be substituted with the mass balance equation according to the reactor type. The energy balance equations for batch and CSTR reactors are shown as follows:

a. For a batch reactor there are no inlet or outlet streams and the mass balance is given by

$$\frac{dn_i}{dt} = \sum_k^{nr} v_{i,k} r_k V \tag{12.122}$$

where $v_{i,k}$ is the stoichiometric coefficient of component i in reaction k. This coefficient is positive when the component is a product and negative when the component is a reactant in a particular reaction; r_k is the rate of reaction k and nr is the total number of reactions taking place in the reactor.

The heat of reaction ($\Delta H_{r,k}$) is defined as

$$\Delta H_{r,k} = \sum_i v_{i,k} \bar{H}_i \tag{12.123}$$

Then substitution of Equations (12.122) and (12.123) into Equation (12.121) gives the energy balance for a batch reactor shown by

$$V \rho C_p \frac{dT}{dt} - \alpha T V \frac{dP}{dt} = -\sum_k \Delta H_{r,k} r_k V + Q \tag{12.124}$$

b. For a CSTR, the mass balance equation in terms of the volumetric flow rate (F) and concentrations (C_i) is given by

$$\frac{dn_i}{dt} = F_{in} C_{in,i} - F_{out} C_{out,i} + \sum_k^{nr} v_{i,k} r_k V \tag{12.125}$$

For homogeneous mixing in CSTR, the concentration and properties of the outlet stream are equal to those of the reaction mixture in the reactor:

$$\sum_i \bar{H}_i \frac{dn_i}{dt} = \sum_i \bar{H}_{out,i} \left[F_{in} C_{in,i} - F_{out} C_{out,i} + \sum_k^{nr} v_{i,k} r_k V \right] \tag{12.126}$$

Then, expressing the energy flow in terms of the volumetric flow (F), concentration and molar enthalpy ($\bar{H}_i$), the term $m_{in} H_{in} - m_{out} H_{out}$ in the energy balance for a CSTR can be written as

$$m_{in} H_{in} - m_{out} H_{out} = \sum_i F_{in} C_{in,i} \bar{H}_{in,i} - \sum_i F_{out} C_{out,i} \bar{H}_{out,i} \tag{12.127}$$

Substituting Equation (12.127) into the generic energy balance Equation (12.121) gives

$$V \rho C_p \frac{dT}{dt} - \alpha T V \frac{dP}{dt} + \sum_i \bar{H}_{out,i} \left[F_{in} C_{in,i} - F_{out} C_{out,i} + \sum_k^{nr} v_{i,k} r_k V \right]$$
$$= \sum_i F_{in} C_{in,i} \bar{H}_{in,i} - \sum_i F_{out} C_{out,i} \bar{H}_{out,i} + Q \tag{12.128}$$

Simplifying the terms and using the definition of heat of reaction gives

$$V \rho C_p \frac{dT}{dt} - \alpha T V \frac{dP}{dt} = \sum_i F_{in} C_{in,i} (\bar{H}_{in,i} - \bar{H}_{out,i}) - \sum_k \Delta H_{r,k} r_k V + Q \tag{12.129}$$

For steady-state assumption the derivatives are equal to zero. Thus, the following equation is applicable for the steady state:

$$\sum_i F_{in}C_{in,i}(\bar{H}_{in,i} - \bar{H}_{out,i}) - \sum_k \Delta H_{r,k}r_kV + Q = 0 \tag{12.130}$$

The enthalpy content of a substance is a function of temperature. Considering the inlet temperature as the reference state, the enthalpy difference at constant pressure can be written as

$$\bar{H}_{in,i} - \bar{H}_{out,i} = C_{p,i}(T_{in} - T) \tag{12.131}$$

where T is the reactor temperature and is equal to T_{out} in the ideal well-mixed reactor or CSTR. For the characteristics of CSTR, refer to the Solution to Exercise 8 in Chapter 11. Assuming an incompressible fluid, constant density and specific heat capacity of the inlet stream (C_p), the first summation term in Equation (12.130) can be expressed as

$$\sum_i F_{in}C_{in,i}C_{p,i}(T_{in} - T_{out}) = F_{in}\rho C_p(T_{in} - T) \tag{12.132}$$

The energy balance equation for a CSTR has the form of

$$\boxed{F_{in}\rho C_p(T_{in} - T) - \sum_k \Delta H_{r,k}r_kV + Q = 0} \tag{12.133}$$

The general energy balance equation for the batch and CSTR reactors in Equations (12.124) and (12.133), respectively, can be used when the assumptions discussed are reasonable for a reaction system. Note that the equations are for single phase. The equations can be further adapted for particular reaction conditions (e.g., gaseous phase, multiple phase, etc.) or reactor types (e.g., PFR).

Exercise 11. Calculation of cooling requirement for an isothermal CSTR with exothermic reaction. Hydroxylated triglycerides are used as monomers for the production of polyurethanes. The first step in the synthesis of these monomers is the epoxidation of vegetable oils. The kinetics of epoxidation of a vegetable oil with hydrogen peroxide is approximated as first order with respect to the concentration of double bonds in the oil C_{DB}

$$r_{epox} = kC_{DB}$$

where C_{DB} is determined from the concentration of the oil C_{oil}, the molar composition of the oil components (x_i) and their number of double bonds ($N_{DB,i}$):

$$C_{DB} = C_{oil} \sum_i^{nc} x_i N_{DB,i}$$

where nc is the number of components in the oil. The reaction constant k (in min^{-1}) varies with the temperature T (in K) according to

$$k = 0.0188e^{[4481.8(1/343 - 1/T)]}$$

A side reaction opening the epoxide rings to various linear, substituted compounds is undesirable. The selectivity (S) of the epoxidized product is correlated to the conversion (X) as

$$S = -1.73X^2 + 2.2927X + 0.1062$$

The epoxidation reaction is highly exothermic with an enthalpy of reaction of -230 kJ mol^{-1} (for each double bond). Answer the following questions using the information given:

a.   Calculate the conversion and selectivity achieved in a CSTR operating at 60 °C processing 100 kmol h^{-1} feed at 25 °C with an oil concentration of 0.353 mol L^{-1}. The oil contains 23% mol of oleic acid (1 double bond), 56% mol of linoleic acid (two double bonds) and 5% mol of linolenic acid (three double bonds).
b.   Determine the heat requirements to keep the reactor at isothermal operation; assume single phase to simplify the energy balance.
c.   Calculate the heat exchanger area required to remove the heat released by the reaction.

$$\rho = 40.26 \text{ mol L}^{-1}$$

$$C_p = 0.996 \text{ kJ mol}^{-1} \text{ °C}^{-1}$$

Reactor volume = 25 m^3 = 25 000 L

Overall heat transfer coefficient $U_h = 41.33$ kJ m^{-2} min^{-1} °C^{-1}

T cooling fluid = 15 °C

Solution to Exercise 11. Refer to *Online Resource Material, Chapter 12 – Additional Exercises and Examples* for the solution to Exercise 11.

Exercise 12. Multiple steady states in a CSTR with exothermic reaction. Calculate the following quantities for various temperatures using the information given in Exercise 11:

$$A = F\rho C_p \left(T_{in} - T \right)$$
$$B = kC_{in,DB}(1 - X)\Delta H_r V$$

Plot the values of A and B against the temperature. Discuss the main findings.

Solution to Exercise 12. Refer to *Online Resource Material, Chapter 12 – Additional Exercises and Examples* for the solution to Exercise 12.

12.5   Synthesis of Unit Operations Combining Reaction and Separation Functionalities

This section discusses two examples of combining reaction and separation functionalities in a unit operation. This leads to *process intensification* by reducing the number of pieces of equipment and floor space required for the production of a chemical. The examples are shown for the case of polyester production but could be adaptable to other polymer and chemical production processes.

12.5.1   Reactive Distillation Column

Polyesters are synthesized by esterification or transesterification between a diol monomer and a diacid monomer using a catalyst. Water is the by-product of esterification reaction, whilst methanol is the by-product of transesterification reaction. The by-products and excess diol are removed by operating at vacuum pressures and distillation.

The production of the polyesters has the common challenge of the removal of by-products at high conversions. Excess diol and water or methanol limits the attainable molar mass of the polyester. A series of reactors operating at increasing temperatures and decreasing pressures are generally used to overcome this limitation. However, high temperatures can result in degradation and subsequent reduction of the molar mass of the polymer. Thus, optimization of process conditions is required. An alternative approach, using process intensification to perform reaction and separation in one reactive distillation column, is shown in the following exercise. Also see Section 11.5.7 for reactive distillation.

Exercise 13. Polyester synthesis in a reactive distillation column. Polyester synthesis reaction is controlled by equilibrium and continuous removal of elimination products such as water and diol is necessary to obtain high conversions. A promising alternative to overcome this challenge and for the intensification of the process is reactive distillation. Describe how a single reactive distillation column can be used for the production of poly(butylene succinate) (PBS) by esterification of 1,4-butanediol (BDO) and succinic acid. Illustrate with diagrams the main functions of the rectifying, reaction and stripping sections. Discuss the advantages and disadvantages of using such a process for the given system. Boiling points of the components are shown in Table 12.9. Consider that some components undergo thermal decomposition at temperatures above 230 °C.

Table 12.9   *Normal boiling points of components in polyester synthesis.*

Component	Normal Boiling Point (at 1 atm) (°C)	Molar mass (g mol^{-1})
Water	100	18.01
Ethylene glycol	197	62.07
1,4-Butanediol (BDO)	230	90.12
Succinic acid	235	118.09
Dimethyl terephthalate	288	194.18
Adipic acid	337	146.14

Solution to Exercise 13.   Figure 12.14 shows the conceptual reactive distillation column scheme for the production of PBS. The column can be divided into three sections, discussed as follows.

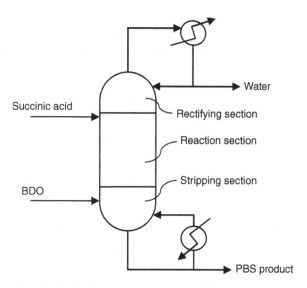

Figure 12.14   *Reactive distillation concept for the production of the biopolymer PBS.*

Reaction section. Appropriate operating conditions can be maintained to ease the reaction between BDO and succinic acid. Succinic acid can be the limiting reactant. By feeding the succinic acid in the top of the reaction zone, the generation of water in the top of this section is low and does not need much energy to be removed from the liquid flow. Also, the temperature here is lower than at the bottom and avoids any possible thermal decomposition of succinic acid.

Rectifying section. Due to the high volatility of water, water can be collected from this section. The boiling points of BDO and succinic acid are much higher than that of water and therefore their contents in the distillate are expected to be minimal.

Stripping section. The polymer product is collected from the bottom of the column and unreacted feed, mainly BDO, may be stripped off. The temperature in the bottom is high and near to the boiling point of BDO (230 °C). Due to its high concentration in the vapor, BDO acts as a stripping agent to remove water.

Advantages. By feeding BDO at the bottom and operating the reboiler at its boiling point, accumulation of BDO in the stripping section is avoided. In this section, the reaction is almost complete and removal of excess BDO from the viscous liquid is done by stripping BDO in the vapor phase. Reactive distillation combines product formation and separation in one apparatus, saves on capital cost, drives equilibrium to the polymer production and thus enhances the purity and molar mass of the polymer product. Exothermic heat of reaction should also give the heat for distillation. Excess BDO, used to drive equilibrium to polymer formation, may not be necessary due to immediate separation of water from the reaction medium.

Disadvantages. Needs favorable volatilities and operating temperatures below the decomposition temperature of the components. Needs thorough pilot scale research and detailed simulation (e.g., rate law based and finite element methods). Needs to make sure good liquid residence time and distribution, minimum catalyst fouling/poisoning/leaching and regeneration. Considering the disadvantages, it may make sense to use excess BDO, recover and recycle BDO through a single-stage flash column and keep succinic acid as a limiting reactant. However, a technoeconomic analysis needs to be performed before final decision making.

Further Questions. Table 12.9 shows the normal boiling point data of monomers used for polyester production in a similar process to the one for PBS. Find other polymers that could be produced from a reactive distillation column as discussed in Exercise 13.

12.5.2   An Example of a Novel Reactor Arrangement

Novel reactor arrangements combine various mixing patterns within the same reactor, which at the same time facilitates separation of reactant, product or by-product to shift the reaction equilibrium. The Vereinfacht Kontinuierliches Rohr (*VK*) *column reactor* is an example used for the polymerization of caprolactam into polyamide 6. Figure 12.15(a) shows a schematic view of a combination of idealized reactor flow patterns consisting of two CSTRs and one PFR. These flow patterns are combined into a VK column reactor as shown in Figure 12.15(b). A mixture of the monomer caprolactam, water and additives is fed at the top of the column. The mixture is heated by internal heat exchangers to about 220–270 °C. At this temperature, a part of the liquid, mainly water, evaporates. The bubbles formed from evaporation cause a vigorous mixing in the top zone of the VK reactor, resembling a CSTR, shown by the first CSTR in Figure 12.15(a). The water vapor and some caprolactam vapor are taken out from the top of the column and condensed in a condenser. The inert gas is vented. A part of the condensate is recycled to the top of the reactor. This is to achieve a countercurrent mass transfer between vapor and liquid. Due to hydrostatic pressure and the polymerization reaction along the top zone, the evaporation becomes difficult. In order to help the evaporation of water, the top zone of the VK column has a higher cross-sectional area. Although water is required as an initiator, the polycondensation reaction produces water and thus water needs to be removed from the top as condensate to drive the polymerization reaction. Further, inert gas is bubbled from the middle zone of the reactor to remove water. The bubbles also enable the generation of turbulence or agitation, creating the flow pattern of the second CSTR, shown in Figure 12.15(a). As the reaction mixture evolves into a viscous phase due to almost complete polymerization, the viscous phase containing the product polymer can be easily separated from the bottom zone of the VK column. The viscous phase is allowed to flow through the bottom zone of the reaction until polymerization is complete and the required molar mass is achieved. This is best achieved by a flow pattern that resembles a plug flow reactor.

Exercise 14. Mass balances in the VK reactor. Develop the mass balance equations for caprolactam and water in the VK reactor shown in Figure 12.15. The reactor can be conceptualized as a combination of two CSTRs and one PFR. Develop the mass balance on an inert gas-free basis.

Solution to Exercise 14. The mass balance for a component in a reactor at steady-state operation has the following form:

Moles of a specie entering the reactor – Moles consumed + Moles generated = Moles leaving the reactor

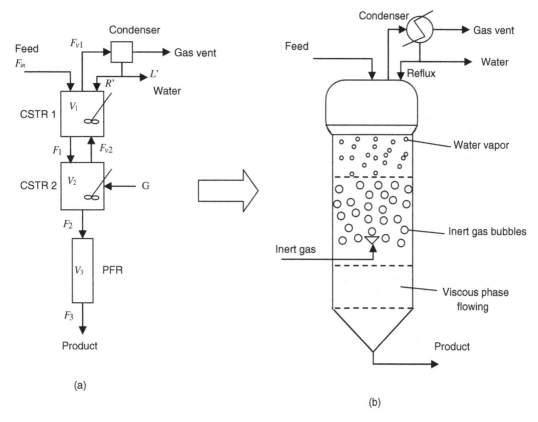

Figure 12.15 (a) Complex reactor arrangement combining two CSTR and one PFR and (b) scheme of a VK reactor combining these flow patterns in one column.

For the CSTR 1, the mass balance for caprolactam is

$$F_{in}C_{in,CL} + R'C_{CL,R'} + F_{v2}y_{CL,2} - r_{CL}V_1 = F_1C_{CL,1} + F_{v1}y_{CL,1}$$

where

F_{in} is the inlet flow rate to the VK column.
R' is the reflux rate from the top condenser.
F_1 is the flow rate from the CSTR 1 to CSTR 2.
F_{v1} is the vapor flow rate from CSTR 1 to the top condenser.
F_{v2} is the vapor flow rate from CSTR 2 to CSTR 1.
$C_{in,CL}$ is the concentration of caprolactam in the feed.
$C_{CL,R'}$ is the concentration of caprolactam in the reflux stream.
$C_{CL,1}$ is the concentration of caprolactam in the stream from CSTR 1 to CSTR 2.
$Y_{CL,1}$ is the caprolactam fraction in the vapor flow from CSTR 1 to the top condenser.
$Y_{CL,2}$ is the caprolactam fraction in the vapor flow from CSTR 2 to CSTR2.
r_{CL} is the rate of consumption of caprolactam.
V_1 is the volume of the first section of the VK column (shown by CSTR 1).

For water:

$$F_{in}C_{in,W} + R'C_{W,R'} + F_{v2}y_{W,2} - r_W V_1 = F_1 C_{W,1} + F_{v1}y_{W,1}$$

where

$C_{in,W}$ is the concentration of water in the inlet stream.
$C_{W,R'}$ is the concentration of water in the reflux stream.
r_W is the reaction rate of water.
$y_{W,1}$ is the caprolactam fraction in the vapor flow from CSTR 1 to the top condenser.
$y_{W,2}$ is the water fraction in the vapor flow from CSTR 2 to CSTR 1.
$C_{W,1}$ is the concentration of water in the outlet from CSTR 1 to CSTR 2.

In the CSTR2, the mass balance for caprolactam is

$$F_1 C_{CL,1} - r_{CL} V_2 = F_2 C_{CL,2} + F_{v2}y_{CL,2}$$

where

$C_{CL,2}$ is the concentration of caprolactam in the outlet from CSTR2.
F_2 is the outlet flow rate from the CSTR 2 to PFR.
V_2 is the volume of the second section of the VK column (shown by CSTR 2).

The mass balance for water is

$$F_1 C_{W,1} - r_W V_2 = F_2 C_{W,2} + F_{v2}y_{W,2}$$

where $C_{W,2}$ is the concentration of water in the outlet from CSTR2 to PFR.
In the PFR section the mass balance for caprolactam is

$$F_2 dC_{CL,3} = -r_{CL} dV_3$$

where

$C_{CL,3}$ is the concentration of caprolactam at a certain point in the PFR section.
V_3 is the volume of the third section of the VK column, the PFR.

The mass balance for water is

$$F_2 dC_{W,3} = -r_W dV_3$$

where $C_{W,3}$ is the concentration of water at a certain point in the PFR section.
Finally, the overall mass balance is (in an inert gas-free basis)

$$F_{in} = F_3 + L'$$

where

L' is the mass flow rate of the liquid from the top condenser.
F_3 is the mass flow rate of the polymer product leaving the PFR.

Note that the previous modeling neglects the possibility of back mixing of the liquid phase from CSTR 2 to CSTR 1.

12.6   Integrated Biopolymer Production in Biorefineries

This section shows potential process pathways for production of biopolymers either as the main product or coproduct, integrated in biorefineries. Pathways for production of polyesters, polyurethanes, polyamides and polycarbonates are discussed. Note that the heuristics discussed in Section 11.6 for integration of bio based chemical synthesis routes in biorefineries apply for integration of biopolymer processes.

12.6.1   Polyesters

Polyesters are produced from a diol and one or more dicarboxylic acids. At least one of these monomers needs to be produced in a biorefinery. Potential pathways for producing polyesters from biorefinery products are shown in Figure 12.16. Replaceable petrochemicals include 1,3-propanediol, 1,4-butanediol, succinic acid and adipic acid. With regard to sustainability and economics, it is important that the monomers are produced at a lower cost and with a lower environmental impact, compared to petrochemical routes.

As shown in Figure 12.16, some polyesters require terephthalic acid or dimethyl terephthalate. Until now, these chemicals could only be produced from petrochemical feedstocks. Options for production of terephthalic acid from biorefinery chemicals or its substitution by other biorefinery products are shown.

Refer to ***Online Resource Material, Chapter 12 – Additional Exercises and Examples*** for process pathways to the monomers for polyester production. Potential pathways for production of terephthalic acid from biomass or the production of poly(ethylene furanoate) as a substitute for PET are also discussed in the online material.

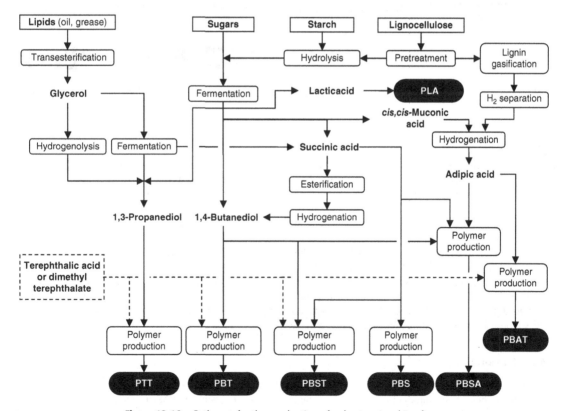

Figure 12.16   *Pathways for the production of polyesters in a biorefinery.*

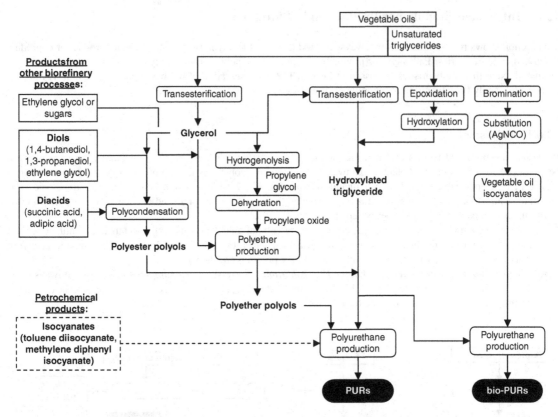

Figure 12.17 *Pathways for the production of polyurethanes (PURs) with a biomass based component or as purely biomass based polymers.*

12.6.2 Polyurethanes

Polyurethanes (PURs) belong to the family of polymers having the functional group urethane (–NHCOO) in the main chain. PURs are produced by addition polymerization between a diol (a molecule having two –OH groups) and a polyol (a molecule having multiple –OH groups) or a polymeric polyol and an isocyanate compound. Polymeric polyols can be polyether polyols, polyesters polyols or hydroxylated vegetable oils. Isocyanate compounds commonly used include toluene diisocyanate ($CH_3C_6H_3(NCO)_2$), methylene diphenyl diisocyanate ($OCNC_6H_4CH_2C_6H_4NCO$) and polymeric isocyanates.

The substitution of petrochemical based polyols by a biorefinery product is possible. Figure 12.17 shows the potential pathways for the production of PURs with a bio based component or a bio-PUR produced from biomass. Vegetable oils and glycerol are the key platforms for the production of PURs in a biorefinery. Although still under research, the substitution of the toxic isocyanate compounds with polyisocyanates produced from vegetable oil is a probable route.

Refer to *Online Resource Material, Chapter 12 – Additional Exercises and Examples* for a discussion on biorefinery pathways for the production of monomers for polyurethanes.

12.6.3 Polyamides

Nylons or polyamides are engineered thermoplastics and long-chain polymers having amide groups [–CO–NH–] in the main chain. Proteins are examples of naturally occurring polyamides. Polyamides (PAs) are synthesized from diamines and dicarboxylic acids, amino acids or lactams. Polyamides are characterized by the number of carbon atoms of the

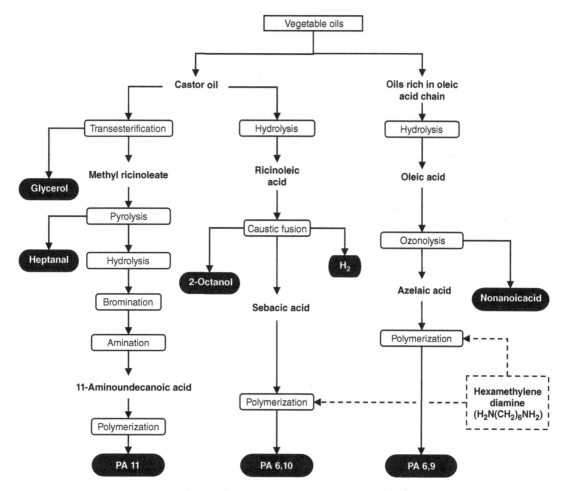

Figure 12.18 *Pathways to polyamides from vegetable oils.*

monomers. For example, PA 6,10 is synthesized from amine with six carbons and a dicarboxilic acid with ten carbons. Potential polyamides from at least one biomass based monomer produced in a biorefinery include:

a. PA 6,6, poly(hexamethylene adipamide), is produced by condensation of hexamethylenediamine ($H_2N(CH_2)_6NH_2$) with adipic acid. This reaction follows a step-growth polymerization. In this case adipic acid can be a biomass based monomer, as discussed in Section 12.6.1.

b. Polycaprolactam (PC) or PA 6 is synthesized by ring-opening polymerization of caprolactam. Caprolactam, the only monomer used, has six carbon atoms ($C_6H_{11}NO$). Caprolactam production from petrochemicals involves several steps and is expensive. Alternative routes to obtain caprolactam from biomass include:

   i. Synthesis from lysine, obtained by fermentation of sugars. Refer to the ***Online Resource Material, Chapter 12 – Additional Exercises and Examples*** for a process flowsheet and process description.

   ii. Synthesis route from HMF. This route is discussed in the ***Online Resource Material, Chapter 11 – Additional Exercises and Examples***.

c. Polyamides from vegetable oils. Figure 12.18 shows pathways to produce various polyamides from vegetable oils. The fatty acid chains present in the triglycerides contained in vegetable oil are first converted into free fatty acid or esters, for example, oleic acid, ricinoleic acid and methyl ricinoleate. Refer to ***Online Resource Material,***

Chapter 12 – Additional Exercises and Examples for a description of biorefinery processes for the monomer production for polyamides from vegetable oils.

As shown in Figure 12.18, the routes to polyamides produce various valuable products such as heptanal (used in cosmetics and perfumes), nonanoic acid (used in the production of herbicides, plasticizers and lacquers), 2-octanol (used as plasticizer, lubricant and in synthesis of amines) and hydrogen.

Note that production of PA 6,10 and PA 6,9 requires hexamethylenediamine. A biomass route to this compound is needed to make polyamides more sustainable.

12.6.4   Polycarbonates

Polycarbonates are polymers containing carbonate groups (–O–CO–O) in the main chain. They are mainly synthesized by condensation of phosgene and aromatic or aliphatic diols.

The synthesis of polycarbonates involves a two-step process, described for an aromatic diol as follows. First, the diol is treated with NaOH to produce a sodium diphenoxide. This intermediate reacts with phosgene ($COCl_2$) to start the polymerization. An alternative route to polycarbonates consists of the transesterification of aromatic diols with diphenyl carbonates.

12.7   Summary

The versatility of biopolymer materials will continue to increase and efficient ways for their production need to be developed. This chapter discussed the possibilities of introducing processes for producing bio based polymers in biorefineries. Reaction and separation pathways can be effectively integrated to make biopolymers part of a biorefinery product portfolio. Models for biopolymer reactors, in particular heterogeneous catalytic reactors, intensified reactors and polymer processing operations are shown with examples. Integrated biorefinery configurations will ensure efficient utilization of carbon and energy in biomass for the production of polymers.

References

1. H.L. Traub, P. Hirt, H. Herlinger, W. Oppermann, Synthesis and properties of fiber-grade poly (trimethylene terephthalate), *Angew. Makromol. Chem.*, **230**, 179–187 (1995).
2. C.E. Carraher, *Polymer Chemistry*, 7th edn., CRC Press, Boca Raton, Florida, 2008.
3. R. Mehtaa, V. Kumar, S.N. Upadhyay, Mathematical modeling of the poly(lactic acid) ring–opening polymerization kinetics, *Polymer-Plastics Technol. Eng.*, **46**, 257–264 (2007).
4. H. Guobin, L. Zuyu, Y. Suling, Y. Rufeng, Study of reaction and kinetics in pyrolysis of methyl ricinoleate, *J. AOCS*, **73**, 1109–1112 (1996).
5. L.A. Belfiore, *Transport Phenomena for Chemical Reactor Design*, John Wiley & Sons, Inc., USA, 2003.
6. O. Levenspiel, *Chemical Reaction Engineering*, John Wiley & Sons, Inc., USA, 1999.
7. M.E. Davis and R.J. Davis, *Fundamentals of Chemical Reaction Engineering*, McGraw-Hill, USA, 2003.

13

Separation Processes: Carbon Capture

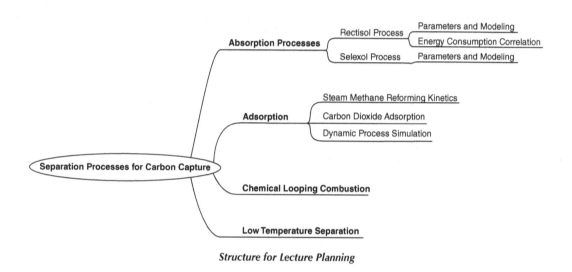

Structure for Lecture Planning

Carbon dioxide is the main contributor to global warming. The main cause to the increasing atmospheric CO_2 concentration is fossil fuel combustion for power generation, transport, industry and domestic use. The reservoir of fossil fuel is fast depleting, needing an alternative source of carbon. Biomass can be a promising replacement. Biomass refers to biological wastes, which also emit CO_2 on combustion and processing. However, biomass consumes CO_2 in their process of growing, hence providing low or neutral CO_2 emission performance. Nevertheless, CO_2 emission is one major concern during the transition phase from fossil resource to renewable resource. CO_2 capture and storage (CCS) from hydrocarbon based combined heat and power (CHP) and hydrogen production processes is generating a lot of interest because it is assumed to be a relatively efficient mid-term option for greenhouse gas reductions. CO_2 can be captured by absorption, adsorption, chemical looping combustion and low temperature processes, discussed in the following sections. These processes need to be integrated with an energy or biorefinery system to reduce emissions. To enable such overall process integration, the following sections discuss insightful and accessible process modeling frameworks. Membrane processes are also of interest, discussed in the context of chemical and biofuel production in Section 11.5.

Biorefineries and Chemical Processes: Design, Integration and Sustainability Analysis, First Edition.
Jhuma Sadhukhan, Kok Siew Ng and Elias Martinez Hernandez.
© 2014 John Wiley & Sons, Ltd. Published 2014 by John Wiley & Sons, Ltd.
Companion Website: http://www.wiley.com/go/sadhukhan/biorefineries

13.1   Absorption

The absorption process is used to absorb components from the gas or vapor phase into a liquid phase of a solvent. If solvent reacts with components to be absorbed, then the process is called a *chemical absorption* or *chemisorption* process. Otherwise, the process involves *physical absorption* or *physisorption*, with no chemical bond formation between absorbed components and solvent. Carbon dioxide can be absorbed by both routes. Amine and alkanolamine such as monoethanolamine and diethanolamine are the solvents for chemisorption, while a mixture of dimethyl ethers and polyethylene glycol is used in Selexol^{TM} processes (physisorption) and refrigerated methanol is used in Rectisol^{TM} processes (physisorption), respectively. The processes can be designed for bulk removal of carbon dioxide and selective absorption of hydrogen sulfide, mercaptans, etc., if present, to ppm or ppb levels as necessary. Due to the higher uptake of solute in chemisorption processes, a lower solvent rate is required, but at the expense of a higher amount of energy input, compared to physisorption processes. Chemical interactions also make the equilibrium correlation between the solute mass (or mole) fraction in gas and the solute mass (or mole) fraction in liquid nonlinear, due to the lack of ideality of the reaction mixture. However, for physical absorption processes, with certain assumptions, the equilibrium correlation can be made linear. The countercurrent operation in an absorber is a more common design feature for absorption processes. The following sections show the shortcut generic mass balance model for the countercurrent absorption processes and rigorous simulation frameworks for Selexol^{TM} and Rectisol^{TM} processes. Figure 13.1 shows a countercurrent absorber column stage-by-stage liquid and gas composition changes and operating and equilibrium lines.

A shortcut generic mass balance model for the countercurrent physisorption processes has the following properties:

G_{in} is the inlet mass (or molar) flow rate of insoluble gas.
G_{out} is the outlet mass (or molar) flow rate of insoluble gas.
L_{in} is the inlet mass (or molar) flow rate of nonvolatile solvent.
L_{out} is the outlet mass (or molar) flow rate of nonvolatile solvent.
X_{in} is the mass (or moles) of solute (absorbed component) in the inlet nonvolatile solvent per inlet mass (or molar) flow rate of nonvolatile solvent.
X_{out} is the mass (or moles) of solute (absorbed component) in the outlet nonvolatile solvent per outlet mass (or molar) flow rate of nonvolatile solvent.

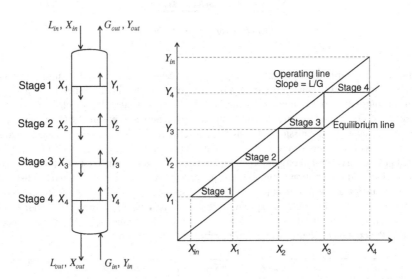

Figure 13.1   *Stage-wise liquid and gas composition changes and operating and equilibrium lines for a countercurrent (physical) absorber column.*

Y_{in} is the mass (or moles) of solute (absorbed component) in the inlet insoluble gas per inlet mass (or molar) flow rate of insoluble gas.

Y_{out} is the mass (or moles) of solute (absorbed component) in the outlet insoluble gas per outlet mass (or molar) flow rate of insoluble gas.

Assuming that the flow rates of the nonvolatile solvent and the insoluble gas do not change (or incompressible fluid),

$$L_{in} = L_{out} = L \text{ and } G_{in} = G_{out} = G.$$

From the overall mass balance,

$$G\left(Y_{in} - Y_{out}\right) = L\left(X_{out} - X_{in}\right) \tag{13.1}$$

At any theoretical stage i, the mass (or molar) compositions in gas and liquid (Y_i and X_i, respectively) are correlated by an assumption of equilibrium (given in the following equation where K is the equilibrium constant, the slope of the equilibrium line in Figure 13.1):

$$Y_i = KX_i \quad \forall i \in \text{Theoretical stages} \tag{13.2}$$

where A is the absorption factor defined in Equation (13.3). Equation (13.4) is to calculate the theoretical number of stages, N, from known inlet and outlet mass (or molar) fractions. Equation (13.5) can be used to calculate the outlet mass (or molar) fractions from the theoretical number of stages and known inlet mass (or molar) fractions. The value of A ranges between 1.2 and 2 and often is at 1.4. High pressure and low temperature favor the absorption process. The equations are also applicable to multicomponent systems. In that case, either a limiting component with the highest equilibrium constant value can be used to represent the system or a group of main solutes to be absorbed can be used to represent the system.

$$\boxed{A = \frac{L}{KG}} \tag{13.3}$$

$$\boxed{N = \frac{\ln\left[\left(\dfrac{A-1}{A}\right)\left(\dfrac{Y_{in} - KX_{in}}{Y_{out} - KX_{in}}\right) + \dfrac{1}{A}\right]}{\ln A}} \tag{13.4}$$

$$\frac{Y_{in} - Y_{out}}{Y_{in} - KX_{in}} = \frac{A^{N+1} - A}{A^{N+1} - 1} \tag{13.5}$$

The actual number of stages is calculated by dividing N by stage efficiency, E, which can be estimated using physical parameters shown in the equation below. MW_L is the molecular weight of the liquid, μ_L is the viscosity of the liquid in kg m^{-1} s^{-1} and ρ_L is the liquid density in kg m^{-3}:

$$-14.333\, E^{0.3854} = \ln\left(\frac{KMW_L\mu_L}{\rho_L}\right) \tag{13.6}$$

Thus, for a value of $KMW_L\mu_L/\rho_L$ of 10^{-5}, $E = 0.566$ and for a value of $KMW_L\mu_L/\rho_L$ of 10^{-1}, $E = 0.009$, respectively. Equation (13.6) is valid for $0.008 < E < 0.6$.

The packing height of a packed bed column is determined by multiplying the number of stages with the height equivalent of a theoretical plate (HETP), with a typical value of the HETP of 1–2 m for absorption (based on the convenience of cleaning).

Exercise 1. A mixture of light hydrocarbons containing methane is to be freed of the methane by absorption in a naphtha stream with an average molar mass of 56. The concentration of methane in the light hydrocarbon gas stream is 4% by volume and the liquid naphtha stream contains 0.2% methane by weight.

The volumetric flow rate of the light hydrocarbon gas stream is 1000 m^3 h^{-1}. The pressure is 1.05 atm and the temperature is 25 °C. The molar mass of methane is 16. Assume that the gas and liquid volumetric flow rates are constant and the fluid system is incompressible (their respective densities are constants). One gram molar volume of gas occupies 22 400 cm^3 at standard conditions. The gas and liquid equilibrium obeys Raoult's law, with their saturation pressure equal to 0.165 atm. If 95% removal of methane by volume is required, estimate the liquid flow rate and the number of theoretical stages. For a liquid viscosity of 0.95 cP (1 poise = 1 g cm^{-1} s^{-1}) and density of 750 kg m^{-3}, calculate the actual number of stages and the height of the column. State the assumptions.

Solution to Exercise 1

$$X_{in} = 0.002 \times \frac{56}{16} = 0.007 \text{ (in mole fractions)}$$

Assuming the ideal gas law and constant volumetric flow rate of the gas, the molar flow rate of the gas from the ideal gas law is as follows:

$$n_2 = \frac{V_2}{V_1} \times \frac{n_1}{\left(\frac{T_2}{T_1}\right)\left(\frac{P_1}{P_2}\right)} \tag{13.7}$$

By replacing state 1 with the standard condition, temperature $T_1 = 0$ °C, pressure $P_1 = 1$ atm, molar volume at the ideal condition $V_1 = 0.0224$ m^3 gmol^{-1} and number of moles at the ideal condition $n_1 = 1/1000$ kmol, the following equation is obtained:

$$G = 1000 \times \frac{1}{0.0224 \times \frac{298}{273} \times \frac{1}{1.05}} \times \frac{1}{1000} = 42.94 \text{ kmol h}^{-1}$$

From Raoult's law,

$$K = \frac{P^{sat}}{P} = \frac{0.165}{1.05} = 0.157 \tag{13.8}$$

Assuming $A = 1.4$, $L = 1.4 \times 0.157 \times 42.94 = 9.44$ kmol h^{-1} (from Equation (13.3)). Assuming G is constant and the ideal gas law is applied, $Y_{out} = 0.04 (1 - 0.95) = 0.002$ and

$$N = \frac{\ln\left[\left(\frac{1.4-1}{1.4}\right)\left(\frac{0.04 - 0.157 \times 0.007}{0.002 - 0.157 \times 0.007}\right) + \frac{1}{1.4}\right]}{\ln 1.4} \approx 8 \text{ theoretical stages (from Equation (13.4))}$$

Substituting the values in Equation (13.6), the stage efficiency, E, can be calculated:

$$-14.333 E^{0.3854} = \ln\left(\frac{0.157 \times 56 \times \left(\frac{0.95}{1000}\right)}{750}\right)$$

$$E = 0.55$$

The actual number of stages = $N/E = 8/0.55 \approx 15$. Taking a typical value of the HETP of 1.1 m for absorption, the height of the column is = $15 \times 1.1 = 16.5$ m.

Exercise 2. A gas stream with a flow rate of 500 kmol h^{-1} having 20% CO_2 and 10% H_2S by volume is to be purified to remove CO_2 and H_2S by at least 90% and 99.99%, respectively, by absorption into a methyldiethanolamine (MDEA) solution with a concentration of 7 kmol m^{-3}. The absorber operates at 40 atm and the temperature at the bottom of the absorber is at 60°C. Solubility of CO_2 and H_2S in the solvent leaving the absorber is 0.75 kmol per kmol MDEA and 0.9 kmol per kmol MDEA, respectively. The relevant parameters of components are given in Table 13.1.

Table 13.1   *Relevant parameters of components.*

Component	Molar Mass (kg kmol^{-1})	Heat of Absorption (MJ kg^{-1})
CO_2	44	2.1
H_2S	34	1.9
	Density (kg m^{-3})	Heat capacity (kJ kg^{-1}°C^{-1})
MDEA	1050	10.8

i.   Calculate the mole fractions CO_2 and H_2S in the pure gas leaving the absorber.
ii. Estimate the liquid flow rate (in m^3 h^{-1}) in the absorber-stripper loop assuming the solvent flow rate to be 1.2 times the minimum.
iii. If the steam consumption is 150 kg m^{-3} of solvent circulated, estimate the steam flow rate (in kg h^{-1}) and the reboiler heat duty (in MW) if the latent heat of the steam is 2100 kJ kg^{-1}.
iv. Estimate the return temperature of the solvent to the absorber to maintain 60°C at the bottom if the main effect of the heat of absorption is to heat the liquid in the absorption column. Explain how this can be achieved.

Solution to Exercise 2

i.   Assume that the gas follows the ideal gas law, so that its volume fraction is equivalent to its mole fraction.
Gas without CO_2 and H_2S = $500 \times (1 - 0.2 - 0.1) = 350$ kmol h^{-1}
CO_2 in the outlet gas = $500 \times 0.2 \times (1 - 0.9) = 10$ kmol h^{-1}
H_2S in the outlet gas = $500 \times 0.1 \times (1 - 0.9999) = 0.005$ kmol h^{-1}
CO_2 mole fraction in the outlet gas = $\frac{10}{350+10+0.005} = 0.028$

H_2S mole fraction in the outlet gas = $\frac{0.005}{350+10+0.005} = 1.39 \times 10^{-5}$

ii. CO_2 in solvent = $100 - 10 = 90$ kmol h^{-1}
H_2S in solvent = $50 - 0.005 = 49.995$ kmol h^{-1}
$X_{CO_2}^{out} = \frac{90}{0.75} = 120$ kmol h^{-1}
$X_{H_2S}^{out} = \frac{49.995}{0.9} = 55.55$ kmol h^{-1}

Actual flow rate of MDEA = $1.2 \times \frac{120+55.55}{7} = 30.09$ m^3 h^{-1}

iii. Steam consumption = $150 \times 30.09 = 4514.143$ kg h^{-1}
Reboiler duty = $\frac{4514.143 \times 2100}{1000 \times 3600} = 2.63$ MW

iv. Heat of absorption = $\frac{(90 \times 44 \times 2.1 + 49.995 \times 34 \times 1.9) \times 1000}{3600} = 3207.133$ kW

Temperature rise = $\frac{3207.133 \times 3600}{10.8 \times 1050 \times 30.09} = 34$ °C

Temperature of the inlet solvent at the top of the absorber must be kept at 26 °C to maintain 60 °C at the bottom of the column.

13.2   Absorption Process Flowsheet Synthesis

Figure 13.2 shows a simplified flowsheet for the absorption process. The main components are two columns, an absorber and a stripper. A stripper column desorbs the solvent. The gas mixture enters the absorber from the bottom of the column and flows upwards under the pressure difference between the bottom and top of the column. During the process, the gas contacts the liquid solvent (mainly recycled and some made-up) entering from the top of the column and the solute is

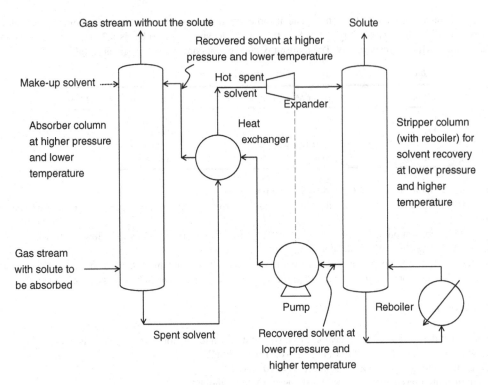

Figure 13.2   *Flowsheet of absorption process.*

transferred from the gaseous phase to the liquid phase. The gaseous stream leaving the absorber is lean in solute content, while the solvent is enriched with the solute. The rich solvent is regenerated by desorbing the solute in a stripper column and then recycled back to the absorber. *Higher pressure and lower temperature favor the absorption process, while the stripper column generally attached to a reboiler operates at a higher temperature and lower pressure.* Hence, there is an opportunity for heat and pressure recovery between the two columns. The recovered solvent from the stripper column at a higher temperature can release the heat to the lean solvent from the absorber column before entering the stripper column. The power generated from an expansion of the lean solvent can be used to operate the pump to pressurize the recovered solvent. This heat and power integration scheme is shown in the figure.

The flowsheets for chemisorption and physisorption processes are very similar except that they work on different principles of separation. In the physical absorption process, solute is transferred from a gas mixture into a solvent by solution, which primarily follows Henry's law (see Chapter 11). Based on that, the *loading capacity* (concentration of solute in solvent, on a solute-free basis) of the physical solvent is proportional to the partial pressure of the solute in the gas mixture. In the chemical absorption process, solute is transferred into a solvent by reaction and the loading capacity is determined by the amount of solvent. Generally, chemical solvents have a higher absorption capacity than physical solvents for the case with a low solute partial pressure, while physical solvents are favored at high acid gas partial pressure conditions.

The most widely used chemical solvents are the amine aqueous solutions. Conventional amine solvents include the primary and secondary amines such as mono- and diethanolamine (MEA and DEA), as well as the tertiary amines such as the methyl-diethanolamine (MDEA). MEA and DEA are very reactive and hence have a fast rate of acid gas removal. However, both MEA and MDEA have a limitation in that the maximum CO_2 loading capacity is lower than 50% based on stoichiometry. On the other hand, the MDEA solvent is relatively slow in the reaction rate with CO_2, but is capable in achieving a 100% approach to equilibrium of the CO_2 loading capacity based on stoichiometry.

The conventional amine separation is very energy intensive. The other problem is the degradation of these conventional amine solvents in the presence of oxygen. The possible solutions are mixed amine solvents and new solvents such as the sterically hindered amines. Mixed amine solvents are primarily a mixture of MEA, DEA and MDEA. The blended solvent has the advantage of both a high reaction rate of the MEA and DEA and a high CO_2 loading capacity of the MDEA. A mixture of aqueous MEA and mixed MEA and MDEA solvents shows a reduction in heat duty, although the chemical stability of the solvent needs to be further tested. Other than the conventional amine solvents, the research on chemical absorption has shown that the sterically hindered amines have a higher CO_2 loading capacity and a lower reaction heat of absorption/regeneration. The representative sterically hindered amines are the proprietary solvents, KS-1, KS-2 and KS-3 developed by Mitsubishi Heavy Industries. KS-1 solvent shows an improvement on conventional amine solvents in energy consumption in regeneration and the tolerance to the existence of oxygen.

Chemical absorption facilities are capital intensive. An estimation targeted to remove 90% of CO_2 in the exhaust gas from a natural gas based power plant generating 500 MW_e, with the chemical absorption unit utilizing MEA as the solvent, shows that six absorber columns with a diameter of 8 meters each, are required for the absorption process, while the stripping process requires 12 columns with a diameter of 6 meters each. To install a chemical absorption unit on such a scale could increase the overall capital cost by 80%.

The two most commonly used physical absorption processes, the Rectisol™ and Selexol™ technologies for the removal of CO_2, H_2S, COS, HCN, NH_3, nickel and iron carbonyls, mercaptans, naphthalene, organic sulfides, etc., to a trace level are discussed as follows.

13.3   The Rectisol™ Technology

The Rectisol™ technology developed by Lurgi uses refrigerated methanol as the solvent for physical absorption or removal of undesired contaminants producing ultra-clean syngas and is widely used in coal gasification plants[1-4]. Rectisol™ provides an excellent option for co-removal of a number of contaminants including H_2S, COS, HCN, NH_3, nickel and iron carbonyls, mercaptans, naphthalene, organic sulfides, etc., to a trace level (e.g., H_2S to less than 0.1 ppm by volume), using one integrated plant, from stringent resources, like coal containing all contaminants more than any other resources. Nearly all coal gasification plants operating in the world today include a Rectisol™ gas purification process for the production of hydrogen, hydrogen rich gases or syngas with hydrogen and carbon monoxide as major constituents. Because of the increasing use of biomass gasification technology in the face of growing interest for chemical production, such as synthesis of ammonia, methanol, Fischer–Tropsch liquids, oxo-alcohols and gaseous products such as hydrogen, syngas, reduction gas and town gas, a steep increase in the application of Rectisol™ processes is expected. A Rectisol™ process needs to be integrated to a low temperature fuel gas, such that minimum cooling is needed to attain the required refrigeration for the Rectisol™ process. A framework for simulation of a Rectisol™ process in Aspen Plus is shown in Figure 13.3. The process specifications for simulation (e.g., in Aspen Plus) is discussed as follows.

The Peng–Robinson property package can be selected to adequately present a Rectisol™ process system. This thermodynamic property package provides liquid and vapor properties for low temperature or cryogenic processes, mixed

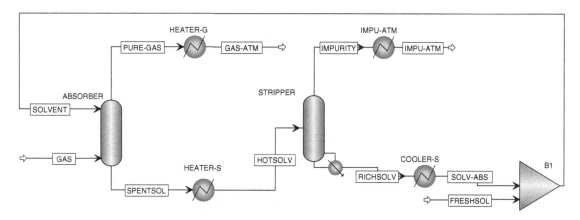

Figure 13.3   *Aspen Plus simulation model for a Rectisol™ process.*

refrigerants, air separation, carbon dioxide freezing and hydrogen systems. The property package can accurately calculate pure, mixture and infinite dilution properties for the system

Columns. For the ABSORBER and the STRIPPER columns shown in Figure 13.3, rigorous two- or three-phase fractionation in a single column, such as the RadFrac column in Aspen Plus, can be used. The ABSORBER column configuration without a condenser and reboiler requires a specification of the number of equilibrium stages. The calculation of the number of equilibrium stages is shown in Exercise 3. The absorber column also requires the feed stream specifications. The SOLVENT stream is entered at the top and the GAS stream is entered at the bottom stages in a counter current ABSORBER column configuration. With increasing ABSORBER column pressure (generally operating between 10 and 30 atm), the amount of solute or impurity absorbed (loading) in the solvent increases. The operating pressure should therefore be adjusted by examining how significant the increase in loading is with respect to the pressure increases. A pressure of 10–12 atm is required in an ADSORBER column for >90% removal of carbon dioxide by mole, from cold dry gas product generated from a biomass gasification process. The stripper column with a reboiler such as of the kettle type requires an additional specification for the distillate mass or molar flow rate, withdrawn from the top of the column. The distillate flow rate is slightly higher than the amount of solute absorbed by the solvent, to allow release of almost all the solute with the distillate stream. The distillate rate is determined iteratively to prevent loss of solvent in the solute rich gas stream (IMPURITY) from the top of the stripper column. As the solvent recovery column is a reboiled stripper column, the spent hot solvent stream, HOTSOLV, must be entered at a top stage of the stripper column.

Heat exchangers. There are various heaters or coolers considered in the flowsheet in Figure 13.3 prior to the heat integration exercise. These are for meeting the conditions required for the streams before entering downstream columns or processes. HEATER-S is to heat up the spent solvent, SPENTSOL, from the ABSORBER column, from −50 °C and 12 atm to 15 °C and 1.2 atm. The hot spent solvent stream, HOTSOLV, then enters the STRIPPER column for solvent recovery. The condition for HEATER-G depends on the downstream process requirement for the pure gas stream (GAS-ATM). For gas turbine applications, keeping the gas at a high temperature and pressure is preferable. Hence the temperature and pressure targets for the GAS-ATM stream are soft values and must be decided based on the whole process site performance, for example, integrated gasification combined cycle (IGCC). The solvent stream recovered from the bottom of the STRIPPER column, RICHSOLV, at 15 °C and nearly atmospheric pressure must be cooled to −50 °C and pressurized to 12 bar before recycling into the ABSORBER column. The COOLER-S is to achieve this condition of SOLV-ABS stream from RICHSOLV stream. IMPURITY stream comprising of CO_2, H_2S and other associated gases stripped off from the solvent is compressed for capture and storage to a pressure of 80–140 atm.

Exercise 3. Determining the number of theoretical equilibrium stages for a Rectisol^{TM} process for the capture of CO_2, H_2S and other associated gases. A gas stream from a biomass gasification process containing 24% by mole of CO_2 and 3% by mole of H_2S and other associated gases is to be freed of 97.7% of CO_2, H_2S and other associated gases by absorption in a refrigerated methanol solvent stream with a total inlet concentration of CO_2, H_2S and other associated gases of 0.4% by mole. The molar ratio of the pure solvent and the gas to be purified is 2:1. Estimate the number of theoretical stages.

Solution to Exercise 3. The basis of the gas flow including impurity is taken to be 100 kmol h^{-1}. Hence, the amount of pure solvent required is 200 kmol h^{-1}.

$$Y_{in} = \frac{24 + 3}{100 - 24 - 3} = 0.37$$

$$X_{in} = \frac{0.4}{100 - 0.4} \approx 0.004$$

$$Y_{out} = \frac{(24 \times 3) \times (1 - 0.977)}{(100 - 24 - 3)} = 0.0085$$

From Equation (13.3),

$$K = \frac{L}{AG}$$

Assuming $A = 1.4$,

$$K = \frac{200}{1.4 \times (100 - 24 - 3)} = 1.96$$

The number of theoretical stages can then be calculated using Equation (13.4):

$$N = \frac{\ln\left[\left(\dfrac{A-1}{A}\right)\left(\dfrac{Y_{in} - KX_{in}}{Y_{out} - KX_{in}}\right) + \dfrac{1}{A}\right]}{\ln A} \tag{13.4}$$

$$N = \frac{\ln\left[\left(\dfrac{1.4-1}{1.4}\right)\left(\dfrac{0.37 - 1.96 \times 0.004}{0.0085 - 1.96 \times 0.004}\right) + \dfrac{1}{1.4}\right]}{\ln 1.4} \approx 15 \text{ theoretical stages}$$

It can be found that the number of theoretical equilibrium stages is very sensitive to the value of Y_{out} and consequently to the value of the percentage removal required. For 97.8% removal of the impurity from the gas, $Y_{out} = 0.008$ and $N = 18$ stages.

A rule of thumb can be followed that:

>97% by mole of CO_2 removal requirement, 14–18 theoretical equilibrium stages are required.
>90% by mole of CO_2 removal requirement, 8 or more theoretical equilibrium stages are required.

13.3.1 Design and Operating Regions of Rectisol™ Process

Table 13.2 shows the stream data for a Rectisol™ process in Figure 13.3 for purification of a biomass gasification derived *gas* (shown as the stream called GAS in Figure 13.3 and Table 13.2). The following specifications are thus required for a simulation model. They include the molar ratio of the solvent to the *gas* to be purified, 2:1, the number of theoretical equilibrium stages for the ABSORBER column, 15, and that for the STRIPPER column, 12. The solvent recovered (with impurity at a mole fraction of 0.0016) at a molar ratio of 1.98:1 (with respect to the *gas*) is recycled with fresh or pure solvent at a molar ratio of 0.02:1 (with respect to the *gas*), to meet the solvent requirement for the *gas* (molar ratio of 2:1). The purity of the solvent recovered can be further enhanced by the application of pressure across the stages in the STRIPPER column. For example, if the solvent is pressurized to the inlet pressure of 12 atm required for the ABSORBER

Table 13.2 *Stream data for a Rectisol™ process simulation in Aspen Plus.*

Stream Name	GAS	SOLVENT	PURE-GAS	SPENTSOL	HOTSOLV	IMPURITY	IMPU-ATM	RICHSOLV	SOLV-ABS
Mole fraction									
Carbon dioxide	0.24	0	0.0139	0.1016	0.1016	0.8130	0.8130	0	0
Carbon monoxide	0.35	0	0.4727	0.0004	0.0004	0.0031	0.0031	0	0
Hydrogen sulfide	0.03	0.0016	0	0.0133	0.0133	0.0954	0.0954	0.0016	0.0016
Hydrogen	0.35	0	0.4736	0	0	0.0007	0.0007	0	0
Methane	0.03	0	0.0398	0.0003	0.0003	0.0023	0.0023	0	0
Methanol	0	1	0.00	0.8844	0.8844	0.0856	0.0856	0.9984	0.9984
Molar flow rate ratio	1	2	0.73	2.26	2.26	0.28	0.28	1.98	1.98
Temperature (°C)	−55	−55	−54	−40	15	14	25	150	−50
Pressure (atm)	12	12	12	12	1.2	1	80	1	12
Vapor fraction	1	0	1	0	0.1102	1	0	0	0
Average molar mass	23	32	15	33	33	42	42	32	32

Table 13.3 *Correlation coefficients and constant for Equation 13.9.*

Coefficient	a	b	c	d	e	f	g
Value	0.6786	−0.0055	5.6441	−0.0586	113.0676	−19.6496	−0.2015

column by increasing 1 atm per stage from top to bottom within the STRIPPER column, pure solvent can be recovered from the bottom of the stripper column. Thus, the minimum make-up solvent required for such an absorption process is zero, but at the expense of an increased pressure gradient and hence energy consumption.

The three most important process variables influencing the loading (solute uptake rate or percentage impurity removal) by a physisorption process are:

1. Operating temperature
2. Operating pressure
3. solvent-to-feed gas molar flow ratio.

As the shortcut modeling approach in Equations (13.1) to (13.6) does not account for these variable changes on the solute loading, it is important to generate a representative set of data using process simulation for incorporation into an overall process design framework. The percentage of carbon dioxide removal or capture (R) is the output variable for such a correlation with respect to the three important independent variables indicated: temperature in Kelvin (T), pressure in atm (P) and solvent-to-feed gas molar flow ratio (S) (Equation 13.9). The correlation coefficients and constant are given in Table 13.3.

The equation is valid for:

1. A >60% carbon dioxide removal (R)
2. Temperature of −45 °C to −55 °C
3. Pressure of 10–30 atm
4. Solvent-to-feed gas molar flow ratio of 1–2

$$R = a \times T + b \times T^2 + c \times P + d \times P^2 + e \times S + f \times S^2 + g \qquad (13.9)$$

In Equation (13.9), R is in molar % of carbon dioxide removal, T in kelvin, P in atm and S in molar ratio. The equation works satisfactorily between 0.1 and 0.25 mole fraction of carbon dioxide in the gas to be purified and for any set of components in the gas, with a maximum of ±3% error in the R estimation. The equation can be applied to determine conditions for purification for a range of gases, such as product gas from gasification consisting of carbon monoxide and hydrogen and biogas rich in methane, etc.

The upper limit of R is attained for the boundary conditions provided in Table 13.4. Alternatively, interpolation between various data points given in Table 13.5 can be carried out for mapping the input operating conditions to achieve a desired value of R. Figure 13.4 shows the bounded regions between operating pressures and CO_2 molar percentage removal. The regions are bounded between $S = 1$ and $S = 2$ for each given temperature, $T = 218$ K, 223 K and 228 K.

Table 13.4 *Boundary conditions for attaining maximum carbon dioxide recovery by the Rectisol^{TM} process.*

Temperature, °C	−55	−55	−45	−45
Pressure, atm	30	12	42	16
Solvent to feed molar ratio*	1	2	1	2

*Here the solvent molar flow rate corresponds to the total of the recycle and make-up solvent molar flow rates. Only 1% of the solvent required constitutes of make-up solvent.

Table 13.5   *Input temperature, pressure and solvent-to-feed molar ratio data mapped for carbon dioxide removal in molar percentage.*

T (K)	P (atm)	S in molar ratio	R in % by mole
223	20	1	62
223	12	1.5	64
218	10	1.5	64
228	25	1	68
218	20	1	71
223	25	1	77
223	12	1.8	78
228	30	1	80
228	20	1.5	85
223	12	2	86
223	30	1	89
223	20	1.5	94
218	12	2	95
228	25	1.45	97
228	42	1	97
218	30	1	98

13.3.2   Energy Consumption of a Rectisol™ Process

It is important to decide the optimum set of conditions for the best trade-off between the loading and the energy consumption of a process. The first option to evaluate minimum energy consumption for a given set of conditions for a Rectisol™ process is the temperature against significant cooling, refrigeration and heating duty curve analyses and the grand composite curve analyses. The grand composite curve analyses show that not all the cooling duties necessarily

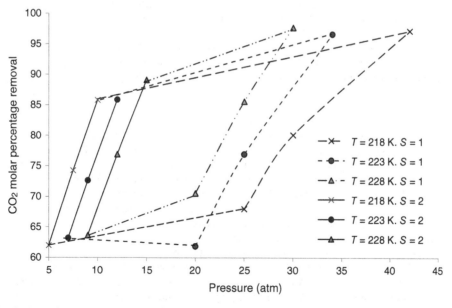

Figure 13.4   *Regions of operating pressure and CO_2 molar percentage removal bounded between $S = 1$ and $S = 2$ for each given temperature, $T = 218$ K, 223 K and 228 K.*

Table 13.6   *Target temperature and pressure conditions of the concerned streams in the Rectisol™ process in Figure 13.3.*

Inlet stream	RICHSOLV	GAS-PDT	SPENTSOL	PURE-GAS
Exchanger side:	HOT	HOT	COLD	COLD
Supply temperature (°C)	149	40	−40	−54
Supply pressure (atm)	1	12	12	12
Inlet vapor fraction	0	1	0	1
Outlet stream:	SOLVENT	GAS	HOTSOLV	GAS-ATM
Target temperature (°C)	−55	−55	15	25
Target pressure (atm)	12	12	1	12
Outlet vapor fraction	0	1	0.1103	1
J per mole of GAS	−47 793	−3024	16 391	1728

require refrigeration duties. In fact, with the help of the grand composite curves, the refrigeration duty can be minimized and shifted to cooling water consumption, which is cheaper. To construct a grand composite curve, the key heating and cooling duties need to be identified. In a gas absorption case, the main cooling duties are required for the feed gas and the solvent to the absorber process. The main heating duties are required for the gas purified for the downstream process and the spent solvent requiring recovery in the stripper column. Table 13.6 shows the temperature and pressure conditions of the concerned streams in the Rectisol™ process in Figure 13.3. The conditions correspond to the simulation instance presented in Table 13.2. RICHSOLV, the recovered solvent at 149 °C and 1 atm from the STRIPPER column, needs to be cooled down to −55 °C before entering the ABSORBER column operating at a pressure of 12 atm. The gasification product gas after cooling, GAS-PDT, at 40 °C and 12 atm also needs to be cooled down to the ABSORBER operating temperature of −55 °C. Both streams are therefore hot streams requiring cooling and some refrigeration. SPENTSOL and PURE-GAS streams from the absorber column require heating for the downstream processes. The outlet stream target temperatures are all soft values. The target temperatures of the hot streams could be higher than the final temperatures to be attained. In that case, some cooling duties are shifted into the ABSORBER. Similarly, the target temperatures of the cold stream can be lower, shifting final heating duties to the respective downstream processes. The target temperature specifications are shown in shaded cells in Table 13.6.

Table 13.6 also shows the enthalpy change during the process of heating or cooling of streams in joules per mole of gas. The negative sign indicates the cooling duties. The net enthalpy change required for these streams is −32 698 J per mole of GAS, indicating a cooling duty requirement. However, this is not yet the true cooling duty requirement for these streams, because the temperature level needs to be considered for feasible heat recovery between hot and cold streams. To enable the feasibility of heat recovery, the use of grand composite curve analyses is recommended. The cooling and heating duty profiles of the hot and cold streams identified with respect to the temperature allow construction of the grand composite curve. The cooling profiles with respect to temperatures of RICHSOLV and GAS-PDT streams from their supply to their target conditions in Table 13.6 are shown in Figures 13.5 and 13.6, while the corresponding heating profiles of SPENTSOL and PURE-GAS streams are shown in Figures 13.7 and 13.8, respectively. The inverse of the slopes of these curves at any point represents the heat capacity flow rate (cf. Section 3.3.1 and Equation (12.118)) of the fluid at that point. Unless there is a phase change, the heat capacity flow rate is fairly constant over the temperature ranges under consideration. Estimating the net enthalpy change over each temperature interval gives rise to the grand composite curve. For constructing the grand composite curve, the temperature of the hot stream is lowered by half of the minimum temperature difference and that of the cold streams is increased by the same amount. The grand composite curve thus created is shown in Figure 13.9. Any x axis value on this line will be the true enthalpy available at that temperature. The pocket indicated in the grand composite curve shows the region of heat recovery between hot and cold streams. The balance of the cooling duty, 32 700 J per mole of GAS, must be met by cooling water to the maximum extent and balance (if any) by refrigeration. This allows over 99% of the cooling duty supply by cooling water (Figure 13.9). The rigorous simulation of the heat exchanger can now follow once the minimum cooling or refrigeration duties are determined from the grand composite curve.

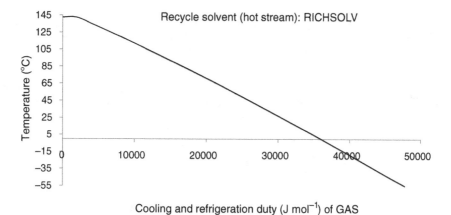

Figure 13.5 *The cooling profile of RICHSOLV with respect to temperature.*

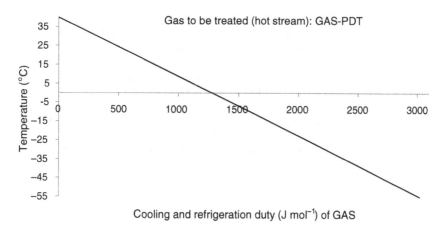

Figure 13.6 *The cooling profile of GAS-PDT with respect to temperature.*

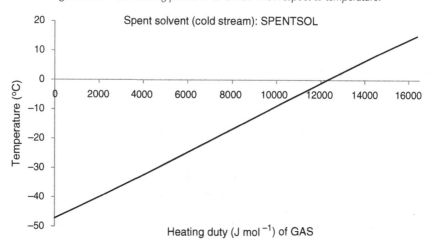

Figure 13.7 *The heating profile of SPENTSOL with respect to temperature.*

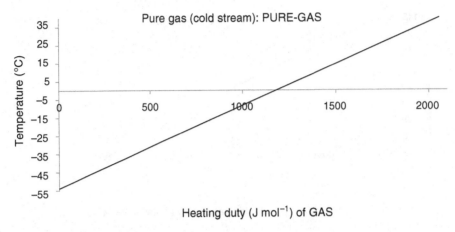

Figure 13.8   *The heating profile of PURE-GAS with respect to temperature.*

Ethane is the recommended refrigerant for a Rectisol™ process. The minimum refrigeration duty determined from the grand composite curve if divided by the specific heat capacity of ethane at a feasible condition provides the product of the amount of ethane and the temperature change, both being independent variables. Hence, one of them must be specified and the other can be determined from this heat balance. Considering a 4 °C temperature difference between the hot streams and the refrigerant ethane, the inlet temperature of ethane in the exchanger is −59 °C. At this condition, the heat capacity of ethane is 348 J mol⁻¹ K⁻¹. If the water temperature is lower than 20 °C, the water is chilled water. Chilled water instead of cooling water may be required to provide the balance of cooling. A 10 °C temperature increase can be assumed for chilled or cooling water after heat exchange.

The total cooling duty is proportional to the % molar recovery of carbon dioxide for a given solvent-to-feed gas molar ratio and is not a strong function of the absorber column temperature and pressure. The difference between the maximum

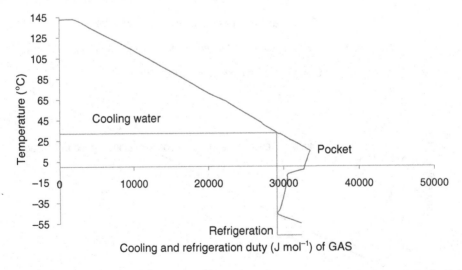

Figure 13.9   *The grand composite curve combining heating and cooling duties of streams shown in Table 13.6 over each temperature interval.*

Table 13.7 *Cooling duty required for a desired % molar recovery of CO₂ for solvent-to-feed GAS molar ratio of (a) 2:1 and (b) 1:1.*

(a)

J per mole of GAS	% molar recovery of CO_2
30 440	62
30 858	63
31 184	64
31 577	69
31 975	76
32 107	80
32 449	86
32 671	89
32 698	95

(b)

14 476	62
14 714	68
14 771	71
15 704	97
15 117	98

and the minimum cooling duties for >95% recovery and < 60% recovery of CO_2 by mole, respectively, is only 7% for a given solvent-to-feed gas molar ratio. The relationship between the % molar recovery of CO_2, solvent-to-feed gas molar ratio and cooling duty is shown in Table 13.7(a) and (b). Interpolation between the data points shown in Table 13.7(a) and (b) can be used to determine the cooling duty required for a desired % molar recovery of CO_2.

Typically, a large fraction of this cooling duty, 0.998, is required only in the form of cooling water, examined from the grand composite curve analysis. The remaining small fraction is the refrigeration duty required for solvent recovery before the absorber column. More solvent means a greater energy need for its recovery. In order to obtain the power requirement for the compressors to supply the refrigeration duty, the refrigerant cycle must be analyzed. Typically, the temperature requirement for ethane refrigerant is −55 °C to −59 °C. Either a single-stage refrigerant cycle with ethane or a cascade or dual-stage refrigerant cycle with ethane and propane as the inner or lower and outer or upper cycle refrigerants, respectively, may be used. The latter increases the coefficient of performance of the refrigerant cycle. The following exercise is on the evaluation of the two typical refrigerant cycles indicated. The multiple stream exchanger that is used to model liquefied natural gas (LNG) exchangers, cold boxes, etc., and perform zone analysis is recommended for heat integration between all process streams indicated in Table 13.6 and the utilities, cooling water and refrigerant ethane.

Exercise 4. Figures 13.10 and 13.11 show conceptual block diagrams for a single-stage refrigerant cycle with ethane and a cascade cycle with ethane and propane as inner and outer cycle refrigerants, respectively, for supplying refrigeration duties to a Rectisol™ process. The conditions at significant points in the cycles on the basis of 1 mole of respective refrigerants are shown in the figures.

i.   Calculate the coefficients of performance of the two cycles for the given data.
ii. Graphically show the difference between the two cycles in terms of work required in temperature versus entropy plot. Points 2 and 4 in Figure 13.10 and points 2' and 4' in Figure 13.11 correspond to the saturation conditions of ethane refrigerant. The entropies of the saturated vapor of ethane are: $-220 \, \text{J mol}^{-1} \, \text{K}^{-1}$ at the saturation temperature

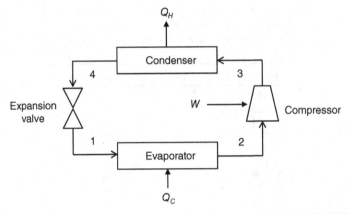

	1	2	3	4
Temperature (K)	215	214	359	305
Pressure (atm)	4	4	47.5	47.5
Vapor fraction	0.80902004	1	1	0
Enthalpy (J mol⁻¹)	−90750	−88209	−82880	−90750
Entropy (J mol⁻¹ K⁻¹)	−214	−202	−201	−225

Figure 13.10   *Conceptual block diagram of a single-stage refrigerant cycle with ethane for supplying the refrigeration duty for a Rectisol^{TM} process (data given for the molar basis of ethane as the refrigerant).*

of 305 K at a pressure of 47.5 atm and −208 J mol⁻¹ K⁻¹ at the saturation temperature of 255 K at 15 atm pressure, respectively.

iii.   Determine the refrigerant requirements in number of moles per mole of the feed gas for various % molar recoveries of CO_2 based on the assumption that 0.2% of the cooling duties provided in Table 13.7(a) and (b) are required in the form of refrigeration duties.

Solution to Exercise 4

i.   The *coefficient of performance of a refrigeration cycle* is the ratio of refrigeration duty performed per unit work required. Hence, for the single-stage refrigerant cycle in Figure 13.10, the coefficient of performance is defined as

$$\frac{Q_C}{W} = \frac{\text{Enthalpy at point 2 − Enthalpy at point 1}}{\text{Enthalpy at point 3 − Enthalpy at point 2}} = \frac{-88\,209 - (-90\,750)}{-82\,880 - (-88\,209)} = 0.48 \qquad (13.10a)$$

For the cascade cycle, the heat duties of the lower cycle condenser and upper cycle evaporator must be equal. This can be achieved by adjusting the molar ratio between the upper and the lower cycle refrigerants. Thus, the molar ratio of propane to ethane is

$$\frac{\text{Molar enthalpy of ethane between point 4' and point 3''}}{\text{Molar enthalpy of propane between point 5 and point 6}} = \frac{-98\,085 - (-85\,513)}{-117\,104 - (-108\,298)} = 1.428$$

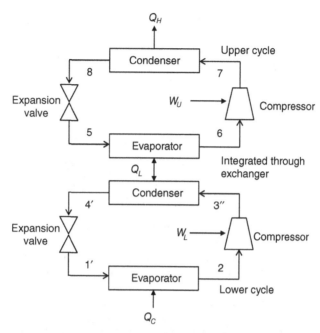

	1'	2	3''	4'	5	6	7	8
Temperature (K)	215	214	286	255	282	250	341	326
Pressure (atm)	4	4	15	15	6	2	18	18
Vapor fraction	0.2622	1	1	0	0.3526	1	1	1
Enthalpy (J mol⁻¹)	−98085	−88209	−85513	−98085	−117104	−108298	−103408	−117104
Entropy (J mol⁻¹ K⁻¹)	−248	−202	−201	−250	−327	−289	−287	−329

Figure 13.11 Conceptual block diagram of a dual-stage refrigerant cycle or a cascade cycle with ethane and propane as lower or inner and upper or outer cycle refrigerants for supplying the refrigeration duty for a Rectisol™ process (data given for the molar basis of the respective refrigerants).

The coefficient of performance of the cascade cycle in Figure 13.11 is defined as

$$\frac{Q_C}{W_L + W_U} = \frac{\text{Enthalpy at point 2} - \text{Enthalpy at point 1}'}{(\text{Enthalpy at point 3}'' - \text{Enthalpy at point 2}) - 1.428\,(\text{Enthalpy at point 7} - \text{Enthalpy at point 6})}$$

$$= \frac{-88\,209 - (-98\,085)}{[-85\,513 - (-88\,209)] + 1.428\,[-103\,408 - (-108\,298)]} \tag{13.10b}$$

$$= 1.02$$

Both the coefficients of performance for the single and cascade cycles are calculated based on 1 molar flow of ethane.

ii. To construct the temperature versus entropy diagrams, the entropy of saturated vapor within a condenser must be accounted for in addition to all other points, as shown in Figure 13.12. This is to reflect the actual area bounded

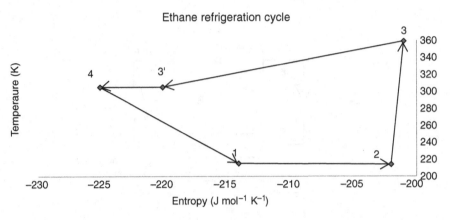

Figure 13.12   *Temperature versus entropy plot of ethane for the given single-stage refrigeration cycle.*

between the curves. Thus, the entropy of the saturated ethane vapor of $-220\,\text{J mol}^{-1}\,\text{K}^{-1}$ at the saturation temperature of 305 K and at a pressure of 47.5 atm is shown by point 3′ to reveal the actual path of the phase change of ethane; 3′–4 is the condensation line of ethane.

Figure 13.12 shows the temperature versus entropy profile of a single stage and Figure 13.13 shows a comparison of the profiles of ethane between the single-stage cycle and inner cycle of the cascade systems. The cascade cycle provides increased refrigeration duty, corresponding to the extended entropy 1′–1, but at a reduced compression due to the reduced pressure level from 3–3′ (47.5 atm) to 3″–3‴ (15 atm).

As a result, the work required for compression of ethane refrigerant is reduced by an area bounded within 3‴, 3″, 3 and 3′. However, this plot only presents the use of ethane as a refrigerant in two given cycle types. The outer cycle with propane as a refrigerant requires some additional work, shown in a pressure versus enthalpy plot of the complete cascade cycle in Figure 13.14.

The inner cycle condenser, 3″–4′, rejects heat to the outer cycle evaporator, 5–6; their heat exchange is in balance through the molar ratio of propane to ethane, 1.428:1. The inner cycle evaporator, 1′–2, provides the process cooling or refrigeration. The outer cycle condenser, 7–8, rejects its heat to chilled or cooling water. The inner and outer cycle compressor flows involved are 2–3″ and 6–7, respectively. Equation (13.10) can be verified from the enthalpy values at various points concerned in Figure 13.14. Thus, the ratio of the coefficients of performance between the cascade and the single-stage refrigerant cycles is 2.14.

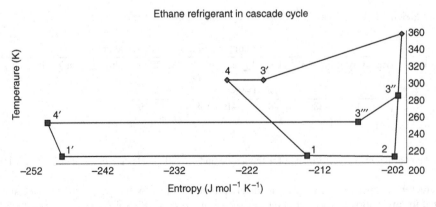

Figure 13.13   *Comparison of temperature versus entropy plot of ethane between the given cascade and single-stage refrigeration cycles.*

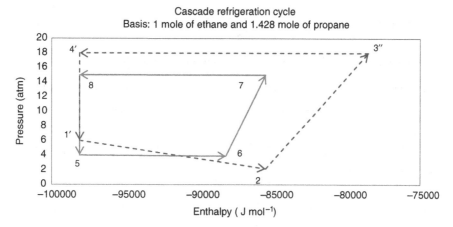

Figure 13.14 *Pressure versus enthalpy plot of ethane (1′–2–3″–4′) and propane (5–6–7–8) refrigerants for the given cascade refrigeration cycle in Figure 13.11.*

iii.   The difference in enthalpy between points 2 and 1 in Figure 13.10 is the heat of evaporation of the refrigerant ethane in the single-stage cycle. This heat of evaporation is the cooling duty available.

Hence, 1 mole of refrigerant ethane at the given temperature and pressure of the single-stage cycle provides $(-88\,209) - (-90\,750) = 2541$ J of cooling or refrigeration duty.

In the cascade refrigerant cycle, 1 mole of refrigerant ethane gives a refrigeration duty of $(-88\,209) - (-98\,085) = 9876$ J at the given temperature and pressure condition of the evaporator (between point 2 and point 1′) in Figure 13.11.

On this basis, the refrigerant requirements in the number of moles per mole of the feed gas for various % molar recoveries of CO_2 in Table 13.7(a) and (b) for the single-stage and the cascade schemes are shown in Table 13.8(a) and (b) and 13.9(a) and (b), respectively, where 0.2% of the cooling duties are required as refrigeration duties.

Exercise 4 helps to determine the ranges of use of refrigerant, ethane and propane. The following section shows how the work or compression duty required for the refrigeration cycle can be systematically minimized.

The *Antoine expression below provides the most common and effective form of correlation between vapor pressure and saturation temperature of gases*, in this case refrigerants, ethane and propane:

$$\boxed{log_{10}P^{sat} = A - \frac{B}{C + T^{sat}}}$$ (13.11)

where A is the correlation constant and B and C are the correlation coefficients and P^{sat} is the saturation vapor pressure at a saturation temperature T^{sat}.

Table 13.10 shows empirically determined values of A, B and C for ethane and propane, when the temperature unit is in K and pressure unit is in atm.

Although there are four levels of pressure to be adjusted for a cascade refrigeration cycle for a Rectisol™ process, the compression work requirement depends on the absorption system pressure and reduces with decreasing system pressure. The compression work is required to pressurize a refrigerant from its saturation vapor pressure for evaporation, P^{evap}, to its saturation vapor pressure for condensing, P^{cond}. As there are two refrigerants, there are four levels of pressure adjustments required based on the process specifications, shown below. Often, the minimum temperature approach in heat exchangers helps to decide these pressure levels. Hence, the Antoine equation (Equation (13.11)) and the equation parameters given in Table 13.10 are used to find the saturation vapor pressure for a given temperature and the saturation temperature for a given pressure.

Table 13.8 *Ethane refrigerant required in single-stage refrigeration cycle for desired %
molar recovery of CO_2 for the solvent to feed the GAS molar ratio of (a) 2:1 and (b) 1:1.
Basis of refrigeration duty: 2541 J per mole of ethane in a single-stage refrigeration cycle.*
(a)

J per mole of GAS	% molar Recovery of CO_2	Mole of Ethane per mole of GAS Required as Refrigerant
30 440	62	0.0240
30 858	63	0.0243
31 184	64	0.0246
31 577	69	0.0249
31 975	76	0.0252
32 107	80	0.0253
32 449	86	0.0256
32 671	89	0.0257
32 698	95	0.0257
(b)		
14 476	62	0.0114
14 714	68	0.0116
14 771	71	0.0116
15 704	97	0.0124
15 117	98	0.0119

Table 13.9 *Ethane and propane refrigerants required in cascade refrigeration cycle for desired
% molar recovery of CO_2 for solvent to feed GAS molar ratio of (a) 2:1 and (b) 1:1. Basis of
refrigeration duty: 9876 J per mole of ethane and propane to ethane molar ratio of 1.428:1 in a
cascade refrigeration cycle.*
(a)

J per mole of GAS	% molar Recovery of CO_2	Mole of Ethane per mole of GAS Required as Refrigerant	Mole of Propane per mole of GAS Required as Refrigerant
30 440	62	0.0062	0.0088
30 858	63	0.0062	0.0089
31 184	64	0.0063	0.0090
31 577	69	0.0064	0.0091
31 975	76	0.0065	0.0092
32 107	80	0.0065	0.0093
32 449	86	0.0066	0.0094
32 671	89	0.0066	0.0094
32 698	95	0.0066	0.0095
(b)			
14 476	62	0.0029	0.0042
14 714	68	0.0030	0.0043
14 771	71	0.0030	0.0043
15 704	97	0.0032	0.0045
15 117	98	0.0031	0.0044

Table 13.10   *A, B and C in the Antoine equation, for ethane and propane.*

	A	B	C
Ethane	4.2122	775.6265	0
Propane	4.2865	987.8092	0

1. Evaporation pressure, P^{evap}, of ethane for the refrigeration performance. The saturation evaporation temperature, T^{evap}, of ethane refrigerant is kept at a temperature 4–5 °C lower than the lowest temperature of the refrigerated solvent methanol (−50 to −55 °C). The Antoine equation (Equation (13.11)) is followed to determine the corresponding saturation evaporation vapor pressure, P^{evap}, of ethane using the correlation constant and coefficients for ethane in Table 13.10. Thus, P^{evap} of ethane = 3.89 atm (at a temperature of −59 °C).

2. Condenser pressure P^{cond} of ethane in the inner cycle is linked with the pressure of evaporation of propane in the outer cycle.

3. Evaporation pressure P^{evap} of propane. T^{evap} of propane refrigerant at this pressure (the Antoine Equation (13.11) and Table 13.10) must be lower than the T^{cond} of ethane refrigerant, corresponding to its P^{cond} by at least 4–5 °C.

   For P^{cond} of ethane = 26 atm, T^{cond} of ethane = 277 K from the Antoine Equation (13.11) and Table 13.10. This sets T^{evap} of propane at 272 K, which corresponds to P^{evap} of propane at 4.56 atm following the Antoine Equation (13.11) and Table 13.10.

   Next the P^{cond} of ethane is lowered until the P^{evap} of propane is reduced to an atmospheric pressure. Thus, the work required to create the vacuum pressure is avoided. This is shown below.

   For P^{cond} of ethane = 9 atm, T^{cond} of ethane = 238 K from the Antoine Equation (13.11) and Table 13.10. This sets T^{evap} of propane at 233 K, which corresponds to P^{evap} of propane at 1.1 atm following the Antoine Equation (13.11) and Table 13.10.

4. Condenser pressure P^{cond} of propane in the outer cycle is chosen to release the heat to cooling water. Thus, P^{cond} of propane = 12−15 atm, resulting in the saturation condensing temperature T^{cond} of propane at 308–318 K.

The above analysis shows that the P^{evap} of ethane and the P^{cond} of propane are pre-set by the refrigerated solvent temperature and the cooling water temperature, respectively. The P^{cond} of ethane and the P^{evap} of propane are linked and must be decided based on minimum compression work requirement. The compression work for a refrigerant is calculated using Equation (13.12) below. Table 13.11 shows the feasible P^{cond} of ethane and P^{evap} of propane based on the Antoine Equation (13.11) and parameters given in Table 13.10 and individual refrigerant compression work requirements using Equation (13.12) for a Rectisol™ process.

$$\text{Compression work requirement in a refrigeration cycle} = \frac{C_v \times T^{evap}}{n_s} \left\{ \left(\frac{P^{cond}}{P^{evap}} \right)^{\frac{\gamma-1}{\gamma}} - 1 \right\} \qquad (13.12)$$

where

P^{cond} = saturation vapor pressure for condensation of a refrigerant
P^{cond} of propane = 12–15 atm
P^{cond} of ethane varies between 9–24 atm

P^{evap} = saturation vapor pressure for evaporation of a refrigerant
P^{evap} of ethane = 3.89 atm
P^{evap} of propane varies between 1–4.56 atm

T^{evap} = saturation temperature for evaporation of a refrigerant at P^{evap}
T^{evap} of ethane = 218 K
T^{evap} of propane varies between 233–272 K

Table 13.11   *Feasible P^{cond} of ethane and P^{evap} of propane and individual refrigerant compression work requirements in a cascade refrigeration cycle for a Rectisol^{TM} process.*

P^{cond} of Ethane (atm)	T^{cond} of Ethane^a (K)	T^{evap} of Propane^{b,c} (K)	P^{evap} of Propane^a (atm)	Compression Work of Ethane^c (kJ mol^{-1})	Compression Work of Propane^c (kJ mol^{-1})
26	277	272	4.56	3.95	3.16
24	274	269	4.10	3.76	3.40
22	270	265	3.65	3.55	3.66
20	266	261	3.22	3.33	3.95
18	262	257	2.80	3.09	4.28
16	258	253	2.40	2.83	4.64
14	253	248	2.01	2.53	5.05
12	248	243	1.64	2.20	5.53
10	241	236	1.29	1.82	6.11
9	238	233	1.12	1.60	6.45

^{a}Calculated using the Antoine equation (Equation (13.11)) and parameters given in Table 13.10.
$^b T^{cond}$ of ethane $- 5 = T^{evap}$ of propane.
^{c}Calculated using Equation (13.12).

n_s = isentropic efficiency of the compressor = 0.9
C_v = heat capacity of the refrigerant at constant volume
C_v = 47.67 J mol^{-1} K^{-1} for ethane (molecular weight = 30)
C_v = 29.86 J mol^{-1} K^{-1} for propane (molecular weight = 44)
$\gamma = C_p/C_v = 1.18$ for ethane
$\gamma = 1.28$ for propane

It is clear that with increasing system pressure the compression work requirement for ethane in a cascade cycle increases. However, increasing the system pressure reduces the difference between P^{cond} and P^{evap} of propane, reducing its compression work requirement. The decrease in compressor work requirement for propane is more than the increase in the compression work requirement for ethane, reducing the total compression work requirement.

The following utility consumptions are established for the Rectisol^{TM} process:

Shaft power: 73.08 kW
LP steam: 323.74 kW
Refrigeration duty: 131.42 kW, respectively, per kmol h^{-1} of sulfur and nitrogen compounds removed

The Rectisol^{TM} process can also be operated at low syngas capacity, ~355 N m^3 h^{-1}.

13.4   The Selexol^{TM} Technology

The UOP Selexol^{TM} process can also be used when a less stringent fuel gas specification is required. The Selexol^{TM} process, unlike the Rectisol^{TM} process, is less energy intensive. A heat integration study shows that excess low pressure (LP) steam can be generated from a Selexol^{TM} process. The operating temperature of the Selexol^{TM} absorber columns is 30–40 °C. The solvent for the Selexol^{TM} process is dimethyl ethers of polyethylene glycol (DMEPEG). The Selexol^{TM} solvent has the formula $[CH_3O(CH_2CH_2O)_x xCH_3]$ where x ranges from 3 to 9 with an average molecular weight of about 272. The Selexol^{TM} process operates at high pressure >20 atm and can operate selectively to recover H_2S and CO_2.

The non-random two liquid (NRTL) thermodynamic property package can be applied for the simulation of the Selexol^{TM} processes for H_2S removal and CO_2 capture in Aspen Plus. The simulation makes use of Tetraglyme (tetraethylene glycol dimethyl ether) to represent the Selexol™ solvent because the Selexol™ solvent is a proprietary solvent of UOP LLC. A comparison of the solubility of CO_2 and H_2S in Selexol™ and Tetraglyme at 21 °C has shown similar results[5]. Also, the Selexol™ solvent experimental data are based on a representative system with Tetraglyme.

Exercise 5. Determine the number of theoretical equilibrium stages for a Selexol™ process for the capture of CO_2, H_2S and other associated gases. A gas stream from a biomass gasification process containing 24% by mole of CO_2 and 3% by mole of H_2S and other associated gases is to be freed of 90% of CO_2, H_2S and other associated gases by a Selexol™ solvent stream with a total inlet concentration of CO_2, H_2S and other associated gases of 0.4% by mole. The molar ratio of the pure solvent (first time charge) and the gas to be purified is 2:1. Estimate the number of theoretical stages.

Solution to Exercise 5. As in Exercise 3, the basis of the gas flow including impurity is taken to be 100 kmol h^{-1}. Hence, the amount of pure solvent required is 200 kmol h^{-1}.

$$Y_{in} = \frac{24 + 3}{100 - 24 - 3} = 0.37$$

$$X_{in} = \frac{0.4}{100 - 0.4} \approx 0.004$$

$$Y_{out} = \frac{(24 + 3) \times (1 - 0.9)}{100 - 24 - 3} = 0.037$$

From Equation (13.3),

$$K = \frac{L}{AG}$$

Assuming $A = 1.4$,

$$K = \frac{200}{1.4 \times (100 - 24 - 3)} = 1.96$$

The number of theoretical stages can be then calculated using Equation (13.4):

$$N = \frac{\ln\left[\left(\frac{A - 1}{A}\right)\left(\frac{Y_{in} - KX_{in}}{Y_{out} - KX_{in}}\right) + \frac{1}{A}\right]}{\ln A}$$

$$N = \frac{\ln\left[\left(\frac{1.4 - 1}{1.4}\right)\left(\frac{0.37 - 1.96 \times 0.004}{0.037 - 1.96 \times 0.004}\right) + \frac{1}{1.4}\right]}{\ln 1.4} \approx 5 \text{ theoretical stages}$$

The difference between the Rectisol™ and the Selexol™ processes is the reduction in removal ratio from >90% in a Rectisol™ to 90% in a Selexol™. Thus, the latter requires a lesser number of stages. Equation (13.4) can only be applied for an initial guess. The equation does not capture any difference due to the operating pressure and temperature and the type of solvent used. N is also insensitive to the value of K or the amount of solvent L for a very small value of X_{in}.

Similar to the Aspen Plus simulation framework for the Rectisol™ process in the previous section, the Aspen Plus simulation model for purification of a biomass gasification derived gas by a Selexol™ process has been discussed. The main difference is the use of a two-outlet flash column with rigorous vapor–liquid equilibrium model in place of the stripper column. This is because the Selexol™ solvent used Tetraglyme, which has a much higher molecular weight than methanol and one equilibrium stage at lower pressure and higher temperature is adequate to flash off the gases from the liquid solvent. Table 13.12 shows the stream data. The naming convention is the same as the Rectisol™ process simulation in Figure 13.3. The shaded cells present the specifications required for the Aspen Plus simulation. The number

Table 13.12 Stream data for the Selexol^TM process simulation in Aspen Plus.

Stream Name	GAS	SOLVENT	PURE-GAS	SPENTSOL	HOTSOLV	IMPURITY	IMPU-ATM	RICHSOLV	SOLV-ABS
Mole fraction									
Carbon dioxide	0.24	0.003	0.0358	0.0940	0.0940	0.7663	0.7663	0.0030	0.0030
Carbon monoxide	0	0	0	0	0	0	0	0	0
Hydrogen sulfide	0.03	0.001	0	0.0131	0.0131	0.1018	0.1018	0.0010	0.0010
Hydrogen	0.35	0	0.4836	0.0008	0.0008	0.0066	0.0066	0	0
Methane	0.03	0	0.0334	0.0026	0.0026	0.0217	0.0217	0	0
Tetraglyme	0	1	0	0.8772	0.8772	0.0010	0.0010	0.9960	0.9960
Molar flow rate ratio	1	2	0.72	2.28	2.28	0.27	0.27	2.01	2.01
Temperature (°C)	25	25	25	27	100	100	25	100	25
Pressure (atm)	30	30	30	30	1	1	80	1	30
Vapor fraction	1	0	1	0	0.1191	1	1	0	0
Average molar mass	23	222	16	200	200	41	41	222	222

of theoretical equilibrium stages for the ABSORBER column is 5. The solvent recovered with impurity at a mole fraction of 0.001 is also adequate for the gas recovery at a molar ratio of 2:1.

The Selexol^TM solvent at given conditions can remove hydrogen sulfide to a ppm level. In the simulation case given, with temperature = 25 °C, pressure = 30 atm and a solvent (first time charge) to-gas-molar ratio = 2:1, hydrogen sulfide is removed to an extent of 38 ppm in the syngas. Hence, for fuel cell applications with a stricter purification requirement at 0.1 ppm – ppb levels, the Selexol^TM process is not suitable. The Rectisol^TM solvent, a refrigerated methanol, on the other hand, has a high affinity towards hydrogen sulfide, removing hydrogen sulfide to a ppb level in the gas. The Selexol^TM process may need additional separation stages for further removal of hydrogen sulfide. With increasing pressure, removal of hydrogen sulfide increases. The following section shows the theoretical extents of removal ratios of most commonly occurring pollutants (e.g., carbon dioxide, hydrogen sulfide and carbonyl sulfide) in gasification-derived product gases, using Aspen Plus based rigorous simulation results.

13.4.1 Selexol^TM Process Parametric Analysis

IGCC product gas contains impurities such as H_2S, COS and a high concentration of capture-ready CO_2. The physical absorption process such as the Selexol^TM process is suitable for IGCC systems because the feed input to gas turbines does not require as stringent specifications as fuel cell, methanol synthesis and Fischer–Tropsch processes and a systematic heat integrated Selexol^TM process can generate excess LP steam from the process. A single-stage Selexol^TM absorption process can achieve 100% COS recovery 99.99% H_2S recovery and >60% CO_2 recovery, by the adjustment of operating pressure of the absorption column and more importantly by increasing the first time charge of the solvent. Figure 13.15 shows the profiles of % recovery of CO_2, H_2S and COS with respect to molar ratio of the first time charge of solvent to carbon dioxide in the feed gas in a single-stage Selexol^TM operation. Thus, to achieve 100% COS recovery, 99.99% H_2S recovery and >60% CO_2 recovery in a single-stage operation, the minimum molar ratio of the first time charge of solvent to carbon dioxide in feed gas is 2:1. Note that the phrase "first time charge of solvent" implies the amount of solvent needed only during a start-up operation. For subsequent operations, solvent is almost completely recovered in the stripping column and recycled back to the absorber column and only a small amount of make-up solvent may be required to sustain the operations. The single-stage process configuration is the same as in Figure 13.2. The operating temperatures of the single-stage absorber and stripper columns are 35–40 °C and 100 °C for the operating pressures of the single-stage absorber and stripper columns of 35–55 bar and atmospheric pressure, respectively.

A dual stage Selexol^TM absorption process can be used for further removal of H_2S to the ppb level and CO_2 by >90%, respectively. Increasing pressure helps in further purification. However, a more sensitive parameter is the first time charge of solvent in the second stage. In order to achieve 100% recovery of H_2S and >90% recovery of CO_2 in the second stage,

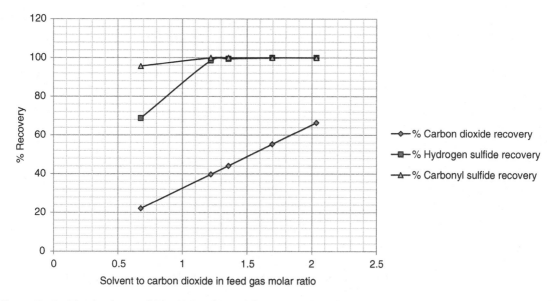

Figure 13.15 *The % recovery of CO_2, H_2S and COS with respect to the molar ratio of the first time charge of solvent to carbon dioxide in feed gas in a single-stage Selexol^{TM} operation.*

after 100% COS, 99.99% H_2S and >60% CO_2 recoveries in the first stage, the minimum molar ratio of the first time charge of solvent in the second stage to carbon dioxide in feed gas is 5:1. The correlation between % overall recovery of CO_2 in a dual-stage absorption system and the molar ratio of the first time charge of solvent to the second stage to carbon dioxide in feed gas to the first stage is shown in Figure 13.16. The pressure of the second stage absorption column is 60 bar and that of the second-stage stripper column is atmospheric. The second-stage absorption column is maintained at 40 °C and the highest temperature at the bottom of the stripper columns is 270–310 °C, sufficient to generate excess LP steam.

A dual-stage absorption process configuration for gas is shown in Figure 13.17. Table 13.13 shows an instance of the Selexol^{TM} process operating conditions based on gasification based product gas.

Exercise 6. Consider the dual stage Selexol^{TM} process operations shown in Table 13.13 and answer to the following questions.

1. Calculate the first stage, second stage and overall removal ratio of CO_2, H_2S and COS for the given process operations. Note that the *gas feed to the second absorption column* is the product gas from the first absorption stage and *gas stream without pollutant* is the final product gas from the second absorption stage (Table 13.13). The removal ratio is calculated based on the pollutant flows in the *gas feed to the first absorption column*.
2. Calculate the molar ratio of the first time solvent charged (approximately equal to the *recycled solvent in the columns*) to each absorption column to carbon dioxide in the *gas feed to the first absorption column*.
3. Calculate the molar ratio of the *make-up solvent* to the *recycled solvent in the columns*.
4. Table 13.13 does not show the recovered pollutant streams. Carry out the component mass balance for the streams shown in Table 13.13 in order to calculate the molar and mass flow rates of the recovered pollutant (or solute) streams from the first- and second-stage stripper columns. Show the mass balance and prove that the molar and mass flow rates of the recovered pollutant (or solute) streams from the first- and second-stage stripper columns are the same as shown in Table 13.14.

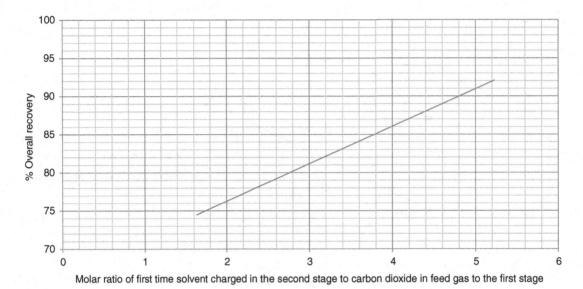

Figure 13.16   *The % overall recovery of CO$_2$ and the molar ratio of first time charge of solvent to the second stage to carbon dioxide in feed gas to the first stage.*

5. From the enthalpy balance, calculate the hot and cold energies available and show their temperature levels.
    Hint. See data extracted in Table 13.15 from the information provided in Table 13.13.
6. Show the composite curves and estimate the excess energy availability for LP generation (in MW) from the dual absorption process system shown in Figure 13.17 and Table 13.13.
    Hint. See Table 13.16 for data points for composite curve construction. Enthalpy change in each temperature interval is shown by these data points. Hot and cold streams from each stage are individually combined. The composite curves are shown in Figure 13.18.

    After the heat exchange between hot and cold streams, hot streams exit at a temperature of 529 K and cold streams at 373 K, respectively. The amount of heat can be exchanged is determined from the heat duty requirement:

    $$(194 + 165) \text{ MW (Table 13.15)} = 359 \text{ MW}$$

    Excess heat available at <529 K or 256 °C is 1480 MW for LP steam generation (this can also be seen from Table 13.15: (847 + 633) MW).

7. Calculate more accurately the excess LP steam generation by taking account of the power requirements by the pumps and compressors. Assume that these power requirements are supplied from the LP steam generated on site. Thus, the net LP steam generation from the dual-stage absorption process will be less compared to the estimated value in the earlier problem. The pump duties can be obtained by subtracting between the enthalpy outlet and the enthalpy inlet across a pump. Show that this enthalpy is negligible compared to the LP steam generation.

Exercise 7. Now consider downstream processing of the purified gas in a combined cycle, capture of carbon dioxide and recovery of sulfur, starting from the process operating instance shown in Table 13.13 and answer to the following questions.

1. Calculate the amount of sulfur recovery in kmol s^{-1} assuming 100% recovery of sulfur in the Claus process. Every mole of H$_2$S or COS provides 1 mole of elemental sulfur from the Claus process.

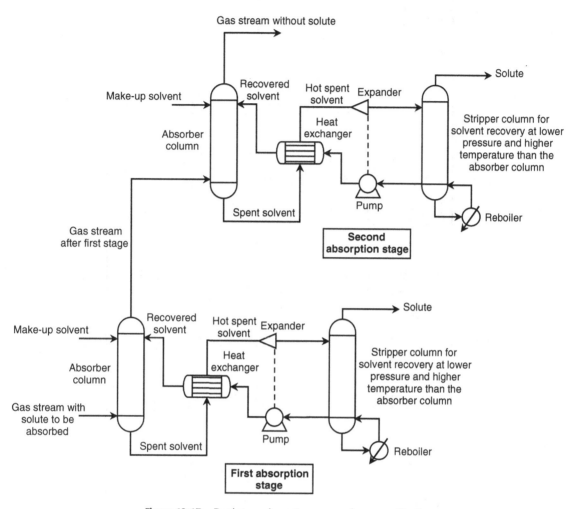

Figure 13.17   *Dual-stage absorption process for gas purification.*

2. Calculate the compression duty requirement in MW for pressurizing captured CO_2 at 1.11 bar to 150 bar. Assume that the correlation for calculating the compression duty ($W^{compression}$) is related to the following parameters in Equation (13.13):

T^{inlet}, inlet temperature in K: take the lowest of the two temperatures of the recovered solute gas streams, 386 K
P^{exit}, exit pressure: 150 bar
P^{inlet}, inlet pressure: 1.11 bar
R is the ideal gas constant: 8.314 J mol^{-1} K^{-1}
γ, heat capacity ratios: $\frac{c_p}{c_v} = 1.4$

The number of moles of CO_2 captured can be assumed from Table 13.14. Assume 94% compression efficiency.

$$W^{compression} = \frac{\text{Number of moles}}{\text{Efficiency}} \times \frac{\gamma}{\gamma - 1} \times R \times T^{inlet} \left(\left(\frac{P^{exit}}{P^{inlet}} \right)^{\frac{\gamma - 1}{\gamma}} - 1 \right) \qquad (13.13)$$

Table 13.13 *Selexol™ process operating conditions based on gasification based product gas: (above) streams in the first absorption stage; (below) streams in the second absorption stage.*

First Absorption Stage	Gas Feed to the First Absorption Column	Make-up Solvent to the First Absorption column	Recycled Solvent to the First Absorption Column	Spent Solvent from the First Absorption Column	Hot Spent Solvent to the First Stripper Column	Hot Recovered Solvent from First Stripper Column	Hot Pressurized Solvent from First Stripper Column
Mole flow (kmol s⁻¹)							
Carbon dioxide	4.10	0	0	2.72	2.72	0	0
Tetraglyme	0	0.00419576	8.33	8.33	8.33	8.33	8.33
Hydrogen	5.72	0	0	0.02373366	0.02373366	0	0
Carbon monoxide	0.08033111	0	0	0.00462014	0.00462014	0	0
Water	0	0	0	0	0	0	0
Carbonyl sulfide	0.00271117	0	0	0.00271117	0.00271117	0	0
Hydrogen sulfide	0.06024833	0	0	0.06024474	0.06024474	0	0
Methane	0.00662731	0	0	0.00090093	0.00090093	0	0
Nitrogen	0.07028972	0	0	0.00377686	0.00377686	0	0
Total flow (kmol s⁻¹)	10.04	0.00419576	8.33	11.15	11.15	8.33	8.33
Total flow (kg s⁻¹)	198.41	0.93264058	1851.52	1974.82	1974.82	1851.52	1851.52
Total flow (m³ s⁻¹)	4.89	0.00095025	1.87	2.07	79.49	2.59	2.61
Temperature (K)	313	325	313	313	373	578	581
Pressure (N m⁻²)	5 350 000	5 500 000	5 500 000	5 370 225	111 457.5	200 000	5 500 000
Vapor fraction	1	0	0	0	0.25	0	0
Liquid fraction	0	1	1	1	0.75	1	1
Enthalpy (J kmol⁻¹)	-161 188 273	-949 476 842	-953 570 216	-808 658 143	-791 251 557	-830 408 903	-828 490 783
Enthalpy (W)	-1 619 000 000	-3 983 777	-7 943 000 000	-9 019 000 000	-8 825 000 000	-6 918 000 000	-6 902 000 000
Entropy (J kmol⁻¹ K⁻¹)	-22 636	-1 392 506	-1 405 416	-1 053 072	-997 499	-1 128 902	-1 125 662
Molar mass	20	222	222	177	177	222	222

(Continued)

Table 13.13 (Continued)

Second Absorption Stage	Gas Feed to the Second Absorption Column	Make-up Solvent to Second Absorption Column	Recycled Solvent to Second Absorption Column	Spent Solvent from the Second Absorption Column	Hot Spent Solvent to the Second Absorption Column	Hot Recovered Solvent from the Second Absorption Column	Hot Pressurized Solvent from the Second Absorption Column	Gas Stream without Pollutant
Mole flow (kmol s⁻¹)								
Carbon dioxide	1.37	0	0	1.05	1.05	0	0	0.32406695
Tetraglyme	0.000001428	0.00149936	7.20	7.20	7.20	7.20	7.20	0.00
Hydrogen	5.70	0	0	0.02722838	0.02722838	0	0	5.67
Carbon monoxide	0.07571096	0	0	0.00502771	0.00502771	0	0	0.07068325
Water	0	0	0	0	0	0	0	0
Carbonyl sulfide	0	0	0	0	0	0	0	0.00
Hydrogen sulfide	0.000003586	0	0	0.000003586	0.000003586	0	0	0.00
Methane	0.00572638	0	0	0.00089966	0.00089966	0	0	0.00482671
Nitrogen	0.06651286	0	0	0.00412599	0.00412599	0	0	0.06238686
Total flow (kmol s⁻¹)	7.22	0.00149936	7.20	8.28	8.28	7.20	7.20	6.13
Total flow (kg s⁻¹)	76.04	0.33328026	1599.37	1646.24	1646.24	1599.37	1599.37	29.50
Total flow (m³ s⁻¹)	3.50	0.00033989	1.61	1.70	31.15	2.14	2.15	2.66
Temperature (K)	313	326	313	313	373	553	556	313
Pressure (N m⁻²)	5 370 225	6 000 000	6 000 000	6 000 000	111 457.5	111 457.5	6 000 000	6 000 000
Vapor fraction	1	0	0	0	0.13	0	0	1
Liquid fraction	0	1	1	1	0.87	1	1	0.00
Enthalpy (J kmol⁻¹)	−75 632 362	−949 097 733	−953 570 216	−878 289 926	−858 452 802	−844 731 009	−842 694 241	−21 680 378
Enthalpy (Watt)	−546 209 859	−1 423 039	−6 861 000 000	−7 276 000 000	−7 111 000 000	−6 078 000 000	−6 063 000 000	−133 000 332
Entropy (J kmol⁻¹ K⁻¹)	−25 110	−1 391 334	−1 405 416	−1 221 924	−1 162 089	−1 153 762	−1 150 152	−28 621
Molar mass	11	222	222	199	199	222	222	5

Table 13.14   *Molar and mass flow rates of the recovered pollutant (or solute) streams from the first- and second-stage stripper columns.*

Mole flow (kmol s^{-1})	Recovered Pollutant or Solute from First Stage	Recovered Pollutant or Solute from Second Stage
Carbon dioxide	2.72	1.05
Tetraglyme	0.00351625	0.00207604
Hydrogen	0.02373366	0.02722838
Carbon monoxide	0.00462014	0.00502771
Water	0	0
Carbonyl sulfide	0.00271117	0
Hydrogen sulfide	0.06024474	0.000003586
Methane	0.00090093	0.00089966
Nitrogen	0.00377686	0.00412599
Total flow (kmol s^{-1})	2.82	1.09
Total flow (kg^{-1})	123.16	47.00

Table 13.15   *Hot and cold energies available and temperature levels of the streams in the dual-stage absorption process shown in Table 13.13: (above) first-stage absorption; (below) second-stage absorption.*

First Absorption Stage	Hot Stream		Cold Stream	
	Hot Pressurized Solvent from the First Stripper Column	Recycled Solvent to the First Absorption Column	Spent Solvent from the First Absorption Column	Hot Spent Solvent to the First Stripper Column
Total flow (kmol s^{-1})	8.33	8.33	11.15	11.15
Total flow (kg s^{-1})	1851.52	1851.52	1974.82	1974.82
Total flow (m^3 s^{-1})	2.61	1.87	2.07	79.49
Temperature (K)	581	313	313	373
Pressure (N m^{-2})	5 500 000	5 500 000	5 370 225	111 457.5
Vapor fraction	0	0	0	0.25
Liquid fraction	1	1	1	0.75
Enthalpy (J kmol^{-1})	−828 490 783	−953 570 216	−808 658 143	−791 251 557
Enthalpy (W)	−6 902 000 000	−7 943 000 000	−9 019 000 000	−8 825 000 000
Heat or cold available (W)	Heat available (W): 1 041 000 000 Excess heat (W): 847 000 000		Cold available (W): 194 000 000	

Second Absorption Stage	Hot Stream		Cold Stream	
	Hot Pressurized Solvent from the Second Absorption Column	Recycled Solvent to the Second Absorption Column	Spent Solvent from the Second Absorption Column	Hot Spent Solvent to the Second Absorption Column
Total flow (kmol s^{-1})	7.20	7.20	8.28	8.28
Total flow (kg s^{-1})	1599.37	1599.37	1646.24	1646.24
Total flow (m^3 s^{-1})	2.15	1.61	1.70	31.15
Temperature (K)	556	313	313	373
Pressure (N m^{-2})	6 000 000	6 000 000	6 000 000	111 457.5
Vapor fraction	0	0	0	0.13
Liquid fraction	1	1	1	0.87
Enthalpy (J kmol^{-1})	−842 694 241	−953 570 216	−878 289 926	−858 452 802
Enthalpy (W)	−6 063 000 000	−6 861 000 000	−7 276 000 000	−7 111 000 000
Heat or cold available (W)	Heat available (W): 798 000 000 Excess heat (W): 633 000 000		Cold available (W): 165 000 000	

Table 13.16 *Data points for composite curve construction combining hot streams and cold streams separately, shown in Table 13.15.*

	Temperature (K)	Heat Available (W)
Hot streams	313	0
	556	1 743 591 325
	581	1 839 000 000
	Temperature (K)	Cold available (W)
Cold streams	313	1 480 000 000
	373	1 839 000 000

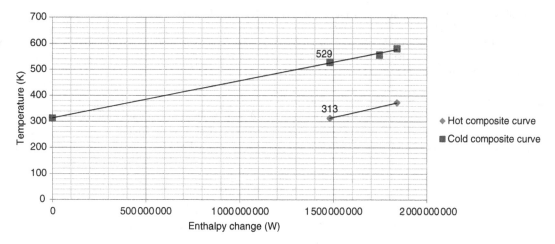

Figure 13.18 *Composite curves using the data shown in Table 13.16.*

3. Calculate the flow rates in kmol s^{-1} of the exhaust gas after complete combustion of the *gas stream without pollutant* from the dual-stage absorption process (Table 13.13) in the presence of a stoichiometric air requirement.

    Calculate the stoichiometric oxygen requirement in kmol s^{-1} for complete combustion of the *gas stream without pollutant*, shown in Table 13.13.

Hint. To calculate the stoichiometric oxygen requirement in kmol s^{-1}, the following need to be added:

0.5 × hydrogen in kmol s^{-1} in *gas stream without pollutant* (Table 13.13)
+ 0.5 × carbon monoxide in kmol s^{-1} in *gas stream without pollutant* (Table 13.13)
+ 3 × methane in kmol s^{-1} in *gas stream without pollutant* (Table 13.13)
= 2.89 kmol s^{-1}

Assume air consists of 79% nitrogen and 21% oxygen, by volume, calculate the flow rates in kmol s^{-1} of the exhaust gas after complete combustion of the *gas stream without pollutant*, as shown in Table 13.17.

Table 13.17 *Exhaust gas composition after complete combustion of the gas stream without pollutant.*

Mole Flow (kmol s^{-1})	
Carbon dioxide	0.40
Water	5.68
Nitrogen	10.92
Total flow (kmol s^{-1})	17.00

4. Calculate the power generation from the gas turbine expander (in MW), expanding the exhaust gas shown in Table 13.17 from 60 bar to atmospheric pressure. Assume that the exit temperature of the exhaust gas from the gas turbine combustor is at 1000 °C. Thus, $T^{inlet} = 1273$ K. Assume that the work generation from the gas turbine ($W^{expansion}$) follows Equation (13.14) below. Assume 75% gas turbine efficiency.

$$W^{expansion} = \text{Number of moles} \times \text{Efficiency} \times \frac{\gamma}{\gamma - 1} \times R \times T^{inlet} \left[\left(\frac{p^{exit}}{p^{inlet}} \right)^{\frac{\gamma - 1}{\gamma}} - 1 \right] \qquad (13.14)$$

Calculate the power consumption by the air compressor for using Equation (13.13), a compression efficiency of 0.5, T^{inlet} of 298 K, p^{inlet} of atmospheric pressure and p^{exit} of 60 bar, where the γ value can be assumed as before.

Calculate the net power generation from the power generation plant by subtracting the air and CO_2 compression power consumptions from the gas turbine expander power generation. The CO_2 compression power consumption can be obtained from part 2 of this Exercise problem.

Exercise 8. Pyrolysis gas produced with mass compositions shown in Table 13.18 from IGCC power plants using three feedstocks, straw, wood and RDF (refuse derived fuel is obtained from de-moisturizing municipal waste) is required to purify using the dual-stage Selexol™ based physical absorption process. Calculate the following using the correlations shown in Figures 13.15 and 13.16 and the parameters and equations shown in Exercises 6 and 7.

Table 13.18   *Primary pyrolysis product distributions in t d^{-1} from each biomass feedstock.*

Component	Straw	Wood	RDF
H_2	0.98	0.81	0.79
CH_4	14.76	12.17	11.93
C_2	75.16	61.97	60.74
CO	132.55	109.29	107.13
CO_2	18.87	15.55	15.25
H_2O	49.25	40.60	39.80
N_2	3.85	0.38	1.66
H_2S	1.00	0	1.18
Total gas	296.42	240.78	238.48

1. Calculate the composition (molar and mass) of the *gas stream without pollutant* from the Selexol™ process.
2. Calculate the composition (molar and mass) of the *recovered pollutant (or solute) streams* from the first- and second-stage stripper columns.
3. Estimate the mass of the first time solvent charged to the first- and second-stage absorption columns for the assumed values of recovery ratios from the graphs shown in Figures 13.15 and 13.16.
4. Estimate the mass of the *make-up solvent* to the first- and second-stage absorption columns. State your assumptions.
5. Estimate the excess LP steam generation using insights from Exercises 6 and 7. State your assumptions.
6. Calculate the mass of sulfur recovery assuming 100% recovery of sulfur in the Claus process.
7. Calculate the compression duty requirement for pressurizing captured CO_2 at 1.11 bar to 150 bar.
8. Calculate the mass flow rates of the components in the exhaust gas after complete combustion of the *gas stream without pollutant* from the dual stage absorption process in the presence of a stoichiometric air requirement.
9. Calculate the stoichiometric oxygen requirement in the mass flow rate for complete combustion of the *gas stream without pollutant* from the dual-stage absorption process.

10. Calculate the power generation from the gas turbine expander expanding the exhaust gas calculated in part 8 of the Exercise problem.
11. Calculate the power consumption by the air compressor for the gas turbine.
12. Calculate the net power generation from the *gas stream without pollutant* from the dual-stage absorption process by subtracting the air and CO_2 compression power consumptions (part 11 and part 7 of the Exercise problem) from the gas turbine expander power generation (part 10 of the Exercise problem).

13.5   Adsorption Process

Adsorption is a separation process by which one or more components from a gaseous mixture can be captured using a solid adsorbent phase. The key advantage of this separation process is that there is no mixing with external material, thus providing very high purity and yield of the desired product. In addition, adsorption can be inherently combined into a reaction process, thus creating advantages of combined reaction separation processes. Catalyst and sorbent materials can be mixed in the form of powders, embedded within a particle and integrated in a bed, etc. When adsorption is combined with a reaction process, high purity desired gaseous or liquid products can be produced by minimizing reverse reaction steps of equilibrium driven reactions, product inhibition and waste production reactions. In these processes integrated and in-situ separation through adsorption of product(s)/stream(s) drives the conversion of desired reactions to completion. Thus, they have the potential to achieve very high purity, selectivity and yields of high value products. Such process intensification strategies have demonstrated high selectivity and productivity, with significant capital and energy saving. The process eliminates downstream reaction and purification steps as well as the need for higher temperature and pressure for reaction processes. The computational tools for designing these processes need to be multiscale integrating surface reaction–adsorption phenomena and reaction–regeneration process systems in a dynamic manner.

This section shows a dynamic simulation framework for modeling and optimization of the (ad)Sorption Enhanced Steam Methane Reforming (SESMR) reaction process for the adsorption of carbon dioxide alongside simultaneous production and recovery of high purity hydrogen[6,7]. Steam methane reforming (SMR) accounts for over 48% of H_2 production globally. However, the efficiency of SMR alone is 65–75% for the best of the commercial productions. Traditional SMR reaction is an equilibrium-limited process. New technologies are required to improve the efficiency and selectivity of the SMR process, such as SESMR.

The conversion and the reaction rate of equilibrium driven reactions can be increased by integrated or in situ separation of one or more components, based on the application of the Le Chatelier's principle. This increases the productivity and purity of the desired product through a more favorable reaction equilibrium. The concept regarding a sorption enhanced reaction process can be described for a generic reaction such as A+B ↔ C+D. A pulse of fluid containing reactants A and B is fed into a reactor adsorbent column. If the adsorbent has a higher affinity for C in comparison to D, the product C is adsorbed and the reaction equilibrium shifts towards the right. Thus, a high conversion of A and B is achieved and pure C and D products are obtained. The processes can be made continuous by changing the location of the feed and reagent stream ports without actual movement of the solid adsorbent and catalyst. Such a simulated moving bed process can be used for continuous production of hydrogen and captured carbon dioxide using SESMR. The earlier work shows its application for biodiesel synthesis and the detailed simulation framework of the simulated moving bed reactor for biodiesel synthesis is available[8].

The reversible reaction steps involved in the SMR process in Figure 13.19 are shown by

$$CH_4 + H_2O \leftrightarrow CO + 3H_2 \qquad \Delta H = +206 \text{ kJ mol}^{-1} \tag{13.15}$$

$$CO + H_2O \leftrightarrow CO_2 + H_2 \qquad \Delta H = -41.1 \text{ kJ mol}^{-1} \tag{13.16}$$

Thus the overall SMR reaction is given as

$$CH_4 + 2H_2O \leftrightarrow CO_2 + 4H_2 \qquad \Delta H = +164.9 \text{ kJ mol}^{-1} \tag{13.17}$$

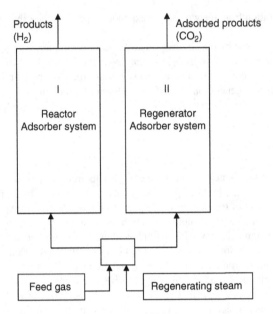

Figure 13.19 *Steam assisted TSA process configuration.*

The process has two sorption enhanced reactors with one reacting and the other regenerating, simultaneously, until the breakthrough point is reached in the reactor, after which the exhausted reactor bed is regenerated and the regenerated bed is brought to reaction operation. This switch in operation between two reactors is attained by switching between feed gas and regenerating steam inlet ports.

The most established catalyst for the SMR reaction is the nickel based catalyst. The various adsorbents for carbon dioxide are K_2CO_3 impregnated on the hydrotalcite with basic materials such as MgO and Al_2O_3, polyethylenimine-modified mesoporous molecular sieve of MCM-41 type, hyperbranched aminosilica material and amine functionalized metal organic frameworks, etc. The feed gas and steam are fed into a tubular reactor operating at 450–550 °C and 1–5 bar pressure. This temperature is much lower than a conventional SMR process, which operates at 800–1000 °C to achieve high conversion of methane into hydrogen.

The reforming catalyst and adsorbent for simultaneous carbon dioxide capture are admixed in the reaction zone. The reactor product is a high purity hydrogen stream with purity over 95% by mole, containing traces of other gases, such as feed methane and product carbon dioxide. When the adsorbent is saturated with CO_2, it is regenerated in situ by using steam assisted temperature swing adsorption (TSA) at the reaction pressure or inert gas, e.g. nitrogen assisted pressure swing adsorption (PSA) at the reaction temperature. The capture-ready CO_2 leaves the regenerator adsorber system either with steam in a TSA process or with an inert gas in a PSA process. Figure 13.19 shows a TSA process configuration. In this case, steam acts as a reactant in the reactor as well as to strip off CO_2 adsorbed in the adsorber column.

13.5.1 Kinetic Modeling of SMR Reactions

The kinetic rate equations of the reactions in Equations (13.15) to (13.17) in the gas phase can be represented in terms of partial pressures of reactants and products. The kinetic rate equations are formed using the reaction stoichiometry. The parameters involved are determined experimentally with Ni based catalyst and hydrotalcite based CO_2 adsorbent, admixed together. Equations (13.18) to (13.20) show the kinetic rate equations for the reactions in Equations (13.15) to (13.17), respectively. The reaction parameters involved are the equilibrium constant keq_j and the pre-exponential factor kit_j and the energy of activation E_j^{act} in the Arrhenius Equation (13.21), for reaction j. The denominator of the

Table 13.19 (a) Equilibrium constants of SMR reactions.
(b) Adsorption constants of species of SMR reactions.
(c) Kinetic constants of SMR reactions.

(a)

keq_1	keq_2	keq_3
9.18×10^6	7.3464	6.75×10^6

(b)

K_{CO}	K_{H_2}	K_{CH_4}	K_{H_2O}
0.0691	1.0496×10^{-4}	0	5.9988×10^{-8}

(c)

kit_1	kit_2	kit_3	E_1^{act}	E_2^{act}	E_3^{act}
0.18	7588	0.02	240.1×10^3	67.1×10^3	243.9×10^3

rate equations below involve adsorption constants K_i of species i. The kinetic parameters determined are shown in Table 13.19(a) to (c)[9].

$$r_1 = \frac{\frac{k_1}{P_{H_2}^{2.5}}\left(P_{CH_4}P_{H_2O} - \frac{P_{H_2}^3 P_{CO}}{keq_1}\right)}{den^2} \tag{13.18}$$

$$r_2 = \frac{\frac{k_2}{P_{H_2}}\left(P_{CO}P_{H_2O} - \frac{P_{H_2}P_{CO_2}}{keq_2}\right)}{den^2} \tag{13.19}$$

$$r_3 = \frac{\frac{k_3}{P_{H_2}^{3.5}}\left(P_{CH_4}P_{H_2O}^2 - \frac{P_{H_2}^4 P_{CO_2}}{keq_3}\right)}{den^2} \tag{13.20}$$

where

$$den = 1 + K_{CO}P_{CO} + K_{H_2}P_{H_2} + K_{CH_4}P_{CH_4} + \frac{K_{H_2O}P_{H_2O}}{P_{H_2}}$$

$$k_j = kit_j \exp\left\{-\frac{E_j^{act}}{R}\left(\frac{1}{T} - \frac{1}{648}\right)\right\} j \in 1,2,3 \tag{13.21}$$

Note that keq_j in Table 13.19(a), K_i in Table 13.19(b) and kit_j and E_j^{act} in Table 13.19(c) are in reaction dependent units when the partial pressures of the species P_i are in Pa, the rate of reactions r_j are in mol kg^{-1} s^{-1}, temperature is in K and the universal gas constant $R = 8.314$ J mol^{-1} K^{-1}. The total pressure $P = 445.7$ kPa.

13.5.2   Adsorption Modeling of Carbon Dioxide

The generic linear driving force model for the rate of adsorption of CO_2 is shown by

$$r_{CO_2}^{ads} = \frac{\partial \bar{q}_{CO_2}}{\partial t} = k_{CO_2}^{ads} \left(q_{CO_2}^* - \bar{q}_{CO_2} \right) \tag{13.22}$$

where $\bar{q}_{CO_2}$ is the concentration of CO_2 in the adsorbed phase, $q_{CO_2}^*$ is the solid phase concentration of CO_2 in equilibrium and $k_{CO_2}^{ads}$ is the *mass transfer coefficient*. The CO_2 sorption parameters involved in Equation (13.22) are given in Equations (13.23 to 13.25) below:

$$q_{CO_2}^* = \frac{m_{CO_2} b_{CO_2} P_{CO_2}}{1 + b_{CO_2} P_{CO_2}} \tag{13.23}$$

where P_{CO_2} is the partial pressure of CO_2 in the gas phase, Pa, and m_{CO_2} is the Langmuir model coefficient for CO_2 adsorption, 0.65 mole of CO_2 per kg of solid adsorbent (in this case hydrotalcite). This is obtained when P_{CO_2} is very large, $b_{CO_2} P_{CO_2} \gg 1$; $q_{CO_2}^* = m_{CO_2}$.

Hence, the Langmuir model coefficient is the solid phase concentration of CO_2 in equilibrium with a gas containing CO_2 predominantly. This parameter is determined from experimental data of CO_2 uptake by the adsorbent estimated using micro balance equipment, with respect to the partial pressure of CO_2 in the gas phase, run at various constant temperatures.

Here b_{CO_2} is the Langmuir model constant for CO_2 adsorption in Pa^{-1}, calculated using

$$b_{CO_2} = 1.36 \times 10^{-4} \exp \left[\frac{10\,000}{R} \left(\frac{1}{T} - \frac{1}{673} \right) \right] \tag{13.24}$$

where T is the temperature of the reactor/adsorber in K and R is the universal gas constant, 8.314 J mol^{-1} K^{-1}. Also,

$$k_{CO_2}^{ads} = \frac{15}{rp^2} \frac{\varepsilon_p D_p}{\varepsilon_p + \rho_p RT \left(\partial q_{CO_2}^* / \partial P_{CO_2} \right)} \tag{13.25}$$

where rp = particle radius, 0.25 mm, D_p = diffusivity inside particle, 3.3×10^{-7} m^2 s^{-1}, ε_p = particle void fraction, 0.24, and ρ_p = density of particle, 139 kg m^{-3}.

For very low P_{CO_2}, $\partial q_{CO_2}^* / \partial P_{CO_2} = m_{CO_2} b_{CO_2}$. This shows the linear correlation between CO_2 uptake by the adsorbent and the partial pressure of CO_2 in the gas phase for low CO_2 partial pressure in the gas phase at a constant temperature (b_{CO_2} is a constant at a constant temperature from Equation (13.24)).

For very low P_{CO_2}, $\partial q_{CO_2}^* / \partial P_{CO_2} = 0$. Hence, to capture CO_2 from a gas predominantly containing CO_2, the adsorbent quickly reaches the equilibrium CO_2 concentration in the solid, which does not change with respect to P_{CO_2}. This is the significance of the Langmuir adsorption isotherm.

13.5.3   Sorption Enhanced Reaction (SER) Process Dynamic Modeling Framework

The dynamic modeling equations of an SER of length L and diameter d_t are shown as follows. It is assumed that the concentration varies only in the axial direction (z) and is constant along the radial direction. The process is considered to be pseudo-homogeneous with uniform voidage and catalyst and adsorbent distributions. With these considerations, the material balance for each specie i in an SER unit k (C_{ik}) is a function of the diffusional transfer (diffusivity D_L), convection (axial velocity u), rates of jth reaction (r_j) and sorption of specie i (r_i^{ads}), which is shown by

$$\frac{\partial C_{ik}}{\partial t} = \frac{D_L}{\varepsilon}\left(\frac{\partial^2 C_{ik}}{\partial z^2}\right) - \frac{u}{\varepsilon}\frac{\partial C_{ik}}{\partial z} - \frac{(1-\varepsilon)}{\varepsilon}\left[\rho_{ads}r_i^{ads} - \rho_{cat}\sum_{j=1}^{nrxn} v_{ij}\eta_j r_j\right]$$ (13.26)

where

$\forall\, i \in \{CO, CO_2, H_2, H_2O\}$
$j \in \{1, 2, 3\}$
$k \in \{1, 2\}$
ε = porosity of the bed, 0.35
u = linear axial velocity, 0.13 m s^{-1}
ρ_{ads} = density of the adsorbent, 139 kg m^{-3}
ρ_{cat} = density of the catalyst, 609 kg m^{-3}
v_{ij} = stoichiometric coefficient of specie i in jth reaction
η_j = effectiveness factor of jth reaction, 1
D_L = axial diffusion coefficient in m^2 s^{-1} is estimated using the physical parameters, shown below

$$D_L = 0.73D_m + \frac{0.5ud_p}{1 + 9.49D_m/(ud_p)}$$ (13.27)

where D_m = molecular diffusivity, 1.6×10^{-5} m^2 s^{-1}, and d_p = diameter of particle, 3.57×10^{-4} m.

Equation (13.28) below shows the energy balance across an element in the bed volume exchanging heat through a wall, for a given gas flow rate through a homogeneous bed of catalyst and sorbent. The rate of change in temperature (T) inside the fixed volume is a function of the conductive heat transfer coefficient (λ_z), convective flow (u), heat transfer coefficient through the wall (U), heat of adsorption (ΔH_i^{ads}) and the heat of reaction (ΔH_j^R), respectively:

$$\frac{\partial T}{\partial t} = \frac{1}{Cp_g\rho_g\varepsilon + Cp_s\rho_b}\left[\begin{array}{l}\frac{\partial}{\partial z}\left(\lambda_z\frac{\partial T}{\partial z}\right) \\ -Cp_g\rho_g u\frac{\partial T}{\partial z} \\ -(1-\varepsilon)\left(1-\varepsilon_p\right)\sum_{i=1}^{nspc}\rho_{ads}\Delta H_i^{ads} r_i^{ads} \\ +4\frac{U}{d_t}\left(T_w - T\right) \\ +(1-\varepsilon)\left(1-\varepsilon_p\right)\sum_{j=1}^{nrxn}\eta_j\Delta H_j^R\rho_{cat}r_j\end{array}\right]$$ (13.28)

where

$\forall\, i \in \{CO, CO_2, H_2, H_2O\}$
$j \in \{1, 2, 3\}$
Cp_g = heat capacity of gas, 42 J mol^{-1} K^{-1}
Cp_s = heat capacity of solid, 850 J mol^{-1} K^{-1}
ρ_g = density of gas, 1 kg m^{-3}
ρ_b = density of bed, 374 kg m^{-3}
λ_z = 0.85 J kg mol^{-1} m^{-1} K^{-1} S^{-1}
U = overall heat transfer coefficient, 71 J kg mol^{-1} m^{-2} K^{-1} s^{-1}
d_t = diameter of tube, 5×10^{-2} m

462   *Biorefineries and Chemical Processes*

T_w = wall temperature, 730 K
ΔH_1^R = heat of reaction in Equation (13.15), 206 kJ mol^{-1}
ΔH_2^R = heat of reaction in Equation (13.16), −41.1 kJ mol^{-1}
ΔH_3^R = heat of reaction in Equation (13.17), 164.9 kJ mol^{-1}
$\Delta H_{CO_2}^{ads}$ = heat of adsorption of CO_2, −178 × 10^3 kJ mol^{-1}.

The boundary conditions for such a continuous process follow Danckwert's boundary conditions. The molar flow into the SER is equal to the diffusion and the convection at the inlet of the SER, shown in Equations (13.29) and (13.30) below. There is no change in the concentration at the outlet of the SER, shown in Equations (13.31) and (13.32):

$$\left(\varepsilon D_L \frac{\partial C_{ik}}{\partial z}\right)_{z=0} = u\left(C_{ik}^f - C_{ik}\right) \tag{13.29}$$

$$\left(\varepsilon \lambda_z \frac{\partial T}{\partial z}\right)_{z=0} = Cp_g u C_{ik}\left(T^f - T\right) \tag{13.30}$$

$$\left(\frac{\partial C_{ik}}{\partial z}\right)_{z=L} = 0 \tag{13.31}$$

$$\left(\frac{\partial T}{\partial z}\right)_{z=L} = 0 \tag{13.32}$$

where C_{ik}^f = feed concentration of specie i in column k, C_{ik} = concentration of specie i in column k, C_i^{ic} = initial concentration of specie i and T^f = feed temperature, 723 K.

$C_{CO_2}^f$	0
$C_{CH_4}^f$	10.3643 mol m^{-3}
$C_{H_2O}^f$	63.6664 mol m^{-3}
C_{CO}^f	0
$C_{H_2}^f$	0
$C_{CO_2}^{ic}$	0
$C_{CH_4}^{ic}$	0
$C_{H_2O}^{ic}$	71.8098 mol m^{-3}
C_{CO}^{ic}	0
$C_{H_2}^{ic}$	2.2209 mol m^{-3}

Exercise 9. A sorption enhanced steam methane reformer (SESMR) is to be designed to produce pure hydrogen with a molar purity above 98% and molar yield above 33 mol m^{-3}. The parameters for the design are set out in Section 13.5. The suggested design and operating ranges are shown in Table 13.20. Develop a dynamic simulation algorithm of the SESMR process MATLAB[9]. Determine the optimum design and operating parameters for achieving the objectives.

Table 13.20   Design and operating ranges of a sorption enhanced reformer.

Length of Reactor (m)	Linear Axial Velocity (m s^{-1})	Switching Time (s)	Temperature (°C)
0.1–0.3	0.13–0.3	100–300	650–850

13.6 Chemical Looping Combustion

The technologies for carbon dioxide capture currently used until now are add-on technologies, adding new process steps to state-of-the-art power plants. Such additional process steps generally result in efficiency penalties and additional costs. The carbon dioxide capture can be done in integrated ways, such as lime enhanced gasification sorption (LEGS) and chemical looping combustion (CLC). These technologies are particularly relevant for pure hydrogen production utilizing coal and biomass.

The LEGS process shown in Figure 13.20 is an integrated CO_2 capture process, in which CaO is employed as a high temperature CO_2 sorbent and carrier between two reactors: a steam gasifier and an oxygen-fired regenerator. Due to the in situ CO_2 capture by CaO sorbent, a hydrogen rich gas is produced in the gasifier. The adsorbed CO_2 is transferred to the regenerator where the sorbent is calcined generating a CO_2 rich stream suitable for storage. The LEGS fuel-to-gas process cycle consists of a LEGS reactor and a regeneration unit. In the LEGS reactor, a hydrogen rich gas is generated from a brown coal feedstock. The relatively low gasification temperatures (<800 °C) require a reactive fuel such as brown coal. The high moisture content of brown coal is advantageous since it provides the required steam for the reforming reaction. The simplified LEGS reaction equation neglecting other side reactions is shown by

$$\text{Brown coal} + a\text{CaO} + H_2O \rightarrow H_2 + (a-z)\,\text{CaCO}_3 + y\text{C} + z\text{CaS} \tag{13.33a}$$

$$\text{CaCO}_3 \rightarrow \text{CaO} + \text{CO}_2 \tag{13.33b}$$

In addition to the main reactions of steam reforming, the water gas shift reaction, CO_2 absorption and the simultaneous capture of sulfur by CaO occur, shown by Equation (13.33).

The CLC is another technique that can be used to carry out simultaneous CO_2 separation. Like the LEGS, the CLC process consists of two reactors, an air reactor and a fuel reactor, shown in Figure 13.21. In the fuel reactor, an oxygen carrier, which is typically a metal oxide, is reduced by the fuel; the reduced metal is circulated back to the air reactor, where it is oxidized back to its original state by air. The stream leaving the fuel reactor consists of only CO_2 and H_2O and once the H_2O is separated by condensation, the CO_2 can be compressed and transported to a suitable storage, such as a depleted oilfield or/and deep salt water reservoirs. The stream leaving the air reactor has mainly nitrogen and small amounts of oxygen and can be released into the atmosphere. As a result, CO_2 can be easily separated in this combustion process. The heat released in the air reactor is equal to the heat of combustion (cf. Section 10.2.1); thus there is a potential to separate CO_2 without any energy loss. The heat should be recovered into process streams or the generation of steam.

One way of performing CLC is by first gasifying a feedstock into the syngas, consisting mainly of CO and H_2 upon clean-up, which could be burned subsequently in the CLC. However, in order to obtain pure CO and H_2 the gasification needs to be carried out with O_2; thus an air separation unit would be needed or allothermal gasifier (discussed in Chapter 16) can be used for biomass feedstocks. Another option is to introduce the solid fuel directly to the fuel reactor

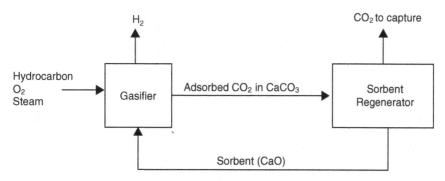

Figure 13.20 *LEGS reactor schematic.*

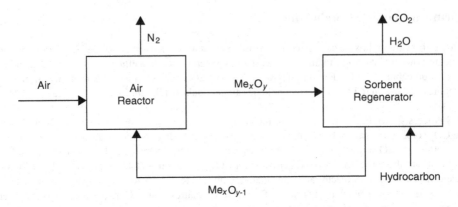

Figure 13.21   *Chemical looping combustion schematic.*

where the oxygen carrier is reduced by the fuel, thus avoiding different gasification and separation steps. However, the solid–solid reaction between coal and a metal oxide is not very likely to occur at an appreciable rate. Instead the solid fuel needs to be gasified using H_2O and then the oxygen carrying particles can react with the produced gas, making the gasification a limiting step in a CLC with a solid fuel process. An important advantage compared to normal gasification is that it takes place in a high concentration of CO_2 and H_2O, which is beneficial for the reaction rates. The main gasification reactions are as follows:

$$C + H_2O \rightarrow CO + H_2 \qquad (13.34a)$$

The water gas shift reaction has been shown in Equation (13.16):

$$CO_2 + C \rightarrow 2CO \qquad (13.34b)$$

The main reactions with the metal oxide are as follows:

$$Me_xO_y + H_2 \rightarrow Me_xO_{y-1} + H_2O \qquad (13.35a)$$

$$Me_xO_y + CO \rightarrow Me_xO_{y-1} + CO_2 \qquad (13.34b)$$

In general, the CLC is a technology in which a gaseous fuel such as natural gas or synthetic gas is burnt with oxygen transported from the combustion air to the fuel by an oxygen carrier. This gives an advantage of an exhaust gas stream containing only CO_2 and H_2O. After H_2O is condensed, almost pure CO_2 is obtained for storage. Recent interest has arisen in the CLC because the lack of nitrogen in its combustion product makes it suitable for CO_2 capture. In comparison to the conventional natural gas combined cycle systems with post-combustion capture, the combined cycle power plants using natural gas CLC show an increase of approximately 5% in energy efficiency.

In the application of CLC to the coal gasification process, the product gas from the gasifier is cleaned up to remove contaminants to a level below the tolerance limit of the oxygen carrier particles used in the process. The cleaned syngas (mostly CO and H_2) is then passed into the fuel reactor or reducer where the main reactions in Equations (13.35a) and (13.35b) occur.

The syngas oxidized by the metal oxide and after heat recovery exits the reactor as primarily CO_2 and steam. The steam is condensed and a concentrated CO_2 stream is obtained, which can be pressurized further and sent for sequestration. The reduced metal oxide particles are then transferred to the air reactor or combustor, where they react with air, shown by

$$Me_xO_{y-1} + 0.5O_2 \rightarrow Me_xO_y \qquad (13.35c)$$

The oxidization reaction in the combustor is highly exothermic and as a result a high temperature and high pressure oxygen depleted exhaust gas stream is generated from the air reactor. This exhaust stream can be used to drive a combined cycle system to generate electricity. The regenerated metal oxide particles are recycled to the fuel reactor to start another cycle.

The main focus areas of research activities are the development of suitable oxygen carriers and the CLC reactor design. The selection of the oxygen carrier particles is one of the most important parts of the CLC process. The oxygen carrier should have a high reactivity in reduction by the fuel and oxidation by oxygen in the air, and have a high conversion of the fuel to CO_2 and H_2O.

A vast amount of oxygen carriers have been investigated with both natural gas and syngas. The most promising materials are oxides of Cu, Fe, Mn and Ni as active oxygen carriers combined with inert materials such as Al_2O_3, SiO_2, TiO_2, $NiAl_2O_4$ and ZrO_2 as support. A fuel conversion efficiency of 99.5% for natural gas or syngas, close to the theoretical maximum conversion, can be obtained. The fuel reactor operates at a temperature of 950 °C. The degree of oxidation in the air reactor can also be 100%. The most common CLC reactor design consists of two interconnected fluidized beds, one for oxidation and the other for reduction, with the oxygen carrier particles transported between the reactors. The high gas/solid heat and mass transfer rates and a stable operating temperature due to the thermal mass of the oxygen carrier particles are the main advantages of this design. Large-scale CLC reactor designs integrated with gas turbine systems are still in a very early phase and this is the main consideration for the system development at present.

Exercise 10. Calculate the molar composition of the product gas from the coal and the biomass feedstocks with their ultimate analyses as shown in Table 13.21, being converted in a steam assisted LEGS process using CaO as a sorbent that is continuously regenerated. Calculate the sorbent-to-feed and the steam-to-feed stoichiometric weight ratios in each case. Assume complete conversion of the LEGS reactions in Equations (13.33a) and (13.33b). The molar masses of various components are shown in Table 13.22.

Table 13.21 Ultimate analysis of coal and biomass feedstocks (values stated in weight %).

Component	Coal	Biomass
C	72.6	48.45
O	15.49	43.78
H	4.11	5.85
N	0.1	0.47
S	0.27	0.01
Cl	0	0.1
Ash	7.43	1.34

Table 13.22 Molar mass of components.

C	O_2	H_2	H_2O	CaS	CaO	N_2	Cl_2
12	32	2	18	72	56	28	71

Solution to Exercise 10

Basis = 100 kg h^{-1} of the feedstock

Assumptions:

1. The carbon present in the feedstock is completely converted into CO_2 that is captured by the sorbent CaO.
2. The generation of CO_2 requires an additional O_2 supply apart from the feedstock itself. The deficit of O_2 that is not present in the feedstock is provided by steam addition. The addition of steam also generates additional H_2 on top of the existing H_2 present in the feedstocks.
3. Any reaction associated with nitrogen and chlorine is neglected.
4. CaO also adsorbs sulfur present in the feedstock, forming CaS.

Based on these assumptions, the flow rates in kmol h^{-1} and mole fraction of each component for individual feedstocks are shown in Table 13.23.

Table 13.23   *Steam gasification with the LEGS case – flow rate and composition of each component in the product on the basis of 100 kg h^{-1} of each feedstock.*

Component	Hypothetical product formation pathway	Coal		Biomass	
		Molar flow rate (kmol h^{-1})	Mole fraction	Molar flow rate (kmol h^{-1})	Mole fraction
CO_2	$C \rightarrow CO_2$ (see Assumption 1)	$\frac{72.6}{12} = 6.05$	0.314	$\frac{48.45}{12} = 4.04$	0.328
H_2	$H \rightarrow H_2$ $H_2O \rightarrow H_2$ (see Assumption 2)	$\frac{4.11}{2} + 11.132 = 13.19$	0.685	$\frac{5.85}{2} + 5.3388 = 8.26$	0.670
N_2	$N \rightarrow N_2$ (see Assumption 3)	$\frac{0.1}{28} = 0.0036$	0.000187	$\frac{0.47}{28} = 0.0168$	0.00136
CaS	$S \rightarrow CaS$ (see Assumption 4)	$\frac{0.27}{32} = 0.0084$	0.000436	$\frac{0.01}{32} = 0.0003$	0.00002
Cl_2	$Cl \rightarrow Cl_2$ (see Assumption 3)	0	0	$\frac{0.1}{71} = 0.0014$	0.0001
Total		**19.25**	**1.000**	**12.32**	**1.000**

Note:
To estimate the amount of H_2 produced, the amount of O_2 required should first be estimated. This is because CO_2 in the product carries O_2 that comes from the feedstock itself and also from the steam added. Hence,

Amount of O_2 contributed by steam = Amount of CO_2 in the product – Amount of O_2 from the feedstock

For the coal LEGS case, the amount of O_2 contributed by steam is

$$\frac{72.6}{12} - \frac{15.49}{32} = 5.566 \text{ kmol h}^{-1}$$

Considering the water dissociation reaction (shown below), it can be deduced that the H_2 in the product contributed by steam is two times the O_2, according to the stoichiometric ratio:

$$H_2O \rightarrow H_2 + \tfrac{1}{2}O_2$$

Hence, 2×5.566 kmol h^{-1} = 11.132 kmol h^{-1}.

The amount of H_2 contributed by steam is added to the existing amount of H_2 from the feedstock, giving the total amount of H_2 present in the product, shown in Table 13.23. The same calculation is applied for the biomass case.

According to Equations (13.33a) and (13.33b), *a* mole of CaO sorbent is required since ($a-z$) moles of CO_2 is generated. For the case of coal as the feedstock (see Table 13.23),

$$(a-z) = 6.05 \text{ kmol h}^{-1}$$
$$z = 0.0084 \text{ (number of mole of CaS)}$$

Hence, the total amount of CaO sorbent required, a,

$$= (6.05 + 0.0084) \text{ kmol h}^{-1}$$
$$= 6.06 \text{ kmol h}^{-1}$$

This is equivalent to 339.27 kg h^{-1} of CaO sorbent and hence a sorbent-to-feed weight ratio = 3.39. The same calculation is applied for the biomass case. The sorbent-to-feed weight ratio = 2.26. The amount of steam added can be deduced from the amount of component carrying O_2 present in the product, that is, CO_2 (cf. water dissociation reaction):

$$\text{Steam required for coal LEGS} = \left(\frac{72.6}{12} - \frac{15.49}{32} \right) \times 2 \times 18 = 200.37 \text{ kg h}^{-1}$$

Hence, the steam-to-feed weight ratio = 2 for the coal LEGS case.

The same calculation is applied for the biomass case. The steam-to-feed weight ratio = 0.96 for the biomass LEGS case.

Exercise 11. Consider Exercise 10 with steam gasification in place of LEGS to produce clean syngas primarily containing carbon monoxide and hydrogen to be fed into a CLC process with nickel oxide (NiO) as an adsorbent. A 77% conversion in the fuel reactor is achieved. Calculate the minimum sorbent-to-feed and the steam-to-feed weight ratios required. Determine the molar composition of the gas from the fuel reactor and the stoichiometric molar ratio of air-to-feedstock. The molar mass of Ni is 58.69. Assume that the sulfur present in the feedstocks is converted into hydrogen sulfide in the steam gasification process. A 100% removal of sulfur by reacting between hydrogen sulfide and NiO forming nickel sulfide (NiS) is achieved in the fuel reactor.

Solution to Exercise 11
Basis = 100 kg h^{-1} of the feedstock
   Refer to Table 13.21 for the ultimate analyses of coal and biomass.
   Assumptions:

1. The carbon present in the feedstock is partially oxidized into carbon monoxide.
2. The partial oxidation requires some more O_2 that not available within the feedstock itself. The deficit of O_2 that is not present in the feedstock is provided by steam addition. This external steam added is also a source of H_2.
3. The net amount of hydrogen present in the product gas from the gasifier is the amount of hydrogen present in the feedstock and the steam, deducted by the amount of hydrogen needed to release the sulfur present in the feedstock as hydrogen sulfide in the gas.
4. Any reaction associated with nitrogen and chlorine is neglected. NiO reacts with CO, H_2 and H_2S as in Equations (13.36a) to (13.36c) below.
5. The 77% conversions are achieved for the reactions in Equations (13.36a) and (13.36b) and 100% conversion is achieved for the reaction in Equation (13.36c).
   The reaction in the fuel reactor is:

$$CO + NiO \rightarrow CO_2 + Ni \tag{13.36a}$$

$$H_2 + NiO \rightarrow H_2O + Ni \tag{13.36b}$$

$$H_2S + NiO \rightarrow H_2O + NiS \tag{13.36c}$$

Based on these assumptions, the flow rates in kmol h^{-1} and mole fraction of each component for individual feedstocks are shown in Table 13.24.

Table 13.24   *Steam gasification with the CLC case – flow rate of each component in the product on the basis of 100 kg h⁻¹ of each feedstock.*

Component	Hypothetical Product Formation Pathway	Molar Flow Rate (kmol h⁻¹) Coal	Molar Flow Rate (kmol h⁻¹) Biomass
CO	$C \rightarrow CO$ (see Assumption 1)	$\dfrac{72.6}{12} = 6.05$	$\dfrac{48.45}{12} = 4.04$
H_2	$H \rightarrow H_2$ $H_2O \rightarrow H_2$ $H_2 \rightarrow H_2S$ (see Assumption 2 and 3)	$\dfrac{4.11}{2} + 5.082 - 0.0084 = 7.13$	$\dfrac{5.85}{2} + 1.30 - 0.0003 = 4.23$
N_2	$N \rightarrow N_2$ (see Assumption 4)	$\dfrac{0.1}{28} = 0.0036$	$\dfrac{0.47}{28} = 0.0168$
H_2S	$S \rightarrow H_2S$ (see Assumption 3)	$\dfrac{0.27}{32} = 0.0084$	$\dfrac{0.01}{32} = 0.0003$
Cl_2	$Cl \rightarrow Cl_2$ (see Assumption 4)	0	$\dfrac{0.1}{71} = 0.0014$
Total		**13.19**	**8.29**

Note:

To estimate the amount of H_2 produced, the amount of O_2 required should first be estimated. This is because CO in the product carries O_2 that comes from the feedstock itself and also from the steam added. Hence,

Amount of O_2 contributed by steam = Amount of CO in the product − (Amount of O_2 from the feedstock)

For coal LEGS case, the amount of O_2 contributed by steam is

$$\frac{72.6}{12 \times 2} - \frac{15.49}{32} = 2.541 \text{ kmol h}^{-1}$$

Considering the water dissociation reaction (shown below), it can be deduced that the H_2 in the product contributed by steam is two times the O_2, according to the stoichiometric ratio:

$$H_2O \rightarrow H_2 \tfrac{1}{2}O_2$$

Hence, 2×2.057 kmol h⁻¹ $= 5.082$ kmol h⁻¹.

The amount of H_2 contributed by steam is added to the existing amount of H_2 from the feedstock, giving the total amount of H_2 present in the product, shown in Table 13.24. The same calculation is applied for the biomass case.

Based on 77% molar conversions of CO and H_2 and 100% conversion of H_2S present in the gas product, the total amount of NiO sorbent required in the fuel reactor for coal as the feedstock is

$$= (6.05 \times 0.77 + 7.1284 \times 0.77 + 0.0084) \text{ kmol h}^{-1}$$
$$= 10.16 \text{ kmol h}^{-1}$$
$$= 758.54 \text{ kg h}^{-1}$$

Sorbent-to-coal feed weight ratio = 7.59

Sorbent-to-biomass feed weight ratio = 4.75

$$\text{Steam required for coal gasification} = \left(\frac{72.6}{12 \times 2} - \frac{15.49}{32}\right) \times 2 \times 18 = 91.5 \text{ kg h}^{-1}$$

Hence, the steam-to-coal feed weight ratio = 0.915 and the steam-to-biomass feed weight ratio = 0.234. The outlet gas from the fuel reactor contains CO_2 generating from 77% molar conversion of CO, H_2O from 77% molar conversion of H_2 and 100% molar conversion of H_2S (Equations (13.36a) to (13.36c)) and unreacted H_2 and CO. The gas will have these components with flow rates in kmol h^{-1} and molar fractions, shown in Table 13.25.

Table 13.25 *Flow rate and molar fraction of the gas products from the fuel reactor.*

	Coal		Biomass	
Component	Molar Flow Rate (kmol h^{-1})	Mole Fraction	Molar Flow Rate (kmol h^{-1})	Mole Fraction
CO_2	4.66	0.354	3.11	0.376
H_2O	5.50	0.417	3.25	0.394
H_2	1.63	0.124	0.97	0.118
CO	1.39	0.106	0.93	0.112
Total	**13.18**	**1.000**	**8.26**	**1.000**

The reactions between air and sorbent, Ni and NiS are shown by

$$Ni + \tfrac{1}{2}O_2 \rightarrow NiO \tag{13.37a}$$

$$NiS + \tfrac{1}{2}O_2 \rightarrow NiO + S \tag{13.37b}$$

For the coal CLC case, the total amount of air (assuming that air contains 21% oxygen by volume) required in the air reactor to regenerate the used-up sorbents, Ni and NiS, from the fuel reactor is

$$= \frac{(6.05 \times 0.77) + (7.13 \times 0.77) + 0.0084}{2} \times \frac{1}{0.21}$$
$$= 24.18 \text{ kmol h}^{-1}$$

The same calculation is applied for the biomass CLC case. Hence, the total amount of air required in the air reactor for the biomass processing = 15.16 kmol h^{-1}. Hence, the stoichiometric molar ratio of air to the feedstock is as follows:

$$\text{For coal} = \frac{24.18}{\frac{72.6}{12} + \frac{15.49}{32} + \frac{4.11}{2} + \frac{0.1}{28} + \frac{0.27}{32}} = 2.8$$

$$\text{For biomass} = \frac{15.16}{\frac{48.45}{12} + \frac{43.78}{32} + \frac{5.85}{2} + \frac{0.47}{28} + \frac{0.01}{32} + \frac{0.1}{71}} = 1.8$$

Exercise 12. Integrated flowsheet configurations with the sorption technologies. Pure hydrogen is to be produced from coal and biomass with the ultimate analysis shown in Table 13.21 using the most energy efficient IGCC configuration, from the following three choices in Figures 13.22 to 13.24. Analyze these alternative designs by carrying out mass and

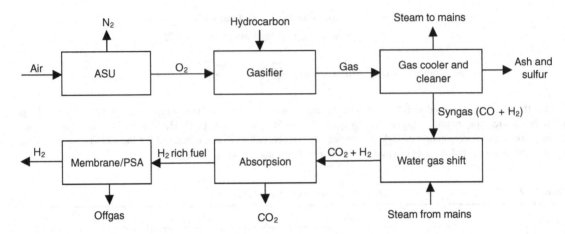

Figure 13.22 *Conventional decarbonized hydrogen production using hydrocarbon gasification. ASU: air separation unit and PSA: pressure swing adsorption.*

energy balance and appropriate selection of absorbents and adsorbents. Remember that the specific chemical information of absorbents and adsorbents is often proprietary to industries. They often only provide feedstock ultimate analysis to evaluate *whole* process performances. Hence, several practical assumptions need to be made to select the design variables, such as purity and yield of products, removal ratios of pollutants as well as absorbents and adsorbents. Restrict the choice of rare earth metals to avoid dependency. Also, market prices and toxicity information are useful to select the materials. For example, Ni is more toxic than Ca. Answer the following questions.

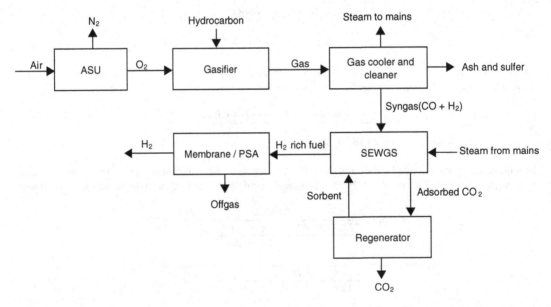

Figure 13.23 *Decarbonized hydrogen production using hydrocarbon gasification followed by a sorption enhanced water gas shift reactor (SEWGS) and regenerator.*

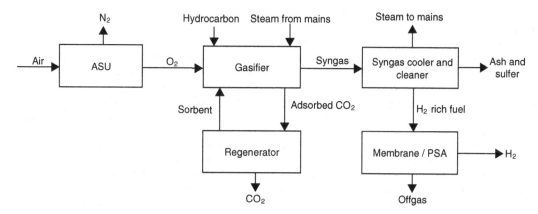

Figure 13.24 *Decarbonized hydrogen production using hydrocarbon gasification in a sorption enhanced reactor and a regenerator.*

1. List the key design variables for each alternative.
2. List the choices of adsorbents and absorbents and benefits (or otherwise) of using them.
3. Name the most capital intensive process units in each design alternative. Comment whether and how capital costs between these alternatives may differ.
4. Name the process units that decide the energy efficiency in each alternative. Comment whether and how the energy efficiency between these alternatives may differ.
5. Perform mass and energy balance calculations. Write the assumptions.
6. List the design parameters that would be most affected by alteration of absorbents in each design alternative.
7. Write the output design objectives and analyze their sensitivities to independent variables.

13.7 Low Temperature Separation

Separation of CO_2 by refrigeration is increasingly used in the industrial community due to the potential of high purity separation between gases. The principle of this separation approach is to reduce the temperature of a H_2 and CO_2 mixture to a low enough level that the CO_2 can be condensed and easily removed from the mixture with phase separation. Two approaches exist. In one approach, the cold energy from liquid natural gas is transferred to refrigerate CO_2, vaporizing the natural gas. In another approach, auto- or self-refrigeration occurs by the process of heat integration or recovery of the refrigeration duty within the process itself. The principles in the latter are similar to air separation.

By reducing the temperature of the gas mixture, the CO_2 component, which has a higher dew point than the other impurity components, can be liquefied first and separated from the gaseous phase. The pressure of the acid gas mixture relates directly to the refrigeration temperature, CO_2 recovery and the scale of the refrigeration facility. For a required CO_2 recovery, a higher pressure of the process stream increases the condensing temperature of CO_2, making the refrigeration separation easier to achieve. The refrigeration separation aims to condense CO_2, but not freeze it. When the refrigeration temperature falls below the liquid area on the CO_2 phase diagram, the CO_2 turns into solid that could easily block and damage the condensing facilities. As a result, the highest CO_2 recovery is determined by the pressure of the gas mixture.

Several approaches are possible to decrease the temperature of the acid gas mixture. The first is the gas expansion, in which the mixture temperature is reduced by the Joule–Thompson effect (gas expansion through throttling valves under an adiabatic condition). The second option is to adopt an external refrigeration cycle, the design of which relates to the operating conditions. Another approach is heat integration. The separated streams of the liquid CO_2 and gaseous hydrogen products are at a low temperature. The cold energy contained in these two streams can be recovered with appropriate heat integration.

In an auto-refrigeration cycle, the gas mixture containing CO_2 is pretreated by a pressure swing adsorption unit or a membrane separator, wherein a portion of hydrogen is separated, leaving the gas mixture more concentrated in CO_2. The pretreatment increases the CO_2 partial pressure so that the downstream CO_2 condensing process could be easier to achieve. Inside the auto-refrigeration system, the gas mixture is recompressed with internal cooling. The reason for the recompression is similar to pretreatment, which is to increase the CO_2 partial pressure so that the required CO_2 recovery can be achieved without freezing. This high pressure gas mixture stream is refrigerated in a coldbox with heat integration with the product streams of hydrogen fuel and liquid CO_2. If the cold energy from the heat integration is not enough to condense the CO_2, an external refrigeration cycle is included. The separation is completed by letting down the high pressure stream through a throttling valve, with which a lower temperature can be reached due to the Joule–Thompson effect. The separated CO_2 is in the liquid state and can easily be pressurized by pumping. With this CO_2 capture technology, the product hydrogen fuel is collected partly from the pretreatment unit and partly from the auto-refrigeration unit. The overall hydrogen fuel must be recompressed to a pressure high enough for use in the downstream power generation section.

There are variations in design and operating conditions that can be achieved by applying the principles of heat integration. This has led to different patents in the field[10].

There are various advantages of the auto-refrigeration process over the other processes for CO_2 capture. The other separation technologies, membrane and pressure swing adsorption, have concerns for scale-up. In comparison, the auto-refrigeration process design concentrates around the design of a compact coldbox.

The pretreatment process involving membrane or pressure swing adsorption can be eliminated leaving the gas feedstock to the auto-refrigeration unit at high pressure. Maintaining a high pressure also eliminates the need for recompression of the product gas. To achieve that, the refrigeration separation must be operated at high pressure.

CO_2 separation by refrigeration is a phase separation process, which provides liquid CO_2 directly. This has the advantage that the pressure of the liquid can easily be increased by pumping. Moreover, the separated hydrogen has the same pressure as the liquefied CO_2. Hence, there is no pressure loss within the CO_2 condensing process.

The dry gas mixture containing mainly hydrogen and CO_2 is refrigerated in a multistream heat exchanger, called a coldbox. Within the coldbox, the temperature of the gas mixture is reduced and the CO_2 content condenses as liquid. The refrigeration temperature required depends on the specification of the CO_2 recovery. It should not be too low to turn the CO_2 content into solid. The gas mixture leaving the coldbox is in two phases, the gas phase mainly with hydrogen and the liquid phase with CO_2. A simple phase separator can be applied to separate the mixture into two product streams. The hydrogen main fuel gas can be reheated within the coldbox to the ambient temperature and sent to power generation. The CO_2 liquid is also reheated within the coldbox after pressurization. This heat integration with product streams provides part of the refrigeration duty and the rest of the refrigeration is supported by an external refrigeration cycle. The only external energy required is for the external refrigeration cycle, which must be minimized, while optimizing the trade-off between product purity and refrigeration cost based on economic parameters. For optimizing the refrigeration cycle, the composition of the refrigerant, a single refrigerant or a mixed refrigerant, must be considered.

The coldbox can be a plate-and-fin heat exchanger, which is widely used in the LNG industry under similar circumstances. The plate-and-fin heat exchanger is capable of accommodating more than 10 streams, with a minimum temperature difference of $1\,°C$ or even less. Also, it should tolerate an operating pressure of up to 100 bar when the aluminium based material is selected. On the other hand, it could tolerate an operating pressure of up to 100 bar in aluminium brazed units. These features make plate-and-fin heat exchangers a candidate for application with refrigeration CO_2 separation.

13.8   Summary

Absorption, adsorption, chemical looping combustion and low temperature separation processes are discussed with respect to carbon dioxide capture. Shortcut modeling equations are shown for estimations of design parameters at the preliminary stage. Rigorous process simulation and, in cases of adsorption, dynamic process simulation are needed. Comprehensive equations and parameters for process simulation are shown.

References

1. J. Sadhukhan, Y. Zhao, N. Shah, N.P. Brandon, Performance analysis of integrated biomass gasification fuel cell (BGFC) and biomass gasification combined cycle (BGCC) systems, *Chem. Eng. Sci.*, **65**(6), 1942–1954 (2010).
2. J. Sadhukhan, Y. Zhao, M. Leach, N.P. Brandon, N. Shah, Energy integration and analysis of solid oxide fuel cell based micro-CHP and other renewable systems using biomass waste derived syngas, *Ind. Eng. Chem. Res.*, **49**(22), 11506–11516 (2010).
3. J. Sadhukhan, K.S. Ng, N. Shah, H.J. Simons, Heat integration strategy for economic production of combined heat and power from biomass waste, *Energy Fuels*, **23**, 5106–5120 (2009).
4. Y. Zhao, J. Sadhukhan, N.P. Brandon, N. Shah, Thermodynamic modelling and optimization analysis of coal syngas fuelled SOFC-GT hybrid systems, *J. Power Sources*, **196**(22), 9516–9527 (2011).
5. K.A. Schmidt, *Solubility of Sulfur Dioxide in Mixed Polyethylene Glycol Dimethyl Ethers*, MSc Dissertation, University of Alberta, 1997.
6. S.A. Bhat and J. Sadhukhan, Process systems engineering aspects for intensifying steam methane reforming: an overview. *AIChE J.*, **55**(2), 408–422 (2009).
7. A. Kapil, S.A. Bhat, J. Sadhukhan, Multi-scale characterisation framework for sorption enhanced reactions processes, *AIChE J.*, **54**(4), 1025–1036 (2008).
8. A. Kapil, S.A. Bhat, J. Sadhukhan, Dynamic simulation of sorption enhanced simulated moving bed reaction processes for high purity biodiesel production, *Ind. Eng. Chem. Res.*, **49**(5), 2326–2335 (2010).
9. A. Kapil, *Multi-Scale Simulation of Sorption Enhanced Reactions in Energy and Biochemical Systems*, PhD Thesis, The University of Manchester, 2009.
10. Y. Lou, *Decarbonisation in Power Production and Process Sites*, PhD Thesis, The University of Manchester, 2009.

Part IV
Biorefinery Systems

14

Bio-Oil Refining I: Fischer–Tropsch Liquid and Methanol Synthesis

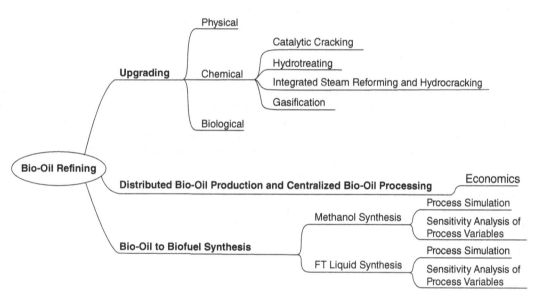

Structure for Lecture Planning

14.1 Introduction

In Chapter 9, the characteristics of bio-oil, the process of transforming biomass into bio-oil via pyrolysis technologies are shown. The chemically unstable nature of bio-oil imposes difficulties in transportation, storage and its application. Therefore, bio-oil needs to be upgraded to eliminate such undesirable behavior. This can be done through physical, chemical and biological upgrading methods, shown in this chapter. Bio-oil can be produced in a centralized site consisting of a number of pyrolysis plants integrated with other processing plants. It can also be produced from various distributed

Biorefineries and Chemical Processes: Design, Integration and Sustainability Analysis, First Edition.
Jhuma Sadhukhan, Kok Siew Ng and Elias Martinez Hernandez.
© 2014 John Wiley & Sons, Ltd. Published 2014 by John Wiley & Sons, Ltd.
Companion Website: http://www.wiley.com/go/sadhukhan/biorefineries

sites and transported to centralized plant for further conversion into useful products. This is shown here covering a range of possible routes for converting bio-oil into fuel production. The emergence of bio-oil has brought an extensive range of opportunities for further research in developing novel technologies in this field, shown in the final section of this chapter.

14.2   Bio-Oil Upgrading

The undesirable characteristics of bio-oil include high acidity, high viscosity, thermal and chemical instability. These characteristics arise due to the high oxygen content in bio-oil. Bio-oil needs to be upgraded to eliminate or reduce such characteristics for various applications: transportation fuels, chemicals and fuels for turbines and boilers. There are two main types of bio-oil upgrading process in practice, that is, physical and chemical catalytic upgrading. Other emerging technology includes the biological upgrading method via fermentation.

14.2.1   Physical Upgrading

Physical means of upgrading involves lowering and controlling the viscosity of bio-oil by addition of water and solvent. This can solve the difficulties during handling and pumping of bio-oil. In addition, the char and solid content can be removed using a physical upgrading method such as hot gas filtration. Char and solid contents have detrimental effects on the processing and storage of bio-oil, causing an increase in viscosity and blockage problems.

14.2.2   Chemical Upgrading

Merely relying on physical upgrading methods is not adequate, since the issues related to the instability of bio-oil and its immiscibility with hydrocarbon fuels have not yet been addressed. Oxygen is chemically bonded within the bio-oil constituents and hence its complete removal needs chemical upgrading of bio-oil. Therefore, chemical catalytic upgrading methods are required to break the bonds with oxygen (deoxygenation). There are two main mechanisms to eliminate oxygen from bio-oil, one by addition of hydrogen followed by formation and removal of water called hydrodeoxygenation (HDO) and the other by removal of carbon dioxide called decarboxylation (DCO). The chemical upgrading method includes catalytic cracking, hydrocracking, hydrotreating, steam reforming, gasification and direct utilization[1]. Another chemical upgrading method involves emulsification of bio-oil with diesel fuel, by adding surfactant to make both the components miscible. However, several drawbacks exist, such as the high cost of surfactants, high energy requirement and high level of corrosion in engines, when a mixture of bio-oil and diesel is used. Partial and full deoxygenations are associated with catalytic cracking, hydrocracking and hydrotreating, where oxygenated components in bio-oil are partially or fully eliminated, hence stabilizing the bio-oil. Application of bio-oil as a transportation fuel requires full deoxygenation.

14.2.2.1   Catalytic Cracking

Catalytic cracking of bio-oil using zeolite catalyst is a mature technology since it has been practiced for many years. The reaction is carried out under atmospheric pressure with temperatures of 350–500 °C. The reactions include dehydration, cracking, polymerization, deoxygenation and aromatization. The oxygen in bio-oil is eliminated and formed into a majority of water, carbon monoxide and carbon dioxide, alongside the main product hydrocarbon, mostly an aromatic compound. The popularity of this method is attributed to its lower operating costs compared to hydrotreating, since no hydrogen is required. However, the main drawback is the low hydrocarbon yield (mostly up to 42%) and high coke deposition, making the process economically unfavorable.

14.2.2.2   Catalytic Hydrotreating

Catalytic hydrotreating has been well developed for rejecting sulfur and nitrogen components in petroleum feedstock. These are known as the hydrodesulfurization and hydrodenitrogenation, which aim to mitigate environmental pollution by reducing SO_x and NO_x generation as well as avoiding catalyst poisoning. Other hydrotreating processes include hydrodemetallization, hydrogenation and HDO.

HDO is rather an uncommon process in petroleum refinery by virtue of the negligible amount of oxygen components presence in petroleum fuel, that is, less than 2 wt% that does not affect the fuel stability. This process is the subject of discussion in the next chapter. In biomass derived fuels, such as bio-oil, particular attention has to be paid to reduce the large proportion of oxygen content in it, that is, up to 50 wt%. For catalytic HDO of bio-oil, severe operating conditions have to be applied, such as high H_2 pressure and consumption. The process is usually carried out at temperatures of 300–600 °C, accompanied by heterogeneous catalyst such as sulfided Ni/Mo and Co/Mo supported on alumina. Consequently, the oxygen in bio-oil is rejected as water. For the production of transportation fuel, full catalytic HDO is preferred rather than catalytic hydrocracking, primarily due to a higher hydrocarbon yield (up to 50%) with less than 0.5% oxygen remaining in bio-oil. Unfortunately, the hydrogen consumption required is very high (700–800 N m^3 t^{-1}) and hence results in expensive production cost. For electricity and heat generation, the HDO process can be compromised with less severe operating conditions, for example, lower hydrogen consumption. In general, this type of partial HDO process can only manage to stabilize the bio-oil, through elimination of the reactive components such as ketones. The following chapter shows the reaction mechanisms of the oxygenated compounds such as phenol, dextrose, furfural, acetic acid, hydroxyacetone and formic acid present in bio-oil in HDO and DCO reactions. It also shows effective ways of sourcing hydrogen from the bio-oil itself.

Besides HDO, DCO is another processing option for removing oxygen in bio-oil. By DCO, CO_2 is rejected instead of H_2O as in HDO. The advantage of DCO is that it occurs favorably under a lower pressure and hence lower hydrogen consumption. However, the optimal yields may often correspond to a combination of the two reaction mechanisms: HDO and DCO.

14.2.2.3 *Integrated Steam Reforming and Hydrocracking Processes*

Hydrogen has been commonly used in various industrial applications such as fuel cell, synthesis of ammonia, upgrading of fossil fuels in refinery and petrochemical plants. Hydrogen can be produced using different feedstocks, namely natural gas, coal as well as renewable resources such as biomass and pyrolysis oil. There are several commercial routes for generating hydrogen, for example, gasification, partial oxidation, steam reforming, photolysis and biophotolysis. Steam reforming of bio-oil to produce hydrogen seems to be a potential route for replacement of fossil-derived hydrogen.

Bio-oil can be separated into two phases, that is, a carbohydrate-derived fraction (aqueous fraction containing aldehydes, ketones, sugars, carboxylic acids, etc.) and a lignin-derived fraction (water insoluble fraction containing mainly phenolic and aromatic compounds), using extraction such as water addition. The aqueous fraction of bio-oil can be reformed into H_2 and CO whereas the lignin fraction that is deficient in hydrogen can be cracked into useful chemicals or fuels such as gasoline and diesel. These processes can be integrated so that H_2 produced from a steam reforming reaction can be availed for hydrogenation reaction in hydrocracking of the lignin fraction of bio-oil, as shown in Figure 14.1. H_2 produced from steam reforming may also be considered for integration with a fuel cell application.

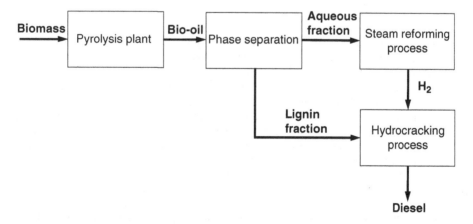

Figure 14.1 *Integrated scheme of steam reforming and hydrocracking processes of bio-oil fractions.*

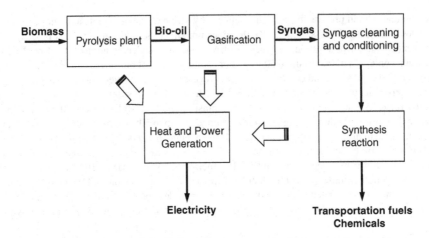

Figure 14.2 *Integrated scheme of bio-oil gasification, synthesis reaction and heat and power generation.*

The following equations describe the reactions that take place in the reformer, that is, steam reforming, water gas shift and overall steam reforming reactions, respectively:

$$C_n H_m O_k + (n - k) H_2O \rightarrow nCO + (n + m/2 - k) H_2 \tag{14.1}$$

$$CO + H_2O \rightleftharpoons CO_2 + H_2 \tag{14.2}$$

$$C_n H_m O_k + (2n - k) H_2O \rightarrow nCO_2 + (2n + m/2 - k) H_2 \tag{14.3}$$

Autothermal steam reforming is another type of reforming technology, wherein O_2 is added to the reformer so that oxidation takes place in combination with the steam reforming reaction, balancing between the exothermic heat and endothermic heat of the two reactions, respectively. This technology offers several advantages such as improvement in reactor temperature control, avoiding formation of hot spot and hindering catalyst deactivation by sintering or carbon deactivation and, most importantly, it consumes less energy compared to steam reforming, since the heat load demand of the reformer is minimized. However, autothermal steam reforming suffers from lower yields of H_2 in comparison to steam reforming as a part of the fuel is exhausted during the oxidation step to supply the heat for the reforming reaction.

14.2.2.4 Gasification

Gasification is principally a partial oxidation reaction, where carbonaceous materials such as coal, heavy oil, biomass and bio-oil, under high temperature, can be converted into syngas upon purification. The syngas containing H_2 and CO as the main constituents has a high heat content. It is thus a favorable and versatile building block for a diverse range of products, such as transportation fuels, chemicals and electricity. Figure 14.2 shows the generic scheme, where bio-oil is gasified into syngas and then converted into either transportation fuels such as Fischer–Tropsch diesel and methanol, electricity, chemicals or a combination of products. Fischer–Tropsch diesel and methanol are among the cleaner and higher quality transportation fuels. By integrating the gasification and synthesis processes with syngas cleaning and water gas shift reaction processes, fuels, chemicals and electricity can be generated efficiently from biomass, mitigating greenhouse gas emission to the atmosphere. Gasification offers many advantages, such as feedstock flexibility, environmentally friendly performance and high efficiency. A detailed discussion on gasification can be found in Chapter 9.

14.2.3 Biological Upgrading

In the earlier section, a thermochemical conversion of bio-oil is shown in Figure 14.1, where the aqueous fraction of bio-oil is steam reformed to generate hydrogen for supporting the hydrocracking of lignin. The aqueous fraction of bio-oil has

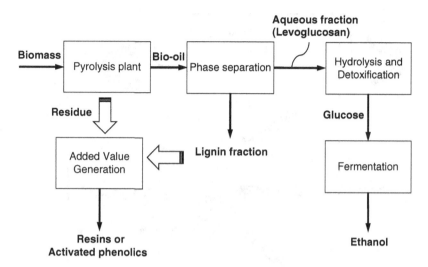

Figure 14.3 *Integrated scheme of ethanol production via pyrolysis and fermentation of aqueous fraction of bio-oil.*

a significant amount of anhydrosugars, normally referring to levoglucosan, a complex substance resulting from breaking the cellulose layers in biomass. Levoglucosan can be hydrolyzed into monosugars, such as glucose, while detoxification also takes place to remove the remaining phenols and acids in the aqueous phase. Glucose is then converted to ethanol via the fermentation process, where enzymatic reaction takes place. The lignin fraction of bio-oil can be used in the utility system to produce heat. Alternative or more effective use of lignin is shown in the Introduction chapter. Integrating these processing options with a thermochemical (pyrolysis), or biological (fermentation) route and lignin value generation utilizing biomass can result in a significantly advanced biorefinery system, shown in Figure 14.3.

14.3 Distributed and Centralized Bio-Oil Processing Concept

14.3.1 The Concept

Bio-oil can be collected from various distributed biomass pyrolysis plants and then transported to a centralized processing plant. In the large centralized processing plant, similar to a petroleum refinery accepting crude oil from different locations within specifications, the bio-oil is converted into different products through a series of processing operations. The distributed–centralized generation concept from biomass to chemicals, fuels, heat and power production is shown in Figure 14.4.

An integrated gasification plant is a good example of a centralized processing plant. An integrated biomass gasification and product synthesis plant is considered as a direct route. In contrast, an indirect route involves two stages: (1) pyrolysis of biomass to produce bio-oil and (2) gasification of bio-oil for further processing into useful products. The indirect route for producing transportation fuel and/or power is a choice for large-scale commercial development soon to be compared to the direct route using solid biomass due to the following advantages it can offer:

- Can be co-fed with other feedstock.
- Handling and storage of liquid bio-oil is easier.
- Bio-oil is cleaner with negligible ash content.
- Bio-oil has more consistent properties.
- Transporting liquid bio-oil is less costly compared to transporting solid biomass.

Distributed processing of bio-oil followed by gasification in large centralized facilities (indirect route) provides a favorable platform for various product generations, such as Fischer–Tropsch (FT) diesel, methanol, heat and power,

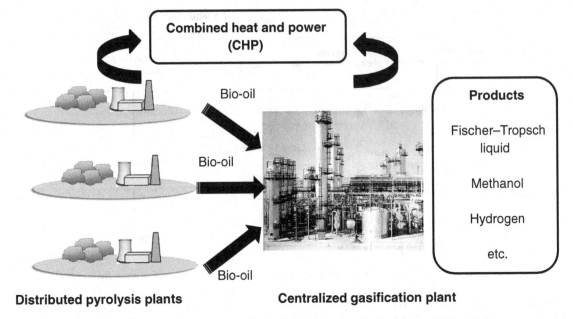

Figure 14.4   *Distributed processing of biomass and centralized processing of bio-oil.*

etc. Such a production strategy offers several benefits over biomass gasification with a synthesis system (direct route):

- Lower production cost.
- Low sensitivity to variation of delivery costs and crop yields.
- Lower transportation cost can be achieved for large scale production.
- Flexibility in choice of biomass.

The economic attractiveness of the indirect route of processing biomass depends on many factors, including the scale, the location and its associated economic and logistic factors, for example, availability and price against competitive fossil resources and industrial practices, etc.

14.3.2   The Economics of Local Distribution of Bio-Oil

The transportation cost of bio-oil is an important factor to analyze, when bio-oil produced from distributed pyrolysis plants needs to be transported to centralized processing plant, such as a bio-oil integrated gasification (BOIG) system. The data and analysis of bio-oil transportation from distributed pyrolysis plants to centralized sites, based on the studies by Bridgwater, Toft and Brammer[2], Rogers and Brammer[3] and Pootakham and Kumar[4], are shown in Table 14.1.

The results shown in Table 14.1 used both fixed and variable costs depending upon the distance or thermal value of bio-oil. (Note that 1 USD = 0.8 euro and 1 GBP = 1.1 euro are assumed.) The zone costing approach uses the number of round trips in a day within the distributed–centralized region as the basis to define a transport zone. Thus, zone 1 is the outermost zone where only one round trip is possible in a day, carrying the highest total cost amongst all zones, 64.2 million euro y^{-1} or 26.42 euro t^{-1}, compared to 8.6 million euro y^{-1} or 3.54 euro t^{-1} for zone 6, which implies 6 round trips in a day. It is hence beneficial to carry out more round trips in a day to reduce the transportation cost. A tanker (truck) with load ranges of 24–44 tonnes and 60 m^3 per truck and pipeline capacity of 560 m^3 d^{-1} are considered for a distance of 100 km between distributed pyrolyzers and a centralized BOIG site.

Table 14.1 Cost of transporting bio-oil from distributed pyrolysis plants to centralized 1350 MW BOIG-methanol system. (Reproduced with permission from Ng and Sadhukhan (2011)[5]. Copyright © 2011, Elsevier.)

Researcher	Bridgwater, Toft and Brammer 2002[2]		Rogers and Brammer, 2009[3]		Pootakham and Kumar, 2010[4]	
Method of transporting bio-oil	Tanker	Tanker	Tanker	Tanker	Tanker	Pipeline
Maximum load or capacity	30.5 t	24.0 t	44.0 t	44.0 t	60 m^3	560 m^3 d^{-1}
Analysis approach Cost estimating Fixed cost	Distance rate Shell UK 4.29 euro t^{-1}	Distance rate Linkman	Zone costing Zone 1 0.66 euro GJ^{-1}	Zone costing Zone 6 0.11 euro GJ^{-1}	Distance rate 4.568 euro m^{-3}	Distance rate 0.03384 euro m^{-3}
Variable cost	0.039 euro t^{-1} km^{-1}	0.043 euro t^{-1} km^{-1}	0.99 euro GJ^{-1}	0.11 euro GJ^{-1}	0.04 euro m^{-3} km^{-1}	0.09608 euro m^{-3} km^{-1}
Distance assumed	100 km	100 km	96–224 km	0–11 km	100 km	100 km
Bio-oil transportation cost (million euro y^{-1})	19.9	10.4	64.2	8.6	19.2	21.6
Bio-oil transportation cost (Euro t^{-1})	8.19	4.28	26.42	3.54	7.90	8.89

Ng and Sadhukhan estimated the cost of transporting 303.75 t h^{-1} of bio-oil (equivalent to 2.43 million t y^{-1} or 2.24 million m^3 d^{-1} or 3.89×10^7 GJ y^{-1}) to 1350 MW BOIG system using the distance rate and zone costing approaches[5]. The resulting transportation cost is 4.28–8.89 euro t^{-1} (or 10.4–21.6 million euro y^{-1} for the 1350 MW case), which reduces the netback of bio-oil from 45.2 euro t^{-1} to 40.9–36.3 euro t^{-1}. In terms of cost of production of bio-oil, transportation adds 5.7–11.9% extra on 75 euro t^{-1}.

14.3.3 The Economics of Importing Bio-Oil from Other Countries

A secure and ample supply of bio-oil is required for a large-scale BOIG system. Importing bio-oil from other countries at an acceptable cost may be a way to fulfil the rising demand for biofuels in the local market. The total delivered cost of bio-oil depends on various aspects, such as loading and discharging, rail and road transportation, labor, taxes, shipping, etc., and is yet not fully understood and estimated. The delivered cost of bio-oil to Rotterdam, The Netherlands, from Canada, Brazil, South Africa, Ukraine and Baltic is estimated to be in the range of 6–10.6 euro GJ^{-1}, using 4700 t tankers (ship) and without a return trip[6].

The shipping cost is the major component of the delivered cost of bio-oil, especially when the distance between the countries is significant. Ng and Sadhukhan estimated that the shipping cost of bio-oil derived from oil palm empty fruit bunches is 5.2 euro GJ^{-1} (equivalent to 83.2 euro t^{-1}) for transporting bio-oil from Port Kelang, Malaysia, to Port of Immingham, UK, over a distance of 15 000 km, assuming a linear relationship between the distance and the shipping cost[7]. For 2.43 million t y^{-1} of bio-oil import to the 1350 MW BOIG system, 202.2 million euro y^{-1} are incurred from shipping of bio-oil, that is four times the original operating cost. However, this cost can be reduced considerably (e.g., by approximately half) by introducing larger tanker for shipment. In the overall context, it is not cost-effective to ship bio-oil across a long distance from other countries rich in biomass resources such as South East Asia to the EU, for example.

14.4 Integrated Thermochemical Processing of Bio-Oil into Fuels

Similar to coal and biomass, bio-oil can be used as a fuel. Bio-oil can be directly used in small stationary applications such as a boiler, gas turbine and diesel engine. In addition, bio-oil can be used in gasification to produce valuable syngas, which can then be converted into transportation fuels through an integrated system.

14.4.1   Synthetic Fuel Production

Integrated System. Gasification followed by Fischer–Tropsch (FT) is a promising route for producing transportation fuel and CHP from lignocellulosic biomass, in a biorefinery fashion. This is an indirect route to produce biofuel. Biomass is first converted into bio-oil via a fast pyrolysis reaction, and further processed in gasification and FT reaction[7]. The conventional route involves the direct use of biomass in the gasification and FT reaction, without undergoing the fast pyrolysis process. This direct route is often known as biomass-to-liquid (BTL), analogous to coal-to-liquid (CTL) and gas-to-liquid (GTL) processes. Gasification of bio-oil (indirect route) is a promising route for producing biofuel compared to gasification of biomass (direct route) since bio-oil is easier to handle (including pumping, atomization and pressurization), lower costs are required for storage and transport, and, most importantly, it can be produced on a small scale and utilized alone from multiple distributed units or as a mixture at a large-scale centralized gasification plant.

Past and Present. FT synthesis was invented by Franz Fischer and Hans Tropsch during the 1920s. The process was used during World War II to produce synthetic fuel from coal in Germany. FT synthesis has been operated commercially by Shell, Malaysia (natural gas based syngas) and Sasol, South Africa (coal based syngas).

Overview of FT Synthesis. FT synthesis produces hydrocarbon with various carbon chain lengths (mainly straight-chain hydrocarbon), used as liquid fuel or further processed into diesel production via a hydrocracking reaction. The diesel (C_9–C_{20}) obtained from FT synthesis has a high cetane number (i.e., a measure of ignition delay of diesel, where a higher cetane number implies that the ignition delay is shorter and thus is more desirable). FT synthesis also produces off-gas containing light hydrocarbons (C_1–C_4), useful for heat and power generation; naphtha (C_5–C_{11}), which can be blended into gasoline, though it has a lower octane number (i.e., resistance of gasoline to detonation in an internal combustion engine) than naphtha obtained from crude oil; and waxes (C_{20+}), to be hydrocracked to form diesel. One of the superiorities of the FT liquid is that it is completely free of sulfur and consists of a minimal amount of aromatics compared to gasoline and diesel, causing less pollution to the environment. This is because the FT synthesis catalyst is sulfur intolerant and sulfur containing compounds such as hydrogen sulfide are removed to the ppb level in the gas feedstock by the gas purification process before FT liquid synthesis. This in turn implies that the feed specifications for FT synthesis are very stringent. The syngas has to be cleaned from all contaminants that would potentially cause fouling in the equipment. Rectisol™, a physical absorption process, is suited to purify the gas. Rectisol™ process modeling is shown in Chapter 13.

FT Reaction. The reactions that occur in FT synthesis are highly exothermic and the main reactions involve paraffin (Equation (14.4)) and olefin (Equation (14.5)) production, as well as a water gas shift reaction (Equation (14.6)). Other side reactions also occur in the FT reactor such as alcohol production, catalyst oxidation and reduction, bulk carbide formation and the Boudouard reaction.

$$n\,CO + (2n+1)\,H_2 \rightarrow C_nH_{2n+2} + n\,H_2O \tag{14.4}$$

$$n\,CO + 2n\,H_2 \rightarrow C_nH_{2n} + n\,H_2O \tag{14.5}$$

$$CO + H_2O \rightleftharpoons CO_2 + H_2 \tag{14.6}$$

The water gas shift reaction (Equation (14.6)) is significant only when iron based catalyst is used. If cobalt based catalyst is used, a separate water gas shift reactor has to be used before the FT reactor, since the water gas shift activity is negligible in the FT reactor. The H_2/CO molar ratio of the syngas is a highly influential parameter for dictating the reaction rate. A higher H_2/CO molar ratio results in a higher selectivity for lighter hydrocarbons due to the higher probability of chain termination. The water gas shift reactor should be used before the FT reactor to adjust the H_2/CO molar ratio. The ideal H_2/CO molar ratio of the FT feed stream is 2, as indicated in the stoichiometric reaction in Equation (14.4). The required H_2/CO ratio in the FT feed stream is 2.15 if cobalt catalyst is used, while the ratio of 1.7 is tolerable for iron catalyst since the water gas shift reaction occurs along with FT synthesis with iron catalyst. Adding steam into the gasifier is also a way to increase the ratio, but steam plays the main role of atomizing the biomass in the gasifier and results in lower energy efficiency. The water gas shift reactor should be operated with a carbon limiting or steam rich condition, so that the largest moles of hydrogen obtained are equal to the moles of carbon monoxide in syngas (water gas shift is an equimolar reaction).

Types of FT Reactors. Now four commercial types of FT reactors are available, that is, the circulating fluidized bed, bubbling fluidized bed, tubular fixed bed and slurry phase reactors. The first two reactors are high temperature FT (HTFT) reactors whilst the latter two reactors are low temperature FT (LTFT) reactors. HTFT operates at about 320–350 °C, whereas LTFT operates at about 220–250 °C. FT reaction has typical operating pressures of 25–60 bar, working with either an iron or cobalt based catalyst. The criterion for choosing the type of reactor configuration is highly dependent on the desired product distribution. For any type of reactor with different operating conditions and catalyst, a wide range of products as aforementioned can be obtained. The weight distribution and selectivity of each product in HTFT and LTFT are modeled using the Anderson–Schulz–Flory distribution model, which describes the relationship between the chain growth probability and product yield[8]. The LTFT favors the production of longer chain paraffinic hydrocarbons, that is, wax that can be further processed into diesel through the hydrocracking reaction. The HTFT is used for the production of olefins, which favor petrol yields.

FT Product Distribution. The fractional conversion of each reaction is estimated based on the weight distribution of each FT product obtained via the Anderson–Schulz–Flory (ASF) distribution model[8]. This model assumes a constant chain growth probability, α. Thus, $(1-\alpha)$ equals the probability of termination of the carbon chain. In general, the FT reaction involves two principal mechanisms (excluding the chain initiation step):

- Chain growth by absorbing CO and H_2 (stepwise addition of CH_2) to form a longer carbon chain length.
- Termination by desorption from the catalyst to form paraffin or olefin.

The ASF model, which relates the weight fraction of the FT product, w, with the chain growth probability α, is shown below, where n denotes the carbon number. Typical values of α fall between 0.7 and 0.9.

$$w_n = \alpha^{n-1}(1-\alpha)^2 n \tag{14.7}$$

14.4.2   Methanol Production

Integrated System. Similar to the FT synthesis reaction, methanol can also be produced from syngas via an integrated gasification system[5].

Overview of Methanol. Methanol serves as a solvent (grade A), a chemical (grade AA) as well as a fuel (fuel grade) and is one of the most important chemicals produced globally. Methanol can be used as a solvent in a gas cleaning process such as Rectisol™ to remove CO_2, H_2S and other impurities. Furthermore, methanol acts as an essential chemical intermediate for producing other chemicals, such as formaldehyde, methyl *tert*-butyl ether (MTBE), acetic acid, etc. Among those, formaldehyde stands as the largest consumer of methanol, which accounts for ~35% of worldwide methanol. As a fuel, methanol can be applied directly to flexible fuel vehicles and fuel cell vehicles. The increasing demand for methanol has driven vigorous production of methanol from a wide range of feedstocks, including natural gas, coal and biomass.

Past and Present. The first large-scale commercial production of synthetic methanol through a synthesis gas platform is dated back to 1923, developed by BASF in Germany. This process required a high pressure (240–300 bar) and temperature 350–400 °C and ZnO/Cr_2O_3 as catalyst. Now, the majority of the methanol is manufactured via the syngas route using natural gas as the feedstock. Natural gas is the preferred feedstock because it has a high hydrogen content, involves less energy consumption, low capital and operating costs. Natural gas has lesser impurities, such as sulfur and metals. Logically, other feedstocks such as coal, biomass and heavy residues can also produce methanol. This broadens the choice of feedstocks for countries with low availability and accessibility to natural gas. Environmental and economic considerations should be taken into account, since gas derived from coal and heavy residues has a substantial amount of impurities needing purification before methanol synthesis. Bio-oil is also suitable for generating syngas and subsequently methanol. Other possible methanol synthesis routes are also available, such as the direct partial oxidation of methane or indirect oxidation via intermediates such as methyl chloride and methyl bisulfate.

Methanol Reaction. The principal reactions involved in a methanol synthesis reactor are shown in the following equations. Other side reactions that produce chemicals such as MTBE and formaldehyde also occur.

$$CO + 2H_2 \rightleftharpoons CH_3OH \qquad \Delta H_R^\circ = -90.6 \text{ kJ mol}^{-1} \qquad (14.8)$$

$$CO_2 + 3H_2 \rightleftharpoons CH_3OH + H_2O \qquad \Delta H_R^\circ = -49.7 \text{ kJ mol}^{-1} \qquad (14.9)$$

$$CO + H_2O \rightleftharpoons H_2 + CO_2 \qquad \Delta H_R^\circ = -41.5 \text{ kJ mol}^{-1} \qquad (14.10)$$

Notably, the methanol synthesis reaction is exothermic and thus favored at low temperature, according to Le Chatelier's principle. Based on the same principle, high pressure is preferred to shift the reaction equilibrium to the production of methanol. The stoichiometric number for methanol synthesis, R (values are in molar terms), is defined as $(H_2 - CO_2)/(CO + CO_2)$. The ideal value of R is 2, but an R value slightly higher than 2 is used in practice due to kinetic reasons and to control the by-product formation. A value of R greater than 2 signifies excess hydrogen in the syngas, whilst an R value lower than 2 means deficient hydrogen in the syngas. The R value should be adjusted for the methanol synthesis reaction to ensure higher conversion and a better quality of methanol. From the definition of R, CO_2 should be in low concentration so that an R value of 2 can be achieved. It has been suggested that a small amount of CO_2 should be present in the feed to the methanol synthesis reactor (i.e., 1–2%), which acts as a promoter of the primary reaction (Equation (14.8)) and helps to maintain catalyst activity.

Types of Methanol Reactors. Several types of commercial methanol synthesis reactors are available. The commercial methanol synthesis reactors can broadly be classified into two categories, that is, gaseous phase and liquid phase reactors. Two gaseous phase reactor types (fixed bed) are widely used for methanol production, that is, ICI and Lurgi low pressure methanol synthesis processes. An ICI reactor operates adiabatically, where cold shot cooling is applied to regulate the temperatures of the reactor. The Lurgi reactor allows circulation of the cooling medium on the shell side in the reactor, so that a near-isothermal operation can be attained. A liquid phase methanol synthesis reactor such as LPMEOH developed by Air Products offers a higher conversion per pass (due to a more effective heat transfer between the solid catalyst and liquid phase), which eliminates the recycle loop. This has the advantage of achieving a higher energy saving.

14.5   Modeling, Integration and Analysis of Thermochemical Processes of Bio-Oil

For designing a highly efficient integrated gasification system for the production of fuels as discussed, a detailed understanding of the effects of operating conditions and process integration on energy efficiency and economic performance is needed. In this section, the design of the integrated system carried out in three steps is shown. First, the overall configuration of the system is synthesized in a simulation environment, in tandem with the consideration of suitable operating conditions for each process. Then, various sensitivity analyses are performed on the major unit operations to explore the interaction of one unit with the other, thus identifying refined operating conditions. The final step involves process enhancement from different aspects such as maximizing energy recovery within the system and improvement of yield through recycling. The methodology for designing the integrated system is shown in Figure 14.5. The bio-oil integrated gasification and Fischer–Tropsch (FT) and methanol synthesis processes are discussed with an example of calculations.

14.5.1   Flowsheet Synthesis and Modeling

Mass and energy balances are the fundamentals to any process synthesis problems. These balances give the basis for later analysis and optimization. Hence, the balances should be solved before analyzing the flowsheets and made as correct as possible to avoid major errors in any further calculation based on the established balances.

In the first attempt to synthesize a flowsheet, one may find that the problem is vaguely defined and little is known about the detail of the processes involved. In principle, two common types of process synthesis problems are found:

1. To find the feedstock. In this case, product specifications such as flow rate and composition are known.
2. To find the product(s). In this case, feed specifications such as flow rate and composition are known.

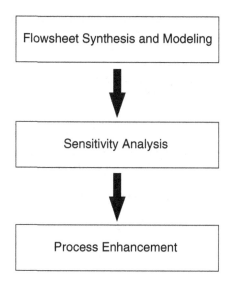

Figure 14.5 *Process design methodology.*

In order to decide upon the objectives, some process operating conditions need to be known. A systematic approach in collecting the information is desirable to reduce the time or extraneous data requirements.

In any flowsheet design and synthesis case, it is always recommended to start creating the flowsheet at a high level and gradually build the individual unit operations at the process level. This gives a preliminary screening of missing information and identifies the level of details required for each important process investigation. A block diagram is the most preferred method in showing the overall system configuration and material and energy flows. The block diagram for an integrated gasification system is shown in Figure 14.6.

The operating conditions (e.g., temperature and pressure) of the gasification process, gas cleaning and conditioning and synthesis reaction are needed to solve the mass and energy balances of the integrated gasification flowsheet, as shown in the block diagram in Figure 14.6. This information can normally be found in literature, technology provider websites, existing technology reports, etc. Mass balances include the flow and composition of feed, gas product, conditioned syngas and product streams. Energy balances include the heat released or consumed within the reactors and separators as well as the heat exchange processes involved between processes.

The next step is detailed process modeling. This can be done using mathematical modeling if detailed reaction equations can be specified. Alternatively, the simulation modeling method can be adopted. Simulation packages such as Aspen Plus, Aspen HYSYS and PRO/II are among the widely used process simulation softwares in industry. They contain well-established unit operation models, thermodynamic property packages, chemical component databases and calculation algorithms that allow process modeling to be carried out efficiently and the generation of reliable results. Nevertheless, it should be remembered that validation of results ought to be performed no matter which modeling method (mathematical or simulation) is adopted.

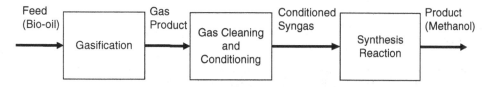

Figure 14.6 *Block diagram showing the integrated gasification system.*

The following general procedures can be adopted for simulation modeling of a process:

Step 1. Set up a flowsheet environment comprising the following:

- Pick the chemical components involved in the whole plant from the component database.
- Choose a suitable thermodynamic property package.

Step 2. Build the flowsheet starting from the main unit operation blocks:

- Select a suitable unit operation model for the process to be modeled. Consider the simplest form of the model from a choice of readily available models. If the results are not satisfactory, then a more rigorous model is to be adopted, including a user-defined model.
- Obtain the operating parameters of the processes to be modeled from the literature.
- Perform simulation.
- Validate the simulation results against the literature results.

Step 3. Create the models for other auxiliary equipment and connect the streams between the processes:

- The auxiliary facility includes devices for increasing or decreasing temperature and pressure, mixer or splitter as well as flash separators for vapor–liquid separation.

Depending on the modeler's preference, Steps 2 and 3 can also be carried out simultaneously by modeling the unit operation blocks in sequential order. This is sometimes more efficient because the intermediate streams between processes can be simulated and used for the later processes. Otherwise, an initial guess has to be made for the feed streams to the main unit operation.

An example using ASPEN Plus simulation modeling is shown for the bio-oil integrated gasification system for the production of methanol. Refer to the ***Online Resource material in the Companion Website: Chapter 14*** for the operating conditions of the processes and their modeling, design and integration using Aspen Plus.

14.5.2   Sensitivity Analysis

14.5.2.1   *Sensitivity Analysis on Gasification Process*

Gasification is the core process of the system. Its operating condition affects the operation, cost, product yields and purities of the downstream processes. For example, if a lower pressure is used in the gasifier, while the downstream process demands a higher pressure, then compression of syngas is needed, thus leading to a higher operating cost. On the other hand, a high pressure gasifer differs from the atmospheric or slightly higher pressure gasifier in terms of material of construction, lining and thickness of the reactor wall. If undesirable components are present in syngas due to an unoptimized operating condition, it would need more downstream cleaning or otherwise it might affect the product yield from the synthesis reactions. Cautious choice of operating parameters of the gasifier is vital, since operating conditions decide the performance of the entire system. The operating parameters that can be manipulated in the gasifier to produce favorable syngas quality are as follows:

- Pressure
- Temperature
- Oxygen flow rate or oxygen-to-feed ratio
- Steam flow rate or steam-to-feed ratio

Steam is normally added as the gasifying medium. Generally, moist biomass has enough moisture to atomize biomass to ease the primary pyrolysis or devolatilization reactions in the gasifier. Since bio-oil is in liquid form and has a significant amount of water, steam is not needed in this case.

Sensitivity analysis should be performed for the variation of the above-mentioned operating parameters. There should be a systematic way to move towards optimal operating conditions. This would give an idea on the range of the correct conditions to generate syngas with the desired quality. There are three main criteria to inspect when carrying out sensitivity analysis for the gasifier:

- The H_2/CO molar ratio
- By-product formation such as CH_4
- No NO_x or SO_x formation by retaining an oxygen lean environment. Oxygen input should just be adequate to convert all carbon present in the biomass into carbon monoxide. Only consider the net oxygen requirement by subtracting the oxygen content in the biomass and associated moisture.

(a) Pressure of Gasifier

Pressure has a negligible impact on the syngas composition. This is primarily due to equimolar stoichiometric gasification reactions and the pressure has less effect in changing the equilibrium composition, according to Le Chatelier's principle.

Although the effect of pressure on the syngas composition is negligible, the downstream operation ought to be taken into account while considering the operating pressure of the gasifier. For example, methanol synthesis operates at high pressure (e.g. 100 bar). Therefore, the pressure of the gasifier is elevated to reduce the compression power. There also exists a trade-off between the cost of operating the gasifier at high pressure and the cost of compression of syngas before the synthesis reaction. For CHP generation, the recommended biomass gasifier pressure is 30–50 bar.

(b) Temperature of Gasifier

A higher temperature in the gasifier is preferred due to the following reasons:

- Higher mole fraction of CO and H_2
- Lower mole fraction CO_2 and CH_4

This is shown in Figure 14.7(a). H_2 and CO are the main constituents of the methanol synthesis reaction; hence higher amounts of these components are desirable. However, the amount of H_2 decreases beyond 1000 °C. The decline of the H_2 component also suggests that the heating value of the syngas is lowered. Lower power is obtained if syngas is utilized for power generation. This can still be improved by placing a water gas shift reaction to increase the proportion of H_2. On the other hand, the H_2/CO molar ratio of the syngas should be monitored so that the syngas can be fed to the downstream process without undergoing rigorous conditioning (i.e., water gas shift). The H_2/CO molar ratio is lowered sharply between temperatures of 500 °C and 700 °C, and the effect is less significant beyond 700 °C. This is shown in Figure 14.7(b).

CO_2 and CH_4 are undesirable products from gasification. CO_2 has a dilution effect on the heating value of syngas and cost is incurred for removing CO_2. The CO_2 content in the syngas at the temperature beyond 1000 °C is acceptable. A relatively higher proportion of CH_4 is helpful if the syngas is used to generate power, but is undesirable for synthesis reactions such as methanol and FT liquids. For CHP generation, a gasifier temperature of 950 °C is recommended.

(c) Variation of oxygen-to-feed ratio

The oxygen input to gasification reaction needs to be properly controlled. This is because excess oxygen input causes more CO_2 rather than CO formation, that is, the combustion reaction dominates the partial oxidation reaction. This is evident from Figure 14.8(a), which shows that increasing the O_2/feed molar ratio leads to undesirable effects such as an increase in CO_2 and a decrease in CO and H_2 formation. Also, a higher H_2/CO molar ratio in syngas can be attained at a lower O_2/feed molar ratio, shown in Figure 14.8(b). Hence, less theoretical oxygen is to be maintained in the gasifier.

The following are the heuristics for gasification process operations:

- Pressure – select a pressure by considering the operating conditions of the downstream processes.
- Temperature – use a temperature close to 1000 °C.
- Oxygen-to-feed molar ratio – use a lower oxygen-to-feed molar ratio, that is, around 0.5.

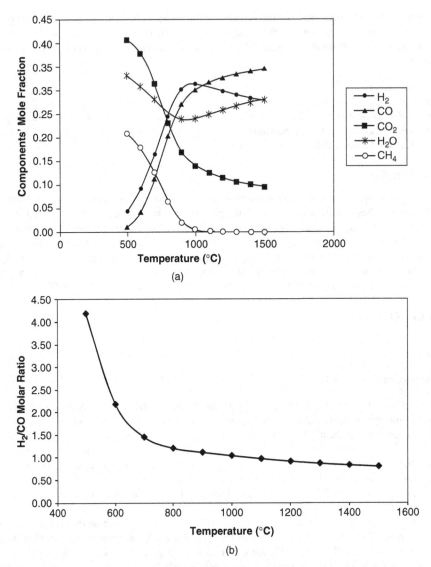

Figure 14.7   *(a) Effect of gasifier temperature on the syngas composition. (b) Effect of gasifier temperature on the H_2/CO molar ratio in the syngas. Pressure of the gasifier is set at 30 bar and the oxygen-to-feed molar ratio at 0.54. (Reproduced with permission from Ng and Sadhukhan (2011)[7]. Copyright © 2011, Elsevier.)*

14.5.2.2   *Sensitivity Analysis on Syngas Composition*

The degree of water gas shift, the degree of water removal as well as the degree of CO_2 capture decide the stoichiometric number, R, and hence the yield of methanol. The stoichiometric number for methanol synthesis, R, defined as $(H_2 - CO_2)/(CO + CO_2)$, should be manipulated to 2. See the ***Online Resource material in the Companion Website: Chapter 14*** for analysis of R and process models.

The water removal unit does not have any effect on the stoichiometric number, but it would have an impact on the methanol conversion. The water gas shift (WGS) reaction is also vital in concentrating CO_2 in the outlet stream so that it can be removed by the CO_2 capture unit (CO2SEP) more economically. A sensitivity analysis approach is required,

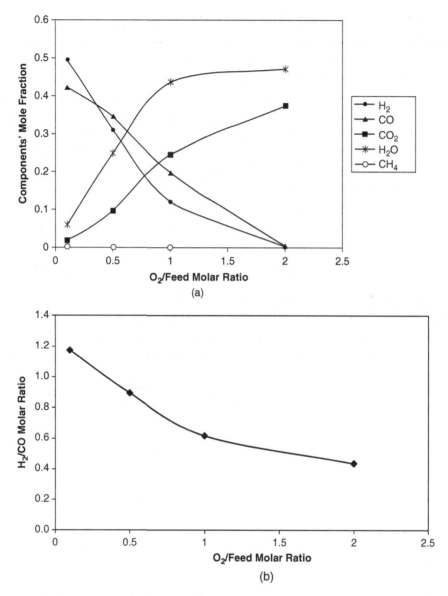

Figure 14.8 *(a) Effect of the oxygen-to-feed molar ratio to gasification on the syngas composition. (b) Effect of the oxygen-to-feed molar ratio to gasification on the H_2/CO molar ratio in the syngas. Pressure of the gasifier is set at 30 bar and temperature at 1300 °C. (Reproduced with permission from Ng and Sadhukhan (2011)[7]. Copyright © 2011, Elsevier.)*

which is to run simulation cases for various temperatures of WGS and split ratios of CO2SEP. The objective is to note the values of R due to the variations in these two input variables. This is shown in Table 14.2.

The syngas generated from the gasifier has an H_2/CO molar ratio of 0.95 and R of 0.41. This ratio needs to be increased to at least 2 for better performance of the methanol synthesis reactor. It can be seen in Table 14.2 that a lower temperature enhances the H_2/CO molar ratio. Pressure does not have any effect on the WGS reaction due to the equimolar stoichiometric reaction. WGS changes the H_2/CO molar ratio but it does not have any effect on the stoichiometric number

Table 14.2 *Sensitivity analysis of operating conditions of WGS and CO2SEP on stoichiometric ratio of reactants in the methanol synthesis reactor. (Reproduced with permission from Ng and Sadhukhan (2011)[5]. Copyright © 2011, Elsevier.)*

	WGS Outlet						CO2SEP Outlet					
	Component Molar Flow Rate (kmol s^{-1})									R		
Temperature of WGS (°C)	Component						Split Ratio of CO2SEP					
	CO	CO$_2$	H$_2$	H$_2$O	R	H$_2$/CO	99%	90%	**85%**	80%	75%	70%
200	0.24	1.13	1.70	0.03	0.41	7.06	6.70	4.49	3.73	3.16	2.71	2.35
250	0.28	1.09	1.66	0.07	0.41	5.95	5.69	4.00	3.39	2.91	2.52	2.21
300	0.33	1.04	1.61	0.12	0.41	4.87	4.69	3.47	3.00	2.62	2.30	2.03
350	0.39	0.98	1.55	0.18	0.41	3.97	3.85	2.99	2.63	2.33	2.07	1.86
400	0.45	0.92	1.48	0.25	0.41	3.28	3.20	2.58	2.30	2.07	1.87	1.69
450	**0.51**	**0.85**	**1.42**	**0.31**	**0.41**	**2.77**	2.71	2.24	**2.03**	1.85	1.69	1.54
500	0.57	0.79	1.36	0.37	0.41	2.38	2.33	1.98	1.81	1.66	1.53	1.41

R. The WGS operating temperature at 450 °C and 85% CO$_2$ removal ratio are able to satisfy the requirement of the methanol synthesis reaction ($R \sim 2$).

Water in the feed stream to the methanol synthesis reactor is undesirable due to its dilution effect and also the water gas shift reaction is hindered, resulting in CO$_2$ instead of methanol. R does not account for the molar ratio of water in the feed. Figure 14.9 shows that the removal of water from 0% to 90% on a molar basis results in an increase in carbon efficiency, defined below, from 38.7% to 72.5%, while R remains unchanged at 2.1. Thus, a higher degree of water removal is desired alongside a higher carbon efficiency. The degree of water removal is controlled by the flash temperature in the separator.

$$\text{Carbon efficiency (\%)} = \frac{\text{Moles of methanol produced}}{\text{Moles of carbon oxides (CO + CO}_2\text{) in feed}} \times 100\% \qquad (14.11)$$

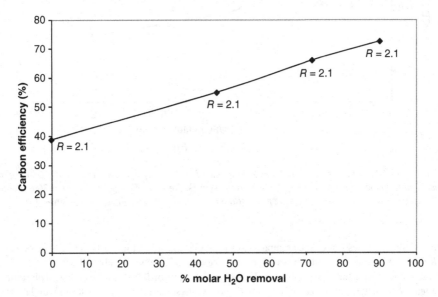

Figure 14.9 *Effect of H$_2$O removal on carbon efficiency. (Reproduced with permission from Ng and Sadhukhan (2011)[5]. Copyright © 2011, Elsevier.)*

14.5.2.3 Sensitivity Analysis on Methanol Synthesis Reaction

The effects of temperature and pressure on the methanol synthesis reaction are shown in Figure 14.10(a) and (b), respectively. It can be observed from these sensitivity studies that a low temperature and high pressure favor higher methanol production, as shown by Le Chatelier's principle. The temperature and pressure of the methanol synthesis reactor not only dictates the methanol yield but also the exothermic heat of reaction available for steam generation. Within the given operating range, the lowest exothermic heat is obtained at 50 bar and 300 °C, whilst the highest exothermic heat is obtained at 150 bar and 210 °C. Figure 14.10(c) shows a comparison between three scenarios of different degrees of exothermic heat released from the methanol synthesis reactor, which operates at different conditions: Case A (50 bar, 300 °C),

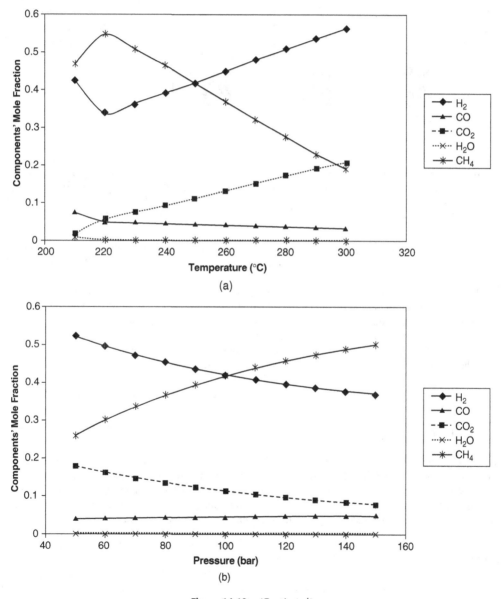

Figure 14.10 (Continued)

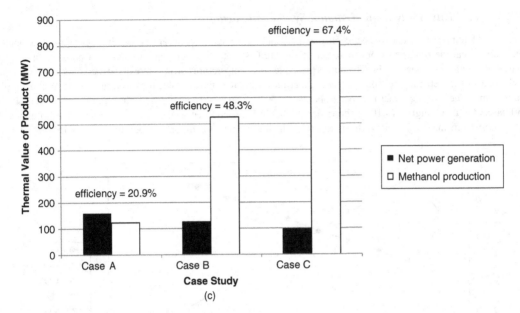

Figure 14.10 *(a) Methanol synthesis product components' mole fractions with temperature (pressure = 100 bar). (b) Methanol synthesis product components' mole fractions with pressure (temperature = 250 °C). (c) Performance of methanol synthesis reactor at different operating conditions: Case A (50 bar, 300 °C), Case B (100 bar, 250 °C) and Case C (150 bar, 210 °C). (Reproduced with permission from Ng and Sadhukhan (2011)[5]. Copyright © 2011, Elsevier.)*

Case B (100 bar, 250 °C) and Case C (150 bar, 210 °C). It is indicated that Cases B and C attain reasonably high efficiency (based on net power generation and methanol production), i.e. 48.3% and 67.4%, respectively, compared to Case A (20.9%).

Although low temperature and high pressure conditions should be fulfilled, it should be noted that a very low temperature lowers the rate of reaction, while a very high pressure necessitates a high compression energy cost. In addition, the allowable temperature for the catalyst should also be taken into consideration.

Using the same analogy as in the methanol process simulation, develop a simulation model of an integrated gasification and Fischer–Tropsch (FT) synthesis system. The current case considers a bio-oil input of 1350 MW.

Refer to the ***Online Resource material in the Companion Website: Chapter 14*** for the simulation model for an integrated gasification and Fischer–Tropsch (FT) synthesis system.

Also see the ***Online Resource material in the Companion Website: Chapter 14*** for further Exercise problems.

14.6   Summary

The undesirable characteristics of bio-oil can be improved through physical, chemical and biological upgrading processes. Bio-oil can be produced in distributed pyrolysis sites and then transported to centralized gasification facilities for synthesizing various products. The economics of local distribution of bio-oil and import of bio-oil from other countries are discussed. The two most important applications of bio-oil in the integrated gasification facilities for the production of FT liquid and methanol are exemplified along with the modeling specifications and process characterizations.

References

1. J. Sadhukhan and K.S. Ng, Economic and European Union environmental sustainability criteria assessment of bio-oil based biofuel systems: refinery integration cases, *Ind. Eng. Chem. Res.*, **50**, 6794–6808 (2011).

2. A.V. Bridgwater, A.J. Toft, J.G. Brammer, A techno-economic comparison of power production by biomass fast pyrolysis with gasification and combustion, *Renew. Sust. Energ. Rev.*, **6**, 181–246 (2002).
3. J.G. Rogers and J.G. Brammer, Analysis of transport costs for energy crops for use in biomass pyrolysis plant networks, *Biomass Bioenergy*, **33**, 1367–1375 (2009).
4. T. Pootakham and A. Kumar, Bio-oil transport by pipeline: a techno-economic assessment, *Bioresource Technol.*, **101**, 7137–7143 (2010).
5. K.S. Ng and J. Sadhukhan, Process integration and economic analysis of bio-oil platform for the production of methanol and combined heat and power, *Biomass Bioenergy*, **35**, 1153–1169 (2011).
6. D. Bradley, *European Market Study for BioOil (Pyrolysis Oil)*, Climate Change Solutions, Canada, 2006.
7. K.S. Ng and J. Sadhukhan, Techno-economic performance analysis of bio-oil based Fischer–Tropsch and CHP synthesis platform, *Biomass Bioenergy*, **35**, 3218–3234 (2011).
8. H. Schulz, Short history and present trends of Fischer–Tropsch synthesis, *Appl. Catal., A.*, **186**, 3–12 (1999).

15

Bio-Oil Refining II: Novel Membrane Reactors

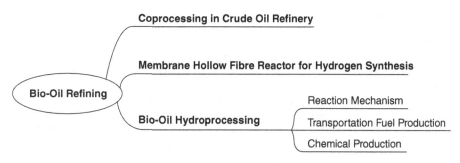

Structure for Lecture Planning

This chapter discusses conceptual designs of coprocessing of bio-oil in a crude oil refinery and refining of bio-oil in a novel membrane based multifunctional reactor. Stable bio-oil from biomass pyrolysis can be coprocessed with refinery middle distillates in a hydrocracker to produce renewable diesel, lowering the CO_2 emission (due to carbon dioxide sequestration during biomass growth) compared to crude oil-derived diesel fuel. With rapidly growing interest in the development of a flexible biomass pyrolysis process, pyrolysis oil or bio-oil can become a promising infrastructure-compatible intermediate to establish biofuels. A number of significant advantages exist in fast pyrolysis as a pretreatment step for converting biomass into liquid bio-oil. Bio-oil has ~8–12 times higher bulk energy density compared to biomass; hence the costs of transporting the liquid to the production and reception sites, handling and storing are lower. Furthermore, char and tar contents in bio-oil are lower compared to biomass[1]. These advantages have led to the concept of integration between decentralized small-scale pyrolyzers utilizing local biomass and a centralized biofuel processing plant.

15.1 Bio-Oil Co-Processing in Crude Oil Refinery

Mixing of biofuels into petroleum-derived transportation fuels has become imperative for the sustainability of today's refineries. Under the UK low carbon scenario, CO_2 emissions are constrained so that they fall 36% below 1990 levels by

Biorefineries and Chemical Processes: Design, Integration and Sustainability Analysis, First Edition.
Jhuma Sadhukhan, Kok Siew Ng and Elias Martinez Hernandez.
© 2014 John Wiley & Sons, Ltd. Published 2014 by John Wiley & Sons, Ltd.
Companion Website: http://www.wiley.com/go/sadhukhan/biorefineries

2025 and 80% by 2050. Given the huge challenge to increase the biofuel mix from currently exploited 5–10% to the target mix within a relatively short time, the most viable way forward is to produce infrastructure-compatible intermediates, such as bio-oil from fast pyrolysis of lignocellulose and wastes, and mix bio-oil with refinery intermediates to produce diesel and gasoline. Alternative to coprocessing of bio-oil in a refinery, biorefineries can be installed to fractionate biomass extensively into renewable biofuels. The latter would take longer to happen because several by-products from biorefineries are yet to find market opportunities.

Bio-oil serves as an intermediate platform for the conversion of biomass into chemicals and fuels. Temperature, pressure and space velocities are the key variables of biomass pyrolysis processes that decide the constituents of the liquid produced. The main constituents of bio-oil are oxygenated organic compounds, such as ketones, alcohols, heterocycles and organic acids. Upon mild to severe hydrodeoxygenation, bio-oil can be converted into transportation fuels, gasoline and diesel and, under specific conditions, chemicals, such as benzene, toluene, xylene, phenol, cyclohexane and various other cycloalkanes, commonly known as naphthenes. The biofuels can be blended to petroleum-derived transportation fuels, while the chemicals, commonly found in petroleum fractions, can be used as an additive to various fractions or produced and sold alongside primary petrochemicals.

Currently, it is not cost-effective to produce fuels and chemicals on a small scale from locally available biomass, while, at the same time, refiners need to mix biofuels into their finished products. New technologies thus can integrate the two infrastructures while taking account of the aims and motives of both infrastructures. Thus, it may be a capital and operating cost-saving way for both refineries and bioindustries to integrate via intermediates such as bio-oil[2]. Bio-oil is characterized by a high oxygen content, to an extent that it is beyond the capacity of hydrogen production and hydrocracking technologies available for today's refineries. Highly energy intensive multiple reaction and separation steps are needed to remove oxygen from bio-oil by reacting with hydrogen. New process technologies must therefore involve energy integrated reaction and separation steps, preferably in a single physical unit operation, to improve productivity and capital and energy savings.

The challenge lies in the efficiency of the hydrogen supply for stabilizing bio-oil by reducing its oxygen content from 35–40 wt% to 2 wt%. The hydrogen requirement is 5% of the weight of bio-oil. This calls for almost an equal amount of bio-oil to fulfill the need for hydrogen, lowering the efficiency to 50%. Some of the lost efficiencies in a bio-oil upgrader can be recovered by infrastructural integration and a share of facilities in a gasification route, for the generation of combined heat and power (CHP) and hydrogen, discussed later. Alternatively, if stable bio-oil and the off-gas (fuel gas) from bio-oil upgrader are transferred to an existing refinery and in return the refinery provides lower-cost hydrogen for the upgrader, the capital investment for the upgrader is reduced by 38% and the production cost of biofuel by 15%[2].

Bio-oil needs to first go through the stabilization process for the reduction of viscosity, oxygen content and molecular weight. The oil produced from the stabilizer is called stable oil, stable bio-oil or simply bio-oil. Stable bio-oil has been used in the text. Refineries can then hydroprocess a blend of such bio-oils from various locations into intermediate or finished products. In this way, small- to medium-scale hydroprocessing technologies can be cost-effectively applied to integrated biorefinery and refinery infrastructures, respectively (Figure 15.1).

Two new process overviews, shown in Figures 15.2 and 15.3, have stabilization and hydrocracking with multiple reaction and separation steps within one physical unit operation, producing heat and pure hydrogen. Both options are

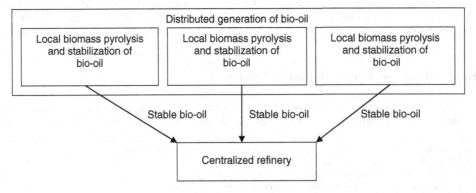

Figure 15.1   Integration between distributed generation of bio-oil and centralized refinery.

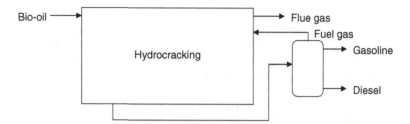

Figure 15.2   *Bio-oil hydrocracking scheme 1.*

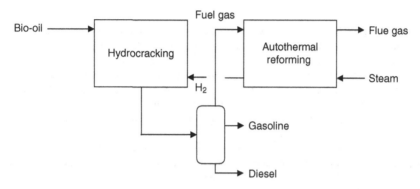

Figure 15.3   *Bio-oil hydrocracking scheme 2.*

aimed at energy and capital savings compared to the conventional routes with reforming and pressure swing adsorption for hydrogen production. Options are considered to be suitable for a coprocessing refinery producing straight-run middle distillates. Overall integrated conceptual flowsheets can be deduced from rigorous kinetic modeling and heat integration, for example, using Aspen Plus simulation. Further application of these intensified hydrogen production technologies to use biogas, derived from anaerobic digestion of biomass waste materials, for the production of transportation fuels, via Fischer–Tropsch, and CHP can also be investigated. See ***Case Study 3 in the companion website*** for biogas use in CHP generation.

15.2   Mixed Ionic Electronic Conducting (MIEC) Membrane for Hydrogen Production and Bio-Oil Hydrotreating and Hydrocracking

Schemes 1 and 2 in Figures 15.2 and 15.3 may exploit the MIEC membrane applying the electrochemical reaction principle illustrated in Figure 15.4. Bio-oil generating from biomass pyrolysis unit is rich in water content ~20 wt% of bio-oil and can be kept at pyrolysis temperature of ~500°C or even higher. This is a good source of steam supply at a high temperature. The MIEC membrane is used for oxygen production selectively from the steam in bio-oil at high temperature of ~650°C. Using electrochemical and green chemistry principles, the membrane can process any hydrocarbon into pure hydrogen production.

Did you know?

Walther Hermann Nernst researched the thermodynamic behavior of matter when approaching absolute zero temperature. The most notable contribution was the calculation of thermochemical affinity of matter that eventually reaffirms that the entropy of a pure substance approaches zero as the absolute temperature approaches zero (third law of thermodynamics). During his many novel investigations, one important observation was made in the ionic conduction of zirconia–yttria solutions in 1899, which later became relevant to oxygen separation. The solution became relevant for conduction of oxygen ions and electrons from air. Nernst's invention had practical uses in the industry. He invented an electric lamp using an incandescent ceramic rod. Nernst won the 1920 Nobel Prize in Chemistry for his work on thermochemistry. However, mixed ionic electronic conducting matter investigations for oxygen separation remained dormant until the last thirty years. A large number of MIEC compounds have been synthesized and characterized since the investigation by Takahashi and coworkers, mainly based on perovskites ($ABO_{3-esi\delta}$ and $A_2BO_{4\pm\delta}$) and fluorites ($A_\delta B_{1-\delta}O_{2-\delta}$ and $A_{2\delta}B_{2-2\delta}O_3$) or dual phases by the introduction of metal or ceramic elements[3,4]. While the research concentrated on improvements of oxygen ionic conduction, recent efforts are motivated to introduce electronic conductivity into materials.

A hollow fiber membrane reactor can be used, with a reactor configuration that offers high surface area per unit volume and is ideally suited for scaled-up applications. The MIEC membrane used is the $La_{0.6}Sr_{0.4}Co_{0.2}Fe_{0.8}O_3$ perovskite (LSCF). The innovative reactor involves the combination, for the first time, of several advanced, but individually proven, membrane technologies with the novel ceramic asymmetric hollow fiber membranes to enable hydrotreating and hydrocracking reactions. Novel methods of upgrading bio-oil into transport-grade fuel are being developed. These are potentially more cost-effective than the hydrogenation routes conventionally used for upgrading bio-oil. Conventionally, steam reforming is used to supply hydrogen. The MIEC reactor provides an electrochemical reaction platform, more environmentally friendly compared to thermochemical reaction routes.

The following basis can be considered for designing an integrated MIEC reactor for bio-oil hydrotreating/hydrocracking (endothermic heat of reactions) and partial oxidation (exothermic heat of reactions). The design goal for the integrated reactor is to reduce the external heat and raw material needs to zero. The heat sinks are the membrane itself and hydrotreating and hydrocracking reactions (HT and HC), while the heat source is the catalytic partial oxidation reaction (POX). The endothermic (HT and HC) and exothermic (POX) reactions can be carried out in the external shell and inside tube, respectively, in a hollow fiber membrane reactor configuration, such that mass and energy input and output flows are in balance and no other external steam or heat is required. The proposed integrated reactor configuration is shown in Figure 15.5.

On the basis of bio-oil throughput of 1000 kg or 1 t (per day) through the shell side of the MIEC membrane reactor, the following flow rates of other reactants are obtained.

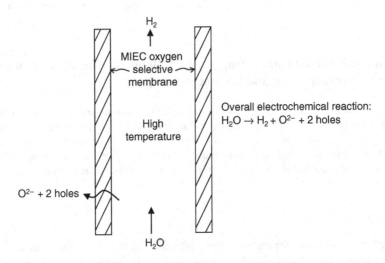

Figure 15.4   *MIEC membrane for oxygen production from steam.*

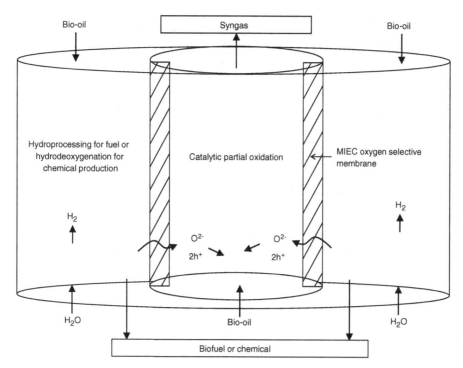

Figure 15.5   *Integrated reactor configuration for biofuel or chemical production from the shell side and hydrogen rich syngas production from the tube side, separated by the MIEC membrane, where h+ represents holes present in the oxygen anion.*

Shell side balance:

Bio-oil throughput through the shell side = 1 t

Steam separated from the bio-oil in the shell side = 212 kg

Hydrogen sourced from the bio-oil in the shell side = $212 \times \frac{2}{18} = 23$ kg

Balance of hydrogen required to stabilize (HT and HC or chemical production) the bio-oil in the shell side = $50 - 23 = 26$ kg

25 kmol of hydrogen per tonne of bio-oil are required to keep the desired partial pressure of hydrogen for hydrodeoxygenation and hydroprocessing reactions

Balance of steam required to stabilize (HT and HC or chemical production) the bio-oil in the shell side = $26 \times \frac{18}{2} = 238$ kg

Water/steam decanted after bio-oil stabilization (HT and HC or chemical production) in the shell side = 450 kg

Steam returned to bio-oil stabilizer in the shell side = 238 kg

Net steam produced from the shell side = 212 kg

Stable bio-oil/biofuel/chemical produced from the shell side = $1000 - 50 \times \frac{18}{2} = 388$ kg

Hence, the shell side is the net producer of steam (212 kg), stable bio-oil/biofuel/chemical (388 kg) and oxygen (400 kg) from 1000 kg bio-oil (and no other input).

Bio-oil (1000 kg) → Stable bio-oil or Biofuel or Chemical (388 kg) + Oxygen (400 kg) + Steam (212 kg)

Tube side balance:

Oxygen supplied from the shell to tube side through the MIEC membrane = 400 kg

The fuel on the tube side can be any hydrocarbon, such as natural gas, biomass, biogas and bio-oil. The tube side energy and mass balance based on bio-oil fuel with the following composition is shown in Table 15.1.

Table 15.1 *Bio-oil composition.*

Chemical Constituent	% Weight of Bio-oil	Chemical Formula
Phenol	33.0	C_6H_6O
Dextrose	27.0	$C_6H_{12}O_6$
Furfural	7.7	$C_5H_4O_2$
Acetic acid	5.6	$C_2H_4O_2$
Hydroxyacetone	4.0	$C_3H_6O_2$
Formic acid	1.5	CH_2O_2
Water	21.2	H_2O

Bio-oil throughput to the tube side can be adjusted to partially oxidize all carbon present into carbon monoxide (with no oxygen and residual carbon left) and release all hydrogen as gas present in the bio-oil. The exit syngas will consist of carbon monoxide and hydrogen.

Table 15.2 shows the carbon, hydrogen and oxygen contents of bio-oil on the basis of 100 kg (see Table 15.1).

Bio-oil throughput through the tube side calculated for no oxygen and residual carbon left and on the basis of 1 t bio-oil throughput through the shell side = 2977 kg.

Syngas constituent:

Carbon monoxide = 113 kmol or 3158 kg
Hydrogen = 109 kmol or 219 kg
Total syngas generated from the tube side = 3377 kg

A summary of the approximate mass and energy balance around the MIEC hollow fiber membrane reactor is shown in Figure 15.6.

15.3 Bio-Oil Hydrotreating and Hydrocracking Reaction Mechanisms and a MIEC Membrane Reactor Based Bio-Oil Upgrader Process Flowsheet

The reactions in the shell side of the MIEC membrane reactor include cracking to olefins and hydrogenation to high octane isoparaffins (mainly in gasoline), ring separation and opening into smaller aromatic compounds and cycloparaffin (mainly in diesel) and side chain hydrocracking and isomerization, which can be regulated to have better control over biofuel product and process performance. Hydrodeoxygenation refers to oxygen removal by hydrogen producing water

Table 15.2 *Carbon, hydrogen and oxygen amounts in bio-oil on 100 kg bio-oil basis.*

Weight, kg	C	H	O	Total
C_6H_6O	25.32	2.11	5.63	33.06
$C_6H_{12}O_6$	10.77	1.80	14.37	26.94
$C_5H_4O_2$	4.79	0.32	2.56	7.67
$C_2H_4O_2$	2.24	0.37	2.98	5.59
$C_3H_6O_2$	1.94	0.32	1.72	3.99
CH_2O_2	0.40	0.07	1.07	1.54
H_2O		2.36	18.85	21.21
Total (kg)	45.47	7.35	47.19	100.00
Total (kmol)	3.79	3.67	1.47	8.93

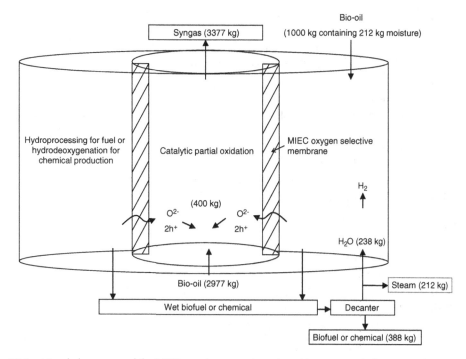

Figure 15.6   *Mass balance around the MIEC membrane reactor, where h$^+$ represents holes present in oxygen anion.*

and decarboxylation indicates carbon dioxide and/or carbon monoxide removal for the generation of olefins, isoparaffins, smaller aromatics and cycloparaffins, etc. These reactions are shown as follows:

Reactions of phenol:

Rxn 1: $C_6H_6O + 4H_2 \rightarrow 0.46C_{13}H_{26} + H_2O$
(Hydrodeoxygenation (HDO) to 1-tridecene)

Rxn 2: $C_6H_6O \rightarrow 0.33C_{18}H_{12} + H_2O$
(Dehydroxygenation (DHO) to chrysene)

Rxn 3: $C_6H_6O + 2.29H_2 \rightarrow 0.43C_{14}H_{20} + H_2O$
(HDO to diamantane)

Rxn 4: $C_6H_6O + 0.5H_2 \rightarrow 0.5C_{12}H_{10} + H_2O$
(HDO to diphenyl)

Rxn 5: $C_6H_6O + 1.33H_2 \rightarrow 0.33C_4H_{10} + 0.33C_{14}H_{10} + H_2O$
(HDO to *i*-butane and phenanthrene)

Rxn 6: $C_6H_6O + 2.5H_2 \rightarrow 0.5C_{12}H_{18} + H_2O$
(HDO to 1,2,4-triethylbenzene)

Rxn 7: $C_6H_6O + 2.83H_2 \rightarrow 0.167C_{13}H_{26} + 0.167C_{14}H_{10} + 0.167C_7H_{16} + 0.167C_2H_6 + H_2O$
(HDO to 1-tridecene, phenanthrene, *n*-heptane and ethane)

Rxn 8: $C_6H_6O + 4H_2 \rightarrow 0.67C_9H_{18} + H_2O$
(HDO to 1-*trans*-3,5-trimethylcyclohexane)

Rxn 9: $C_6H_6O + 4H_2 \rightarrow 0.67C_9H_{18} + 6H_2O$
(HDO to *n*-propylcyclohexane)

Rxn 10: $C_6H_6O + 1.75H_2 \rightarrow 0.75C_8H_{10} + H_2O$
           (HDO to *p*-xylene)

Rxn 11: $C_6H_6O + 3.33H_2 \rightarrow 0.33C_{13}H_{26} + 0.33C_4H_{10} + 0.33CO_2 + 0.33H_2O$
           (HDO and decarboxylation (DCO) to 1-tridecene and *i*-butane)

Rxn 12: $C_6H_6O + 3.5H_2 \rightarrow 0.5C_{12}H_{22} + H_2O$
           (HDO to bicyclohexyl)

Rxn 13: $C_6H_6O + H_2 \rightarrow 0.75C_8H_{10}O + 0.25H_2O$
           (HDO to 2,5-xylenol)

Rxn 14: $C_6H_{12}O_6 + 2H_2 \rightarrow 0.67C_9H_{12} + H_2O$
           (HDO to 1,2,3-trimethylbenzene)

Rxn 15: $C_6H_6O + 5.5H_2 \rightarrow 0.5C_7H_{16} + 0.5C_4H_{10} + 0.5CH_4 + H_2O$
           (HDO to *n*-heptane, *i*-butane and methane)

Reactions of dextrose:

Rxn 16: $C_6H_{12}O_6 + 3H_2 \rightarrow 0.75C_8H_{10}O + 5.25H_2O$
           (HDO to 2,5-xylenol)

Rxn 17: $C_6H_{12}O_6 + 0.67H_2 \rightarrow 0.33C_{14}H_{20} + 1.33CO_2 + 3.33H_2O$
           (HDO and DCO to diamantane)

Rxn 18: $C_6H_{12}O_6 + 0.33H_2 \rightarrow O.33CH_4 + CO_2 + 4H_2O$
           (HDO and DCO and methane production)

Rxn 19: $C_6H_{12}O_6 + 6H_2 \rightarrow 0.67C_9H_{18} + 6H_2O$
           (HDO to 1-*trans*-3,5-trimethylcyclohexane)

Rxn 20: $C_6H_{12}O_6 + 6H_2 \rightarrow 0.67C_9H_{18} + 6H_2O$
           (HDO to *n*-propylcyclohexane)

Rxn 21: $C_6H_{12}O_6 + 2H_2 \rightarrow 0.5C_9H_{12} + 0.5CH_4 + CO_2 + 4H_2O$
           (HDO and DCO to 1,2,3-trimethylbenzene)

Rxn 22: $C_6H_{12}O_6 + 4.5H_2 \rightarrow 0.5C_{10}H_{18} + 0.5CH_4 + 0.5CO_2 + 5H_2O$
           (HDO and DCO to *cis*-decalin)

Rxn 23: $C_6H_{12}O_6 + 2.33H_2 \rightarrow 0.33C_{13}H_{26} + 0.33CH_4 + 1.33\ CO_2 + 3.33H_2O$
           (HDO and DCO to 1-tridecene)

Rxn 24: $C_6H_{12}O_6 + 5H_2 \rightarrow 0.5C_{10}H_2O + 0.5CH_4 + 0.5CO_2 + 5H_2O$
           (HDO and DCO to *n*-butylcyclohexane)

Rxn 25: $C_6H_{12}O_6 + 5.5H_2 \rightarrow 0.5C_{12}H_{22} + 6H_2O$
           (HDO to bicyclohexyl)

Reactions of furfural:

Rxn 26: $C_5H_4O_2 + 4.5H_2 \rightarrow 0.5C_{10}H_{18} + 2H_2O$
           (HDO to *cis*-decalin)

Rxn 27: $C_5H_4O_2 + 5.5H_2 \rightarrow 0.5C_{10}H_{22} + 2H_2O$
           (HDO to 3,3,5-trimethylheptane)

Rxn 28: $C_5H_4O_2 + 3.5H_2 \rightarrow 0.5C_{10}H_{14} + 2H_2O$
           (HDO to 1,2-dimethyl-3-ethylbenzene)

Rxn 29: $C_5H_4O_2 + 3.5H_2 \rightarrow 5CH_4 + 2H_2O$
           (HDO to methane)

Rxn 30: $C_5H_4O_2 + 4H_2 \rightarrow C_3H_8 + CH_4 + CO_2$
            (DCO to Propane and Methane)

Rxn 31: $C_5H_4O_2 + 4H_2 \rightarrow 0.5C_{10}H_{20} + 2H_2O$
            (HDO to *n*-butylcyclohexane)

Reactions of acetic acid:

Rxn 32: $C_2H_4O_2 + 1.5H_2 \rightarrow 0.5C_2H_6 + 0.5CH_4 + 0.5CO_2 + H_2O$
            (HDO and DCO to ethane and methane)

Rxn 33: $C_2H_4O_2 + 2.22H_2 \rightarrow 0.11C_7H_{16} + 0.11C_4H_{10} + 0.11C_3H_8 + 0.11C_2H_6 + 0.11CH_4 + 0.11CO_2 + 1.78H_2O$
            (HDO and DCO to alkanes and isoparaffins)

Reactions of hydroxyacetone:

Rxn 34: $C_3H_6O_2 + 2H_2 \rightarrow 0.25C_7H_{16} + 0.25C_3H_8 + 0.25CH_4 + 0.25CO_2 + 1.5H_2O$.
            (HDO and DCO to alkanes)

Rxn 35: $C_3H_6O_2 + 5H_2 \rightarrow 3CH_4 + 2H_2O$
            (HDO to alkane)

Reactions of formic acid:

Rxn 36: $CH_2O_2 + 1.75H_2 \rightarrow 0.25C_2H_6 + 0.25CH_4 + 0.25CO_2 + 1.5H_2O$
            (HDO and DCO to alkanes)

Rxn 37: $CH_2O_2 + H_2 \rightarrow 0.5CH_4 + 0.5CO_2 + H_2O$
            (HDO and DCO to alkane)

Reactions of intermediate reactants (if severe hydrocracking of the diesel fraction takes place followed by diesel separation after the MIEC membrane reaction):

Rxn 38: $C_{14}H_{10} + 9H_2 \rightarrow 2C_7H_{14}$
            (Hydrogenation of phenanthrene to cyclohexane: methylcyclohexane)

Rxn 39: $C_{18}H_{12} + 12H_2 \rightarrow 2C_9H_{18}$
            (Hydrogenation of chrysene to cyclohexane: 1-*trans*-3,5-trimethylcyclohexane)

Rxn 40: $2C_7H_{14} \rightarrow C_{10}H_{12} + C_4H_{10} + 3H_2$
            (Methylcyclohexane to 1,2,3,4-tetrahydronaphthalene and *n*-butane)

The process flowsheet for the synthesis of biofuels from bio-oil using the MIEC hollow fiber membrane reactor is shown in Figure 15.7. After the MIEC membrane reactor's shell side reactions Rxn 1–37 followed by the decanting of the water are undertaken, the stable bio-oil can be taken to a gas separator, a debutanizer, a naphtha splitter and a diesel separator, respectively, all of which can be integrated within one reboiler distillation column with pressure varying between 1.1 and 3.5 bar. The process will generate gas that can be used for CHP generation, gasoline and diesel for further coprocessing with crude oil refinery fractions in hydrocracker or for blending with refinery gasoline and diesel, respectively.

Figure 15.8 shows the bio-oil upgrader flow scheme with further hydrocracking of the diesel fraction from the bottom of the diesel separator of the reboiler distillation column. The two new units added are the pressure swing adsorption (PSA) or membrane process for hydrogen separation and the hydrocracking reactor using this hydrogen. The naphtha fraction from the naphtha splitter and the diesel fraction from the diesel separator of the reboiler distillation column and the hydrocracked diesel are blended together to generate bio-oil-derived diesel from the upgrader. The off-gas from the PSA or membrane process and the syngas from the MIEC membrane reactor tube side can be used for CHP generation in excess from the upgrader. The gasoline fraction from the debutanizer forms the gasoline product from the upgrader.

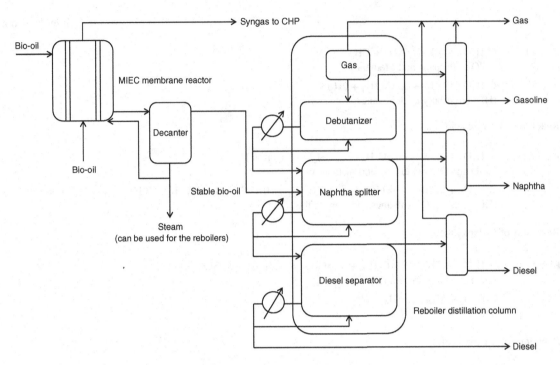

Figure 15.7 *Integrated MIEC membrane reactor and separator flowsheet for bio-oil processing into gasoline and diesel.*

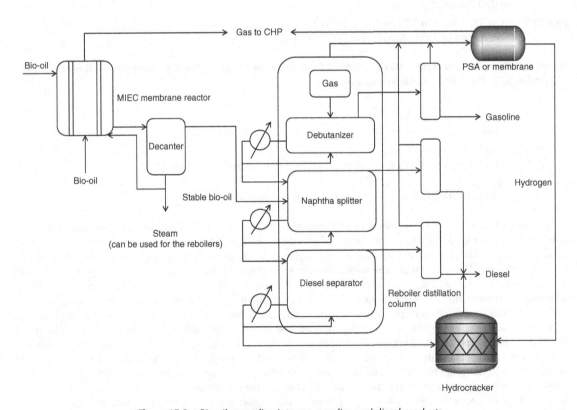

Figure 15.8 *Bio-oil upgrading into gas, gasoline and diesel products.*

Table 15.3 *% Molar conversions of reference chemicals in the reactions Rxn 1–40 to obtain the diesel and gasoline compositions and properties shown in Table 15.4.*

Rxn no.	Reference Chemical	% Molar Conversion
1	Phenol	15.7
2	Phenol	15.3
3	Phenol	13.6
4	Phenol	10.7
5	Phenol	10.4
6	Phenol	7.4
7	Phenol	7.2
8	Phenol	6.9
9	Phenol	6.9
10	Phenol	4.1
11	Phenol	0.5
12	Phenol	0.4
13	Phenol	0.4
14	Phenol	0.2
15	Phenol	0.2
16	Dextrose	37.6
17	Dextrose	19.6
18	Dextrose	12.4
19	Dextrose	9.4
20	Dextrose	9.4
21	Dextrose	4.1
22	Dextrose	3.6
23	Dextrose	2.0
24	Dextrose	1.2
25	Dextrose	0.8
26	Furfural	31.3
27	Furfural	23.2
28	Furfural	20.4
29	Furfural	13.9
30	Furfural	10.8
31	Furfural	0.5
32	Acetic acid	61.6
33	Acetic acid	38.4
34	Hydroxyacetone	96.6
35	Hydroxyacetone	3.4
36	Formic acid	95.4
37	Formic acid	4.6
38	Phenanthrene	1 (in the case of further hydrocracking in Figure 15.8)
39	Chrysene	1 (in the case of further hydrocracking in Figure 15.8)
40	Methylcyclohexane	0.1 (in the case of further hydrocracking in Figure 15.8)

Table 15.3 shows the % molar extent conversions of reference chemicals in the reactions Rxn 1–40 obtained from linear regression for gasoline and diesel production with desired compositions, shown in Table 15.4. The compositions and properties of the products, gasoline and diesel thus derived from the new upgrader process flow scheme developed in Figure 15.8, are shown in Table 15.4. Table 15.4 shows that the properties of the diesel fuel obtained from the shell side bio-oil reactions closely match within those of a typical middle distillate to heavy feedstock to a hydrocracker in a crude oil refinery. This shows the feasibility of coprocessing of bio-oil-derived diesel and refinery fractions through a hydrocracker into a final diesel fuel with desired properties.

Table 15.4 *Gasoline and diesel mass compositions in % and physical properties (relevant for crude oil refinery) from the MIEC membrane reactor's shell side bio-oil reactions.*

Chemical Components	Gasoline	Diesel	
n-Heptane	47.8	0.5	
Isobutane	2.4	0.0	
2,5-Xylenol	2.0	12.8	
1-*trans*-3,5-Trimethylcyclohexane	17.0	5.8	
3,3,5-trimethylheptane	4.2	2.8	
n-Propylcyclohexane	9.7	6.9	
1,2,3-Trimethylbenzene	0.7	1.0	
n-butylcyclohexane	0.2	0.4	
1,2-Dimethyl-3-ethylbenzene	0.8	2.7	
cis-Decalin	1.4	5.1	
1-Tridecene	1.2	14.4	
1,2,4-Triethylbenzene	0.7	5.3	
Bicyclohexyl	0.0	0.5	
Diphenyl	0.3	7.3	
Diamantane	0.9	14.3	
Phenanthrene	0.0	10.2	
Chrysene	0.0	3.9	
p-Xylene	10.6	1.8	
1-*trans*-3,5-Trimethylcyclohexane	0.0	4.4	
Property	**Gasoline**	**Diesel**	**Refinery Hydrocracker Feed**
Flow rate (weight % of bio-oil)	2.12	36.68	
Specific gravity	0.737	0.873	0.87–0.97
API gravity	60.6	30.6	14–30
Volumetric average boiling point (°C)	115	225	200–450
Reid vapor pressure (bar)	1.1	0.1	
Flash point (°C)	−38	47	50–150
Aniline point (°C)	45	28	25–65
Cetane number (°C)	26	29	29–32

A refinery hydrocracker unit to coprocess the refinery fractions and stablize bio-oil after hydrotreating and the combustion of fuels are analyzed for an estimation of the environmental drivers, for example, global warming or greenhouse gas (GHG) impact reduction, and crude oil saving and barriers, for example, product yield loss and land use. Figure 15.9 shows the empirical correlations of avoided CO_2 emission and crude oil saving; and, as well as the diesel and gasoline yield loss and land use with respect to the percentage of stable bio-oil coprocessing in a refinery hydrocracker. These correlations shown in the plot are valid for the configuration of the bio-oil upgrader shown in Figure 15.8. The analysis shows a whole range of environmental incentives from those without biofuel coprocessing (no incentive case) to those with complete replacement by biofuel (greatest incentive). The saving in (or avoided) carbon dioxide emission and crude oil, relative to the without coprocessing case, increases with the increasing percentage of biofuel coproduction: for example, the emission saving increases from 313 kg of CO_2 per tonne of hydrocracker feed (the basis is 1 tonne of hydrocracker feed) in the 10% biofuel coproduction case to 1876 kg t^{-1} in the 60% biofuel coproduction case and a crude oil saving from 141 to 845 kg t^{-1}, respectively. Henceforth, the bio-oil upgrader technology can provide a 31.3% carbon dioxide saving for 10% biofuel blending and over 60% carbon dioxide saving for 20% biofuel blending, respectively. However, the diesel and gasoline fuel yields are reduced due to low energy density of the upgrader products. The transportation fuel yield from a refinery hydrocracker is 98.9%, while that from a bio-oil upgrader is 70.6%, on a weight basis. The projection of land use with respect to increasing the biofuel blend is also shown for 13.5 t ha^{-1} growth of a short rotation coppice. Aspen Plus modeling is recommended for analysing the process for other biomass feedstocks. Figure 15.10 shows the bio-oil upgrader for integration to the crude oil refinery and energy market. For economic feasibility analysis and data for business models, please refer to Sadhukhan and Ng (2011)[2].

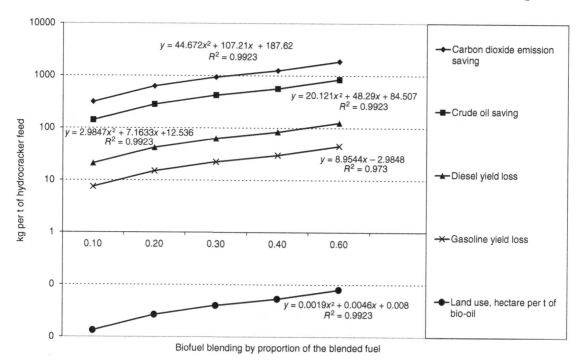

Figure 15.9 *Correlations of avoided CO$_2$ emission and crude oil saving, as well as diesel and gasoline yield loss and land use with respect to the percentage of stable bio-oil coprocessing in a refinery hydrocracker.*

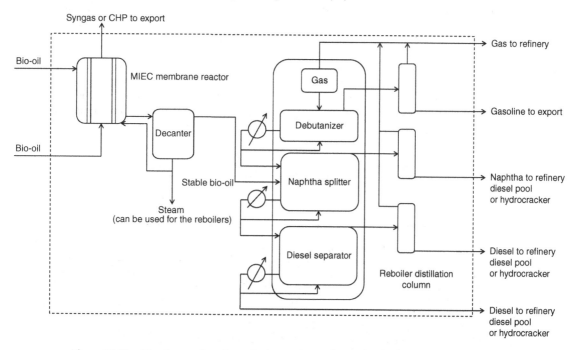

Figure 15.10 *Bio-oil upgrader scheme for integration to the crude oil refinery and energy market.*

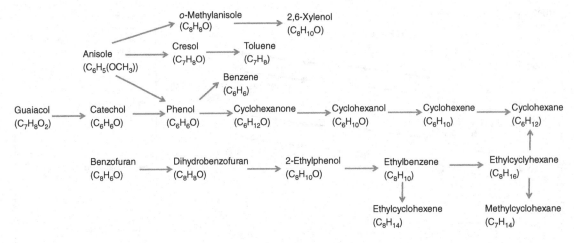

Figure 15.11   *Reaction pathways of guaiacol.*

The bio-oil can also be used in the MIEC hollow fiber membrane reactor to produce chemicals such as those in the following section, requiring further research to optimize operating conditions for each production route. Here a coursework problem has been given to help formulations for other chemical production problems utilizing bio-oil. As an example of chemical productions from bio-oil the guaiacol reaction pathways are shown in Figure 15.11.

15.4   A Coursework Problem

The world has diverse biomass feedstocks, which can be used to provide building block chemicals. The building block chemicals can then be converted into high value chemicals, the majority of which are now produced from fossil fuels. In this coursework, the aim is to investigate technical feasibility and economic potential of two such building block chemicals, benzene and cyclohexane, from biomass feedstocks.

Bio-oil derived from biomass pyrolysis and high pressure liquefaction processes is often a complex mixture of valuable chemicals, which can be converted into building block chemicals. The reaction pathways from model components of bio-oils, guaiacol and phenol (an analysis of biomass is shown in Table 15.5) to benzene and cyclohexane are shown in Table 15.6. The two reaction pathways from phenol to benzene and cyclohexane are parallel and competing, depending

Table 15.5   *Biomass composition analysis.*

Bio-Oil Model Components	Bio-Oil 1 (kg d^{-1})	Bio-Oil 2 (kg d^{-1})
Guaiacol	5986.6	5986.6
Phenol	40691.5	40691.5
Inert	0	35330

Table 15.6   *Reaction pathways from guaiacol and phenol to benzene and cyclohexane.*

Guaiacol to catechol (1,2-benzenediol)	$C_6H_4(OH)(OCH_3) + H_2 \rightarrow C_6H_4(OH)(OH) + CH_4$
Catechol to phenol	$C_6H_4(OH)(OH) + H_2 \rightarrow C_6H_5(OH) + H_2O$
Phenol to benzene	$C_6H_5(OH) + H_2 \rightarrow C_6H_6 + H_2O$
Phenols to cyclohexanone	$C_6H_5(OH) + 2H_2 \rightarrow C_6H_{10}O$
Cyclohexanone to cyclohexanol	$C_6H_{10}O + H_2 \rightarrow C_6H_{11}OH$
Cyclohexanol to cyclohexene	$C_6H_{11}(OH) \rightarrow C_6H_{10} + H_2O$
Cyclohexene to cyclohexane	$C_6H_{10} + H_2 \rightarrow C_6H_{12}$

Table 15.7 A hypothetical product gas composition and two sets of price data.

Components	Product Composition (wt% on Dry Basis)	Minimum Price ($ kg^{-1})	Maximum Price ($ kg^{-1})
Methane	1.9		
1,2-Benzenediol	Trace		
Phenol	Trace		
Benzene	56.6	900.0	1305.3
Cyclohexanone	4.9		
Cyclohexanol	3.4		
Cyclohexene	9.0		
Cyclohexane	24.1	1075.0	1559.0
Price of bio-oil 1		75	300
Price of hydrogen		700	1400

upon the amount of hydrogen reactant and temperature. The targeted final product distribution depends upon market demands, economics and process efficiency. Your aim is to conceptually design a commercial-scale process flowsheet for the production of the given products and assess the sensitivity of process and market variability on decision making.

Answer the following questions. These will help you in decision making at various design stages, from conceptual analysis to simulation of the flowsheet.

1. Basis: bio-oil 1 in Table 15.5 and a hypothetical product gas composition in wt% (on dry basis) from the process in Table 15.7.
    a. Perform the material balance across the reaction pathways shown in Table 15.5, and for the given hypothetical product gas composition in Table 15.7.
    b. Estimate the amount of hydrogen required.
    c. Calculate the conversion for each reaction step in Table 15.6.
    d. Calculate the following selectivity and yield on a mass basis:
        i. Benzene with respect to guaiacol
        ii. Benzene with respect to phenol
        iii. Cyclohexane with respect to guaiacol
        iv. Cyclohexane with respect to phenol
    e. Estimate the preliminary economic potential for the given hypothetical product gas composition and two sets of price data in Table 15.7.
    f. Discuss how economic potential can be affected by the presence of an inert substance in the bio-oil feedstock by taking an example of bio-oil 2 in Table 15.7.
2. Basis: bio-oil 1 in Table 15.5. Simulate the process flowsheet in Aspen Plus. Select two equilibrium reactors, the first one for converting bio-oil into a mixture of phenol, catechol and methane and the second one for converting the product from the first reactor into benzene and cyclohexane. The first reactor is kept at 135 bar and 250 °C, whilst the second reactor is maintained at 150 bar and 400 °C.
    a. Use a stoichiometric amount of hydrogen reactant in the first reactor. Calculate the final product composition in both wt% and mol% and amount of product in kg per day and kmol per day for two flow rates of hydrogen reactant (assume pure hydrogen): 400 kmol per day and 1200 kmol per day.
    b. Calculate the following selectivity and yield on a mass basis for the two simulation cases in 2a:
        i. Benzene with respect to guaiacol
        ii. Benzene with respect to phenol
        iii. Cyclohexane with respect to guaiacol
        iv. Cyclohexane with respect to phenol

c. Estimate the heat of reactions of the two reactors for the two simulation cases in 2a (assume reactants enter at the same temperatures as the reactors). A simulation package such as Aspen Plus may be used.

d. Design heat recovery schemes for both the cases and check the feasibility of exchangers (e.g., there should not be any temperature crossover between hot and cold streams within the exchangers). Show the net heat demand (or availability) by the site after heat recovery for the two simulation cases in 2a.

e. Estimate the economic potential based on the list of two sets of prices in Table 15.7 and a price of heat of $0.05 per kWh for the two simulation cases in 2a.

f. Calculate the ratio between the total energy cost and the raw material cost for the two simulation cases in 2a.

g. Based on the preliminary economic potential predicted in 2e, calculate the largest allowable capital and other fixed process costs.

h. Discuss how economic potential (obtained from 2e) can be affected by the presence of an inert substance in the bio-oil feedstock by taking an example of bio-oil 2 in Table 15.5.

3. Basis: bio-oil 1 in Table 15.5. Examine the sensitivity of temperature (250 °C to 700 °C in the second reactor) on the product shift. Use adequate hydrogen in the second reactor.

a. Calculate the final product composition in both wt% and mol% and total amount of product in kg and kmol per day for at least two operating temperatures in the second reactor and show the switching between products as a function of the reactor operating temperature.

b. Comment on the phase in the two reactors for the two simulation cases in 3a. Select types of reactors in each case and justify your answer.

c. Estimate the economic potential based on the list of two sets of prices in Table 15.7 and a price of heat of $0.05 per kWh for the two simulation cases in 3a.

d. Calculate the ratio between the total energy cost and the raw material cost for the two simulation cases in 3a.

e. Based on the preliminary economic potential predicted in 3c, calculate the largest allowable capital and other fixed process costs.

f. An alternative to purchasing hydrogen from the market is to produce hydrogen from steam using an oxygen selective membrane (Figure 15.12). Estimate the allowable cost for this membrane for the given purchasing prices of hydrogen (Table 15.7) and the cost of electricity of $0.05 per kWh. This oxygen selective membrane needs to be operated at 650 °C. Identify the product that is energetically more favorable to produce if this membrane is integrated within the second reactor. Perform the energy balance around the second reactor for the energetically more favorable production. How much oxygen can be produced from such integration? Comment on the usage of this oxygen within the site.

4. Pervaporation is an effective technique to separate benzene (in permeate) and cyclohexane products[5]. Select two simulation cases (basis: bio-oil 1 in Table 15.5) resulting in a 40–80% mole fraction of benzene in the product gas from the second reactor. Experimental results for three types of membrane materials are available, shown in Figures 15.13 to 15.15.

a. Calculate the selectivity, compositions of permeate and retentate and total flux for both simulation cases chose above and for all three types of membrane in Figures 15.13 to 15.15.

b. For the given capacity of bio-oil 1 in Table 15.5, estimate the areas required for three types of membranes in Figures 15.13 to 15.15.

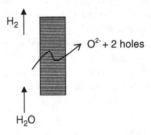

Figure 15.12   *Mechanism of dissociation of steam through an oxygen selective membrane.*

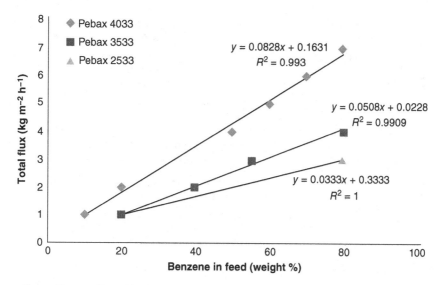

Figure 15.13 *Effect of feed concentration on flux for three types of membrane materials.*

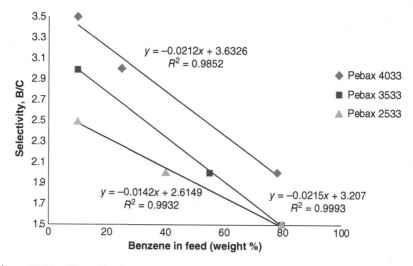

Figure 15.14 *Effect of feed concentration on selectivity for three types of membrane materials.*

c. Show the trade-offs between the heat demand by the reactors and membrane area required for the separation. Discuss your results.

d. Discuss the basis of selection of a membrane type from the given three membranes. How may this selection be affected if bio-oil 2 from Table 15.5 is processed?

15.5 Summary

A novel MIEC hollow fiber membrane reactor design for bio-oil conversion into transportation fuels and chemicals is discussed to incentivize efficient process innovations for biorefineries. This highly intensified reactor integrates

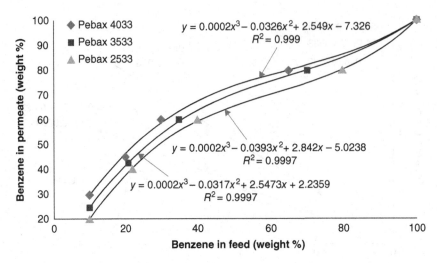

Figure 15.15 *Feed mixture–permeate equilibrium diagram for three types of membrane materials.*

hydrogen production, hydrotreating and hydrocracking reactions, oxygen separation and oxidation within one physical unit operation, where all materials and energy are sourced from bio-oil. Integrated bio-oil and crude oil refining reduces carbon dioxide emission and crude oil depletion. The design methodology can be applied to commercialize MIEC hollow fiber membrane reactor processes for biorefining. Coursework problems have been provided to help in designing highly intensified and efficient reactors for chemical production.

References

1. A.V. Bridgwater, *Technical and economic assessment of thermal processes for biofuels. Life cycle and techno-economic assessment of the Northeast biomass to liquid projects.* NNFCC Project 08/018, COPE Ltd., 2009.
2. J. Sadhukhan and K.S. Ng, Economic and European Union environmental sustainability criteria assessment of bio-oil based biofuel systems: refinery integration cases, *Ind. Eng. Chem. Res.*, **50**(11), 6794–6808 (2011).
3. J. Sunarso, S. Baumann, J.M. Serra, W.A. Meulenberg, S. Liu, Y.S. Lin, J.C. Diniz da Costa, Mixed ionic–electronic conducting (MIEC) ceramic-based membranes for oxygen separation, *J. Membrane Sci.*, **320**, 13–41 (2008).
4. T. Takahashi, T. Esaka, H. Iwahara, Electrical conduction in the sintered oxides of the system Bi_2O_3–BaO, *J. Solid State Chem.*, **16**, 317–323 (1976).
5. A.E. Yildirim, N.D. Hilmioglu, S. Tulbentci, Separation of benzene/cyclohexane mixtures by pervaporation using PEBA membranes, *Desalination*, **219**, 14–25 (2008).

16

Fuel Cells and Other Renewables

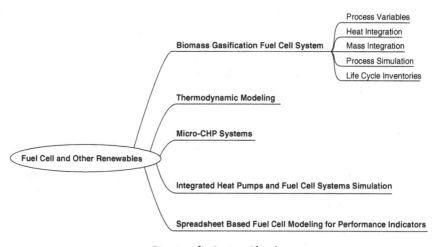

Structure for Lecture Planning

For combined heat and power (CHP) generation from biomass derived fuel gas such as syngas, biogas, etc., gas turbines and combined cycles are more commonly used as the main CHP generation technology. However, with the development of electrochemistry and fuel cells, a biomass integrated gasification fuel cell (BGFC) system can be designed to enhance the CHP generation efficiency compared to the biomass integrated gasification combined cycle (BIGCC) using gas turbines. The energy efficiency is defined as a dimensionless ratio between net total output energy (e.g., CHP) generation and total input biomass calorific value. Solid oxide fuel cells (SOFC) are ideally placed for integration with the high temperature gasification process. This chapter shows the modeling equations and parameters for design of integrated BGFC systems for micro to decentralized generations, to enhance energy efficiency and economic productions and reduce process emissions.

Fuel cells are the reverse of electrolysis processes. In fuel cells, hydrogen and oxygen are combined to produce usable electricity, heat and water. Sir William Grove constructed the first fuel cell in 1839 using platinum electrodes and sulfuric acid as the electrolyte[1]. These early devices, however, had very low current densities. It was not until the 1930s that Francis Bacon developed a fuel cell with the capability of producing a current density of 1 A cm^{-2} at 0.8 V. Bacon

Biorefineries and Chemical Processes: Design, Integration and Sustainability Analysis, First Edition.
Jhuma Sadhukhan, Kok Siew Ng and Elias Martinez Hernandez.
© 2014 John Wiley & Sons, Ltd. Published 2014 by John Wiley & Sons, Ltd.
Companion Website: http://www.wiley.com/go/sadhukhan/biorefineries

substituted the acid electrolyte of the earlier fuel cells with an alkaline electrolyte, referred to as the "Bacon cell" or now more commonly known as the alkaline fuel cell (AFC)[2]. In the 1960s, NASA chose the AFC for the power supply of the Apollo lunar missions, with the fuel cells being designed, developed and manufactured by Connecticut based International Fuel Cells (now UTC Fuel Cells). The late 1950s saw the first development of the proton exchange membrane (PEM) fuel cell by General Electric in the US for use by NASA to provide power for the Gemini space project. After GE's early work, the development of PEM fuel cells became dormant, but was reactivated by Ballard in the late 1980s, with other companies also starting their own development programs. It was during the 1960s that other electrolytes were developed and formed the basis of the high temperature types of fuel cells, which are in an advanced stage of development today. The development work during the period up to the end of the 1980s was largely conducted in government and independent laboratories, universities and a relatively small number of commercial companies. However, the 1990s saw an explosion of activity with a large number of companies presently involved in the industry.

In an electrochemical cell making use of gaseous reactants, the anodic and cathodic reactants are fed into their respective chambers and an electrolyte layer is situated between the two electrodes. The half-cell reaction at the anode yields electrons, which are transported through the external circuit to reach the cathode. These electrons are then transferred to the cathodic reactants. The circuit is completed by the transport of ions from one electrode to the other through the electrolyte. By the process of transport of electrons through the external circuit in a complete circuit, electricity is generated. Fuel cells are named after the type of electrolytes used. The electrode reactions of a hydrogen–oxygen fuel cell in an acid or alkaline electrolyte are shown in Figure 16.1.

The electrolyte in Figure 16.1 is a solid, mixed ionic oxygen conducting membrane, used in SOFC. This membrane transports the oxygen atom and electrons from the cathode to the anode. The fuel electrochemical combustion reaction occurs in the anode, while the air separation occurs in the cathode. SOFC is flexible in terms of fuel intake. Hydrogen rich fuel, for example, biomass-derived syngas can be used, as long as the gas is ultra-clean. This is required to avoid catalyst poisoning, which is usually platinum.

The balanced electrochemical reactions based on one mole of water production are as follows. One hydrogen molecule in contact with the anode relieves two electrons to flow through an external circuit for the generation of electricity. After relieving two electrons, one hydrogen molecule is converted into two hydrogen cations or protons. Two electrons and an atom of oxygen are transported through the solid electrolyte membrane from the cathode to the anode, where an oxygen anion is formed after combining with two electrons. In the anode chamber, two hydrogen cations and one oxygen anion electrochemically react to form one mole of water.

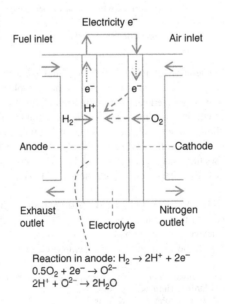

Reaction in anode: $H_2 \rightarrow 2H^+ + 2e^-$
$0.5O_2 + 2e^- \rightarrow O^{2-}$
$2H^+ + O^{2-} \rightarrow 2H_2O$

Figure 16.1   *SOFC schematic.*

16.1 Biomass Integrated Gasification Fuel Cell (BGFC) System Modeling for Design, Integration and Analysis

Most of the biomass gasifiers operating for power generation today are combined either with a gas engine or with gas turbine based combined cycles. The energy efficiency of a biomass gasification distributed system can be greatly enhanced if coupled with high efficiency power generation systems, such as the SOFC.

The SOFC has the potential to become an energy technology, due to its inherently clean and efficient operation combining green chemistry and electrochemistry. The SOFC can be used for community and district level generation of electricity and heat, for example, a few hundred kilowatts to a few megawatts of electricity.

Based on green electrochemical principles, the SOFC works on the reverse electrolysis process, oxidizing gaseous fuels such as hydrogen, syngas, etc., in the anode in the presence of an oxidant (air) in the cathode. Significant integration synergies in terms of process operating conditions and material and heat exchange, such as described below, exist between SOFC and biomass gasification processes, which can enhance the overall BGFC system's energy efficiency.

There are various ways to improve heat recovery, including waste heat, energy efficiency and cleaner operation of BGFC systems. The BGFC system considers integration of the exhaust gas from the SOFC as a source of steam and unreacted fuel to the steam gasifier, utilizing biomass volatile gases and tars, which is separately carried out from the combustion of the remaining char of the biomass in a char combustor in the presence of depleted air from the SOFC. The high grade process heat is utilized to directly heat the process streams. The product gas from the steam gasifier, after cooling, condensation and ultra-cleaning with the Rectisol^{TM} process, generates the clean syngas. This syngas feed to the SOFC is heated up using the hot product gas from the steam gasifier. Also, the air to the SOFC is heated up using exhaust gas from the char combustor. The medium to low grade process heat that is available in excess is extracted into excess steam and hot water generation from the BGFC site.

Operating conditions. Both gasifiers and SOFCs operate effectively at elevated temperatures of around 500–1000 °C and can be operated at atmospheric as well as elevated pressures. The SOFC for higher power generation >10 kW can be operated at a higher pressure than 1 atm, while pressurized gasifiers are commonplace. This provides opportunities for process integration.

Material integration. The nitrogen rich depleted air (from the cathode side) and the exhaust gas (from the anode side) from an SOFC are a good source of high temperature oxygen and steam, respectively. These are the two essential oxidizing agents used in gasification processes. The exhaust gas from the SOFC is a source of steam and unreacted fuel, for the steam gasifier, utilizing biomass volatile gases and tars. Biomass char combustion in a separate char combustor in an *allothermal gasifier* (Figure 16.2) can use the depleted air from the SOFC as a combustion agent. Additionally, the SOFC has fuel flexibility, in which hydrogen, hydrocarbons and syngas can be used as feedstocks. Even greater environmental benefits can be gained if gaseous fuels, such as syngas, which is a good source of hydrogen, from biomass waste can be used as a fuel to the SOFC.

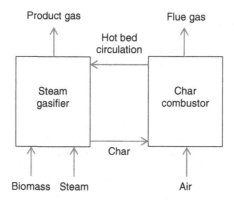

Figure 16.2 *Allothermal gasifier schematic.*

Heat integration. To maximize the heat recovery from the product gas from the gasification process, a hot gas clean-up strategy, followed by heat recovery from a high to low temperature below the gas's dew point can be adopted. This leaves the gas dry (as water in the gas is condensed out) and high in heating value, and hence is an ideal fuel for the SOFC. The cold and dry product gas after ultra-cleaning to a trace level removal of contaminants, discussed later, can itself be used to extract the heat from the hot syngas generating from the gasifier, before entering the SOFC. Preheating the syngas and air fed to the SOFC to thermodynamically maximum achievable temperatures ensures maximum power generation efficiency from the SOFC. Preheating of feed gases facilitates endothermic reforming reactions and increases the net exothermic heat generation (due to combustion) from the SOFC. There are several other high temperature heat sources, such as exit gases from the SOFC. The excess heat from a highly integrated BGFC site can be recovered into high pressure superheated steam, which can further be utilized to generate additional power from the site.

A major hindrance to the commercialization of a BGFC system is the stringent tolerance limits on the contaminants required for the SOFC feedstock. The most common contaminant in the syngas feed to an SOFC is H_2S, which originates from the raw materials used in gasification. It acts as a poison to the reforming and anode catalysts used in the SOFC. A tolerance limit as stringent as 0.1 ppm for H_2S in the SOFC feedstock is required to ensure thousands of hours of trouble free operation. The Rectisol™ technology developed by Lurgi that uses refrigerated methanol as the solvent for physical absorption or removal of undesired contaminants producing ultra-clean syngas is widely used in coal gasification plants. Rectisol™ provides an excellent option for co-removal of a number of contaminants, including H_2S, COS, HCN, NH_3, nickel and iron carbonyls, mercaptans, naphthalene, organic sulfides, etc., to a trace level (for example, H_2S to less than 0.1 ppm by volume), using one integrated plant from stringent resources, like coal. Nearly each of the coal gasification units for the production of hydrogen or hydrogen rich gases and syngas with hydrogen and carbon monoxide as major constituents is equipped with a Rectisol™ gas purification system. Because of the increasing use of biomass gasification technology in the face growing interest for chemical production, such as the synthesis of ammonia, methanol, Fisher–Tropsch liquids, oxo-alcohols and gaseous products such as hydrogen, syngas, reduction gas and town gas, a steep increase in the application of Rectisol™ processes is expected. A Rectisol™ process needs to be integrated to a low temperature fuel gas, such that minimum cooling is needed to attain the required refrigeration for the Rectisol™ process. Rectisol™ process modeling, design and energy integration are shown in Section 13.3. The one-by-one process integration challenges are now discussed in greater detail.

There are synergies for simultaneous heat and material integration between the gasification and the SOFC systems, shown in Figure 16.3.

1. Design of the gasifier. The gasification process under consideration consists of two interconnected fluidized beds, a char combustor, combusting char in the presence of air, and a steam gasifier, gasifying biomass volatilized gases and tars (Figure 16.2). Such gasifiers are called allothermal gasifiers. Direct contact between the steam gasifier and the char combustor is avoided; the heat required for steam gasification is achieved by means of the circulation of a hot bed generally made up of sand. It is in a loop with an end-to-end configuration composed of a circulating fluidized bed char combustor, a cyclone and a bubbling fluidized bed steam gasifier. This scheme avoids dilution of the gasification product gas with nitrogen whilst avoiding the use of an oxygen plant (air separation unit) for supplying pure oxygen to the gasifier. A biomass allothermal fluidized bed gasifier comprising a steam gasifier, tar cracker and combustor can be operated at around 800 °C to produce a medium calorific value gas mixture, rich in H_2, CO and CH_4, which are fuel species for the SOFC. In view of the thoroughness of mixing and good gas–solid contact, the use of fluidized bed gasifiers is recommended. These designs have various advantages, such as a relatively simple construction, greater tolerance to the particle size range than fixed beds, good temperature control and high reaction rates, high carbon conversion, high specific capacity, high conversion efficiency, limited turndown and a very good scale-up potential.
2. Integration of syngas. The syngas rich in hydrogen and carbon monoxide from the steam gasifier, followed by cooling–condensation and ultra-cleaning using Rectisol™, is an excellent feedstock to the SOFC.
3. Syngas cleaning and heat recovery. The hot product gas clean-up and cooling comprise hot gas filtration for the removal of particulates, cooling or heat recovery and cleaning of contaminants to a trace level using the Rectisol™ technology. The particulate free hot gas is cooled down to preheat clean and dry syngas feed to the SOFC and to further generate superheated steam. These two heat recovery exercises can be done in parallel or series or combined within

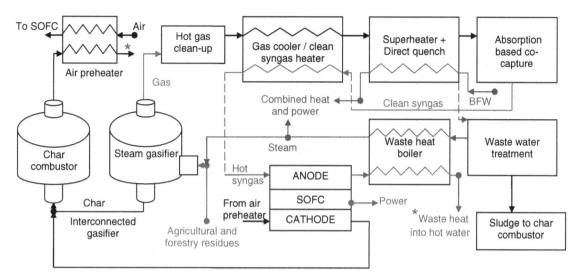

Figure 16.3 *BGFC schematic. (Reproduced with permission from Sadhukhan et al. (2010)[3]. Copyright © 2010, Elsevier.)*

one heat exchanger unit. However, the hot gas coolers inherently require high maintenance; thus, introducing further complexity in the operation and maintenance in the latter case is not desirable. Figure 16.3 shows the preheating of the clean and dry syngas feed to the SOFC and the generation of superheated steam in series. This is followed by a direct quench of the cold gas with cooling water below its dew point, so as to allow the separation of water and tar condensables from the remaining dry syngas. After sulfur and all other contaminants are removed to a trace level using the Rectisol™ process, the ultra-clean and high heating value syngas rich in hydrogen is fed to the SOFC via heating in the hot gas cooler. The water generated, after waste water treatment and purge of sludge, can be fed back as boiler feed water (BFW) for steam generation within the BGFC system.

4. Steam from the SOFC. A SOFC produces high temperature steam, after electrochemically oxidizing hydrogen present in the syngas, while the gasification process requires such high grade steam, over and above that present as moisture in a biomass feedstock, thus adding more to the hydrogen concentration in the syngas feed to the SOFC. Steam gasification is essential to reform gas and tar and consequently reduce the tars. The tar can be reformed catalytically in the steam gasifier. Steam is also known to reduce the concentration of other forms of oxygenates including condensable ones. Thus, a part of the exhaust gas generated from the SOFC (anode) containing steam can be routed to the steam gasifier (Figure 16.3). This also helps gasify any unreacted fuel from the SOFC. The total exhaust gas generating from the SOFC is divided between that emitted to the atmosphere (after heat recovery, discussed later) to balance the carbon across a BGFC system and as a source of steam to the steam gasifier.

5. Steam generation from the SOFC exhaust gas cooler. The amount of the exhaust from the SOFC at a high temperature that is not fed back as a source of steam to the steam gasifier can be cooled (heat extracted) to generate superheated steam (from the waste heat boiler in Figure 16.3). The water recovered from the product gas from the steam gasifier via the effluent treatment plant can be reused to recover this heat into superheated steam. A part of this steam can be routed to the steam gasifier to fulfill the balance of its minimum steam requirement. The rest of the steam is available as an excess steam from the BGFC site.

6. Supply of air to the SOFC and gasifier processes. Both gasification and SOFC processes require oxygen (air). Oxygen needs to be added selectively at various gasification stages, such as in the secondary zones of a pyrolysis-cracker reactor, in order to preferentially oxidize tars. Its main role is to supply heat to the steam gasifier by combusting char in the char combustor. In a perfectly energy balanced BGFC flowsheet, the heat from the char combustor must satisfy the heat requirements of the steam gasifier after integration of the exhaust gas from the SOFC. The resulting depleted air from the SOFC cathode can thus be utilized as a source of oxygen in the char combustor. The amount of air to the SOFC can be adjusted so as to maximize syngas fuel utilization efficiency in the SOFC and consequently combust

char in order to fulfill the heat requirement of the steam gasifier. The steam gasifier as such would not require any additional oxygen. Thus the char combustion and dilution of the char combustor exhaust gas with nitrogen sourced from air can be completely isolated from the steam gasifier, resulting in the highest calorific value syngas generation and use within the system.

7. Feed preheating. The air and syngas feedstocks need preheating before entering the SOFC, so as to avoid thermal shock of the ceramic components and such that the sensible heat in them can be made available for power generation through an electrochemical process from the SOFC. Either of the exhaust gases from the char combustor and/or the SOFC can be used to preheat air. From a heat integration point of view, a heat exchange between the exhaust gas from the char combustor and air is preferred, based on a closer match of the heat capacities between the two streams (see Chapter 3 for heat integration concepts and heuristics). The hot gas directly from the gasifier cannot be fed to the SOFC at the gasifier temperature without a thorough quench with cooling water. In order to avoid this heat loss, heat from the hot and moist gas from the gasifier is recovered into preheating the clean and dry syngas product, fed to the SOFC, shown in Figure 16.3.

8. Excess steam. An overall BGFC site can be a net generator of heat. Excess heat in the form of superheated steam can be generated by utilizing a hot product gas from the steam gasifier in the superheater and a part of the hot exhaust gas from the SOFC in the waste heat boiler, respectively.

9. Low grade heat. The various sources of low grade heat include the heat of condensation of the SOFC exhaust gas, hot water recovered via condensation of the SOFC exhaust gas and low temperature sensible heat from the exhaust gas from the char combustor. Whether the low grade heat recovery is cost-effective or not depends on the amount of low grade heat generation, which obviously is more justifiable for higher capacity BGFC sites.

16.2   Simulation of Integrated BGFC Flowsheets

Simulation of an integrated BGFC flowsheet, shown in Figure 16.3 in such a way as to improve energy efficiency, heat recovery and cleaner operation, can be undertaken in a process simulator such as Aspen Plus, with modeling specifications shown in Table 16.1.

To generate 652.61 kW of electricity from the SOFC, 9.13 t d^{-1} of clean syngas feed needs to be produced from 5.44 t d^{-1} of straw slurry, the ultimate analysis of which is shown in Table 16.2. The primary pyrolysis product distribution calculation can be done based on the mass ratios or transfer functions shown in Table 16.3.

Figure 16.4 shows the Aspen Plus simulation flowsheet and results (compositional and thermal) obtained for the BGFC flowsheet in Figure 16.3. The SOFC for a range of power generation from 100 kW to a few MW can be operated at around 5 bar. The same or slightly lower pressure can also be maintained in the biomass gasifier. The ambient temperature is assumed to be 25 °C.

The two reactors in the interconnected circulating fluidized bed gasifier shown in Figure 16.2 can be simulated as RGibbs reactors, STGASIFY, fed with gas (GASIN) and tar (TARIN), and CHAR-RCT, fed with char (CHARIN) and ash (ASH), respectively (Figure 16.4). These streams are derived from primary pyrolysis of a biomass feedstock, provided in Tables 16.2 and 16.3. The primary pyrolysis occurs as soon as a biomass feedstock comes in contact with the hot bed within a gasifier before any mixing, mass transfer and chemical reaction with other reactants, steam and oxygen, take place in a gasifier. The composition of GASIN and the amounts of GASIN, TARIN and CHARIN, shown in Figure 16.4, are predicted using the ultimate analysis of straws in Table 16.2 and correlations provided in Table 16.3. The yield correlations of primary pyrolysis products in Table 16.3 are generic for biomass applications.

TARIN can be modeled as phenol, while CHARIN and ASH can be modeled as nonconventional components. The compositions of GASIN, TARIN, CHARIN and ASH are thus determined to balance with the C, H, N, O, S, ash and moisture contents in a biomass (Tables 16.2 and 16.3 and Figure 16.4). The nonrandom two liquid (NRTL) thermodynamic package can be used for property estimation. The SOFC cathode (CATHODE) and anode (ANODE) can be modeled as a two-component separator (Sep2) with 95% efficiency of oxygen separation from air on a molar basis and a Gibbs reactor (RGibbs), respectively.

The materials to construct the cathode and anode are typically nickel oxide and yttria stabilized zirconia (YSZ) and graphite powders, and lanthanum strontium manganate (LSM: La$_{1-x}$Sr$_x$MnO$_3$), YSZ and graphite powders, respectively. The materials inventories for the manufacture of the SOFC unit are shown in the ***Online Resource material, Chapter 16 – Additional Exercises and Examples***.

Table 16.1   *Basis for BGFC process flowsheet simulation in Aspen Plus. (Reproduced with permission from Sadhukhan et al. (2010)[3]. Copyright © 2010, Elsevier.)*

Unit Names	Modeling Framework	Process Specification
Gasification	Estimate pyrolysis or devolatilization product yield using EXCEL spreadsheet based model shown in Table 16.2. Use RGibbs reactor in Aspen Plus for the gasification of pyrolysis product into product gas.	Temperature = 900–950 °C. The pressure of the desired product (syngas) decides the system and the gasification operating pressure. Pressure is kept at less than or equal to 5 bar.
Air compressor	Compressor Isentropic model in Aspen Plus.	Desired product decides the system and compressors' operating pressures. Isentropic efficiency = 75%.
Gas cooler and heat recovery steam generator (HRSG)	Cooler in Aspen Plus.	Temperature is just above the dew point of the gas at the gasification pressure, such that single-phase gas still leaves the cooler/HRSG without requiring special design of the gas cooler/HRSG.
Gas or exhaust condenser	Flash2, two-phase flash separator in Aspen Plus.	Temperature at or lower than the dew point of the gas, so as to dry the gas from the water present in the gas. After purification in the effluent treatment plant (ETP) and 10% purge, the rest of the water is recovered as BFW.
Gas clean-up and carbon capture and storage; air separation unit	Sep2: two-outlet component separator based on component purity, flow, etc., for the site flowsheet simulation in Aspen Plus. Individual processes (such as Rectisol™) can be simulated in detail.	Specify the mole fraction of one of the components to be separated in the outlet pure gas stream.
SOFC anode	RGibbs in Aspen Plus.	85% fuel utilization efficiency.
SOFC cathode	Sep2: two-outlet component separator based on component purity, flow, etc. for the site flowsheet simulation in Aspen Plus.	Pure oxygen recovery by 95% by mole.
Steam gasifier in allothermal gasification	RGibbs reactor in Aspen Plus. Gas and tar yields from the spreadsheet based pyrolysis or devolatilization product yield modeling (Table 16.2) are to be entered as feedstock to the steam gasifier model in Aspen Plus.	Temperature = 900–950 °C. Desired product decides the system and the gasification operating pressure. Pressure less than equal to 5 bar.
Char combustor cooler	Cooler in Aspen Plus.	Temperature is above the dew point of the flue gas at atmospheric pressure, such that single-phase gas still leaves the cooler without requiring special design of the cooler.
Direct quench	Flash2, two-phase flash separator in Aspen Plus.	Temperature at or lower than the dew point of the gas, so as to dry the gas from the water present in the gas. After effluent treatment and 10% purge, the rest of the water is recovered as BFW.
Char combustor in allothermal gasification	RGibbs reactor in Aspen Plus. Char yield from the spreadsheet based pyrolysis or devolatilization product yield model (Table 16.2) entered as feedstock to the char combustor model in Aspen Plus.	About 50 °C higher temperature than steam gasifier to maintain a temperature gradient and supply exothermic heat of combustion reaction to the steam gasifier.

Table 16.2   *Ultimate analysis of straw in weight%. (Reproduced with permission from Sadhukhan et al. (2010)[3]. Copyright © 2010, Elsevier.)*

wt%	Straw
C	36.57
H	4.91
N	0.57
O	40.70
S	0.14
Ash	8.61
Moisture	8.50
Lower heating value (MJ kg^{-1})	14.60

The cathode works to split air into a pure oxygen stream and nitrogen rich exit stream from the cathode compartment. The molar recovery ratio of oxygen can be assumed at 95%: moles of oxygen separated in the cathode/moles of oxygen present in the inlet air stream. This stream of pure oxygen enters the anode chamber via the ceramic oxygen selective membrane. The remaining 5% oxygen along with the nitrogen inlet exits the cathode chamber of the SOFC and enters the char combustor.

The anode of the SOFC drives the reactions shown in Figure 16.1 in the forward direction according to thermodynamic optimality, to minimize the Gibbs free energy of the overall reactions (cf Section 10.3.2). Continuous removal of the exhaust or flue gas drives the reactions in the forward direction. The reactions are fundamentally combustion reactions (powered by electrons) of the combustible components of the fuel gas, for example, hydrogen, carbon monoxide, methane and any other hydrocarbons present in the syngas. The combustion reactions take place by electrochemical driving forces. Hence the reactions occur between the cations of the combustible gases and oxygen anions, so that the electrons released from the combustible gases can flow through an external circuit from the anode, to the cathode, generating electricity. The electrochemical reactions process in the anode can be perfectly modeled for a given temperature using a Gibbs reactor in Aspen Plus. The Gibbs reactor in Aspen Plus does not require reaction scheme specifications and works to minimize the difference in the Gibbs energy of formation between products and reactants (see Section 10.3.2 for the Gibbs free energy minimization calculations for reaction thermodynamic optimality). The model allows accounting of all components participating in the reactions. The simulation results shown in Figure 16.4 will help to generate design specifications of the equipment involved in a BGFC flowsheet.

A BGFC site can be a good source of heat with both the gasifier and SOFC operating at high temperatures. It can be observed that with increasing temperature and at a lower pressure of the gasifier, the concentration of hydrogen in the syngas increases, and hence the heating value of the syngas increases. However, a higher temperature, >1000 °C, may cause operational difficulties and maintenance problems, while increasing pressure is associated with increased

Table 16.3   *Correlations for the yields of products: gas, tar and char from biomass primary pyrolysis or devolatilization. (Reproduced with permission from Sadhukhan et al. (2010)[3]. Copyright © 2010, Elsevier.)*

Component	kg per kg Biomass
Total devolatilization	0.96
Total gas as follows	0.48
H_2	0.00
CH_4	0.02
C_2	0.12
CO	0.22
CO_2	0.03
H_2O	0.08
Tar	Total devolatilization – Total gas
Char	1 – Total devolatilization

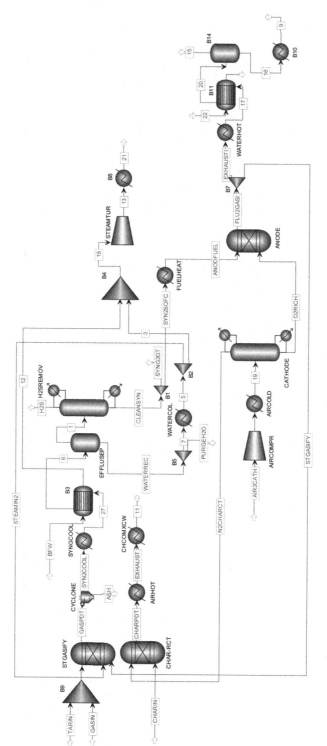

(Continued)

Stream Names	GASIN	TARIN	CHARIN	ASH	GASPDT	SYN2COOL	27	8	1
Mole Fraction									
H_2O	0.228	0.056	0.000	0.000	0.314	0.314	0.314	0.314	0.032
N_2	0.011	0.000	0.000	0.000	0.005	0.005	0.005	0.005	0.006
O_2	0.000	0.763	0.000	0.000	0.000	0.000	0.000	0.000	0.000
H_2	0.041	0.000	1.000	0.000	0.232	0.232	0.232	0.232	0.329
C	0.000	0.000	0.000	0.000	0.000	0.000	0.000	0.000	0.000
CO	0.395	0.000	0.000	0.000	0.232	0.232	0.232	0.232	0.328
CO_2	0.036	0.000	0.000	0.000	0.216	0.216	0.216	0.216	0.304
H_2S	0.002	0.000	0.000	0.000	0.000	0.000	0.000	0.000	0.001
CH_4	0.077	0.000	0.000	0.000	0.000	0.000	0.000	0.000	0.000
Ethane	0.209	0.000	0.000	0.000	0.000	0.000	0.000	0.000	0.000
Phenol	0.000	0.181	0.000	0.000	0.000	0.000	0.000	0.000	0.000
Total Flow (kmol h^{-1})	4.018	2.296	0.685		22.659	22.659	22.659	22.659	16.006
Total Flow (t d^{-1})	2.388	2.341	0.197	0.469	12.125	12.125	12.125	12.125	9.228
Total Flow (m^3 h^{-1})	76.481	42.498			460.868	459.732	1362.370	646.900	396.778
Temperature (°C)	25	25	25		950	947	450	70	25
Pressure (bar)	1.034	1.034	1.034		5.000	5.000	1.000	1.000	1.000
Vapor Fraction	0.794	0.772			1.000	1.000	1.000	1.000	1.000
Liquid Fraction	0.206	0.228			0.000	0.000	0.000	0.000	0.000
Solid Fraction	0.000	0.000	1.000	1.000	0.000	0.000	0.000	0.000	0.000
Enthalpy (kJ mol^{-1})	-145.298	-43.762			-152.570	-152.566	-172.083	-185.352	-163.693
Entropy (J mol^{-1} K^{-1})	-32.225	-70.560			56.783	56.660	49.603	23.766	39.235
Density (kg m^{-3})	1.301	2.295	187.330	3486.884	1.096	1.099	0.371	0.781	0.969

(Continued)

Stream Names	CLEANSYN	SYN2SOFC	ANODFUEL	FLU2GASI	STGASIFY	EXHAUSTI	17	20	15
Mole Fraction									
H_2O	0.032	0.032	0.032	0.292	0.292	0.292	0.292	0.292	0.123
N_2	0.006	0.006	0.006	0.009	0.009	0.009	0.009	0.009	0.012
O_2	0.000	0.000	0.000	0.051	0.051	0.051	0.051	0.051	0.068
H_2	0.329	0.329	0.329	0.049	0.049	0.049	0.049	0.049	0.000
C	0.000	0.000	0.000	0.000	0.000	0.000	0.000	0.000	0.000
CO	0.328	0.328	0.328	0.049	0.049	0.049	0.049	0.049	0.000
CO_2	0.305	0.305	0.305	0.550	0.550	0.550	0.550	0.550	0.798
H_2S	0.000	0.000	0.000	0.000	0.000	0.000	0.000	0.000	0.000
CH_4	0.000	0.000	0.000	0.000	0.000	0.000	0.000	0.000	0.000
Ethane	0.000	0.000	0.000	0.000	0.000	0.000	0.000	0.000	0.000
Phenol	0.000	0.000	0.000	0.000	0.000	0.000	0.000	0.000	0.000
Total Flow (kmol h^{-1})	15.997	15.837	15.837	16.731	6.542	10.189	10.189	10.189	7.627
Total Flow (t d^{-1})	9.221	9.129	9.129	13.809	5.399	8.410	8.410	8.410	7.290
Total Flow (m^3 h^{-1})	396.545	392.580	282.613	298.563	116.738	181.825	303.409	288.662	204.914
Temperature (°C)	25	25	800	800	800	800	85	72	50
Pressure (bar)	1.000	1.000	5.000	5.000	5.000	5.000	1.000	1.000	1.000
Vapor Fraction	1.000	1.000	1.000	1.000	1.000	1.000	1.000	0.988	1.000
Liquid Fraction	0.000	0.000	0.000	0.000	0.000	0.000	0.000	0.012	0.000
Solid Fraction	0.000	0.000	0.000	0.000	0.000	0.000	0.000	0.000	0.000
Enthalpy (kJ mol^{-1})	−163.775	−163.775	−135.869	−284.609	−284.609	−284.609	−316.211	−317.201	−342.845
Entropy (J mol^{-1} K^{-1})	39.192	39.192	70.907	34.501	34.501	34.501	0.578	−2.265	5.511
Density (kg m^{-3})	0.969	0.969	1.346	1.927	1.927	1.927	1.155	1.214	1.482

(Continued)

Stream Names	16	AIR2CATH	2	19	O2RICH	N2CHARCT	CHARPDT	EXHAUST	11
Mole Fraction									
H_2O	0.993	0.000	0.000	0.000	0.000	0.000	0.000	0.000	0.000
N_2	0.000	0.790	0.790	0.790	0.007	0.973	0.973	0.973	0.973
O_2	0.000	0.210	0.210	0.210	0.993	0.027	0.001	0.001	0.001
H_2	0.000	0.000	0.000	0.000	0.000	0.000	0.000	0.000	0.000
C	0.000	0.000	0.000	0.000	0.000	0.000	0.000	0.000	0.000
CO	0.000	0.000	0.000	0.000	0.000	0.000	0.000	0.000	0.000
CO_2	0.007	0.000	0.000	0.000	0.000	0.000	0.026	0.026	0.026
H_2S	0.000	0.000	0.000	0.000	0.000	0.000	0.000	0.000	0.000
CH_4	0.000	0.000	0.000	0.000	0.000	0.000	0.000	0.000	0.000
Ethane	0.000	0.000	0.000	0.000	0.000	0.000	0.000	0.000	0.000
Phenol	0.000	0.000	0.000	0.000	0.000	0.000	0.000	0.000	0.000
Total Flow (kmol h^{-1})	2.562	32.200	32.200	32.200	6.100	26.100	26.100	26.100	26.100
Total Flow (t d^{-1})	1.120	22.296	22.296	22.296	4.681	17.615	17.812	17.812	17.812
Total Flow (m^3 h^{-1})	0.048	760.201	277.442	574.611	108.855	465.756	530.857	1200.358	722.949
Temperature (°C)	50	25	245	800	800	800	950	280	60
Pressure (bar)	1.000	1.050	5.000	5.000	5.000	5.000	5.000	1.000	1.000
Vapor Fraction	0.000	1.000	1.000	1.000	1.000	1.000	1.000	1.000	1.000
Liquid Fraction	1.000	0.000	0.000	0.000	0.000	0.000	0.000	0.000	1.000
Solid Fraction	0.000	0.000	0.000	0.000	0.000	0.000	0.000	0.000	0.000
Enthalpy (kJ mol^{-1})	-284.679	0.000	6.490	24.161	25.249	23.906	19.010	-2.747	-9.296
Entropy (J mol^{-1} K^{-1})	-155.733	3.977	7.285	30.306	27.949	26.698	31.745	19.582	4.508
Density (kg m^{-3})	972.297	1.222	3.348	1.617	1.792	1.576	1.398	0.618	1.027

(Continued)

Stream Names	WATERREC	PURGEH2O	7	5	STEAMIN2	3	BFW	12	18	13	22	25
Mole Fraction												
H_2O	0.995	0.995	0.995	0.995	0.995	0.995	1.000	1.000	0.999	0.999	1.000	1.000
N_2	0.000	0.000	0.000	0.000	0.000	0.000	0.000	0.000	0.000	0.000	0.000	0.000
O_2	0.000	0.000	0.000	0.000	0.000	0.000	0.000	0.000	0.000	0.000	0.000	0.000
H_2	0.000	0.000	0.000	0.000	0.000	0.000	0.000	0.000	0.000	0.000	0.000	0.000
C	0.000	0.000	0.000	0.000	0.000	0.000	0.000	0.000	0.000	0.000	0.000	0.000
CO	0.000	0.000	0.000	0.000	0.000	0.000	0.000	0.000	0.000	0.000	0.000	0.000
CO_2	0.005	0.005	0.005	0.005	0.005	0.005	0.000	0.000	0.001	0.001	0.000	0.000
H_2S	0.000	0.000	0.000	0.000	0.000	0.000	0.000	0.000	0.000	0.000	0.000	0.000
CH_4	0.000	0.000	0.000	0.000	0.000	0.000	0.000	0.000	0.000	0.000	0.000	0.000
Ethane	0.000	0.000	0.000	0.000	0.000	0.000	0.000	0.000	0.000	0.000	0.000	0.000
Phenol	0.000	0.000	0.000	0.000	0.000	0.000	0.000	0.000	0.000	0.000	0.000	0.000
Total Flow (kmol h^{-1})	6.653	0.665	5.988	5.988	4.580	1.408	5.630	5.630	7.038	7.038	3.500	3.500
Total Flow (t d^{-1})	2.897	0.290	2.607	2.607	1.994	0.613	2.434	2.434	3.047	3.047	1.513	1.513
Total Flow (m^3 h^{-1})	0.121	0.012	0.109	59.057	45.171	13.886	0.102	53.673	67.561	257.931	0.063	0.066
Temperature (°C)	25	25	25	320	320	320	25	300	304	168	25	65
Pressure (bar)	1.000	1.000	1.000	5.000	5.000	5.000	5.000	5.000	5.000	1.000	1.000	1.000
Vapor Fraction	0.000	0.000	0.000	1.000	1.000	1.000	0.000	1.000	1.000	1.000	0.000	0.000
Liquid Fraction	1.000	1.000	1.000	0.000	0.000	0.000	1.000	0.000	0.000	0.000	1.000	1.000
Solid Fraction	0.000	0.000	0.000	0.000	0.000	0.000	0.000	0.000	0.000	0.000	0.000	0.000
Enthalpy (kJ mol^{-1})	−286.138	−286.138	−286.138	−232.207	−232.207	−232.207	−285.683	−232.280	−232.265	−237.090	−285.683	−282.801
Entropy (J mol^{-1} K^{-1})	−161.793	−161.793	−161.793	−33.248	−33.248	−33.248	−162.687	−35.059	−34.679	−30.824	−162.687	−153.682
Density (kg m^{-3})	995.576	995.576	995.576	1.839	1.839	1.839	993.515	1.890	1.879	0.492	993.515	954.064

Figure 16.4 BGFC flowsheet in Aspen Plus simulation results. (Reproduced with permission from Sadhukhan et al. (2010)[3]. Copyright © 2011, Elsevier.)

power generation from gas turbines in the case of the BGFC system. SOFCs operate at elevated temperatures of around 500–1000 °C and therefore can be a good source of high grade heat. The results shown here are for the gasifier and SOFC temperatures at 950 °C and 800 °C, respectively. The pressure in the SOFC may be varied from atmospheric to ~10 bar for 1 kW to a few MW power generation, respectively.

Figure 16.4 shows the gas clean-up processes that comprise the hot gas filtration (CYCLONE) for the removal of particulates, flash separator (EFFLUSEP) for condensation of water and other condensates (e.g., tar) (WATERREC), from the gas by cooling the gas below its dew point and the Rectisol™ process, modeled as a two-component separation unit (H2SREMOV) operating at 99% efficiency, respectively. The following utility consumptions are established for the Rectisol™ process: shaft power: 73.08 kW, LP steam: 323.74 kW and refrigeration duty: 131.42 kW, respectively, per kmol h^{-1} (per hour) of sulfur and nitrogen compounds removed, where the total kmol h^{-1} sulfur and nitrogen compounds removed were calculated from the inlet and the outlet stream analyses, 1 and SYN2SOFC, respectively (Figure 16.4). Alternatively, the rigorous process simulation model can be used for the Rectisol™ process discussed in Section 13.3. The Rectisol™ process can also be operated at low syngas capacity, such as in this case ~390 Nm^3 h^{-1}. The clean and dry syngas, SYN2SOFC (and after preheating, ANODFUEL), has almost equal molar compositions of hydrogen and carbon monoxide (32.8% each), both of which are combustible in the SOFC. Based on 85% of the total enthalpy change from SYN2SOFC and O2RICH, fed to the SOFC anode, to FLU2GASI, produced from the SOFC anode, the electricity generation shown in Figure 16.4 is 653 kW. Alternatively, 85% of the heat of combustion of hydrogen and carbon monoxide can also be taken into account to estimate the electricity generation from the SOFC.

A key energy efficiency exercise consists of heat integration between hot–cold process streams or coolers–heaters, governed by thermodynamic optimality or maximum heat recovery strategy and identification of basic processing chains to aid with process decision making and establish mass and energy balance, as discussed below.

16.3 Heat Integration of BGFC Flowsheets

Once the temperature, pressure and stream compositions are decided for the major process units (e.g., reactors and separators), coolers and heaters are placed on the hot and cold streams, respectively, to achieve their respective target temperatures (Figure 16.4).

Two heat recovery strategies can be straightaway adopted from conventional integrated gasification combined cycle (IGCC) systems: syngas cooling and heat recovery from the exhaust gases, into the generation (economizer and evaporation) and superheating of steam. Additionally, the site has a major high temperature heat requirement for preheating SYN2SOFC from 25 °C up to the operating temperature of the SOFC, 800 °C, which can only be satisfied by the high temperature (950 °C) heat available in SYN2COOL (Figure 16.4). Hence, a heat exchange between SYNGCOOL (cooler) and FUELHEAT (heater) is an obvious choice. After preheating SYN2SOFC to 800 °C, the medium grade heat remained in SYN2COOL (450 °C to the dew point of the gas) can then be utilized to generate superheated steam, stream 12 at 300 °C and 5 bar, from BFW, in unit B3 (Figure 16.4).

Similar to the syngas feed to the SOFC, air also needs preheating, after compression (AIRCOMPR), up to a maximum temperature of the SOFC, 800 °C. The two sources of exhaust gas are from the SOFC and the char combustor, EXHAUSTI (at 800 °C) from ANODE and CHARPDT (at 950 °C) from CHAR-RCT, respectively, from which the high grade heat can be recovered into superheated steam and/or to preheat the air. As CHARPDT provides a feasible temperature driving force for maximum preheating of air (up to 800 °C) and is also based on a closer match of the heat capacities between the CHARPDT and AIR2CATH-19 streams, thermal integration between AIRHOT (cooler) and AIRCOLD (heater) is thermodynamically more favorable. Figure 16.5 shows the enthalpy change of the hot stream (CHARPDT) and the cold stream (2) over the temperature range in the AIRHOT-AIRCOLD exchanger and ensures that there is no temperature crossover between the two. The inlet temperature of the air from the air compressor (with AIRCOMPR operating at 75% efficiency, Table 16.1) to AIRCOLD is 245 °C. A minimum temperature approach of 35 °C occurs at the inlet of the air and outlet of EXHAUST, in a countercurrent exchanger, for achieving the maximum air preheating. The WATERHOT and WATERCOL streams can be combined for exhaust gas heat recovery (down to 85 °C or the dew point of the exhaust gas, stream 17) to superheat steam, stream 5 (at 320 °C and 5 bar from 25 °C) in Figure 16.4, following the strategy adopted in conventional IGCC processes. Thus, these pairs show the major exchangers for the high grade process to process heat recovery in a BGFC site.

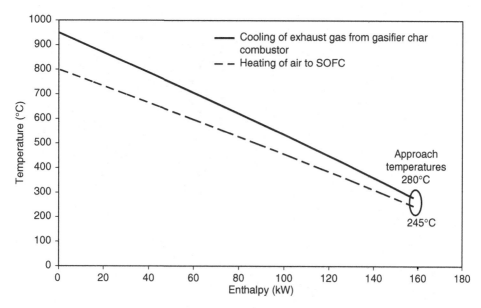

Figure 16.5   *Heat balance in the air preheater or char combustor exhaust cooler. (Reproduced with permission from Sadhukhan et al. (2010)[3]. Copyright © 2010, Elsevier.)*

16.4   Analysis of Processing Chains in BGFC Flowsheets

Four main processing pathways need to be balanced for integrated BGFC system design and equipment specifications. These four routes are integrated with the heat recovery schemes. Hence, the energy efficiency and material integration are interdependent. Refer to Figure 16.4 to track the streams of processing pathways.

1. TARIN-GASIN-GASPDT-SYN2COOL-27-8-1-CLEANSYN-SYN2SOFC-ANODFUEL-FLU2GASI-EXHAUSTI -17-20-15-STGASIFY or the syngas production to exhaust emission routes, shown in Figure 16.6. The independent variable is the mole fraction of the exhaust gas resulting from the syngas combusted in the SOFC anode, used to recover the heat into BFW preheating: the molar ratio between EXHAUSTI and FLU2GASI. The carbon intake through the primary pyrolysis gas and tar, GASIN and TARIN, to the steam gasifier (STGASIFY) must be equal to the carbon released to the atmosphere via stream 15 (the part of the SOFC exhaust gas finally released to the atmosphere). The system consideration for the carbon balance is shown in Figure 16.6. For the given set of simulation results, a molar split ratio of EXHAUSTI of 60.9% of FLU2GASI in B7 has been shown to maintain the carbon balance in the processing routes.

   The steam content in the recycle stream, STGASIFY, enhances the hydrogen concentration in the syngas, while its unreacted hydrogen and carbon monoxide can provide the balance between the heating requirement of the endothermic steam gasification (STGASIFY) and the cooling requirement of the exothermic char combustion (CHAR-RCT), resulting in an overall thermally neutral gasification process operation.

2. WATERREC-7-PURGEH2O-5-STEAMIN2-3 or the BFW to steam cycles, shown in Figure 16.7. The carbon balance route in Figure 16.6 fixes the amount of STGASIFY recycled to the steam gasifier. The main decision making involved in the BFW to steam cycles is the condition of the steam to be generated and any additional amount of steam needed to be recycled to the STGASIFY process to meet the recommended steam-to-biomass weight ratio of 0.6 for a good mixing and conversion in the steam gasifier. The flow rate of STEAMIN2 should thus be decided to achieve this recommended steam-to-biomass ratio. With respect to the BFW balance, after a 10% purge due to waste water sludge removal, the rest is recovered as BFW (stream 7). This BFW is economized, evaporated and superheated using the

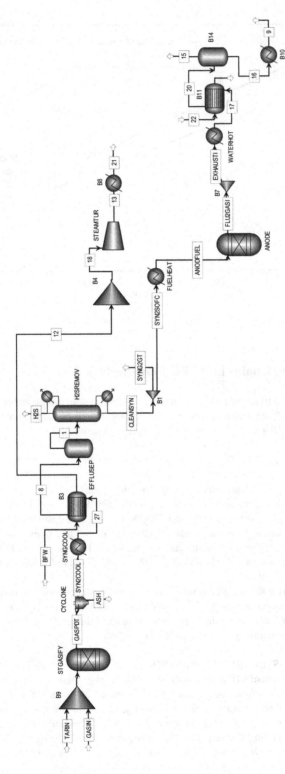

Figure 16.6 *Syngas production to exhaust emission routes are shown for the carbon balance between the inlet and outlet streams, GASIN, TARIN and stream 15, respectively.*

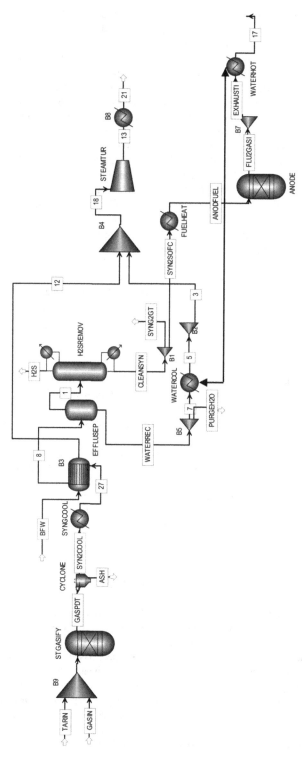

Figure 16.7 BFW production through to steam generation routes are shown for the heat balance between WATERCOL and WATERHOT and for meeting the steam (STEAMIN-2)-to-biomass weight ratio requirement of 0.6 in the steam gasifier.

heat recovered from the fraction of the exhaust gas from the SOFC anode decided for carbon balance, EXHAUSTI. A connection between WATERHOT and WATERCOL is shown for this heat balance. After sending STEAMIN2 to STGASIFY, the excess superheated steam 3 can be utilized to generate power using a steam turbine. The steam temperature and pressure conditions of STEAMIN2 and 3 are variables to adjust against the excess amount of steam generation. Steam conditions with lower than 300 °C temperature were not recommended in order to avoid any thermal shock in the steam gasifier, STGASIFY. The results shown are for excess superheated steam generation at 320 °C.

The two processing chains so far discussed are closely interlinked via the steam balance. The pairing of WATERHOT-WATERCOLD determines the conditions of the steam generation. Additionally, the flow rates of the excess steam 3 and the recycle stream STGASIFY are interdependent, and any change in either of their flow rates affects the conditions of the steam entering the STGASIFY unit, which thus gives the hydrogen and carbon monoxide concentrations in the syngas feed to the SOFC and overall performance of the SOFC. Therefore, boundary conditions need to be further investigated for equipment design specifications.

3. AIR2CATH-2-19-O2RICH-N2CHARCT-CHARPDT-EXHAUST-11 or the air-to-char combustion exhaust route shown in Figure 16.8. The total air requirement for complete combustion of char in the char reactor and syngas in the SOFC anode is calculated. The heat integration between AIRCOLD and AIRHOT is shown in Figures 16.5 and 16.8. The low grade heat available from the char combustor waste heat recovery unit, CHCOMXCW, remains unchanged, irrespective of any modification in the first two processing chains shown, due to the fixed biomass or char flow rate.

4. BFW-12-3-18-13 or the utilization of the excess steam to generate CHP shown in Figure 16.7. The excess steam can generate power only in a few kW range (microturbine). The microturbines provide a great research challenge and their investment can only be justified for the higher range of power generation. An alternative is to produce the entire superheated steam, as a source of high grade heat (stream 18 at 304 °C and 5 bar in Figure 16.4) to export to other industrial processes.

Further Challenges

1. Determine the feasibility to balance the steam requirement in a BGFC site so that there is no excess steam generation from the site for the given feedstock with the ultimate analysis shown in Table 16.2.

2. Determine the biomass specification in terms of its carbon, hydrogen and oxygen contents in weight percent required to balance between steam sources and sinks in a BGFC site, so that there is no excess steam generation from the site.

A BGFC site is also a source of low grade heat, shown in Table 16.4. The low grade heat can be obtained from the cooling of the char combustor exhaust gas (CHCOMXCW) and condensation of the SOFC exhaust gas below the dew point (B11) using cooling water, hot water or condensate recovered from the SOFC exhaust (stream 16 from the bottom of the Flash separator, B14) and the low pressure steam extracted from the steam turbine (stream 13 from STEAMTUR), shown in Figure 16.4. An energetic analysis of the BGFC site is shown in Table 16.4.

Further challenge questions on biomass gasification combined cycle[3] and LCA are provided in Online Resource Material, Chapter 16 – Additional Exercises and Examples.

16.5   SOFC Gibbs Free Energy Minimization Modeling

In order to predict the amount of power generation from an SOFC, a steady-state irreversible model in Equations (16.1) to (16.4) later in this section can be implemented in an Excel spreadsheet[4]. In a fuel cell, chemical reaction energy is converted into electrical energy by the reaction between fuel and oxidant. An Aspen Plus simulation of the SOFC can be used to establish the kinetic reaction parameters to feed to the spreadsheet based SOFC electrochemical model, which predicts the power generation from the SOFC. The SOFC process simulation parameters include the composition of fuel, standard molar enthalpy and entropy changes of the system. An example is shown in Table 16.5. In the example, the syngas produced and cleaned from biomass gasification has been used as the fuel to the SOFC anode, which is oxidized with the oxygen obtained from the air separation in the SOFC cathode. The flow of air to the cathode can be adjusted so that the stoichiometric amount of pure oxygen can flow from the cathode to the anode for complete oxidation of the fuel in the anode. Thus, the flue gas generating from the oxidation of the fuel gas in the anode consists of steam, nitrogen and

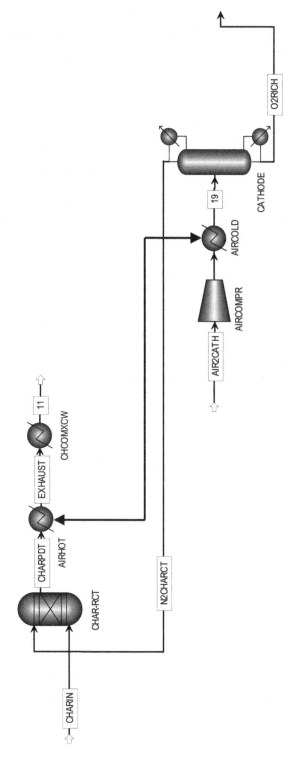

Figure 16.8 Heat recovery from char combustor exhaust gas into preheating air to the SOFC cathode or heat balance between AIRHOT and AIRCOLD.

Table 16.4 Summary of energetic analysis of the BGFC flowsheet. (Reproduced with permission from Sadhukhan et al. (2010)[3]. Copyright © 2010, Elsevier.)

Energy Generation or Consumption	kW
Power generation from SOFC based on 85% fuel efficiency	652.61
Power generation from steam turbines	9.43
Power consumption by compressors	58.05
Sources of low grade heat	
Waste heat recovery (WHR) from char combustor exhaust (CHCOMXCW in Figure 16.8)	47.48
WHR from SOFC or HRSG exhaust (WATERHOT and B11 in Figure 16.6)	2.8 (sensible heat) 31.18 (condensation heat)

carbon dioxide and can be a source of pure water on condensation of the flue gas below its dew point. The fuel gas (in this case syngas) needs to be free from contaminants in order to be a good source of pure water. For the anode of the SOFC, it is essential to avoid contaminants in the fuel. There are two sets of flow rates chosen in an Aspen Plus simulation and the enthalpy and entropy changes are noted. The parameters used in the spreadsheet based SOFC electrochemical model (Equations (16.1) to (16.4)) are shown in Table 16.6.

The electrochemical model adopted for the SOFC is an irreversible and steady-state model, in which the irreversibilities resulting from the electrochemical reaction, electrical resistance and heat transfer to the environment are taken into account. In addition, the following assumptions/approximations are usually made without any loss of accuracy: fuel in the anode inlet consists of hydrogen, carbon monoxide, methane, nitrogen, carbon dioxide and water, amongst which the first three species participate in the reaction, although the oxidation of hydrogen is much faster than the other two (Table 16.6). In the model given, all the reacting species in the place of partial pressure of hydrogen and nonreacting species are taken into account to replace the partial pressure of P_{H_2} and P_{H_2O}, respectively. The enthalpy and entropy changes are extracted from the Aspen Plus simulation results in Table 16.5. The SOFC output voltage is less than the reversible cell voltage because there are some voltage drops across the cell caused by irreversible losses. The electrochemical model takes account of these losses, due to overpotential, or polarization, including three main sources: activation, ohmic and concentration overpotential. In terms of thermodynamic descriptors of the SOFC system, the Gibbs free energy and enthalpy changes are developed for electrochemical reactions, which accounted for both the electrical as well as the chemical parameters (Tables 16.5 and 16.6).

It can be noted that the enthalpy and entropy changes obtained from the Aspen Plus simulation (Table 16.5) reflect the nature of the chemical reactions occurring in the anode chamber of the SOFC system at the given temperature and

Table 16.5 Input parameters to the SOFC electrochemical model generated from the Aspen Plus simulation. (Reproduced with permission from Sadhukhan et al. (2010)[3]. Copyright © 2010, Elsevier.)

Mole Fraction	Syngas to Anode	Oxygen to Anode	Flue Gas from Anode	Change
H_2O	0.01	0.00	0.46	
N_2	0.17	0.00	0.17	
O_2	0.00	1.00	0.00	
H_2	0.44	0.00	0.00	
CO	0.22	0.00	0.00	
CO_2	0.15	0.00	0.38	
CH_4	0.01	0.00	0.00	
Option 1				
Total Flow (kmol h^{-1})	0.024	0.00875	0.0248	
Enthalpy (J mol^{-1})	−61 443.909	19 187.489	−223 371.943	$\Delta h°$ (−170 691.07)
Entropy (J mol^{-1} °C^{-1})	43.359	34.492	40.313	$\Delta s°$ (−13.809)
Option 2				
Total Flow (kmol h^{-1})	0.0258	0.0091	0.0263	
Enthalpy (J mol^{-1})	−70 426.953	12 101.013	−226 505.027	$\Delta h°$ (−161 604.03)
Entropy (J mol^{-1} °C^{-1})	33.534	25.505	39.782	$\Delta s°$ (−1.939)

Table 16.6 An example of SOFC operating conditions and performance-related parameters. (Reproduced with permission from Sadhukhan et al. (2010)[3]. Copyright © 2010, Elsevier.)

Parameters	Value
Operating pressure, p_0 (bar)	1
Fuel composition, p_{H_2}; p_{H_2O}	From Aspen Plus*
Air molar composition, p_{O_2}; p_{N_2}	0.21; 0.79
Charge-transfer coefficient, β	0.5
Number of electrons, n_e	2
Pre-factor for anode exchange current density, γ_a (A m^{-2})	5.5×10^8
Activation energy of anode, $E_{act,a}$ (J mol^{-1})	1.0×10^5
Pre-factor for cathode exchange current density, γ_c (A m^{-2})	7.0×10^8
Activation energy of cathode, $E_{act,c}$ (J mol^{-1})	1.2×10^5
Electrolyte thickness, L_{el} (μm)	20
Activation energy of O^{2-}, E_{el} (J mol^{-1})	8.0×10^4
Pre-factor of O^{2-}, σ_0 (S m^{-1})	3.6×10^7
Ratio of the internal resistance to the leakage resistance, k	1/100
Anode limiting current density, $i_{L,a}$ (A m^{-2})	2.99×10^4
Cathode limiting current density, $i_{L,c}$ (A m^{-2})	2.16×10^4
Faraday constant, F (C mol^{-1})	96 485
Universal gas constant, R (J mol^{-1} K^{-1})	8.314
Standard molar enthalpy change at 1073 K, $\Delta h°$ (J mol^{-1})	From Aspen Plus*
Standard molar entropy change at 1073 K, $\Delta s°$ (J mol^{-1} K^{-1})	From Aspen Plus*

*Obtain from Table 16.5.

pressure conditions and given compositions of the syngas and air inputs to the SOFC. The total enthalpy change of the SOFC is divided into electrical and thermal energies. As long as the enthalpy change is more negative than the Gibbs free energy change of the reaction, a part of the total energy is released as heat. This part is a measure of the entropy loss or irreversibility of the SOFC system. Once all these entropy losses (henceforth power losses due to irreversibilities) are calculated, the net power generation is predicted from the difference between the theoretical or thermodynamically maximum power generation and the power losses due to irreversibilities. The final equations derived in this way to predict the power generation (in watts) and efficiency of a SOFC are

$$P_{fc} = \frac{iA}{n_e F}\left(m - \frac{k}{RTd_1}m^2\right) \tag{16.1}$$

$$\eta_{fc} = \frac{P_{fc}}{-\dfrac{iA}{n_e F}\Delta h°} = \frac{1}{-\Delta h°}\left(m - \frac{k}{RTd_1}m^2\right) \tag{16.2}$$

where

$$d_1 = 2n_e \sinh^{-1}\left(\frac{i}{2i_{0,a}}\right) + 2n_e \sinh^{-1}\left(\frac{i}{2i_{0,c}}\right) - \ln\left(1 - \frac{i}{i_{L,a}}\right) - \ln\left(1 - \frac{i}{i_{L,c}}\right) + \frac{in_e F L_{el}}{\sigma_0 R}\exp\left(\frac{E_{el}}{RT}\right)$$

$$k = R_{int}/R_{leak}$$

and

$$m = -\Delta h° + T\Delta s° + RT\ln\left(\frac{p_{H_2}p_{O_2}^{1/2}}{p_{H_2O}}\right) - RTd_1$$

The terms in d_1 from the left hand side to the right hand side indicate the voltage drops across the anode and cathode caused by activation overpotential (the first two terms), concentration overpotential (the second two terms) and ohmic overpotential (the last term), respectively; i is the current density, $F = 96\,485$ C mol^{-1} is Faraday's constant,

$R = 8.314$ J mol^{-1} K^{-1} is the universal gas constant, n_e is the number of electrons participating in the reaction (e.g., on a per mol of product basis), A is the surface area of the fuel cell polar plate, $\Delta h°$ and $\Delta s°$ are the standard molar enthalpy and entropy changes of the reaction at temperature T, respectively, k denotes the ratio of the equivalent leakage resistance to the internal resistance of the fuel cell and is assumed to be a constant, L_{el} is the thickness of the electrolyte, E_{el} represents the activation energy for ion transport, σ_0 is the reference ionic conductivity, p_0 is the ambient pressure, p_{H_2}, p_{O_2} and p_{H_2O} are the partial pressures of reactants, O_2 and nonreacting components, respectively, and $i_{L,a}$ and $i_{L,c}$ are the limiting current densities of the anode and cathode, respectively. The anode and cathode exchange current densities, $i_{0,a}$ and $i_{0,c}$, are calculated using Equations (16.3) and (16.4), respectively, and the rest of the parameters present in Equations (16.1) and (16.2) are either fixed values or obtained from the Aspen Plus simulation, shown in Tables 16.5 and 16.6. Table 16.6 also shows the nomenclature and the values of the parameters used.

$$i_{0,a} = \gamma_a \left(\frac{p_{H_2}}{p_0}\right)\left(\frac{p_{H_2O}}{p_0}\right)\exp\left(-\frac{E_{act,a}}{RT}\right) \tag{16.3}$$

$$i_{0,c} = \gamma_c \left(\frac{p_{O_2}}{p_0}\right)^{1/4}\exp\left(-\frac{E_{act,c}}{RT}\right) \tag{16.4}$$

where γ_a and γ_c are pre-exponential coefficients for the anode and cathode, respectively. $E_{act,a}$ and $E_{act,c}$ are the activation energies for the anode and cathode, respectively.

16.6　Design of SOFC Based Micro-CHP Systems

Figure 16.9 shows the heat integrated SOFC based micro CHP-generation scheme that utilizes countercurrent preheating of air and syngas to SOFC by nitrogen rich depleted air and flue gas, respectively. Preheating of syngas and air to SOFC can

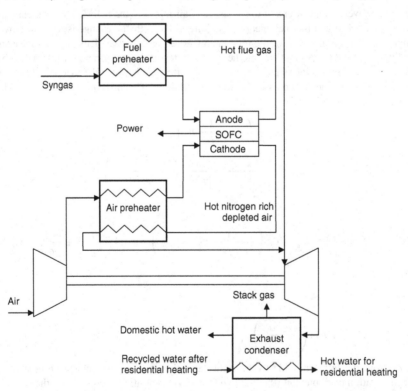

Figure 16.9　*SOFC scheme with countercurrent preheating of air and syngas to SOFC by nitrogen rich depleted air and flue gas, respectively.*

be achieved by exchanging heat with the outlet gases from SOFC, flue gas and nitrogen rich depleted air from the anode and cathode, respectively, based on the thermodynamic matching rule between hot and cold streams with closer specific heat values (see Section 3.5.2). Preheating of feed gases facilitates endothermic reforming reactions and increases the net exothermic heat generation (due to combustion) from the SOFC. Since the Gibbs free energy change of an electrochemical reaction is a measure of the maximum electrical energy obtainable as work from the reaction, preheating of feed gases to a maximum feasible temperature (allowing 20 °C minimum temperature approach between the hot and cold streams) ensures maximum energy output and hence maximum power generation from the SOFC. However, maximum preheating or a power generation strategy is associated with the requirements of the maximum heat exchanger areas. Additionally, there is no high grade heat available from the flue gas and nitrogen rich depleted air from the SOFC in this option, which can be used for space heating. Henceforth, a co-current preheating option can also be explored, which keeps a better balance between the heat-to-power ratio obtainable from the SOFC and the heat exchanger area required (Option 2). Option 3, without any preheating, provides the maximum heat-to-power ratio with a minimal exchanger area required.

In all three pressurized SOFC schemes, condensation of the exhaust gas from the SOFC after expansion, below its dew point, has been considered. After combining the outlet nitrogen rich depleted air and flue gas from the cathode and anode respectively, the resulting exhaust gas contains the following constituents: water, nitrogen and carbon dioxide. Hence condensation of the exhaust gas below its dew point by cooling water supplied at 20 °C results into a gas free water phase and a dry gas phase. Cooling water after recovering heat from the SOFC exhaust gas at ~60 °C could be used for space heating and returned back to the exhaust gas condenser after providing heat. The water phase recovered from the SOFC exhaust gas (at the dew point ~58 °C) can be used as the hot water. Thus, the hot water resulting from the hydrogen content in the SOFC feed gas can also be utilized for space heating. The remaining dry stack gas is exhausted into the atmosphere.

A comparison of energetic analysis among the three SOFC design options is shown in Table 16.7. The heat transfer related results, such as heat duties of exchangers, outlet temperatures of either of the hot or cold streams, log mean temperature differences (LMTDs), overall heat transfer coefficients (U) and heat exchanger areas, as shown in Table 16.7, can be used for future scenario evaluations. The power generation from the combined air compressor and exhaust gas expander for the three SOFC design options is shown in Table 16.7. The total residential heat consists of the heat recovered from the exhaust condenser as well as the condensate recovered as domestic hot water from the remaining stack gas. The heat recovered into the cooling water through the condensation of the exhaust gas from the SOFC and the hot water produced from the exhaust gas are the same irrespective of the SOFC schemes, 0.382 kW and 0.24 kW, respectively. Hence, thermal results for the cooler part of the exhaust condenser to 60 °C are only shown in Table 16.7 for a comparison between the three options.

A ground source heat pump (GSHP) can provide effective solutions to heating or cooling applications (including space heating as well as hot water) for all types of buildings and displays a unique integration synergy with an SOFC process. The GSHP works on the heat pump or refrigeration principle, in which the solar energy stored in the ground is absorbed via evaporation or heating of a refrigerant at low pressure, followed by compression of the refrigerant vapor. The high pressure and high temperature refrigerant vapor is then condensed or cooled, providing residential heating, expanded and sent back to extract heat from the ground. Figure 16.10 shows a GSHP refrigeration cycle in its simplest form and modified GSHP loop after coupling with the SOFC. The cycle can also be operated to provide cooling (in summer months) instead of heating (in winter months). The technology exploits the fact that the temperature of the earth at a moderate depth is slightly higher in winter than the air temperature at the surface (and in summer slightly lower). A pressurized SOFC and GSHP have a thermodynamic integration synergy between them to provide space heating and hot water. The high pressure and high temperature condensation or cooling part of a GSHP cycle can be replaced by an SOFC, as also shown in Figure 16.10. In general, a water and antifreeze mixture is used as the refrigerant in a GSHP. The exhaust gas from an SOFC using syngas as fuel typically has water (20% by mole), nitrogen (63% by mole) and carbon dioxide (16% by mole), respectively, and inherently exhibits properties of the refrigerant.

The *process simulation framework for an integrated fuel cell and GSHP system design is provided in Online Resource material, Chapter 16 – Additional Exercises and Examples.*

16.7 Fuel Cell and SOFC Design Parameterization Suitable for Spreadsheet Implementation

The term fuel cell used in this section implies any fuel cell. The model provided in this section is suitable for implementation and parameterization in an Excel spreadsheet.

Table 16.7   *Comparison of energetic analysis among three design options of an SOFC based micro-CHP system.*

Heat exchanger units		Air preheater	SOFC fuel gas preheater	Exhaust cooler
Hot stream		N_2 rich depleted air from SOFC cathode	Flue gas from SOFC anode	Exhaust from GT
Cold stream		Air to SOFC cathode	Fuel gas to SOFC anode	Cooling water for residential heat
Option 1		Minimum temperature driving force/maximum heat recovery in preheaters/maximum SOFC power generation		
Hot side temperatures	(°C)	Hot stream/cold stream: 800/660.9	800/775	85.6/60[b]
Cold side temperatures	(°C)	228.8/208.8	244.2/60	60/20
Heat duty	(kW)	0.5696	0.5739	0.0737[a]
Area	(m²)	0.0109	0.0085	0.0027
U	(W m⁻² K⁻¹)	850.00	850.00	850.01
LMTD	(°C)	61.41	79.71	32.27
Total area	(m²)		0.0221	
SOFC/GT power	(kW)		3.94/0.059	
SOFC current density	(A)		29.70	
Total residential heat	(kW)		0.48	
Power-to-heat ratio	(%)		89.29	
Option 2		Cocurrent preheaters with minimum temperature approach		
Hot side temperatures	(°C)	Hot stream/cold stream: 800/208.8	800/60	277.6/60[b] (countercurrent)
Cold side temperatures	(°C)	487.9/467.9	497.6/477.6	60/20 (countercurrent)
Heat duty	(kW)	0.3195	0.3249	0.4286[a]
Area	(m²)	0.0012	0.0010	0.0048
U	(W m⁻² K⁻¹)	850.00	850.00	850.01
LMTD	(°C)	304.83	377.07	104.85
Total area	(m²)		0.0071	
SOFC/GT power	(kW)		3.44/0.204	
SOFC current density	(A)		26.30	
Total residential heat	(kW)		0.83	
Power-to-heat ratio	(%)		81.36	
Option 3		No preheating of air and fuel gas to SOFC/maximum residential heat generation		
Hot side temperatures	(°C)			322.4/60[b]
Cold side temperatures	(°C)			60/20
Heat duty	(kW)			0.5142[a]
Area	(m²)			0.0051
U	(W m⁻² K⁻¹)			850.01
LMTD	(°C)			118.24
Total area	(m²)		0.0051	
SOFC/GT power	(kW)		3.32/0.237	
SOFC current density	(A)		25.80	
Total residential heat	(kW)		0.92	
Power-to-heat ratio	(%)		79.45	

[a]Contributing to residential heat.
[b]Just above the dew point.

Hydrogen is a more common fuel for fuel cells, but other gases such as syngas with equal volumes of hydrogen and carbon monoxide and biogas rich in methane are also becoming prominent as feedstock to fuel cells, as long as the gas is free from impurity or at the most contains trace amounts of impurities below 1 ppb. Though any mechanistic studies have not been carried out for fuel cell reactions with gas mixtures, the reactions are commonly assumed to be reforming reactions (of methane or low molecular weight alkanes), water gas shift reactions (of carbon monoxide) and hydrogen oxidation within the cells. The outlet compositions from the fuel cells suggest that almost complete conversions are possible. A single fuel cell unit may be repeated to reach a technologically and economically viable power level (hence reaction completion). The usual chemical reaction potentials are powered by electrochemical reaction potentials. The three step reactions mentioned are shown below:

$$\text{Reforming: } CH_4 + H_2O \rightarrow CO + 3H_2$$

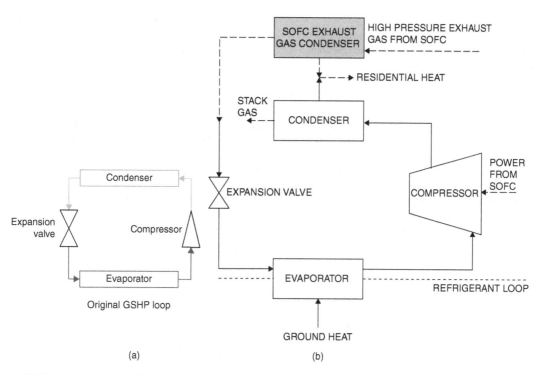

Figure 16.10 (a) In the original GSHP cycle, a condenser is connected to an expansion valve; (b) an SOFC exhaust gas condenser is connected to the expansion valve in the proposed integrated scheme and the modifications are shown in the shaded area and by dotted arrows.

The CO formed by the reforming reaction also undergoes the water gas shift reaction, in addition to the carbon monoxide present in the fuel gas:

$$\text{Water gas shift: } CO + H_2O \rightarrow CO_2 + H_2$$

$$\text{Hydrogen oxidation: } H_2 + 0.5O_2 \rightarrow H_2O$$

A complete model of the SOFC should have the models for reforming, water gas shift and hydrogen oxidation reactions, electrochemical polarizations, mass transfer, heat transfer and overall energy balance. The model can then readily evaluate cell performance related quantities, such as stack voltage and stack power.

16.7.1 Mass Balance

In an electrochemical reaction, the fuel consumed is related to the current generated by Faraday's law:

$$n_{in}(j) = n_{in}x_j = \frac{iAx_j}{n_eFU_fx_{fc}} \quad \forall j \in \{H_2, CO, CH_4\} \tag{16.5}$$

where n_{in} is the molar flow rate of fuel intake to the cell. The reacting components of the fuel in the given model are hydrogen, carbon monoxide and methane; hence $j \in \{H_2, CO, CH_4\}$ in Equation (16.5). The reacting components can be extended to include hydrocarbons, such as methanol, ethanol, etc.:

$n_{in}(j)$ is the molar flow rate of fuel j to the cell, $j \in \{H_2, CO, CH_4, H_2O, CO_2\}$.
x_j is the molar fraction of fuel j intake to the cell, $j \in \{H_2, CO, CH_4, H_2O, CO_2\}$.

i is the current density.

A is the area of the cell for ionic and electronic transfer.

n_e is the number of electrons transferred in reactions.

Faraday's constant, $F = 96\,485$ C mol^{-1} (C is coulomb).

U_f is the fuel utilization factor on a molar basis.

$x_{fc} = x_{H_2} + x_{CO} + 4x_{CH_4} =$ moles of hydrogen produced per mole of the fuel.

In addition, steam is supplied to carry out the reforming and water gas shift reactions. Double the stoichiometric amount of steam required is supplied in order to avoid carbon deposition. The molar flow rate of steam needed is given by

$$n_{H_2O} = 2\{n_{in}(CO) + 2n_{in}(CH_4)\} \tag{16.6}$$

The fuel cell can be designed to complete the steam reforming of methane and water gas shift of carbon monoxide reactions, thus leaving no methane and carbon monoxide in the effluent gas from the fuel cell. The exit gas consists of carbon dioxide formed (Equation (16.7)), excess hydrogen if any (Equation (16.8)), steam (Equation (16.9)), nitrogen from the air supply (Equation (16.10)) and remaining oxygen (Equation (16.11)):

$$n_{out}(CO_2) = n_{in}(CO_2) + n_{in}(CO) + n_{in}(CH_4) \tag{16.7}$$

$$n_{out}(H_2) = n_{H_2} \times (1 - U_f) \tag{16.8}$$

$$n_{out}(H_2O) = \frac{n_{H_2O}}{2} + n_{H_2} \times U_f \tag{16.9}$$

$$n_{out}(N_2) = n_{air-in}(N_2) = n_{air-in}(O_2) \times \frac{0.79}{0.21} \tag{16.10}$$

$$n_{out}(O_2) = n_{air-in}(O_2) \times (1 - U_{air}) \tag{16.11}$$

where

$n_{out}(j)$ is the molar outlet flow rate of component j from the cell, $j \in \{CO_2, H_2, H_2O, N_2, O_2\}$.

n_{air-in} is the inlet molar flow rate of air to the cell.

Inlet air is assumed to have 79% nitrogen and 21% oxygen by volume.

U_{air} is the utilization factor of inlet oxygen on a molar basis.

16.7.2   Electrochemical Descriptions

The theoretical maximum voltage corresponding to the minimum Gibbs free energy change across the cell is given by the *Nernst equation*. The Gibbs free energy change is related to the process operating conditions, in this case, cell temperature and pressure. The Nernst equation and the equation for Gibbs free energy change are shown, respectively, by

$$E = \frac{-\Delta g(T, P)}{n_e F} \tag{16.12}$$

$$-\Delta g(T, P) = \Delta g^o(T) - RT \ln \left(\frac{P_{H_2} P_{O_2}^{0.5}}{P_{H_2O}} \right) \tag{16.13}$$

where

$\Delta g(T, P)$ is the molar Gibbs free energy change as a function of temperature (T) and pressure (P).

$\Delta g^o(T)$ is the molar Gibbs free energy change as a function of T at a pressure of 1 atm.

R is the universal gas constant, 8.314 J mol^{-1} K^{-1}.

P_j is the partial pressure of component j.

Equation (16.13) does not take account for the partial pressures of other chemicals in the fuel feedstock, carbon monoxide and methane. This assumption is reasonable as hydrogen oxidation is the main contributor towards the Gibbs free energy change.

The voltage calculated by the Nernst Equation (16.12) is the theoretical maximum voltage obtained corresponding to the reversible process that results in the minimum Gibbs free energy change. However, the actual voltage is less than the theoretical voltage due to the resistances and overpotential losses. Three types of polarization effects were considered: activation, ohmic and concentration, shown by Equations (16.14) to (16.19).

Activation overpotentials are a result of resistance to electrochemical reaction kinetics occurring in the anode and cathode, shown, respectively, by

$$V_{activation,anode} = \frac{2RT}{n_e F} \sinh^{-1}\left(\frac{i}{2i_{0,anode}}\right) \tag{16.14}$$

$$V_{activation,cathode} = \frac{2RT}{n_e F} \sinh^{-1}\left(\frac{i}{2i_{0,cathode}}\right) \tag{16.15}$$

where

$V_{activation,anode/cathode}$ is the voltage loss due to activation overpotentials in anode/cathode.
$i_{0,anode/cathode}$ is the anode/cathode exchange current density.

Ohmic overpotentials (V_{ohmic}) are due to the resistance to conduction of ions through the electrolyte, electrons through the electrodes and current collectors and by contact resistance between cell components, shown by

$$V_{ohmic} = i\left(\frac{L_{electrolyte}}{\sigma_{electrolyte}} + \frac{L_{anode}}{\sigma_{anode}} + \frac{L_{cathode}}{\sigma_{cathode}} + \frac{L_{interconnect}}{\sigma_{interconnect}}\right) \tag{16.16}$$

where L_k is the thickness, $\forall k \in \{electrolyte, anode, cathode, interconnect\}$; σ_k is the electronic or ionic conductivity in k, $\forall k \in \{electrolyte, anode, cathode, interconnect\}$, shown by the following equations:

$$\sigma_k = \frac{C1_k}{T} \exp\left(\frac{C2_k}{T}\right) \quad \forall k \in \{anode, cathode, interconnect\} \tag{16.17a}$$

$$\sigma_k = C1_k \exp\left(\frac{C2_k}{T}\right) \quad \forall k \in \{electrolyte\} \tag{16.17b}$$

where $C1_k$ and $C2_k$ are the constants shown in Table 16.8. Concentration overpotentials caused by mass transfer limitations from the gas phase and through the electrode result in a further voltage loss, shown by

$$V_{concentration,anode} = -\frac{RT}{n_e F} \ln\left(1 - \frac{i}{i_{l,anode}}\right) \tag{16.18}$$

$$V_{concentration,cathode} = -\frac{RT}{n_e F} \ln\left(1 - \frac{i}{i_{l,cathode}}\right) \tag{16.19}$$

where $V_{concentration,anode/cathode}$ is the voltage loss due to concentration overpotentials in the anode/cathode and $i_{l,anode/cathode}$ is the limiting current density of the anode/cathode, assumed constant, shown in Table 16.8.

Table 16.8 *SOFC modeling parameters in Equations (16.5) to (16.21). (Reproduced with permission from Zhao et al. (2011)[5]. Copyright © 2011, Elsevier.)*

Parameter	Symbol	Value	Unit
Ambient temperature	(required for the calculation of $\Delta g°$)	298	K
Ambient pressure		1	atm
Fuel utilization	U_f	0.8	Molar fraction
Oxygen utilization	U_{air}	0.2	Molar fraction
Number of electrons	n_e	2	(based on hydrogen mole)
Anode exchange current density	$i_{0,anode}$	6500	A m^{-2}
Cathode exchange current density	$i_{0,cathode}$	2500	A m^{-2}
Limiting current density, anode/cathode	$i_{l,anode/cathode}$	9000	A m^{-2}
Anode thickness	L_{anode}	500	μm
Cathode thickness	$L_{cathode}$	50	μm
Electrolyte thickness	$L_{electrolyte}$	10	μm
Interconnect thickness	$L_{interconnect}$	0.3	cm
Anode conductivity constants	$C1_{anode}, C2_{anode}$	$95 \times 10^6, -1150$	
Cathode conductivity constants	$C1_{cathode}, C2_{cathode}$	$42 \times 10^6, -1200$	
Electrolyte conductivity constants	$C1_{electrolyte}, C2_{electrolyte}$	$3.34 \times 10^4, -10\,300$	
Interconnect conductivity constants	$C1_{interconnect}, C2_{interconnect}$	$9.3 \times 10^6, -1100$	
Air blower power consumption factor	$\omega_{air\ blower}$	0.1	

The net voltage thus obtained is the Nernst theoretical voltage (Equation (16.12)) minus the voltage losses (Equations (16.14) to (16.19)), shown by

$$V = E - V_{activation} - V_{ohmic} - V_{concentration} = E - \frac{RT}{n_e F} d_1 \qquad (16.20)$$

where E is shown in Equation (16.12) and

$$d_1 = 2\sinh^{-1}\left(\frac{i}{2i_{0,anode}}\right) + 2\sinh^{-1}\left(\frac{i}{2i_{0,cathode}}\right)$$

$$+ \frac{i n_e F}{RT}\left(\frac{L_{electrolyte}}{\sigma_{electrolyte}} + \frac{L_{anode}}{\sigma_{anode}} + \frac{L_{cathode}}{\sigma_{cathode}} + \frac{L_{interconnect}}{\sigma_{interconnect}}\right)$$

$$- \ln\left(1 - \frac{i}{i_{l,anode}}\right) - \ln\left(1 - \frac{i}{i_{l,cathode}}\right)$$

16.7.3 An air Blower Power Consumption

An air blower is required to supply air to overcome the pressure drop across the cell component. The power consumption by the air blower ($W_{air\ blower}$) of a fuel cell is about 10% of the power generated by the fuel cell ($W_{fuel\ cell}$). Hence, the net power generation from the fuel cell module will be 10% less than the power generation from the fuel cell itself, taking account of the power consumption by the air blower. The air blower power consumption relation with fuel cell power generation is shown by

$$W_{air\ blower} = W_{fuel\ cell} \times \omega_{air\ blower} \qquad (16.21)$$

where $\omega_{air\ blower}$ is the ratio of the power consumption by the air blower to power generation by the fuel cell, usually at 0.1.

16.7.4   Combustor Modeling

A combustor may be needed after the fuel cell to ensure complete oxidation of the exhaust gas from the fuel cell. Hence, the exhaust gas from the combustor does not contain any combustible component, such as hydrogen. The exhaust gas from the combustor thus has steam (Equation (16.22)), carbon dioxide (Equation (16.23)), nitrogen (Equation (16.24)) and some remaining oxygen (Equation (16.25)). It is assumed that the outlet gas from the fuel cell contains an adequate amount of oxygen required for its complete combustion or, in other words, air containing slightly over a stoichiometric amount of oxygen of the fuel gas was supplied by the air compressor to the fuel cell.

$$n_{exhaust}(H_2O) = n_{out}(H_2O) + n_{out}(H_2) \tag{16.22}$$

where $n_{out}(H_2O)$ and $n_{out}(H_2)$ are the steam and hydrogen molar flow rates from the fuel cell, obtained from Equations (16.8) and (16.9), respectively, and $n_{exhaust}(j)$ is the molar flow rate of component j in the exhaust gas from the combustor; hence, here $\forall j \in \{H_2O, CO_2, N_2, O_2\}$.

$$n_{exhaust}(CO_2) = n_{out}(CO_2) \tag{16.23}$$

where $n_{out}(CO_2)$ is the carbon dioxide molar flow rate from the fuel cell, obtained from Equation (16.7).'

$$n_{exhaust}(N_2) = n_{out}(N_2) \tag{16.24}$$

where $n_{out}(N_2)$ is the nitrogen molar flow rate from the fuel cell, obtained from Equation (16.10).

$$n_{exhaust}(O_2) = n_{out}(O_2) - 0.5\, n_{out}(H_2) \tag{16.25}$$

where $n_{out}(O_2)$ is the carbon dioxide molar flow rate from the fuel cell, obtained from Equation (16.11).

16.7.5   Energy Balance

The total energy balance must take account of the heat generation from the combustor, heat recovery from the exhaust gas (or the amount of sensible heat available in the exhaust gas above its dew point) and the enthalpy change in the fuel cell process. Any preheating duty of the fuel gas involved must be subtracted from the total enthalpy output to calculate the net energy output.

   Thus, there are various sources/sinks of enthalpy from the interactions between the fuel cell and the environment, shown below:

1.   Feed preheating (sink)
2.   Reforming and water gas shift reactions for methane or lower hydrocarbons (sink, though the water gas shift reaction is slightly exothermic)
3.   Enthalpy from the fuel cell electrochemical reaction (source)
4.   Exhaust gas cooling (source).

Thus, the total absolute values of the required enthalpies by the sinks are subtracted from the total values of the available enthalpies from the sources to obtain the net energy output from the fuel cell system to the environment. Figure 16.11 shows the conceptual block diagrams for the analysis of the SOFC enthalpy output and sensible heat available from the exhaust gas. The difference in the net enthalpy outputs between the two block diagrams is the sensible heat available from the exhaust gas. The enthalpy balance across the upper block diagram shows the net SOFC enthalpy output, with integrated feed preheating and reforming, water gas shift and electrochemical reactions.

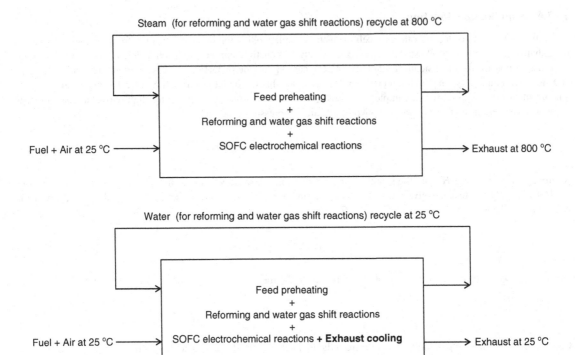

Figure 16.11   *Conceptual block diagrams for the analysis of the SOFC enthalpy output and sensible heat available from the exhaust gas.*

The sensible heat available from the exhaust gas and the net SOFC enthalpy output (Gibbs free energy change from electrochemical reactions – (enthalpy for feed preheating + enthalpy for reforming – enthalpy for water gas shift)) are analyzed separately with respect to the molar fraction of methane in the fuel, assuming equal moles of hydrogen and carbon monoxide comprising the balance of the fuel, as shown in Figure 16.12. The stoichiometric molar flow rate of air is assumed. The temperature of the SOFC is at 800 °C. It can be seen that with the SOFC temperature variation, individual sensible heat available from the exhaust gas and the net SOFC enthalpy output will vary, but the total energy output from the SOFC system to the environment will remain the same for the ambient or basis temperature of 25 °C. The thermodynamic property package used for deriving the linear correlations shown is the Peng–Robinson property package in Aspen Plus.

Assume that y is the enthalpy in joule per mole and x is the methane mole fraction in the fuel. With increasing methane fraction in the fuel gas, the overall enthalpy output from the SOFC system increases, in spite of increases in reforming reactions. This is because one mole of methane provides two moles of hydrogen. Two extreme cases can be generated using the correlations shown in Figure 16.12: no methane in the fuel and above 90% molar fraction of methane (e.g., biogas contains 60-90% mole fraction of methane depending on the process configuration). The correlations are representative as long as the fuel gas constituents are methane and (equimolar fractions of) hydrogen and carbon monoxide. An Aspen Plus simulation is recommended for different fuel compositions.

Power generation from SOFC is the net power generation and the power requirement for driving the air blower. This is equal to the Gibbs free energy change subtracted by all the voltage losses. The deduction is shown in Equations (16.26) and (16.27):

$$T_oS = I(V_{activation} + V_{ohmic} + V_{concentration}) \tag{16.26}$$

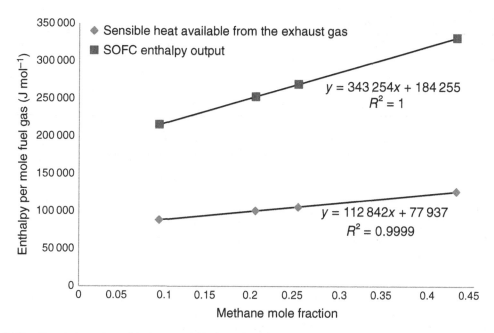

Figure 16.12 *Correlations of sensible heat available from the exhaust gas and SOFC enthalpy output with respect to the methane mole fraction in the fuel gas.*

where S is the total entropy at T_o or ambient temperature.

$$W_{fuel\ cell} + W_{air\ blower} = -\Delta G - T_o S = \frac{iA}{n_e F}(-\Delta g - RTd_1)$$

$$W_{fuel\ cell}(1 + \omega_{air\ blower}) = \frac{iA}{n_e F}(-\Delta g - RTd_1)$$

$$W_{fuel\ cell} = \frac{iA(-\Delta g - RTd_1)}{n_e F(1 + \omega_{air\ blower})} \qquad (16.27)$$

where $W_{fuel\ cell}$ is the net power generation from the SOFC. Calculate d_1 from the parametric values shown in Table 16.8 (Equation (16.20)). The absolute value of the Gibbs free energy change $(-\Delta g)$ can be approximated from the SOFC enthalpy output correlation $(-\Delta g = 343\,254x_{CH_4} + 184\,255)$, shown in Figure 16.12.

In Equation (16.27),

T is the temperature in K, 1073 K.
R is the ideal gas constant, 8.314 J mol^{-1} K^{-1}.
A is the cell area in m^2.
n_e is the number of electrons transferred, 2.
F is Faraday's constant, 96 485 C mol^{-1}.
$\omega_{air\ blower}$ is 0.1.
d_1, calculated from the parametric values shown in Table 16.8 and using Equation (16.20), 1.44161.
The power density (i) can be assumed at 1564 W per m^2 for the SOFC.

Table 16.9 *Fuel mole fractions to the SOFC.*

Fuel Component	Composition Set 1	Composition Set 2	Composition Set 3	Composition Set 4	Composition Set 5
Hydrogen	0.4	0.3	0.2	0.1	0.05
Carbon monoxide	0.4	0.3	0.2	0.1	0.05
Methane	0.2	0.4	0.6	0.8	0.9

Thus, substituting these values in Equation (16.27), an empirical model for given specifications for SOFC power generation is obtained in terms of the methane mole fraction in fuel gas to the SOFC, shown by

$$W_{fuel\ cell}(\text{W per m}^2) = 2529\, x_{CH_4} + 1262.85 \qquad (16.28)$$

The model in Equations (16.5) to (16.28) allows changes in parameters and parameterization of any variable in it. The fuel cell current density (i), temperature (T), etc., can be varied. The fuel utilization factor (U_f) and $\omega_{air\ blower}$ can also be varied. However, their values shown in Table 16.8 are more practical and state of the art to use. The fuel utilization factor of above 0.9 is not possible, due to safety concerns.

Further challenges:

1. Calculate the outlet exhaust gas composition and power generation from the SOFC for the fuel compositions to the SOFC, shown in Table 16.9, using the parameters given in Table 16.8, correlations in Figure 16.12 and the model shown in Equations (16.5) to (16.28). The power density (i) is 1564 W per m^2.

16.8 Summary

This chapter shows comprehensive process modeling of fuel cells for biorefinery integration. Three types of SOFC model are shown:

1. Aspen Plus simulation based using a Gibbs reactor model for the SOFC anode and the separator model for the SOFC cathode.
2. Gibbs free energy minimization based modeling to estimate SOFC power generation.
3. Spreadsheet based SOFC model correlating fuel composition with power generation and energy and mass balance.

There exist significant mass (fuel and water) and energy integration potentials between biomass gasification and the SOFC. Syngas and biogas can be important fuels for fuel cells. The process synthesis approach is shown to design high efficiency BGFC processes. Both distributed and microgeneration options using fuel cells are discussed alongside process modeling equations, design parameters and simulation framework.

References

1. W.R. Grove, On voltaic series and the combination of gases by platinum, *Phil Mag*, **14**, 127–130 (1839).
2. F.T. Bacon, *BEAMA J.*, **6**, 61–67 (1954).
3. J. Sadhukhan, Y. Zhao, N. Shah, N.P. Brandon, Performance analysis of integrated biomass gasification fuel cell (BGFC) and biomass gasification combined cycle (BGCC) systems, *Chem. Eng. Sci.*, **65**(6), 1942–1954 (2010).
4. J. Sadhukhan, Y. Zhao, M. Leach, N.P. Brandon, N. Shah, Energy integration and analysis of solid oxide fuel cell based micro-CHP and other renewable systems using biomass waste derived syngas, *Ind. Eng. Chem. Res.*, **49**(22), 11506–11516 (2010).
5. Y. Zhao, J. Sadhukhan, N.P. Brandon, N. Shah, Thermodynamic modeling and optimization analysis of coal syngas fuelled SOFC-GT hybrid systems, *J. Power Sources*, **196**(22), 9516–9527 (2011).

17

Algae Biorefineries

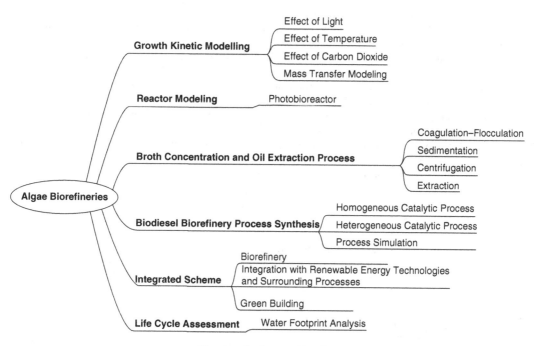

Structure for Lecture Planning

Biodiesel is a mixture of fatty acid methyl esters produced from transesterification of triglycerides. Vegetable oils are the main sources of triglycerides. Oily crop residues are more desired to reduce environmental impacts due to the use of waste to produce biofuel. However, the oily crops are more established. The use of oily crops is dependent on the local availability based on weather conditions. Soybean oil is the main feedstock in the US, rapeseed oil is more available in the EU and *Jatropha* is a tropical plant becoming common in regions such as India, Malaysia, Africa and Latin America. Other feedstocks include sunflower and palm oil. Direct land competition between food and energy crops is one of the

Biorefineries and Chemical Processes: Design, Integration and Sustainability Analysis, First Edition.
Jhuma Sadhukhan, Kok Siew Ng and Elias Martinez Hernandez.
© 2014 John Wiley & Sons, Ltd. Published 2014 by John Wiley & Sons, Ltd.
Companion Website: http://www.wiley.com/go/sadhukhan/biorefineries

Table 17.1 Oil yields from different feedstocks used to produce biodiesel.

Feedstock	Oil Yield (L ha^{-1})
Corn	170
Soybean	450
Sunflower	950
Rapeseed	1190
Jatropha	1890
Palm oil	5940
Algae (10 g m^{-2} d^{-1} at 15% oil)	11 225
Algae (g m^{-2} d^{-1} at 50% oil)	93 540

main challenges faced by crop based biofuels. A more viable alternative to overcome this challenge is the production of biodiesel from algae oil, called *algal biodiesel*.

Algae are photosynthetic organisms that grow in water and in the presence of sunlight and CO_2. While growing, algae produce and accumulate oil suitable for biodiesel production. Hence, growing algae as a source of oil for biodiesel production is an alternative solution for CO_2 capture. Table 17.1 shows oil yields from algae and other common feedstocks. Algae are especially attractive because of their capacity to produce large amounts of oil per unit of land area compared to other feedstocks. Apart from the extracted oil, proteins and the residual biomass can be recovered from algae. Proteins can be used for animal feed and the residual biomass can be used as fertilizer. Furthermore, nutrient requirements, such as nitrogen and phosphorus, can be supplied by wastewater. Thus, the simultaneous treatment of organic waste along with biodiesel production by growing algae could be possible.

Biofuels from algae, however, incur high operating and capital costs during algae cultivation and oil extraction. The optimization of algae production systems involves experimental studies combined with process modeling and economical and environmental impact assessment. These aspects of algal biodiesel are discussed here.

17.1 Algae Cultivation

Algae are a diverse group of unicellular and multicellular organisms living in aquatic ecosystems under different environmental conditions. *Microalgae* are unicellular or have a simple multicellular structure. Many species of microalgae can be cultivated for the production of oil feedstock for biofuels. Oil content in the microalgae cell mass guides the selection of algal species. Table 17.2 shows some known marine and freshwater species of microalgae and their oil content and productivity. Under typical growing conditions, the species have a similar oil content. The effects of changes in factors such as the availability of nutrients, environmental conditions, harvesting and processing methods need to be considered while selecting the species of microalgae.

The two main systems for large-scale cultivation are *open ponds* and *photobioreactors* (PBRs).

17.1.1 Open Pond Cultivation

This type of cultivation system consists of artificial open ponds using natural sunlight for microalgae photosynthesis and growth. As shown in Figure 17.1, *raceway open ponds* consist of oval-shaped recirculation channels. Constant circulation and mixing is required to stabilize algae growth and productivity. This is performed by a motor-operated paddle wheel which also divides the pond into two sides. The algal broth is introduced on one side of the paddle wheel and is circulated along the pond. After growing and circulation, algae biomass is collected at the harvest point on the other side of the paddle wheel. Carbon dioxide is supplied either by the surrounding air or by waste streams from other industries (e.g., flue gas).

The costs and the energy requirements for the production of algae biomass in open pond systems are lower than for other algae cultivation systems. However, they have lower biomass productivity than PBRs and can be easily contaminated with other microorganisms or substances that could kill the growing algae. In algae pond cultivation, it is difficult to control the cultivation conditions (salinity, pH and nutrient availability and distribution) to sustain the growth of the algae

Table 17.2 *Marine and freshwater species of microalgae and their oil content and productivity. (Reproduced with permission from Pienkos (2007)[1].)*

Species	Lipid Content (% dry weight biomass)	Oil Productivity (mg L^{-1} d^{-1})	Land Productivity of Biomass (g m^{-2} d^{-1})
Chlorella emersonii	25.0–63.0	10.3–50.0	0.91–0.97
Chlorella vulgaris	5.0–58.0	11.2–40.0	0.57–0.95
Chlorella sp.	10.0–48.0	42.1	1.61–16.47
Chlorella	18.0–57.0	18.7	3.50–13.90
Dunaliella salina	6.0–25.0	116	1.6–3.5
Nannochloropsis sp.	12.0–53.0	37.6–90.0	1.9–5.3
Phaeodactylum tricornutum	18.0–57.0	44.8	2.4–21
Porphyridium cruentum	9.0–18.8	34.8	25
Scenedesmus sp.	19.6–21.1	40.8–53.9	2.43–13.52
Spirulina maxima	4.0–9.0		25
Tetraselmis suecica	8.5–23.0	27.0–36.4	19

species. Algae biomass is generally kept as a suspension. Nutrient distribution depends on the flow rate and depth of the raceway as well as the mixing imparted by the paddle wheel. CO_2 can be bubbled to promote growth and provide further mixing. Temperature, sunlight irradiation and pond geometry are also important parameters for algae biomass production.

The kinetics of algae growth can be expressed by the product of functions describing the influence of light intensity, temperature, nutrients and pH, as shown by

$$\mu = \mu_{max} f(N)f(I)f(T)f(CO_2) \qquad (17.1)$$

where

μ is the specific growth rate of algae, d^{-1}.
μ_{max} is the maximum algal production under optimum conditions, d^{-1}.
$f(N)$ is a function describing the influence of nutrient concentration.
$f(I)$ is a function describing the influence of light irradiation.
$f(T)$ is a function describing the influence of the temperature.
$f(CO_2)$ is a function describing the influence of CO_2.

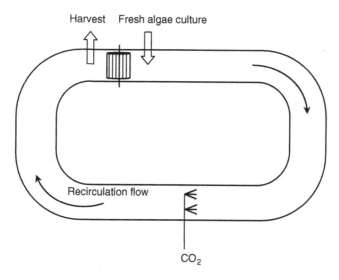

Figure 17.1 *Raceway open pond.*

The functions of the various effects are expressed as efficiency factors. Then, these factors do not have units and have a value ranging between 0 and 1. The supply of nutrients can be controlled as required. Therefore, it can be assumed that there is no variation in nutrient concentration, so that $f(N) = 1$.

17.1.1.1   *Influence of Light*

Algae biomass production increases with light intensity until an optimum point is reached. To describe the effect of light, the simplest approach is to use the Steele equation. This equation is an exponential function of the optimum irradiation per unit of area (*irradiance*), shown by

$$f(I) = \frac{I}{I_{op}} e^{\left(1 - \frac{I}{I_{op}}\right)}$$

(17.2)

where

I is the light irradiance on the liquid surface, W m^{-2}.
I_{op} is the optimum irradiance for maximum algae growth, W m^{-2}.

The response to light depends on the light source and the algal species. When sunlight is used as the light source, the light irradiation is intermittent. If data on the hourly irradiation received in the location of the pond is known, it is possible to track the variations in algae growth and biomass production during the day and throughout the year. When continuous light is supplied, the algae growth rate decreases after the optimum irradiation point is reached. A model for continuous light irradiation can account for the intermittency of sunlight during a day using a factor called the *fractional day length*. This factor varies between 0 and 1.

A more detailed modeling approach considers the effect of the irradiance on the *algae photosynthesis* process within the cells. The growth rate in the following equation is related to the carbon fixation per amount of irradiation during photosynthesis, the ratio of chlorophyll to carbon and the cell respiration rate[2]:

$$\mu_{(light)} = P_{C,max} \left(1 - e^{-\frac{aI\theta}{P_{C,max}}}\right) - r_m$$

(17.3)

where

$\mu_{(light)}$ is the specific growth rate of algae, d^{-1}.
$P_{C,max}$ is the maximum carbon-specific photosynthesis rate at light saturation, d^{-1}.
a is a parameter relating the photon flux with the carbon or C fixation per amount of chlorophyll, g C (mol^{-1} photons) m^2 g^{-1} chlorophyll.
I is the photon flux measured, (mol photons) m^{-2} d^{-1}.
θ is the mass ratio of chlorophyll to carbon, g chlorophyll g^{-1} C.
r_m is the maintenance metabolic coefficient equal to the respiration rate at $\mu = 0$, d^{-1}.

Equation (17.3) is valid when the rest of the influencing factors (nutrients, temperature, etc.) are optimal. Irradiance here is measured as the photon flux. It is possible to convert this measure to units of W m^{-2} but this depends on the light source and its characteristic range of wavelength. For the sunlight wavelength of 400–700 nm, the factor for the conversion from irradiance in photon flux as μmol m^{-2} s^{-1} to irradiance in W m^{-2} is 0.219. The value of θ is found from[2]

$$\theta = \theta_{max} \left(\frac{1}{1 + (\theta_{max} aI)/(2P_{C,max})}\right)$$

(17.4)

where θ_{max} is the maximum mass ratio of chlorophyll to carbon, g chlorophyll g^{-1} C.

Exercise 1. Calculate the specific growth rate of an algae species X in a location with sun irradiation during a summer day of 150 W m^{-2}. The following parameters have been determined:

$a = 5$ g C (mol^{-1} photons) m^2 g^{-1} chlorophyll
$P_{C,max} = 1$ d^{-1}
$r_m = 0.05$ d^{-1}
$\theta_{max} = 0.09$ g chlorophyll g^{-1} C

Solution to Exercise 1. First, the value of the irradiance is transformed to the photon flux units (mol photons m^{-2} d^{-1}) using the conversion factor given (1 μmol m^{-2} s^{-1} = 0.219 W m^{-2}) as

$$I = \frac{150}{0.219} \times \frac{3600 \times 24}{1\,000\,000} = 59.2 \text{ mol photons m}^{-2} \text{ d}^{-1}$$

The value of the mass ratio of chlorophyll to carbon (g chlorophyll g^{-1} C), θ, is determined from Equation (17.4):

$$\theta = 0.09 \left(\frac{1}{1 + (0.09 \times 5 \times 59.2)/(2 \times 1)} \right) = 0.0063$$

The specific growth rate as a function of the sunlight irradiation is calculated using Equation (17.3):

$$\mu_{(light)} = 1 \times \left(1 - e^{-\frac{5 \times 59.2 \times 0.0063}{1}} \right) - 0.05 = 0.8 \text{ d}^{-1}$$

17.1.1.2   *Influence of Temperature*

The algae growth rate increases with temperature. A maximum rate is achieved at an optimal value of temperature. If the temperature increases beyond that value, the growth rate declines. The effect of temperature can be approximated using an exponential function of the form shown by

$$f(T) = e^{-K(T - T_{op})^2} \tag{17.5}$$

where

T is the liquid temperature, °C.
T_{op} is the optimal temperature for maximum algae growth, °C.
K is an empirical constant, °C^{-2}.

In reality the influences of light irradiation and temperature are not independent of each other, since light irradiation affects the temperature of the liquid. Developing an accurate correlation between the growth rate and temperature is difficult for an open pond since temperature can vary with the channel length and depth, while measurements are taken at few specific points. Another model takes into account the optimum temperature and the temperature limit at which algae can not grow. This correlation is shown as follows[3]:

$$\mu_{(T)} = \mu_{(T),max} \left(\frac{T_d - T}{T_d - T_{op}} \right)^\beta e^{-\beta \left(\frac{T_d - T}{T_d - T_{op}} - 1 \right)} \tag{17.6}$$

where

$\mu_{(T),max}$ is the maximum specific growth rate at optimum temperature, d^{-1}.
T is the liquid temperature, °C.
T_{op} is the optimal temperature for maximum algae growth, °C.
T_d is the temperature limit for algae growth, °C.
β is a correlation parameter.

If the effect of light irradiation and temperature are combined, the effective specific growth rate can be expressed as

$$\mu = \mu_{(light)} \left(\frac{T_d - T}{T_d - T_{op}} \right)^{\beta} e^{-\beta \left(\frac{T_d - T}{T_d - T_{op}} - 1 \right)}$$ (17.7)

17.1.1.3 *Influence of CO$_2$*

The effect of CO$_2$ depends on the CO$_2$ dissolved from air and the CO$_2$ released to the liquid medium during transpiration of algae. When CO$_2$ is dissolved in water, carbonic acid is formed by the reaction

$$H_2CO_3 \leftrightarrows HCO_3^- + H^+$$ (17.8)

To simplify, the concentration of CO$_2$ is assumed to be equal to the concentration of carbonic acid. CO$_2$ concentration also depends on the pH and temperature. The equation below shows the influence of CO$_2$ on algae growth following a Monod kinetic equation when CO$_2$ is a limiting resource for algae growth[4]:

$$f(CO_2) = \frac{[CO_2]}{K_{CO_2} + [CO_2]}$$ (17.9)

where

$[CO_2]$ is the concentration of CO$_2$ dissolved in the liquid medium, kg m^{-3}.
K_{CO_2} is the CO$_2$ saturation constant, kg m^{-3}.

For modeling CO$_2$ concentration considering the carbonic acid equilibrium and the effect of pH, see Equation (11.12). When CO$_2$ is supplied using a gas stream at the optimum rate, the effect of CO$_2$ can be neglected in the modeling.

17.1.1.4 *Modeling of Algae Pond Cultivation*

After modeling the effect of the various factors affecting algae growth, the overall mass balance for algae biomass production can be formulated assuming that: (a) no algae biomass enters into the pond (or is negligible compared to the final concentration) and (b) the system is well-mixed and the pond circulation rate is constant (similar to the assumption for a CSTR discussed in Section 11.4). With these assumptions, for a constant pond volume, the unsteady state mass balance can be developed, as shown by

$$V_{pond} \frac{dC_{alg}}{dt} = \mu C_{alg} V_{pond} - F_c C_{alg}$$ (17.10)

where

V_{pond} is the pond volume, m^3.
C_{alg} is the algae biomass concentration, kg m^{-3}.
F_c is the pond circulation flow rate, m^3 d^{-1}.
t is time, d.

Dividing Equation (17.10) by the pond volume, the rate of change of algae biomass concentration is obtained:

$$\frac{dC_{alg}}{dt} = \mu C_{alg} - \frac{F_c}{V_{pond}} C_{alg}$$
(17.11)

where F_c/V_{pond} is the dilution factor D, d^{-1}. Then Equation (17.11) can be rewritten for the algae biomass production as

$$\frac{dC_{alg}}{dt} = (\mu - D)\, C_{alg}$$
(17.12)

At steady state, the algae biomass concentration is constant and the variation in concentration with time is equal to zero. The dilution rate is then equal to the specific biomass growth rate. If the effect of CO_2 concentration in water is neglected, the algae biomass yield per land area used (Y_{alg}) can be calculated from the pond depth (z) using

$$Y_{alg} = \mu C_{alg} h z$$
(17.13)

where h is the fraction of area actually occupied by water (e.g., 0.8). The rest is occupied by the construction materials of the pond and other additional facilities. Equation (17.12) considers that light irradiation is continuous along the pond depth, but this can be an oversimplification. Consideration of light reflection and shadowing between algae at different depth levels is shown after the following exercise.

Exercise 2. Estimate the annual algae biomass production per unit of land for the species X in Exercise 1. The steady-state concentration is 0.5 kg m^{-3} and the pond depth is $z = 0.3$ m. The actual water surface is 80% of the total area, $h = 0.8$. Assume algae cultivation at average pond water temperature and sunlight. Neglect the effect of CO_2 and variations in concentration with depth. The monthly average irradiation for the pond location and average pond water temperatures are shown in Table 17.3.

Table 17.3 *Monthly average irradiation for the pond location and average pond water temperature.*

Month	Irradiation (W m^{-2})	Water Temperature (°C)
January	30	3
February	50	5
March	80	8
April	100	12
May	120	15
June	150	20
July	145	21
August	130	18
September	100	16
October	75	13
November	40	11
December	35	5

Calculate the annual oil yield if the average oil content in the algae biomass cell is 30%.

$T_{op} = 22 \,^{\circ}\text{C}$

$T_d = 35 \,^{\circ}\text{C}$

$\beta = 1.5$

Oil density $= 0.87$ kg L^{-1}

Solution to Exercise 2. The specific growth is first calculated for each data point in Table 17.3, as shown in Exercise 1. For example, for January the specific growth as a function of irradiation has been determined as $\mu_{(light)} = 0.716$ d^{-1}. Then, using Equation (17.7), the effective specific growth rate is

$$\mu = 0.716 \times \left(\frac{35 - 3}{35 - 22} \right)^{1.5} e^{-1.5 \left(\frac{35-3}{35-22} - 1 \right)} = 0.3 \text{ d}^{-1}$$

The yield is calculated using Equation (17.13). Thus, the average algae biomass yield in January is

$$Y_{alg} = 0.8 \times 0.3 \times 0.5 \times 0.3 = 0.036 \text{ kg m}^{-2} \text{ d}^{-1}$$

The total yield for the month, which has 31 days, is then

$$0.038 \times 31 = 1.1 \text{ kg m}^{-2}$$

Similar calculations are performed for all months. The results are shown in Figure 17.2. The total annual yield per amount of land is then the sum of the average monthly yield, which is about 25 kg m^{-2}. To compare against other oil feedstocks, it is convenient to express the yield in tonnes per hectare (1 ha = 10 000 m^2). The yield is then equal to 250 t ha^{-1}.

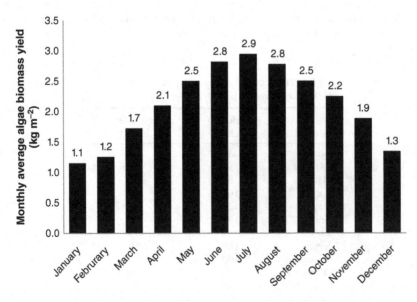

Figure 17.2   *Monthly average algae biomass yield.*

For the oil content of 30%, the annual oil yield is 75 t ha^{-1}. Using the oil density of 0.87 kg L^{-1}, the volumetric yield is about 86 200 L ha^{-1}. Compared to palm oil tree, which produces about 5900 L ha^{-1}, the algae species X produces up to 14.6 times more oil per amount of land used.

Figure 17.3 shows how the light intensity or irradiance changes along the pond depth due the algae shadowing effect. The algae grown at levels close to the water surface can shadow the algae in the sublevels, thus reducing available irradiation and reducing algae growth and oil production. A model representing the light intensity distribution along the pond depth uses *Beer–Lambert's law*[4]:

$$I(z) = \frac{1}{z} \int_0^z I_s e^{(-K_e z)} dz \qquad (17.14)$$

where

$I(z)$ is the light irradiance as a function of pond depth, W m^{-2}.
I_s is the light irradiance on the pond liquid surface, W m^{-2}.
K_e is the light extinction coefficient, m^{-1}.
z is the pond depth, m.

The light extinction coefficient is correlated to algal concentration, as shown by[5]

$$K_e = K_{e1} + K_{e2}C_{alg} \qquad (17.15)$$

where K_{e1}(m^{-1}) and K_{e2} (m^2 kg^{-1}) are the correlation constants.

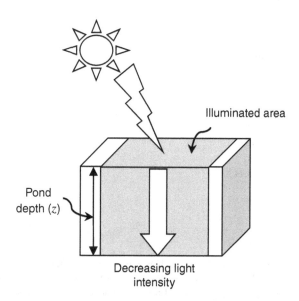

Figure 17.3 *Decreasing light irradiance or intensity along the pond depth due to the algae biomass shadowing effect.*

Exercise 3. Develop a model for the algae biomass yield per amount of land used (Y_{alg}) expressed as a function of the pond depth (z) considering the algae shadowing effect.

a.   For $K_{e1} = 0.3$ m^{-1} and $K_{e2} = 30$ m^2 kg^{-1}, evaluate the annual algae and oil yields for data in Exercise 2 and compare the results. Carry out a sensitivity analysis for the results obtained using different values of the parameters K_{e1} and K_{e2}.

b.   Evaluate the yields for total pond depths between 0 and 1 m. Is there any optimum depth?

Solution to Exercise 3. In a model for the influence of light intensity (e.g., Equations (17.2) or (17.3)), I can be expressed as a function of z (Equation (17.14)). Thus, to develop the model required, substitute first Equation (17.15) into Equation (17.14). Then substitute the expression for specific growth as a function of light intensity (e.g., Equation (17.4) → Equation (17.3) → Equation (17.7)). The biomass growth as a function of z, that is, $\mu = \mu(z)$, must result. Thus, the specific biomass growth in Equation (17.13) is not a constant. A distribution of growth rate will result, leading to a distribution of biomass and oil production. To determine the total production, integration along the pond depth is required, as expressed in Equation (17.11). Solve Equation (17.16) by numerical integration between the bounds $z = 0$ and $z =$ *pond depth Z*, using MATLAB. For a well-mixed pond, C_{alg} can be assumed constant:

$$Y_{alg} = h \int_{0}^{z=Z} \mu C_{alg} dz \qquad (17.16)$$

Exercise 4. An energy company is exploring carbon capture using algae open ponds. The company has 100 ha of land available adjacent to the power plant. The average CO_2 content in the flue gas is 12% by volume with a total gas flow rate of 1500 m^3 h^{-1}. Facilities for desulfurization of the flue gas already exist in the plant and thus the gas is available at a sulfur level tolerable by the algae. Using the information provided:

a.   Determine if the algae carbon capture alternative is competitive with carbon capture and storage (CCS) technologies which have a range of costs of \$30–330 t^{-1} CO_2.

b.   Determine the percentage of CO_2 in the flue gas that can be captured in the available land.

The following data are available:

Algae biomass production = 1.1 kg m^{-2}
Carbon content in algae biomass = 40% (mass basis)
Maximum pond size = 15 ha
Pond construction cost = \$98 000 ha^{-1}
Nutrients costs = \$28 t^{-1} algae biomass
Harvest costs = \$23 t^{-1} algae biomass
Drying costs = \$0.025 t^{-1} algae biomass
Electricity costs = \$8 t^{-1} algae biomass
Other costs = 15% of total operating costs
Capital charge ratio = 12%
Project lifetime = 15 years

17.1.2   Photobioreactors (PBRs)

PBRs are closed reactors made of transparent material such as glass or plastic. Figure 17.4 shows the more common PBR designs including flat plate, bubble columns and tubular. PBRs improve algae and oil yields and allow a more controlled cultivation environment and stable production compared to open ponds. The risk of contamination is also minimized. Agitation and mixing is performed by recirculating the liquid culture medium using a mechanical pump or by an airlift system. The CO_2 and O_2 transfer rate is also enhanced.

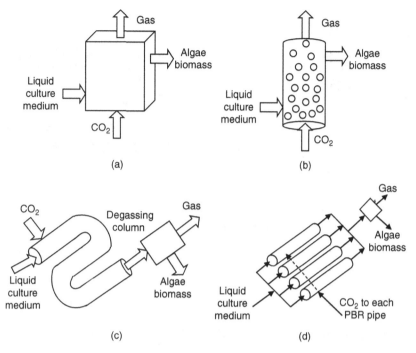

Figure 17.4 *PBRs for algae cultivation: (a) plate PBR, (b) bubble column PBR, (c) interconnected tubular PBR system, (d) parallel tubular PBR system.*

PBRs can be modified to use a combination of sunlight and artificial lights. For example, they can be illuminated using halogen lamps. However, this option requires higher investment and operating costs due to additional energy demand. Higher GHG emissions are incurred if electricity is imported to the algae cultivation plant. The major issue limiting the use of PBRs is the higher operating and capital costs in comparison to open pond systems.

17.1.2.1 Modeling of Tubular PBRs

Tubular reactors are suitable for outdoor cultivation and are comparatively cheaper than the other types of PBRs. One of the main limitations of *tubular PBRs* is that they usually require a large area to achieve good biomass productivity. Thus, the performance of PBRs needs to be optimized to decrease the cost of production of biomass and thereby the cost of production of biodiesel.

Mass balance models need to be developed for these continuous reactors in order to analyze and optimize their performances. The mass balance for a tubular PBR assumes uniform concentration in the radial direction but varying concentration in the axial direction (see the mass balance of a PFR in Section 12.4 and the solution to Exercise 8 in Chapter 11). Hence, C_{alg} changes with the length of the PBR. For an element of volume ΔV in the tubular PBR (Figure 17.5), the mass balance is shown by

$$\frac{d(C_{alg}\Delta V)}{dt} = C_{alg}F\Big|_z - C_{alg}F\Big|_{z+\Delta z} + r_{alg}\Delta V \tag{17.17}$$

where

C_{alg} is the algae biomass concentration at time t, kg m^{-3}.

$C_{alg}F\Big|_z$ is the biomass flow rate entering the element of volume ΔV in m^3 at length z, kg h^{-1}.

$C_{alg}F\Big|_{z+\Delta z}$ is the biomass flow rate leaving the element of volume ΔV at length $z + \Delta z$, kg h^{-1}.

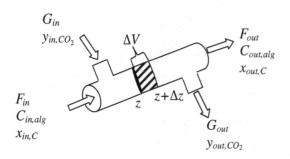

F_{in}, F_{out}: inlet and outlet liquid flow rate, respectively (m^3 h^{-1})

$C_{in,alg}$, $C_{out,alg}$: algae concentration in the inlet and outlet liquid stream, respectively (kg m^{-3})

$x_{in,C}$, $x_{out,C}$: moles of carbon in the algae biomass in the inlet and outlet stream, respectively (mol C kg^{-1} biomass)

G_{in}, G_{out}: inlet and outlet gas flow rate, respectively (mol h^{-1})

y_{in,CO_2}, y_{out,CO_2}: CO_2 mole fraction in the inlet and outlet gas stream, respectively

Figure 17.5   *Idealized tubular PBR.*

r_{alg} is the algae biomass production rate, kg m^{-3} h^{-1}.
F is the volumetric flow rate across the element of volume ΔV, m^3 h^{-1}.

Dividing by ΔV and taking the limit as ΔV goes to zero (this is to get the definition of the derivative $d(C_{alg}F)/dV$) allows Equation (17.17) to be rewritten as

$$\frac{dC_{alg}}{dt} = -\frac{d(C_{alg}F)}{dV} + r_{alg} \tag{17.18}$$

At steady state, the accumulation term is equal to zero and thus Equation (17.18) becomes

$$\frac{d(C_{alg}F)}{dV} = r_{alg} \tag{17.19}$$

For a constant flow rate, $F = F_{in} = F_{out}$, the PBR mass balance equation is given by

$$F_{in}dC_{alg} = r_{alg}dV \tag{17.20}$$

where r_{alg} is determined from the algae growth kinetics as

$$r_{alg} = \mu C_{alg} \tag{17.21}$$

The specific growth rate can also follow a Monod kinetics form (Equation (11.1)), but in this case, with light irradiance I as the limiting factor:

$$\mu = \mu_{max}\frac{I}{I + K_I} \tag{17.22}$$

where K_I is the saturation constant for light irradiance.

To determine the algae biomass production, Equations (17.21) and (17.22) can be combined and then introduced as an expression for the production rate r_{alg} into the PBR design equation (Equation (17.20)). The design equation then becomes as shown by

$$F_{in}dC_{alg} = \mu_{max}\frac{I}{I + K_I}C_{alg}dV \tag{17.23}$$

Separation of variables gives

$$\frac{dC_{alg}}{C_{alg}} = \mu_{max}\frac{I}{I + K_I}\frac{dV}{F_{in}} \tag{17.24}$$

The residence time τ (similar to space–time in Equation (11.42)) is defined by

$$\tau = \frac{V}{F_{in}} \tag{17.25}$$

Then, the rate of change of the algae biomass concentration in terms of the residence time can be expressed as

$$\frac{dC_{alg}}{C_{alg}} = \mu_{max}\frac{I}{I+K_I}d\tau \tag{17.26}$$

Integration from $C_{in,alg}$ to $C_{out,alg}$ for $\tau=0$ to τ gives the solution as

$$\ln\frac{C_{out,alg}}{C_{in,alg}} = \mu_{max}\frac{I}{I+K_I}\tau \tag{17.27}$$

Upon rearranging, the value of the biomass concentration in the outlet stream is given by

$$C_{out,alg} = C_{in,alg}e^{\mu_{max}\frac{I}{I+K_I}\tau} \tag{17.28}$$

Exercise 5. A new algae production development has been planned in a location with low sunlight irradiance. Halogen lamps are going to be installed to irradiate tubular PBRs in a parallel arrangement. The effective continuous irradiance provided by these lamps is 500 µmol m^{-2} s^{-1}. The initial biomass concentration in the liquid feed is 1 kg m^{-3}. Calculate:

a. The algae biomass productivity in a tubular PBR with a volume of 1.7 m^3 for a liquid feed flow rate of 10 m^3 h^{-1}.
b. The number of PBRs of the given volume capacity and the land used for an oil production of 10 000 t y^{-1} assuming an oil content of 35% by mass in the algae biomass. Assume 300 operating days a year.

$\mu_{max} = 0.1$ h^{-1}
$K_I = 70$ µmol m^{-2} s^{-1}

Floor area occupied by one PBR = 18.6 m^2.

Solution to Exercise 5.

a. The residence time for the flow rate of 10 m^3 h^{-1} and a PBR of volume $V_{PBR} = 1.7$ m^3 is calculated from Equation (17.25):

$$\tau = \frac{1.7}{10} = 0.17 \text{ h}$$

Then the biomass concentration is calculated using Equation (17.28):

$$C_{out,alg} = 1 \times e^{0.1\frac{500}{500+70}0.17} = 1.02 \text{ kg m}^{-3}$$

The net biomass production is

$$(C_{out,alg} - C_{in,alg}) \times F = (1.02 - 1) \times 10 = 0.2 \text{ kg h}^{-1}$$

b.   The algae biomass production for one reactor was found to be 0.2 kg h^{-1}. For 300 operating days per year and assuming an oil content in the algae biomass of 0.35, the annual oil production of one PBR of the given volume is

$$\frac{0.2 \times 24 \times 300 \times 0.35}{1000} = 0.5\ t\ y^{-1}\ oil$$

Thus, the number of PBRs for the production of 10 000 t y^{-1} of oil is

$$\frac{10\,000}{0.5} = 20\,000\ PBRs$$

If the area occupied by one reactor is 18.6 m^2, the total land required is

$$20\,000 \times 18.6 = 372\,000\ m^2 = 37.2\ ha$$

The average land productivity is

$$\frac{10\,000}{37.2} = 269\ t\ ha^{-1}\ y^{-1}$$

The carbon capture achieved by an algae cultivation system is important. The previous models assumed a CO_2 supply at an optimum rate. Algae growth is influenced by CO_2 levels. Cultivation of algae species that tolerate and capture high concentrations of CO_2 is an enabling *CO_2 capture* technology (for other carbon capture technologies see Chapter 13). The kinetics of the conversion of inorganic carbon depends on the form of carbon supplied and on the equilibrium established between CO_2, carbonic acid and its dissociated species. The supply of inorganic carbon can be in the form of a carbonate salt added to the culture medium. In PBRs, the most common supply method is bubbling of CO_2 into the liquid. CO_2 bubbling provides an optimal supply of CO_2 for algae growth. Under this condition, light irradiation is the only limiting resource for biomass growth. Figure 17.5 shows an idealized tubular PBR for the analysis of the carbon balance and algae biomass production. In a continuous PBR, the carbon input comes from the culture medium feed and the input gas stream supplying CO_2. The total carbon input must be equal to the carbon leaving the reactor with the algae biomass produced and the depleted gas stream. Since there is one mole of C in one mole of CO_2, the carbon balance can be developed for CO_2 when this is the main carbon supply to the algae. Then, the carbon balance is formulated as

$$y_{in,CO_2} G_{in} + x_{in,C} C_{in,alg} F_{in} = y_{out,CO_2} G_{out} + x_{out,C} C_{out,alg} F_{out} \qquad (17.29)$$

where

F_{in} is the inlet liquid flow rate, m^3 h^{-1}.
F_{out} is the outlet liquid flow rate, m^3 h^{-1}.
$C_{in,alg}$ is the algae concentration in the inlet liquid stream, kg m^{-3}.
$C_{out,alg}$ is the algae concentration in the outlet liquid stream, kg m^{-3}.
$x_{in,C}$ is the carbon in the algae biomass in the inlet stream, mol C kg^{-1} biomass.
$x_{out,C}$ is the carbon in the algae biomass in the outlet stream, mol C kg^{-1} biomass.
G_{in} is the inlet gas flow rate, mol h^{-1}.
G_{out} is the outlet gas flow rate, mol h^{-1}.
y_{in,CO_2} is the CO_2 mole fraction in the inlet gas stream.
y_{out,CO_2} is the CO_2 mole fraction in the outlet gas stream.

The CO_2 in the system is converted into carbon stock in the algae cell biomass. The molar balance of carbon in the algae biomass is given by

$$x_{out,C} C_{out,alg} F_{out} = x_{in,C} C_{in,alg} F_{in} + x_{out,C} r_{alg} V \qquad (17.30)$$

where V is the PBR volume.

Substituting Equation (17.30) into Equation (17.29) gives

$$y_{in,CO_2} G_{in} = y_{out,CO_2} G_{out} + x_{out,C} r_{alg} V \qquad (17.31)$$

Exercise 6. In the system studied in Exercise 5, CO_2 is supplied by bubbling 200 mol h^{-1} of a flue gas stream containing 10% by mole of CO_2.

a. Calculate the CO_2 capture rate and percentage in relation to the CO_2 input in the gas stream.
b. Develop a correlation between irradiance intensity, residence time of PBR and carbon capture.

Oil density $= 0.87$ kg L^{-1}
Algae biomass carbon fraction $= 0.44$ kg C kg$^{-1} = 37$ mol C kg^{-1}

Solution to Exercise 6.

a. Equation (17.29) can be used to determine the rate of CO_2 capture, which is the carbon fixation in algae biomass. Remember that there is one mole of carbon in one mole of CO_2. For a constant gas flow rate, constant liquid flow rate and constant carbon content in the algae biomass (x_C), the CO_2 capture can be expressed as

$$CO_2 \text{ capture} = G_{in}(y_{in,CO_2} - y_{out,CO_2}) = x_C F_{in}(C_{out,alg} - C_{in,alg}) \qquad (17.32)$$

Substituting the values from the previous exercise ($F_{in} = 10$ m^3 h^{-1}, $C_{in,alg} = 1.0$ kg m^{-3}, $C_{out,alg} = 1.02$ kg m^{-3}) and the carbon molar content provided ($x_C = 37$ mol kg^{-1}):

$$CO_2 \text{ capture} = 37 \times 10 \times (1.02 - 1.0) = 7.4 \text{ mol } CO_2 \text{ h}^{-1}$$

To calculate the percentage of carbon CO_2 capture, CO_2 in the input gas stream is determined as

$$CO_2 \text{ input} = G_{in} \times y_{in,CO_2} = 200 \times 0.1 = 20 \text{ mol } CO_2 \text{ h}^{-1}$$

Thus, the percentage of CO_2 captured is

$$\frac{7.4}{20} = 37\%$$

b. Substituting Equation (17.28) in Equation (17.32) gives Equation (17.33) for CO_2 capture:

$$CO_2 \text{ capture} = x_C F_{in} C_{in,alg} \left(e^{\mu_{max} \frac{I}{I+K_I} \tau} - 1 \right) \qquad (17.33)$$

This equation can be used to compare carbon dioxide capture capacities between different tubular PBR systems growing the same algae specie or between different species grown in the same PBR system by changing the residence time and kinetic parameters. A more generic formulation, so that μ can be expressed as a function of light irradiance by the model, that is, $\mu = \mu(I)$, is shown by

$$\boxed{CO_2 \text{ capture} = 44 x_C F_{in} C_{in,alg} \left(e^{\mu(I)\tau} - 1 \right)} \qquad (17.34)$$

where the factor 44 is the molar mass of CO_2 to convert into a mass unit. Hence, CO_2 capture will be obtained in the mass flow rate.

Exercise 7. Develop the model of algae biomass production by using *flat plate PBRs* (Figure 17.4(a)).

Solution to Exercise 7. You can conceptualize plate PBR as a CSTR reactor (see Section 11.4).

17.2   Algae Harvesting and Oil Extraction

17.2.1   Harvesting

Harvesting of algae biomass is one of the bottlenecks for a successful and wide commercial emergence of algae as a feedstock for biofuels. The small size of the algae cells (3–30 μm of diameter) makes a stable suspension that is difficult to separate. Algae are generally produced at dilute concentrations (0.02–0.06%, mass basis), which means that high volumes of algae broth need to be handled by the process equipment. This also means that a lot of energy is required for concentrating and drying the biomass for further processing. The harvesting method depends on various factors such as algae species, biomass concentration, conditions of growth medium and economics.

Harvesting of algae biomass consists of recovery and concentration of it. The first stage involves the separation of algae cells from the bulk suspension by coagulation–flocculation and flotation or gravity sedimentation. After this step, the biomass concentration in the resulting effluent is about 2–7% by mass. The second stage consists of dewatering of the slurry in order to concentrate the algae biomass using centrifugation, filtration, evaporation, etc. This stage brings the algae biomass to a concentration of up to 25% (mass basis). More recent developments in algae harvesting include ultrasonic aggregation, ion exchange, electrophoresis, electrocoagulation, magnetic separation and others.

The more important methods used in algae harvesting are coagulation–flocculation and sedimentation, discussed as follows.

1. *Coagulation–flocculation.* Aggregation of microalgae cells is not spontaneous due to the electrostatic repulsion of negatively charged cells. Thus, microalgae suspension is highly stable. Coagulation and flocculation need to occur in a sequence to achieve a separation of particles in a suspension. *Coagulation* refers to the destabilization of the particles by a chemical agent (*coagulant*) so that the particles can start to agglomerate. A coagulant must have a charge opposite to that of the suspended particles and is added to the suspension to neutralize the charges and destabilize the suspension. After destabilization, the particles start to form agglomerates. These initial agglomerates, called *microfloccules*, are of microscopic scale. They are needed for the subsequent process of flocculation. *Flocculation* refers to further agglomeration of the microfloccules into particles of a bigger size by collisions between them and the coagulation agent. A flocculation agent can also be added to generate floccules of a proper size with a binding strength enough to prevent them from tearing apart. Coagulation involves vigorous mixing to ensure the maximum contact between the coagulation agent and the particles. However, flocculation requires slow agitation to avoid breaking down of the floccules but enough to ensure collision between microfloccules and coagulation and flocculation agents. The coagulation step required for flocculation to occur is similar to the nucleation step required for crystallization (see Section 11.5.8). Thus, the mechanisms of coagulation–flocculation are equally complex and difficult to model.

   Algae cells are negatively charged particles, requiring cationic coagulation–flocculation agents to form the floccules. Multivalent cations or cationic polymers are used. Metal salts providing multivalent cations include ferric chloride ($FeCl_3$), aluminum sulfate ($Al_2(SO_4)_3$) and ferric sulfate ($Fe_2(SO_4)_3$). Long-chain and high molecular weight polymers are widely used in combination with metal coagulants. Low-to-medium weight cationic polymers can also be used alone or in combination with aluminum or ferrous coagulants. Cationic polymer molecules attract the suspended solids for neutralization and can serve as bridges between particles highly dispersed in a liquid medium, as in the case of algae biomass. Polyacrylamide and the biopolymer chitosan (produced from shrimp shells) are common polymers used for algae flocculation. Anionic polymers can also be used in combination with metal coagulants.

   Table 17.4 compares the advantages and challenges for algae biomass separation between main coagulation–flocculation agents. A third type of agent is also shown in this table: the alkalis. Alkalis promote autocoagulation due to an increase in pH. The choice of agent depends upon the nature of the suspended solid to be removed, the pH and salinity of the liquid medium, the cost of equipment and the doses and costs of the chemicals. Final selection should consider the required separation efficiency and the downstream processing.

Table 17.4 *Comparison between types of coagulation–flocculation agents.*

Type	Inorganic Salts	Polymers	Alkalis
Examples	$FeCl_3$, $Al_2(SO_4)_3$, $Fe_2(SO_4)_3$, $Na_2Al_2O_4$	Polyacrylamide Polyethyleneimine Chitosan	NaOH, CaOH, MgOH, etc.
Advantages	Low cost. Widely available. Highly efficient.	Effective over a wide range of pH. Low doses required. Produce more concentrated floccules. Floccules more resistant to agitation and mixing.	Avoids the use of other chemical agents.
Disadvantages	Modify pH of the liquid medium. Require special materials of construction for equipment.	Expensive due to low production available.	Depend on algae species. Extremely high pH may cause breaking down of the cells.

2. *Sedimentation.* Gravity sedimentation is a common unit operation in algae harvesting. It is also used in conjunction with coagulation–flocculation to sediment the floccules. Sedimentation refers to the separation of destabilized suspended solids by the force of gravity. The settling characteristics of the suspended solids are determined by the sedimentation velocity and the density and radius of the particles. While the sedimentation of floccules follows a more complex mechanism, direct sedimentation of the algae biomass can be considered to follow Stoke's law for discrete particles. A *discrete particle* has a constant size, shape and specific gravity during the sedimentation time. As floccules are aggregating molecules, the aforementioned particle characteristics may not remain constant during the sedimentation time.

   Stoke's law for discrete particles results from the balance of three forces acting upon a particle immersed in a fluid medium: gravity, friction or drag force and the uplift force of the fluid. When the forces are in balance at equilibrium, the acceleration at which a particle travels down to the bottom of a container is zero and the *settling velocity* is constant. Then, the settling velocity can be expressed as

$$v_s = \sqrt{\frac{2g\left(\rho_p - \rho_f\right)V_p}{C_D \rho_f A_p}} \tag{17.35}$$

where

v_s is the settling velocity of the particle.
ρ_p is the density of the particle.
V_p is the volume of the particle.
A_p is the cross-sectional area of the particle in the direction of particle flow.
ρ_f is the density of the fluid medium (generally water).
C_D is the drag coefficient.
g is the gravitational acceleration.

   The drag coefficient is a function of the particle shape and the flow regime of the fluid around the particle. Assuming that algae biomass particles are spherical, the drag coefficient can be defined as

$$C_D = \frac{24}{Re} + \frac{24}{\sqrt{Re}} + 0.34 \tag{17.36}$$

where Re is the *Reynolds number* (dimensionless) defined by

$$Re = \frac{v_s d_p \rho_f}{\mu_f} \tag{17.37}$$

where

μ_f is the viscosity of the fluid.
d_p is the diameter of the particle.

For a turbulent flow regime ($Re > 10^4$), a drag coefficient in the range of 0.34–0.4 is commonly used.

Exercise 8. Develop the expression to calculate the settling velocity of an algae biomass particle of spherical shape in a fluid in the laminar flow regime ($Re < 1$).

Solution to Exercise 8. For a small value of Re, the last two terms in Equation (17.36) can be neglected and then $C_D = 24/Re$. Therefore, Equation (17.36) becomes

$$C_D = \frac{24}{Re} = \frac{24\mu_f}{v_s d_p \rho_f}$$

The ratio of the volume of a spherical particle (with radius r) and its cross-sectional area is

$$\frac{V_p}{A_p} = \frac{\frac{4}{3}\pi r^3}{\pi r^2} = \frac{4}{3}r = \frac{2d_p}{3}$$

Substituting the above two equations into Equation (17.35) gives the following equation for the settling velocity under the laminar flow regime known as Stoke's law:

$$v_s = \frac{g\left(\rho_p - \rho_f\right) d_p^{\,2}}{18\mu_f} \tag{17.38}$$

The particles to separate can be present in the fluid at a wide range of sizes. Thus, particles of different sizes have different settling velocities. The size distribution and the resulting velocity distribution determine the percentage of particles separated by a sedimentation process. Another important variable is the *critical velocity*, which is defined as the tank depth over the residence time. The depth and volume (V) are related by the area. The critical velocity (v_c) can be calculated from the volumetric feed flow rate (F) and the surface area (A) of sedimentation, as shown by

$$v_c = \frac{\text{Depth}}{V/F} = \frac{\text{Depth}}{(\text{Depth} \times A)/F} = \frac{F}{A} \tag{17.39}$$

The critical velocity corresponds to the settling velocity of the slowest-settling particles that are 100% removed.

To determine the overall recovery percentage of the particles for a given design flow rate, the settling velocity distribution for the suspension must be determined by sedimentation experiments. Sedimentation is carried out in a small vessel of depth z. The concentration of particles in the liquid is measured at different times. The fraction between the concentration at time t and the initial concentration ($t = 0$) is the fraction of particles that remain in the liquid. The sedimentation velocity is calculated using the known depth of the vessel and the time at which each sample is taken.

From the experimental results, a plot of the cumulative fraction of particles remaining in the fluid versus settling velocities is generated. A typical distribution diagram is shown in Figure 17.6, where x_c is the fraction of particles that remains in the fluid at the critical velocity (v_c). The total fraction of particles that settles and can be recovered (s) is the sum of:

a.   The fraction of particles with velocity greater than v_c: $(1 - x_c)$. Since v_p(particle velocity) $= v_s$(settling velocity) $> v_c$(critical velocity), these particles are 100% recovered.

b.   The fraction of particles with velocity less than v_c. This fraction is present in the shaded area above the distribution curve between the origin and the coordinate (v_c, x_c), as shown in Figure 17.6.

    Thus the total fraction recovered can be expressed as

$$s = (1 - x_c) + \frac{1}{v_c} \int_0^{x_c} v_p \, dx$$
(17.40)

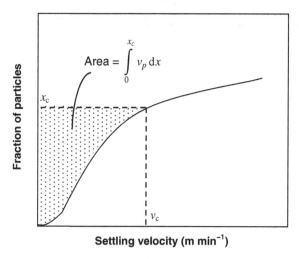

Figure 17.6 *Settling velocity distribution diagram.*

Exercise 9. The data obtained from sedimentation experiments with an algae broth from a raceway pond are shown in Table 17.5. Determine the fraction of biomass recovery achieved by sedimentation if the critical velocity is 0.5 m min^{-1} and the depth of the experimental sedimentation basin is 2 m.

Table 17.5 *Experimental data for algae sedimentation.*

Time (min)	Concentration (mg L^{-1})
0	200
3	184
5	157
10	120
20	56
40	32
60	16

Solution to Exercise 9. The settling velocity and fraction of particles that remains in water can be calculated for each data point, as shown in Table 17.6.

    The calculated fractions versus settling velocities are plotted to find the value of x_c and the area above the curve related to the fraction of particles with $v_s < v_c$ that settle. The resulting settling velocity diagram is shown in Figure 17.7. For $v_c = 0.5$ m min^{-1}, $x_c = 0.83$.

Table 17.6 Calculations of settling velocity and fraction of biomass remaining in water.

Time (min)	Concentration (mg L^{-1})	Settling Velocity v_s = depth/time (m min^{-1})	Fraction Remaining in Water
0	200	–	–
3	184	2/3 = 0.667	(200−184)/200 = 0.92
5	157	2/5 = 0.4	(200−157)/200 = 0.785
10	120	2/10 = 0.2	(200−120)/200 = 0.6
20	56	2/20 = 0.1	(200−56)/200 = 0.28
40	32	2/20 = 0.05	(200−32)/200 = 0.16
60	16	2/60 = 0.033	(200−16)/200 = 0.08

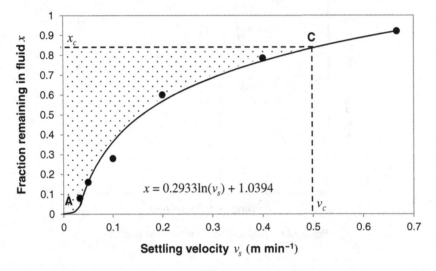

Figure 17.7 *Settling velocity distribution diagram for data in Exercise 9.*

To determine the area above the curve, a function could be fitted to the data as shown in Figure 17.7. Then, the area above the curve is equal to the area of the rectangle A–x_c–C–v_c minus the area under the curve. Thus, the area above the curve in Figure 17.7 is

$$\int_0^{x_c} v_p dx = x_c \times v_c - \int_0^{v_c} (0.2933 \ln v_s + 1.0394) dv_s$$

Integration and substitution of values gives

$$\int_0^{x_c} v_p dx = 0.83 \times 0.5 - (0.2933(v_c \times \ln v_c - v_c) + 1.0394 v_c)$$

$$= 0.83 \times 0.5 - (0.2933 (0.5 \times \ln 0.5 - 0.5) + 1.0394 (0.5)) = 0.14 \text{ m min}^{-1}$$

Substituting the above results into Equation (17.40), the fraction of algae biomass recovered is

$$s = (1 - 0.83) + \frac{1}{0.5} (0.14) = 0.45$$

Thus, 45% of the algae biomass could be recovered at $v_c = 0.5$ m min^{-1}. To increase the recovery, a lower value of critical velocity is required. According to Equation 17.39, this means that the flow rate must be decreased for a constant sedimentation area. Alternatively, a basin must be designed with a larger sedimentation area for a given flow rate. Thus, a higher recovery requires a higher area and capital cost; the trade-off between recovery and capital cost of the basin must be evaluated from more experimental data points.

3. *Centrifugation.* Centrifugal force is applied to a fluid to increase the settling velocity of particles in the fluid to enhance separation. Centrifugation is suitable to most of the microalgae species but it is only preferred for recovery of high value products from the algae biomass. This is because centrifugation incurs higher energy costs and maintenance than other harvesting and concentration methods. The advantage of centrifugation is that the algae biomass can be obtained at a high concentration of up to 25% (mass basis) with a harvesting efficiency of greater than 95%. The energy requirements are higher at lower concentrations of the feed. Thus, centrifugation can be used after an upstream recovery process that pre-concentrates the feed such as sedimentation or coagulation–flocculation.

Exercise 10. The process in Figure 17.8 consisting of coagulation–flocculation and centrifugation is being evaluated for harvesting algae from a raceway pond. Estimate the minimum cost per kg of algae paste produced. Aluminum sulfate is used as the coagulation–flocculation agent.

Algae biomass recovery in centrifugation, $\%R_{CFG} = 90\%$
Price of aluminum sulfate, $P_{AS} = \$0.5$ kg^{-1}
Price of electricity, $P_E = \$0.03$ MJ^{-1}

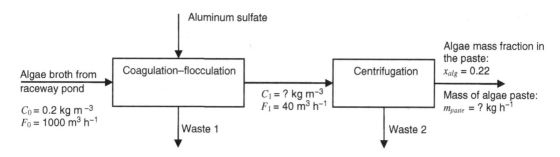

Figure 17.8   *Algae harvesting process for evaluation in Exercise 10.*

Coagulation–flocculation performance curve relating the recovery of biomass ($\%R_{C-F}$) and the dose of aluminum sulfate (dose range: 40–200 mg L^{-1}):

$$\%R_{C-F} = \frac{96.5754}{1 + e^{[-0.06857(\text{dose}-67.84)]}}$$

Specific energy consumption (MJ kg^{-1}, algae mass basis) of centrifuge as a function of the algae biomass concentration in the feed (C_1):

$$E_s = \frac{1}{0.007114 + 0.01331C_1}$$

Solution to Exercise 10.   The balance in Figure 17.8 needs to be completed for various values of the chemical dose applied to the coagulation–flocculation process. The mass of the stream going to centrifugation is related to the biomass in the feed by the biomass recovery fraction in the coagulation–flocculation unit:

$$C_1F_1 = \frac{\%R_{C-F}}{100} \times C_0F_0$$

Thus, if $\%R_{C-F}$ is known for different values of chemical dose, the concentration of the feed to the centrifugation unit can be found as

$$C_1 = \frac{\dfrac{\%R_{C-F}}{100} \times C_0 F_0}{F_1}$$

Then the energy consumption per kg of biomass in the algae paste product can be calculated for each known value of C_1 using the correlation provided.

The algae biomass flow rate (kg h^{-1}) from the centrifugation unit in Figure 17.8 is

$$m_{paste} x_{alg} = C_1 F_1 \times \frac{\%R_{CFG}}{100}$$

Then the total mass flow rate (kg h^{-1}) of the algae paste from centrifugation is

$$m_{paste} = \frac{C_1 F_1 \times \dfrac{\%R_{CFG}}{100}}{x_{alg}}$$

The total operating costs are the sum of the costs of aluminum sulfate (AS) required for coagulation–flocculation and the costs of electricity for the centrifuge. The costs of AS are

$$\text{Cost of AS} = \frac{\text{Dose} \times F_0}{1000} \times P_{AS}$$

The cost of electricity for the production of paste with x_{alg} content of algae biomass is

$$\text{Cost of electricity} = E_s \times m_{paste} x_{alg} \times P_E$$

The sum of the costs per kg of wet algae paste ($CWAP$) is then

$$CWAP = \frac{\left[\dfrac{\text{Dose} \times F_0}{1000} \times P_{AS} + E_s \times m_{paste} x_{alg} \times P_E\right]}{m_{paste}} = \left[\frac{\text{Dose} \times F_0}{1000 m_{paste}} \times P_{AS} + E_s \times x_{alg} \times P_E\right]$$

Substitution of the expression for m_{paste} gives

$$CWAP = \left[\frac{\text{Dose} \times F_0}{1000} \times \frac{x_{alg}}{C_1 F_1 \times \frac{\%R_{CFG}}{100}} \times P_{AS} + E_s \times x_{alg} \times P_E\right]$$

Substituting E_s as expressed earlier:

$$CWAP = \left[\frac{\text{Dose} \times F_0}{1000} \times \frac{x_{alg}}{C_1 F_1 \times \frac{\%R_{CFG}}{100}} \times P_{AS} + \frac{1}{0.007114 + 0.01331 C_1} \times x_{alg} \times P_E\right]$$

C_1 is a function of $\%R_{C-F}$ as given in the problem statement. In turn, $\%R_{C-F}$ is a function of the dose of aluminum sulfate according to the equation provided. Thus, the whole expression for $CWAP$ is a function of the aluminum sulfate dose when the rest of the parameters are fixed.

The calculation for the dose of AS of 40 mg L^{-1} is shown. First the biomass recovery in the coagulation–centrifugation unit is calculated as

$$\%R_{C-F} = \frac{96.5754}{1 + e^{[-0.06857(40-67.84)]}} = 12.47\%$$

The value of the concentration in the effluent is

$$C_1 = \frac{\dfrac{12.47}{100} \times 0.2 \times 1000}{40} = 0.623 \text{ kg m}^{-3}$$

Thus, the sum of the costs per kg of wet algae paste is

$$CWAP = \left[\frac{40 \times 1000}{1000} \times \frac{0.22}{0.623 \times 40 \times \dfrac{90}{100}} \times 0.5 + \frac{1}{0.007114 + 0.01331 \times 0.623} \times 0.22 \times 0.03 \right]$$

$$CWAP = 0.196 + 0.428 = \$\, 0.624 \text{ kg}^{-1}$$

where the coagulation–flocculation (indicated as C-F in Figure 17.9) chemical costs are \$0.196 kg^{-1} and the centrifugation electricity costs are \$0.428 kg^{-1}. Similar calculations can be done for a range of AS doses of 40–200 mg L^{-1} with a step size of 20 mg L^{-1}. The results for chemical costs, centrifugation electricity costs and the sum of these costs can be plotted against the dose of AS. The plot is then used to find out if there is an optimum value of dose of AS for which the sum of the costs is minimum. Figure 17.9 shows the resulting plot. As the dose of AS is increased, the cost (per kg of algae paste) of adding the chemical reduces up to a minimum of ~90 mg L^{-1}, after which the cost (per kg of algae paste) of adding the chemical increases. Centrifugation costs have a decreasing trend with the increasing dose of AS. This is due to an increased amount of biomass recovered in the coagulation–flocculation process, which leads to a higher concentration in the feed to the centrifuge for a constant volumetric flow rate F_1. Thus, even with the increasing cost of chemical addition with increasing doses of AS, greater than 90 mg L^{-1}, the sum of costs still decreases until the optimum point appears at a dose of about 108 mg L^{-1}. Doing the calculations for $CWAP$, as in the example above, gives a minimum cost per amount of wet algae paste (22% mass content) of \$0.17 kg^{-1}.

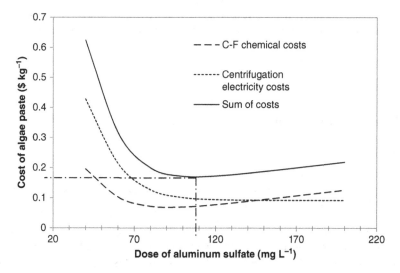

Figure 17.9 *Coagulation–flocculation (C-F) chemical costs, centrifugation electricity costs and the sum of these costs versus dose of aluminum sulfate.*

At the optimum point, the following values can be calculated to complete the mass balance in Figure 17.8:

$$C_1 = 4.54 \text{ kg m}^{-3}$$

$$m_{paste} = 743 \text{ kg h}^{-1}$$

Further Questions

a. Determine the amount of waste streams shown in Figure 17.8 if aluminum sulfate is supplied from a solution of 0.1 mol L^{-1}. Assume that the percentage of aluminum sulfate that goes in the floccules with the algae is equal to the algae biomass recovery.
b. Aluminum sulfate may need to be removed from the biomass after flocculation. Suggest a process to achieve such a separation.
c. The wet algae paste obtained by the process still contains high amounts of water and may need to be dried before going to oil extraction. Calculate the amount of energy required to dry the algae paste up to a moisture content of 10% by mass.

Another technique that can be used for concentrating algae biomass is filtration. Conventional filtration can be used for relatively large (>70 μm) algal species. For smaller species with cell sizes <30 μm, membrane microfiltration or ultra-filtration are more suitable. However, these processes consume energy due to the high operating pressure. In addition, fouling problems make the replacement of membranes more often, increasing the costs.

Harvesting costs can be up to 30% of the total cost of algae biomass production. Process optimization using data from experimental trials or commercial facilities exploring the parameters influencing the performance of harvesting process units should aim to decrease the costs.

17.2.2 Extraction

Algae *oil extraction* can be achieved by mechanical or chemical methods. The expeller press is a common mechanical method, whilst *Soxhlet extraction* with hexane is the most common chemical method. Each method has its own limitations. The expeller press requires the algae biomass to be dried first, making the process energy intensive. The use of solvents in the chemical methods presents health and safety risks and adds an extraneous material requiring secondary solvents for the primary solvent recovery, etc. A higher oil recovery is obtained by the chemical extraction methods (up to 99%) compared to the expeller press (up to 75%). The main challenge in oil extraction is the breaking down of the algae cell wall to release the oil content. More recent developments include *sonication* (ultrasonic waves) and magnetic fields for breaking the cells. Another chemical method is the *supercritical extraction* using CO_2, remark discussed in the Introduction chapter (Section 1.6.1).

17.3 Algae Biodiesel Production

17.3.1 Biodiesel Process

Similar to the oil extraction from plants, the main components of algal oil are *triglycerides*. Triglycerides are molecules having the backbone of a glycerol molecule substituted by chains of fatty acids. The oil can also contain *free fatty acids* (FFA), that is, not forming part of triglycerides. FFA undergo esterification reactions in the biodiesel production process (see also reactions in Chapter 18). Table 17.7 shows the names and chemical formulae for fatty acids and triglycerides commonly found in vegetable and algae oils. The fatty acid chains are denoted by the number of carbons and the number of double bonds in the chain. For example, oleic acid is a fatty acid having 18 carbons and 1 double bond and is denoted as [C18:1]. Biodiesel production involves chemical conversion of triglycerides into fatty acid alkyl esters. The properties of these biodiesel components depend on the participating fatty acid chains and FFA content. Thus, an oily feedstock for

Table 17.7 *Common fatty acids and triglycerides in vegetable and algae oils.*

Notation	Fatty Acid	Formula	Triglyceride	Formula
[C12:0]	Lauric acid	$C_{12}H_{24}O_2$	Trilaurin	$C_{39}H_{74}O_6$
[C14:0]	Myristic acid	$C_{14}H_{28}O_2$	Trimyristin	$C_{45}H_{86}O_6$
[C16:0]	Palmitic acid	$C_{16}H_{32}O_2$	Tripalmitin	$C_{51}H_{98}O_6$
[C16:1]	Palmitoleic acid	$C_{16}H_{30}O_2$	Tripalmitolein	$C_{51}H_{92}O_6$
[C18:0]	Stearic acid	$C_{18}H_{36}O_2$	Tristearin	$C_{57}H_{110}O_6$
[C18:1]	Oleic acid	$C_{18}H_{34}O_2$	Triolein	$C_{57}H_{104}O_6$
[C18:2]	Linoleic acid	$C_{18}H_{32}O_2$	Trilinolein	$C_{57}H_{98}O_6$
[C18:3]	Linolenic acid	$C_{18}H_{30}O_2$	Trilinolenin	$C_{57}H_{92}O_6$

biodiesel production is characterized by its fatty acid profile. The fatty acid profile accounts for the percentage of fatty acid chains present in the oil.

The conversion of oil into biodiesel is carried out by a *transesterification reaction* of the triglycerides with an alcohol to form fatty acid alkyl esters (Figure 17.10). The triglycerides are first converted into diglycerides, then to monoglycerides and finally to glycerol. More details about the mechanism of the reaction are shown in Section 18.1.1. The transesterification reaction is an equilibrium reaction and requires three moles of alcohol per mole of triglycerides. The alcohol is generally in excess depending on the oil composition and the catalyst used. The most commonly used alcohols are methanol and ethanol. Methanol is preferred due to its wide availability and low cost. Methanol being the lightest of all alcohols is most suited to dilute highly viscous triglycerides. When methanol is used, the transesterification products are known as *fatty acid methyl esters*, or FAMEs. These are formed from the fatty acid chains and their number of carbons is within the range of the number of carbons of the molecules present in petroleum diesel.

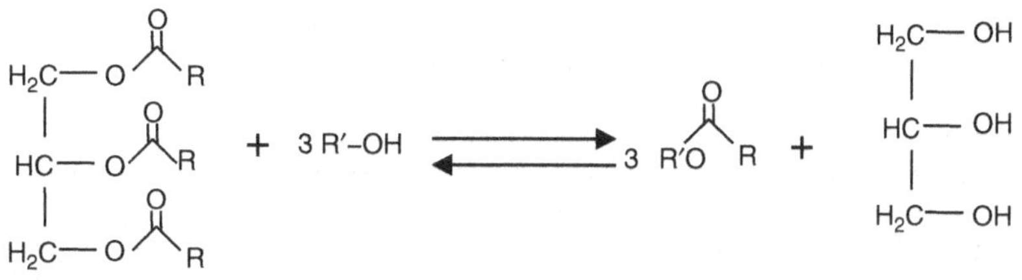

Figure 17.10 *Transesterification reaction of triglyceride with alcohol. R = alkyl chain in triglyceride and R' = alkyl chain in alcohol.*

As shown in the reaction in Figure 17.10, glycerol is produced as a by-product to biodiesel. Glycerol can be sold or refined for the pharmaceutical industry. The increasing biodiesel production has increased the availability of glycerol and decreased its value. Various routes are available for the exploitation and conversion of glycerol into high added value products; see Section 11.6 and ***Online Resource Material, Chapter 17 – Additional Exercises and Examples***.

The transesterification reaction requires a catalyst to achieve a high reaction rate and high conversion. Alkalis such as NaOH, KOH and methoxides such as $NaOCH_3$ are the most commonly used homogeneous catalysts. Typical catalyst concentration is about 1% by weight of oil. The reaction is carried out at 60 °C under atmospheric pressure. The yield of methyl esters may exceed 98% on a weight basis. Despite being widely established, the homogeneously catalyzed reaction makes downstream processing more complex: (i) requires additional reaction to neutralize the alkali, which generates a salt waste; (ii) the glycerol produced is of low quality and requires further purification; (iii) a water washing step is required to remove glycerol and catalyst for biodiesel purification.

17.3.2   Heterogeneous Catalysts for Transesterification

Heterogeneously catalyzed transesterification facilitates downstream processing, thus enhancing operation efficiency and reducing production costs and environmental impacts. Solid heterogeneous catalysts require higher alcohol-to-oil molar ratios and higher operating temperatures than a homogeneous catalyst, but they are easily separated from the product or can be fixed in a catalyst bed within the reactor. This avoids the need for the neutralization and water washing steps. In addition, the glycerol produced has a higher purity. Common heterogeneous catalysts include metal oxides such as CaO, ZnO, MgO, metal carbonates, zeolite type catalysts, polymeric resins and enzymes. Zirconium and titanium based catalysts have also been tested. Heterogeneous catalysts could also achieve complete conversion of free fatty acids and eliminate soap formation (a common issue in homogeneous alkali catalysts). This capability provides higher feedstock flexibility to the biodiesel process.

The *Online Resource Material, Chapter 11 – Additional Exercises and Examples* shows the process simulation of biodiesel production from algal oil to estimate overall biodiesel process yield, energy requirements and fuel properties of the biodiesel product. The simulation framework can serve as the basis for the analysis of processes using other oil feedstocks. Refer to Section 12.4.3 for heterogeneous catalytic reactor modeling and design.

Exercise 11. Kinetic studies of a new heterogeneous catalyst for transesterification of triolein have shown a consumption rate of 8.51×10^{-5} mol s^{-1} g^{-1} catalyst[6]. The concentration of the triolein is 0.0693 mol L^{-1}. The catalyst particles are spherical with a diameter of 0.3 cm and have a density of 0.9 g cm^{-3}. The effective diffusivity is 0.005 cm^2 s^{-1}. The reaction has been determined to be second order with respect to oil concentration.

a.   Calculate the reaction rate constant and the effectiveness of the catalyst particles. Discuss the regime governing the catalytic process.
b.   Determine the diameter of the particle that allows elimination of the intraparticle diffusional limitation.

Solution to Exercise 11. Refer to Section 12.4.3 to relate the reaction rate constant, Thiele modulus and effectiveness with the physical properties of the catalyst. Notice that in this case the order of the reaction is two. Once you determine the unknown parameters, you can change the particle diameter to investigate if the decrease or increase in size allows elimination of the intraparticle diffusional limitation. Follow the derivation of Equations (12.71) to (12.96).

Exercise 12. A promising feedstock for biodiesel production in tropical countries is *Jatropha Curcas*. The seeds of this perennial plant are also rich in triglycerides. Table 17.8 shows an average composition of the *Jatropha* oil. Perform a process simulation as shown for algae oil conversion into biodiesel production using a heterogeneous catalyst. Use the process specifications shown in the *Online Resource Material, Chapter 11 – Additional Exercises and Examples*. Compare the results in terms of overall product yields, biodiesel properties and utility requirements between *Jatropha* oil and algal oil as feedstocks.

Table 17.8   *Composition of Jatropha oil for Exercise 12.*

Component	Mass Fraction
Triolein	0.45
Trilinolein	0.403
Tripalmitin	0.114
FFA (as oleic acid)	0.0103
Tristearin	0.0227

17.4   Algae Biorefinery Integration

Algae are also a source of proteins, carbohydrates, nucleic acids and other important molecules such as vitamins, amino acids, antioxidants and pigments. The proportions of these components are variables depending on the microalgae species

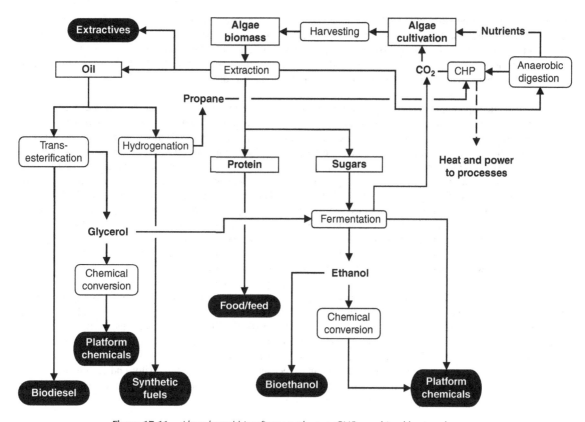

Figure 17.11 *Algae based biorefinery pathways. CHP, combined heat and power.*

and the nutritional conditions during cultivation. Under a rich nitrogen and phosphorus medium, microalgae generally produce high amount of proteins. Under a limited supply of such nutrients, algae store energy either as carbohydrates or lipids. The full potential of algae could be realized in integrated biorefineries. Biorefinery pathways for full algae biomass exploitation are shown in Figure 17.11. Algae biomass is first fractionated by using solvent or supercritical fluid extraction into high value added extractives, oil, protein, sugars and substrate for anaerobic digestion. The interesting feature is the possibility of capture of CO_2 by algae cultivation resulting from energy production or fermentation. There is also recycling of nutrients, water and energy by using anaerobic digestion of biorefinery waste streams.

Exercise 13. A company is planning to invest on an algae based biorefinery to use its algae biomass production from a raceway pond system. Table 17.9 shows the composition of the biomass produced. Carry out spreadsheet calculations for

Table 17.9 *Mass content of the various algae biomass fractions.*

Algae Biomass Fraction	Mass Content (%)
Oil	35
Protein	25
Starch	20
Nitrogen	5
Phosphorus	1
Others	14

the mass balance of a possible integrated algae based biorefinery flowsheet and report the relevant values for the mass balance. The selected product portfolio may include biodiesel, protein meal and succinic acid. The integration should be aimed to supply the energy and hydrogen demand by on site generations. Determine if it is possible to produce all three products or if there is any limiting co-product that is to be spent to meet energy demands. Make assumptions on the processes shown here, as appropriate. Use 100 kg h^{-1} of algae biomass as the calculation basis.

Other synergetic opportunities for algae based biorefineries can exploit interactions with natural resources and industrial wastes. Natural resources include, for example, wind and solar irradiation, which can be used by *renewable energy technologies* (e.g., solar panels and wind turbines) to supply energy for algae production. Industrial wastes include wastewater from the food industry, power plants, buildings, etc. An example of this type of integration is shown in Figure 17.12. Algae cultivation by either open pond or artificially illuminated PBRs is used as part of the treatment process of residential wastewater. The energy for the paddle wheels in open ponds or the illumination lamps in PBRs is provided by electricity from solar panels and wind turbines. The residual biomass after oil extraction is digested to produce biogas. The biogas is used in a combined heat and power (CHP) plant to provide energy for households. Nutrients and water are recycled from the anaerobic digestion plant while CO_2 from biogas combustion in a CHP plant is used for algae growth. One challenge is the regeneration of water for recycling and reuse in households in order to avoid losses during algae biomass processing. The energy production from the residual algae biomass via anaerobic digestion offers flexibility to cope with the intermittency in the energy supply from solar panels and wind turbines. Thus, part of the electricity generated from a CHP plant can be used as backup for periods of low sunlight irradiance and low wind speed.

Another possibility is the algae cultivation as part of the architecture of buildings to produce various energy carriers. Flat panel PBRs form the façade producing algae biomass from solar irradiation, thus making the buildings "green." Furthermore, the wastewater generated in the building can be treated by the algae growing in the façade PBRs. The algae biomass produced is converted into biogas by anaerobic digestion. Biogas can be used as fuel to provide energy to the building or can be upgraded to high purity methane for the gas grid. In addition, solar thermal energy is also captured within the liquid culture medium. The associated heat can be recovered and stored geothermally, underneath the building, or by the use of heat pumps. The recovered heat is then used for thermal comfort inside the building and to supply hot water.

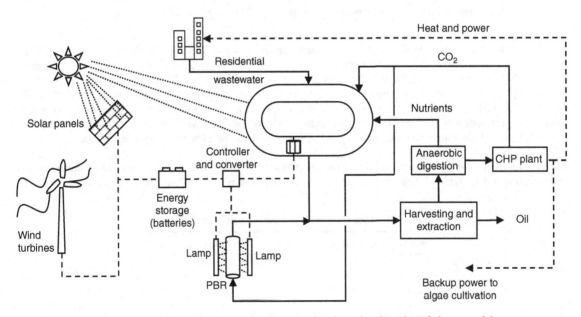

Figure 17.12   *Integration of algae cultivation with other natural, industrial and residential elements of the ecosystem.*

Exercise 13. For the integrated "green building" discussed here draw a process block diagram. Develop mass and energy balance models to determine if a planned office building can be designed to be self-sufficient in energy by using the algae based system. The planned office building has the following characteristics:

Dimensions = width: 15 m, length: 15 m, height: 22 m
Electricity requirements = 100 kW h m^{-2} y^{-1}
Space heating requirement = 70 kW h m^{-2} y^{-1}
Hot water requirement = 280 L h^{-1}
Wastewater generation = 50 L h^{-1}

The algae system has the following characteristics:

PBR panel thickness = 25 cm
Algae biomass digestion efficiency = 55%
Algae biomass production rate = 2.25 g m^{-2} h^{-1}
Biogas yield = 0.45 m^3 kg^{-1} biomass
Biogas heating value = 21 MJ m^{-3}
Biogas energy conversion into electricity generation efficiency = 50%
Biogas energy conversion into waste heat generation efficiency = 35%
Collected solar thermal heat = 30 kW h m^{-2} y^{-1}

Make rational assumptions. If there is excess heat available, consider geothermal storage. Assume that water is previously clarified in an adjacent plant and the algae system is used for nutrient disposal from wastewater (via anaerobic digestion) to avoid eutrophication (for the eutrophication potential impact refer to Chapter 4, Life Cycle Assessment).

17.5   Life Cycle Assessment of Algae Biorefineries

The resulting environmental benefits or detrimental effects of algae based production systems need to be assessed to determine the overall sustainability. Despite being an effective solution for CO_2 capture and avoidance of land for food (by using nonagricultural land and producing protein and other nutritional components), algae need some kind of fertilization. Cultivation using clean water could also cause depletion of local water resources. In addition, the environmental impact from the supply of construction materials can also be important. This is more relevant in PBR cultivation systems which may need special materials for their fabrication.

The various stages involved in algae cultivation have been shown earlier. These stages form the basis for life cycle assessment of algae based biorefineries. Figure 17.13 shows a life cycle flowsheet for a biorefinery producing biodiesel, glycerol and protein meal from algae biomass. Alternatives for algae cultivation and harvesting systems are included. The extraction cake can either be sold as protein meal or converted into biogas for CHP generation. As remarked before, the energy produced from biogas can be used in various processing stages and CO_2 and nutrients can be recycled to the algae cultivation stage. Oxygen is generated during algae photosynthesis and is released in a degassing column prior to biomass recovery. Note that not all the CO_2 injected to the algae culture medium is captured. Mass and energy balance calculations are needed to determine the process energy that can be supplied from anaerobic digestion of cake for biogas and CHP generation as well as the mass of nutrients recovered and recycled.

Another important issue in algae based biorefineries is the use of fresh water. Fresh water consumption is reduced by growing algae in sea water or wastewater. Sea water could avoid the need for additional nutrients except for phosphorus. Some wastewaters (e.g., residential wastewater) also provide nutrients for algae growth. In algae systems that use wastewater and convert extraction cake for biogas generation, the nutrients recovered from the wastewater sludge can be used as agricultural fertilizer. With the controlled release of nutrients, the accumulation of nutrients is avoided.

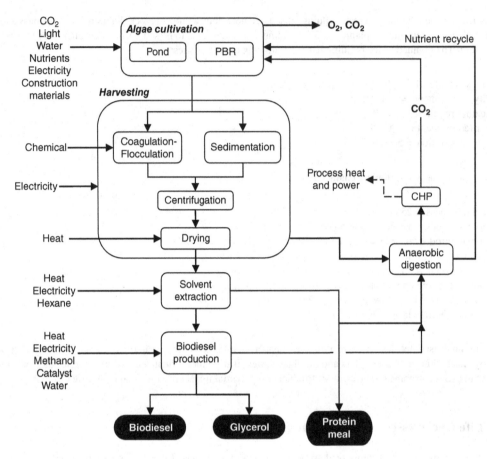

Figure 17.13   *Biorefinery system flowchart producing biodiesel, glycerol and protein meal from algae oil.*

Figure 17.14 shows the water flows in a highly integrated biorefinery. Most of the water is kept within the life cycle by means of regeneration and recycling of streams. The net water inputs are the water streams used in cultivation, as make-up water for CHP generation and to compensate for any loss due to evaporation. Rain can be considered an additional input for open ponds. The net water outputs are the water generated from biogas combustion in CHP plant and the losses by evaporation in open ponds and in the cooling towers. The cooling towers are used to cool down condensates and hot cooling water. Note that the water streams generated in algae harvesting processes (coagulation–flocculation, centrifugation) can be regenerated in an auxiliary water treatment process and recycled to algae cultivation. Process water is also used and regenerated in the case of homogeneously catalyzed transesterification processes. Clarified water from sedimentation without the use of chemical additives may be reused as produced.

Figure 17.14 can be analyzed for water footprints from biodiesel production. Water footprint analysis, also discussed in Web Chapter 3, refers to (i) the total volume of fresh water used in a process or activity or (ii) the net volume of fresh water consumption and the potential impacts on water availability and quality in certain regions or ecosystems. The first approach is a measure of the quantity of water used or consumed and could be called *water accounting*. This measure is useful in increasing awareness of the amounts of water that are being used to obtain certain products or services and can guide the management of regional water resources. However, the second approach is closer to the definition of a footprint as it aims to get an indicator of the real impacts caused by water use. This idea is more in line with the life cycle assessment (LCA) methodologies (see Chapters 4 and 5). Water footprint analysis in this chapter refers to this more evolved approach. A series of impact indicators are under development with some of them taken from LCAs such as

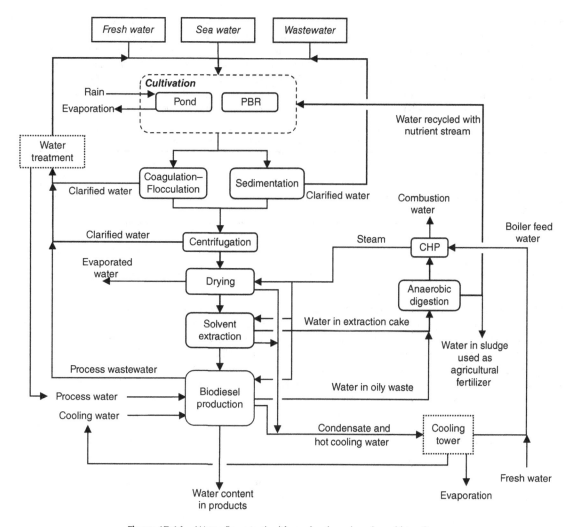

Figure 17.14 *Water flows in the life cycle of an algae based biorefinery.*

acidification, eutrophication and water toxicity. From a water accounting point of view, for example, all the water flow inputs in Figure 17.14 are to be accounted for in water inventories. Consideration of the regeneration, reuse and recycling as well as the type of water used for cultivation determines the net consumption. Scarcity or availability of freshwater for human activities is a major concern of society. Therefore, the net consumption may be focused on the amount of fresh water consumed in a system that potentially impacts the availability and quality of this resource. A water footprint analysis considers this by using a *water stress index* (*WSI*), as shown by[7]

$$WSI = \frac{\text{Withdrawal or extraction of water resources (m}^3)}{\text{Total avilability of water resources (m}^3)} \qquad (17.41)$$

The availability of water resources varies with the location and climatic characteristics of such locations, such as precipitation (which increases availability) and solar energy irradiation (which decreases availability due to evaporation produced by heating surface water bodies). These variations confer dynamism to a water footprint analysis, since the inventories vary with both time and space.

Exercise 14. AlgDiesel is a company producing biodiesel from algae oil and is planning a new facility in a tropical country. The government of such a country is promoting the production of biofuels but is also concerned about the scarcity of fresh water in some regions. The company has to assess the impact on fresh water that an algae based biorefinery producing biodiesel could have in the potential locations of the facility. The company has estimated the water inventories of the biorefinery in three potential locations, shown in Table 17.10. Only a small fraction of the fresh water used is recycled within the process. Determine the water stress index to guide decision making about the locations of the facility with the lowest potential impact on water availability. At this stage, static average values can be used.

Table 17.10 *Water inventories for an algae based biorefinery in three different locations.*

ID	Location	A	B	C
1	Estimated fresh water use for algae cultivation	$20\,000$ m^3 y^{-1}	$30\,000$ m^3 y^{-1}	$18\,000$ m^3 y^{-1}
2	Fresh water for processing and utilities	$10\,000$ m^3 y^{-1}	$10\,000$ m^3 y^{-1}	$10\,000$ m^3 y^{-1}
3	Available wastewater for algae cultivation	8000 m^3 y^{-1}	5000 m^3 y^{-1}	$15\,000$ m^3 y^{-1}
4	Net water loss in ponds (evaporation minus precipitation)	3000 m^3 y^{-1}	2300 m^3 y^{-1}	2000 m^3 y^{-1}
5	Other losses	2500 m^3 y^{-1}	2000 m^3 y^{-1}	1800 m^3 y^{-1}
6	Treated water discharged to a river	$25\,500$ m^3 y^{-1}	$20\,500$ m^3 y^{-1}	$25\,500$ m^3 y^{-1}
7	Average annual availability in reservoir	$35\,000$ m^3 y^{-1}	$50\,000$ m^3 y^{-1}	$30\,000$ m^3 y^{-1}

Solution to Exercise 14. Water use by the biorefinery is the summation of fresh water and wastewater for cultivation, fresh water for processing and utilities and the water required to balance the water losses (i.e., inventories 1 to 5 in Table 17.10). The net consumption is the summation of treated water discharged to the river, fresh water required for algae cultivation and fresh water used to balance the water losses (i.e., inventories 4 to 6 in Table 17.10). This is because a fraction of fresh water is recycled within the process and the wastewater input does not come from the fresh water reservoir. Although the water discharged is treated, it is degraded in quality and is discharged into another water body. Thus, it must be counted as consumed fresh water. The results are shown in Figure 17.15. This figure shows that location B has the lowest water stress index of the three options evaluated. Although the water use is higher, the net water consumption is the lowest and the availability of fresh water in the underground reservoir is higher. Thus, the biorefinery will cause less water stress impact if it is located in B.

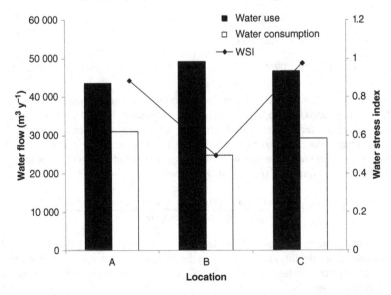

Figure 17.15 *Comparison of water use, water consumption and water stress index in three options for a biorefinery location.*

17.6   Summary

Algae show promising opportunities for the production of biofuel, energy and value-added coproducts. At the same time, various challenges remain unresolved and prevent the successful emergence of commercial and large-scale production. There is scope for improvement in the algae cultivation and harvesting to reduce production costs. This chapter discusses comprehensive models to gain insights into the performance of algae cultivation in open ponds and photobioreactors. Approaches for modeling of cultivation systems have been shown using idealized reactor models such as a continuously stirred tank reactor, batch reactor and plug flow reactors. The models can be used to estimate reactor size, the CO_2 capture rate and algae biomass yield. Important aspects of algae harvesting processes are also discussed in this chapter with an emphasis on the process synthesis and optimization. This is required to reduce capital and energy costs, which are preventing successful establishment of commercial facilities. Although production of oil for biodiesel has been the main driver for algae cultivation, a wide range of products is accessible from algae biomass using the biorefinery concept. Integrated production of biofuel, protein meal, platform chemical and high value extractive needs to be explored to utilize whole algae biomass. Furthermore, algae based biorefineries can become more sustainable not only by internal process integration but also by integration with renewable energy technologies, surrounding processes, buildings and regional systems. Integration of renewable energy (solar and wind power) and wastewater for algae growth enhances the environmental performance of algae based production systems. Opportunities for improvement in downstream processing include the use of heterogeneous catalysts for oil conversion into biodiesel. Process simulation strategies are discussed in *Online Resource Material, Chapter 11 – Additional Exercises and Examples* for an estimation of physical properties for the biodiesel production process. As with other biorefinery systems, life cycle assessment is crucial for the development of sustainable algae based biorefineries. Particular attention is paid to the water footprint analysis due to the inherent relationship of algae cultivation with this vital resource.

Algae biomass production and its exploitation in biorefineries and "green buildings" are still evolving. However, process engineers must be prepared to invent new and sustainable biorefining systems. The concepts shown in this book form a basis for such preparation.

References

1. P.T. Pienkos, *The Potential for Biofuels from Algae*, Algae Biomass Summit, NREL, San, Francisco, CA, 2007.
2. R.J. Geider, H.L. MacIntyre, T.M. Kana, A dynamic model of photoadaptation in phytoplankton, *Limnol. Oceanogr*, **41**, 1–15 (1996).
3. G.F. Blanchard, J.M. Guarini, R. Richard, Ph. Gros, F. Mornet, Quantifying the short-term temperature effect on lightsaturated photosynthesis of intertidal microphytobenthos, *Mar. Ecol. Progr. Ser.*, **134**, 309–313 (1996).
4. A. Yang, Modeling and evaluation of CO_2 supply and utilization in algal ponds, *Ind. Eng. Chem. Res.*, **50**, 11181–11192 (2011).
5. H. Jupsin, E. Praet, J.L. Vasel, Dynamic mathematical model of high rate algal ponds (HRAP), *Water Sci. Technol.*, **48**, 197–204 (2003).
6. Z. Huaping, W. Zongbin, C. Yuanxiong, Z. Ping, D. Shijie, L. Xiaohua, M. Zongqiang, Preparation of biodiesel catalyzed by solid super base of calcium oxide and its refining process, *Chinese J. Catal.*, **27**, 391–396 (2006).
7. S. Pfister, A. Koehler, S. Hellweg, Assessing the environmental impacts of freshwater consumption in LCA, *Environ. Sci. Technol.*, **43**, 4098–4104 (2009).

18

Heterogeneously Catalyzed Reaction Kinetics and Diffusion Modeling: Example of Biodiesel

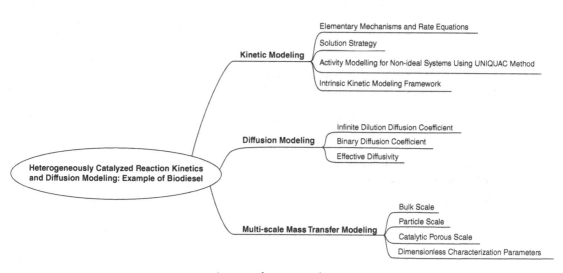

Structure for Lecture Planning

Modeling to understand the kinetic mechanism and diffusive transport is important for designing heterogeneously catalyzed reaction systems. Models of these mechanisms play an important role in identifying important parameters and their optimal values for design and control of heterogeneous systems. A heterogeneous system is a liquid or gas phase reaction on a solid catalyst.

The modeling and simulation frameworks are for determining the intrinsic kinetic mechanism on to nonporous catalysts and effective diffusivities of species for a scaling-up reactor with porous catalysts. Two key effects, reaction kinetics and diffusion transport, determine the productivity and purity of desired products. Thus, these two modeling frameworks are shown in detail, with an example of a biodiesel production system.

Biorefineries and Chemical Processes: Design, Integration and Sustainability Analysis, First Edition.
Jhuma Sadhukhan, Kok Siew Ng and Elias Martinez Hernandez.
© 2014 John Wiley & Sons, Ltd. Published 2014 by John Wiley & Sons, Ltd.
Companion Website: http://www.wiley.com/go/sadhukhan/biorefineries

Biodiesel is fast becoming one of the key transport fuels as the world endeavors to reduce its carbon footprint and find viable alternatives to oil-derived fuels. Biodiesel is made from nonedible oils that undergo a series of transesterification reactions to produce fatty acid methyl esters (FAMEs) (the main constituents of biodiesel). The use of solid catalysis in the esterification and transesterification process has started to gain attention over homogeneous catalysts, due to elimination of the neutralization and quench steps, reduced energy requirements for downstream separation processes and recovery of by-product glycerol (if any) at a sufficient purity that can be sold.

18.1   Intrinsic Kinetic Modeling

An *intrinsic kinetic model* is the set of rate equations derived from elementary adsorption–reaction steps, without accounting for diffusion transport. The three most widely used elementary solid and liquid or gas reaction mechanism models are:

Eley–Rideal (ER)
Langmuir–Hinshelwood–Hougen–Watson (LHHW)
Hattori

The ER model assumes that one of the reactants bonds with an active site on a solid catalyst, proceeding to react with the other reactant, present in the bulk phase (liquid or gas). This forms one of the products bonded with the active site and the other in the bulk phase.

The LHHW model assumes both reactants adsorb on to adjoining active sites, reacting to form products bonded to the active sites.

The Hattori mechanism is a combination of the ER and LHHW models. The reactants are first adsorbed on to adjacent active sites on a solid catalyst. The reactants then react to form an intermediate molecule bonded to only one of the sites. This intermediate molecule splits to form the products, one bonded on to the active site and the other released in the bulk phase.

Amongst all mechanistic models, the ER type is the simplest form, involving the least number of kinetic parameters and thereby providing statistically the most reliable model for a given set of experimental data.

18.1.1   Elementary Reaction Mechanism and Intrinsic Kinetic Modeling of the Biodiesel Production System

Biodiesel or FAME (fatty acid methyl ester) is produced by the transesterification reactions of nonedible oils. The transesterification reaction occurs in three stages, the first reaction producing a diglyceride and the first of the methyl esters. The second and third reactions produce a monoglyceride and glycerol, respectively, along with a further two methyl ester molecules.

The ER reaction mechanism for the production of FAME involves adsorption of methanol (CH_3OH) on empty catalyst sites (*) and the reactions between adsorbed methanol (CH_3OH^*) with triglyceride, T, diglyceride, D, and monoglyceride, M, in the bulk phase that form adsorbed diglyceride (D^*), monoglyceride (M^*) and glycerol (G^*), respectively, along with FAME. Equation set (18.1) shows the reversible elementary steps involved in the ER reaction mechanism.

Equation set (18.1) (ER reaction steps):

$$* + CH_3OH \Leftrightarrow CH_3OH^*$$

$$CH_3OH^* + T \Leftrightarrow D^* + FAME$$

$$CH_3OH^* + D \Leftrightarrow M^* + FAME$$

$$CH_3OH^* + M \Leftrightarrow G^* + FAME$$

$$D^* \Leftrightarrow D + {}^*$$

$$M^* \Leftrightarrow M + {}^*$$

$$G^* \Leftrightarrow G + {}^*$$

To develop the rates of generation or consumption of the species, one or more or all of the elementary steps shown can be considered as the *rate limiting steps* (or the slowest steps that control the rates). The rates of generation or consumption of the species (generating differential equations with respect to time) are then equated to the kinetic rate expressions derived based on the *rate limiting step* assumptions.

The differential equations are solved to generate the concentration or number of moles per unit volume profiles of reactant and product species with respect to time, up to a batch reaction time. These profiles can be closely converged with those obtained from the laboratory experimentations by adjusting kinetic rate equation parameters, using mathematical programming. The combination of the reaction mechanism (e.g., ER or LHHW or Hattori) and elementary step assumptions providing the best fit with experimental data, presents the likely elementary mechanism for the given reaction system.

The lab based experimental design should be such that the effect of the transport of species through catalytic pores can be neglected at this stage and thus the rates of generation or consumption of the species would primarily be dependent on the intrinsic kinetics of the reaction system.

The intrinsic kinetic reaction condition can be attained by carrying out the reaction using a nonporous solid catalyst in the laboratory-scale reactor.

There are various assumptions of elementary reaction steps as follows. These assumptions can be applied to a reaction mechanism hypothesis. One or more of the following assumptions along with a chosen mechanism can then be used for experimental data validation.

1. Assuming a pseudo-steady state, all elementary steps in a mechanism are the rate limiting steps.
2. Only the surface reactions are quasi-equilibrated. The concentrations of the surface species are calculated assuming their pseudo-steady state.
3. The adsorption steps are rate limiting steps. This could be a valid assumption because active sites on a solid catalyst may be difficult to be accessible to liquid/gaseous reactant species.

The differential rate of generation or consumption of species are then written as a function of the activities of the reaction components or species (rather than concentrations of species due to the consideration of nonideality discussed later), the reaction rate coefficients and the adsorption and surface reaction equilibrium coefficients of the elementary steps in a mechanism.

Assuming that the surface of the catalyst is quasi-equilibrated or at a pseudo-homogeneous state with respect to the surface species, such that the concentrations of the catalyst surface species remain constant with respect to time (assumption 2), the rates of the consumption or generation of the various species in the ER mechanism of FAME production system in Equation set (18.1) can be expressed as in Equation set (18.2).

The backward reaction rate constants are neglected. k_1–k_7 represent the kinetic rate constants of the forward reactions in Equation set (18.1) in the ER mechanism. The activities of species $i = T, D, M, G, CH_3OH$ and $FAME$ in the bulk phase are represented as "$[i]$."

Equation set (18.2) (ER mechanism with the assumption of all elementary steps as the rate limiting steps):

Rate of consumption of T:

$$r_T = \left(\frac{k_2[T]}{1 + \left(\dfrac{k_2[T] + k_3[D] + k_4[M]}{k_1[CH_3OH]} \right) + \dfrac{k_2[D]}{k_6} + \dfrac{k_2[T]}{k_5} + \dfrac{k_4[M]}{k_7}} \right)$$

Rate of consumption of D:

$$r_D = \left(\frac{k_3[D] - k_2[T]}{1 + \left(\dfrac{k_2[T] + k_3[D] + k_4[M]}{k_1[CH_3OH]} \right) + \dfrac{k_2[D]}{k_6} + \dfrac{k_2[T]}{k_5} + \dfrac{k_4[M]}{k_7}} \right)$$

Rate of consumption of M:

$$r_M = \left(\cfrac{k_4\,[M] - k_3\,[D]}{1 + \left(\cfrac{k_2[T] + k_3[D] + k_4[M]}{k_1[CH_3OH]} \right) + \cfrac{k_2[D]}{k_6} + \cfrac{k_2[T]}{k_5} + \cfrac{k_4[M]}{k_7}} \right)$$

Rate of consumption of CH_3OH:

$$r_{CH_3OH} = \left(\cfrac{-\left(k_5\,[D] + k_3\,[T] + k_7\,[M]\right)}{1 + \left(\cfrac{k_2[T] + k_3[D] + k_4[M]}{k_1[CH_3OH]} \right) + \cfrac{k_2[D]}{k_6} + \cfrac{k_2[T]}{k_5} + \cfrac{k_4[M]}{k_7}} \right)$$

Rate of change of G:

$$r_G = \left(\cfrac{k_7 k_1 [CH_3OH]}{k_3[T] + k_5[D] + k_7[M]} \right)$$

Rate of generation of FAME:

$$r_{FAME} = \left(\cfrac{\left(k_5\,[D] + k_3\,[T] + k_7\,[M]\right)}{1 + \left(\cfrac{k_2[T] + k_3[D] + k_4[M]}{k_1[CH_3OH]} \right) + \cfrac{k_2[D]}{k_6} + \cfrac{k_2[T]}{k_5} + \cfrac{k_4[M]}{k_7}} \right)$$

There are nine elementary reaction steps in the LHHW mechanism shown in Equation set (18.3). The corresponding expressions for the equilibrium rate constants $K_1 - K_9$ are shown in Equation set (18.4). The model in Equation set (18.4) assumes that all nine elementary reaction steps are rate limiting. The first elementary reaction step in the LHHW kinetic mechanism is the adsorption of methanol, as in the ER mechanism. The main difference between the ER and LHHW mechanisms is the adsorption of triglyceride on the surface of the catalyst. The adsorbed methanol and triglyceride react with each other if they are adjacent, to produce adsorbed diglyceride and FAME. Subsequently the adsorbed methanol reacts with adsorbed diglyceride or monoglyceride to form adsorbed monoglyceride and glycerol, along with FAME.

Equation set (18.3) (LHHW reaction steps):

$$* + CH_3OH \Leftrightarrow CH_3OH^*$$

$$T + {}^* \Leftrightarrow T^*$$

$$CH_3OH^* + T^* \Leftrightarrow D^* + FAME^*$$

$$CH_3OH^* + D^* \Leftrightarrow M^* + FAME^*$$

$$CH_3OH^* + M^* \Leftrightarrow G^* + FAME^*$$

$$FAME + {}^* \Leftrightarrow FAME^*$$

$$D + {}^* \Leftrightarrow D^*$$

$$M + {}^* \Leftrightarrow M^*$$

$$G + {}^* \Leftrightarrow G^*$$

Equation set (18.4) (LHHW mechanism with the assumption of all elementary steps as the rate limiting steps):

$$K_1 = \frac{[CH_3OH^*]}{[*][CH_3OH]}$$

$$K_2 = \frac{[T^*]}{[*][T]}$$

$$K_3 = \frac{[D^*][FAME^*]}{[CH_3OH^*][T^*]}$$

$$K_4 = \frac{[M^*][FAME^*]}{[CH_3OH^*][D^*]}$$

$$K_5 = \frac{[G^*][FAME^*]}{[CH_3OH^*][M^*]}$$

$$K_6 = \frac{[FAME^*]}{[*][FAME]}$$

$$K_7 = \frac{[D^*]}{[*][D]}$$

$$K_8 = \frac{[M^*]}{[*][M]}$$

$$K_9 = \frac{[G^*]}{[*][G]}$$

When the surface reaction steps are assumed to be quasi-equilibrated, the LHHW reaction steps generate the kinetic rate expressions for individual reaction steps, shown in Equation set (18.5). In addition to the adsorbed species defined in the ER mechanism, * and T^* are used to represent the empty site and adsorbed triglyceride on the catalyst surface, respectively; k_j indicates the forward reaction kinetic rate constants for nine rate determining steps and K_j represents the equilibrium constants of reaction j, respectively.

Equation set (18.5) (LHHW mechanism with the assumption that the surface reaction steps are quasi-equilibrated):

$$r_1 = \frac{k_1\left([CH_3OH] - {}^1/_{K_1}\frac{K_6 K_7}{K_3 K_2}\frac{[D][FAME]}{[T]}\right)}{\left(1 + \frac{K_6 K_7}{K_3 K_2}\frac{[D][FAME]}{[T]} + K_6[FAME] + K_2[T] + K_7[D] + K_8[M] + K_9[G]\right)}$$

$$r_2 = \frac{k_2\left([T] - {}^1/_{K_2}\frac{K_6 K_7}{K_3 K_1}\frac{[D][FAME]}{[CH_3OH]}\right)}{\left(1 + K_1[CH_3OH] + K_6[FAME] + \frac{K_6 K_7}{K_3 K_1}\frac{[D][FAME]}{[CH_3OH]} + K_7[D] + K_8[M] + K_9[G]\right)}$$

$$r_3 = \frac{k_3\left(K_1 K_2[CH_3OH][T] - {}^1/_{K_3}K_7 K_6[D][FAME]\right)}{\left(1 + K_1[CH_3OH] + K_6[FAME] + K_2[T] + K_7[D] + K_8[M] + K_9[G]\right)^2}$$

$$r_4 = \frac{k_4\left(K_1 K_7[CH_3OH][D] - {}^1/_{K_4}K_8 K_6[M][FAME]\right)}{\left(1 + K_1[CH_3OH] + K_6[FAME] + K_2[T] + K_7[D] + K_8[M] + K_9[G]\right)^2}.$$

$$r_5 = \frac{k_5 \left(K_1 K_8 [CH_3OH] [M] - \frac{1}{K_5} K_9 K_6 [G] [FAME] \right)}{\left(1 + K_1 [CH_3OH] + K_6 [FAME] + K_2 [T] + K_7 [D] + K_8 [M] + K_9 [G] \right)^2}$$

$$r_6 = \frac{k_6 \left(\dfrac{K_3 K_2 K_1 [T] [CH_3OH]}{K_7 [D]} - K_6 [FAME] \right)}{\left(1 + K_1 [CH_3OH] + \dfrac{K_3 K_2 K_1 [T] [CH_3OH]}{K_7 [D]} + K_2 [T] + K_7 [D] + K_8 [M] + K_9 [G] \right)}$$

$$r_7 = \frac{k_7 \left(\dfrac{K_3 K_2 K_1 [T] [CH_3OH]}{K_6 [FAME]} - K_7 [D] \right)}{\left(1 + K_1 [CH_3OH] + K_6 [FAME] + K_2 [T] + \dfrac{K_3 K_2 K_1 [T] [CH_3OH]}{K_6 [FAME]} + K_8 [M] + K_9 [G] \right)}$$

$$r_8 = \frac{k_8 \left(\dfrac{K_4 K_1 K_7 [D] [CH_3OH]}{K_6 [FAME]} - K_8 [M] \right)}{\left(1 + K_1 [CH_3OH] + K_6 [FAME] + K_2 [T] + K_7 [D] + \dfrac{K_4 K_1 K_7 [D] [CH_3OH]}{K_6 [FAME]} + K_9 [G] \right)}$$

$$r_9 = \frac{k_9 \left(\dfrac{K_5 K_8 K_1 [M] [CH_3OH]}{K_6 [FAME]} - K_9 [G] \right)}{\left(1 + K_1 [CH_3OH] + K_6 [FAME] + K_2 [T] + K_7 [D] + K_8 [M] + \dfrac{K_5 K_8 K_1 [M] [CH_3OH]}{K_6 [FAME]} \right)}$$

The elementary steps in the Hattori kinetic mechanism are shown in Equation set (18.6). Similar to LHHW, The Hattori mechanism also considers the adsorption of triglyceride on the surface of the catalyst as a rate determining step. The Hattori mechanism differs from the LHHW mechanism, by the consideration of intermediate specie formation by the reactions between adsorbed methanol and adsorbed triglyceride, diglyceride and monoglyceride.

Equation set (18.6) (Hattori mechanism reaction steps):

$$^* + CH_3OH \Leftrightarrow CH_3OH^*$$

$$T + ^* \Leftrightarrow T^*$$

$$CH_3OH^* + T^* \Leftrightarrow [TsCH_3OH]^* + ^*$$

$$[TsCH_3OH]^* \Leftrightarrow D^* + FAME$$

$$CH_3OH^* + D^* \Leftrightarrow [DsCH_3OH]^* + ^*$$

$$[DsCH_3OH]^* \Leftrightarrow M^* + FAME$$

$$CH_3OH^* + M^* \Leftrightarrow [MsCH_3OH]^* + ^*$$

$$[MsCH_3OH]^* \Leftrightarrow G^* + FAME$$

$$D + ^* \Leftrightarrow D^*$$

$$M + ^* \Leftrightarrow M^*$$

$$G + ^* \Leftrightarrow G^*$$

The adsorbed methanol and triglyceride react to form adsorbed intermediate ($TsCH_3OH^*$) and an empty site. Subsequently, the adsorbed intermediate ($TsCH_3OH^*$) decomposes into the production of adsorbed diglyceride and FAME in the bulk phase, respectively. Adsorbed diglyceride, monoglyceride and glycerol thereafter desorb from the catalyst surface to the bulk phase. $TsCH_3OH^*$, $DsCH_3OH^*$ and $MsCH_3OH^*$ represent the intermediates from the reactions between adsorbed methanol and adsorbed T^*, D^* and M^*, respectively.

The equilibrium rate constants for these eleven rate limiting reaction steps in the Hattori mechanism are correlated to the activities of species, as shown in Equation set (18.7); k_j indicates the forward reaction rate constants for eleven rate determining Hattori mechanism steps and K_j represents the equilibrium constant of reaction j, respectively.

Equation set (18.7) (Hattori mechanism with the assumption of all elementary steps as the rate limiting steps):

$$K_1 = \frac{[CH_3OH^*]}{[*][CH_3OH]}$$

$$K_2 = \frac{[T^*]}{[*][T]}$$

$$K_3 = \frac{[*]\left[TsCH_3OH^*\right]}{\left[CH_3OH^*\right][T^*]}$$

$$K_4 = \frac{[D^*][FAME]}{\left[TsCH_3OH^*\right]}$$

$$K_5 = \frac{[*]\left[DsCH_3OH^*\right]}{\left[CH_3OH^*\right][D^*]}$$

$$K_6 = \frac{[M^*][FAME]}{\left[DsCH_3OH^*\right]}$$

$$K_7 = \frac{[*]\left[MsCH_3OH^*\right]}{\left[CH_3OH^*\right][M^*]}$$

$$K_8 = \frac{[G^*][FAME]}{\left[MsCH_3OH^*\right]}$$

$$K_9 = \frac{[D^*]}{[*][D]}$$

$$K_{10} = \frac{[M^*]}{[*][M]}$$

$$K_{11} = \frac{[G^*]}{[*][G]}$$

The elementary reaction expressions in Equation set (18.6) result in the kinetic rate expressions in Equation set (18.8), when methanol adsorption is assumed as the slowest or rate limiting step.

Equation set (18.8) (Hattori mechanism with the assumption of methanol adsorption as the rate limiting step):

$$r_1 = \frac{k_1\left([CH_3OH] - {}^1\!/\!K_1 \frac{K_9}{K_4K_3K_2}\frac{[D][FAME]}{[T]}\right)}{\left(1 + \frac{K_9}{K_4K_3K_2}\frac{[D][FAME]}{[T]} + \frac{K_9[D][FAME]}{K_4} + \frac{K_{10}[M][FAME]}{K_6} + \frac{K_{11}[G][FAME]}{K_8} + K_2[T] + K_9[D] + K_{10}[M] + K_{11}[G]\right)}$$

$$r_2 = \frac{k_2\left([T] - {}^1\!/\!K_2 \frac{K_9}{K_4K_3K_1}\frac{[D][FAME]}{[CH_3OH]}\right)}{\left(1 + K_1[CH_3OH] + \frac{K_9[D][FAME]}{K_4} + \frac{K_{10}[M][FAME]}{K_6} + \frac{K_{11}[G][FAME]}{K_8} + \frac{K_9}{K_4K_3K_1}\frac{[D][FAME]}{[CH_3OH]} + K_9[D] + K_{10}[M] + K_{11}[G]\right)}$$

$$r_3 = \frac{k_3\left(K_1 K_2 [CH_3OH][T] - \frac{1}{K_3}\frac{K_9}{K_4}[D][FAME]\right)}{\left(1 + K_1[CH_3OH] + \frac{K_9[D][FAME]}{K_4} + \frac{K_{10}[M][FAME]}{K_6} + \frac{K_{11}[G][FAME]}{K_8} + K_2[T] + K_9[D] + K_{10}[M] + K_{11}[G]\right)^2}$$

$$r_4 = \frac{k_4\left(K_3 K_2 K_1 [CH_3OH][T] - \frac{1}{K_4}K_9[D][FAME]\right)}{\left(1 + K_1[CH_3OH] + K_3 K_2 K_1[CH_3OH][T] + \frac{K_{10}[M][FAME]}{K_6} + \frac{K_{11}[G][FAME]}{K_8} + K_2[T] + K_9[D] + K_{10}[M] + K_{11}[G]\right)}$$

$$r_5 = \frac{k_5\left(K_1 K_9 [CH_3OH][D] - \frac{1}{K_5}\frac{K_{10}}{K_6}[M][FAME]\right)}{\left(1 + K_1[CH_3OH] + \frac{K_9[D][FAME]}{K_4} + \frac{K_{10}[M][FAME]}{K_6} + \frac{K_{11}[G][FAME]}{K_8} + K_2[T] + K_9[D] + K_{10}[M] + K_{11}[G]\right)^2}$$

$$r_6 = \frac{k_6\left(K_5 K_9 K_1 [CH_3OH][D] - \frac{1}{K_6}K_{10}[M][FAME]\right)}{\left(1 + K_1[CH_3OH] + \frac{K_9[D][FAME]}{K_4} + K_5 K_9 K_1[CH_3OH][D] + \frac{K_{11}[G][FAME]}{K_8} + K_2[T] + K_9[D] + K_{10}[M] + K_{11}[G]\right)}$$

$$r_7 = \frac{k_7\left(K_1 K_{10} [CH_3OH][M] - \frac{1}{K_7}\frac{K_{11}}{K_8}[G][FAME]\right)}{\left(1 + K_1[CH_3OH] + \frac{K_9[D][FAME]}{K_4} + \frac{K_{10}[M][FAME]}{K_6} + \frac{K_{11}[G][FAME]}{K_8} + K_2[T] + K_9[D] + K_{10}[M] + K_{11}[G]\right)^2}$$

$$r_8 = \frac{k_8\left(K_7 K_{10} K_1 [CH_3OH][M] - \frac{1}{K_8}K_{11}[G][FAME]\right)}{\left(1 + K_1[CH_3OH] + \frac{K_9[D][FAME]}{K_4} + \frac{K_{10}[M][FAME]}{K_6} + K_7 K_{10} K_1[CH_3OH][M] + K_2[T] + K_9[D] + K_{10}[M] + K_{11}[G]\right)}$$

$$r_9 = \frac{k_9\left(\frac{K_{10}[M][FAME]}{K_5 K_6 K_1[CH_3OH]} - K_9[D]\right)}{\left(1 + K_1[CH_3OH] + \frac{K_9[D][FAME]}{K_4} + \frac{K_{10}[M][FAME]}{K_6} + \frac{K_{11}[G][FAME]}{K_8} + K_2[T] + \frac{K_{10}[M][FAME]}{K_5 K_6 K_1[CH_3OH]} + K_{10}[M] + K_{11}[G]\right)}$$

$$r_{10} = \frac{k_{10}\left(\frac{K_{11}[G][FAME]}{K_7 K_8 K_1[CH_3OH]} - K_{10}[M]\right)}{\left(1 + K_1[CH_3OH] + \frac{K_9[D][FAME]}{K_4} + \frac{K_{10}[M][FAME]}{K_6} + \frac{K_{11}[G][FAME]}{K_8} + K_2[T] + K_9[D] + \frac{K_{11}[G][FAME]}{K_7 K_8 K_1[CH_3OH]} + K_{11}[G]\right)}$$

$$r_{11} = \frac{k_{11}\left(K_8 K_7 K_1 K_{10}\frac{[CH_3OH][M]}{[FAME]} - K_{11}[G]\right)}{\left(1 + K_1[CH_3OH] + \frac{K_9[D][FAME]}{K_4} + \frac{K_{10}[M][FAME]}{K_6} + \frac{K_{11}[G][FAME]}{K_8} + K_2[T] + K_9[D] + K_{10}[M] + K_8 K_7 K_1 K_{10}\frac{[CH_3OH][M]}{[FAME]}\right)}$$

From the discussion earlier, there is rationality behind the assumption of liquid/gaseous reactant species adsorption on solid catalyst active sites as the rate limiting steps. When some of the reactant species are lighter than the others in a system, the lighter species inevitably find less accessible sites, due to the presence of bulkier species. Consequently, higher concentrations of lighter species are maintained for the reaction to proceed.

FAME production is a reaction system with bulky triglyceride and lighter alcohol (e.g., methanol) reactant species. Therefore, methanol adsorption as a rate determining step in the ER mechanism is a valid assumption. In order to improve accessibility of catalytic active sites by methanol, its concentration is maintained in excess. The methanol-to-triglyceride molar ratio is maintained at 10:1–30:1 in order to achieve a higher conversion of triglyceride into FAME production.

The ER mechanism involving the least number of kinetic parameters is the simplest form amongst all mechanisms and is thus statistically the most reliable form of the mechanistic representations. The resulting ER mechanism with methanol adsorption as the rate determining step can be further simplified by the assumption of equal rate constants of all the adsorption equilibrium steps. The resulting rate expressions are shown in Equation set (18.9).

The equilibrium constant for the adsorption of methanol (K_{eq}) is assumed to be constant and equal to unity. The kinetic rate expression is reduced to two parameters, k_f (rate of forward reaction for adsorption of methanol) and K_A (adsorption equilibrium constant for diglyceride, monoglyceride, glycerol and FAME).

Equation set (18.9) (ER mechanism with methanol adsorption as the rate limiting step):

Rate of consumption of T:

$$r_T = \frac{k_f \left([CH_3OH] - {}^1\!/\!_{K_{eq}} \frac{[D][FAME]}{[T]} \right)}{\left(1 + \frac{K_A}{K_{eq}} \frac{[D][FAME]}{[T]} + K_A[FAME] + K_A[T] + K_A[D] + K_A[M] + K_A[G] \right)}$$

Rate of consumption of D:

$$r_D = \frac{k_f \left([CH_3OH] - {}^1\!/\!_{K_{eq}} \frac{[M][FAME]}{[D]} \right)}{\left(1 + \frac{K_A}{K_{eq}} \frac{[M][FAME]}{[D]} + K_A[FAME] + K_A[T] + K_A[D] + K_A[M] + K_A[G] \right)}$$

Rate of consumption of M:

$$r_M = \frac{k_f \left([CH_3OH] - {}^1\!/\!_{K_{eq}} \frac{[G][FAME]}{[M]} \right)}{\left(1 + \frac{K_A}{K_{eq}} \frac{[G][FAME]}{[M]} + K_A[FAME] + K_A[T] + K_A[D] + K_A[M] + K_A[G] \right)}$$

Exercise 1. Write the rate equations of consumption of triglyceride, diglyceride and monoglyceride for the LHHW and Hattori mechanisms with methanol adsorption as the rate determining step. To simplify the equations, assume a unit equilibrium constant for the adsorption of methanol (K_{eq}) and equal equilibrium constants for all other adsorption equilibrium steps. How many kinetic parameters are involved in the rate equations?

Exercise 2. FAME is produced using esterification reactions between fatty acids (FA) and methanol as follows:

$$FA + CH_3OH \rightleftharpoons FAME + H_2O \tag{18.10}$$

Write the elementary reaction steps in the LHHW mechanism for the esterification reaction shown in Equation (18.10).

Deduce the expressions for the rates of consumption of triglyceride, diglyceride and monoglyceride for the LHHW mechanism based on the assumption that the surface reaction between adsorbed methanol and fatty acid, CH_3OH^* and FA^*, is the rate determining step. All other steps are in equilibrium.

Solution to Exercise 2. The elementary steps of the reaction in Equation (18.10) involved in LHHW mechanism are shown in Equation set (18.11).

Equation set (18.11) (LHHW mechanism for the esterification reaction in Equation (18.10)):

$$* + CH_3OH \Leftrightarrow CH_3OH^*$$

$$* + FA \Leftrightarrow FA^*$$

$$CH_3OH^* + FA^* \Leftrightarrow FAME^* + H_2O^*$$

$$FAME^* \Leftrightarrow FAME + *$$

$$H_2O^* \Leftrightarrow H_2O + *$$

The overall reaction rate expression for the LHHW mechanism with an assumption that the surface reaction between adsorbed methanol and fatty acid, CH_3OH^* and FA^*, is the rate determining step, is shown in Equation (18.12).

All other adsorption steps are in equilibrium. K_1–K_5 represent the equilibrium rate constants for five reaction steps, respectively, shown in Equation set (18.11); k_f represents the forward surface reaction rate constant for the reaction between adsorbed methanol and fatty acid, CH_3OH^* and FA^*.

$$r = \frac{k_f \left(K_1 K_2 \, [FA] \, [CH_3OH] - \frac{K_4 K_5}{K_3} \, [FAME] \, [H_2O] \right)}{\left(1 + K_1 \, [FA] + K_2 [CH_3OH] + K_4 \, [FAME] + K_5 [H_2O] \right)^2} \tag{18.12}$$

18.1.2   Solution Strategy for the Rate Equations Resulting from the Elementary Reaction Mechanism

The rate of change in the bulk concentration (C_i) of species i with respect to time is a function of the kinetic rates (r_i) in

$$\frac{dC_i}{dt} = r_i \tag{18.13}$$

The differential equations are solved to closely match with the concentration versus time profiles obtained from the laboratory experimentations, using the linear regression technique.

The expressions for the reaction rate (r_i) are shown in terms of activities of species of the FAME production system for various combinations of reaction mechanisms (e.g., ER, LHHW and Hattori, etc.) and rate limiting step or quasi-steady state assumptions. The lab based reactor system using a nonporous solid catalyst ensures that only intrinsic kinetics, as discussed so far, influence the rates of generation or consumptions of species and there is a negligible effect of transport on these rate expressions. Thus, the solution of these differential equations provides an intrinsic kinetic model for a reaction system under consideration.

The differential equations are solved to find appropriate values of the parameters, equilibrium constants and rate constants of surface reactions involved in a mechanistic model that minimize the *residual sum of square of errors (RSSQ)* between the experimentally observed and the model predicted concentrations of species ($i = 1$ to *nspc*) at subsequent time points ($j = 1$ to *ntime*):

$$RSSQ = \sum_{j=1}^{ntime} \sum_{i=1}^{nspc} \left(C_i \, (j) - C_i^{epxerimental} \, (j) \right)^2 \tag{18.14}$$

This set of ordinary differential equations (ODEs) can be solved by an ODE solver such as in MATLAB to derive the concentration of component as a function of time. The kinetic parameters, rate constants and equilibrium constants can also be obtained by the curve fitting function 'lsqcurvefit' in MATLAB. The RSSQ is defined as the objective function to minimize the overall difference between an experimentally observed concentration profile and a profile predicted by a model.

More than one solution in terms of the values of the kinetic parameters may be obtained depending on the starting values or initial guesses of the kinetic parameters in the solution space. To obtain a more robust solution in terms of the values of kinetic parameters in a given mechanistic model, the genetic algorithm based stochastic search can be applied[1]. The iterations are repeated until a satisfactory value of *RSSQ* is obtained and there is no change in the values of the kinetic parameters over a specified number of iterations.

18.1.3 Correlation between Concentration and Activity of Species Using the UNIQUAC Contribution Method

To account for the *nonideality of a reaction mixture*, the correlation between the *activity of species* [i] and its *concentration* C_i needs to be accounted for. The *UNIQUAC* contribution method outlined here is used to calculate the activity coefficient γ_i of species i:

$$[i] = C_i\gamma_i \quad \forall i \in nspc \tag{18.15}$$

Using the UNIQUAC method, the *activity coefficients* of species in a multicomponent fluid mixture are correlated to interactions between structural groups. The usefulness of this methodology lies in the suitability of application of interaction parameters between pairs of structural groups deduced from one nonelectrolyte system to another. The UNIQUAC method often provides satisfactory representation of activity coefficients of liquid–liquid equilibrium for multicomponent mixtures containing a wide variety of hydrocarbons, such as ketones, esters, water, alcohols, etc.

The molecular activity coefficient (γ_i) of species i comprises two parts, one part known as the combinatorial part (γ_i^C) is contributed due to the effect of molecular size and shape predicted from constituent group sizes and shapes and the other part known as the residual part (γ_i^R) reflects the effect of molecular interactions (Equation (18.16)). The total number of species is presented by *nspc*.

$$\ln\gamma_i = \ln\gamma_i^C + \ln\gamma_i^R \quad \forall i \in nspc \tag{18.16}$$

Using the mole fraction x_i, area fraction θ_i and volume or segment fraction Φ_i of species i, γ_i^C is evaluated (Equation (18.17)). The molecular van der Waals area and volume parameters, q_i and s_i, are calculated as the sum of the group (p) van der Waals surface area and volume, Q_p and S_p, experimentally determined, respectively (Equations (18.18) and (18.19); $v_p^{(i)}$ is the number of p group present in molecule i and is always an integer. Equation (18.20) presents a group of parameters l_i used in Equation (18.17). The total number of groups is represented by g.

$$\ln\gamma_i^C = \ln\frac{\Phi_i}{x_i} + 5q_i\ln\frac{\theta_i}{\Phi_i} + l_i - \frac{\Phi_i}{x_i}\sum_{i'}^{nspc} x_{i'}l_{i'} \quad \forall i \in nspc \tag{18.17}$$

$$\theta_i = \frac{x_iq_i}{\sum_{i'}^{nspc} x_{i'}q_{i'}} \qquad \Phi_i = \frac{x_is_i}{\sum_{i'}^{nspc} x_{i'}s_{i'}} \quad \forall i \in nspc \tag{18.18}$$

$$q_i = \sum_p^g v_p^{(i)}Q_p \qquad s_i = \sum_p^g v_p^{(i)}S_p \quad \forall i \in nspc \tag{18.19}$$

$$l_i = 5(s_i - q_i) - (s_i - 1) \quad \forall i \in nspc \tag{18.20}$$

Equation (18.21) for the estimation of γ_i^R also requires a group residual activity coefficient Γ_p and another similar type in a reference solution containing only molecules of type i, $\Gamma_p^{(i)}$. Γ_p and $\Gamma_p^{(i)}$ are related to the corresponding area parameters in Equations (18.22) to (18.25).

$$\ln\gamma_i^R = \sum_p^g v_p^{(i)}\left(\ln\Gamma_p - \ln\Gamma_p^{(i)}\right) \quad \forall i \in nspc \tag{18.21}$$

$$\ln\Gamma_p = Q_p\left\{1 - \ln\left(\sum_m^g \theta_m\Psi_{m,p}\right) - \sum_m^g \frac{\theta_m\Psi_{p,m}}{\sum_n^g \theta_n\Psi_{n,m}}\right\} \quad \forall m,n,p \in g \tag{18.22}$$

$$\theta_p = \frac{Q_p X_p}{\sum_p Q_p X_p} \tag{18.23}$$

$$X_p = \frac{\sum_i x_i}{\sum_i x_i \sum_p v_p^{(i)}} \tag{18.24}$$

$$\ln \Gamma_p^{(i)} = Q_p \left\{ 1 - \ln \left(\sum_m^g \theta_m^{(i)} \Psi_{m,p} \right) - \sum_m^g \frac{\theta_m^{(i)} \Psi_{p,m}}{\sum_n^g \theta_n^{(i)} \Psi_{n,m}} \right\} \quad \forall m, n, p \in g \text{ and } \forall i \in nspc \tag{18.25}$$

where θ_p is the area fraction relating to a particular group p over the sum of all different groups in given species (Equation (18.26)). It is calculated in the same manner as $\theta_p^{(i)}$, except that $X_p^{(i)}$ is the fraction of a group p amongst all groups within species i.

$$\theta_p^{(i)} = \frac{Q_p X_p^{(i)}}{\sum_p Q_p X_p^{(i)}} \tag{18.26}$$

The group interaction parameter $\Psi_{m,n}$ in species i is related to $a_{m,n}$, which is indicative of the energy of interaction between the main groups m and n and the temperature of the mixture, Te (Equation (18.27)). An extensive set of experimentally determined values of $a_{m,n}$ is provided by Reid, Praustnitz and Poling (1987)[2].

$$\Psi_{m,n} = \exp \left(-\frac{a_{m,n}}{Te} \right) \tag{18.27}$$

18.1.4   An Example of EXCEL Spreadsheet Based UNIQUAC Calculation for a Biodiesel Production System is Shown in Detail for Implementation in Online Resource Material, Chapter 18 – Additional Exercises and Examples

18.1.5   Intrinsic Kinetic Modeling Framework

Intrinsic kinetic modeling is needed to identify a kinetic mechanism to best represent experimentally obtained concentration profiles of a reaction system. A well-mixed batch reactor with nonporous catalysts (so as to neglect the effect of diffusion limitation) generates the concentration or mole fraction versus time profiles of components in the reaction system. The concentration–time profile then represents the effect of intrinsic kinetics only. In addition, various characterization techniques are applied to identify catalyst structure–concentration of species patterns. Experimental observations are interpreted to develop a hypothesis on the kinetic mechanism. The rate of consumption or production of components developed based on the elementary reaction mechanism and assumptions selected is correlated to the time derivative of the concentration of components in the system (Equation (18.13)). The UNIQUAC calculations are used to correlate the activities used in the rate equations with the concentration of species. Analytical integration can be undertaken for the solution of Equation (18.13). The *RSSQ* between the experimentally and computationally derived concentration or mole fraction–time profiles of species is minimized in order to arrive at a set of kinetic parameters. The gradient based optimization can be used, however, although not recommended, as the mechanistic intrinsic kinetic rate equations are highly nonlinear.

The stochastic search techniques, such as the genetic algorithm (GA) can be applied to increase the solution space and versatility and generate robust solution. The solution is obtained in terms of the optimal values of the kinetic parameters involved in the rate equations to provide the best representation of the experimental data.

Figure 18.1 shows the intrinsic kinetic modeling framework.

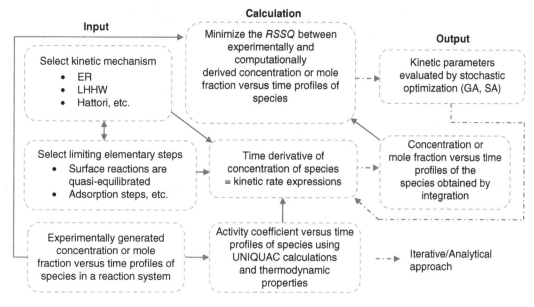

Figure 18.1   *Intrinsic kinetic modeling framework.*

The following exercise is on the calculation of the activity coefficients from the mole fraction–time profile of a biodiesel system.

Exercise 3. Estimate the activity coefficients of the components, triglyceride, diglyceride, monoglyceride, methanol and FAME, in a multicomponent FAME production system from their mole fraction–time profiles given in Table 18.1. Use the UNIQUAC method and the structural property input data shown in the *Online Resource Material, Chapter 18 – Additional Exercises and Examples*. Follow the calculation procedure shown in the companion website for the various mole fractions versus time data points in Table 18.1. Note that the calculation procedure for the data point at 1440 min is shown in the companion website.

Table 18.1   *Mole fraction–time profiles of species in a FAME production system.*

Time (min)	Triglyceride	Diglyceride	Monoglyceride	Methanol	FAME
0	0.032258	1×10^{-10}	1×10^{-10}	0.967742	1×10^{-10}
30	0.028709	0.003799	1×10^{-10}	0.964846	0.002605
60	0.024962	0.006187	0.000535	0.95997	0.00522
120	0.022846	0.00955	0.001471	0.959563	0.009416
180	0.021034	0.011444	0.002521	0.957932	0.013331
240	0.018602	0.011712	0.00294	0.952991	0.015241
300	0.017174	0.012271	0.003616	0.950638	0.017308
360	0.016149	0.013166	0.004657	0.949523	0.020342
1440	0.006337	0.010573	0.008725	0.922622	0.039365

Solution to Exercise 3.   Table 18.2 shows the results in terms of the activity of species for the mole fraction–time profiles in Table 18.1. Glycerol values have not been included, because glycerol concentration is negligibly small.

Table 18.2　*Activity of species for the mole fraction–time profiles in Table 18.1.*

Triglyceride	Diglyceride	Monoglyceride	Methanol	FAME
0.0949	0	0	0.9750	0
0.0849	0.0031	0	0.9719	0.0076
0.0771	0.0052	0.0003	0.9656	0.0154
0.0657	0.0078	0.0007	0.9675	0.0271
0.0579	0.0091	0.0012	0.9673	0.0377
0.0542	0.0096	0.0014	0.9604	0.0440
0.0504	0.0101	0.0018	0.9578	0.0501
0.0456	0.0106	0.0022	0.9579	0.0581
0.0209	0.0093	0.0043	0.9263	0.1189

Consider a generic reactant concentration versus time profile in Figure 18.2. The various points in terms of time points and concentration at those time points, $(t1, C1)$, $(t2, C2)$, $(t3, C3)$, $(t4, C4)$,..., etc., are predicted from the analytical integration of Equation (18.13) (Taylor's series first-order expansion). C presents concentration of a reactant and t presents the time.

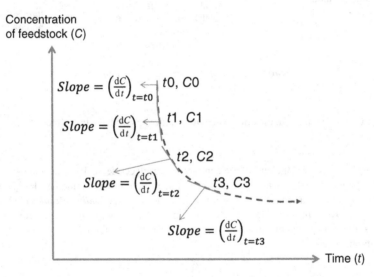

Figure 18.2　*A generic reactant concentration versus time profile.*

Start the computation at $(t0, C0)$ and proceed with time. The activity coefficient at $(t1, C1)$ is estimated using the UNIQUAC method, shown in the previous section. The value of the rate of reaction of the given component at $t = t2$ is then calculated from the activity coefficient at $t = t2$ and the initial guesses of the kinetic parameters. Using the backward derivative method, $C2$ at $t2$ is predicted in Equation (18.28a). The forward derivative method is shown in Equation (18.28b).

Backward derivative method:

$$C2 = C1 + (t2 - t1)\left(\frac{dC}{dt}\right)_{t=t2} = C1 + (t2 - t1) \times r_{t=t2} \tag{18.28a}$$

$r_{t=t2} = f$ (activity coefficient at $t2$ estimated from $C2$ and the kinetic parameters to be found by optimization)

Forward derivative method:

$$C2 = C1 + (t2 - t1)\left(\frac{dC}{dt}\right)_{t=t1} = C1 + (t2 - t1) \times r_{t=t1} \tag{18.28b}$$

$r_{t=t1} = f$ (activity coefficient at $t1$ estimated from $C1$ and the kinetic parameters to be found by optimization)

Note that the value of the reaction rate r of a reactant is negative (as the concentration decreases with the reaction time) and the value of the formation rate r of a product is positive (as the concentration increases with the reaction time).

The kinetic rate for FAME is a summation of the kinetic rate of triglyceride, diglyceride and monoglyceride shown in Equation (18.29), where the subscripts T, D and M represent triglyceride, diglyceride and monoglyceride, respectively. The species are consumed, thus the negative sign is used to show that FAME is produced.

$$r_{FAME} = -(r_T + r_D + r_M) \tag{18.29}$$

The backward calculation method shown in Equation (18.28a) can be followed to obtain the subsequent concentration points, for example, $C3$ from $C2$: $C3 = C2 + (t3 - t2) \times r_{t=t3}$.

Once all the concentration–time data points relevant to the experimental profile are predicted using the initial guesses of the kinetic parameters, the $RSSQ$ between the experimentally (e.g., $C1_{exp}$) and the computationally derived concentration (e.g., $C1_{calc}$), shown below, can be minimized using an optimization algorithm to obtain the optimal values of the kinetic parameters:

$$RSSQ = (C1_{exp} - C1_{calc})^2 + (C2_{exp} - C2_{calc})^2 + (C3_{exp} - C3_{calc})^2 + \cdots \tag{18.30}$$

For a robust optimization approach, the eigenvalues (λ_i) can be introduced as multipliers for the $RSSQ$ of individual species. The optimization problem is thus defined as a constrained nonlinear programming problem, with the $RSSQ$ as the objective function in Equation (18.31) and the constraints in Equations (18.32) and (18.33):

$$RSSQ = \sum_i^{species} \lambda_i \sum_{time} \left(C_{exp}^i - C_{calc}^i \right) \tag{18.31}$$

$$\sum \lambda_i = 1 \tag{18.32}$$

$$0 \leq \lambda_i \leq 1 \tag{18.33}$$

A kinetic modeling framework shown so far can be implemented in a spreadsheet environment. The minimization of $RSSQ$ can be done using an EXCEL solver algorithm.

18.2 Diffusion Modeling

Transport modeling of liquid/gas in solid systems is best represented by molecular type diffusion, potentially using some transition region modeling for the micropores. To check this assumption, mean free path calculations must be performed for the larger molecules in a range of pore sizes. For the scale-up from using short chain length triglycerides to the longer chain length triglycerides that make up the bulk of plant oils, these mean free path calculations will have to be repeated.

The mean free path of larger components can be calculated using Equation (18.34). The ratio between the mean free path length and the pore diameter is the indicator of which diffusion regime, the system is operating in. If this ratio is less than one, Knudsen diffusion is the controlling regime whereas if it is much more than one then bulk diffusion is the controlling regime. When the ratio is about one then there is a transition region encompassing traits of both Knudsen and bulk diffusion.

$$\text{Mean free path length} = \frac{1}{\pi d^2 \eta_v} \tag{18.34}$$

where
d = pore diameter
η_v = number of molecules per unit volume

First the *infinite dilution diffusion coefficients* are calculated, which represent the diffusion coefficient for one molecule of solute in pure solvent. This can be calculated for all binary combinations of components in the system. The Wilke and Chang correlation in Equation (18.35) is the most effective method for estimating the infinite dilution diffusion coefficient[3]. This equation relies on a few and commonly used physical properties, such as the molar mass, molar volume and viscosity of the components, to estimate the infinite dilution diffusion coefficient:

$$\mathcal{D}^\theta_{12} = 7.4 \times 10^{-8} \frac{(\phi_2 M_2)^{0.5} T}{\mu_2 V_1^{0.6}} \tag{18.35}$$

where $\mathcal{D}^\theta_{12}$ is the infinite dilution diffusion coefficient of solute 1 in solvent 2 (cm^2 s^{-1}), ϕ_2 is an association constant, M_2 is the molar mass (g mol^{-1}), T is the temperature (K), μ_2 is the viscosity (mPa s) and V_1 is the molar volume (cm^3 mol^{-1}).

The *binary diffusion coefficients* can be estimated using the mole fractions and the infinite dilution coefficients of components, shown in Equation (18.36). Note that this equation provides equal diffusion coefficients for component i moving through component j and vice versa.

$$\mathcal{D}_{ij} = (\mathcal{D}^\theta_{ij})^{(1+x_j-x_i)/2} (\mathcal{D}^\theta_{ji})^{(1+x_i-x_j)/2} \tag{18.36}$$

where $\mathcal{D}_{ij}$ is the diffusion coefficient of component i against component j (m^2 s^{-1}) and x_i is the mole fraction of component i in the reaction mixture.

Equation (18.37) is applicable at scales where the size of the molecule is comparable to the diameter of the pore. This change of diffusion coefficients is due to the viscous 'squeezing' of large molecules through a pore already full of molecules.

$$\frac{D}{D_O} = 1 + \frac{9}{8} \lambda \ln \lambda - 1.54 \lambda \tag{18.37}$$

where

$$\lambda = \frac{2R_O}{d}$$

R_O is the radius of the molecule and d is the pore diameter, D_O is the diffusion coefficient in large pores and D is the 'squeezed' diffusion coefficient.

A square matrix [B] is defined to evaluate the inverse of diffusive fluxes, for example, in s per m^2, in terms of mole fractions (x_i, x_j) and the binary diffusion coefficients (calculated using Equations (18.35) to (18.37)) in

$$B_{ii} = \sum_{j \neq i}^{nspc} \frac{x_j}{\mathcal{D}_{ij}} \tag{18.38a}$$

$$B_{ij} = \frac{x_i}{\mathcal{D}_{ij}} \tag{18.38b}$$

$\mathcal{D}^{eff}_i$, the *effective diffusivity* of species i for an average mole fraction in a given time interval (t_1 to t_2), can be calculated by the ratio between the change in mole fractions in the time interval and summation of B elements for the species for a given average mole fraction:

$$\mathcal{D}^{eff}_i = \frac{x_i^{t=t_2} - x_i^{t=t_1}}{\sum\limits_{j=1}^{nspc} B_{ij}} \tag{18.39}$$

The intrinsic diffusion modeling framework only taking account of diffusion is shown in Figure 18.3.

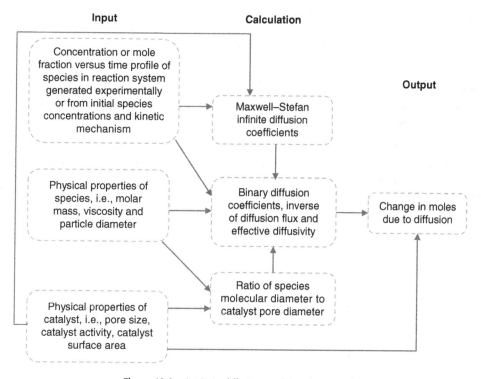

Figure 18.3 *Intrinsic diffusion modeling framework.*

Exercise 4. Calculate the effective diffusivity of species with the physical properties given in Table 18.3 and the mole fractions given in Table 18.4 in a biodiesel reaction mixture. The six species, triglyceride (tributyrin), diglyceride (dibutyrin), monoglyceride (monobutyrin), methanol, glycerol and fatty acid methyl ester (FAME known as biodiesel) (methyl butyrate), are denoted by species $i = 1$–6, respectively.

Table 18.3 *Physical properties of species used in the diffusivity calculation.*

Species	1	2	3	4	5	6
Molar mass	302	232.3	162.2	32.04	92.09	102.13
Viscosity (10^{-2} g cm^{-1} s^{-1})	14.338	19.321	23.835	0.310	47.442	2.615
Molar volume (cm^3 mol^{-1})	328	257	144	41	73	121
Particle diameter (nm)	0.9847	0.88	0.778	0.39	0.47	0.712

Table 18.4 *Mole fraction x_i of species i for a sample calculation of diffusivity.*

Species	1	2	3	4	5	6
x_i	0.02684	0.00499	0.00027	0.96241	0.00158	0.00391
Δx_i	−0.00375	0.00239	0.00053	−0.00488	0.00309	0.00261
λ	0.3961	0.3539	0.3129	0.1569	0.1890	0.2864
D/D_O	−0.0226	0.0413	0.1091	0.4315	0.3546	0.1561

Solution to Exercise 4. Refer to the ***Online Resource Material, Chapter 18 – Additional Exercises and Examples*** for the solution of this Exercise.

18.3   Multi-scale Mass Transfer Modeling

It is possible to develop an analytical solution for multiscale diffusion and reaction models, shown in the following section.

The kinetic modeling shown in Section 18.1 and the physical property based effective diffusivity modeling shown in Section 18.2 can be put together to predict the production profile of FAME for the scaled-up reactor, where heterogeneous catalysts cannot be nonporous and catalytic pores are needed to help diffusion of the species to access active sites for production on a large scale.

In order to estimate the change in the number of moles of species i from time t_1 to t_2 due to diffusion, multiscale mass transfer phenomena should be considered. Simplifications are introduced while going along the various levels. Figure 18.4 shows the multiscale phenomena and mass transfer modeling framework. Note that this dynamic simulation framework does not include an unsteady state energy transfer model, because the reactions are undertaken at the isothermal condition. See Sections 11.3.1 and 12.4.4 and Web Chapter 2 for temperature control of reactors and energy transfer models. The equations are based on mass balance or mass flux models. Mass (or energy or momentum) transfer equations feature the $\partial/\partial t$ operator.

Nomenclature

C_i	concentration of species i in the bulk, with the unit of moles per volume
C_i^p	concentration of species i inside the catalyst particle
C_i^{pore}	concentration of species i inside the pore, inside a catalyst particle
D_L	axial diffusion coefficient, cm^2 s^{-1}
d_i^{eff}	diffusivity of species i inside the particle, cm^2 s^{-1}
$Ð_i^{eff}$	effective molecular diffusivity of species i, cm^2 s^{-1}
$k(j)$	forward rate of reaction j
$keq(j)$	equilibrium constant for reaction j
L	characteristic length of a pore
$N_{p,i}^*$	mass flux of species i through the catalyst particle
N_i	number of moles of species i consumption (of reactants) or formation (of products)
$(N_i)^{intrinsic\ kinetics}$	number of moles of species i consumption (of reactants) or formation (of products) due to intrinsic reaction kinetics
R	universal gas constant, 8.314 J mol^{-1} K^{-1}
$\bar{r}$	dimensionless catalyst particle radius
r_p	particle radius
r	scale for the radius of catalyst particles
t	time, s
T	temperature, K
u	velocity
z	axial length

Greek letters

ε	bed void fraction
ε_p	void fraction inside particle
τ	tortuosity
σ	constriction factor
μ	viscosity

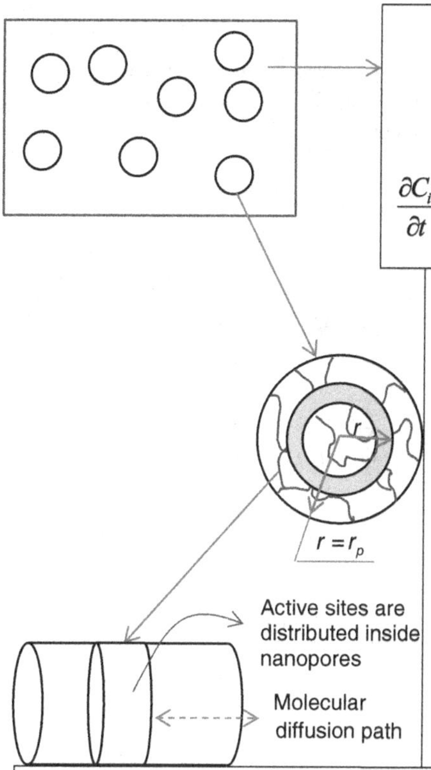

Species mass transfer through the bulk phase in a reactor:

The bulk phase species i balance equation is a function of bulk diffusion D_L, axial velocity u and mass flux at the surface of particle $N_{p,i}^*$.

$$\frac{\partial C_i}{\partial t} = D_L\left(\frac{\partial^2 C_i}{\partial z^2}\right) - \frac{u}{\varepsilon}\left(\frac{\partial C_i}{\partial z}\right) + N_{p,i}^* \frac{3}{r_p}\frac{(1-\varepsilon)}{\varepsilon} \tag{I}$$

Mass flux through spherical catalyst particles:

$$N_{p,i}^* = -d_i^{eff}\left.\frac{\partial C_i^p}{\partial r}\right|_{r=r_p} \tag{II}$$

The component balance inside a particle is given by

$$\frac{\partial C_i^p}{\partial t} = \frac{d_i^{eff}}{\varepsilon_p}\left(\frac{1}{r^2}\frac{\partial^2 (r^2 C_i^p)}{\partial r^2}\right) - \frac{(1-\varepsilon_p)}{\varepsilon_p} \times \text{Rate of reaction 1} = 0 \tag{III}$$

It is assumed that there are no mass transfer limitations at the surface of particles and the concentration inside a particle is symmetric with the radius. The boundary conditions for particles are given as

At $r = r_p$: $C_i^p(r = r_p, z, t) = C_i(z, t)$

$$r = 0: -d_i^{eff}\frac{\partial C_i^p}{\partial r} = 0 \tag{IV}$$

The rate of reaction equations are shown in Section 18.1.

$r = r_p$

Active sites are distributed inside nanopores

Molecular diffusion path

Cylindrical nanopores inside particles:

$$\mathcal{D}_i^{eff} = d_i^{eff}\frac{\varepsilon_p \sigma}{\tau} \tag{V}$$

$\mathcal{D}_i^{eff}$ is deduced from Equations (18.35) to (18.39).

ε_p : Porosity

σ : Constriction factor ~0.8

τ : Tortuosity

Porosity is ~0.416 and tortuosity is 1.56 for a powder packed reaction in a column. At steady state, the mass flux neglecting axial velocity within a pore is of the following form:

$$\mathcal{D}_i^{eff}\frac{\partial^2 C_i^{pore}}{\partial z^2} - \text{Rate of reaction} = 0 \tag{VI}$$

Assumptions:
  Bulk diffusivity and axial velocity are neglected.
  Thus, combining Equations (I), (II) and (V):

$$\frac{\partial C_i}{\partial t} = -\frac{3}{r_p}\frac{(1-\varepsilon)}{\varepsilon} \times \frac{\mathcal{D}_i^{eff}}{0.2133}\left.\frac{\partial C_i^p}{\partial r}\right|_{r=r_p} \tag{VII}$$

Within a time interval, the following equations are sequentially solved and concentrations are updated from nano through meso to bulk scale.

Equations (VI) → (III) → (VII)

Figure 18.4 *Multiscale phenomena and mass transfer modeling framework for a heterogeneously catalyzed reaction system.*

Did you know?

In early 1800s, G.G. Stokes in England and M. Navier in France came up with mass, momentum and energy unsteady state or transfer equations for fluid flow in three-dimensional space. These equations are universally valid and are yet to be evidenced by mathematical proof. The equations are of the following forms:

Mass continuity:

$$\frac{\partial \rho}{\partial t} + \frac{\partial (\rho u_x)}{\partial x} + \frac{\partial (\rho u_y)}{\partial y} + \frac{\partial (\rho u_z)}{\partial z} = 0$$

where ρ = density, t = time and u = linear velocity, which can be in the x, y and z dimensions.

Momentum transfer:

$$\frac{\partial (\rho u_i)}{\partial t} + \sum_{j=x,y,z} \frac{\partial (\rho u_i u_j)}{\partial j} = -\frac{\partial P}{\partial i} + \frac{1}{Re} \sum_{j=x,y,z} \frac{\partial \tau_{ij}}{\partial j}$$

$$i \in x, y, z$$

where
P = pressure of the fluid system

$$Re = \text{Reynolds number (dimensionless)} = \frac{\text{Characteristic length} \times \text{Velocity} \times \text{Density}}{\text{Viscosity}}$$

τ = stress

Energy transfer:

$$\frac{\partial E}{\partial t} + \sum_{i=x,y,z} \frac{\partial (E u_i)}{\partial i} + \sum_{i=x,y,z} \frac{\partial (P u_i)}{\partial i} + \frac{1}{Re}\frac{1}{Pr} \sum_{i=x,y,z} \frac{\partial (q_i)}{\partial i}$$

$$= \frac{1}{Re} \sum_{i=x,y,z} \frac{\partial \left(\sum_{j=x,y,z} u_j \tau_{ij} \right)}{\partial i}$$

$$i \in x, y, z$$

where
E = total energy, q = heat flux

$$Pr = \text{Prandtl number (dimensionless)} = \frac{\text{Viscous diffusion rate}}{\text{Thermal diffusion rate}} = \frac{\text{Heat capacity} \times \text{Density}}{\text{Thermal conductivity}}$$

Navier–Stokes equation solutions are taught in Fluid Mechanics.

Did you know?

The mass conservation equation of elementary particulates follows the form of the Navier–Stokes equation. Paul A, M. Dirac and Erwin Schrödinger received Nobel Prize in Physics in 1933 for their work on wave equations for nonrelativistic and relativistic particulates.

For elementary particulates such as electrons and positrons, the mass conservation equation at the atomic scale consists of time domain and three-dimensional operators $h\frac{\partial}{\partial t}$, $-ih\frac{\partial}{\partial x}$, $-ih\frac{\partial}{\partial y}$ and $-ih\frac{\partial}{\partial z}$.

Schrödinger provided a wave equation for nonrelativistic particulates (e.g., a proton) as follows:

$$E = \frac{p^2}{2m} \rightarrow ih\frac{\partial \vartheta}{\partial t} = -\frac{h^2}{2m}\nabla^2 \vartheta$$

where E = energy, p = momentum and m = mass of the particulate.

Dirac provided mass conservation of relativistic particulates (e.g., the electron) based on a first-order wave equation in quantum theory, as follows:

$$ih\frac{\partial \vartheta}{\partial t} + ih\left(\sigma_x\frac{\partial \vartheta}{\partial x} + \sigma_y\frac{\partial \vartheta}{\partial y} + \sigma_z\frac{\partial \vartheta}{\partial z}\right) + \sigma_m m\vartheta = 0$$

where σ_x, σ_y, σ_z and σ_m are the "anticommuting matrices, whose squares are unity:"

$$\sigma_x = \begin{bmatrix} 0 & 1 \\ 1 & 0 \end{bmatrix}; \quad \sigma_y = \begin{bmatrix} 0 & -i \\ i & 0 \end{bmatrix}; \quad \sigma_z = \begin{bmatrix} i & 0 \\ 0 & i \end{bmatrix}; \quad \sigma_m = \begin{bmatrix} 1 & 0 \\ 0 & -1 \end{bmatrix}$$

Can you think of multiscale problems that require interactive simulation of atomistic scale quantum mechanistic and fluid mechanistic models to solve the problems?

There are analytical and numerical ways to solve partial differential equations. A numerical way of solving the partial differential equations (PDEs) is to convert them into ordinary differential equations (ODEs) using the method of orthogonal collocation. For a change in effective diffusivity, there is a large variation and even instability in the solution (stiff equations). Therefore a stiff equation solver needs to be used to solve these equations. These coupled sets of ODEs can be solved simultaneously using the MATLAB function ode15s, appropriate for a system with stiff equations.

The other way of solving these equations is by representation of concentration profiles (expected results of PDE solutions) by *hyperbolic functions (analytical ways)* (cf. Section 12.4.3). The two basic hyperbolic functions, $\sinh(x)$ (shown in Figure 18.5) and $\cosh(x)$ and their ratios, $\tanh(x)$, universally present generic trends of reaction concentration profiles when dimensionless variables in 0 to 1 scale are used:

$$\sinh(x) = \frac{e^x - e^{-x}}{2}$$

$$\cosh(x) = \frac{e^x + e^{-x}}{2}$$

$\tanh(x)$, $\coth(x)$ and $\operatorname{sech}(x)$ have formulae accordingly.

A generic form of PDE of $\partial^2 y/\partial x^2 - \theta y = 0$ has a solution in algebraic form as

$$y(x) = a1\sinh(\theta x) + a2\cosh(\theta x) \tag{18.40}$$

Hence, solutions of PDEs (VI) and (III) (Figure 18.4) have the generic form of Equation (18.40). Physical properties grouped together are evaluated. Unknown constants ($a1$ and $a2$) are evaluated from boundary conditions, based on rational assumptions.

Now consider the solution strategy for PDE (VI) in Figure 18.4 at the porous scale within spherical catalytic particles. For a cylindrical pore of length $z = 0$ to L, inside a spherical catalyst particle, with $z = 0$ at the centre of the particle and $z = L = r_p$ at the surface of the particle (or radius of the particle) (as in Figure 18.6), the following assumptions about the boundary conditions can be made. As the PDE is of second order, two boundary conditions are needed to solve the PDE.

1. Boundary condition 1. At $z = 0$, a symmetry around z axis implies

$$\frac{dC_i^{pore}}{dz} = 0 \tag{18.41}$$

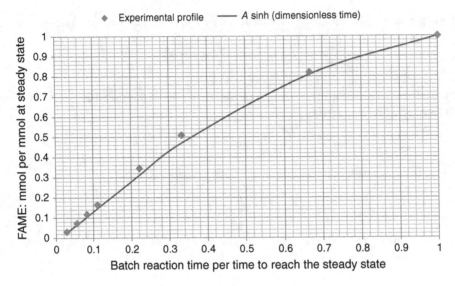

Figure 18.5 *Hyperbolic algebraic functions represent concentration profiles at the dimensionless scale. "A" is a coefficient of the sinh function, which can be deduced analytically such as in Equations (18.46) and (18.53).*

2. Boundary condition 2. At $z = L$, there is no mass transfer limitation at the surface of particles and hence the concentration at the surface of a particle is equal to the bulk concentration, as shown by

$$C_i^{pore}(z = L = r_p) = C_i^p(r = r_p) = C_i(t) \tag{18.42}$$

The solution of PDE (VI) in Figure 18.4 takes the following form based on Equation (18.40):

$$\frac{d^2 C_i^{pore}}{dz^2} - \frac{\phi L}{\mathcal{D}_i^{eff}} \frac{C_i^{pore}}{L} = 0 \tag{18.43}$$

where $\phi L/\mathcal{D}_i^{eff}$ is the dimensionless Thiele modulus: $\varphi = $ Rate of reaction/Diffusion rate.

The Thiele modulus quantifies the ratio between the reaction and diffusion rates inside a heterogeneous catalyst (cf. Section 12.4.3). Note that the definition or physical meaning of the Thiele modulus remains the same, but its expression

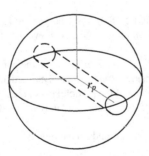

Figure 18.6 *A cylindrical pore inside a spherical catalyst particle, with $z = 0$ at the centre of the particle and $z = r_p$ at the surface of the particle ($r_p = $ radius of the particle).*

may change depending on design considerations and convenience of presentation of results. The modulus should always be dimensionless.

The concentration and length of the pore are made dimensionless by the following ratios in order to apply a hyperbolic algebraic solution of the form of Equation (18.40):

$$\overline{C}_i^{pore} = \frac{C_i^{pore}}{C_i(t)}; \quad \overline{z} = \frac{z}{r_p}$$

where $C_i(t)$ is a constant concentration at the surface of the catalyst particle at a given time.

Equation (18.43) thus becomes

$$\frac{d^2 \overline{C}_i^{pore}}{d\overline{z}^2} - \varphi \overline{C}_i^{pore} = 0 \tag{18.44}$$

Applying Equation (18.40) as the algebraic equation form of the PDE, Equation (18.44) results in

$$\overline{C}_i^{pore} = a1 \sinh(\varphi \overline{z}) + a2 \cosh(\varphi \overline{z}) \tag{18.45}$$

Applying the boundary condition in Equation (18.41) gives $a2 = 0$ and applying the boundary condition in Equation (18.42) gives $a1 = 1/\sinh \varphi$.

Thus, Equation (18.45) has the final form of

$$\overline{C}_i^{pore} = \frac{\sinh(\varphi \overline{z})}{\sinh(\varphi)} \tag{18.46}$$

Remember that Equation (18.46) is only valid for the following design variables:

1. Cylindrical nanopores inside a spherical catalytic particle extending from the surface to the centre of the particle.
2. Catalytic active sites are on the cylindrical surface of the catalyst pores.
3. Axial velocity is neglected and only the molecular diffusion effect is considered, which is valid for nanopores.

Any of the above three design variable changes, Equations (18.41) to (18.46) are to be deduced.

Exercise 5. This is to show the solution of PDE III in Figure 18.4 for spherical catalytic particles. Now take the example of the particle scale mass conservation Equation III in Figure 18.4. Deduce the algebraic solution of the PDE at a given time using the approach shown in Equations (18.41) to (18.46). Make rational assumptions along the way of solutions and list them.

Solution to Exercise 5. At any given time, the concentration within a catalyst particle varies only with the geometric position. Thus, Equation III in Figure 18.4 reduces to the following form, similar to Equation (18.43). The difference between the two forms of equations is due to the geometric shapes: Equation (18.43) is for a cylindrical shape and Equation (18.47) is for spherical shape, respectively.

$$d_i^{eff} \left(\frac{1}{r^2} \frac{\partial^2 (r^2 C_i^p)}{\partial r^2} \right) - (1 - \varepsilon_p) \times \text{Rate of reaction 1} = 0 \tag{18.47}$$

The boundary conditions in the following equations, similar to Equations (18.41) and (18.42), are for two locations, at the centre and at the surface of the sphere, based on the following rational assumptions:

$$\text{At } r = 0, \text{ a symmetry around } r \text{ implies } \frac{dC_i^p}{dr} = 0 \tag{18.48}$$

At $r = r_p$ (radius of the catalyst spherical particle), there is no mass transfer limitation at the surface of the particles and hence the concentration at the surface of the particle is equal to the bulk concentration, as shown by

$$C_i^p(r = r_p) = C_i(t) \qquad (18.49)$$

The concentration and radius of the spherical particle are made dimensionless by the following ratios in order to apply a hyperbolic algebraic solution of the form of Equation (18.40) to the PDE in Equation (18.47):

$$\overline{C}_i^p = \frac{C_i^p}{C_i(t)}; \quad \overline{r} = \frac{r}{r_p}$$

where $C_i(t)$ is a constant concentration at the surface of the catalyst particle at a given time.

Rearrangement of Equation (18.47) and incorporation of dimensionless quantities results in

$$\frac{\partial^2 (\overline{r}^2 \overline{C}_i^p)}{\partial \overline{r}^2} - \frac{(1 - \varepsilon_p) r_p^2 \phi 1}{d_i^{eff}} (\overline{r}^2 \overline{C}_i^p) = 0 \qquad (18.50)$$

where $(1 - \varepsilon_p) r_p^2 \phi 1 / d_i^{eff}$ is the dimensionless Thiele modulus: $\varphi 1$ = *Rate of reaction/Diffusion rate*. In this case, the Thiele modulus quantifies the ratio between reaction and diffusion rates inside the catalyst particle.

Equation (18.50) thus takes the form of Equation (18.40):

$$\frac{\partial^2 \left(\overline{r}^2 \overline{C}_i^p\right)}{\partial \overline{r}^2} - \varphi 1 \left(\overline{r}^2 \overline{C}_i^p\right) = 0 \qquad (18.51)$$

Applying Equation (18.40) as the algebraic equation form of the PDE, Equation (18.51) results in

$$\overline{r}^2 \overline{C}_i^p = a1 \sinh(\varphi 1 \overline{r}) + a2 \cosh(\varphi 1 \overline{r}) \qquad (18.52)$$

Applying the boundary condition in Equation (18.48) gives $a2 = 0$ and applying the boundary condition in Equation (18.49) gives $a1 = 1/\sinh \varphi 1$.

Thus, Equation (18.52) takes a final form of

$$\overline{C}_i^p = \frac{\sinh(\varphi 1 \overline{r})}{\overline{r}^2 \sinh(\varphi 1)} \qquad (18.53)$$

Remember that Equation (18.53) is only valid for the following design variables (list of assumptions):

1. Spherical catalytic particle.
2. Uniform concentration at a given r between 0 (centre of particle) and r_p (radius of particle).
3. Axial velocity is neglected and only the diffusion effect is considered, which is valid at the meso scale.

Any of the above three design variable changes, Equations (18.47) to (18.53) are to be deduced, for example, mass transfer through a bundle of tubes in a rectangular slab (Figure 18.7).

The key question now is how to integrate the nano-scale porous level mass transfer model (Equation (18.46)) and the meso-scale particle level mass transfer model (Equation (18.53)). Inhomogeneity and faster change in concentration are expected at the smaller scale. Thus, within a given time change at the bulk scale, there will be a number of meso-scale

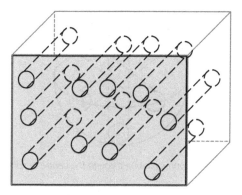

Figure 18.7 *A bundle of tubes in a rectangular slab.*

phenomena happening at a smaller time scale that need to be integrated to update the concentration at the bulk scale. Similarly, within a small domain at the meso scale, there will be many nano-scale porous phenomena that need to be integrated to decide their overall effect at the meso-scale particle. A multi-scale model thus needs to adopt a strategy to send information from the bulk through the meso to the nano scale and update results in the reverse order through the scales, as shown in Figure 18.8. The approach of data accumulation from the lower level into the compressed data set at a higher level is called coarse graining. There are two possible ways of coarse graining, in the time domain and in the space domain. In time domain coarse graining, for one slow event at the top level, there are multiple implementations of fast events at the bottom level and the compressed time domain events are sent upward from the bottom to the top level. For coarse graining in the space domain, for a smaller area at the top level, there are many heterogeneity data at the nano scale that need to be accounted for, compressed and retained at the top level. Hence, as a top-down analysis is done, a

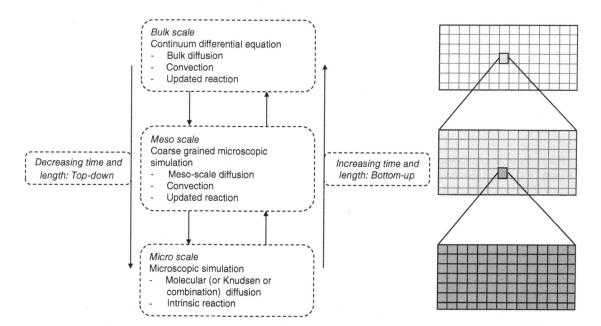

Figure 18.8 *Multiscale integration strategy.*

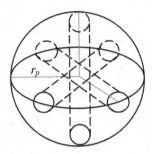

Figure 18.9 *Cylindrical uniform pores passing through the centre of spherical catalytic particles.*

more detailed design is accounted for and as a bottom-up analysis is done, the detailed analyses from the bottom level are grouped or compressed to update the information at the top level.

The concept can be applied to various problems, fuel cells, chemical vapor depositions, heterogeneous reactions, nanotechnology, drug delivery, crustal growth and medical fields. Depending upon a specific problem, models can be simplified for better data management.

The basic transfer equations in Figure 18.4 assume a time domain change at the bulk scale, but only length domain heterogeneity at the meso and nano scales. Hence, within a small time change, all the events happening at the nano-scale spatial domain can be accounted for and the total concentration change can be updated in the bulk scale. Hence, within a small time scale, Equation (18.46) is applied first, Equation (18.53) is updated with important changes and parameters from Equation (18.46) and finally Equation (VII) (Figure 18.4) is updated with a parameter and variables from Equation (18.53). The steps are repeated over the entire reaction time to generate the complete concentration versus time profile.

To generate two limiting conditions, by which either of Equations (18.46) or (18.53) can be used, two sets of assumptions are shown:

1. Cylindrical uniform pores passing through the centre of spherical catalytic particles (Figure 18.9). In this case, Equation (18.53) does not need to be considered and the concentration obtained at the particle surface (to be derived from Equation (18.46)) can be multiplied by the total number of pores in a given reactor volume (number of particles × number of pores in each particle) and average porous volume to obtain the molar change of species at a given time interval. The molar change of individual species at a given time interval is calculated is and added to that obtained from the previous time interval to update the number of moles of species in the bulk scale.
2. Uniform pores and uniform concentrations within particles at a given $r = 0$ to r_p (particle radius). In this case, Equation (18.46) is not considered. Equation (18.53) is used to calculate the concentration obtained at the particle surface to incorporate in Equation (VII) in Figure 18.4, which is solved analytically to obtain the bulk-scale concentration profile with time.

For any other situation, when there is spatial heterogeneity (not uniform), all three Equations (18.46), (18.53) and (VII) (Figure 18.4) should be modified and solved alongside boundary conditions.

18.3.1 Dimensionless Physical Parameter Groups

Two dimensionless groups of physical parameters used in Equations (18.46) and (18.53) are as follows:

$$\varphi = \frac{\phi L}{\mathcal{D}_i^{\mathit{eff}}} \tag{18.54}$$

The effective diffusivity has a unit of length^2 time^{-1}. The characteristic length has a unit of length, while ϕ is the rate of reaction with the unit of length time^{-1}. This makes the Thiele modulus φ dimensionless. The characteristic length of reaction of a cylindrical pore is pore radius/2.

Hence,

$$\phi = \text{Reaction rate} \times \text{Characteristic length} = \text{Reaction rate} \times \frac{\text{Cylindrical pore radius}}{2}$$

For accurate modeling, the rate of reaction is strictly the intrinsic rate of reaction and can be of any form, such as ER, with methanol adsorption as the rate limiting step (Equation set (18.9)). For estimating the FAME rate of production, use Equation (18.29). Equations (18.15) to (18.27) are used to address the nonideality of the reaction mixture through activity calculations, while Equations (18.35) to (18.39) are used to estimate $\mathcal{D}_i^{eff}$.

$$\varphi 1 = \frac{(1 - \varepsilon_p) r_p^2 \phi 1}{d_i^{eff}} \tag{18.55}$$

where $\phi 1$ is the rate of reaction with the unit of time^{-1}, to make $\varphi 1$ dimensionless, and d_i^{eff} is shown in Equation (V) in Figure 18.4.

Exercise 6. This is to show how the bulk concentration profile with respect to the batch reaction time can be deduced from estimation of the intrinsic kinetic rate of reactions and effective diffusivities.

1. Take Exercise 3 as an example. Determine the intrinsic kinetic rates of reactions of species with respect to time from the activity values shown in Table 18.2. Use Equation set (18.9) for the ER mechanism with methanol adsorption as the rate limiting step to estimate the intrinsic kinetic rates of reactions of triglyceride, diglyceride and monoglyceride at different times and Equation (18.29) to estimate the intrinsic kinetic rates of reactions of FAME. The total number of moles of 310 mmol in 16 cm^3 reaction mixture does not change with the reaction time.
2. Calculate the dimensionless Thiele modulus, $\varphi = \phi L / \mathcal{D}_i^{eff}$ of FAME, for a cylindrical pore radius of 4 nanometers (nm). The effective diffusivity of species can be calculated using Equations (18.35) to (18.39). Assume $L = 5$ mm. Tables 18.5 and 18.6 show the average mole fractions and difference in mole fractions in each time interval, respectively.

Table 18.5 Average mole fraction with respect to time interval.

Time Interval (min) ($t2 - t1$)	0–30	30–60	60–120	120–180	180–240	240–300	300–360
Triglyceride	0.03048	0.02684	0.02390	0.02194	0.01982	0.01789	0.01666
Diglyceride	0.00190	0.00499	0.00787	0.01050	0.01158	0.01199	0.01272
Monoglyceride	0	0.00027	0.00100	0.00200	0.00273	0.00328	0.00414
Methanol	0.96629	0.96241	0.95977	0.95875	0.95546	0.95181	0.95008
Glycerol	0.00002	0.00158	0.00014	0	0	0	0
FAME	0.00130	0.00391	0.00732	0.01137	0.01429	0.01627	0.01882

Table 18.6 Difference in mole fractions with respect to time interval.

Time Interval (min) ($t2 - t1$)	0–30	30–60	60–120	120–180	180–240	240–300	300–360
Triglyceride	−0.00355	−0.00375	−0.00212	−0.00181	−0.00243	−0.00143	−0.00103
Diglyceride	0.00380	0.00239	0.00336	0.00189	0.00027	0.00056	0.00090
Monoglyceride	0.00000	0.00053	0.00094	0.00105	0.00042	0.00068	0.00104
Methanol	−0.00290	−0.00488	−0.00041	−0.00163	−0.00494	−0.00235	−0.00111
Glycerol	0.00004	0.00309	−0.00597	−0.00342	0.00478	0.00048	−0.00283
FAME	0.00261	0.00261	0.00420	0.00392	0.00191	0.00207	0.00303

3. Calculate $\partial C_i^p / \partial r|_{r=r_p}$ at different time intervals and estimate the bulk concentration profile with respect to the batch reaction time (Equation (VII) in Figure 18.4), following an analytical procedure similar to that shown in Equation (18.28). Assume a uniform porosity of 0.8.

Solution to Exercise 6

1. The intrinsic kinetic rates of reactions of species with respect to time are calculated, in Table 18.7, from the activity values in Table 18.2 using Equation set (18.9). Further, the formation of number of moles of FAME is calculated as follows, which is similar to Equation (18.28), with $\Delta t_2 = t_2 - t_1 > 0$:

$$\text{(Number of moles)}_{\Delta t_2} = \text{(Number of moles)}_{\Delta t_1} \pm \text{(Rate of reaction)}_{\Delta t_2} \times (t_2 - t_1) \qquad (18.56)$$

The above form of equation follows the Taylor series expansion for first order.

The number of moles of reactants reduces with time and that of products increases with time. Accordingly, the plus or minus sign is selected in Equation (18.56). Following this procedure, the FAME number of mole formations is shown in Table 18.7.

Table 18.7 *Intrinsic kinetic rates of reactions of species with respect to time estimated using the mechanistic model, ER, with methanol adsorption as the rate limiting step.*

Time (min)	Triglyceride	Diglyceride	Monoglyceride	FAME	FAME Number of Mole Formations (mmol)
		Rate of Reactions			
0	0.0000093	0.0000000	0.0000000	−0.0000093	0
30	0.0000093	0.0000093	0.0000000	−0.0000185	0.17243984
60	0.0000092	0.0000092	0.0000053	−0.0000236	0.39226347
120	0.0000091	0.0000092	0.0000122	−0.0000305	0.95894219
180	0.0000090	0.0000091	0.0000151	−0.0000332	1.57665788
240	0.0000089	0.0000089	0.0000103	−0.0000281	2.09960782
300	0.0000088	0.0000088	0.0000098	−0.0000274	2.60976415
360	0.0000087	0.0000087	0.0000117	−0.0000292	3.15232831
1440	0.0000072	0.0000072	0.0000024	−0.0000168	8.78958983

2. The effective diffusivity of FAME is calculated from the physical properties given in Table 18.3 with a pore radius of 4 nanometers. The effective diffusivity is multiplied with the total number of moles (= 310 mmol) in the system to find out the change in number of moles of FAME due to diffusion. The Thiele modulus, $\varphi = \phi L / D_i^{eff}$ of FAME, for a pore length of $L = 5$ mm and radius of 4 nm, is then calculated, for example, at time 60 min:

$$0.0000236 \,(\text{reaction rate}) \times (2 \times 10^{-7}) \left(\frac{\text{Pore radius}}{2} \text{ or Characteristic length of a cylindrical pore in cm} \right)$$

$$\times \, (0.5) \,(\text{pore length in cm}) / 0.0000178 \,(\text{effective diffusivity in cm}^2 \text{ s}^{-1})$$

The values calculated are shown in Table 18.8.

Table 18.8 *Effective diffusivity and Thiele modulus, $\varphi = \frac{\phi L}{D_i^{eff}}$ of FAME.*

Time (min)	Reaction Rate (1)	Effective Diffusivity (2)	Ratio of (1) to (2)	φ
30	0.0000185	0.0000534	0.3475	3.4755×10^{-8}
60	0.0000236	0.0000178	1.3282	1.3282×10^{-8}
120	0.0000305	0.0000153	1.9964	1.9964×10^{-8}
180	0.0000332	0.0000092	3.6217	3.6217×10^{-8}
240	0.0000281	0.0000040	6.9759	6.9759×10^{-8}
300	0.0000274	0.0000034	8.1299	8.1299×10^{-8}
360	0.0000292	0.0000043	6.8169	6.8169×10^{-8}

3. $\partial C_i^p / \partial r \big|_{r=r_p}$ can be shown from the first derivative of Equation (18.46) for $\bar{z} = 1$. Assume a uniform porous and particle distribution, such that:

$$\bar{C}_i^p = \bar{C}_i^{pore} = \frac{\sinh(\varphi\bar{z})}{\sinh(\varphi)}$$

$$\frac{\partial \bar{C}_i^p}{\partial \bar{r}}\bigg|_{r=r_p} = \varphi\frac{\cosh(\varphi)}{\sinh(\varphi)} \qquad (18.57)$$

$$\frac{\partial C_i^p}{\partial r}\bigg|_{r=r_p} = \frac{C_i}{r_p} \times \frac{\varphi}{\tanh\varphi}$$

Thus, Equation (VII) in Figure 18.4 is deduced to give the following form:

$$\Delta C_i = -\frac{3}{r_p}\frac{(1-\varepsilon)}{\varepsilon} \times \frac{D_i^{eff}}{0.2133} \times \frac{C_i}{r_p} \times (t_2 - t_1) \times \frac{\varphi}{\tanh\varphi}$$

By analytical integration with Taylor's series of first order:

$$C_i \text{ (at } t = t_2) = C_i \text{ (at } t = t_1) \pm \frac{3}{r_p}\frac{(1-\varepsilon)}{\varepsilon} \times \frac{D_i^{eff}}{0.2133} \times \frac{C_i}{r_p} \times (t_2 - t_1) \times \frac{\varphi}{\tanh\varphi}$$

or in terms of the number of moles:

$$N_i \text{ (at } t = t_2) = N_i \text{ (at } t = t_1) \pm \frac{3}{r_p}\frac{(1-\varepsilon)}{\varepsilon} \times \frac{D_i^{eff}}{0.2133} \times \frac{(N_i)^{intrinsic\,kinetics}}{r_p} \times (t_2 - t_1) \times \frac{\varphi}{\tanh\varphi} \qquad (18.58)$$

At time $t = 30$ min, N_i of FAME is calculated as follows

$(N_i)^{intrinsic\,kinetics}$ value of FAME $= 0.17244$ mmol (Table 18.7)

N_i (at $t = 30$ min)

$$= \frac{3}{0.5}\frac{1-0.8}{0.8} \times \frac{0.0000534}{0.2133} \times \left\{ \frac{0.17244}{(0.5)} \times (30 \times 60) \times \frac{3.4755 \times 10^{-8}}{\tanh(3.4755 \times 10^{-8})} \right\}$$

$$= 0.5456206$$

N_i (at $t = 60$ min)

$$= 0.5456206 + \frac{3}{0.5}\frac{1-0.8}{0.8} \times \frac{0.0000178}{0.2133} \times \left\{ \frac{0.39226347\,(\text{Table 18.7})}{(0.5)} \times (30 \times 60) \times \frac{1.3282 \times 10^{-8}}{\tanh(1.3282 \times 10^{-8})} \right\}$$

$$= 0.7276193$$

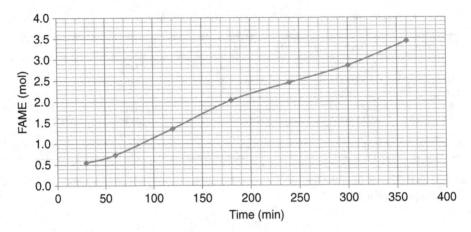

Figure 18.10   *FAME profile in mmol with respect to time using kinetic and diffusion modeling at catalytic pores.*

N_i (at $t = 120$ min)

$$= 0.7276193 + \frac{3}{0.5}\frac{1-0.8}{0.8} \times \frac{0.0000153}{0.2133} \times \left\{ \frac{0.95894219 \,(\text{Table } 18.7)}{(0.5)} \times (60 \times 60) \times \frac{1.9964 \times 10^{-8}}{\tanh(1.9964 \times 10^{-8})} \right\}$$

and so on.

Figure 18.10 shows the FAME profile in mmol with respect to time obtained from the above calculations.

Exercise 7. Consider the isopropyl palmitate (IPA) production system using palmitic acid (PA) and isopropanol (IPP) as reactants at a temperature of 130 °C. The by-product is water. PA, IPP, IPA and water are denoted by species $i =$ 1–4, respectively. The seven groups involved are CH_3, CH_2, CH, OH, H_2O, COOH and COO and their secondary group numbers are $p = 1, 2, 3, 14, 16, 42$ and 77, respectively[2]. The group parameters are shown in Tables 18.9 and 18.10.

Table 18.9 $v_p^{(i)}$ matrix of p group present in species i.

	Species			
	1	2	3	4
Group: 1	1	2	3	0
2	14	0	14	0
3	0	1	1	0
14	0	1	0	0
16	0	0	0	1
42	1	0	0	0
77	0	0	1	0

Table 18.10   *Group (p) van der Waals surface area and volume parameters Q_p and S_p.*

	Groups						
	1	2	3	14	16	42	77
Q_p	0.848	0.54	0.228	1.2	1.4	1.224	1.2
S_p	0.9011	0.6744	0.4469	1.1	0.92	1.3013	1.38

Calculate the activity coefficients for the mole fraction at $t = 240$ min shown in Table 18.11. The binary energy of interaction between two groups is shown in Table 18.12.

Table 18.11 Mole fraction x_i of species i.

Species	1	2	3	4
x_i	0.0691	0.7281	0.1014	0.1014

Table 18.12 Binary energy of interaction between two groups.

Group	$n = 1$	2	3	14	16	42	77
$m = 1$	0	0	0	986.5	1318	663.5	387.1
2	0	0	0	986.5	1318	663.5	387.1
3	0	0	0	986.5	1318	663.5	387.1
14	156.4	156.4	156.4	0	353.5	199	190.3
16	300	300	300	−229.1	0	−14.09	−197.5
42	315.3	315.3	315.3	−151	−66.17	0	−337
77	529	529	529	88.63	284.4	1179	0

Calculate the kinetic parameters for the ER quasi-steady state mechanism.

Calculate the effective diffusivity for the system, with physical properties shown in Table 18.13. Examine its variation with respect to the pore size changes from 5 to 10 nanometers. The following equation can be used to estimate the viscosity at one temperature (T in K) from a known value at another temperature (T_0 in K):

$$\mu_T^{-0.2661} = \mu_{T_0}^{-0.2661} + \frac{T - T_0}{233} \tag{18.59}$$

Table 18.13 Physical properties of species in the IPA production system.

	Palmitic Acid	Isopropanol	Isopropyl Palmitate	Water
Molecular weight	256.43	60.1	298.51	18.02
Viscosity (10^{-2} g cm^{-1} s^{-1})	7.8 at 70 °C	1.33 at 40 °C	2.5 at 60 °C	0.404 at 70 °C
Molar volume (m^3 kmol^{-1})	0.3006	0.0765	0.3502	0.0184
Particle diameter (nm)	0.7	0.41	0.66	0.3
Specific density (cm^3 g^{-1})	1.1723	1.2723	1.1730	1.0235

Table 18.14 Mole fraction against time profiles of species in the isopropyl palmitate production system.

Time (min)	Palmitic Acid	Isopropanol	Isopropyl Palmitate	Water
0	0.1651	0.8349	0	0
30	0.1506	0.8146	0.0174	0.0174
40	0.1394	0.8026	0.029	0.029
50	0.1272	0.7896	0.0416	0.0416
60	0.1166	0.7832	0.0501	0.0501
90	0.0985	0.7625	0.0695	0.0695
120	0.0835	0.7541	0.0812	0.0812
150	0.0807	0.7439	0.0877	0.0877
180	0.0787	0.7407	0.0903	0.0903
210	0.0714	0.7327	0.0979	0.0979
240	0.0691	0.7281	0.1014	0.1014

Assuming that the effect of nanopores facilitates the diffusional mass transfer and thereby the reaction conversion process, calculate the concentrations of the various species with time, taking account of the additive kinetic and diffusive effects.

The experimentally determined mole fraction–time profiles of the species in the isopropyl palmitate production system are shown in Table 18.14[4].

18.4   Summary

This chapter shows material pertaining to various aspects of the multiscale modeling of heterogeneously catalyzed reaction systems. Modeling of the intrinsic kinetics has been shown for ER, LHHW and Hattori mechanisms with assumptions of rate limiting steps. The UNIQUAC model for activity and concentration correlations for a nonideal reaction system has been shown with calculations for transesterification reactions between triglyceride and methanol for FAME production. Analytical integration by Taylor's series first-order expansion can be done to estimate concentration versus time profiles of species. Modeling of multicomponent diffusion coefficients and effective diffusivities of species has been shown to calculate the fluxes of the components at a given composition and diffusion driving force. A simulation framework for implementation of a multiscale diffusion–reaction model has been provided. The mass transfer models are discussed. The calculations are shown for the FAME production system, but could be adapted to model other heterogeneously catalyzed reactive diffusive systems that incorporate pore networks. The models at every step are discussed with practical example problems.

References

1. A. Kapil, A.F. Lee, K. Wilson, J. Sadhukhan, Kinetic modelling studies of heterogeneously catalyzed biodiesel synthesis reactions, *Ind. Eng. Chem. Res.*, **50**(9), 4818–4830 (2011).
2. R.C. Reid, J.M. Praustnitz, B.E. Poling, *The Properties of Gases and Liquids*, 4th edn, McGraw-Hill, New York, 1987.
3. R. Krishna and R. Taylor, *Multicomponent Mass Transfer*, 1st edn, John Wiley & Sons, Ltd, Chichester, UK, 1993.
4. S.Y. Chin, A.L. Ahmad, A.R. Mohamed, S. Bhatia, Characterization and activity of zinc acetate complex supported over functionalized silica as a catalyst for the production of isopropyl palmitate, *Applied Catalysis A: General*, **297**, 8–17 (2006).

Index

Biorefineries and Chemical Processes: Design, Integration and Sustainability Analysis, First Edition.
Jhuma Sadhukhan, Kok Siew Ng and Elias Martinez Hernandez.
© 2014 John Wiley & Sons, Ltd. Published 2014 by John Wiley & Sons, Ltd.
Companion Website: http://www.wiley.com/go/sadhukhan/biorefineries